SOYBEAN PHYSIOLOGY AND BIOCHEMISTRY

Edited by **Hany A. El-Shemy**

INTECHOPEN.COM

Soybean Physiology and Biochemistry
Edited by Hany A. El-Shemy

Published by InTech

Janeza Trdine 9, 51000 Rijeka, Croatia

Edition 2014

Publishing Process Manager Petra Nenadic

Technical Editor Teodora Smiljanic

Cover Designer InTech Design Team

Image Copyright Steve Collender, 2011. Used under license from Shutterstock.com

Additional hard copies can be obtained from orders@intechopen.com

Soybean Physiology and Biochemistry, Edited by Hany A. El-Shemy
p. cm.
ISBN 978-953-307-534-1

INTECH

open science | open minds

Contents

Preface

Worldwide, soybean seed proteins represent a major source of amino acids for human and animal nutrition. Soybean seeds are an important and economical source of protein in the diet of many developed and developing countries. Soy is a complete protein and soyfoods are rich in vitamins and minerals.

Soybean protein provides all the essential amino acids in the amounts needed for human health.

Recent research suggests that soy may also lower risk of prostate, colon and breast cancers as well as osteoporosis and other bone health problems and alleviate hot flashes associated with menopause. This volume is expected to be useful for student, researchers and public who are interested in soybean.

Hany A. El-Shemy, Ph.D.
Professor
Director of Biotechnology Labs
FARP
Biochemistry and Molecular Biology
Biochemistry Department
Faculty of Agriculture
Cairo University
Giza
Egypt

1

Soybean Yield Formation: What Controls It and How It Can Be Improved

James E. Board and Charanjit S. Kahlon
School of Plant, Environmental, and Soil Sciences
Louisiana State University Agricultural Center
US

1. Introduction

Soybean [*Glycine max* (L.) Merr.; family leguminosae, sub family Papilionoideae; tribe Phaseoleae] is the most important oilseed crop grown in the world (56% of world oil seed production) (US Soybean Export Council, 2008). Major producers are the US (33% of world production), followed closely by Brazil (28%) and Argentina (21%). Remaining producers are China, India, and a few other countries. Currently, soybean is grown on about 90.5 million hectares throughout the world with total production of nearly 220 million metric tons (US Soybean Export Council, 2008). At current prices, total value of the world's soybean crop is about $100 billion. Soybean is used as human food in East Asia, but is predominately crushed into meal and oil in the US, Argentina, and Brazil; and then used for human food (as cooking oil, margarine, etc.) or livestock feed (Wilcox, 2004). These uses are derived from the crop's high oil (18%) and protein (38%) content. Soybean meal is a preferred livestock feed because of its high protein content (50%) and low fiber content. Soybean oil is mainly used by food processors in baked and fried food products or bottled into cooking oil. Other uses are biodiesel products and industrial uses. Global demand for soybean has been increasing over the last several years because of rapid economic growth in the developing world and depreciation of the US dollar (US Soybean Export Council, 2008).

In response to this demand, world production has been increasing through a combination of increased production area and greater yield. Among major producers, most of this increase in Argentina and Brazil has come from increased production area, whereas in the US it has come from increased yield (US Soybean Export Council, 2008). However, over the last 10 years US soybean yields have been increasing by only 66 kg ha^{-1} yr^{-1} compared to 396 kg ha^{-1} yr^{-1} for corn (USDA, 2007). An even greater problem is the disparity in yield between the three main producing countries [US, Argentina, and Brazil (2,800 kg ha^{-1})] and that in the remainder of the world (1,510 kg ha^{-1}) (US Soybean Export Council, 2008). Because of the limited potential for increasing production area, it is very important that yield be accelerated in order to meet increasing global demand. Our objective is to describe the basic processes affecting yield formation in soybean and to apply this information to development of management and genetic strategies for increasing soybean yield. First, we will outline potential yield gains possible with management modifications in soybean. Secondly, the main abiotic and biotic stresses will be detailed describing their modes of action on yield.

This will be followed by development of a paradigm integrating how these stresses act on crop growth dynamics and yield component formation to affect final yield. This paradigm will be applied to examples of everyday problems faced by soybean farmers in coping with environmental stresses such as determination of stress-prone developmental periods, identification of stress problems affecting yield, determining the efficacy for modified management practices, and predicting yield potential of a field. Once environmental parameters have been discussed, a similar analysis will be applied to genetic strategies for yield improvement. Our objective here is to identify which plant factors explain yield improvement during cultivar development. Such factors may serve as indirect selection criteria for increasing the efficiency of cultivar development breeding programs.

2. Enviromental stress and soybean yield

Recent yield increases for soybean production in the US (66 kg ha^{-1} yr^{-1}) can be attributed to both a genetic and environmental component (USDA, 2007). Comparison of old and new US soybean cultivars have shown a range of genetic gain from cultivar development of 10 to 30 kg ha^{-1} yr^{-1} (Boerma, 1979; Specht and William, 1984; Specht et al., 1999; Wilcox, 2001). More recent research has indicated gains towards the higher end of this range (Kahlon et al., 2011). Thus, it can be approximated that recent yield gains within the US are about 50% due to cultivar genetic improvement and 50% to improved cultural practices. Potential gains from improved cultural practices for any given locale are usually determined by comparing farmer yields with those done using recommended practices (Foulkes et al., 2009). In the US, many states conduct these studies within farmer fields in which one area of a field receives typical practices and an adjacent area receives recommended practices (Louisiana Agric. Ext. Serv., 2009). In Louisiana, the typical soybean farmer produces an average yield 70% of that expected if recommended production practices were followed. Similar yield potential studies in other parts of the world show yields ranging from 60 to 80% of the optimal level (Foulkes et al., 2009). This yield gap is attributed to a suboptimal physical environment (i.e. inadequate solar radiation, temperature, photoperiod, water, soil factors) coupled with inadequate application of fertilizer and pest control. Thus, improvement of cultural practices can be expected to increase yield anywhere from 25 to 66%. Yield increases for countries outside the US, Brazil, and Argentina would be even greater, since their yield levels are substantially below those of the major producers (1510 vs. 2800 kg ha^{-1} , US Soybean Export Council, 2008).

The inability of a soybean farmer to achieve optimal yield, when adapted cultivars are grown, is caused by environmental stress. We define environmental stress as a deficiency or excess of some factor large enough to significantly reduce yield and/or impair crop quality. Environmental stresses are divided into two kinds, abiotic and biotic. Abiotic stresses are non-living stresses which can be divided into atmospheric factors (e.g. solar radiation, air temperature, humidity, and rainfall) and soil factors (eg. fertility, pH, compaction, waterlogging, soil structure, saline intrusion). Biotic stresses are living factors which are generally referred to as pests (weeds, insects, diseases, and nematodes). Although environmental stresses can initially affect crops by several physiological mechanisms, in most cases the final effect on yield occurs by reducing the canopy photosynthetic rate [uptake of CO_2 m^{-2} (land area) d^{-1}] (Fageria et al., 2006). Canopy photosynthesis combines the plant's basic genetic photosynthetic capacity per unit leaf area (leaf photosynthesis) with leaf area index (LAI, leaf area/ground area ratio) and canopy architecture to give a comprehensive picture of the crop's ability to obtain CO_2 from the atmosphere. The importance of the photosynthetic reactions in crop growth and yield formation cannot be overestimated. It is estimated that 75 to 95% of crop dry weight is derived from CO_2 fixed through photosynthesis (Imsande, 1989; Fageria et al., 2006). Photosynthesis produces the basic carbohydrates used for producing more complex carbohydrates, proteins, and lipids, all of which contribute to dry matter (Loomis and Connor, 1992a). It also supplies the chemical energy for metabolism. Because of this close linkage between canopy photosynthesis and dry matter accumulation, seasonal crop patterns of canopy photosynthetic activity and crop growth rate [CGR, dry matter accumulation per day per m^2 [g m^{-2} (land area) d^{-1}] parallel one another (Imsande, 1989). For the remainder of the chapter, CGR will be used synonymously with canopy photosynthetic rate.

Both parameters increase slowly after emergence and then increase exponentially until early reproductive development (Fig. 1) [R1-R3, stages according to Fehr and Caviness (1977) (see Table 1 for definitions and descriptions)] (Imsande, 1989). Plateau rates are maintained until R5 and then fall as the seed filling period progresses. Seasonal total dry matter (TDM) curves reflect these patterns for CGR and canopy photosynthetic rate (Fig. 2, Carpenter and Board, 1997). The first period of seasonal dry matter accumulation is called the exponential phase. Growth is initially slow, but increases exponentially with plant size until maximal light interception is achieved. At this point, maximal CGR is achieved and the crop enters the linear growth phase where CGR is relatively constant (subject to stress-induced decreases). As senescence nears and leaf fall commences, the CGR slows until reaching zero. This last period is called the senescent phase. Crop growth rate is an example of a growth dynamic parameter. Growth dynamic parameters are rates and levels of total dry matter (TDM), dry matter partitioning (e.g. harvest index), leaf area index (LAI), light interception (LI), and radiation use efficiency that characterize soybean's seasonal growing pattern (Loomis and Connor, 1992a). Canopy photosynthetic rate and CGR are important to study because they directly control TDM production. Final yield is a function of TDM produced and the percentage of dry matter transferred into the seed (i.e. harvest index) (Loomis and Connor, 1992a). Crop growth rate, in turn, is regulated by the level of ambient light and the percentage of this light intercepted by the crop [the two terms combined will be called light interception (LI)]. The importance of LI in controlling CGR is derived from its use as an energy source to produce ATP and NADPH for fixation of CO_2 into carbohydrates. The effect of LI on CGR and TDM is measured by radiation use efficiency (dry matter/intercepted light; g MJ^{-1}). Optimal radiation use efficiency depends on the absence of any stress reducing the effect of LI on TDM. Light interception and radiation use efficiency are controlled by LAI and net assimilation rate [dry matter produced per unit leaf area; g m^{-2}(leaf area) d^{-1}]. Crop growth rate is maximized when LAI is large enough to intercept 95% of the sun's light [3-4 for narrow rows; 5-6 for wide rows (Board et al., 1990a)], sunlight is not blocked by clouds, and no stress factors are present to interfere with the ability of intercepted light to stimulate net assimilation rate and CGR (as measured by radiation use efficiency). For example, a crop can be maximizing LI, but if drought stress is present and the stomata are closed so CO_2 cannot enter the leaf, net assimilation would fall, reducing CGR and TDM. This effect would be reflected in reduced radiation use efficiency.

Developmental Stages	Descriptions of Developmental Stages
Vegetative Stages	
VE	Emergence - cotyledons have been pulled through the soil surface.
V1	Completely unrolled leaf at the unifoliate node.
V2	Completely unrolled leaf at the first node above the unifoliate leaf.
V5	Completely unrolled leaf at the fifth node on the main stem beginning with the unifoliate node.
Reproductive stages	
R1	First flower: One flower at any node on the plant.
R3	Pod initiation: Pod 0.5 cm (1/4") long at one of the four uppermost nodes on the main stem with a fully developed leaf.
R4	Pod elongation: Pod 2 cm (3/4") long at one of the four uppermost main stem nodes with a fully developed leaf.
R5	Seed Initiation: Seed within one of the pods at the four uppermost main stem nodes having a fully developed leaf that is 0.3 cm long (1/8").
R6	Full seed stage: Pod at one of the four uppermost main stem nodes having a fully developed leaf that has at least one seed that has extended to the length and width of the pod locule.
R7	Physiological maturity: Presence of one pod anywhere on the plant having the mature brown color. 50% or more of leaves are yellow.

Table 1. Descriptions of the vegetative and reproductive developmental stages of soybean during the typical growing season.

Dry matter accumulation is important in yield formation because yield components recognized as important in controlling yield on the environmental level [node m^{-2}, reproductive node m^{-2} (node bearing a viable pod), pod m^{-2}, and seed m^{-2}] are responsive to TDM accumulation (Egli and Yu, 1991; Board and Modali, 2005). Yield components are morphological characteristics whose formation is critical to yield. For soybean, yield components which have potential to influence yield are seed number per area (seed m^{-2}), seed size (g per seed), seed per pod (no.), pod number per area (pod m^{-2}), pod per reproductive node (no.), reproductive node number per area (reproductive node m^{-2}), percent reproductive nodes (%; percentage of nodes becoming reproductive), and node number per area (node m^{-2}). Yield components in soybean can be organized into a sequential series of causative relationships where: yield is controlled by primary yield components seed size and seed m^{-2}; seed m^{-2} is controlled by secondary yield components seed per pod and pod m^{-2}; pod m^{-2} is controlled by tertiary yield components pod per reproductive node and reproductive node m^{-2}; and reproductive node m^{-2} is controlled by quaternary yield components node m^{-2} and percent reproductive nodes. Thus, yield components are the vehicle through which canopy photosynthetic rate and CGR affect yield.

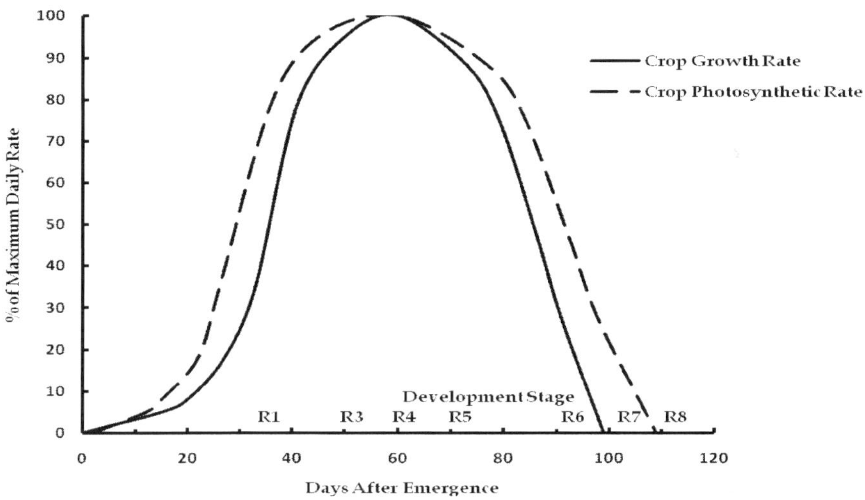

Fig. 1. Temporal profiles of the relative daily rates of plant growth and canopy CO_2 exchange. Profiles for dry matter accumulation and canopy CO_2 exchange were derived by curve fitting. For each of these two parameters several sets of published data, obtained with field grown plants, were plotted and the best-fit curves were generated. Curve presented in Imsande (1989).

Development and growth of soybean during the growing season are summarized in Fig. 3. Soybean development is separated into the vegetative development period (emergence to R1) and reproductive development period (R1 to R7). However, vegetative growth (leaves, stems, and nodes) extends from emergence to R5 (Egli and Leggett, 1973). The reproductive development period is separated into the flowering/pod formation period (R1 to R6) and the seed filling period (R5 to R7). The seed filling period, in turn, is divided into the initial lag period of slow seed filling (R5-R6) and the rapid seed filling period (R6-R7) when seed growth rate is maximal (Egli and Crafts-Brandner, 1996). Pod and seed numbers are determined by R6 (Board and Tan, 1995), before rapid seed filling starts. The linkage of environmental stress with canopy photosynthetic activity, CGR, yield component formation, and yield can be illustrated by examining the effects of the three most common abiotic stresses for soybean production: temperature extremes, drought, and canopy light interception (Hollinger and Angel, 2009).

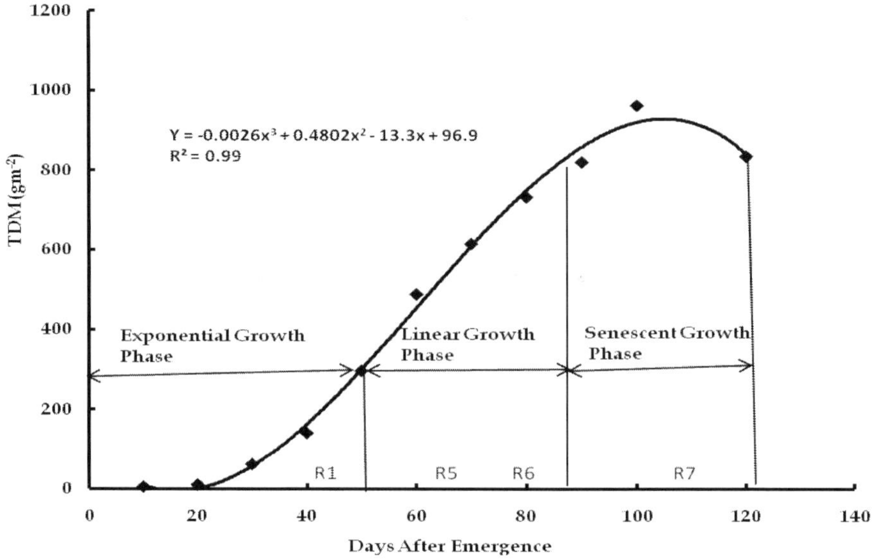

Fig. 2. Seasonal growth curve for a typical soybean crop showing the progression of total dry matter (TDM) accumulation across the exponential, linear, and senescent growth phases. Data adapted from Carpenter and Board (1997).

2.1 Temperature extremes and soybean yield

Temperature stress in soybean is manifested through effects on photosynthesis and CGR (Paulsen, 1994), reproductive abnormalities (Salem et al., 2007), and phenological events (Huxley and Summerfield, 1974). Among these factors, the effect on canopy photosynthesis and CGR has the greatest effect on yield. Temperatures above 35^0 C can inhibit pollen germination and pollen tube growth (Salem et al., 2007; Koti et al., 2004). However, since anther dehiscence occurs at 8 to 10 A.M., temperatures in most soybean growing areas would not be above the critical level during these events. The effect of warmer temperature interacting with shorter photoperiod to hasten phenological development (Hadley et al., 1984) can result in small plants having insufficient light interception for optimal canopy photosynthesis and crop growth rate (Board et al., 1996a). Thus, temperature effects on phenology indirectly affect yield through the same processes as direct temperature effects on canopy photosynthesis and CGR. Determination of heat units for soybean developmental timing uses a base temperature of 7^0 C, minimum optimum temperature of 30^0 C, maximum optimum temperature of 35^0 C, and an upper limit of 45^0 C (Boote et al., 1998).

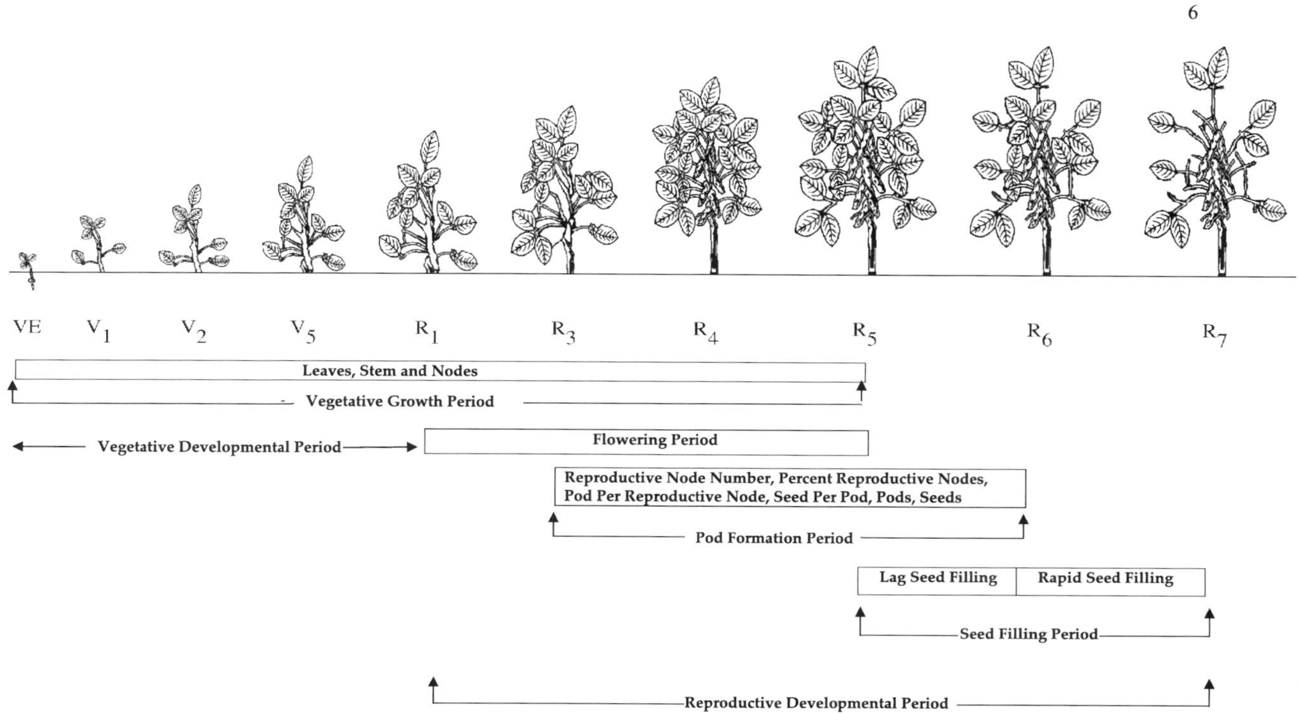

Fig. 3. Progression of vegetative organs and yield components across the developmental and growth periods of Soybean. Definitions and description of stages are in Table 1. Stages according to Fehr & Caviness (1977).

Effects of temperature on canopy photosynthesis and CGR are characterized by an optimal temperature response range falling between minimal and maximal optimal temperatures, and suboptimal and supraoptimal temperatures falling below and above the optimal range, respectively (Hollinger and Angel, 2009). The most sensitive part of the photosynthetic apparatus to heat stress is photosystem II. Specifically, the splitting of water to provide electrons to the light reactions is inhibited (Paulsen, 1994). Temperatures falling below the minimal optimal level reduce canopy photosynthesis and CGR through reduced reaction rates and/or enzyme inactivation. Studies conducted under constant day time temperatures (12-16 hours per day) across an extended period generally have reported an optimal temperature range for photosynthesis of 25-35^0 C (Jeffers and Shibles, 1969; Campbell et al., 1990; Jones et al., 1985; Gesch et al., 2001; Vu et al., 1997). However, under natural growing conditions, maximal daily temperature usually occurs for only 1-2 hours (Louisiana Agric. Exp. Stn., 2010). When heat stress studies are conducted under more realistic conditions of short-term stress, temperature had to be raised to 42-43^0 C to have a deleterious effect on soybean photosynthesis (Ferris et al., 1998). These results are corroborated by Fitter and Hay (1987) who stated that for plants from most climatic regions, temperatures of 45-55^0 C for 30 minutes were sufficient to cause irreversible damage to the photosynthetic apparatus. In conclusion, under typical growing conditions, the optimal temperature range for soybean canopy photosynthetic rate appears to be 25-40^0 C. A similar optimal temperature range of 26 to slightly above 36^0 C for crop growth rate has also been reported (Sato and Ikeda, 1979; Raper and Kramer, 1987; Sionit et al., 1987; Baker et al., 1989, Hofstra, 1972). Adverse effects on yield were entirely due to high day time temperatures rather than night time temperatures (Hewitt et al., 1985; Raper and Kramer, 1987; Gibson and Mullen, 1996).

At the crop level, heat-stress induced reductions in canopy photosynthesis affect yield components being formed at the time of the stress. Stresses occurring during flowering and pod formation (R1-R5) affect seed number, whereas stress during seed filling (R5-R7) reduces seed size (Gibson and Mullen, 1996). Both reductions were linked with lower photosynthetic rates. Concomitant with these reductions in canopy photosynthesis and yield components are decreased TDM and plant size. Soybean yield was as sensitive to heat stress during flowering/pod formation (R1-R5) as during seed filling (R5-R7). A summary for heat stress effects on yield formation is shown in Table 2.

Similar to heat stress, cold stress also adversely affects canopy photosynthesis when temperatures fall below 25^0 C. This results in less LAI, TDM, seed production, and yield (Baker et al., 1989). However, when research is conducted under cold temperature regimes similar to field conditions (intermittent nightly cold temperature or short-term cold treatments) adverse effects do not occur until temperature drops to 10^0 C (Seddigh and Jolliff, 1984 a,b; Musser et al., 1986). Seddigh and Jolliff (1984 a,b) showed that nightly cold temperature of 10^0 C vs. 16^0 C or 24^0 C slowed CGR during the vegetative and early reproductive periods. However, pod and seed numbers were not reduced, because the cooler temperatures extended the period to R5, thus allowing vegetative TDM accumulation to equilibrate across nightly temperature treatments. The 24% yield loss caused by reducing nightly temperature from 16^0 C to 10^0 C was entirely due to reduced seed size. Musser et al. (1986) reported that 1-wk chilling treatments (10^0 C) during the late vegetative and early reproductive period did not reduce early pod production. Chilling stress is a unique cold effect to plants where temperature at 10-12^0 C or below causes a cell membrance phase transition from liquid-crystalline to solid-gel form (Bramlage et al., 1978). Consequently, cell metabolism is disrupted resulting in potential adverse effects on yield. In the case of the Bramlage et al. (1978) study, pod numbers equilibrated after return to normal conditions resulting in no effect on yield. Thus, under natural growing conditions, soybean yield is resilient to cold temperatures that fall to as low as 15^0 C. However, temperatures below this level pose a significant risk for reducing yield, especially when they fall to 10^0 C. Yield loss is assured with even short term exposure to freezing temperatures [2 hr a night for 1 wk (Saliba et al., 1982)]. Effects of freezing injury are irreversible. Thus, freezing temperatures during flowering/pod formation (R1-R5) cause much greater yield losses (70% loss) compared with freezing at R6 (25% loss). A summary of cold stress effects on yield formation is shown in Table 3.

Physiological Disruptions	Affected Canopy Level Growth Processes	Affected Yield Components	Temperature Parameters
Impairment of photosystem II	Reduced canopy photosynthesis and CGR	Reduction of seed number or seed size depending on timing of stress.	Short-term exposure to temp.>40^{0}C
Enzyme denaturation and deactivation	Reduced canopy photosynthesis and CGR	Reduction of seed number or seed size depending on timing of stress.	Short-term exposure to temp >40^{0}C
Increased development rate	Reduced canopy photosynthesis and CGR by shortening emergence-R5 period	Reduction of seed number.	Under short days development rate increases with degree days [Base temp=7^{0}C Min. optimum temp=30^{0}C Max. optimum temp= 35^{0}C Upper limit=45^{0}C]. Developmental stage sensitivity to heat stress not clearly defined.

Table 2. Summary of heat stress effects on soybean physiology, growth, and yield components.

2.2 Drought stress and yield

Drought stress (i.e. soil water too low for optimal yield) is recognized as the most damaging abiotic stress for soybean production in the US (Heatherly, 2009). However, only about 8% of the entire hectarage is irrigated. In the main part of the Midwestern US soybean region east of the Mississippi River, little irrigation is done. For example, in Illinois, the nation's largest soybean producing state, most areas receive sufficient rainfall for optimal yield (Cooke, 2009). Soybean water relations are aided by the state's deep soils that allow greater water extraction relative to shallow claypan soils in the Southeastern US. Irrigated areas are concentrated in the drier parts of the soybean growing region (western Midwest or Great Plains states) such as Nebraska where 46% of soybean hectarage is irrigated (Pore, 2009). Irrigation is also common in some Southeastern states where shallow-rooted soils combined with erratic rainfall make drought stress a threat. Currently, about 75% of soybean hectarage in Arkansas is irrigated and the figure for Mississippi is 25-30%. Increased irrigation in the Great Plains and Southeastern states has been stimulated by research showing large yield increases of over 1,000 kg ha^{-1} under irrigated vs. nonirrigated conditions (Specht et al., 1999; Heatherly and Elmore, 1986).

Drought stress is a complicated agronomic problem that is conditioned not only by lack of rain but by evapotranspiration from the soil/plant system, rooting depth and proliferation,

Physiological Disruption	Affected Canopy Level Growth Processes	Affected Yield Components	Temperature Parameters
Reduced metabolic reaction rates	Reduced canopy photosynthesis and CGR	Reduced seed number or seed size depending on timing of stress.	Although 25^{0}C required for optimal canopy photo. and CGR, cold effects under natural growing conditions usually affect yield only <15^0 C.
Chilling stress (Membrane malfunctioning, enzyme inactivation, ion leakage)	Reduced canopy photosynthesis and CGR	Reduced seed number or seed size depending on timing of stress.	0-10/12 ^{0}C; Yield more affected by chilling stress during seed filling than flowering/pod formation period.
Freezing (cell death) tissue damage	Reduced canopy photosynthesis and CGR	Reduced seed number or seed size depending on timing of stress.	0^{0}C Yield more affected by freezing during flowering/pod formation than seed filling.

Table 3. Summary of cold stress effects on soybean physiology, growth, and yield components.

and how much rainfall gets into and stays in the rooting zone (Loomis and Connor, 1992b). Thus, in addition to rainfall, other factors that influence occurrence of drought stress are: tillage systems (conservation vs. conventional tillage), plant genetics (rooting characteristics, stomatal control, leaf reflectance, osmotic adjustments, leaf orientation and size, etc.), climatic factors (relative humidity, temperature, and wind), and soil factors (soil texture and structure, compaction, hardpans, pH, and slope).

Drought stress occurs when loss of water from leaves exceeds that supplied from the roots to such a degree that water potential in those leaves falls to levels resulting in physiological disruptions that eventually reduce CGR and yield (Loomis and Connor, 1992b). Another aspect of drought stress is low water potential in root nodules which reduces nitrogen fixation. Consequently, the crop may become N deficient which can also contribute to reduced CGR and yield (Purcell and Specht, 2004). Although there are many physiological processes potentially affected by drought stress, the main factors which are most important in yield loss are seed germination and seedling establishment, cell expansion, photosynthesis, and nitrogen fixation (Raper and Kramer, 1987). Water entrance and loss from a crop is controlled by water potential, the energy of water measured as a force in bars or pascals (1 bar=0.1 MPa) (Loomis and Connor, 1992b). Water potential differences between components of a system describe the direction of water flow, since water will always flow from a greater to a lesser water potential. Pure water has the highest water potential (0 MPa) and water potential of natural systems will have negative values below that for pure water. In plants, water potential is mainly controlled by solute potential (increased concentration makes water potential lower or more negative) and turgor pressure (positive hydrostatic pressure against the cell wall makes water potential greater or less negative). In soil, solute concentration also affects water potential. However, matric potential (adhesion of water onto soil particles) is also an important component of soil water potential. Water is lost from the leaves by transpiration to the atmosphere. For this water to be replaced, root water potential must be lower than soil water potential to create water inflow from soil to root.

When a soil is initially at field capacity (maximal water a soil will hold after natural drainage), soil water potential is at about -0.02 MPa (Loomis and Connor, 1992b). This corresponds to volumetric water contents (volume of water per volume of soil) of 0.6 and 0.35 for clay and sand soils, respectively. At night, water potentials for soil, roots, and leaves are in equilibrium. During the day, water loss from the leaves depresses leaf water potential below root water potential resulting in movement of water from root to leaves in the xylem. Consequently, root water potential falls below soil water potential resulting in water flowing into the root. As water is withheld from the crop for successive days, the water potential for soil, roots, and leaves steadily drops. When midday leaf water potential falls to -1.5 MPa, stomata will close to conserve water. Meanwhile, as the soil dries the conductance of water from soil to root drops making it difficult to resupply the plant with water. Continued drought past this point will cause leaf water potential to fall below -1.5 MPa resulting in possible death. Eventually, soil water potential may fall to -1.5 MPa at which point water no longer enters the root from the soil (wilting point). Plant available water is defined as the soil water content between field capacity and the wilting point. Irrigation to avoid drought stress is usually recommended when plant available water falls to 50%, a level indicated by a soil water potential of -0.05 to -0.06 MPa for a silt loam or clay soil and -0.04 to -0.05 for a sandy soil. (Univ. of Arkansas Coop. Ext., 2006). This corresponds to a volumetric water content of 0.4 and 0.23 for clay and sand soils, respectively.

Once injurious soil water potential levels are reached, physiological disruptions occur which adversely affect CGR, yield component formation, and yield. Because of the large amount of water the soybean seed must imbibe for successful germination (50% of fresh weight), adequate moisture at planting is an important agronomic problem. Helms et al. (1996) cautioned that stand establishment could be difficult when soil water is sufficient to cause seed imbibition but not germination. Seed planted into a soil having a gravimetric water content (water wgt./soil wgt.) of 0.07 kg kg^{-1} was great enough for imbibition, but too low for root emergence. Increasing water content to 0.09 kg kg^{-1} allowed successful germination and emergence. Drought stress during the seedling emergence and stand establishment period can result in a suboptimal plant population for optimal yield. Because of low plant population, LAI and LI are inadequate to create a CGR that optimizes yield.

Once successful stand establishment is achieved, one of the most sensitive physiological processes to drought stress is reduced cell expansion resulting from decreased turgor pressure (Raper and Kramer, 1987). As leaf water potential falls, cell and leaf expansion are affected before photosynthesis. Bunce (1977) reported a linear relationship between soybean leaf elongation rate and turgor pressure. Decreasing leaf water potential to -0.80 MPa reduced leaf elongation rate by 40% relative to greater values. Consequently, leaf area and plant dry matter were reduced 60% and 65%, respectively. These results were subsequently confirmed in field experiments (Muchow et al., 1986). Thus, occurrence of drought stress during vegetative growth (emergence to R5) can reduce LAI and LI to levels insufficient for optimal CGR and yield. Decreased photosynthetic rate is not initiated until leaf water potential falls into the range of -1.0 to -1.2 MPa (Raper and Kramer, 1987). The rate starts declining more rapidly as water potential falls below -1.2 MPa. Plants suffering this level of drought would have greater reductions of CGR and yield because not only would LAI be reduced, but the net assimilation rate (photosynthetic rate per unit LAI) would also be reduced. Drought stress effects on photosynthesis become irreversible once water potential falls below -1.6 MPa.

Another physiological process sensitive to drought stress is nitrogen fixation (Purcell and Specht, 2004). Decreased nitrogen fixation starts when water potential of root nodules starts falling below -0.2 to -0.4 MPa (Pankhurst and Sprent, 1975). Because of the high protein content of its seed, soybean has a greater demand for nitrogen compared with other crops (Sinclair and de Wit, 1976). Soybean obtains nitrogen from fixation and directly from the soil. During seed filling, much of seed nitrogen demand is met by remobilization from the leaves. The contribution of nitrogen fixation to the plant's nitrogen supply varies inversely with soil nitrogen availability (Harper, 1987). In the Midwestern US which has soils of relatively high residual NO_3, about 25-50% of total plant nitrogen comes from fixation. In contrast, in soils having low nitrogen, fixation can contribute up to 80-94% of the plant's nitrogen. Thus, any stress (drought or other) that restricts nitrogen fixation can result in a nitrogen deficiency (leaf nitrogen falling below 4%, Jones, 1998) which can reduce net assimilation rate and CGR. Ample evidence indicates that nitrogen fixation is more sensitive to drought than photosynthesis, TDM accumulation, transpiration, or soil nitrogen uptake (Purcell and Specht, 2004).

Because of its effects on CGR, drought creates changes in certain growth dynamic and yield component parameters. In general, drought stress during the vegetative growth period (emergence to R5) has adverse effects on LAI, TDM, CGR, and plant height (Scott and Batchelor, 1979; Taylor et al., 1982; Muchow, 1985; Meckel et al., 1984; Desclaux et al., 2000; Pandey et al., 1984; Ramseur et al., 1985; Cox and Jolliff, 1986; Constable and Hearn, 1980; Hoogenboom et al., 1987; Cox and Jolliff, 1987). In a dry growing season, nonirrigated vs. irrigated soybean will begin showing diminished TDM accumulation by the late vegetative or early reproductive period (Scott and Batchelor, 1979). By R3, LAI differences between irrigated vs. drought-stressed soybeans will be obvious (Cox and Jolliff, 1987), with concomitant effects on LI and CGR (Muchow, 1985; Taylor et al., 1982; Ramseur et al., 1985; Pandey et al., 1984). Among vegetative growth indicators of drought stress, reduced internode length and plant height are the most sensitive (Desclaux et al., 2000). The effect of drought stress on plant height is reflected in rooting depth (Mayaki et al., 1976b). During the emergence-R5.5 period, rooting depth is twice the plant height. Thus, occurrence of early-season drought impairs the plant's future potential for obtaining water. If a fortuitous rainfall interrupts this impaired growth dynamic process, TDM levels may return to normal without yield being affected (Hoogenboom et al., 1987). However, continuation of drought will accentuate TDM differences between irrigated and nonirrigated soybean. Decreased TDM and yield are closely correlated in such a condition (Cox and Jolliff, 1986; Meckel et al., 1984). In cases where drought stress occurs during the seed filling period, growth characteristics are of course different. Since plant height and vegetative TDM have already been determined, no effect on these parameters is seen. Drought during seed filling accelerates the senescence process by increasing the rate of chlorophyll and protein degradation. This shortens the seed filling period causing reduced seed size and yield (De Souza et al., 1997).

When soybean faces seasonal drought or drought initiated by R1, yield loss results predominately from reduced pod and seed numbers and seed size is relatively unaffected (Sionit and Kramer, 1977; Ramseur et al., 1984; Pandey et al., 1984; Meckel et al., 1984; Cox and Jolliff, 1986; Constable and Hearn, 1980; Lawn, 1982; Ball et al., 2000). Thus, when confronted with drought stress, soybean reduces seed m^{-2} so that normal seed size can be maintained. Although some have reported mild adverse effects of drought on seed per pod (Ramseur et al., 1984; Pandey et al., 1984), others have shown no effect (Lawn, 1982; Elmore et al., 1988). In contrast, consistent reports have shown pod m^{-2} is reduced by drought during the R1-R6 seed formation period (Sionit and Kramer, 1977; Ramseur et al., 1984; Pandey et al., 1984; Snyder et al., 1982; Neyshabouri and Hatfield, 1986; Cox and Jolliff, 1986; Ball et al., 2000). Based on these results, we conclude that reduced seed m^{-2} from drought stress is derived predominately from reduced pod m^{-2} rather than seed per pod. Because pod per node is not severely affected by drought (Elmore et al., 1988), reduced pod and seed m^{-2} caused by drought results mainly from decreased node m^{-2}, mainly resulting from reduced branch development (Taylor et al., 1982; Snyder et al., 1982; Frederick et al., 2001). In addition to reduced node m^{-2}, drought stress during the flowering period retards early ovary expansion because of reduced photosynthetic supply (Westgate and Peterson, 1993; Liu et al., 2004; Kokubun et al., 2001). The period from 10 days before R1 to 10 days after R1 is the critical period.

Drought stress occurring at the start of linear seed filling (R6) can also reduce seed number, but the main effect of drought initiated at this time or later is on reduced seed size (Sionit and Kramer, 1977; De Souza et al., 1997; Brevedan and Egli, 2003; Doss and Thurlow, 1974). In cases where drought stress is similar at different developmental periods, yield loss is generally twice as great for the R1-R6 vs. R6-R7 periods (Kadhem et al., 1985; Korte et al., 1983b; Shaw and Laing, 1966; Eck et al., 1987; Brown et al., 1985; Hoogenboom et al., 1987; Korte et al., 1983b). Some studies show that within the R1-R6 period, the most drought sensitive phase is R3-R5 (Kadhem et al., 1985; Korte et al., 1983a). This explains why most irrigation studies have identified parts or all of the seed formation period as the most drought prone period (Heatherly and Spurlock, 1993; Elmore et al., 1988; Kadhem et al., 1985; Hoogenboom et al., 1987; Eck et al., 1987; Korte et al., 1983a; Korte et al., 1983b; Brown et al., 1985; Morrison et al., 2006). These studies far outweigh early studies indicating that seed filling had the same or greater sensitivity to drought as the seed formation period (Shaw and Laing, 1966; Snyder et al., 1982; Sionit and Kramer, 1977). Irrigation during the vegetative period has consistently proven unnecessary for alleviating drought stress (Heatherly and Spurlock, 1993; Neyshabouri and Hatfield, 1986). Lack of irrigation response during the vegetative period is likely due to the limited water use during that period (Reicosky and Heatherly, 1990). In conclusion, based on yield component responses, the most drought prone period during soybean development is R1-R6. Drought effects on soybean yield formation are summarized in Table 4.

2.3 Light interception and yield

Because of the importance of canopy photosynthesis and CGR in affecting yield, the level of intercepted photosynthetically active radiation (commonly referred to as light) is one of the most important stresses affecting soybean yield (Loomis and Connor, 1992a). Although a very complicated process, photosynthesis can be simplified by viewing it as three basic parts: 1) Movement of CO_2 from the atmosphere to the chloroplasts; 2) Light reactions in which absorption of specific wavelengths of radiation (red and blue light) cause ionization (photoelectric effect) and result in production of the high-energy compounds ATP and NADPH; and 3) Carbon fixation reactions in which the ATP and NADPH produced in the light reactions is used to fix CO_2 into organic compounds (Fageria et al., 2006). The major environmental factors affecting canopy photosynthetic rate and CGR are atmospheric [CO_2], temperature, water availability, and light level absorbed by the canopy. An understanding of how light affects canopy photosynthesis is critical for analyzing the effect of environmental stress on yield.

Physiological Disruptions	Affected Canopy Level Growth Processes	Affected Yield Components	Drought Parameters
Reduced cell expansion	Reduced LAI and LI. Reduced canopy photosynthesis and CGR.	Reduced seed m^{-2} or seed size depending on timing of stress	Decrease of leaf water potential to -0.80 MPa or less reduces turgor pressure and cell expansion.
Reduced nitrogen fixation	Reduced canopy photosynthesis and CGR.	Reduced seed m^{-2} or seed size depending on timing of stress	Decline starts at -0.2 to -0.4 MPa.
Reduced net assimilation rate	Reduced CGR	Reduced seed m^{-2} or seed size depending on timing of stress.	Water potential below -1.2 MPa
		Seed m^{-2} reduction mainly due to reduced node m^{-2} and pod m^{-2}. Reduced seed size due to reduced effective filling period.	Most drought prone period is the R1-R6 seed formation period. Irrigation recommended when soil at 50% available water. Drought sensitivity of rapid seed filling (R6-R7) is less than half that for R1-R6 period.

Table 4. Summary of drought stress effects on soybean physiology, growth, and yield components.

For soybean, as well as other C3 crop species, photosynthetic rates of individual leaves increase asymptotically to a light intensity of 500 micro moles m^{-2} s^{-1} (or 100 W m^{-2}) (Hay and Porter, 2006); an intensity equivalent to about 25% of full sun in many soybean-growing regions. However, this relationship does not transfer to the canopy level; largely because of uneven shading for leaves in the mid and lower canopy levels which do not receive saturating light intensities. Although top leaves do not increase their photosynthetic rates as light intensity increases above 25% of full sun, mid and lower canopy leaves would receive increased light within the responsive range; thus resulting in an overall increase in canopy photosynthetic rate (Hay and Porter, 2006). In cases of crops having erect leaves with low canopy light extinction coefficients such as ryegrass, canopy photosynthetic rate increases linearly with increasing intensity to the full-sun level (Hay and Porter, 2006). Although soybean canopies having LAI<4.0 [canopy cover (95%) (Shibles and Weber, 1965)] saturate the canopy photosynthetic rate at intensity levels less than full sun, those having LAI >4.0 show continual increase up to full-sun conditions (Shibles et al., 1987). The increased canopy photosynthesis responds to increased light intensity in an asymptotic rather than linear fashion (Jeffers and Shibles, 1969). At any given time, light intercepted by the canopy depends on LAI and the intensity of ambient light. Prior to canopy closure (LAI of 3.0 to 5.0 depending on row spacing), CGR primarily is influenced by LAI (Shibles and Weber, 1965), whereas ambient light level mainly affects CGR after canopy closure. Major research aims have been to determine yield response to reduced LI across different developmental periods; to assess yield losses related to specific reductions in LI; and to determine if different stresses reducing LI (e.g. shade, nonoptimal row spacing, subnormal plant population, and defoliation) affect yield by similar mechanisms. In the current discussion, we will examine the effects of shade, row spacing, plant population, and defoliation on yield.

2.3.1 Light interception and yield: Shade stress

Studies with heavy shade treatment (63%) demonstrated that the flowering/pod formation period (R1 to R6) was more sensitive to reduced LI than the period of linear seed filling (R6 to R7; rapid seed filling period) (Jiang and Egli, 1995; Egli, 1997). Application of shade during the seed determination period reduced yield by 52% (Jiang and Egli, 1995), whereas the same light interception reduction during rapid seed filling reduced yield by only 24% (Egli, 1997). Thus, within the reproductive period, the flowering/pod formation period was twice as sensitive to reduced LI as compared with the rapid seed filling period. Within the flowering (R1-R4) and pod formation (R4-R6) periods, yield responses to shade were similar (Jiang and Egli, 1993, 1995). Yield loss can occur with as few as 9 continuous days of heavy shade (80%) at any time during the flowering/pod formation period (Egli, 2010).

When shade stress is applied continuously across the reproductive period, yield losses occur with as little as 30% shade (22-31% yield loss) (Egli and Yu, 1991). Increasing shade stress to 50% resulted in a 55% yield loss. Yield losses were entirely due to reduced seed number rather than seed size. When faced with a reduced crop growth rate induced by shade stress starting at first flowering, soybean reduces its seed number so that when seed filling commences, seed size is unaffected. In such cases, yield is said to be "source restricted" during flowering and pod formation (i.e. yield reduction occurred due to lower CGR); whereas during seed filling yield was "sink restricted" (i.e. yield reduction occurred due to reduced seed number and was unaffected by changes in CGR). A summary of shade effects on yield is shown in Table 5.

2.3.2 Light interception and yield: Row spacing and plant population

Early studies which altered LI through row spacing and plant population demonstrated that optimizing light during the reproductive period (R1 to R7) was more important than during the vegetative period (emergence to R1) (Brun, 1978; Christy and Porter, 1982; Johnson, 1987; Tanner and Hume, 1978; Shibles and Weber, 1965). More recent studies suggest that reduced LI during the vegetative period can reduce yield if it results in a suboptimal CGR during the subsequent flowering/pod formation period (Board et al., 1992; Board and Harville, 1996). Row spacing and plant population have similar effects on LI, CGR, TDM, and yield component formation as do the aforementioned shade studies. Reducing row spacing from 100 to 50 cm increases LI and accelerates CGR during the vegetative, flowering/pod formation, and seed filling periods (Board et al., 1990). Greater CGR in narrow vs. wide rows was evident as early as 16 days after emergence (Board and Harville, 1996). During most of the vegetative and flowering/pod formation periods, accelerated CGR was due more to increased LAI than to net assimilation rate (Board et al.,

1990b). However, initial increases in CGR in narrow vs. wide rows during the vegetative period were influenced as much by increased net assimilation rate as increased LAI. This probably occurred due to greater interception of light per unit LAI in narrow vs. wide rows at this time (Board and Harville, 1992). Increased yield in narrow vs. wide rows is more evident in short-season soybean production, such as in late vs. normal planting dates or growing early vs. late maturing cultivars (Board et al., 1990a; Boerma and Ashley, 1982; Carter and Boerma, 1979).

Physiological Disruption from Shade Stress	Affected Canopy Level Growth Processes	Affected Yield Components	Shade Parameters
Reduced photosynthetic light reactions	Reduced canopy photosynthesis and CGR	Reduced seed number if shade applied during R1-R6 period. Reduced seed size if shade applied during R6-R7 period.	Most sensitive stress period is R1-R6. Reduced yield occurs (24% yield loss) with as few as 9 d of heavy shade (83%). Shade decreases yield as it decreases CGR < 16 $gm^{-2}d^{-1}$ during R1-R5 period. Moderate shade (30%) during R1-R6 period reduces yield 22-31%. Shade stress during linear seed filling period (R6-R7) has half the effect on yield vs. the R1-R6 period.

Table 5. Summary of shade effects on soybean physiology, growth, yield components, and yield.

Although narrow vs. wide culture enhances CGR at all three developmental periods, yield increases result entirely from increased pod and seed production (Egli and Yu, 1991; Board et al., 1990b, 1992). Seed per pod and seed size, yield components formed during the seed filling period (R5 to R7) were not affected by reduced row spacing. The dominant yield components controlling pod and seed production were node m^{-2} and reproductive node m^{-2}, which are formed during the vegetative period and part of the flowering/pod formation periods (emergence to R5) (Board et al., 1990b; 1992). Thus, greater LI and CGR in narrow vs. wide rows has its beneficial effect on yield between emergence and R6, with the main effect occurring from emergence to R5. In cases where wide rows achieve 95% light interception by first flowering, no yield loss occurs (Board et al., 1990a). Reduced yield in wide vs. narrow rows starts occurring when average LI across the R1 to R5 period is reduced by 14% (Board et al., 1992). In summary, changes in row spacing affected yield by a mechanism very similar to that reported for shade treatments applied throughout the reproductive period (Egli and Yu, 1991); i.e. pod and seed numbers produced during the emergence to R6 period were reduced by the lower CGR so that seed size (produced during the R5 to R7 seed filling period) could remain constant.

Plant population studies conducted under short-season conditions also have outlined a yield-control mechanism very similar to those described for narrow vs. wide row spacing and shade (Ball et al., 2000, 2001; Purcell, 2002). Increasing plant population above the normal recommendation of 25-35 plant m^{-2} increased LI early in the vegetative period [similar to the findings for narrow vs. wide row spacing (Board and Harville, 1996)] resulting in an accelerated CGR during the R1 to R5 period, greater dry matter accumulation, and yield (Ball et al., 2000). Purcell et al. (2002) determined that increased yield responded linearly to increased photosynthetically active radiation accumulated across the emergence to R5 period. Thus, the period during which increased LI benefitted yield in high vs. normal plant population was the same as that described for narrow vs. wide rows (Board et al., 1990a; Board et al., 1992). Yield increases were shown to be caused by increased node m^{-2} and pod m^{-2} (Ball et al., 2001), similar to findings by Board et al. (1990b, 1992) for narrow vs. wide row spacing. Data indicate that subnormal plant populations can achieve yields similar to those of normal populations if average light interception across the R1 to R5 period is 90% (Carpenter and Board, 1997). Yield losses started occurring when average light interception across this period falls 14% below that for full-coverage canopies. This yield response to reduced light interception corresponds very closely to that shown by Board et al. (1992) for wide vs. narrow row spacing. Row spacing and plant population effects on yield, growth and yield components are summarized in Table 6.

2.3.3 Light interception and yield: Defoliation

Several biotic and abiotic stresses such as hail, insect leaf feeders, and diseases affect yield through defoliation. Potential physiological responses to defoliation include effects on canopy photosynthesis, TDM, altered partitioning of TDM to plant parts, leaf abscission, delayed leaf senescence, delayed crop maturity, changes in leaf specific weight, and reduced nitrogen fixation, as well as others (Welter, 1993). Convincing evidence has shown that insect defoliation reduces yield through LI effects on canopy photosynthetic activity and/or CGR. Ingram et al. (1981) infested soybean with velvetbean caterpillar during the reproductive period to study effects on physiological processes and yield. The treatments resulted in a 50% reduction in LAI resulting in LI falling to 83% of the control level during the seed filling period. Corresponding to reduced LI, canopy photosynthetic activity declined to 85% of control and yield was reduced to 86% of control. Yield loss occurred through reduced seed size caused by reduced seed growth rate which was entirely attributed to decreased photosynthetic supply. Similar results were found by Board and Harville (1993) for partial defoliation treatments made to create a LI gradient during the reproductive period. Yield loss occurred only when defoliation was severe enough to reduce LAI below 3.0 and light interception below 95% for extended periods. Although these studies involved manual defoliation, rather than insect defoliation, research has shown that yield responses from either manual or insect defoliation are similar if applied during the same growth period and if leaf removal rates are similar (Higgins, et al., 1983; Turnipseed and Kogan, 1987). The connection between LI and soybean yield response to defoliation has been reinforced by research showing that photosynthetic rates in leaves damaged by defoliation are similar to undamaged controls (Peterson and Higley, 1996). Thus, leaves remaining after defoliation cannot compensate photosynthetically for lost leaf material and the reduced LI directly decreases the photosynthetic rate and yield. Browde et al. (1994) using a combination of defoliating insects, nematodes, and herbicide damage, concluded that light interception was the "unifying explanation for yield losses". Similar conclusions were reached by Board et al. (1997) who reported a linear relationship between yield and LI at the temporal midpoint of the seed filling period.

Previous defoliation studies have indicated that yield response is affected not only by the severity of insect infestation, but also the timing of the attacks. Defoliation during the vegetative period (emergence to first flowering) usually has shown little effect on yield, largely due to leaf regrowth potential at this time. Since defoliation during the vegetative period usually does not have a long-term depressing effect on LI and CGR, little effect on yield has been reported (Haile et al., 1998a,b; Weber, 1955). These results are similar to those of Jiang and Egli (1995) where shade during the vegetative period did not reduce yield if crop growth rate during the R1 to R5 period was unaffected. Fifty percent defoliation between appearance of the first trifoliate leaf and full flowering had little effect on yield (Weber, 1955). Significant yield losses (20%) occurred in this study only when 100% defoliation was applied during this period. Pickle and Caviness (1984) reported no yield loss when soybean received 100% defoliation at the fifth leaf stage.

Greater yield responses to defoliation have been reported during the reproductive period with greatest effect near the start of seed filling (R5). Yield losses from 100% defoliation at R2 were only 25%, but rose sharply as defoliation was delayed to R3, R4, and R5 (see Table 1 for definitions of R stages) (Fehr et al., 1977). Increased defoliation tolerance at early reproductive stages (near first flower) were later determined to be caused by rapid leaf regrowth (Haile et al., 1998a,b). Greatest yield loss (75-88%) occurred at R5 (Fehr et al., 1977), a finding substantiated by later studies (Fehr et al., 1981; Gazzoni and Moscardi, 1998; Goli and Weaver, 1986). Delay of total defoliation to R6.6 resulted in only a 20% yield loss (Board et al., 1994), supporting the view of greater tolerance to defoliation as seed filling progresses.

Partial defoliation treatments initiated at R1 and terminated at R3, R4, R5, and R6.5 resulted in significant yield loss (approximately 15%) when average LI during the R1 to R5 period was reduced at least by 17-20% (Board and Harville, 1993; Board and Tan, 1995). Although these partial defoliations resulted in decreased LI during seed filling, yield losses were almost entirely due to reduced pod m^{-2} and seed m^{-2} rather than seed size. These results are similar to shade responses shown by Egli and Yu (1991). In summary, shade, defoliation, wide row spacing, and subnormal plant population affect yield through reduced CGR during all or part of the period between emergence and R6. Stresses that operate during the entire emergence to R6 period (wide row spacing, subnormal plant population), cause these reductions to pod and seed numbers through lower production of node m^{-2} and reproductive node m^{-2}. However, in cases where CGR is reduced only during the flowering/pod formation period (e.g. defoliation stress initiated at R1), lower pod and seed numbers can also be affected by decreased pod per reproductive node.

As defoliation is delayed past the start of initial seed filling (R5), yield losses attenuate and yield components causing the yield loss change. By the time seed number is determined and soybean starts rapid seed filling (R6), yield losses from 100% defoliation are half that compared with 100% defoliation at R5 (Goli and Weaver, 1986). Thus, similar to findings with shade stress (Egli and Yu, 1991), yield was twice as sensitive to defoliation stress during the flowering/pod formation period compared with the rapid seed filling period. Defoliation during seed filling affects yield mainly through reduced seed size, although seed number is also affected if defoliation occurred at or before R6 (Board et al., 2010). Every 0.1 unit delay in developmental stage from R5 to R7 (e.g. 5.4 to 5.5 or 6.2 to 6.3) resulted in a 5% reduction in yield loss caused by 100% defoliation. Throughout early and mid seed filling (R5 to R6.2), defoliation had to be sufficient to reduce light interception by about 20% to decrease yield (Board et al., 2010; Ingram et al., 1981; Board et al., 1997). Once soybean

Physiological Disruptions from Wide vs. Narrow Row Spacing	Affected Canopy Level Growth Processes	Affected Yield Components	Row Spacing Parameters
Reduced LAI and LI efficiency results in lower canopy LI.	Growing at nonoptimal wide row spacing reduces canopy photosynthesis and CGR during emergence-R6 period.	Reduced node and reproductive node numbers, pods and seeds.	Sensitive stress period is emergence to R6. Wide vs. narrow rows reduces yield whenever LI falls enough to reduce average CGR (R1-R5) below 15 gm^{-2} d^{-1}. Seed filling period is unaffected by LI in wide vs narrow rows .
Physiological Disruption From Subnormal Plant Population	**Affected Canopy Level Growth Processes**	**Affected Yield Components**	**Plant Population Parameters**
Reduced LAI results in lower canopy LI	Reduced canopy photosynthesis and CGR during emergence-R6 Period	Reduced node and reproductive node numbers, pods and seeds.	Sensitive period is emergence-R6. Yield losses occur when average CGR (R1-R5) falls below 15 g m^2 d^{-1}. Seed filling period is unaffected.

Table 6. Summary of row spacing and plant population effects on soybean physiology, growth, yield components, and yield.

passes into the last half of the seed filling period, defoliation must be at or close to 100% (resulting in a 50% relative LI reduction) to cause yield loss (Board et al., 2010; Board et al., 1997). The effects of defoliation stress on yield formation are summarized in Table 7.

Physiological Disruption from Defoliation	Affected Canopy Level Growth Processes	Affected Yield Components	Defoliation Parameters
Reduced LAI and canopy LI	Reduced canopy photosynthesis and CGR.	Defolilation during R1-R6 period reduces node and reproductive node numbers, pod per reproductive node, pods and seed. Defoliation during R6-R7 reduces seed size.	Vegetative period is not sensitive to defoliation stress unless at 100% level. Period most sensitive to defoliation stress is R1-R6.2. Significant yield losses start occurring when light interception across this period falls 17-20% and CGR falls below 15 g m^{-2} d^{-1}. During R6.2-R7 period must have total defoliation to get significant yield loss; i.e. 50% reduction in relative LI. Thus, yield is half as sensitive to defoliation during R6.2- R7 as during R1-R6.2.

Table 7. Summary of defoliation effects on soybean physiology, growth, yield components, and yield.

3. A general mechanism for explaining stress effects on yield

Our discussion on temperature, drought, and light interception has outlined a paradigm of how these factors cause yield loss (Fig. 4). Despite differences in initial physiological disruptions, environmental stress first affected canopy photosynthesis and CGR. Coupled with length of the emergence to R5 period (related to maturity group), these growth dynamic rates influence TDM(R5), the dry matter level at which vegetative TDM, node m^{-2}, reproductive node m^{-2}, and pod per reproductive node are maximized (Board and Harville, 1993; Board and Tan, 1995) (Fig. 3). These yield components, in turn, regulate pod m^{-2} and seed m^{-2} which mediate stress effects on yield. Thus, TDM(R5) serves as a benchmark indicator for yield potential. Because yield component production per unit dry matter (yield component production efficiency) differs with environmental (Board and Maricherla, 2008) and genotypic factors (Kahlon and Board, 2011), final yield component number is also affected by this factor (Fig.4). Seed size usually plays a much smaller role in explaining environmental influences on yield. Support for this paradigm can be seen in data for a single cultivar grown across a wide environmental range (Fig. 5). Yield is highly correlated with seed m^{-2} (R^2=0.83), but shows no relationship with seed size. Because of the paramount importance of nodes, pods, and seeds in regulating environmental effects on yield, the most stress-prone period is between emergence and R5, the predominant period in which these yield components are formed.

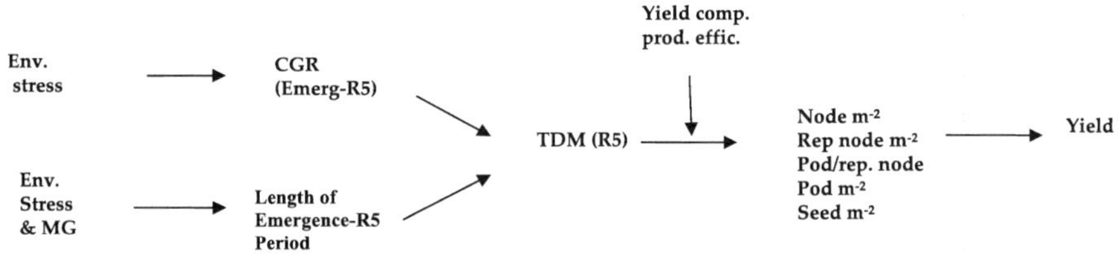

Fig. 4. Paradigm for explaining how environmental stresses affect growth, yield components, and yield. MG=Maturity Group.

We acknowledge that some abiotic and biotic stresses do not follow the paradigm outlined above. Stresses that directly impair reproductive structures (i.e. flowers, pods, and seeds) without acting through CGR fall into this category. Examples are the southern green stink bug [*Nezara viridula* (L.)] which punctures the soybean seed; temperatures that are sufficiently hot or cold during or near to fertilization to disrupt pod development (Salem et al., 2007; Koti et al., 2004); and diseases such as pod and stem blight [*Diaporthe phaeseolorum* (var. sojae)] which enter pods through abrasions, cracks, or other injuries (Athow and Laviolette, 1973). Although these exceptions exist, analyses of many environmental stresses indicates that the mode of action for yield reduction at the canopy level is similar to that described for temperature extremes, drought, and reduced light interception; and that such stresses affect yield through the paradigm explained in Fig. 4. Although it is impossible to cover all the possible abiotic and biotic environmental stresses affecting soybean in a single chapter, a few of them will be described.

Nitrogen deficiency is a common limiting factor for soybean yield (Tolley-Henry and Raper, 1986). Optimal growth and yield of soybeans, as well as other crops, requires a greater input of N than any other nutrient. Soybean obtains its N either directly from the soil or from symbiotic N_2 fixation by the bacteria *Bradyrhizobium japonicum*. Deficiency symptoms

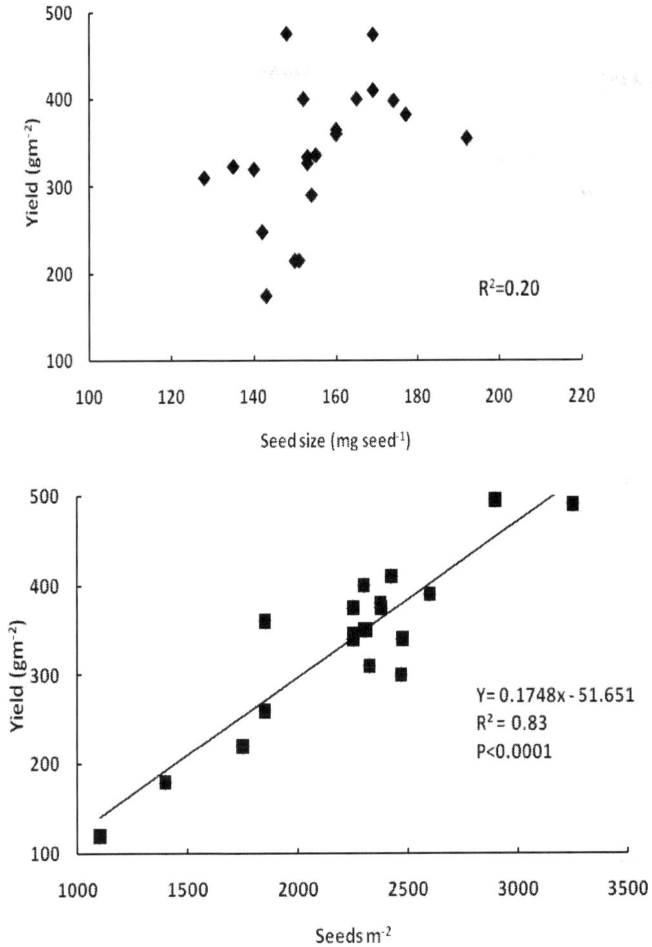

Fig. 5. The relationship of soybean yield with seed m^{-2} and seed size for cultivar Iroquois grown across 21 locations in the Midwestern US, 1996 (USDA, Unpublished data).

are manifested as decreased photosynthetic rate, reduced initiation and expansion of leaves, and lower growth rates for stems and roots (Tolley-Henry and Raper, 1986). Approximately 50% of soybean's leaf N is in rubisco, the enzyme involved in CO_2 carboxylation onto Ribulose di Phosphate (Sinclair, 2004). This enzyme is recognized as the rate-limiting step for photosynthesis. Thus, when N becomes deficient, the entire photosynthetic cycle declines, as evidenced by high correlation of leaf photosynthetic rate with [N] (Tolley-Henry et al., 1992) and soluble protein (Ford and Shibles, 1988; Sung and Chen, 1989). Thus, any stress which adversely affects N_2 fixation (e.g. inadequate inoculation, low pH soils, drought, etc.) can create N deficiency and yield loss in soybean. When a N deficiency results in leaf [N] falling below 5%, photosynthetic rate starts declining (Tolley-Henry et al., 1992).

Unavailability of N for 10 or more d results in cessation of dry matter accumulation (Tolley-Henry and Raper, 1986). Associated with this, leaf initiation and expansion stops. Consequently, LAI, LI, and CGR are greatly reduced during the emergence-R5 period, resulting in decreased seed m^{-2} and yield (Koutroubas et al., 1998). Thus, on the canopy level, N deficiency affects yield in a manner similar to that shown for temperature extremes, drought, and deficient light interception.

Several biotic stresses of soybean show a similar mechanism of yield loss. Among biotic stresses, farmers in the Southeastern US spend the greatest amount of money for weed control. Weeds reduce yield through competition with soybeans for water, light, and nutrients (Hoeft et al., 2000). Depending on weed species, weed population, and environmental conditions, there is a "critical period" early in soybean development when weeds must be controlled to maintain yield (Hoeft et al., 2000). Failure to control weeds in the critical period results in reduced soybean vegetative TDM(R5) and yield (Hagood et al., 1980, 1981). As with drought, reduced light interception and N deficiency, yield loss occurred through reduced pod and seed numbers.

4. Development of yield-loss prediction tools for diagnosing environmental stress problems

A major barrier to improved yield is correct identification of environmental stresses causing yield losses. During any given growing season, a soybean crop can be faced with a series of potential yield-limiting stresses. For example, an early-season drought stress may have slowed CGR during the vegetative period. This might be followed by a waterlogging stress during the flowering/pod formation period (R1-R6) which left standing water on the field for 2-3 d (sufficient to slow CGR, Scott et al., 1989). Finally, a late-season attack of defoliating insects during rapid seed filling (R6-R7) may have decreased LAI enough to cause significant yield loss. Correct identification of which factor(s) caused the yield loss aids in devising remedial strategies to improve yield. If the entire yield loss was due to early-season drought stress, then the farmer may consider irrigation when a similar future stress occurs. On the other hand, if the early-season drought stress was shown not to play a role in yield loss, the farmer would know that his crop could tolerate such drought periods without suffering yield loss. If waterlogging was identified as the causative factor of yield loss, then the farmer may wish to consider planting on raised beds or sloping the field in a given direction so that water runs off the field rather than ponding. If the yield loss was caused by the late-season insect defoliation, the farmer should consider more vigilant monitoring and control of whatever pest was infesting the field.

Using the paradigm outlined in Fig. 4 for explaining environmental stress effects on yield, yield-loss prediction tools can be identified which aid farmers in making decisions such as those described above. Because CGR during the emergence to R5 period plays a critical role in stress effects, TDM levels at developmental stages that are easily identifiable could be used as putative yield-loss prediction tools. Since vegetative growth ends near R5 (Egli and Leggett, 1973), TDM(R5) serves as an integrative measure of growing conditions during the emergence to R5 period. Total dry matter (R5) also has value in predicting yield (Fig. 6). The R5 stage is easy to identify by the appearance of fully-elongated pods at the top four main stem nodes. Total dry matter at R1 (also an easily identifiable developmental stage) could be used to indicate growing conditions at an intermediate stage of the vegetative growth period. Based on an analyses of studies conducted across 1987-1996 near Baton Rouge, LA involving a wide range of environmental conditions (years, planting dates, row spacings, plant populations, and waterlogging stress) achievement of optimal yield was shown to be associated with a TDM(R1) level of 200 g m^{-2} and a TDM(R5) level of 600 g m^{-2} (Fig. 6) (Board and Modali, 2005). Dry matter levels below these resulted in a curvilinear decline in yield, while increases above this level gave only small insignificant yield increases. Yield components identified as important for yield formation (seed m^{-2}, pod m^{-2}, reproductive node m^{-2}, and node m^{-2}) demonstrated similar curvilinear responses to TDM(R1) and TDM(R5) as did yield.

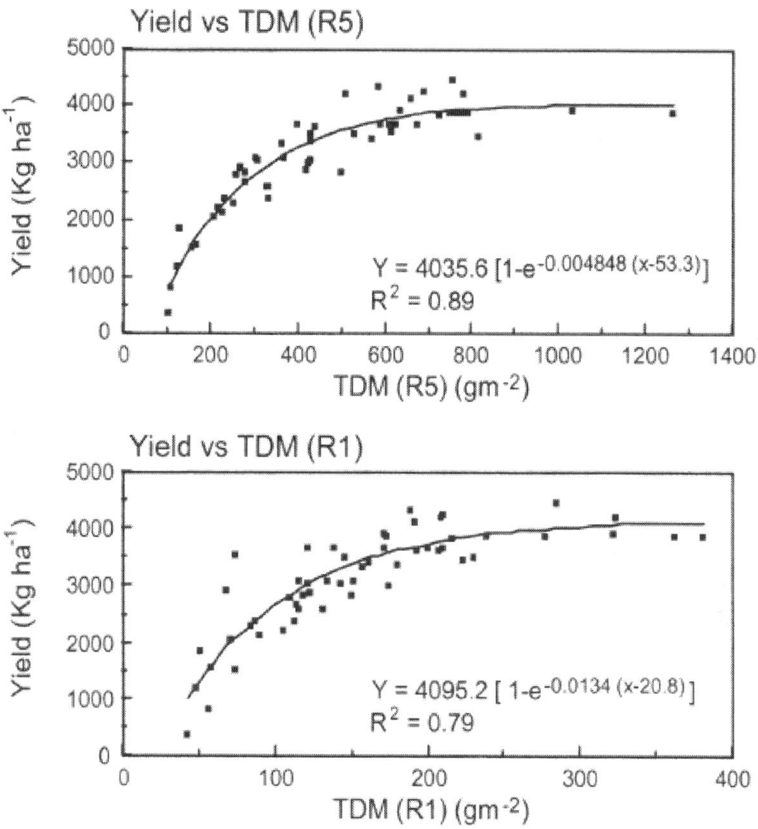

Fig. 6. Yield response to total dry matter at R1 [TDM (R1)] and total dry matter at R5 [TDM (R5)] for soybean grown across a range of environment conditions near Baton Rouge, LA, 1987 through 1996.

Use of TDM(R1) and TDM(R5) as yield-loss prediction tools can be illustrated by analyzing the aforementioned case of decreased yield resulting from three possible stresses across the growing season: drought during the vegetative period, waterlogging during the flowering/pod formation period, and insect defoliation during rapid seed filling.

Determination that TDM(R1) was optimal (200 g m^{-2} or greater), but TDM(R5) was suboptimal (<600 g m^{-2}) would indicate that the waterlogging stress contributed to the yield loss, but the drought stress did not. Such a result would be manifested in a reduction in seed m^{-2}. If TDM levels at R1 and R5 were both optimal, then the yield loss probably resulted from the insect defoliation. This would be reflected by a reduction in seed size, but no reduction in seed m^{-2}. Seed size can be determined from a field by random sampling of 100-seed samples. Seed m^{-2} can then be easily calculated by dividing seed size into seed yield (as dry matter). Thus, a knowledge of when stresses occur, developmental stage timing, TDM(R1) and TDM(R5), and seed size and seed m^{-2} data, greatly aid in diagnosing yield-limiting stresses.

Because of the large size of many commercial soybean farms, it is not practical to determine TDM(R1) and TDM(R5) by conventional sampling methods. However, simple regression methods have been developed that allow easy, rapid, and accurate determination of these parameters. Total dry matter (R5) can be predicted from a multiple regression equation using canopy closure (CC) date (achievement of 95% light interception) and days to R5 (R5days) [TDM(R5)= -20.1-(5.9 x CC)+(13.7 x R5days)] (R^2=0.81). The regression model was verified using independent data (R^2=0.90). Both canopy closure date and days to R5 are parameters that can easily, rapidly, and accurately be determined in commercial soybean fields.

Because of the relationship between TDM, LAI, and LI (Loomis and Connor, 1992a), TDM(R1) can be predicted from LI. Determination of light interception under field conditions can now be done rapidly and accurately for commercial soybean farms using digital photographic methods developed by Purcell (2000). When grown in narrow-row culture (50 cm or less), a light interception of 92% at R1 is associated with a dry matter level of 200 g m^{-2} (Board et al., 1992; Board and Harville, 1996). In the case of wide rows (75-100 cm), light interception of 68% at R1 is associated with a dry matter of 200 g m^{-2}. The greater light interception value for narrow rows occurs because LAI in narrow rows intercepts more light per unit LAI (Board and Harville, 1992). In conclusion, TDM(R1) and TDM(R5) are robust yield-loss prediction tools that can be used in conjunction with seed m^{-2} and seed size data to efficiently analyze environmental stress problems in soybean.

5. Genetic strategies for yield improvement

Across a 60-year period, cultivar development efforts by soybean breeders have resulted in a 21-31 kg $ha^{-1}yr^{-1}$ increase in soybean yield (Wilcox, 2001). Selection for yield during this process has been done through empirical yield trials across a range of different environments (Fehr, 1987; Frederick and Hesketh, 1994). Desirable lines are selected as future cultivars based on high and stable yields across years and locations. Thus, factors responsible for this yield improvement have not been clearly identified. In an effort to identify indirect yield criteria for streamlining cultivar development, scientists have endeavored to determine the pertinent factors related to genetically-induced yield enhancement in the cultivar development process.

Several studies have sought to explain yield improvement in the cultivar development process through greater production of specific yield components. However, results have been mixed. Boerma (1979) reported that yield improvement was attributed to greater pod production, although this was apparent only in maturity group VIII cultivars, and not in maturity group VI and VII. Frederick et al. (1991) also demonstrated that increased yield in new compared with old cultivars was related to increased pod number. In contrast, Specht and Williams (1984) demonstrated a small increase in seed size averaging 0.1 g/year. Other research indicated that the relative importance of seed number and seed size in explaining greater yield in the cultivar development process may depend on cultivar comparisons being made. Gay et al. (1980) demonstrated that within indeterminate maturity group III cultivars, the newer cultivar Williams yielded more than the older cultivar Lincoln because of greater seed size. On the other hand, in comparing determinate maturity group V cultivars, the newer cultivar Essex yielded more than the older cultivar Dorman because of greater seed number. More recent studies comparing old and new Midwestern cultivars clearly indicated that yield improvement was more strongly related to seed m^{-2} than seed size (De Bruin and Pedersen, 2009). The authors also stated that greater seed m^{-2} appeared to be related to greater seed per pod, although other yield components were not examined. Comprehensive research from China involving determinate and indeterminate soybeans in four areas of the country showed that greater yield occurred through differential increases of pods per plant, seed per pod, and seed size (Cui and Yu, 2005). Based on the diversity of results from different researchers, countries, and germplasms, it appears that yield improvement with cultivar development can occur through different yield component mechanisms. However, for the Southeastern and Midwestern US soybean-growing regions, most studies conclude that cultivar yield improvement in new vs. old cultivars has been more controlled by changes in seed m^{-2} than seed size. Recent studies involving Southeastern US cultivars indicated that genetic differences in new vs. old cultivars were sequentially controlled by node m^{-2}, reproductive node m^{-2}, pod m^{-2}, and seed m^{-2} (Kahlon et al., 2011).

Because of its importance in crop production, researchers have also tried to determine if leaf photosynthetic rate plays a role in explaining yield improvement during cultivar development. This objective has been studied by comparing carbon exchange rates (CER) per unit leaf area in new vs. old cultivars and also between parents and progeny in a breeding program. Results have been mixed. Early studies by Larson et al. (1981) involving cultivars released between 1927 to 1973 found no correlation between yield and leaf photosynthetic rate. Gay et al. (1980) also found little change in CER between two new and two old cultivars. Similar results were reported by Frederick et al. (1989). In contrast, Dornhoff and Shibles (1970) compared 20 cultivars released across time and demonstrated a general trend between CER and yield, although exceptions occurred. More recent studies by Morrison et al. (2000) with new and old Canadian cultivars did report a 0.52 % per yr increase in the photosynthetic rate, a level very similar to the annual yield increase shown by these cultivars. However, an inverse relation of photosynthetic rate per leaf with LAI may have negated some of the positive effect of increased photosynthetic rate. The increase in photosynthetic rate was related to an increase in stomatal conductance.

Results of studies looking at CER in progeny of a breeding program have also been mixed. Buttery and Buzzell (1972) determined that over 60% of cultivars developed from breeding programs had CER greater than their parent cultivars. Ojima (1972) also was successful in demonstrating increased CER in early progeny lines vs. parental cultivars. However, other research has not demonstrated positive results. Wiebold et al. (1981) crossed two parental cultivars with contrasting high and low CER and could not find improved CER by the F_3 and F_4 generations. Ford et al. (1983) found similar disappointing results. The current general consensus is that using CER as an indirect selection criterion in a breeding program has limited value (Frederick and Hesketh, 1994).

Measurement of photosynthesis on the canopy level (canopy apparent photosynthesis, CAP) has shown greater association with final yield compared with CER (Harrison et al., 1981;

Wells et al., 1982). However, the degree of correlation was not high (r=0.5). Using cultivars and plant introductions differing in CAP and seed filling period, Boerma and Ashley (1988) showed positive partial correlations of yield with CAP (averaged during the reproductive period) (r=0.63) and seed filling period (r=0.54). The product of CAP x seed filling period was even more closely related to yield (r=0.78). However, the inherent difficulties involved in measuring CAP (variable light and temperature conditions; tedious equipment set-up) preclude its use as an indirect selection tool in a breeding program.

The roles of TDM accumulation and harvest index in explaining yield improvement during cultivar development have also shown mixed results. Salado-Navarro et al. (1993) examined 18 Southeastern cultivars released from 1945 to 1982, but found no relationships between improved yield with either TDM or harvest index. Gay et al. (1980) explained yield differences between new and old cultivars as governed more by increased harvest index rather than TDM accumulation. More recent studies involving new vs. old cultivars in Canada (Morrison et al., 1999) and Japan (Shiraiwa and Hashikawa, 1995) have also supported the importance of harvest index for explaining greater yield. In the case of the Canadian study, no differences in TDM were shown between new and old cultivars. These results are supported by Chinese studies which reported a greater role for harvest index vs. TDM accumulation for explaining yield improvement in cultivar development programs (Cui and Yu, 2005).

In contrast, Frederick et al. (1991) (US cultivars) reported little role for harvest index in explaining genetic improvement in soybean and attributed greater importance to TDM accumulation. Cregan and Yaklich (1986) reported similar findings. These results were supported by Kumudini et al. (2001) who showed that TDM accumulation contributed 78% to greater yield in new vs. old cultivars, whereas harvest index contributed only 22%. Greater TDM accumulation occurred entirely during the seed filling period and was supported by the longer leaf area duration (leaf area index integrated over time) for the new cultivars. De Bruin and Pedersen (2009) supported Kumudini's findings and attributed yield enhancement in new vs. old Midwestern cultivars as entirely due to dry matter and not harvest index. However, this more recent study differed from Kumudini in concluding that the greater dry matter accumulation was partly due to greater crop growth rate (R1-R5.5) prior to seed filling.

6. Summary and conclusion

Because of soybean's importance in meeting world food needs, increased demand for agricultural commodities fueled by global economic development, and the limited potential for expansion of arable land, it is imperative that strategies be developed for coping with the effects of environmental stress on crop yields. Accurate identification and correction for environmental stress problems potentially can increase yield from 25-66%, with increases being greater in the developing compared with developed world. Environmental stresses can be divided into either abiotic stresses (atmospheric and soil factors) or biotic stresses (pest problems). Because such a high proportion of crop dry matter is derived from either current or previous photosynthesis, the vast majority of environmental stresses affect yield through the canopy photosynthetic rate and CGR. The majority of soybean research has conclusively demonstrated that environmental stress affects yield through control of seed m^{-2}, which, in turn, is controlled by sequential formation and growth of node m^{-2}, reproductive node m^{-2}, and pod m^{-2}. Since formation of these yield components occurs across the emergence to R6 period, this is the period where stresses depressing crop growth rate have their greatest effect on yield. Although yield is less sensitive to stress during the rapid seed filling period (R6-R7), stresses during this period can also reduce yield if sufficiently severe. Correct advice to soybean farmers concerning correction of environmental stresses depends on accurate identification of which potentially damaging biotic and abiotic factors occurring in any growing season significantly reduce yield (i.e. act as stresses). Development of TDM levels at R1 and R5 as yield-loss prediction tools facilitates this process. Both developmental stages are easy to identify and yield has shown robust asymptotic relationships of TDM(R1) and TDM(R5) with yield reaching plateau levels at 200 g m^{-2} TDM(R1) and 600 g m^{-2} TDM(R5). Accurate and rapid regression methods were outlined for indirect calculation of these parameters. Thus a farmer having knowledge of TDM(R1) and TDM(R5), the timing of potential stress events, and knowledge of seed m^{-2} and seed size, would be able to identify which potential stresses actually cause yield loss.

On the genetic level, the majority of yield formation studies indicate that seed m^{-2} plays a larger role in yield improvement than seed size. However, exceptions to this exist and it must be realized that alternative mechanisms of yield improvement are possible between different germplasm pools and geographic regions. Although little research has been done beyond the primary yield component level, studies that have been conducted indicate that genetic influences on seed m^{-2} are mediated by node m^{-2}, reproductive node m^{-2}, and pod m^{-2}. Genetic studies involving old and new soybean cultivars indicate that both TDM accumulation and harvest index play roles in explaining yield improvement. However, the evidence is so conflicting at this point in time that definitive statements are not possible. Although much research has been done on the subject, there is little evidence to suggest that improved yield has resulted from improved photosynthetic rate per unit leaf area, canopy photosynthesis, or CGR.

7. Acknowledgements

The authors wish to acknowledge financial support from the Louisiana Soybean Promotion Board, the USDA Risk Management Agency, the southern Soybean Reseach Program, and the Kentucky Soybean Board.

8. References

Athow, K.L., & Laviolette, F.A. (1973). Pod protection effects on soybean seed germination and infection with Diaporthe Phaeseolorum var. sojae and other microorganisms. *Phytopathology* Vol., 63 pp. 1021-1023.

Baker J.T, Allen, L.H.J. & Boote, K.J. (1989). Response of soybean to air temperature and carbon dioxide concentration. *Crop Sci.*, Vol. 29, pp. 98-105.

Ball, R.A., Purcell, L.C., & Vories, E.D. (2000). Short-season soybean yield compensation in response to population and water regime. *Crop Sci.*, Vol. 40, pp. 1070 - 1078

Ball, R.A., Mcnew, R.W., Vories, E.D., Keisling, T.C. & Purcell, L.C. (2001). Path analysis of population density effects on short season soybean yield. *Agron. J.*, Vol. 93, pp. 187-195.

Board, J.E., Harville, B.G. & Saxton, A.M. (1990a). Narrow-row seed-yield enhancement in determinate soybean. *Agron. J.*, Vol. 82, pp. 64–68.

Board, J.E., Harville, B.G., & Saxton, A.M. (1990b). Growth dynamics of determinate soybean in narrow and wide rows at late planting dates. *Field Crops Res.* Vol. 25, pp. 203-213.

Board, J.E., Harville, B.G., and Saxton, A.M. (1990c). Branch dry weight in relation to yield increase in narrow row soybean. *Agron. J.* Vol. 82, pp. 540-544.

Board, J.E & Harville, B. G. (1992). Explanations for greater light interception in narrow vs wide row soybean. *Crop Sci.*, Vol. 32, pp. 198-202.

Board, J.E., Kamal, M. & Harville, B.G. (1992). Temporal importance of greater light interception to increased yield in narrow-row soybean. *Agron. J.*, Vol. 84, pp. 575-579.

Board, J.E. and Harville, B.G. (1993). Soybean yield component responses to a light interception gradient during the reproductive period. *Crop Sci.*, Vol. 33, pp. 772-777.

Board, J.E., Wier, A.T. & Boethel, D.J. (1994). Soybean yield reductions caused by defoliation during mid to late seed filling. *Agron. J.*, Vol. 86, pp. 1074-1079.

Board, J.E. & Tan, Q. (1995). Assimilatory capacity effects on soybean yield components and pod formation. *Crop Sci.*, Vol. 35, pp. 846–851.

Board, J.E., Zhang, W. & Harville, B.G. (1996). Yield rankings for soybean cultivars grown in narrow and wide rows with late planting dates. *Agron. J.*, Vol. 88, pp. 240–245.

Board J.E., & Harville B.G. (1996). Growth dynamics during the vegetative period affects yield of narrow-row late planted soybean. *Agron. J.*, Vol. 88, pp. 567-572.

Board, J.E., Wier, A.T. & Boethel, D.J. (1997). Critical light interception during seed filling for insecticide application and optimum soybean grain yield. *Agron. J.*, Vol. 89, pp. 369-374.

Board, J.E. & Modali, H. (2005). Dry matter accumulation predictors for optimal yield in soybean. *Crop Sci.*, Vol. 45, pp. 1790-1799.

Board, J.E. & Maricherla, D. (2008). Explanations for decreased harvest index with increased yield in soybean. *Crop Sci.* Vol. 48, pp. 1995-2002.

Board, J.E., Kumudini, S., Omielan, J., Prior, E. & Kahlon, C. S. (2010). Yield response of soybean to partial and total defoliation during the seed-filling period. *Crop Sci.*, Vol. 50, pp. 703-712.

Boerma, H.R. (1979). Comparison of past and recently developed soybean cultivars in maturity groups VI, VII, and VIII. *Crop Sci.* Vol. 19, pp. 611-613.

Boerma, J.R., and Ashley, D.A.. (1982). Irrigation, row spacing, and genotype effects on late and ultra-late planted soybeans. *Agron. J.*, Vol. 74, pp. 995-999.

Boerma, H.R. & Ashley, D.A. (1988). Canopy photosynthesis and seed-fill duration in recently developed soybean cultivars and selected plant introductions. *Crop Sci.*, Vol. 28, pp. 137–140.

Boote, K.I., Jones, J.W. & Hoogenboom, G. (1998). Simulation of crop growth: CROPGRO model. In *Agricultural systems modeling and simulation,* Peart, R.M. & Curry, R.B. (Eds.),pp. 651-692, Marcel Dekker, New York.

Bramlage, W.J., Leopold, A.C. & Parrish, D.J. (1978). Chilling stress to soybeans during imbibitions. *Plant Physiol.*, Vol.6, pp. 525-529.

Brevedan, R.E & Egli, D.B. (2003). Short periods of water stress during seed filling, leaf senescence, and yield of Soybean. *Crop Sci.*, Vol. 43, pp. 2083-2088.

Browde, J A., Pedigo, L. P., Owen, M. D. K.& Tylka, G. L. (1994). Soybean yield and pest management as influenced by nematodes, herbicides, and defoliating insects. *Agron. J.*, Vol. 86, pp. 601-608.

Brown, E.A., Caviness, C. E. & Brown, D. A. (1985). Response of selected soybean cultivars to soil moisture deficit. *Agron. J.*, Vol. 77, pp. 274-278.

Brun, W.A. (1978). Assimilation. In: *Soybean: Physiology, Agronomy and Utilization*, Norman, A.G. (Ed.), pp 45–73, Academic Press, New York.

Bunce, J. A. (1977). Leaf elongation in relation to leaf water potential in soybean. *Exp. Bot.*, Vol. 28, pp. 158-63.

Buttery, B.R. & Buzzell, R.I. (1972). Some differences between soybean cultivars observed by growth analysis. *Can. J. Plant Sci.*, Vol. 52, pp.13-20.

Campbell, W.J., Allen, L.H.J. & Bowes, G. (1990). Response of soybean canopy photosynthesis to CO_2 concentration, light, and temperature. *J. Exp. Bot.*, Vol. 41, pp. 427–433.

Carpenter, A.C. & Board, J.E. (1997). Growth dynamic factors controlling soybean yield stability across plant populations. *Crop Sci.* 37, pp. 1520-1526.

Carter, T.E. & Boerma, H.R. (1979). Implicaitons of genotype X planting date and row spacing interactions in double-cropped soybean cultivar development. *Crop Sci.*, Vol. 19, pp. 607-610.

Christy, A.L. & Porter, C.A. (1982). Canopy photosynthesis and yield in soybean In: *Photosynthesis: Vol. II, Development, carbon metabolism, and plant productivity*, Govindjee (Ed.), pp. 499-511, Academic Press, New York.

Constable, G. A. & Hearn, A. B. (1980). Irrigation for crops in a sub-humid environment. I. The effect of irrigation on the growth and yield of soybean. *Irrigation Sci.*, Vol. 2, pp. 1-12.

Cooke, R. (2009). Water Management. In: *Illinois Agronomy Handbook.*, Nafziger, E. (Ed)., pp. 143-152, University of Illinois 24th Edition.

Cox, W.J. & Jolliff, G.D. (1986). Growth and yield of sunflower and soybean under soil water deficits. *Agron. J.*, Vol. 78, pp..226-230.

Cox, W.J. & Jolliff, G.D. (1987). Crop-water relations of sunflower and soybean under irrigated and dryland conditions. *Crop Sci.*, Vol. 27, pp. 553-557.

Cregan, P.B. & Yaklich, R.W. (1986). Dry matter and nitrogen accumulation and partitioning in selected soybean genotypes of different derivation. *Theor. Appl. Genet*, Vol. 72, pp. 782-786.

Cui, Y.S. & Yu., D.Y. (2005). Estimates of relative contribution of biomass, harvest index, and yield components to soybean yield improvements in China. *Plant Breed.*, Vol. 5, pp. 473-476.

De Bruin, J.L. & Pedersen, P.(2009). Growth, yield, and yield component changes among old and new soybean cultivars. *Agron. J.*, Vol. 101, pp.123–130.

De Souza, P.I., Egli, D. B. & Bruening, W.P. (1997) Water stress during seed filling and leaf senescence in soybean. *Agron . J.*, Vol. 89, pp. 807-812.

Desclaux, D., Huynh, T.T. & Roumet, P. (2000). Identification of soybean plant characteristics that indicate the timing of drought stress. *Crop Sci.*, Vol 40, pp. 716 - 722.

Dornhoff, G.M., & Shibles, R.M. (1970). Varietal differences in net photosynthesis of soybean leaves. *Crop Sci.*, Vol. 10, pp. 42-45.

Doss, B. D. & Thurlow, D. L. (1974). Irrigation, row width, and plant population in relation to growth characteristics of two soybean carieties. *Agron. J.*, Vol. 66, pp. 620-623.

Eck, H. V., Mathers, A. C. & Musick, J.T. (1987). Plant water stress at various growth stages and growth and yield of soybeans. *Field Crops Res.*, Vol. 17, pp.1-16.

Egli, D.B. & Leggett, J.E. (1973). Dry matter accumulation patterns in determinate and indeterminate soybeans. *Crop Sci.*, Vol. 13, pp. 220-222.

Egli, D.B. & Yu, Z.W. (1991). Crop growth rate and seeds per unit area in soybean. *Crop Sci.* Vol. 31, pp. 439–442.

Egli, D.B., & Crafts-Brandner, S.J. (1996). Soybean. In *Photoassimilate distribution in plants and crops. Source-sink relationships,* Zamski, E. and Schaffer, A.A. (Eds.), pp 595-623, Marcel Dekker, New York.

Egli, D.B. (1997). Cultivar maturity and response of soybean to shade stress during seed filling. *Field Crops Res.*, Vol. 52, pp. 1-8.

Egli, D.B. (2010). Soybean reproductive sink size and short-term reductions in photosynthesis during flowering and pod set. *Crop Sci.*, Vol. 50, pp. 1971-1977.

Elmore, R. W., Eisenhauer, D. E. , Specht, J. E. & Williams, J. H.(1988). Soybean yield and yield component response to limited capacity sprinkler irrigation systems. *J. of Prod. Agric.*, Vol. 1, pp. 196-201.

Fageria, N.K., Baligar, V.C. & Clark, R.B. (2006). Photosynthesis and crop yield. In *Physiology of Crop Production*, pp. 95-116, Food Products Press, New York.

Fehr, W.R.& Caviness, C.E. (1977). *Stage of soybean development.* Iowa State University, Cooperative Extension Service, 11 p. (Special report, 80), Ames, IA.

Fehr, W.R., Caviness, C.E. & Vorst, J.J. (1977). Response of indeterminate and determinate soybean cultivars to defoliation and half-plant cut-off. *Crop Sci..*, Vol. 17, pp. 913-917.

Fehr, W.R., Lawrence, B.K., & Thompson, T.A. (1981). Critical stages of development for defoliation of soybean. *Crop Sci.*, Vol. 21, pp. 259-262.

Fehr, W.R. (1987). *Principles of cultivar development, Vol. 1.* Macmillan, New York.

Ferris, R., Wheeler, T. R., Hadley, P. & Ellis, R.H. (1998). Recovery of photosynthesis after environmental stress in soybean grown under elevated CO_2. *Crop Sci.*, Vol. 38, pp. 948–955.

Fitter, A. H. & Hay, R.K.M. (1987). Temperature In *Environmental Physiology of Plants,* pp. 187-224, Academic Press, New York.

Ford, D.M., Shibles, R. & Green, D.E. (1983). Growth and yield of soybean lines selected for divergent leaf photosynthetic ability. *Crop Sci.*, Vol. 23, pp. 517–520.

Ford, D.M. & Shibles, R. (1988). Photosynthesis and other traits in relation to chloroplast number during soybean leaf senescence, *Plant Physiol.* Vol. 86, pp. 108-111.

Foulkes, M.J., Reynolds, M.P. & Sylvester-Bradley, R. (2009). Genetic improvement of grain crops: yield potential. In *Crop Physiology: Applications for Genetic Improvement and Agronomy.* Sadras, V.O. & Calderini, D.F. (Eds.), pp. 355-385 Elsevier, Burlington, MA.

Frederick, J.R., Aim, D.M. & Hesketh, J.D. (1989). Leaf photosyntyhetic rates, stomatal resistances, and internal CO_2 concentrations of soybean cultivars under drought stress, *Photosynthetica*, Vol. 23, pp. 575-584.

Frederick, J.R., Woolley, J.T., Hesketh, J.D., & Peters, D.B. (1991). Water deficit development in old and new soybean cultivars, *Agron. J.*, Vol. 82, pp. 76-81.

Frederick, J.R. & Hesketh, J.D. (1994). Genetic improvement, In *Soybean: Physiological Attributes*, Slafer, G.A. (ed.), pp. 237-286, Marcel Dekker, New York.

Frederick, J.R., Camp, C.R. & Bauer, P.J. (2001). Drought-stress effects on branch and mainstem seed yield and yield components of determinate soybean. *Crop Sci.*, Vol. 41, pp. 759 - 763.

Gardner, F.P., Pearce, R.B. &Mitchell, R.L. (1985). Carbon fixation by crop canopies. In *Physiology of crop plants*, pp. 31-57, Iowa State Univ. Press, Ames, IA.

Gay, S., Egli, D.B., & Reicosky, D.A. (1980). Physiological basis of yield improvement in soybeans. *Agron. J.*, Vol. 72, pp. 387-391.

Gazzoni, D.L. & Moscardi, F. (1998). Effect of defoliation levels on recovery of leaf area, yield and agronomic traits of soybeans. *Pesq. Agropec. Bras., Brasilia,* Vol. 33, pp. 411-424.

Gesch, R.W., Vu, J.C.V, Allen, L.H.J., & Boote, K.J. (2001). Photosynthetic responses of rice and soybean to elevated CO_2 and temperature. *Recent Res. Devel. Plant Physiol.*, Vol. 2, pp. 125-137.

Gibson, L.R. & Mullen, R.E. (1996). Soybean seed quality reductions by high day and night temperature. *Crop Sci.*, Vol.36, pp. 1615–1619.

Goli, A. & Weaver, D.B. (1986). Defoliation responses of determinate and indeterminate late-planted soybeans. *Crop Sci.*, Vol. 26, pp. 156-159.

Hadley, P., Roberts, E.H., Summerfield, R.J. & Minchin, F.R. (1984). Effects of temperature and photoperiod on flowering in soybean [*Glycine max* (L.) Merrill]: A quantitative model. *Ann. Bot.*, Vol. 53, pp.669–681.

Hagood Jr., E.S., Bauman, T.T. Williams Jr., J.L., &. Schreiber, M.M. (1980). Growth analysis of soybeans (Glycine max) in competition with velvetleaf (*Abutilon theophrasti*). *Weed Sci. Vol.* 28, pp 729-734.

Hagood Jr., E.S., Bauman, T.T., Williams Jr., J.L., & Schreiber, M.M. (1981). Growth analysis of soybeans (*Glycine max)* in competition with Jimsonweed (*Datura stramonium). Weed Sci.*, Vol. 29, pp. 500-504.

Haile, F.J, Higley, L.G, and Specht, J.E, and Spomer, S.M. (1998a) Soybean leaf morphology and defoliation tolerance. *Agron. J.*, Vol. 90, pp. 353-362.

Haile, F.J., Higley, L.G., & Specht, J.E. (1998b). Soybean cultivars and insect defoliation: yield loss and economic injury levels. *Agron. J.*, Vol. 90, pp. 344-352.

Harper, J.E. (1987). Nitrogen metabolism. In: *Soybeans: Improvement, production, and uses,* Wilcox, J.R. (Ed.), pp. 498–533, Amer. Soc. of Agron., Madison, WI.

Harrison, S.A., Boerma, H.R., & Ashley, D.A. (1981). Heritability of canopy apparent photosynthesis and its relationship to seed yield in soybean. *Crop Sci.*, Vol. 21, pp. 222-226.

Hartwig, E.E. & Edwards, C.J. (1970). Effects of morphological characteristics upon seed yield in soybeans. *Agron. J.*, Vol. 62, pp. 64-65.

Hartwig, E.E. (1973). Varietal development. In *Soybeans: Improvement, production, and uses,* Caldwell, B.E. (Ed.), pp. 187-210, Amer. Soc. of Agron., Madison, WI.

Hay, R. & Porter, J. (2006). *The physiology of crop yield,* 2nd Ed., Blackwell Publishing , Singapore.

Heatherly, L. G. & Spurlock, S. R. (1993). Timing of farrow irrigation termination for determinate soybean on clay soil. *Agron. J.*, Vol. 85, pp. 1103-1108.

Heatherly, L.G. & Elmore, C.D. (1986). Irrigation and planting date effects on soybean grown on clay soil. *Agron. J.*, Vol. 78, pp. 576-580.

Heatherly, L.G. (2009). U.S. Soybean Production. United Soybean Board, St. Louis, Missouri.

Helms, T.C., Deckard, E., Goos, R.J. & Enz, J.W. (1996). Soybean seedling emergence influenced by days of soil water stress and soil temperature. *Agron. J., Vol.* 88, pp. 657–661.

Hewitt, J.D., Casey, L.L. &. Zobel, R.W. (1985). Effect of day length and night temperature on starch accumulation and degradation in soybean. *Ann. of Botany*, Vol. 56, pp. 513-522.

Higgins, R. A., Pedigo, L. P. & Staniforth, D. W. (1983). Selected preharvest morphological characteristics of soybeans stresses by simulated green cloverworm (Lepidoptera : Noctuidae) defoliation and velvetleaf competition. *J. of Econ. Ent.*, Vol. 76, pp. 484-491.

Hoeft, R.G., Nafziger, E.D., Johnson, R.R. & Aldrich, S.R. (2000). Weed management. In *Modern Corn and Soyben Production*, pp. 173-183, MCSP publications, Champaign, IL.

Hofstra, G. (1972). Response of soybeans to temperature under high light intensities. *Can. J. Plant Sci.*, Vol. 52, pp. 535–543.

Hollinger, S.E. & Angel, J.R. (2009). Weather and crops In: *Illinois Agronomy Handbook 24th edition*, Nafziger, E. (Ed.), Illinois Agr. Ext. Pub. C1394, Champaign, IL.

Hoogenboom, G, Peterson, C.M. & Huck, M. G. (1987). Shoot growth rate of soybean as affected by drought stress. *Agron. J.*, Vol. 79, pp. 598-607.

Huxley, P. A. & Summerfield, J.R. (1974). Effect of night temperature and photoperiod on the reproductive ontogeny of cultivars of cowpea and of soybeans selected for the wet tropics. *Plant Sci. Letter*, Vol. 3, pp. 11-17.

Imsande, J. (1989). Rapid dinitrogen fixation during soybean pod filling enhances net photosynthetic output and seed yield: a new perspective. *Agron. J.* Vol., 81, pp. 549-556.

Ingram, K.T., Herzog, D.C., Boote, K.J., Jones, J.W. & Barfield, C.S. (1981). Effects of defoliating pests on soybean canopy CO_2 exchange and rep. growth. *Crop Sci.*, Vol. 21, pp. 961-968.

Jeffers, D.L. & Shibles, R.M. (1969). Some effects of leaf area, solar radiation, air temperature, and variety on net photosynthesis in field-grown soybeans. *Crop Sci.*, Vol. 9, pp. 762–764.

Jiang, H. & Egli, D.B. (1993). Shade induced changes in flower and pod number and flower and fruit abscission in soybean. *Agron. J.*, Vol. 85, pp. 221-225.

Jiang, H. & Egli, D.B. (1995). Soybean seed number and crop growth rate during flowering. *Agron. J.*, Vol. 87, pp. 264-267.

Johnson, R.R.(1987). Crop management. In: *Soybeans: Improvement, production, and uses.* Wilcox, J.R. (Ed.), pp. 355-390, Amer. Soc. of Agron., Madison, WI.

Jones Jr., J.B. (1998). *Plant nutrition manual.* CRC Press, Boca Raton, FL.

Jones, P., Jones, J.W. & Allen, L.H. (1985). Carbon dioxide effects on photosynthesis and transpiration during vegetative growth in soybean. *Soil and Crop Sci. Soc. of Florida*, Vol. 44, pp. 129-134.

Kadhem, F. A., Specht, J. E. & Williams, J. H. (1985). Soybean irrigation serially timed during stages R1 to R6. II. Yield component responses. *Agron. J.*, Vol. 77, pp. 299-304.

Kahlon, C.S., Board, J.E., and Kang, M.S. (2011). An analysis of yield component changes for new vs. old soybean cultivars. *Agron. J.*, Vol. 103, pp. 13-22.

Korte, L. L.,. Williams, J.H., Specht, J. E. & Sorensen, R. C. (1983a). Irrigation of soybean genotypes during reproductive ontogeny. I. Agronomic responses. *Crop Sci.*, Vol. 23, pp. 521.

Korte, L.L., Specht, J.E., Williams, J.H. & Sorensen, R.C. (1983b). Irrigation of soybean genotypes during reproductive ontogeny. II. Yield component responses. *Crop Sci.*, Vol. 23, pp. 528-533.

Koti, S., Reddy, K.R., Reddy, V.R., Kakani, V.G., & Zhao, D. 2004. Interactive effects of carbon dioxide, temperature, and ultraviolet-B radiation on soybean (*Glycine max L.)* flower and pollen morphology, pollen production, germination and tube lengths. *J. Exper. Bot.* Vol. 56, pp. 725-736.

Koutroubas, S.D., Papakosta, D.K., and Gagianas, A.A. (1998). The importance of early dry matter and nitrogen accumulation in soybean yield. *European Jr. of Agron.*, Vol. 9, pp. 1-10.

Kumudini, S., Hume, D.J., & Chu, G. (2001). Genetic improvement in short-season soybeans: dry matter accumulation, partitioning and leaf area duration. *Crop Sci.*, Vol. 41, pp. 391-398.

Larson, E.M., Hesketh, J.D., Wooley, J.T. & Peters, D.B. (1981). Seasonal variation in apparent photosynthesis among plant stands of different soybean cultivars. *Photosyntheis Res.*, Vol. 2, pp. 3-20.

Lawn, R.J. (1982). Response of four grain legumes to water stress in south-eastern Queensland. III. Dry matter production, yield and water use efficiency. *Aus. J. of Agric. Res.* Vol. 33, pp. 511-521.

Liu, F., Jensen, C.R. & Andersen, M.N. (2004). Drought stress effect on carbohydrate concentration in soybean leaves and pods during early reproductive development: its implication in altering pod set. *Field Crops Res.*, Vol. 86, pp. 1-13.

Loomis, R.S., & Connor D.J. (1992a). Community concepts. In: *Crop Ecology: Productivity and management in agricultural systems.* pp. 32–39, Cambridge Univ. Press, Cambridge, England..

Loomis, R.S. & Connor, D.J. (1992b). Water concepts. In: *Crop Ecology: Productivity and management in agricultural systems.* pp. 224-256, Cambridge Univ. Press, Cambridge, England.

Louisiana Agric. Ext. Serv. (2009). Louisiana soybean verification program. <http://www.lsuagcenter.com/en/crops_livestock/crops/soybeans/lsrvp/>.

Louisiana Agric. Exp. Stn. (2010). Louisiana climatic data. http://weather.lsuagcenter.com/reports.aspx.

Mayaki, W.C, Stone, L.R., & Teare, I.D. (1976a). Irrigated and nonirrigated soybean, corn, and grain sorghum root systems. *Agron. J.*, Vol. 68, pp. 532-534

Mayaki, W.C, Teare, I.D. & Stone, L.R. (1976b). Top and root growth of irrigated and nonirrigated soybeans. *Crop Sci.*, Vol. 16, pp. 92-94.

Meckel, L., Egli, D.B., Phillips, R.E., Radcliffe, D., & Leggett, J.E. (1984). Effect of moisture stress on seed growth in soybeans. *Agron. J.*, Vol. 76, pp. 647-650.

Morrison, M.J., Voldeng, H.D., & Cober, E.R. (1999). Physiological changes form 58 years of genetic improvement of short-season soybean cultivars in Canada. *Agron. J.*, Vol. 91, pp. 685-689.

Morrison, M.J., Voldeng, H.D., & Cober, E.R. (2000). Agronomic changes from 58 years of genetic improvement of short-season soybean cultivars in Canada. *Agron. J.*, Vol. 92, pp. 780-784.

Morrison, M. J., McLaughlin, N.B., Cober, E.R. & Butler, G.M. (2006). When is short-season soybean most susceptible to water stress? *Can. J. Plant. Sci.*, Vol. 86, pp. 1327-1331.

Muchow, R. C. (1985). Phenology, seed yield and water use of grain legumes grown under different soil water regimes in a semi-arid tropical environment. *Field Crops Res.*, Vol. 11, pp. 81-97.

Muchow, R.C., Sinclair, T.R. , Bennett, J.M. & Hammond, L.C. (1986). Response of leaf growth, leaf nitrogen, and stomatal conductance to water deficits during vegetative growth of field-grown soybean. *Crop Sci.*, Vol. **26**, pp. 1190–1195.

Musser, R.L., Kramer, P. J. & Thomas, J. F. (1986). Periods of shoot chilling sensitivity in soybean flower development, and compensation in yield after chilling. *Ann. of Bot.*, Vol. 57, pp. 317-329.

Neyshabouri, M. R., & Hatfield, J. L. (1986). Soil water deficit effects on semi-determinate and determinate soybean growth and yield. *Field Crops Res.*, Vol. 15, pp. 73-84.

Ojima, M. (1972). Improvement of leaf photosynthesis in soybean varieties. *Bull. Natl. Inst. Agric. Sci. Ser. D.*, Vol. 23, pp. 97-154.

Pandey, R. K., Harrera, W. A. T. & Pendleton, J. W. (1984). Drought responses of grain legumes under an irrigation gradient. I. Yield and yield components. *Agron. J.*, Vol. 76, pp. 549-553

Pankhurst, C.E., & Sprent, J.I. (1975). Effects of water stress on respiratory and nitrogen-fixing activity of soybean root nodules. *J. Exper. Bot.*, Vol. 26, pp. 287-304.

Paulsen, G.M. (1994). High temperature responses of crop plants. In: *Physiology and determination of crop yield.* Boote, K.J., Sinclair, T.R., & Paulsen, G.M. (Eds.), pp. 365-389, American Society of Agronomy, Madison, WI.

Peterson, R. K. D. & Higley, L. G. (1996). Temporal changes in soybean gas exchange following simulated insect defoliation. *Agron. J.*, Vol. 88, pp. 550-554.

Pickle, C. S. & Caviness, C.E. (1984). Yield reduction from defoliation and plant cutoff of determinate and semideterminate soybean. *Agron J.*, Vol. 76, pp. 474-476.

Pore, R. (2009). Climate change threatens state's irrigation industry. The Independent (local newspaper).

Purcell, L.C. (2000). Soybean canopy coverage and light interception measurements using digital imagery. *Crop Sci.*, Vol. 40, pp. 834-837.

Purcell, L.C. & Specht, J.E. (2004). Physiological traits for ameliorating drought stress. In: *Soybeans: Improvement, production, and uses (third ed.),* Boerma, H.R., & Specht, J.E. (Eds.), American Society of Agronomy. Madison, WI.

Purcell, L.C., Ball, R.A. , Reaper, J.D. & Vories, E.D.(2002). Radiation use efficiency and biomass production in soybean at different plant population densities. *Crop Sci.* Vol. 42, pp. 172 - 177.

Ramseur, E.L., Quisenberry, V.L., Wallace, S.U., and Palmer, J.H. (1984). Yield and yield components of 'Braxton' soybeans as influenced by irrigation and intrarow spacing. *Agron. J.*, Vol. 76, pp. 442-446.

Ramseur, E. L., Wallace, S. U. & Quisenberry, V. L. (1985). Growth of 'Braxton' soybeans as influenced by irrigation and intrarow spacing. *Agron. J.*, Vol. 77, pp. 163-168.

Raper, C.D. & Kramer, P.J. (1987). Stress Physiology. In: *Soybeans: Improvement, production, and uses,* Wilcox, J.R. (Ed.)., pp. 589-641. American Society of Agronomy, Madison, WI.

Reicosky, D.C, & Heatherly, L.G.. (1990). Soybean. In *Irrigation of Agricultural Crops,* Stewart, B.A. & Nelson, D.R. (Eds.), pp .659-674, Amer. Soc. of Agron., Madison,WI.

Salado-Navarro, L.R., Sinclair, T.R., & Hinson, K. (1993). Changes in yield and seed growth traits in soybean cultivars released in the Southern USA from 1945-1983. *Crop Sci.*, Vol. 33, pp. 1204-1209.

Salem, M.A., Kakani, V. G., Koti, S. & Reddy, K.R. (2007). Pollen-based screening of soybean genotypes for high temperatures. *Crop Sci.*, Vol.47, pp. 219–231.

Saliba, M.R., Schrader, L.E., Hirano, S.S. & Upper, C. D. (1982). Effects of freezing field-grown soybean plants at various stages of podfill on yield and seed quality. *Crop Sci.*, Vol. 22, pp. 73-78.

Sato, K. and Ikeda, T. (1979). The growth responses of soybean to photoperiod and temperature. IV. The effect of temperature during the ripening period on the yield and characters of seeds. *Jpn. J. Crop Sci.*, Vol. 48, pp. 283–290.

Scott, H. D. & Batchelor, J. T. (1979). Dry weight and leaf area production rates of irrigated determinate soybeans. *Agron. J.*, Vol. 71, pp. 776-782.

Scott, H.D., De Angula, J., Daniels, M.B. & Wood, L.S. (1989). Flood duration effects on soybean growth and yield. *Agron. J.*, Vol. 83, pp. 631-636.

Seddigh, M. & Jolliff, G.D. (1984a). Night temperature effects on morphology, phenology, yield and yield components of indeterminate field-grown soybean. *Agron. J.*, Vol. 76, pp. 824-828.

Seddigh, M. & Jolliff, G.D. (1984b). Effects of night temperature on dry matter partitioning and seed growth of indeterminate field-grown soybean. *Crop Sci.*, Vol. 24, pp. 704-710.

Shaw, R. H., & Laing, D.R.. (1966) Moisture stress and plant response. In:, *Plant environment and efficient water use,* Pierre, W.H., Kirkham, D., Pesek, J. & Shaw, R. (Eds.), pp. 73–94, Amer. Soc. of Agron. Madison, WI.

Shibles, R.M. & Weber, C.R. (1965). Leaf area, solar radiation interception, and dry matter production by soybeans. *Crop Sci.*, Vol. 5, pp. 575–578.

Shibles, R., Secor, J. & Ford, D.M. (1987). Carbon assimilation and metabolism. In: *Soybeans: Improvement, production, and uses,* Wilcox, J.R. (Ed.), pp 535–588, Amer. Soc. of Agron., Madison, WI.

Shiraiwa, T., & Hashikawa, U. (1995). Accumulation and partitioning of nitrogen during seed filling in old and modern soybean cultivars in relation to seed production. *Jpn. J. Crop Sci.*, Vol. 64, pp. 754-759.

Sinclair, T.R. & de Wit, C.T. (1976). Analysis of the carbon and nitrogen limitations to soybean yield. *Agron. J., Vol.* 68, pp. 319–324.

Sinclair, T.R. (2004). Improved carbon and nitrogen assimilation for increased yield. In *Soybeans: Improvement, production, and uses.* Boerma, H.R. & Specth, J.E. (Eds.), pp. 537-568, Amer. Soc. of Agron., Madison, WI.

Sionit, N. & Kramer, P. (1977). Effect of water stress during different stages of growth of soybean. *Agron. J.*, vol. 69, pp. 274-278.

Sionit, N., Strain, B.R. & Flint, E. (1987). Interaction of temperature and CO_2 enrichment on soybean photosynthesis and seed yield. *Can. J. Plant Sci.*, Vol. 67, pp. 629–636.

Snyder, R. L., Carlson, R. E., & Shaw, R. H. (1982). Yield of indeterminate soybeans in response to multiple periods of soil-water stress during reproduction. *Agron. J.*, Vol. 74, pp. 55-859.

Specht, J.E. & Williams J.H. (1984). Contribution of genetic technology to soybean productivity-retrospect and prospect.. In *Genetic contributions to yield gains of five major crop plants*, Fehr, W.R. (Ed.), pp. 49-74, Amer. Soc. of Agron. Madison.WI

Specht, J.E., Hume, D.J. & Kumudini, S.V. (1999). Soybean yield potential—a genetic and physiological perspective. *Crop Sci.*, Vol. 39, pp. 1560-1570.

Sung, F.J.M., & Chen, J.J. (1989). Changes in photosynthesis and other chloroplast traits in lanceolate leaflet isolines of soybean. *Plant Physiol.*, Vol. 90, pp. 773-777.

Tanner, J.W. & Hume, D.J. (1978). Management and production. In Norman, G. (Ed.) *Soybean physiology, agronomy and utilization,* pp. 157-217, Academic Press, New York.

Taylor, H. M., Mason, W. K., Bennie, A. T. P. & Rouse, H. R. (1982). Responses of soybeans to two row spacings and two soil water levels. I. An analysis of biomass accumulation, canopy development, solar radiation interception and components of seed yield. *Field Crops Res.*, Vol. 5, pp. 1-14.

Tolley-Henry, L. & Raper Jr., C.D. (1986). Nitrogen and dry matter partitioning in soybean plants during onset of and recovery from nitrogen stress. *Bot. Gaz.*, Vol. 147, pp. 392-399.

Tolley-Henry, L., & Raper Jr., C.D. (1992). Onset of and recovery from nitrogen stress during reproductive growth of soybean. *Int. Jr. Plant Sci.*, Vol. 153, pp. 178-185.

Turnipseed, S.G. & Kogan, M. (1987). Integrated control of insect pests. In *Soybean: Improvement, production, and uses,* Wilcox, J.R. (Ed.), pp. 780-817, Amer. Soc. of Agron., Madison, WI.

US Soybean Export Council. (2008). How the global oilseed and grain trade works. Soyatech, LLC, Southwest Harbor, Maine, US.

USDA (2007). The Census of Agriculture. Natl. Agric. Stat. Serv. USDA. Washington, D.C.,

Univ. of Arkansas. (2006). Soil and water management. In *Soybeans-crop irrigation,* Univ. of Arkansas Coop. Ext. Bul., Little Rock, AR.

Vu, J.C.V., Allen Jr, L.H., Boote, K.J. & Bowes, G. (1997). Effects of elevated CO_2 and temperature on photosynthesis and Rubisco in rice and soybean. *Plant Cell. Environ.*, Vol. 20, pp. 68–76.

Weber, C. R. (1955). Effects of defoliation and topping simulating hail injury to soybeans. *Agron. J.*, Vol., 47, pp. 262-266.

Wells, R., Schulze, L.L., Boerma, H.R., & Brown, R.H. (1982). Cultivar differences in canopy apparent photosynthesis and their relationship to seed yield in soybeans. *Crop Sci.*, Vol. 22, pp. 886-890.

Welter, S.C. (1993). Responses of plants to insects: eco-physiological insights. In *International Crop Science I.*, Buxton, D.R.. (Ed.), p. 773-778, Crop Sci. Soc. of Amer., Madison, WI.

Westgate, M. E. & Peterson, C. M. (1993). Flower and pod development in water deficient soybeans (*Glycine max. L. Merr.*). *J. of Exp. Bot.*, Vol. 44, pp. 109-117.

Wiebold, W.J., Shibles, T. & Green, D.E. (1981). Selection for apparent photosynthesis and related leaf traits in early generations of soybeans. *Crop Sci.*, Vol. 21, pp. 969-973.

Wilcox, J.R. (2001). Sixty years of improvement in publicly developed elite soybean lines. *Crop Sci.*, Vol. 41, pp. 1711-1716.

Wilcox, J.R. (2004). World distribution and trade of soybean. In *Soybean: Improvement, production, and uses,* Boerma, H.R. & Specht, J.E. (Eds.), pp. 1-14, Amer. Soc. of Agron., Madison, WI..

2

Use of Climate Forecasts to Soybean Yield Estimates

Andrea de Oliveira Cardoso[1], Ana Maria Heuminski de Avila[2], Hilton Silveira Pinto[3] and Eduardo Delgado Assad[4]
[1]*CECS, UFABC, Santo André - SP*
[2]*CEPAGRI, UNICAMP, Campinas - SP*
[3]*IB and CEPAGRI, UNICAMP, Campinas - SP*
[4]*Embrapa Informática Agropecuária, Campinas - SP*
Brazil

1. Introduction

The soybean is an annual legum that have many industrial, human, and agricultural uses. United States are the main producing and exporters of soybean grain, ranking as the first highest agricultural commodity of this specific agricultural cultural (FAO, 2008). Considering the total production of soybeans by the 20 highest producing countries in 2008 was 35% from US, 26% from Brazil and 20% from Argentine, having equivalent agricultural commodities values.

Some studies in agronomic experimental stations suggest that this culture was initially introduced in Bahia State, northeastern of Brazil, 1882. However, only after the 40's in southern of Brazil, the soybean crop became commercial in the country. Nowadays, this culture is considered the most important agricultural commodity in Brazil, and one of its main export products (Esquerdo, 2001).

The production of soybean has a great importance for the economy of Brazil. Historical data of soybean harvest for Brazil (IBGE, 2008) show a high correlation between soybean production economic value and productivity of this culture (Cardoso et al., 2010). According to IBGE, the Brazilian production in 2007 was 58 million tons, with Mato Grosso, Paraná and Rio Grande do Sul States adding higher crop production.

Soybean cultivars can be classified according to the duration of your cycle, being early (75 to 115 days), semi-early (116 to 125 days), medium (126 to 137 days), late medium (138 to 150 days) and late (over 150 days), according to Farias et al. (2000).

According to Camargo (1994), the climate is the main factor responsible by annual fluctuations in grain production in Brazil. The occurrence of drought is the main cause of harms (71% of cases), followed by excessive rainfall (22% of cases), hail, frost, pests and diseases (Göpfert et al., 1993).

The observations of weather conditions applied to the crop forecast models are useful to provide the most accurate crop simulations, and the importance of solar radiation, precipitation and air temperature variables is stood out (Hoogenboom, 2000).

Research has been conducted with the goal of exploring the climate patterns to improve the yield of this crop agriculture. The soybean production can be significantly affected by water conditions, according to the intensity of water deficit (Thomas & Costa, 1994; Confalone & Dujmovich, 1999). The water necessity of the soybean crop will increase with plant development, reaching a maximum during the flowering and grain filling phases, decreasing after this period (EMBRAPA, 2004). Significant water deficits during flowering and grain filling, causing physiological changes in the plant, such as stomatal closure and winding sheets, consequently cause premature leaf drop and flower and pod abortion, resulting in decrement grain yield. Thus, precipitation over the planted area is an important determiner of crop yield, considering the high cost of effective irrigation being.

Studies developed by IPEA (Institute of Applied Economic Research) in 1992 indicated that 95% of Brazilian agricultural losses were due to events of drought or heavy rain. Based on these data, was established, in 1996, the zoning program in Brazil for climate risks, being the public policy currently adopted by Brazilian Ministry of Agriculture of Agrarian Development to direct credit and agricultural insurance in the country. The zoning established, statistically, the risk levels of the regions studied for various types of culture, assuming crop losses of up to 20%. This indirectly increased agricultural productivity. The agroclimatic zoning of risks is a tool that indicates what to plant, where to plant and when to plant according to the climatic region (Assad et al., 2008).

According to Farias et al. (1997), in modern agriculture, increases in income and reductions in costs and risks of failure ever more dependent of judicious use of resources. In this case, the farmer must make decisions based on available production factors and involved risk levels involving your activity, looking for in order to reach greater prosperity. Among the risk factors can be considered as the main the market uncertainties and the unpredictable climate conditions.

A way that can minimize the effects of drought is to plant only varieties adapted to the region, in appropriate period and soil condition. Farias et al. (2001) delimited areas with fewer risks for the soybean crop in Brazil, based on: sowing dates; water availability in each region; water consumption in the different stages of development of the soybean crop; soil type; and cultivar cycle. Results are presented in a map that represents the drought risk classification of different areas of the state for a given sowing date, as a function of the soil type and cultivar cycle.

Assad et al. (2007) evaluated the performance of soybean yield forecast system for Brazil that is based on the conceptual model proposed by Doorenbos & Kassam (1979), including some empirical adjustments for each Brazilian region. Statistical analysis was performed to evaluate the estimated soybean yield for harvests from 2000/2001 to 2005/2006. According to the results of correlation were not significant differences between the estimates and official data. Additionally it was observed that such system has a good performance in the soybean yield forecast in the Brazilian States of Mato Grosso, Paraná, São Paulo, Minas Gerais, Tocantins and Goiás, and a low performance in the soybean yield forecast in the Brazilian States of Rio Grande do Sul, Santa Catarina, Mato Grosso do Sul, Maranhão, Piauí and Bahia.

There are evidences that the accuracy of yield forecast models increases when the meteorological forecast information is used (Challinor et al., 2003; Cantelaube & Terres, 2005). Reliable meteorological forecasts having considerable lag may contribute to anticipate the productivity estimates and give good estimates of crop yield losses. Using the ensemble forecast - where the initial condition (IC) uncertainties are explored by making a certain number of IC disturbed forecasts - had a positive impact on increasing the predictability (Gneiting & Raftery, 2005; Sivillo et al., 1997). Mendonça & Bonatti (2004) compared the performance of ensemble weather forecasts from the Center for Weather and Climate Prediction of National Institute for Space Research (CPTEC/INPE) for the period from October 2001 to September 2003 and found out that the average ensemble performance is higher than that of the control forecast.

Analyses of soybean yield estimate models, showing that water is the factor that has greater influence on soybean grain yield could be incorporated into programs forecasting the crop harvest (Fontana et al., 2001).

Recent studies of Cardoso et al. (2010) show that the use of accurate meteorological forecasts can be useful to improve the productivity prediction and consequently contribute to agricultural planning. According to the results the use of up to 15 day meteorological forecasts lead to more reliable crop productivity estimates than those generated using only climatological information. The combination of precipitation forecasts by the CPTEC ensemble system combined with climatology date after the end of the forecast cycle already show significant improvement of the final productivity forecast compared to estimates solely based on past observed data and climatology. Highlighting the importance to turn meteorological forecasts available for periods as longer as possible in real time, primarily in periods when the crop is more sensitive to water deficit.

1.1 The interannual variability

El Niño-Southern Oscillation (ENSO) is a phenomenon of the coupled atmospheric-ocean that forms the link with the anomalous global climate patterns, being a dominant source of interannual climate variability. The atmospheric component tied to El Niño is termed the Southern Oscillation. El Niño corresponds to the warm phase of ENSO, consisting of a basinwide warming of the tropical Pacific. The term La Niña is applied to the cold phase of ENSO, associated with a cooling of the tropical Pacific (Trenberth, 1997).

The La Niña and El Niño phenomenons influence the precipitation over some regions of the South America such as Northeast Brazil, eastern Amazônia, Southern Brazil, Uruguay and NE Argentina (Ropelewski & Halpert, 1987, 1989). Studies have suggested that sea surface temperature (SST) positive anomalies in the equatorial Pacific, related El Niño events, favours to increase of precipitation in the South of Brazil (Grimm et al.,1998; Coelho et al., 2002) and decrease of precipitation in the Brazilian Northeast (Rao & Hada, 1990; Moura & Shukla, 1981). There is a general behavior towards opposite signals in the precipitation anomalies over southern South America during almost the same periods of the El Niño and La Niña cycles, indicating a large degree of linearity in the precipitation response to these events.

Peak rainfall in central-east Brazil during part of spring holds a significant inverse correlation with rainfall in peak summer monsoon, especially during ENSO years (Grimm et al., 2007). As shown by the latter paper, a surface-atmosphere feedback hypothesis is proposed to explain this relationship: low spring precipitation leads to low spring soil moisture and high late spring surface temperature; this induces a topographically enhanced low-level anomalous convergence and cyclonic circulation over southeast Brazil that enhances the moisture flux from northern and central South America into central-east Brazil, setting up favorable conditions for excess rainfall. Antecedent wet conditions in spring lead to opposite anomalies.

Marques et al. (2005) observed that part of the variability of rainfall and air temperature in the state of Rio Grande do Sul (southern Brazil) is associated with the variability of Sea Surface Temperature (SST) in the Pacific and Atlantic oceans. This knowledge is of great relevance, given the importance of these elements on vegetation growth. Also was verified the existence of the association between SST in the Pacific and Atlantic oceans and NDVI (Normalized Difference Vegetation Index) in the Rio Grande do Sul State, which is dependent on season and region of the state. NDVI is correlated to SST of the Pacific Ocean during the summer, while for the winter period the SST of the Atlantic Ocean shows greater correlation.

Berlato & Cordeiro (2005) studied the variability of soybean yield in Rio Grande do Sul State, southern Brazil, noting that the variability in yields coincides with higher rainfall variability. They observed that the increasing trend of yields between 1990 and 2000 years coincides with the increasing trend of rainfall from October to March, caused by the merger of El Niño events that cause positive anomalies of precipitation in spring and early summer, in Rio Grande do Sul State. The risk of El Niño be prejudicial to non-irrigated summer crops is restricted to "rebound" phenomenon in the fall of the second year of the event, especially if the months of April and May are wet anomalous can harmful the final maturation and harvest, as was the case of the large El Niño of 1982/1983.

This increasing trend of soybean yields is also present in the productivity historical data of Brazil (Figure 1). When compared the productivity historical data of the productivity with El Niño and La Niña events occurrences the relationship is unclear, except for the El Niño episodes of 1988, 1992, 2003 and 2007 that indicate a positive relation. This can see by Figure 1 that presents the Oceanic Nino Index (ONI), based 3 month running mean of sea surface temperature anomalies in the Niño 3.4 region (5°N-5°S, 120°-170°W) to quarterly november-december-january (NDJ). Several studies discuss the impact of El Niño and La Niña phenomena on rainfall patterns in some regions of Brazil, however it is observed that the response on soybean yield in years of these events is not linear, probably due the distribution of rain throughout the development of the crop, that varies in different events.

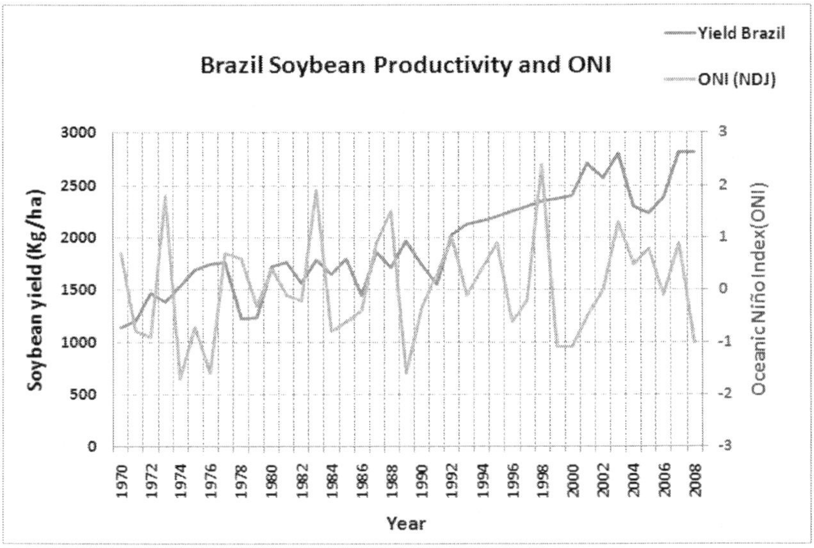

Fig. 1. Data yearly productivity (kg/ha) for Brazil for the period 1970 to 2008 (Source: IBGE) and data quarterly (november-december-january) Oceanic Niño Index (Source: NOAA).

1.2 Climate forecasts for estimative of soybean productivity

The climatic condition in the critical phases of the plant vegetative development influences the crop productivity, thus being a basic parameter for crop forecast. It is important to evaluate the possibility to use of climate forecasts in estimating the productivity given that rainfall patterns on Brazil are influenced by climate variability, as in other areas of the globe. The Center for Weather and Climate Prediction of the National Institute for Space Research (CPTEC/INPE) develops, produces and disseminates weather forecasts as well as seasonal climate forecasts since early 1995. This Center is part of the research network of the Ministry of Science and Technology of Brazil. The model used for seasonal climate predictions is the COLA-CPTEC Atmospheric Global Circulation Model (AGCM) which was originally derived from the National Center for Environmental Prediction (NCEP) model by COLA. The same model is used at CPTEC for medium-range.

This model is able to simulate the main features of the global climate, and the results are consistent with analyses of other AGCMs. The seasonal cycle is reproduced well in all analyzed variables, and systematic errors occur at the same regions in different seasons. The Southern Hemisphere convergence zones are simulated reasonably well, although the model overestimates precipitation in the southern portions and underestimates it in the northern portions of these systems. The high and low level main circulation features such as the subtropical highs, subtropical jet streams, and storm tracks are depicted well by the model, albeit with different intensities from the reanalysis (Cavalcanti et al., 2002).

The CPTEC-COLA AGCM simulates the broad aspects of the observed El Nino Southern Oscillation (ENSO) variations reasonably well, as may be expected since these variations are primarily driven by prescribed SST. Some regions, such as Northeast Brazil, Amazonia, southern Brazil-Uruguay exhibit better predictability due to the large skill of the AGCM in reproducing interannual variability of climate in those regions (Marengo et al., 2000).

Occurs monthly at CPTEC a forum to develop a consensus seasonal forecast. The participants were climate experts and operational forecasters, who reached a consensus forecast for the coming 3-month season (Berri et al., 2005). In general the consensus forecast shows better results than the operational purely forecasts of the CPTEC-COLA AGCM (Camargo Jr. et al., 2004). This type of prediction can be useful in estimating crop productivity.

This chapter present an study about the investigate of possible contribution of climate forecasts for soybean productivity forecast, considering quarterly rain forecasts updated monthly, due the importance of this variable to the yield of soybean. The improvement of estimated soybean productivity may give a contribution to agribusiness sector, in order to turn more realistic expectations available and assist on the strategic planning.

2. Investigating the possibility of improving crop forecast

Several studies indicate that Southern Region of Brazil is the main area of the country affected by interannual climate variability, as commented above. As the climate predictability tends to be better in ENSO years over regions affected by this phenomenon, it is important to investigate the potential role of precipitation climate forecast as a partial substitute for the usual climatological data in crop forecast models.

The Southern Region of Brazil is included in the areas with low climatic risk for soybean production (AGRITEMPO, 2007), participating with 32% of the total production of soybeans in Brazil (IBGE, 2009). The Rio Grande do Sul State (RS) is the second main producer state of this region, being its rainfall regime strongly affected by impacts ENOS (Grimm et al.,1998; Coelho et al., 2002; Berlato & Cordeiro, 2005). Thus, were studied productivity cases on a municipality of RS in three years corresponding to different phases of ENSO: 2005/2006 harvest (neutral year), 2006/2007 harvest (El Niño year) and 2007/2008 harvest (La Niña year). The municipality evaluated is located in the interest area and has longs historical data series, being a reliable reference for studies on agricultural productivity.

2.1 Model description

The FAO model proposed by Doorenbos & Kassam (1979) was applied to estimate the crop agricultural productivity. This is an empirical model that includes the following components: soil, with its water balance; plant, with its development, growth and yield processes; and atmosphere, with its thermal regime, rainfall and evaporative demand. This model correlates the relative yield drop to the relative evapotranspiration deficit, being formulated by the Equation (1). Therefore, it is necessary to first estimate the potential productivity (Yp) that represents the maximum crop yield in suitable conditions and then estimate YR accounting the relative water deficiency that is weighed by a crop sensitivity factor for the water deficit.

$$YR/YP = 1 - ky \cdot (1 - ER/EP), \qquad (1)$$

being: YR the actual productivity;
YP the potential productivity;
ER the actual crop evapotranspiration;
EP the potential crop evapotranspiration;
ky the productivity penalization coefficient per water deficit, variable with the crop phenological stage.

The actual crop evapotranspiration (ER) is determined by the sequential water balance based on daily temperature and precipitation data. The potential evapotranspiration (EP), or maximum crop evapotranspiration, is given by the product between the reference evapotranspiration (ETo), and the crop coefficient (Kc) for each phenological stage, as recommended by FAO, considering temperature information in its estimate. The Thornthwaite equation is a simpler method for estimating ETo, since it just requires mean temperature data. As there are limitations of this method to some climatic conditions, Camargo et al. (1999) proposed an adjust of Thornthwaite method using the concept of an effective temperature, which is a function of the local thermal amplitude. In this work was used the adjusted Thornthwaite's method was used, also considering the effective temperature corrected by photoperiod (Pereira et al., 2004). The penalization coefficient ky is an empirical adjustment factor that is specific for each crop, each phenological stage and each region of Brazil, considering the regional particularities of the varieties and the used production systems, according to values recommended by FAO (Assad et al., 2007). The potential productivity Yp represents the maximum value that can be obtained in each region and presupposes in its estimate that the phyto-sanitary, nutritional and water crop requirements are met and that the productivity is conditioned only by crop characteristics and the environmental conditions that are represented by solar radiation, photoperiod and air temperature.

The actual productivity (YR) calculation by the Equation 1 is normally made with the daily data obtained in surface stations or by climatology data. In the studied cases, one considered the values listed in the Table 1 for the ky and Kc coefficients were considered. Using extended weather forecast data may allow the actual productivity (YR) estimates to be made with the same anticipation and accuracy of meteorological models. The more accurate and anticipated will the productivity estimate be, more useful and strategic it will be.

	Duration (days)	Kc	ky
1 - Establishment	10	0.30	0.10
2 - Vegetative Development	35	0.70	0.20
3 - Flowering	40	1.10	0.80
4 - Fructification	30	0.70	1.00
5 - Maturation	10	0.40	0.10

Table 1. Values of ky and Kc coefficients per phenological stage that are considered in calculating the actual productivity.

2.2 Data

In this study, we used forecast and observed precipitation daily data in Passo Fundo/RS (28.23°S; 52.31°W) and observed temperature daily data, both from October 20 to February 21 of years: 2005/2006, 2006/2007, 2007/2008, related harvest periods according drought risk sowing date recommended (AGRITEMPO, 2007). This is a municipality of Rio Grande do Sul State (RS), is located in southern Brazil, presenting humid subtropical local climate (Cfa) according to the Koppen's classification.

The climatologic values of daily precipitation that were calculated for each year's month, on the basis of a 40-year observation period (1961 to 2000). Were also analyzed precipitation historical data to found the precipitation thresholds associated with the range that each precipitation tercile for the coming 3-month season. These values was used to represent qualitatively the precipitation climate forecast.

The climate forecasts of precipitation were obtained by consensus seasonal forecast developed monthly at CPTEC, based in forecasts of CPTEC-COLA AGCM compared with results of other models climate, being presented by maps on the CPTEC web. As the precipitation seasonal forecast are available by tercile maps displays the probability of occur above normal, normal and below normal precipitation. The use of tercile probabilities provides both the direction of the forecast relative to climatology, as well as the uncertainty of the forecast. The probability that any of the three outcomes will occur is one-third, or 33.3%. Recall that for each location and season, the tercile correspond to actual precipitation ranges, based on the set of historical observations. Thus, were used the values of observed precipitation thresholds to represent the forecasts precipitation of the category forecasts most likely. Based on values of seasonal precipitation were obtained forecast daily precipitation. These values were updated monthly to each new climate forecast, as well as is possible on real-time situation.

The observed soybean productivity data was obtained by the Brazilian Geographical and Statistical Institute (IBGE) for harvests studied.

2.3 Simulations

The simulations were developed considering the possibility of its real-time replicating. Thus to evaluate the possible contribution of precipitation climate forecasts, the actual productivity was estimated in three different ways, changing only the precipitation data set, as follows:

i. A suitable model simulation for the productivity estimate, using the observed precipitation and temperature data (October to February);
ii. Productivity estimates using series that are composed of climatologic and observed precipitation - containing observed precipitation values from the first cycle day - on different periods (ends extended at each 1 day) that are completed by climatologic values until the crop cycle ends;
iii. Productivity estimates considering precipitation series that are respectively composed by observation, climate forecast (3-month season) and climatology on different periods of the crop cycle that are extended at each 1 day in the same way as in the previous case. In this case, the values corresponding values of climatic forecasts were updated monthly.

Thus, to accomplish the second and third types of simulation was necessary to process the crop forecast model 125 times, that corresponds to the cycle day number, since the observed data is updated daily.

3. Results and discussion

In regard to climatology precipitation in the studied region is well known that the Rio Grande do Sul State presents a double peak in the wet season: from summer to spring and then to late winter, presenting a phase discontinuity (Grimm et al, 1998). In the location of Passo Fundo, at the northern part of the state of Rio Grande do Sul, the peak rainy season is the austral spring, with the largest volume in September (206 mm) and the lowest in April (118 mm). In general, the precipitation in Passo Fundo is well distributed throughout the year. There is a marked seasonal cycle of temperatures in Passo Fundo, with maximum temperatures around 28 °C (17 °C) in January (July) and minimum temperatures near 17 °C (7 °C) in January (July).

Whereas, that the suitable productivity estimate is reached by using the data observed throughout the period, that is, simulating a yield forecast model with the data observed throughout the crop cycle period, this type of simulation was used as basis comparison. The estimated soybean productivity for Passo Fundo using observed meteorological data was of 2588 Kg/ha in 2005/2006, 2525 Kg/ha in 2006/2007 and 2652 Kg/ha in 2007/2008. The verified soybean productivity was of 2500 kg/ha, 3000 kg/ha and 2450 kg/ha for the harvest of 2005/2006, 2006/2007 and 2007/2008, respectively (IBGE, 2008). The estimate productivity using data observed was better adjusted to 2005/2006 and 2007/2008 harvest than 2006/2007 period.

It is important to estimate the productivity estimate gain at different forecast periods, using whatever precipitation forecasts are available in a real-time application. To develop real-time productivity estimates in different periods of the crop cycle, there are observed data until the day of estimative, 3-month season of climate forecast and climatology for the remainder. Thus, various soybean productivity estimates were made in Passo Fundo, assuming that such estimates had been made in different crop periods, considering that there were observed data up to the beginning of the process, completed by forecasts and climatology from the end of the precipitation forecast period until harvest. Results are presented in form of graphics in Figures 2 to 4.

When comparing the results of the productivity estimated by the observed precipitation with that based on the precipitation climate forecasts throughout the crop cycle and with the climatological precipitation, it is verified that the estimate based on the precipitation forecasts is closer to the observed productivity than the estimate based on the climatological rainfall to

2005/2006 and 2007/2008 harvest (Figure 2 and Figure 4), in the period of between 40th and 70th day of the cycle. This is a period that of plant is most affected by water necessity is higher, being the estimative of productivity sensitive to variations of precipitation. This demonstrates the importance of using precipitation climate forecasts accurate to attain the productivity estimates, main in this cycle periods. Cardoso et al. (2010) found similar results by use of up to 15 day wheather forecasts to improve the soybean productivity prediction.

For the cases of 2005/2006 harvest (neutral year) and 2007/2008 harvest (La Niña year) was verified gain when using precipitation climate forecasts, because the climate forecast hit the category of precipitation occurred between November and January, periods when the crop is more sensitive to water deficit. In these two years was verified below normal precipitation in November and December, persisting until February in case of La Niña year.

There were no differences between estimate productivity using forecasts precipitation and only climatology information to 2006/2007 harvest, because although it is an El Niño year the climate forecast indicated normal to all period, being observed above normal precipitation from November by the end of the period. Maybe the forecast climate wrong because it was a weak El Niño, making more difficult the estimation of their impacts. However the error of climate forecast no harmful the estimate crop, because was forecast normal precipitation, ie, climatology that is data used when there is not climate forecast.

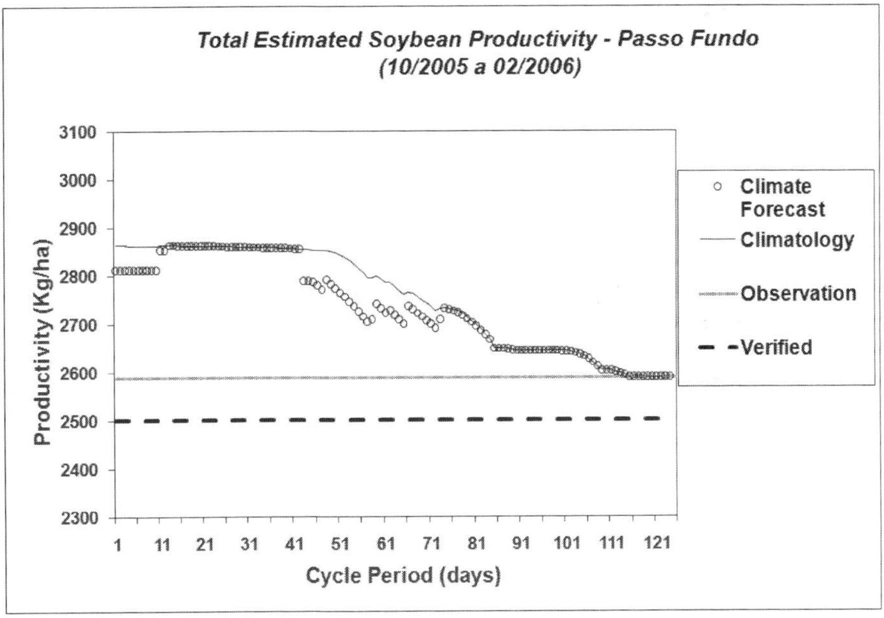

Fig. 2. Values of the total estimated productivity of soybean in Passo Fundo, 2005/2006 harvest, from the data observed throughout the cycle period (black dotted line) and from the series composed respectively by observation-climatology (black line) and observation-forecast-climatology (circles). It is highlighted that these composed series contain observed precipitation values from the first cycle day in different periods (extended at each 1 day). This also includes the value of the verified productivity (black thick line) published by the IBGE.

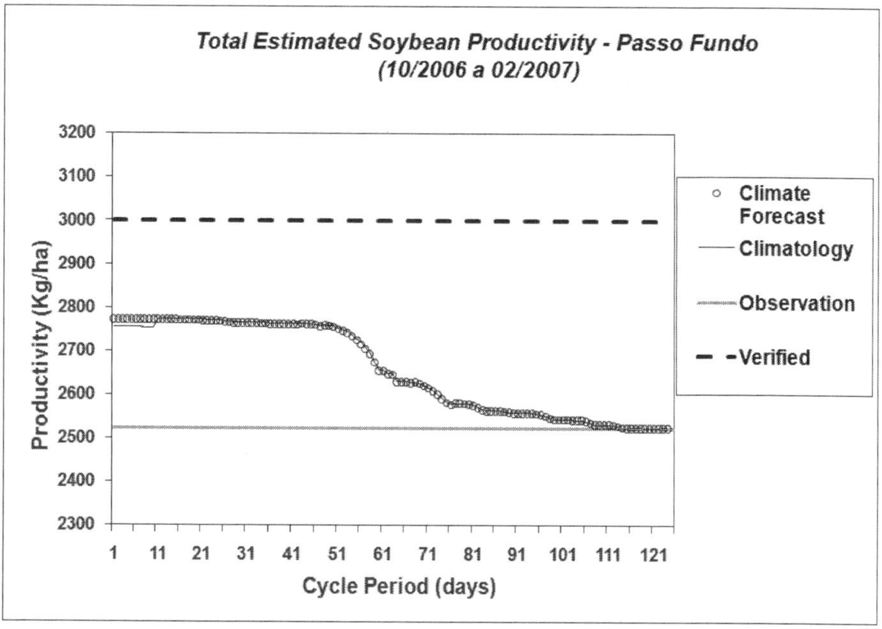

Fig. 3. Values of the total estimated productivity of soybean in Passo Fundo, 2006/2007 harvest, from the data observed throughout the cycle period (black dotted line) and from the series composed respectively by observation-climatology (black line) and observation-forecast-climatology (circles). It is highlighted that these composed series contain observed precipitation values from the first cycle day in different periods (extended at each 1 day). This also includes the value of the verified productivity (black thick line) published by the IBGE.

4. Conclusion

This chapter approached the importance of using climate forecasts to etimative of agricultural productivity and presented case studies of soybean productivity estimative, evaluating the possible contribution this type of information.

Was verified that in general the precipitation climate forecasts contribution to the improvement of estimated soybean productivity, primarily in periods when the crop is more sensitive to water deficit. For this period is important that the category of forecast precipitation be the same of observed precipitation. Thus, to achieve a gain by the use of climate forecast is necessary to know the skill of climate model used, preferring to apply this type of information in periods of greater reliability.

The improvement of estimated soybean productivity may give a contribution to agribusiness sector, in order to turn more realistic expectations available and assist on the strategic planning. This demonstrates the importance develop research that aim at better understanding the potential use of climate forecasts to estimate agricultural productivity, over the globe.

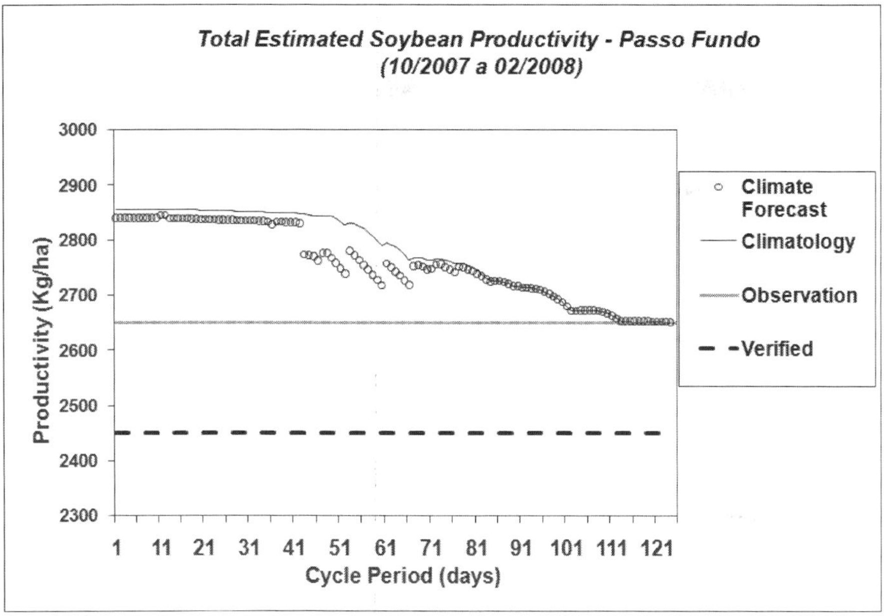

Fig. 4. Values of the total estimated productivity of soybean in Passo Fundo, 2007/2008 harvest, from the data observed throughout the cycle period (black dotted line) and from the series composed respectively by observation-climatology (black line) and observation-forecast-climatology (circles). It is highlighted that these composed series contain observed precipitation values from the first cycle day in different periods (extended at each 1 day). This also includes the value of the verified productivity (black thick line) published by the IBGE.

5. References

AGRITEMPO (2007). Zoneamento Mapas. http://www.agritempo.gov.br. Accessed 1 November 2007.

Assad, E.D., Marin, F.R., Evangelista, S.R., Pilau, F.G., Farias, J.R.B., Pinto, H.S., Zullo Júnior, J. (2007). Sistema de previsão da safra de soja para o Brasil. *Pesquisa Agropecuária Brasileira,* Vol. 42, No. 5, pp. 615-625.

Assad, E.D., Zullo Junior, J. Pinto, H.S. (2008). Zoneamento Agricola de riscos climáticos. In: Ana Christina Agebin Albuquerque; Aliomar Gabriel da Silva. (Org.). *Agricultura Tropical Quatro décadas de inovações técnológicas, institucionais e políticas. 1 ed. Brasília: Embrapa Informação técnológica,* 2008, Vol. 1, pp. 1291-1318.

Berri, G., Antico, P., Goddard, L. (2005). Evaluation of the Climate Outlook Forums' seasonal precipitation forecasts of southeast South America during 1998-2002. *Int. J. Climatol.*, Vol. 25, pp. 365-377.

Belato, M. A., Cordeiro, A. P. A. (2005). Variabilidade climática e agricultura do Rio Grande do Sul. In: Federação dos Clubes de Integração e Troca de Experiênci-

FEDERACITEa. (Org.). *As Estiagens e as Perdas na Agricultura: Fenômeno Natural ou Imprevidência?*. 1ª ed. Porto Alegre: Ideograf Editora Gráfica, pp. 43-59.

Camargo Jr., HC; Marengo, JA; Preste, ACA (2004). Skill da previsão sazonal de clima e avaliação da previsão de consenso do CPTEC. In: *XIII Congresso Brasileiro de Meteorologia*, Fortaleza, Brasil. Anais.

Camargo, M. B. P. (1994). Exigências bioclimáticas e estimativa da produtividade para quatro cultivares de soja no Estado de São Paulo. Piracicaba, SP (Tese de mestrado).

Cantelaube, P; Terres, J (2005) Seasonal weather forecasts for crop yield modelling in Europe. Tellus, Vol. 57, No. 3, pp. 476-487.

Cavalcanti, I. F. A., Marengo, J., Satyamurty, P., Nobre, C., Trosnikov, I., Bonatti, J., Manzi, A., Tarasova, T., Pezzi, L., D'Almeida, C., Sampaio, G., Castro, C., Sanches, M., Camargo, H. (2002). Global Climatological Features in a Simulation Using the CPTEC-COLA AGCM. *J. Climate*, 15(21): p. 2965-2988.

Challinor A. J., Slingo J. M., Wheeler T. R., Craufurd P. Q. , Grimes, D. I. F. (2003). Toward a Combined Seasonal Weather and Crop Productivity Forecasting System: Determination of the Working Spatial Scale. *J of Appl Meteorology*, Vol. 42, pp. 175-192.

Cardoso, A. O.; Pinto, H. S.; Silva Dias, P. L.; Ávila, A. M. H; Marin, F. R.; Pilau, F. (2010). Extended time weather forecasts contributes to agricultural productivity estimates. *Theoretical and Applied Climatology*, Vol. 102, No. 3-4, p. 343-350.

Coelho, C. A. S., Uvo, C. B. and Ambrizzi, T. (2002). Exploring the impacts of the tropical Pacific SST on the precipitation patterns over South America during ENSO periods. *Theoretical and Applied Climatology*, Vol. 71, pp. 185-197.

Confalone, A. E. & Dujmovich, M. N. (1999). Influência do déficit hídrico sobre o desenvolvimento e rendimento da soja. *Revista Brasileira de Agrometeorologia*, Vol. 6, No. 2, pp. 165-169.

Doorenbos, J. & Kassam, A. H. (1979). Yields response to water. Rome: FAO, 306p. (FAO: Irrigation and Drainage Paper, 33).

EMBRAPA (2004). Empresa Brasileira de Pesquisa Agropecuária. Tecnologias de produção de soja na região central do Brasil 2004. Embrapa Soja. Sistema de Produção. Exigências Climáticas. http://www.cnpso.embrapa.br/producaosoja/SojanoBrasil.htm. Accessed 5 June 2007.

Esquerdo, J. C. D. M. (2007). Utilização de perfis multi-temporais do NDVI/AVHRR no acompanhamento da safra de soja no Oeste do Paraná.Campinas, 168p. Tese - Universidade Estadual de Campinas, 2007.

Farias, J. R. B., Assad, E. D., Almeida, I. R. et al. (1997). Identificação de regiões com riscos de déficit hídrico à cultura da soja. In: *X CONGRESSO BRASILEIRO DE AGROMETEOROLOGIA*, 1997, Piracicaba. Anais.

Farias, J. R. B., Nepomuceno, A. F., Neumaier, N., Oya, T. (2000). ECOFISIOLOGIA. In: *Empresa Brasileira de Pesquisa Agropecuária (EMBRAPA). A cultura da soja no Brasil.* Londrina: Embrapa Soja, 2000.

Farias, J. R. B., Assad, E. D., Almeida, I. R. et al (2001). Caracterização de risco de déficit hídrico nas regiões produtoras de soja no Brasil. *Revista Brasileira de Agrometeorologia*, Vol. 9, No. 3, pp. 415-421.

FAO (2008). Food and Agriculture Organization of the United Nations. Statistical databases: Faostat: agriculture, 2008. http://www.fao.org. Accessed 10 Nov 2010.

Fontana, D. C., Berlato, M. A., Lauschner, M. H., Mello, R. W. (2001). Modelo de estimativa de rendimento de soja no Estado do Rio Grande do Sul. *Pesquisa Agropecuária Brasileira*, Vol. 33, No. 3, pp.339-403.

Gneiting, T., Raftery, A. E. (2005). Weather Forecasting with Ensemble Methods. *Science 310,* Vol. 5746, pp. 248-249.

Göpfert, H., Rossetti, L. A., SOUZA, J. (1993). *Eventos generalizados e securidade agrícola.* Brasília: IPEA, Ministério do Planejamento, 78p.

Grimm, A. M., Ferraz, S. T, Gomes, J. (1998). Precipitaition Anomalies in Southern Brazil Associated with El Niño and La Niña Events. *J. of Climate*, Vol. 11, pp. 2863-2880.

Grimm, A. M., PAL, J. ; Giorgi, F. (2007). Connection between Spring Conditions and Peak Summer Monsoon Rainfall in South America: Role of Soil Moisture, Surface Temperature, and Topography in Eastern Brazil. *J. Climate*, Vol. 20, p.5929-5945.

Hoogenboom, G. (2000). Contribution of agrometeorology to the simulation of crop production and its application. *Agricultural and Forest Meteorology*, Vol. 103, pp. 137–157.

IBGE (2008). Instituto Brasileiro de Geografia e Estatística. Banco de Dados Agregados. Sistema IBGE de Recuperação Automática - SIDRA. http://www.sidra.ibge.gov.br. Accessed 15 December 2008.

IBGE (2009). Instituto Brasileiro de Geografia e Estatística. Levantamento Sistemático da Produção Agrícola, 21 (9): 1-80. http://www.ibge.gov.br. Accessed 7 November 2009.

Marengo, J. A., Cavalcanti, I. F. A., Satyamurty, P., Bonatti, J., Nobre, C., Sampaio, G., D´Almeida, C., Camargo, H., Cunningham, C., Sanches, M., Pezzi, L. (2000). Ensemble Simulation of Interannual Climate Variability using the CPTEC/COLA AGCM for the período 1982-1991. In: *XI Congresso Brasileiro de Meteorologia*, 2000, Anais.

Marques, J. R., Fontana, D., Mello, R. W. (2005). Estudo da correlação entre a temperatura da superfície dos oceanos Atlântico e Pacífico e o NDVI, no Rio Grande do Sul. *Revista Brasileira de Engenharia Agrícola e Ambiental*, Vol. 9, No. 4, pp.520-526.

Mendonça, A. M., Bonatti, J. P. (2004). Avaliação Objetiva do Sistema de Previsão de Tempo Global por Ensemble do Cptec e Relação entre o Espalhamento e o Desempenho do Ensemble Médio. In: *XIII Congresso Brasileiro de Meteorologia, Sociedade Brasileira de Meteorologia*. Anais.

Moura, A. D. & Shukla, J. (1981). On the dynamics of droughts in northeast Brazil: observations, theory and numerical experiments with a general circulation model. *J. Atmos. Sci.*, Vol. 38, pp. 2653 - 2675.

Pereira, A. R., Pruitt, W. O. (2004). Adaptation of the Thornthwaite scheme for estimating daily reference evapotranspiration. *Agric. Water. Manage*, Vol. 66, pp. 251–257.

Rao, V. B. & Hada, K. (1990). Characteristics of Rainfall over Brazil: Annual Variations and Connections with the Sourthern Oscillations. *Theor. Appl. Climatol.*, Vol. 42, pp. 81-91.

Ropelewski, C. H. & Halpert, S. (1987). Global and regional scale precipitation patterns associated with the El Niño/Southern Oscillation. *Mon. Wea. Rev.*, 115 ,1606-1626.

Ropelewski, C. H. & Halpert, S. (1989). Precipitation patterns associated with the high index phase of the Southern Oscillation. *J. Climate*, Vol. 2, pp. 268-284.

Sivillo, J. K.; Ahlquist, J. E.; Toth, Z. (1997). An Ensemble Forecasting Primer. *Weather and Forecasting*, Vol. 12, pp. 809-818.

Thomas, A. L.; Costa, J. A. (1994). Influência do déficit hídrico sobre o rendimento da soja. *Pesquisa Agropecuária Brasileira*, Vol. 29, No. 9, pp. 1389-1396.

Trenberth, K. E. (1997). The Definition of El Niño. *Bull. Amer. Met. Soc.*, Vol. 78, pp. 2771-2777.

3

Properties of Soybean for Best Postharvest Options

Seth I. Manuwa
Department of Agricultural Engineering, School of Engineering and Engineering Technology, The Federal University of Technology, Akure Nigeria

1. Introduction

Soybean is considered as one very important grain grown commercially in more than 35 countries of the world and the leading producer is the USA (41%) followed by Brazil (23%), Argentina (16%) and China (9%), (F A O 1988).
Soybean contains 40% protein, 35% total carbohydrate and 20% cholesterol-free oil (Deshpande et al., 1993). Mineral content of whole soybean is about 1.7% for potassium, 0.3% for Magnesium, 110 ppm iron, 50 ppm zinc and 20 ppm copper (Smith and Circle, 1972). Soybean is the world leading vegetable oil and accounts for about 20 to 24% of all fats and oil in the world. Soybean is becoming increasingly important in agriculture because it is a food source in human and animal nutrition
So many varieties have been developed around the world considering desired traits. The properties of the developed cultivars could be considered to vary from one cultivar to the other. Sometimes, such variations in properties (especially physical properties) are easily observable, especially in the size and shape of such cultivars. Other properties would have to be measured to know them or to see how they vary from one cultivar to another. By extension, the properties (physical, mechanical and chemical) of a cultivar affect the post harvest options to which a cultivar may be subjected. The challenge of post harvest processing of soybean into animal and human food is increasing by the day. This is so because, the world's population is increasing and the challenge of eradicating hunger and producing quality food on the surface of the earth is staring.
Manuwa (2000, 2007), Manuwa et al.(2004, 2005) reported on similar improved varieties of Soybean that were developed in Nigeria. The major improvements made on soybean varieties from 1987 through 1992 at IITA were to increase grain yield by about 20%, improve resistance to pod shattering and to maintain the level of all other traits constant. In order to design equipment for threshing, winnowing, separation, grading, sorting, size reduction, storage, and other secondary processing of soybean, especially the new improved cultivars, the physical properties should be determined.

2. Varieties of soybeans

So many varieties of soybean have been developed around the world so that it is a major task to know all of them. The main aim of developing varieties (cultivars) was to improve desired traits such as:

- Early maturity,
- Disease resistance e.g phytophthora root rot resistant
- High grain yield,
- Shattering and lodging resistant,
- Intact seed coat and some weathering tolerance,
- Seed quality that meets culinary market standards, for example a light hila culinary type.

A number of varieties have been reported in literature (Tables 1, 2, 3).

3. Harvesting and utilisation of soybean

Needless to say that before soybean can be utilised as food for either man or animal, it must first of all be harvested from the field. However, harvest management is a crucial skill for the specialty soybean producer, simply because the physical appearance of the beans is so important to the buyer. Small-seeded soybeans tend to thresh well, but air adjustments may have to be fine-tuned to remove chaff without blowing the small seeds out the back of the combine. Large-seeded soybeans are extremely prone to mechanical damage during threshing operations, which can knock off the seed coat and/or split the embryo into its cotyledonal halves. The combine's cylinder speeds will have to be slowed considerably to avoid this, and the crop may require harvesting at somewhat higher moisture content.

Prompt harvesting will always be a must, as field deterioration of the seed affecting appearance can commence soon after the moisture content of the physiologically mature seed drops to 14%.. If storage is necessary, the producer will have to ensure that storage facilities are clean, dry, and free from any materials that may be toxic to humans. The conditions under which beans are stored greatly influence the quality of the processed product. Moisture content of 13% or less will prevent mold growth. However, very dry beans tend to split when being transferred, and the splitting lowers the quality.

Soybeans can be used for oil, livestock feeds and for preparing various dishes. A number of traditional foods have been produced from soybeans: *Tofu. Miso, Natto, Tempeh, Soymilk, Soyflour, Soyoil, soy milk* (Bschmann, 2001). According to the report, the size of the seed is often crucial, and may be either smaller or larger than average soybean cultivars. For example, small seeds are sought out for *natto,* while large seeds are preferred for *tofu.* Perfectly round seeds are generally prized, while oblong or kidney-shaped soybeans are usually avoided

4. Post harvest options

Post harvest options are generally all the activities that can be carried out after the harvesting of crops in order to convert it to use by man and animal. It can be classified into primary and secondary processing.

Primary processing: This includes threshing, winnowing, cleaning, separation, grading, sorting, packaging, transportation, marketing, storage and so on.

Grains or seeds from harvesters are not directly suitable for its final use such as re-sowing, animal feed or human consumption. The standards of seeds in the three categories have risen in the last few decades to date. Reasons, especially for re-sowing seeds include the need to achieve international marketing standard, and secondly the uniform, high germination product required in precision drilling.

COUNTRY	VARIETY	YIELD (Kg/ha)	SOURCE
USA	Jim	-	www.ag.ndsu.nodak.edu/aginfo/variety/soybean.htm
	Traill	-	
	RG200RR	-	
	Walsh	-	
	MN0201	-	
	MN0302	-	
	Barnes	-	
	Normatto	-	
	Nannonatto	-	
	Norpro	-	
	SD1081RR	-	
	Sargent	-	
	Surge	-	
	SD1091RR	-	
AUSTRALIA	Arunta	3.81	Adapted from: www.ag.ndsu.nodak.edu/aginfo/variety/ Soybean.
	Stephens	3.80	
	Bowyer	3.55	
	Curringa	3.73	
	Djakal (BAF 212)	3.93	
SLOVENIA	Aldama	1791	Acko and Trdan (2009)
	Borostyan	1242	
	Essor	2757	
	Ika	3138	
	Kador	3702	
	Major	2342	
	Nawiko	2748	
	Olna	2272	
	Tarna	3381	
	Tisa	4216	
NIGERIA	Samsoy 2	1745	Manuwa, 2005; 2007
	TGx 923-2E	1736	
	TGx992-22E	1642	
	TGx 1440-1E	1629	
	TGx 1448-2E	1558	
	TGx 1660- 19F	2134	
	TGx 1489-1D	2071	
	TGx 1447-2D	1970	
	TGx 1437-1D	1877	
	TGx 1455- 2E	1660	
	TGx 849-313D	1524	

Table 1. Some Soybean cultivars from USA, Australia, Slovenia & Nigeria

COUNTRY	VARIETY	YIELD (Kg/ha)	OIL CONTENT (%)
INDIA	Alankar	2200	-
	Ankur	2300	-
	Clark - 63	1800	-
	PK-1042	3300	-
	PK-262	2800	-
	PK-308	2600	20-23
	PK-327	2300	-
	PK-416	3200-3800	41-56
	PK-564	3000	-
	Shilajeeth	2200	-
	Bragg	1800	-
	Calitur	1800	-
	Durga	2100	-
	Gaurav	2200	-
	Indira Soya -9	2300	-
	JS-2	1800	-
	JS-71-05	2000-2400	41
	JS-75-46	1600-3100	-
	JS-76-205	1600-2000	-
	JS-79-81	2800	-
	JS-80-21	2500-3000	-
	JS-90-41	2500-3000	-
	JS-335	2500-3000	17-19
	MACS-13	2700	15-22
	MACS-58	2000-2500	-
	MAUS-47 (Parbhani ona)	2500-3000	20
	MS-335	2800	-
	NRC-12(Ahilya-2)	2800	-
	NRC-2(Ahilya-1)	3500-4000	21
	NRC-7(Ahilya-3)	3200	-
	PK-472	3300	-
	PUSA-16	2800	-
	PUSA-22	2600	-
	PUSA-37	2800	-
	TYPE-49	2200	-
	MACS-57	2800	-
	MACS-450	2500	20
	MAUS-2	2450	-
	MAUS-1	2800	-
	MAUS-32(Prasad)	3000-3500	19
	KB-79(Sneha)	1700	-
	MACS-124	2500-3200	-
	PUSA-40	2600	-

Source: http://agmarknet.nic.in/soybean-profile.pdf

Table 2. Some Soybean cultivars from India

BR 16	BR 36	BRS 153	BRS 155	Embrapa 1
Embrapa 48	FT 106 I	FT 109 I	FT 2	FT 20 (Jau)
FT 4	FT 7 (Taroba)	FT 9 (Inae)	FT Manaca	FT Seriema
IAC 13	IAC 15	IAC 15-1	IAC 16	IAC 4
IAC Foscarin-31	IAC/Holambra twart-1	KI-S 601	KI-S 602 RCH	MS/BR 34 (Empaer 10)
Ocepar 10	Ocepar 16	Ocepar 4 (Iguaçu)	Ocepar 7 (Brilhante)	Ocepar 8
RB 502	RS 9 (Itaúba)	BRS 156	IAC 11	Paraná
BRS 157	BRSMS Apaiari	CEP 12 (Cambará)	Cobb	FT 103
FT 104	FT 2000	IAS 4	Ivorá	Ocepar 17
Ocepar 5 (Piquiri)	RS 5 (Esmeralda)	BRS 134	BRS 136	BRS 138
BRS 65	BRS 66	BRSMA Sambaíba	BRSMA Seridó RCH	BRSMG Confiança
BRSMS Piapara	BRSMS Piracanjuba	CEP 20 (Guajuvira)	DM Nobre	Embrapa 30 (V. R Doce)
Embrapa 62	Emgopa 313 (Anhang.)	Emgopa 316 (Rio Verde)	FT 101	FT 19 (Macacha)
GO/BR 25 (Aruanã)	IAC 100	IAC 12	MS/BR 19 (Pequi)	Ocepar 14
Santa Rosa	BR 28 (Seridó)	BR 38	BRS Carla	RB 603
RB 604	DM 247	DM 339	1 BR 6 (Nova Bragg) Bragg(3)	Bragg
BRS 137	BRS 154	BRS Celeste	BRSMG Garantia	BRSMG Robusta
BRSMG Segurança	BRSMG Virtuosa	BRSMS Mandi	Embrapa 20	(Doko RC)
Embrapa 63 (Mirador)	Emgopa 315 (R. Verm.)	FT 10 (Princesa)	FT 18 (Xavante	FT 6 (Veneza
FT Cometa	IAC 18	IAC 22	MG/BR48 (Gar. RCH)	
UFV 19 FT	UFV/ITM-1	BR 30	BRS 135	BRS Milena
BRSMS Carandá	BRSMS Lambari	BRSMS Piraputanga	BRSMS Taquari	BRSMS Tuiuiú
DM Soberana	Embrapa 64 (Ponta Porã)	Emgopa 301	FT 14 (Piracema)	FT 5 (Formosa)
FT Abyara	FT Maracajú	FT Saray	Fundacep 33	Ocepar 12
UFV 10 (Uberaba)	Bossier	BR IAC 21	BRSMA Parnaíba	BRSMG 68 (Vencedora)
BRSMG Liderança	BRSMG Renascença	BRSMS Bacuri	BRSMS Surubi	DM Vitória
FT 11 (Alvorada)	FT Guaira	IAC 17	IAC 8	IAC 8-2
KI-S 702	KI-S 801	MG/BR-46 (Conquista)	Ocepar 3 (Primavera)	UFV 18 (Patos de Minas)
BRSGO Goiatuba	BR 4	BR 9 (Savana)	BRSMA Pati	Embrapa 4
Embrapa 46	Embrapa 47	Emgopa 304 (Campeira)	Emgopa 309 (Goiana)	FT 8 (Araucária)
FT Bahia	FT Cristalina	FT Cristalina CH	FT Estrela	FT Iramaia
FT Líder	Ivaí	MT/BR 50 (Parecis)	MT/BR 51 (Xingu)	MT/BR 53 (Tucano)
Planalto	UFV 5	BRSMT Crixás	CAC-1	CS 301
CS 303	DM 118	Dourados	FEPAGRO-RS 10	FT 102
IAC 20	M-SOY 2002	BRSGO Catalão	Campos Gerais	Embrapa 9 (Bays)
Emgopa 308 (S.Dourada)	FT 100	FT 45263	FT Canarana	FT Eureka
IAC 14	Invicta	Ipagro 21	RS 7 (Jacuí)	Emgopa 303

Adapted from: Glass et al.(2006)

Table 3. Some Soybean cultivars from Alabama, USA

Requirements for seed cleaning:

- To obtain graded lots of seed which will meet home and international testing standards for the variety under consideration, in terms of purity, viability, vigour and size variation
- To remove completely any seeds, the sale of which in a batch may contravene the Noxious Weeds Act, of some countries.
- To avoid any loss of good seeds in the cleaning process.
- To avoid excessive wear that may be due to sorting machines.
- To remove all contaminants that is capable of damaging subsequent processing machinery such as size reduction machines. Typical contaminants include weed seeds, straws, leaves, stones and soil particles.

Principles of separation:

It is important to identify differences in the physical properties of the seeds and the contaminants that will enable the machine (to be designed) make them flow in different directions. Such properties include the following:

- Seed dimensions: length, width, thickness, geometric mean diameter
- Specific gravity
- Falling rate (float)
- Surface texture, friction
- Colour
- Resilience (ability to bounce)
- Electrical conductance

Typically, most processing machines identify differences in properties between good seeds and contaminants. For example a sieve identifies size while other machines identify a combination of properties such as specific gravity table. The shaking table for example identifies friction, size and density of the seeds. The air-screen cleaner for example make use of differences in size, shape and density of the seeds and such machine range from a small, one fan, single screen machine to the large multi-fan eight screen machine with several air columns. Other machines that are used for primary processing include threshing machine, from simple hand operated threshers to high capacity multi cop threshers, combine harvesters, winnowers, air-screen separators (oscillating or vibrating), graders (band, spiral), separators (spiral, table, magnetic, electrostatic, colour, pneumatic, and so on).

Secondary crop processing: It involves processing of food for direct consumption. This requires grinding, milling, oil extraction and so on. To accomplish these, machines are used such as size reduction machines such as milling machines, dehullers, grinding machines, oil press and so on.

5. Methodology for evaluating soybean properties

Sample preparation: Dry mature Soybeans [Glycine max.] are normally used for all the experiments. Before the experiments, the grains were further cleaned by removing those that were physically bad, unhealthy or broken. The moisture content of the grain would be determined using a standard method. Physical properties were determined at the initial moisture content. Thereafter, grain sample of the desired moisture levels were prepared by adding calculated amount of distilled water and sealed in separated polythene bags. The samples would be kept at about 278 ^{0}K in a refrigerator for 1 week to enable the moisture to distribute uniformly throughout the sample. Before the commencement of a test, the required quantity of the grain was taken out of the refrigerator (if kept there to cool), and allowed to warm up to room temperature at about 305 ^{0}K.

Physical properties: The physical properties of Soybean to be determined include linear dimensions, mass, bulk density, seed density, volume, surface area, sphericity, porosity, coefficient of static friction on structural surfaces and angle of repose, angle of internal friction, terminal velocity. Experiments were conducted at five levels of moisture content in the desired range and replicated five times. Average values were normally reported. The choice of the range of moisture content was due to the fact that the lower limit was the safe storage moisture content, and the upper range, the maximum moisture content obtainable after the seeds were soaked overnight.

Linear dimensions and geometric mean diameter: To determine the size of the grain, 10 sub samples each consisting of 100 grains were randomly taken. From each sub sample, 10 grains were taken and their three linear dimensions namely, length (L), width (W) and thickness (T) were measured with a venier calipers having accuracy of 0.01mm. The geometric mean diameter (D_{GM}) of the grain was calculated by using the following relationship (Sreenarayanan et al 1985, Sharma et al 1985).

$$D_{GM} = (LWT)^{1/3} \qquad (1)$$

Test weight: Sub samples of One, one hundred and one thousand soybean grains from each sample were randomly selected and weighed. The averages of the replicated values are usually reported.

Bulk and seed density: A method similar to that reported by Shephered and Bhardwaj (1986) can be used to determine the bulk density at each moisture level: a 180 ml cylinder was filled continuously from a height of about 15 cm. Tapping during filling was done to obtain uniform packing and to minimize the wall effect, if any. The filled sample was weighed and the bulk density of the material filling the cylinder was computed (Shephered and Bhardwaj, 1986; Deshpande and Ali, 1988; Mohsenin, 1970). The seed density of the grain can be determined by the liquid displacement method to determine the seed volume similar to that reported (Shephered and Bhardwaj, 1986; Deshpande and Ali, 1988).

Sphericity and porosity: According to Mohsenin (1970), sphericity ф, was calculated using the formula.

$$\phi = \frac{(LWT)^{1/3}}{L} \qquad (2)$$

Fractional porosity is defined as the fraction of space in the bulk grain which is not occupied by grain. Thompson and Isaacs (1967) gave the following relationship for fractional porosity.

$$\varepsilon = \frac{(1-\rho_b)}{\rho_s} x100 \qquad (3)$$

where,

ε = fractional porosity

ρ_b = bulk density of the seed

ρ_s = Seed density

Angle of repose: The emptying angle of repose θ is normally determined at the moisture levels using the pipe method (Henderson 1982, Jha 1999). A pipe of 40 cm height and 106 mm internal diameter was kept on the floor vertically and filled with the sample, Tapping during filling was done to obtain uniform packing. The tube was slowly raised above the floor so that the whole material could slide and form a heap. The height above the floor H and the diameter of the heap D at its base were measured with a measuring scale and the angle of reposes θ of the soybean computer using the equation;

$$\theta = Arc\tan(2H / D) \quad (4)$$

Surface area: The surface area of the grain can be found by analogy with a sphere of geometric mean diameter for the different levels given by (McCabe et al., 1986)

$$S = \pi D_{GM}^2 \quad (5)$$

Coefficient of static friction: The coefficient of static friction for seed grain can be determined against structural surfaces such as plywood (with grain parallel to direction of motion and then with grain perpendicular to direction of motion), galvanized steel (GS), glass, concrete and so on. A bottom less wooden box of 150 mm x 150 mm x 40 mm was constructed for this purpose. This was similar to that reported by (Oje, 1994). The box shall be filled with soybean grains on an adjustable tilting surface. The surface would be raised gradually using a screw device until the box started to slide down and the angle of inclination read on a graduated scale.

Terminal velocity: The terminal velocity of soybean at different moisture content can be determined using an air column (Polat et al., 2006). For each test, a seed was dropped from the top of a 75 mm diameter, 1 m long glass tube. The air was made to flow upwards in the tube from bottom to the top and the air velocity at which the sample seed was suspended was noted with an anemometer having at least 0.1 m/s sensitivity.

Angle of internal friction: To determine the angle of internal friction of soybean at different moisture contents, the direct shear method can be used according to Uzuner (1996), Zou and Brucewitz (2001), Molenda et al.(2002) and Mani et al.(2004). Typical velocity to be used during the experiment is 0.7 mm/min (Kibar and Ozturk, 2008) and the angle of internal friction can be calculated using the following equations:

$$\sigma = \frac{N}{A}100 \quad (6)$$

Where: σ - normal stress (kPa), N - load applied over sample (kg), A - cellular area (cm2),

$$\tau = \frac{T_s}{A}100 \quad (7)$$

Where: τ – stress of cutting (kPa), Ts – strength of cutting (kg),

$$\tau = (C + \sigma tg\phi) \quad (8)$$

Where: C- cohesion

6. Rupture force and rupture energy

To determine the rupture force and rupture energy, a Universal Testing Machine (UTM) can be used such as Instron Universal Testing Machine reported by Tavakoli et al. (2009). It was equipped with a 500 kg compression load cell and integrator. The measurement accuracy was

0.001 N in force and 0.001 mm in deformation. The individual grain was loaded between two parallel plates of the machine and compressed along with thickness until rupture occurred as is denoted by a rupture point in the force-deformation curve. The rupture point is a point on the force-deformation curve at which the loaded specimen shows a visible or invisible failure in the form of breaks or cracks. According to them the point was detected by a continuous decrease of the load in the force-deformation diagram. The loading rate of 5 mm/min was used according to ASAE (2006a). The energy absorbed by the sample at rupture was determined by calculating the area under the force-deformation curve from the relationship:

$$E_a = \frac{F_r D_r}{2} \tag{9}$$

Where E_a is the rupture energy in mJ, F_r is the rupture force in N and D_r is the deformation at rupture point (Braga et al., 1999).

Cultivars	MC range %(db)	Dimensions (mm)	Mass (g)	Reference
JS- 7244	8.7- 25.0	L: 6.32 to 6.75 W: 5.23 to 5.55 T: 3.99 to 4.45 GMD: 5.09 to 5.51	- - 1000 grains: 110 to 127	Deshpande et al., 1993
TGX 1440-1E	10.5- 34.1	L: 8.58 to 10.02 W: 6.51 to 7.22 T: 5.43 to 5.69 GMD: 6.71 to 7.44	1 grain: 0.11 to 0.21 100 grains: 14.67 to 19.98 1000 grains: 139.18 to 190.6	Manuwa, 2000
TGX 1871- 5E	7.1- 43.7	L: 7.52- 9.11 W: 6.47- 7.05 T: 5.49- 5.05 GMD: 6.44- 7.29	1 grain: 0.136 to 0.206 100 grains: 12.3 to 16.59 1000 grains: 119.17 to 153.15	Manuwa and Afuye, 2004
TGX 1019-2EB	6.7- 47.1	L: 7.37 to 9.96 W: 6.48 to 7.45 T: 5.33 to 5.54 GMD: 6.33 to 7.39	1 grain: 0.178 to 0.218 100 grains: 13.78 to 18.79 1000 grains: 130.67 to 180.21	Manuwa and Odubanjo, 2005
Unspecified	6.7- 15.3	L: 7.41 to 9.57 W: 5.34 to 6.75 T: 4.5 to 5.17 GMD: 5.62 to 6.94	1000 grains: 121.76 to 223.65	Polat et al., 2006
TGX 1448- 2E	9.9 to 39.6	L: 8.3 to 10.4 W: 6.4 to 7.5 T: 5.4 to 5.8 GMD: 6.6 to 7.6	1 grain: 0.19 to 0.24 100 grains: 15.6 to 19.4 1000 grains: 154.2 to 185.6	Manuwa, 2007
Unspecified	8- 16	L: 7.24 to 8.19 W: 6.79 to 7.12 T: 5.78 to 6.23 GMD: 6.57 to 7.14	NAV	Kibar and Ozturk, 2008
Unspecified	6.92- 21.19	L: 7.27 to 8.23 W: 6.48 to 6.97 T: 5.41 to 5.94 GMD: 6.34 to 6.98	1000 grains: 171.5 to 219.04	Tavakoli et al., 2009

MC= moisture content, NAV= not available

Table 4. Effect of moisture content on mass and dimensional properties of some soybean cultivars

Cultivars	MC range %(db)	Seed density	Bulk density	Sphericity (%)	Porosity (%)	V_t (m/s)	Reference
JS- 7244	8.7- 25.0	1216 - 1124	735 - 708	80.6 – 81.6	40 - 37	NAV	Deshpande et al., 1993
TGX 1440-1E	10.5- 34.1	1184 - 1076	720 - 631	79 – 73.3	23.6 – 34.2	NAV	Manuwa, 2000
TGX 1871- 5E	7.1- 43.7	1222.3 – 935.7	686.5 – 616.7	85.87- 78.23	25.64 – 40.96	NAV	Manuwa and Afuye, 2004
TGX 1019-2EB	6.7- 47.1	1157 - 952	728.5 – 608.4	86 – 74.9	23.46 – 42.33	NAV	Manuwa and Odubanjo, 2005
Unspecified	6.7- 15.3	1062.6 to 1086.5	804.8 to 689.3	75 to 72	51 to 44.2	7.13 to 9.24	Polat et al., 2006
TGX 1448- 2E	9.9 to 39.6	1465 - 1074	714 - 638	79.1 – 72.7	19.5 – 33.7	NAV	Manuwa, 2007
Unspecified	8- 16	983.33 – 905.67	766.12 – 719.00	91- 87	22.58 – 20.61	NAV	Kibar and Ozturk, 2008
Unspecified	6.92- 21.19	1147.86 to 1126.43	650.95 to 625.36	87.25 to 84.75	43.29- 44.48	NAV	Tavakoli et al., 2009

MC= moisture content, NAV= not available
Vt = terminal velocity

Table 5. Effect of moisture content on density, sphericity, porosity and terminal velocity of some soybean cultivars

Cultivars	MC range %(db)	Angle of repose (degree)	Coefficient of static friction				Reference
			Galvanised steel	PWLG	PWDG	Glass	
JS- 7244	8.7- 25.0						Desshpande et al., 1993
TGX 1440-1E	10.5-34.1	24.1 – 31.5	0.344 – 0.509	0.446 – 0.600	0.481 – 0.653	-	Manuwa, 2000
TGX 1871- 5E	7.1- 43.7	23.43 – 32.23	0.434 – 0.679	0.4245 – 0.601	0.4243 – 0.6789	-	Manuwa and Afuye, 2004
TGX 1019-2EB	6.7- 47.1	25.87 – 32.45	0.3839 – 0.5774	0.4877 – 0.6249	0.4922 – 0.6876	NAV	Manuwa and Odubanjo, 2005
Unspecified	6.7- 15.3		0.21 – 0.34	0.22 – 0.35*		0.19 – 0.33	Polat et al., 2006
TGX 1448- 2E	9.9 to 39.6	24.2 – 30.2	0.391 – 0.510	0.466 – 0.601		-	Manuwa, 2007
Unspecified	8- 16		0.164 - 0.286				Kibar and Ozturk, 2008
Unspecified	6.92-21.19	24.56 – 29.93	0.28 – 0.326	0.287 – 0.361		0.262 – 0.307	Tavakoli et al., 2009

MC = moisture content, NAV= not available, PWLG = plywood parallel to grain, PWDG = plywood perpendicular to grain *PLWD = plywood

Table 6. Effect of moisture content on angle of repose and coefficient of static friction of some soybean cultivars

7. Estimated values of soybean properties

Some typical values and models of physical, mechanical and aerodynamic properties of soybean cultivars are reported in this section (Tables 4 to 6). Table 4 shows the effect of moisture content on mass and dimensional properties of some soybean cultivars. Table 5 shows the effect of moisture content on density, sphericity, porosity and terminal velocity of some soybean cultivars. Table 6 shows the effect of moisture content on angle of repose and coefficient of static friction of some soybean cultivars.

8. General comments

It can be seen that the number of soybean cultivars that have been developed around the world is numerous and can be better imagined. However, it appears that very little has been reported in literature concerning physical and engineering properties of such soybean cultivars. Nevertheless, it is obvious that post harvest options or technology are *sine qua non* in order to convert soybean seeds into quality food for human and animal in view of the quality of food nutrition available in the seeds.

9. References

ASAE.(2006a) Compression tests of food materials of convex shape. S368.4, 609- 616

Braga G.C., Couto S. M., Hara T., J.T.P.A. Neto (1999). Mechanical behaviour of macadamia nut under compression loading. Journal of Agricultural Engineering Research, 72: 239- 245

Carman K (1996). Some physical properties of lentil seeds. Jour Agric Engng Res 63: 87-92

Carman K, Ogut H (1991). The determination of porosity rate on different moisture content of several crops. Agric Fac, Univ of Selcuk, Kenya, Turkey 1: 55-62

Chung J H, Verma L R (1989). Determination of friction coefficients of beans and peanuts. Trans ASAE 32: 745-750

Deshpande S D, N Ali (1988). Effect of harvest moisture on some engineering properties of 'Wheat'. Int AgroPhys 4: 83-91

Deshpande S D, Bal S, Ojha T P (1993). Physical properties of soybean seeds. J Agric Eng Res 56: 89-92

Dutta S K, Nema V K, Bhardwaj R K (1988). Physical properties of gram. J Agric Eng Res 39: 259-268

F A O (1988). Production yearbook. Food and agricultural organization of the united nations, Rome, Italy

Fraser B M, Verma S S, Muir W E (1978). Some physical properties of fababeans. J Agric Eng Res 22: 53-57

Glass K.M., Delaney D.P., E.V. Santen (2006). Performance of Soybean Varieties in Alabama, USA. Agronomy and Soils Department Series No. 279. Alabama Agricultural Experimental Station

Henderson M E (1982). Agricultural processes engineering 3rd edn. AVI Publ Co. Inc, Westport, Connecticut

Jha S W (1999). Physical hygroscopic properties of makhana. J Agric Engng Res 72: 145-150

Josh D C, Das S K, Mukherjee R K (1993). Physical properties of pumpkin seed. J Agric Eng Res 54: 219-229

Kibar H. And T. OOzturk (2008). Physical and mechanical properties of soybean. Int. Agrophysics, 22, 239- 244

Liu B Y, Nearing M A, Baffaut C, Ascongh J C (1997). The watershed model: III. Comparisons to measured data from small watersheds. Trans ASAE 40: 945-952

Mani S., Tabil L. G., and Sokhansanj S. (2004). Mechanical properties of corn stover grind. Transactions of the ASAE, 47, 1983- 1990.

Manuwa S I (2000). Properties of soybean 'TGX1440-1E' Annals of Agric Sci. 1 (3) 85-93.

Manuwa S I and Afuye G. (2004). Moisture-dependent physical properties of soybean (var TGX1871-5E) Nigerian Journal of Industrial and System Studies , Vol3 (2): 45-53.

Manuwa S I and Odubanjo O.O. (2005). Physical properties of soybean (var TGX 1019-2EB) LAUTECH Journal of Engineering and Technology, Vol 3 (2): 88-92.

Manuwa S I (2007). Moisture-dependent physical properties of improved soybean 'var TGX1448-2E' J. of Food Sci and Technol, 44 (4): 371- 374.

McCabe W L, Smith J C, Harriot P (1986).Unit operation of chemical engineering McGraw Hill Book of Co. New York

Mohsenin N N (1970). Physical properties of plant and animal material. Gordon and Breeach, New York

Molenda M., Montross M D., Horabik j., and Rose I.J. (2002). Mechanical properties of corn and soybean meal. Transactions of the ASAE, 45, 1929- 1936.

Ogut H (1998). Some physical properties of white lupin. J Agric. Eng Res 69: 273-277

Ogut H and K Carmam (1991). Determination of coefficient of friction on different surfaces of small grain crops. National Symp Mechanization in Agriculture, Konya, Turkey 471-480

Oje K (1994). Moisture dependence of some physical properties of cowpea. Ife J Technol 4 (2): 23-27

Polat R., Atay U., and C. Saglam (2006). Some physical and aerodynamic properties of soybean. Journal of Agronomy 5(1): 74= 78

Sharma S K, R K Dubey, C K Techchandani (1985). Engineering properties of black gram, soybean and green gram. Proc Indian Society of Agricultural Engineers, 3: 181-185

Shephered H, Bhardwaj R K (1986). Moisture-dependent physical properties of pigeon pea. J Agric. Eng Res 35: 227-234

Smith A K, Circle S J (1972). Chemical composition of the seed. *In* soybeans: chemistry and technology, AVI publ Co., Smith A K, Circle S J (eds), Westport, Connecticut,1: 61-92

Sreenarayanan V V, Subramainan V, Visvanthan R (1985). Physical and thermal properties of soybean. Proc Indian Society of Agricultural Engineers 3: 161-167

Tavakoli H., Rajabipour A., S.S. Mohtasebi (2009). Moisture-dependent some Engineering properties o Soybean grains. Agricultural Engineering International: the CIGR Ejournal. Manuscript 1110. Vol. XI. February, 2009

Thompson R A, Isaacs B W (1967). Porosity determinations of grains and seeds with comparison Pyonometer. Trans ASAE 10: 693-696

Uzuner, B A (1996). Soil mechanics. Technique Press, Ankara, Turkey

Zou Y. and G H. Brucewitz (2001). Angle of internal friction and cohesion of consolidated gound marigold petals. Transactions of the ASAE, 44, 1255- 1259.

4

Fuzzy Logic System Modeling Soybean Rust Monocyclic Process

Marcelo de Carvalho Alves[1], Edson Ampélio Pozza[2], Luiz Gonsaga de Carvalho[2] and Luciana Sanches[1]
[1]*Federal University of Mato Grosso*
[2]*Federal University of Lavras*
Brazil

1. Introduction

The soybean rust (*Phakopsora pachyrhizi* H. Sydow & P. Sydow) was reported in soybean (*Glycine max* L. Merrill) in many tropical and subtropical regions, causing significant reductions in productivity and quality of seeds (Bromfield, 1984, Hartman et al., 2005; Kawuki et al., 2004; McGee, 1992, Medina et al., 2006, Sinclair & Backman, 1989; Vale, 1985, Yang et al., 1990, Yang et al., 1991; Yorinori & Lazzarotto, 2004), with losses of up to 70% in production (Bromfield, 1976). The rust occurs in almost all soybean fields in Brazil. The states with high occurrence of the disease in 2003/04 were Mato Grosso, Goias, Minas Gerais and São Paulo. Considering Brazilian states in 2002/03, soybean rust caused losses of 4.011 million of megagrams or the equivalent of US$ 884.25 million, while in 2004, the losses were approximately US$ 2.28 billion (Yorinori & Lazzarotto, 2004).
The success of pathogen infection depends on the sequence of events determined by spore germination, appressoria formation and penetration. Each of these events, the subsequent colonization and sporulation, are influenced by biotic factors such as pathogen-host and abiotic environment. Among abiotic factors, temperature and leaf wetness play a crucial role, especially in the monocyclic germination, infection and colonization of *P. pachyrhizi* in soybeans. Thus, several studies were conducted to model the effects of temperature and humidity on the disease progress for Brazilian cultivars (Vale, 1984, Vale et al., 1990) and for different cultivars adapted to other countries (Batchelor et al., 1997, Kim et al., 2005, Marchetti et al. 1975; Melching et al. 1989; Pivonia & Yang, 2004, Reis et al., 2004). According to Sinclair & Backman (1989), the range of optimum temperature for infection is 20 °C to 25 °C. Under these conditions, with the availability of free water on the leaf surface, the infection starts after 6 hours of the deposition of the spore (Marchetti et al., 1975; Melching et al. 1989; Vale et al., 1990). However, after 12 hours (Marchetti et al. 1975; Melching et al., 1989) up to 24 hours of leaf wetness (Vale et al., 1990) was more successful in establishing infection (Sinclair & Backman, 1989). Therefore, such studies are important for estimating the potential occurrence and formulate strategies to control disease in geographic regions not yet reported (Pivonia & Yang, 2005) and to investigate the potential of spreading in major producing regions throughout the months of the year (Alves et al., 2006; Pivonia & Yang, 2004).

Linear regression approaches (Vale et al., 1990), nonlinear regression (Reis et al., 2004), artificial intelligence techniques, such as neural networks (Batchelor et al., 1997, Pinto et al., 2002) and fuzzy logic (Kim et al., 2005), were used to model the influence of abiotic variables on the disease progress. However, in the case of using regression and neural networks, there is a need to perform data collection for the best fitting models (Reis et al., 2004) and network training (Batchelor et al., 1997). On the other hand, considering fuzzy logic technique, quantitative measures are no longer urgently needed to develop a model (Kim et al., 2005), notwithstanding the choice of these observations are used in the modeling process (Mouzouris & Mendel, 1997). In this context, fuzzy logic was applied to model physical, chemical and biological process, with uncertainty and ambiguous nature (Kim et al., 2005, Massad et al. 2003; Schermer, 2000; Uren et al., 2001).

Other features that justify the application of fuzzy logic systems (FLS) are related to the flexibility of the technique, ease of understanding the concepts, ability to model complex nonlinear functions, development based on the expertise of specialists, integration with other automation techniques and finding support in the natural language used by humans (Cox, 1994; Tanaka, 1997).

Likewise, there is no precise measurements of the influence of other variables such as soil fertility, resistant cultivars, climatic variables, management practices in the progress of the disease, being necessary to create a subjective measure to assess the potential progress of the disease.

Considering the importance of the soybean crop in Brazil, as well as the risk caused by the rust and the losses due to its occurrence, it is necessary to know epidemiological aspects of the disease in Brazilian cultivars in order to enable disease intensity prediction. Therefore, the objective of this work was to study the effects of temperature and leaf wetness on the monocyclic process of soybean rust in cultivars Conquista, Savana and Suprema, based on a fuzzy logic system and nonlinear regression models.

2. Material and methods

The phases of problem selection, development, evaluation and implementation were used to develop the FLS.

2.1 Problem selection

As criteria to study the application of a FLS for estimating soybean rust, there were considered the selection of the problem, seasonal occurrence, the existence of experts and literature in the area, the soybean crop importance and the ease of acquiring information. In the prototype development phase, information from the literature about the epidemiology of the disease and experts in the field were consulted (Batchelor et al., 1997; Bromfield, 1984, Kim et al., 2005, Marchetti et al., 1975; Melching et al., 1989; Pivonia & Yang, 2004, Reis et al., 2004; Valley, 1984, Vale et al., 1990). Some important aspects were considered in the design, such as simplicity to facilitate its subsequent implementation, to be based on knowledge and experience of experts in order to produce accurate and flexible results and the possibility to incorporate new variables (Von Altrock, 1995; Zadeh, 1965).

2.2 Development

In the early stage of development, membership functions were defined into five categories related to the variables temperature, leaf wetness, and area under the disease progress curve, classified as very low, low, medium, high and very high, in order to constitute the fuzzy sets. It was specified a set of if-then rules, with the input and output variables to form the inference mechanism (Tanaka, 1997). The system used the implication operator Min of Mamdani, because it was intuitive and widely accepted to translate the human experience (Driankov et al., 1993), and the limited sum composition method (Cox, 1994), chosen due to the nature of the rules, as each one defined an increase or decrease in the occurrence of rust (Vargens et al., 2003). When compared to the operator max, which considers only the maximum value of relevance, the limited summation method was more suitable, similar to that found in the study of Vargens et al. (2003). At the final stage of development, corrections were made to confirm the internal logic and its full operation based on expert knowledge, references in the area, fuzzyfication, inference and defuzzification processes (Figure 1).

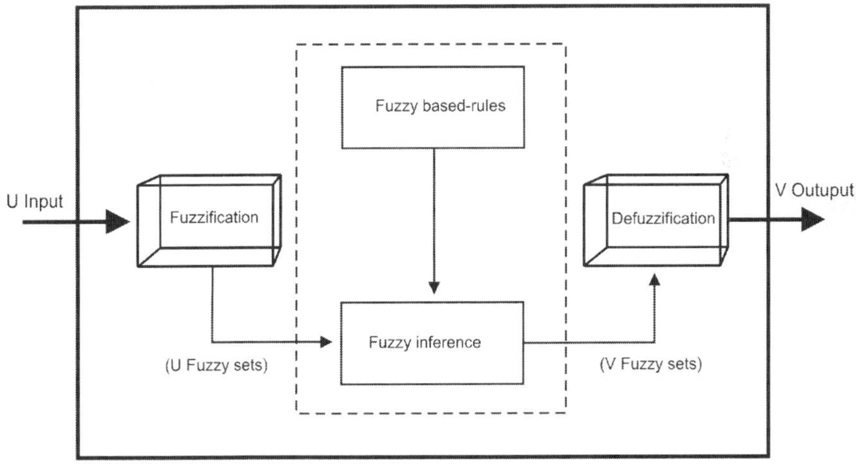

Fig. 1. Structure of a logic fuzzy system (Adapted from Mouzouris & Mendel, 1997).

2.3 Validation

Data collection for the system validation was obtained through experiments conducted in growth chambers at the Laboratory of Epidemiology and Management of the Department of Plant Pathology, Federal University of Lavras (UFLA), Brazil. The experimental design was in blocks at random arrangement of 4 x 5 factorial treatments, with three replications, four temperatures (15 °C, 20 °C, 25 °C and 30 °C) and five leaf wetness periods (0, 6, 12 , 18 and 24 hours). After designing the layout, soybean cultivars Conquista, Savana and Suprema were planted in pots containing 5 kg of soil mixture, sand and organic matter (manure) in the proportion 2:1:0.5. Thinning was performed 15 days after planting, leaving two plants per pot. The plants were kept at green house until V3 vegetative stage, according to the soybean phenological scale proposed by Ritchie et al. (1982). The inoculum of the fungus was obtained by collecting *P. pachyrhizi* uredospores directly from Conquista diseased plants, in a greenhouse, at the UFLA experimental campus, and stored in liquid nitrogen (-180 °C). Test was performed to verify the viability of the inoculum before the inoculation, which presented 89% germination.

The inoculation was done by spraying all the leaves with a suspension at a concentration of 10^4 uredospore of *P. pachyrhizi*.mL^{-1} until runoff. For the different periods of leaf wetness, the plants recently sprayed with the suspension of uredospore were kept in a moist chamber for the duration of each treatment, wrapped in clear plastic bags. In the treatment of zero hours of leaf wetness, the plants were taken to the growth chambers without moist step, allowing the rapid drying of the sprayed suspension. During the experiment, irrigation was accomplished by depositing water directly in the lap of the plants. From the 6th day after inoculation, there were four disease severity (% leaf area with lesions) and incidence (% of leaflets core of all trifoliate leaves of plants) every three days, depending on the onset of signs. The severity and incidence of rust were recorded in the central leaflet of all trifoliate leaves of each plant. The severity was obtained using Bromfield (1984) scale: where score 0 = 0%, 1 = 0.15%, 2 = 1.0%, 3 = 2.5%, 4 = 8.0%, 5 = 13.0%. By having the data of disease severity, the area under the curve of progress of disease incidence (AUDPCI) and severity (AUDPCS) was calculated, according to Campbell & Madden (1990), for each combination of temperature and leaf wetness inside of each cultivar susceptible to disease (Zambenedetti, 2005).

After obtaining the data, it was proceeded the analysis of variance for AUDPCI and AUDPCS, according to a factorial design between temperature and leaf wetness. The significant variables in the F test were subjected to analysis of nonlinear regression to obtain equations to represent the effects of the interaction of temperature and leaf wetness duration on the rust intensity (Figure 2). It is noteworthy that the dependent variable in the case of FLS was named as area under the disease progress curve (AUDPC), since in this case, both results of AUDPCI and AUDPCS were considered for the FLS development. The FLS was validated using Pearson correlation coefficients and linear regression between estimated and observed values of diseased plants, comparatively with the nonlinear regression models.

2.4 Implementation

After the validation phase, the implementation phase was proceeded with the use of a geographic information system and geostatistics (Burrough & McDonnell, 1998). Thus, the FLS was used to estimate the disease based on observations of mean monthly temperature of 39 weather INMET stations, referring to Climatological Normals (1961-1990) (BRASIL, 1992) for the month of January, simulating the occurrence of leaf wetness for 12 hours at all considered stations, because there is no historical data of this variable (BRASIL, 1992). As the number of weather stations available in Minas Gerais and surrounding regions are scarce, the co-kriging technique (Isaaks & Srivastava, 1989) was used to improve the quality of the data interpolation and to increase the spatial resolution of the estimates, through a database of altitude, latitude and longitude, in a regular 1 km grid within the boundaries of the Minas Gerais state, considering the digital elevation model of the surface with a spatial resolution of 90m (NASA, 2005). After, co-kriging was used to map the potential spatial progress of the disease (Figure 3). Co-kriging technique was chosen to explore the known influence of altitude, latitude and longitude in the variation of temperature (Sediyama Mello Jr., 1998), as well as in the occurrence of disease (Yang & Feng, 2001), and to improve the spatial resolution of the estimates.

After mapping rust, the same co-kriging procedure was applied to characterize the climate of Minas Gerais, in order to verify the relationship between the intensity of rust and moisture annual Thornthwaite index (Iu), as well as the annual potential

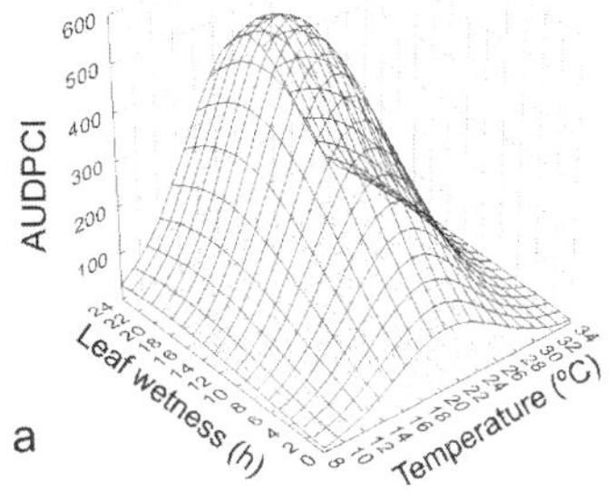

$AUDPCI=557.182*exp(-3.8363*(((T-20.5104)/15.0491)^2+((M-17.3542)/28.1683)^2)\ R^2=0.77**$

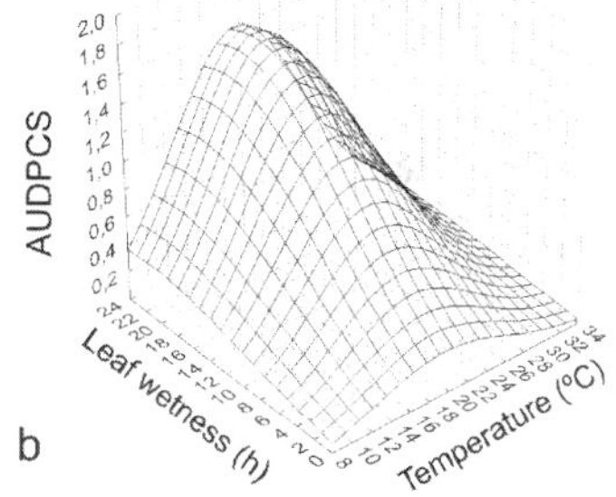

$AUDPCS=1.65469*exp(-6.84246*(((T-19.4871)/24.6295)^2+((M-20.9021)/42.1387)^2)\ R^2=0.50*$

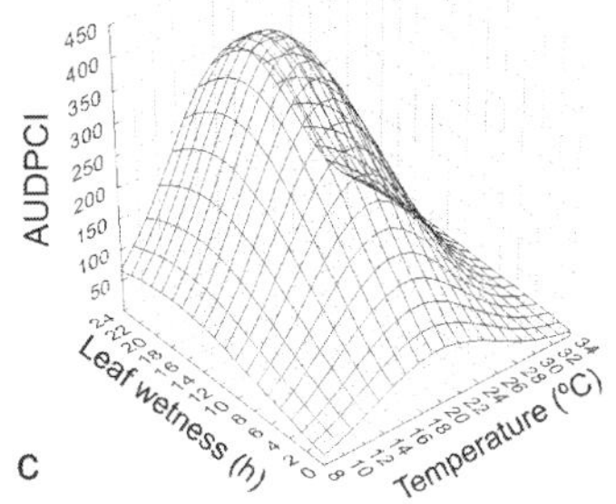

$AUDPCI=408.431*exp(-6.71891*(((T-19.4741)/23.2664)^2+((M-17.8745)/37.4345)^2)\ R^2=0.67**$

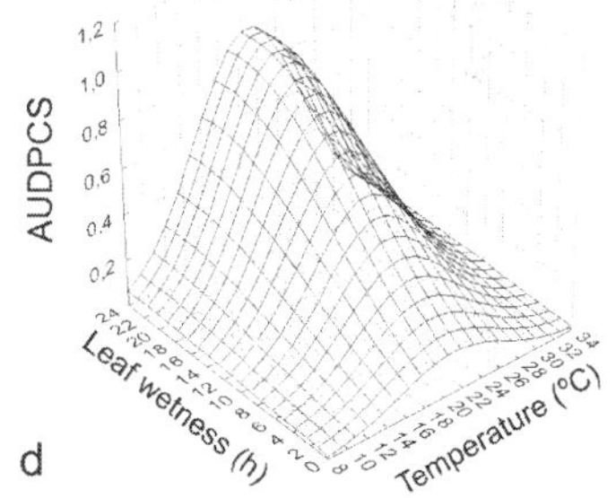

$AUDPCS=0.967155*exp(-19.2256*(((T-20.9475)/34.5077)^2+((M-22.4356)/75.4584)^2)\ R^2=0.57**$

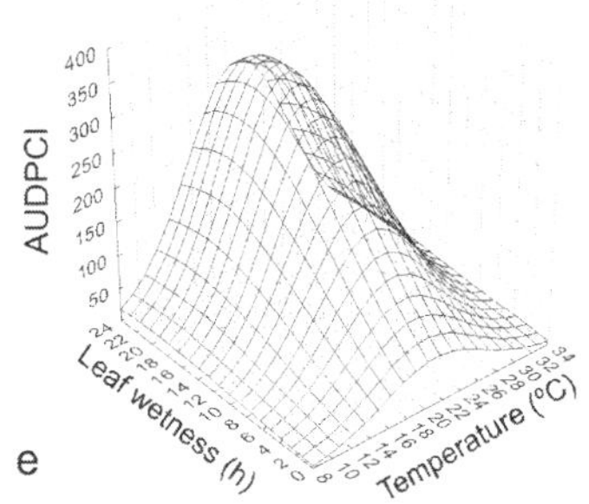

$AUDPCI=340.711*exp(-1.32273*(((T-21.1179)/9.00065)^2+((M-19.2224)/17.5206)^2)\ R^2=0.76**$

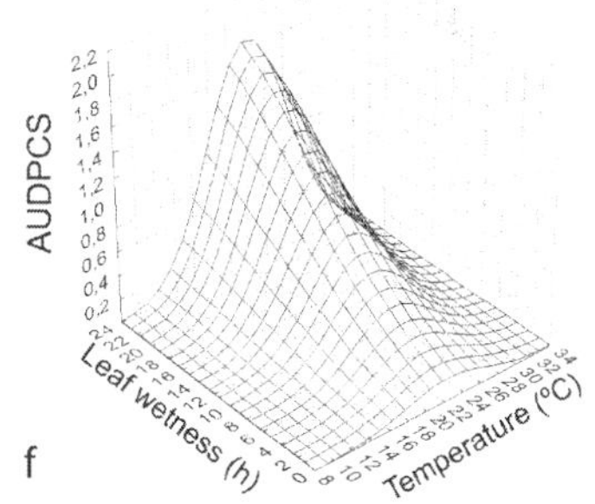

$AUDPCS=1.99399*exp(-61.473*(((T-22.1362)/48.3311)^2+((M-29.0547)/147.917)^2)\ R^2=0.58**$

Fig. 2. Nonlinear regression models used to represent the incidence progress (AUDPCI) (a, c, e) and severity curves (AUDPCS) (b, d, f) of soybean rust in cultivars Conquista (a, b), Savana (c, d) and Suprema (e, f).

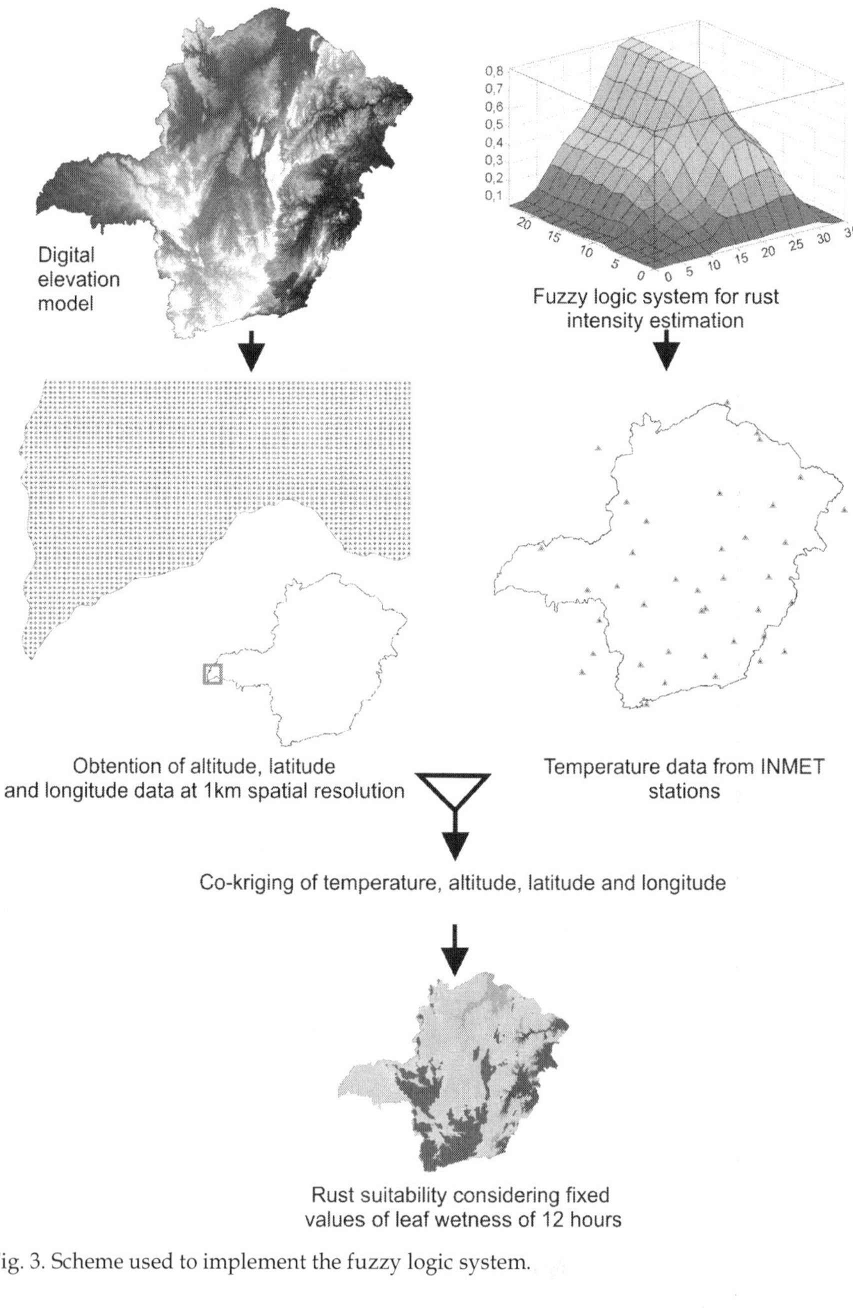

Fig. 3. Scheme used to implement the fuzzy logic system.

evapotranspiration (ETp) (Thornthwaite, 1948; Thornthwaite and Mather, 1955). The climatic characterization was based on climatological data of temperature and rainfall referring to 32 locations INMET (BRAZIL, 1992). For this, the ETp was estimated by Thornthwaite method based on average monthly values of air temperature and, thereafter, in possession of rainfall and considering the storage capacity of soil water equivalent to 100 mm (mean value for most crops), the climatic water balance was calculated. Based on values obtained from the excess and deficit water balance, it was possible to estimate the water index and index of aridity, in order to obtain the Iu, for each location. It is noteworthy that the method of ordinary kriging was used in a comparative manner with co-kriging to estimate areas favorable to rust in Minas Gerais in order to compare the quality of the estimates of both methodologies.

3. Results and discussion

In the construction of the FLS, the input and output variables were divided into five categories, according to information from experts, and were classified according to the proximity of the universe of discourse. For example, in a position of fuzzy sets in the universe of high temperatures, and fuzzy sets in the universe of low duration of leaf wetness, implied unfavorable conditions to the progress of soybean rust, characterized by membership functions (Figure 4). Then, it was specified a set of rules based on expert knowledge, according to the influence of temperature and leaf wetness on disease (Table 1), to form, together with the fuzzy sets, the inference system (Figure 5). Then, a response surface of the FLS was generated for the input and output variables (Figure 6). At the end of the development phase, tests were performed with data in order to verify full operation of the FLS, according to an appropriate structure to process input data of temperature and leaf wetness, giving a response concerning the area under the disease progress curve consistent with the literature (Batchelor et al., 1997; Bromfield, 1984; Kim et al., 2005; Marchetti et al., 1975; Melching et al., 1989; Pivonia & Yang, 2004; Valley, 1984, Vale et al., 1990).

Subsequently, it was proceeded the model validation based on data from the experiment carried out under controlled conditions. In this case, models of nonlinear regression were fitted to data of rust incidence and severity in the cultivars Conquista, Savana and Suprema, to compare with the developed FLS. Thus, it could be observed higher correlation with observed estimates of FLS than the nonlinear regression models used to estimate the monocyclic process of rust in all the progress curves of incidence and severity, except for the severity variable of the Suprema cultivar (Figures 7, 8 and Table 2). This probably occurred because, in this particular case, the leaf wetness duration tended to increase until the period of 29 h, unlike the progress curves of the disease of the Savana and Conquista cultivars, which showed response of leaf wetness between 17 h and 23 h.

Similar to this study, Kim et al. (2005) developed an FLS for estimating the infection rate of apparent severity of soybean rust considering the results of 73 field experiments in Taiwan. However, in this case, the model was developed based on the average night temperature, maximum and minimum temperatures of the day, associated with biological criteria relating to the disease, in order to explain 85% of the progress and severity of the disease in TK 5 and G8587 cultivars, especially in the epidemic early.

Castañeda-Miranda et al. (2006) also developed an FLS to control the environment inside a greenhouse with meteorological variables, however, after validating step, the system was implemented in an electronic circuit integrated with FLS.

Similarly to the work of Castañeda-Miranda et al. (2006), it is expected to develop an electronic circuit to integrate the FLS developed in this study, with automated weather stations, to assist the decision making of farmers on the most appropriate time to conduct the integrated management of soybean rust.

After estimating the potential progress of the disease in INMET weather stations, co-Kriging was used to map potential suitability areas for disease occurence, considering better application of co-kriging method when compared to ordinary kriging (Table 2).

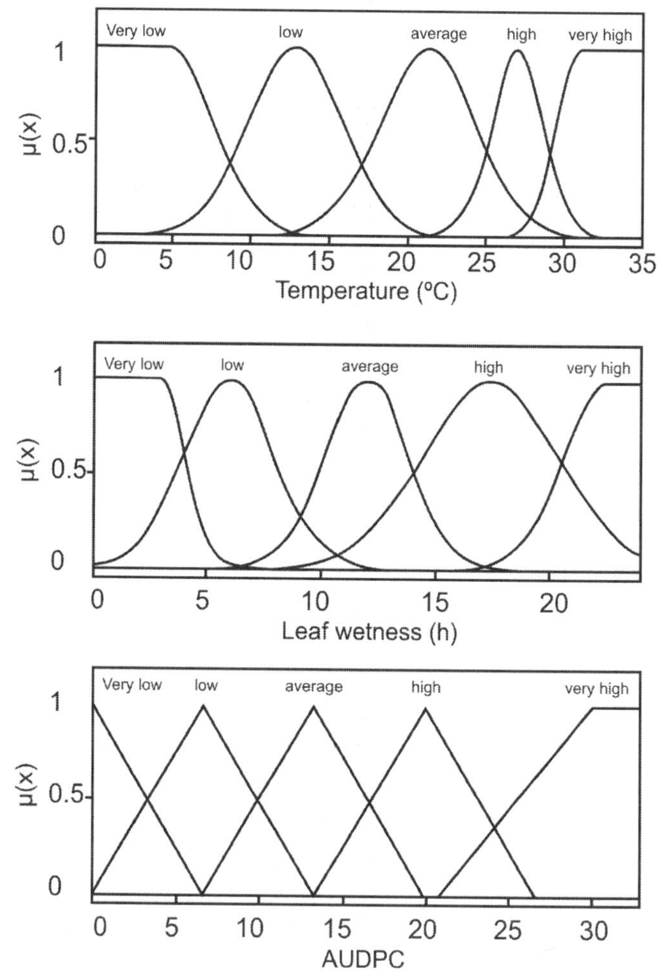

Fig. 4. Membership functions of temperature, leaf wetness, and area under the progress curve of soybean rust severity (AUDPC).

Rule N°	If (Temperature - °C)	and (Leaf wetness - hours)	Then (AUDPC)
1	-	Very low	Very low
2	Very high	Low	Very low
3	Very low	Average	Very low
4	Very low	High	Very low
5	Very low	Very high	Very low
6	Low	Low	Low
7	Low	Average	Average
8	Low	High	Average
9	Low	Very high	Average
10	Average	Low	High
11	Average	Average	Very high
12	Average	High	Very high
13	Average	Very high	Very high
14	High	Low	Average
15	High	Average	Average
16	High	High	Average
17	High	Very high	Average
18	Very high	Low	Low
19	Very high	Average	Low
20	Very high	High	Low
21	Very high	Very high	Very low

Table 1. Rules used to develop the FLS to characterize the monocyclic process of soybean rust

Other studies had applied the co-kriging to improve estimates based on covariates. For example, Desbarats et al. (2002) also used the co-kriging to estimate the water table of the aquifer Oak Ridges Moraine, in Ontario, Canada, in an area of 250 km^2, considering altitude as covariate. According to the authors, areas with higher water table occurred in areas of higher altitude.

Thus, it was observed the most favorable areas for disease in regions of higher altitude and less favorable areas, with the blue color, especially in the east and north of Minas Gerais (Figure 10). Based on a comparison of the ranges used for classifying the disease as high or low, relative to other rust forecasting models previously developed, there was consistency of the results according to the available literature. Therefore, to Sinclair (1975) and Bromfield (1981), the optimum temperature for infection by *P. pachyrhizi* was in the range of 18 °C to 21 °C if the leaf remain wet for at least 16 hours. Vale (1985), studying the cultivar Paraná, cited the value of 20 °C and relative humidity above 90%, while Casey (1980), in Australia, determined temperature of 18 °C to 26 °C and extended periods of leaf wetness, approximately 10 hours per day, required to occur epidemics with high rates of progress and severity. In another review, Sinclair & Backman (1989) cited the optimum range of temperature for infection by *P. pachyrhizi* on soybeans from 20 °C to 25 °C, ie, all these authors observed temperatures around 20 °C, although in some cases close to 25 °C as the optimum to occur higher intensity of the disease, with extended periods of leaf wetness . These differences may be related to the cultivars, as discussed earlier. Regarding the limiting temperatures, Casey (1980) quoted values above 30 °C and below 15 °C, in dry conditions, ie with fewer hours of leaf wetness, as responsible for delaying the progress of the rust, while Bromfield (1981), quoted temperatures below 20 °C or above 30 °C. According to Vale et al. (1990), temperature and leaf wetness can be determinant for sporulation and reduction of the latent period of the disease in cultivar Paraná with 20 °C of temperature and 12 h to 24 h of leaf wetness, similar to the present study, with Conquista, Savana and Suprema cultivars. Marchetti et al. (1975) already studied the effect of rust in cultivar Wayne and observed that plants incubated at 27.5 °C showed no infection regardless of the leaf wetness. Likewise, Melching et al. (1989), studying the effects of duration, frequency and temperature of leaf wetness periods on soybean rust in Taiwan, Wayne cultivar, found that after 8 hours of dew period between 18 °C and 26.5 °C, intensities of Injuries were 10 times higher than those in the 6 hours corresponding temperatures, despite the increased of leaf wetness from 12 to 16 hours did not result in significant increase in the rust intensity, even in favorable temperatures between 18 °C and 26.5 °C. There was no appearance of lesions at 9 °C and 28.5 °C even in wet periods of 20 hours. Thus, because the Wayne cultivar and the rust race being probably adapted to conditions of latitude, longitude, different from Lavras, Minas Gerais, where *P. pachyrhizi* was first reported in Brazil (Bromfield, 1984), probably under conditions of temperatures above 28 °C, there was no disease infection in Wayne cultivar, deviating from this study.

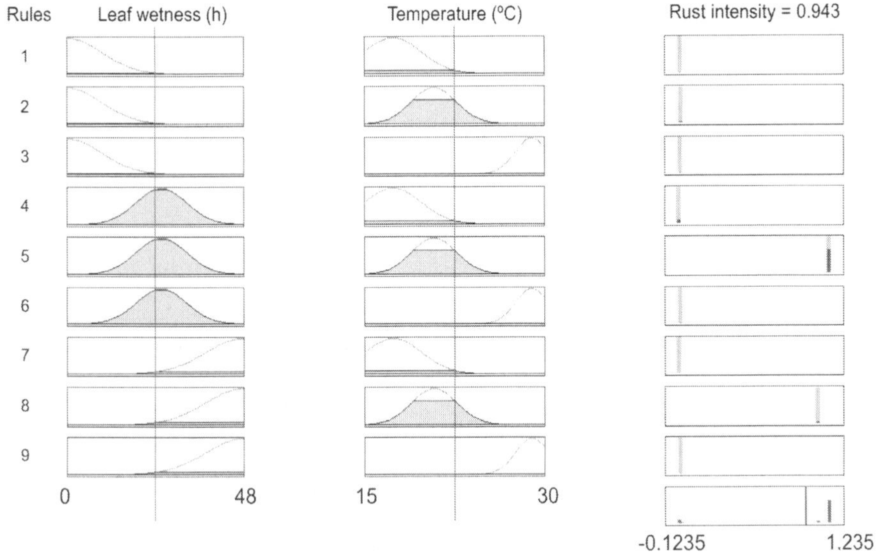

Fig. 5. Inference fuzzy diagram used to estimate the monocyclic process of soybean rust.

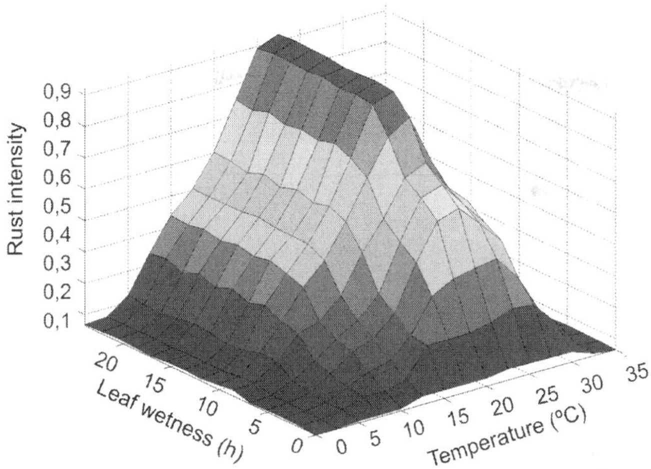

Fig. 6. Three-dimensional representation of the FLS model for estimating the monocyclic process of soybean rust, based on temperature and leaf wetness.

However, the climate model to predict soybean rust in soybeans in Brazil, Reis et al. (2004), based on data from Melching et al. (1989) with cultivar Wayne, suggested daily value of the probability of infection of uredospore with occurrence of infection even at temperatures around 29 °C with 16 hours of leaf wetness, and at lower temperatures of 9 °C with 11 hours of leaf wetness, after nonlinear regression model fit to the observed data, disagreeing with the results of themselves Melching et al. (1989), but similar to situations found in the present study, with Brazilian cultivars adapted to the region of Lavras, Minas Gerais. Thus, despite having been reported in the literature similar responses of the disease with respect to temperature variation and leaf wetness duration, in some situations, differences in the intervals for disease suitability probably occurred due to host characteristics, differences between genotypes, vegetative stage, soil and plant nutrition, in order to justify the development of a subjective measure for evaluating the monocyclic disease process, as in the case of the present FLS.

After spatialize rust using the co-kriging technique, the same procedure was applied to characterize the climate of Minas Gerais, in order to verify the relationship between the intensity of rust with the moisture annual index of Thornthwaite (Iu), as well as the annual potential evapotranspiration (ETp) (Thornthwaite, 1948; Thornthwaite & Mather, 1955). Therefore, comparing the maps of disease severity (Figure 10) with those of ETp and Iu (Figure 11), it could be seen correspondence between areas of high rust intensity with lowest values of ETp and highest values of Iu. This relationship was also verified by the linear relationship of disease intensity with ETp and Iu, in the 39 INMET evaluated localities (Figure 12) and the negative correlation between the intensity of rust with the ETp ($r = -0.86457$, $p < 0.0001$) and positively with Iu ($r = 0.76682$, $P < 0.0001$). Another finding was the better application of co-kriging method when compared to kriging method, for detailing the spatial resolution of a database of macroclimatic variable scale from a database of covariates on mesoclimatic scale (Table 3). Likewise, based on climatic zoning, the planning and implementation of various areas such as industry, agriculture, transport, architecture, biology, medicine (Vianello & Alves, 1991), could be supported, in a sustainable manner (Mitchell et al., 2004), in order to minimize risks and impacts as well as negative effects of climate on natural resources (Machado, 1995; Hansen, 2002), based on appropriated decision-making.

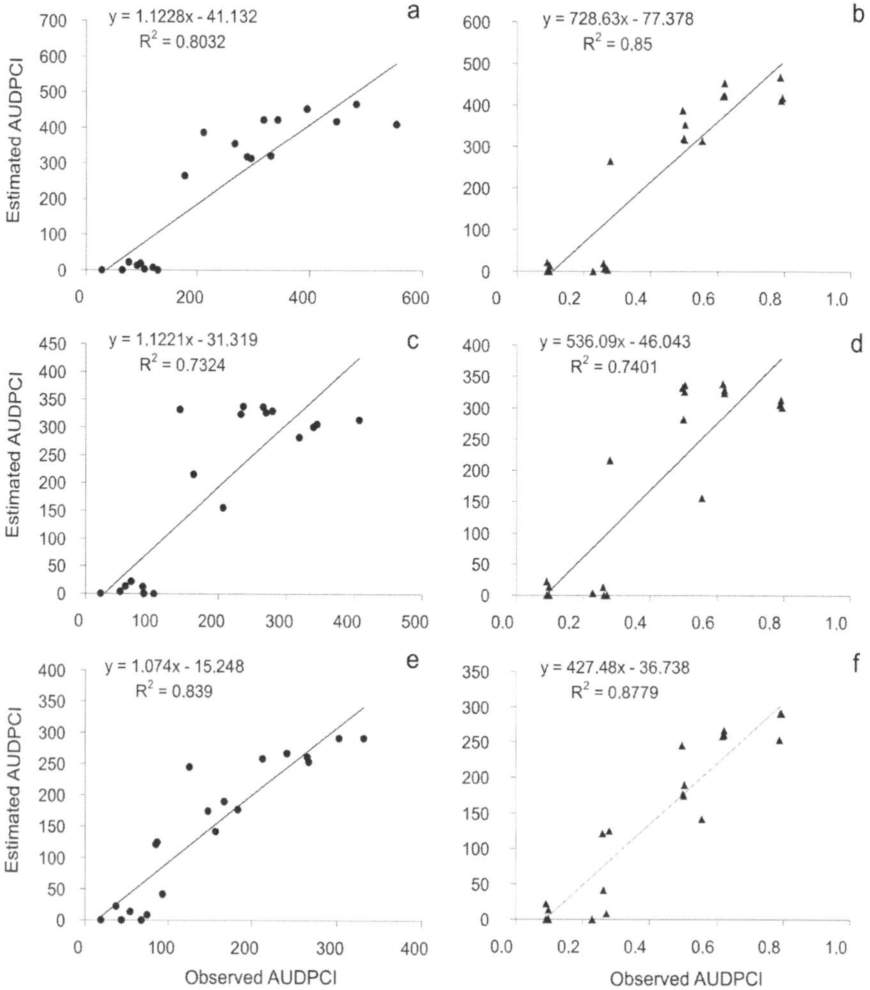

Fig. 7. Linear relationship between observed and predicted values of the area under the curve of incidence progress (AUDPCI) of soybean rust through models of nonlinear regression (a, c, e) and FLS (b , d, f) on the Conquista (a, b), Savana (c, d) and Suprema (e, f) cultivars.

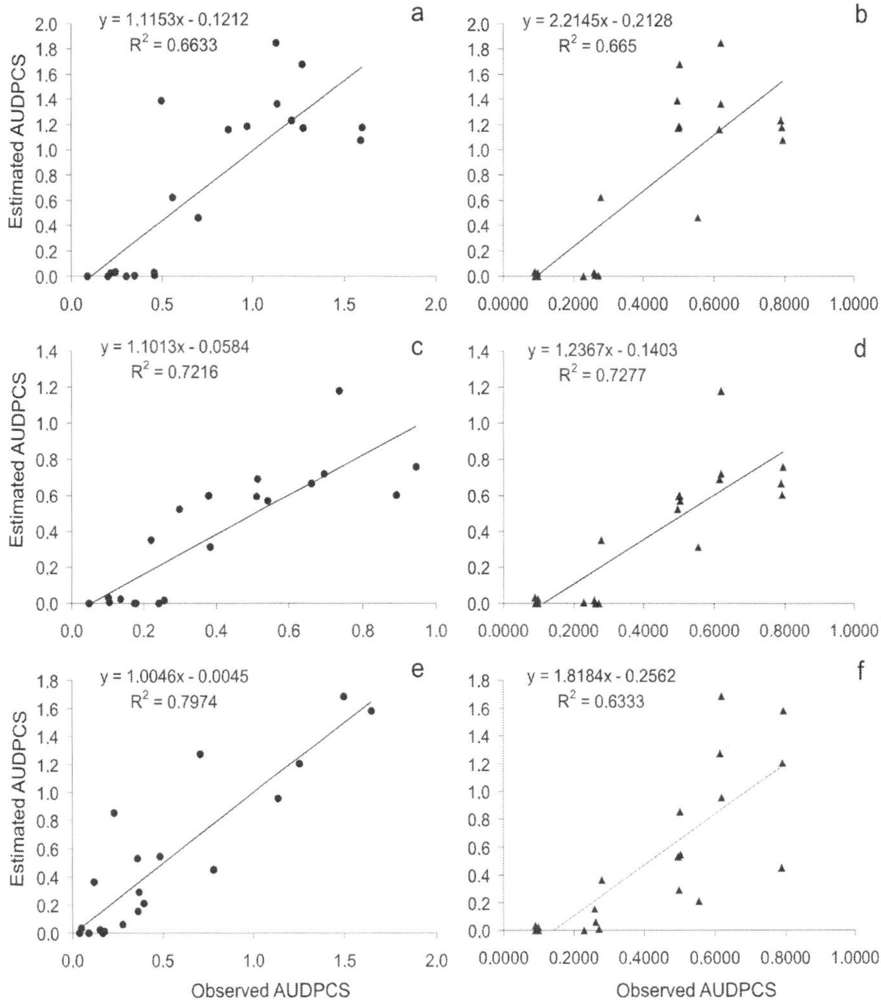

Fig. 8. Linear relationship between observed and predicted values of the area under the curve of severity progress (AUDPCS) of soybean rust through models of nonlinear regression (a, c, e) and FLS (b , d, f) on the cultivar Conquista (a, b), Savana (c, d) and Suprema (e, f) cultivars.

Method	AUDPCI observed			AUDPCS observed		
	Conquista	Savana	Suprema	Conquista	Savana	Suprema
RNL	0.8962*	0.85583*	0.91599*	0.81441*	0.84947*	0.89295*
FLS	0.92195*	0.8603*	0.93697*	0.81548*	0.85303*	0.7958*

*1% significant.

Table 2. Pearson correlation coefficients (r) for the observed values of the area under the curve of incidence progress (AUDPCI) and severity (AUDPCS) of soybean rust and the models estimated by nonlinear regression (RNL) and fuzzy logic system (FLS)

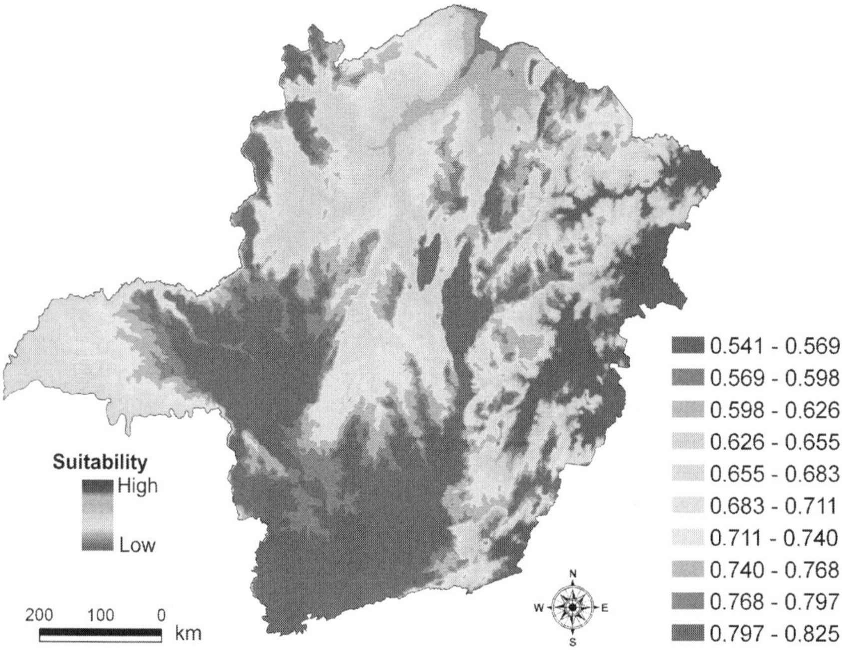

Fig. 9. Intensity of soybean rust in Minas Gerais, estimated by FLS, for the period of 1961 to 1990, based on observations of average monthly temperature in January of 39 weather INMET stations, with the leaf wetness period fixed at 12 hours, using altitude, latitude and longitude as covariates.

Variable	Ordinary kriging		Co-kriging	
	RMSE	Standard error	RMSE	Standard error
Rust intensity	0.07526	0.07746	0.05497	0.03604

Table 3. Coefficients of the estimate quality of the methods of ordinary kriging and co-kriging.

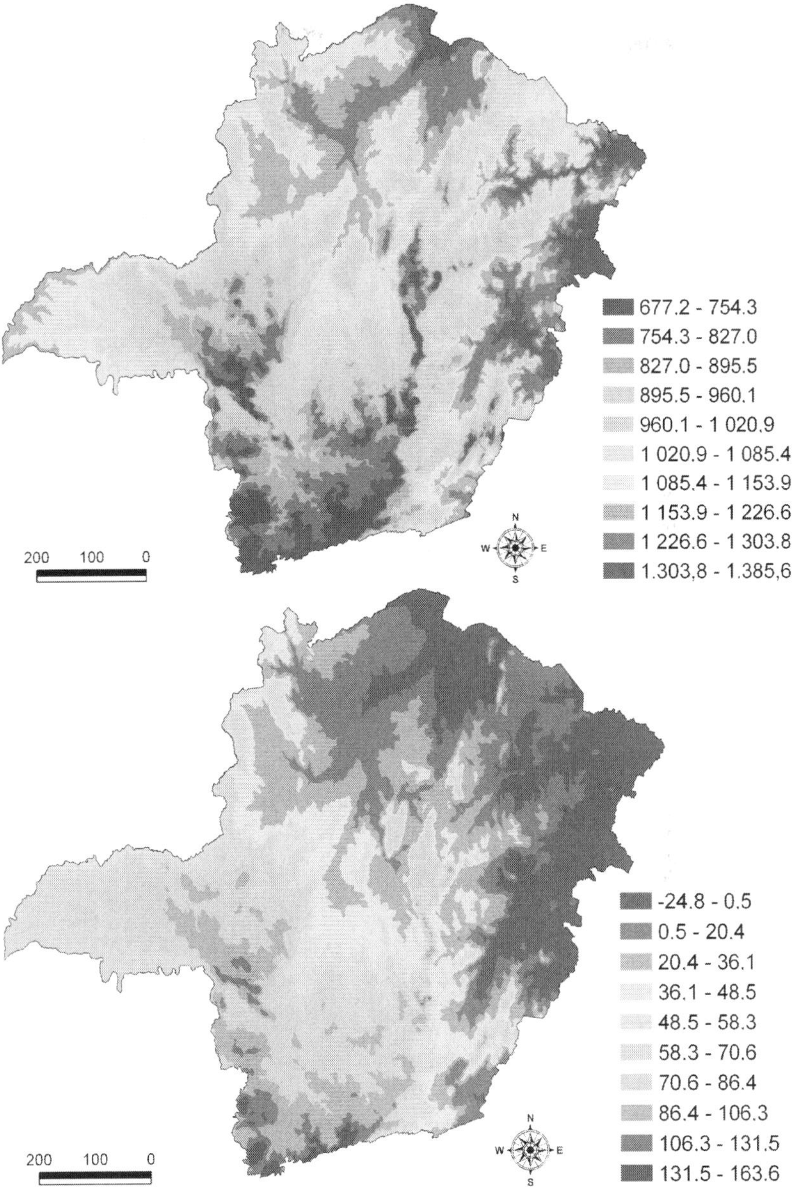

Fig. 10. Annual potential evapotranspiration (ETp) estimated by Thornthwaite (TW) (a) and annual moisture index (Iu) estimated by TW (b) in Minas Gerais, based on of 39 meteorological INMET stations, using co-kriging with altitude, latitude and longitude covariates.

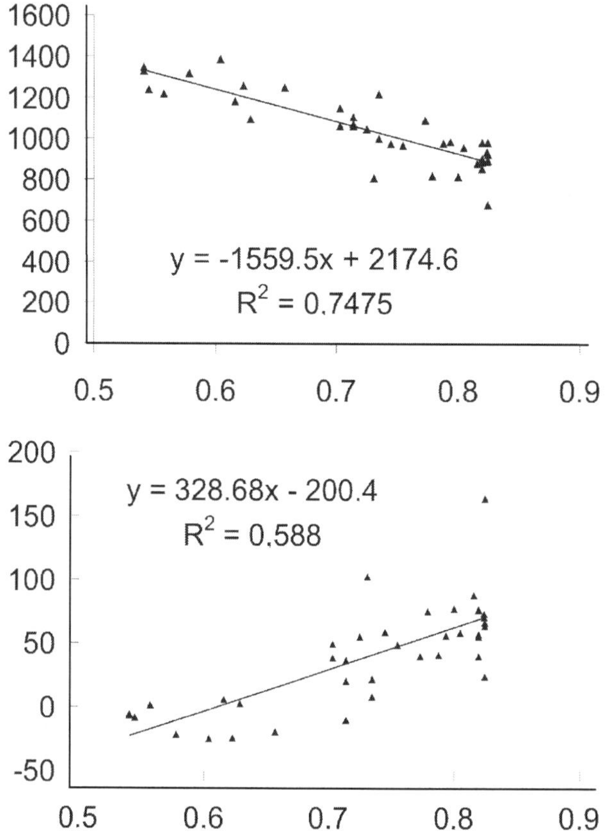

Fig. 11. Linear relationship between annual potential evapotranspiration (ETp) (Y axis) estimated by Thornthwaite (TW) (top) and annual moisture index (Iu) (Y axis) estimated by TW (down), with the potential intensity of soybean rust estimated by FLS (X axis), for the observations of monthly average temperature in January, at 39 INMET weather stations in Minas Gerais and surrounding states, with the leaf wetness period fixed at 12 hours.

Similarly, Morales & Jones (2004) used GIS to study the ecology and epidemiology of whitefly (*Bemisia tabaci* Gennadius 1889), transmitting geminiviruses in tropical crops in Latin America, at 304 georeferenced locations, where the whitefly and geminiviruses have caused significant damage. For this, it was developed a mathematical model including two climatic variables, temperature and precipitation, to map the probability of occurrence of favorable areas for pests. Later, using the Köeppen climatic classification, it was possible to verify that 55% of the localities affected by geminiviruses were located in the tropical wet-dry, 22% in humid-dry tropical regions, subtropical and local remnants of humid equatorial climates, with frequent coastal winds. According to the authors, based on the results, it was possible to understand the epidemic of whiteflies and geminiviruses, in order to assist the sustainable integrated pest management and disease in the studied regions. Vale et al. (2004) also reported the influence of climate on the inoculum survival, both between crop seasons and within the crop season. According to the authors, the survival of inoculum between cropping seasons is lower in temperate regions with arid or semi-dry summer, because under these characteristics, there is destruction of the survival structures, limiting the pathogen infection. Once inside the growing season for disease caused by polycyclic fungi and bacteria, the inoculum survival was higher in temperate regions, with low temperatures, low solar radiation and longer duration of leaf wetness. According to these authors, the temperature interfered with plant physiological processes, such as evapotranspiration, however, according to the results of this study, this variable may also be related to processes of infection, colonization, sporulation and survival of pathogens.

In this context, it became possible to develop, validate and implement a FLS for soybean rust, based on temperature and leaf wetness, for the cultivars Conquista, Savana and Suprema. Other important features on the FLS may be related to the system's simplicity, ease of implementing in field conditions, and the flexibility of the used method to incorporate other variables.

4. Conclusion

It was possible to develop, validate and implement a fuzzy logic system to estimate the monocyclic process of soybean rust, regarding Conquista, Savana and Suprema cultivars, based on temperature, leaf wetness and area under disease progress curve. The co-kriging method was more accurate and precise than the ordinary kriging method for mapping rust intensity.

FLS was better applied then non linear regression models to estimate the potential disease spatial progress.

The moisture index and potential evapotranspiration of Thornthwaite were significantly correlated with the estimates of the soybean rust intensity.

Leaf wetness up to 12 hours and temperatures around 20 °C, determined higher rust intensity. Temperatures above 30 °C and 15 °C as well as leaf wetness below 6 hours, reduced the rust intensity.

5. References

Alves, M.C.; Pozza, E.A.; Carvalho, L.G.; Oliveira, M.S.; Carvalho, L.M.T.; Machado, J.C.; Souza, P.E. (2006). Sistema de Informação Geográfica, Geoestatística e Estatística aplicados ao zoneamento ecológico potencial da ferrugem asiática da soja. *Fitopatologia Brasileira*, Brasília, v. 31, p. 181, 2006 Supl. CONGRESSO BRASILEIRO DE FITOPATOLOGIA, 39, Salvador.

Batchelor, W.D.; Yang, X.B.; Tschanz, A.T. (1997). Development of a neural network for soybean rust epidemics. *Transactions of the ASAE*, St. Joseph, v. 40, n. 1, p. 247-252.

BRASIL. (1992). Ministério da Agricultura e Reforma Agrária. Secretaria Nacional de Irrigação. Departamento Nacional de Meteorologia. *Normais climatológicas* (1961-1990). Brasília, 84 p.

Bromfield, K.R. (1981). Differential reaction of some soybean accessions to *Phakopsora pachyrhizi*. *Soybean Rust Newsletter*, Shanhua, v. 4, n. 2.

Bromfield, K.R. (1984). *Soybean rust.* St. Paul: American Phytopathological Society, 64 p. (Monograph, 11).

Bromfield, K.R. (1976). World soybean rust situation. In: Hill, L. D. *World Soybean Research*: proceedings of the world soybean research conference. Danville: The Interstate Printers and Publichers, p. 491-500.

Burrough, P.A.; McDonnell, R.A. (1998). *Principles of geographical information systems: spatial information systems and geostatistics*. 2. ed. Oxford: Oxford University Press,. 333 p.

Campbell, C.L.; Madden, L.V. (1990). *Introducion to plant disease epidemiology*. New York: Jonhn Wiley. 532 p.

Casey, P. S. (1980). The epidemiology of soybean rust - *Phakopsora pachyrhizi* Syd. *Soybean Rust Newsletter*, Shanhua, v. 4, n. 1, p. 3-5.

Castañeda-Miranda, R.; Ventura-Ramos Jr., E.; Peniche-Vera, R.R.; Herrera-Ruiz, G. (2006). Fuzzy Greenhouse Climate Control System based on a Field Programmable Gate Array. *Biosystems Engineering*, San Diego, v. 94, n. 2, p. 165-177.

Cox, E. (1994). *The fuzzy systems*: handbook a practitioner's guide to building, using, and maintaining fuzzy systems. London: Academic Press. 625 p.

Desbarats, A.J.; Logan, C.E.; Hinton, M.J.; Sharpe, D.R. (2002). On the kriging of water table elevations using collateral information from a digital elevation model. *Journal of Hydrology*, Amsterdam, v. 255, n. 1/4, p. 25-38.

Driankov, D.; Hellendoorn, H.; Reinfrank, M. (1993). *An introduction to fuzzy control*. New York: Springer-Verlag, 316 p.

Hansen, J.W. (2002). Realizing the potential benefits of climate prediction to agriculture: issues, approaches, challenges. *Agricultural Systems*, Oxford, v. 74, n. 3 p. 309-330.

Hartman, G.L.; Miles, M.R.; Frederick, R.D. (2005). Breeding for resistance to soybean rust. *Plant Disease*, St Paul, v. 89, n. 6, p. 664-666.

Isaaks, E.H.; Srivastava, R.M. (1989). *Applied geostatistics*. New York: Oxford University Press. 561 p.

Kawuki, R.S.; Tukamuhabwa, P.; Adipala, E. (2004). Soybean rust severity, rate of rust development, and tolerance as influenced by maturity period and season. *Crop Protection*, Oxford, v. 3, n, 5, p. 447-455.

Kim, K.S.; Wang, T.C.; Yang, X.B. (2005). Simulation of apparent infection rate to predict severity of soybean rust using a fuzzy logic system. *Phytopathology*, St Paul, v. 95, n. 10, p. 1122-1131.

Klir, G.J.; Yuan, B. (1995). *Fuzzy sets and fuzzy logic*: theory and applications. New Jersey: Prentice Hall. 574 p.

Machado, M.A.M. (1995). *Caracterização e avaliação climática da estação de crescimento de cultivos agrícolas para o estado de Minas Gerais*. 61 p. Dissertação (Mestrado) - Universidade Federal de Viçosa, Viçosa, MG.

Marchetti, M.A.; Uecker, F.A.; Bromfield, K.R. (1975). Uredial development of *Phakopsora pachyrhizi* in soybean. *Phytopathology*, St. Paul, v. 65, n. 7, p. 822-823.

Massad, E.; Ortega, N.R.S.; Struchiner, C.J.; Burattini, M.N. (2003). Fuzzy epidemics. *Artificial Intelligence in Medicine*, Amsterdam, v. 29, n. 3, p. 241-259.

McGee, D.C. (1992). *Soybean diseases*: a reference source for seed technologists. St. Paul: The American Phytopathological Society. 151 p.

Medina, P.F.; Wutke, E.B.; Miranda, M.A.C.; Braga, N.R.; Ito, M.F.; Barreto, M.; Harakawa, R. (2006). Qualidade de sementes de soja de cultivares IAC, produzidas diante da ocorrência natural a campo de *Phakopsora pachyrhizi,* agente causal da ferrugem asiática. *Informativo Abrates*, Londrina, v. 16, n. 1/3, p. 24.

Melching, J.S.; Dowler, W.M.; Koogle, D.L.; Royer, M.H. (1989). Effects of duration, frequency, and temperature of leaf wetness periods on soybean rust. *Plant Disease*, St. Paul, v. 73, n. 2, p. 117-122.

Mitchell, N.; Espie, P.; Hankin, R. (2004) Rational landscape decision-making: the use of meso-scale climatic analysis to promote sustainable land management. *Landscape and Urban Planning*, Amsterdam, v. 67, n. 1/4, p. 131-140.

Morales, F.J.; Jones, P.G. (2004). The ecology and epidemiology of whitefly-transmitted viruses in Latin America. *Virus Research*, Amsterdam, v. 100, n. 1, p. 57-65.

Mouzouris, G.C.; Mendel, J.M. (1997). Dynamic Non-Singleton Fuzzy Logic Systems for Nonlinear Modeling. *IEEE Transactions on Fuzzy Systems*, New York, v. 5, n. 2, p. 199-208.

NASA. (2005) *Shuttle Radar Topography Mission (SRTM) 2000*. Land Information Worldwide Mapping, LLC. Raster, 1:50000.

Pinto, A.C.S.; Pozza, E.A.; Souza, P.E.; Pozza, A.A.A.; Talamini, V.; Boldini, J.M.; Santos, F.S. (2002). Descrição da epidemia da ferrugem do cafeeiro com redes neurais. *Fitopatologia Brasileira*, Brasilia, v. 27, n. 5, p. 517-524.

Pivonia, S.; Yang, X.B. (2005). Assessment of epidemic potential of soybean rust in the United States. *Plant Disease*, St Paul, v. 89, n. 6, p. 678-682.

Pivonia, S.; Yang, X.B. (2004). Assessment of the potential year-round establishment of soybean rust throughout the world. *Plant Disease*, St. Paul, v. 88, n. 5, p. 523-529.

Reis, E.M.; Sartori, A.F.; Câmara, R.K. (2004). Modelo climático para previsão da ferrugem da soja. *Summa Phytopathologica*, Botucatu, v. 30, n. 2, p. 290-292.

Ritchie, S.; Hanway, J.J.; Thompson, H.E. (1982). *How a soybean plant develops*. Ames: Iowa State University of Science and Technology, Cooperative Extension Service, 20 p. (Special Report, 53).

Scherm, H. (2000). Simulating uncertainty in climate-pest models with fuzzy numbers. *Environmental Pollution*, Oxford, v. 108, n. 3, p. 373-379.

Sediyama, G.; Mello Jr., J.C. (1998). Modelos para estimativas das temperaturas normais mensais médias, máximas, mínimas e anual no estado de Minas Gerais. *Engenharia na Agricultura*. Viçosa, v. 6, n. 1, p. 57-61.

Sinclair, J.B. (Ed.). (1975). *Compedium of soybean diseases*. Minesota: American Phytopathological Society, 69 p.

Sinclair, J.B.; Backman, P.A. (Ed.). (1989). *Compendium of soybean diseases*. 3. ed. St. Paul: APS Press, p. 24-27.

Tanaka, K. (1997). *An introduction to fuzzy logic for pratical applications*. 138 p.

Thornthwaite, C.W. (1948). An approach towards a rational classification of climate. *Geographycal Review*, London, v. 38, n. 1, p. 55-94.

Thornthwaite, C.W.; Mather, J.R. (1955). *The water balance*. Centerton, NJ: Drexel Institute of Technology - Laboratory of Climatology. 104 p. (Publications in Climatology, v. 8, n. 1).

Urenã, R.; Rodríguez, F.; Berenguel, M.A. (2001). Machine vision system for seeds germination quality evaluation using fuzzy logic. (2001). *Computers and Electronics in Agriculture,* Oxford, v. 32, n. 1, p. 1-20.

Vale, F.X.R. (1985). *Aspectos epidemiológicos da ferrugem (Phakopsora pachyrhizi* Sydow) da soja (*Glycine max* L. Merrill). 104 p. Tese (Doutorado em Fitopatologia) - Universidade Federal de Viçosa, Viçosa, MG.

Vale, F.X.R.; Zambolim, L.; Chaves, G.M. (1990). Efeito do binômio temperatura-duração do molhamento foliar sobre a infecção por *Phakopsora pachyrhizi* em soja. *Fitopatologia Brasileira,* Brasília, v. 15, n. 3, p. 200-202.

Vale, F.X.R.; Zambolim, L; Costa, L.C.; Liberato, J.R.; Dias, A.P.S. (2004). Influência do clima no desenvolvimento de doenças de plantas. In: VALE, F. X. R.; JESUS JUNIOR, W. C.; ZAMBOLIM, L. *Epidemiologia aplicada ao manejo de doenças de plantas.* Belo Horizonte: Editora Perffil, p. 47-87.

Vargens, J.M.; Tanscheit, R.; Vellasco, M.M.B.R. (2003). Previsão de produção agrícola baseada em regras lingüísticas e lógica fuzzy. *Revista Brasileira de Controle & Automação,* Campinas, v. 2, n. 14, p. 114-120.

Vianello, R.L.; Alves, A.R. (1991). *Meteorologia básica e aplicações.* Viçosa: Imprensa Universitária/UFV. 449 p.

Von Altrock, C. (1995). *Fuzzy Logic and NeuroFuzzy Applications Explained.* USA: Prentice Hall. 384 p.

Yang, X.B.; Feng, F.(2001). Ranges and diversity of soybean fungal diseases in North America. *Phytopathology,* St Paul, v. 91, n. 8, p. 769-775.

Yang, X.B.; Royer, M.H.; Tschanz, A.T.; Tsai, B.Y. (1990). Analysis and quantification of soybean rust epidemics from 73 sequential planting experiments. *Phytopathology,* St. Paul, v. 80, n. 12, p. 1421-1427.

Yang, X.B.; Tschanz, A.T.; Dowler, W.M.; Wang, T.C. (1991). Development of yield loss models in relation to reductions of components of soybean infected with *Phakopsora pachyrhizi. Phytopathology,* St. Paul, v. 81, n. 11, p. 1420-1426.

Yorinori, J.T.; Lazzarotto, J.J. (2004). *Situação da ferrugem asiática da soja no Brasil e na América do Sul.* Londrina: Embrapa Soja. 30 p. (Documentos, 236). Disponível em <http://www. cnpso. embrapa. br> em:< dez. 2004.

Zambenedetti,E.B. *Preservação de Phakopsora pachyrhizi Sydow & Sydow e aspectos epidemiológicos e ultra-estruturais da sua interação com a soja (Glycine max* (L.). Merril). (2005). 92 p. Dissertação (Mestrado) - Universidade Federal de Lavras, Lavras, MG.

Zadeh, L.A. (1965). Fuzzy sets. *Information and Control,* San Diego, v. 8, n. 3, p. 338-353, 1965.

The Effectiveness of FeEDDHA Chelates in Mending and Preventing Iron Chlorosis in Soil-Grown Soybean Plants

W. D. C. Schenkeveld and E. J. M. Temminghoff
Wageningen University
The Netherlands

1. Introduction

1.1 Iron deficiency – The problem

Iron (Fe) is an essential micronutrient for plants, humans and other animals. An adequate uptake of Fe is needed to ensure proper growth and development, as well as good health of organisms (Marschner, 1995; Vasconcelos and Grusak, 2007). When provided with insufficient quantities of Fe, organisms will suffer from Fe deficiency symptoms.

Fe deficiency is a worldwide problem in crop production, affecting yield both qualitatively and quantitatively (Mortvedt, 1991); plants do not reach their full growth potential, and the nutritional value is compromised, leading to economic losses and limitations in crop selection (Chaney, 1984). In extreme cases, Fe deficiency may result in complete crop failure (Chen and Barak, 1982). The list of plant species affected is vast and includes apple, citrus, grapevine, peanut, dryland rice, sorghum and soybean (Marschner, 1995).

Fe deficiency is typically found in crops grown on calcareous or alkaline soils, in arid and semi-arid regions of the world; these soils cover over 30% of the earths' land surface (Figure 1) (Alvarez-Fernandez, et al., 2006; Chen and Barak, 1982; Hansen, et al., 2006; Mortvedt, 1991). Fe is abundantly present in all soils including calcareous ones; in mineral soils the average Fe content is approximately 2% (20,000 μg/g) (Marschner, 1995; Mengel and Kirkby, 2001). Most agricultural crops require less than 0.5 μg/g in the plough layer (Lindsay, 1974). The occurrence of Fe deficiency in plants grown on calcareous soils, despite the excessive soil-Fe pool, is caused by a limited bioavailability of Fe in such soils.

1.2 Symptoms of Fe deficiency

Fe deficiency in plants typically causes chlorosis of leaf tissue because of inadequate chlorophyll synthesis; the leaves become pale green to yellow (Figure 2), often with darker coloured veins. In case of severe chlorosis, leaves can also become necrotic (Figure 2). Due to the reduction in photosynthetic capacity, carbon fixation by plants also becomes reduced, leading to slower growth rates and yield losses (Figure 2) (Alvarez-Fernandez, et al., 2006).

Fe chlorosis develops most strongly in young leaves, because growing plant parts (also fruits, buds and storage organs) have incomplete xylem structures. As a result, Fe is not directly transported from the roots to these sites with the highest demand, but remobilized from older plant parts and secondarily transported through the phloem (Grusak, et al., 1999;

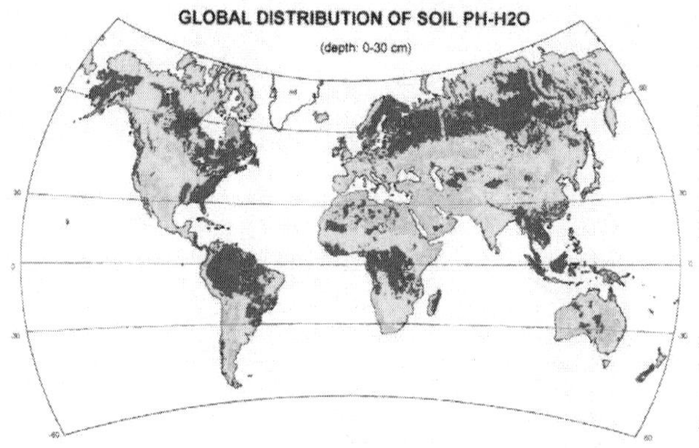

Fig. 1. Global pH-map of the top soil (0-30 cm); red indicates pH < 5.5; yellow indicates 5.5 < pH < 7.3; green indicates pH > 7.3. Calcareous soils are to be found in the green areas. Source: ISRIC, 1995, derived from the WISE- database.

Zhang, et al., 1995). It has been observed that chlorotic leaves can have comparable or even higher Fe contents than green leaves (the "chlorosis paradox"). This phenomenon has been attributed to impaired expansion growth, leading to diminished dilution of the high Fe concentration in young leaves (Römheld, 2000). Fe deficiency also causes morphological changes in the roots: inhibition of root elongation, increase in diameter of apical rootzone, abundant root hair formation (Römheld and Marschner, 1981) and formation of rhizodermal transfer cells.

1.3 Causes of Fe deficiency

Two related soil characteristics are principally responsible for the low Fe availability in calcareous soils: 1) the relatively high pH (7 - 8.5) (Figure 1.1), and 2) the presence of a bicarbonate pH-buffer in soil solution (Boxma, 1972; Chaney, 1984; Lucena, 2000; Marschner, 1995; Mengel, et al., 1984; Mengel and Kirkby, 2001).

In order for soil-Fe to be taken up, it needs to be transported through the soil solution to the root surface. The solubility of soil Fe(hydr)oxides is a function of pH and the type of Fe(hydr)oxide. The concentration of inorganic Fe species in solution reaches a minimum around pH 7.5 - 8.5: in the order of 10^{-10} M (Figure 3); the free Fe^{3+} concentration is around 10^{-21} M (Lindsay and Schwab, 1982). For optimal growth, plants require an Fe concentration in soil solution in the order of 10^{-6} to 10^{-5} M (Marschner, 1995). Complexation by dissolved organic substances, like humic acids, fulvic acids and siderophores can increase the total Fe concentrations in soil solution by orders of magnitude in comparison to the inorganic Fe concentration (O'Conner, et al., 1971), but not always sufficiently to prevent Fe deficiency.

The bicarbonate pH-buffer prevents plants from adapting the rhizosphere pH and causes impairment of Fe deficiency stress response mechanisms (except in grasses). Although the pH-buffer capacity of calcareous soils is largely determined by the lime content, the dissolution of carbonate minerals is relatively slow in comparison to bicarbonate diffusion. Therefore, on the short term, the bicarbonate concentration in soil solution is more

Fig. 2. Examples of Fe deficiency symptoms in soybean plants. *Upper:* from left to right - decreasing degree of chlorosis; *Lower left*: necrosis in the leaves; *Lower right:* reduced growth.

important for maintaining a high rhizosphere pH (Lucena, 2000). In addition to the role of bicarbonate as pH-buffer in soil solution, there has been much debate on bicarbonate uptake leading to Fe immobilization inside plants (Gruber and Kosegarten, 2002; Mengel, 1994; Nikolic and Romheld, 2002; Römheld, 2000).

1.4 Prevention and remediation of Fe deficiency

When Fe stress response mechanisms of plants prove inadequate, techniques to prevent or remedy Fe deficiency need to be applied to avoid yield losses. Breeding and genetically modifying plants for a more efficient Fe uptake mechanism is a promising approach. Developing new cultivars should however be done carefully and requires much time. Once crops are in the field, application of Fe fertilizer is the most certain and efficient treatment to ensure that plants do not suffer from Fe deficiency.

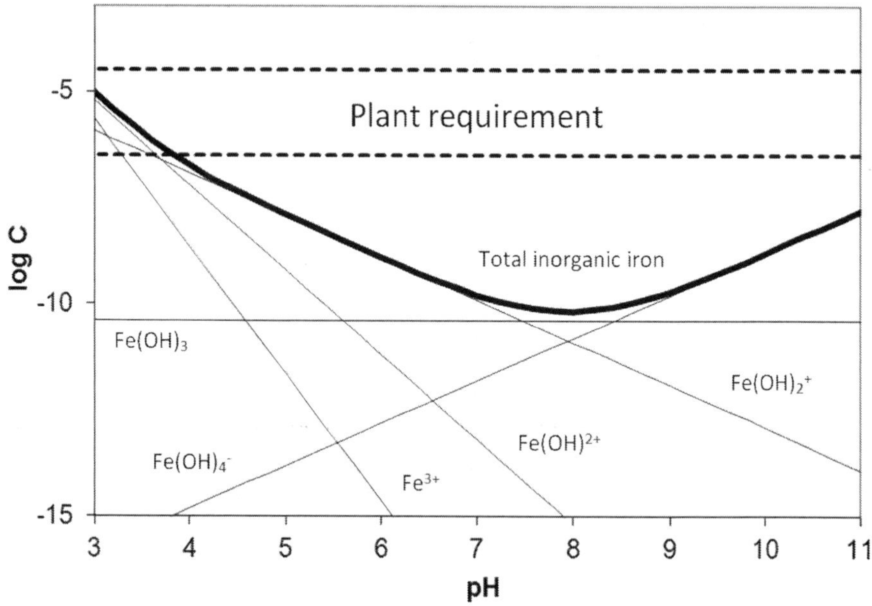

Fig. 3. Hydrolysis species of Fe(3+) in equilibrium with soil-Fe (pK_{sol} = 39.3; I = 0.03 M), after Lindsay (1979).

Fe fertilizers can be administered through trunk injection, foliar application, and soil application. Trunk injection is expensive and only suitable for trees. Foliar application does not provide full control of Fe chlorosis, but can be useful as complementary technique next to soil application (Alvarez-Fernandez, et al., 2004). Soil application is the most common technique to manage Fe deficiency in soil grown crops (Lucena, 2006). The technique is based on increasing the Fe concentration in soil solution. On calcareous soils, soil application of Fe fertilizers based on organic Fe salts, Fe complexes of lignosulfonates, citrates, gluconates, and synthetic Fe chelates of limited stability (e.g. FeEDTA, FeDTPA and FeHEDTA) has limited or no result, because these fertilizers are not able to maintain Fe in soil solution. Only Fe chelates of higher stability (FeEDDHA and derivatives, with phenolic functional groups) are effective and provide the most efficient treatment to control Fe deficiency (Lucena, 2006).

1.5 Fe deficiency in soybean

Fe deficiency chlorosis is a persistent, yield-limiting condition for soybean (*Glycine max* (L.) Merr.) production in regions with calcareous soils (Inskeep and Bloom, 1986). In the North Central U.S., Fe deficiency is responsible for an estimated loss in soybean grain production of $120 million per year (Hansen et al., 2004). Foliar Fe treatments and soil application of Fe chelates can be efficient in alleviating Fe deficiency chlorosis in soybean. However, in agricultural practice, these methods are only economically feasible for high-value crops and not for soybean (Fairbanks 2000).

Although soybean is not a target species for application of synthetic Fe chelates, it is an attractive test species due to the availability of soybean cultivars with a high susceptibility to Fe deficiency, the ease in handling of the plants, and the relatively short growth cycle in comparison to many of the target species (e.g citrus trees and grape vines). There is much experience with soybean in Fe chlorosis research; in nutrient solutions, in pot cultures and in the field (e.g. Garcia-Marco et al. 2006; Goos et al. 2004; Goos and Johnson 2000; Heitholt et al. 2003; Wallace and Cha 1986).

1.6 FeEDDHA based fertilizers

FeEDDHA is the iron(3+) complex of the chelating agent EDDHA, which is an acronym for ethylene diamine di(hydroxy phenyl acetic acid). EDDHA is also referred to as EHPG (ethylenebis-(hydroxy phenyl glycine)). This chelating agent was first synthesized by Kroll, introduced in 1955, but only fully described in 1957 (Kroll, 1957; Kroll, et al., 1957; Wallace, 1966). FeEDDHA was quickly recognized as very effective in correcting Fe chlorosis under soil conditions, also in comparison to other chelating agents (Wallace, et al., 1955; Wallace, 1962). The Fe^{3+} ion is bound by 2 carboxylate groups, 2 phenolate groups and 2 secondary amine groups in an octahedral complex of high stability with an intense red colour at neutral pH. The FeEDDHA complex owes its high stability in comparison to FeEDTA or FeDTPA complexes to the Fe-O (phenolate) bonds.

The current synthesis pathway for manufacturing EDDHA on an industrial scale is a Mannich-like reaction between phenol, ethylenediamine and glyoxylic acid. This reaction produces a mixture of 1) positional isomers, 2) diastereomers and 3) polycondensates, because 1) the reaction pathway allows for aromatic substitution in (o) ortho and (p) para position, 2) two chiral centers are introduced into the molecule leading to (R,R); (R,S); (S,R) and (S,S) configurations, and 3) undesired addition reactions take place between reactants and half products. The composition of the mixture of reaction products can be steered. After the reaction is terminated, an Fe salt is added to the reaction products to form Fe chelates.

Commercial FeEDDHA formulations can be operationally divided into 4 groups of compounds:

1. racemic o,o-FeEDDHA (Figure 4a); referring to the (R,R) and (S,S) configurations of o,o-FeEDDHA (iron (3+) ethylene diamine-N,N'-bis(2-hydroxy phenyl acetic acid) complex). These configurations are mirror images, but identical in most physical and chemical properties, including binding strength.
2. meso o,o-FeEDDHA (Figure 4b); referring to the (S,R) = (R,S) configuration of o,o-FeEDDHA. Due to the internal mirror plane of the chelate, the (S,R) and (R,S) configurations are identical.
3. o,p-FeEDDHA (Figure 4c); referring to the 4 configurations of o,p-FeEDDHA (iron (3+) ethylene diamine-N-(2-hydroxy phenyl acetic acid)-N'-(4-hydroxy phenyl acetic acid) complex). The o,p-FeEDDHA configurations are not identical in physical and chemical properties.
4. rest-FeEDDHA; referring to the 3 configurations of p,p-FeEDDHA (iron (3+) ethylene diamine-N,N'-bis(4-hydroxy phenyl acetic acid) and a variety of polycondensates and half products. An example of a polycondensate is depicted in Figure 4d.

In this chapter, these 4 groups will be referred to as the FeEDDHA components. In commercial FeEDDHA formulations, the sum of the racemic and meso o,o-FeEDDHA content is referred to as the o,o-FeEDDHA content of the product. Generally racemic and meso o,o-FeEDDHA are synthesized in a ratio close to 1.

Racemic and meso o,o-FeEDDHA are diastereomers; the chelated Fe is bound by the same functional groups, but the geometry of the chelate differs: in racemic o,o-FeEDDHA, both phenolic rings are in equatorial position, while in meso o,o-FeEDDHA one phenolic ring is in equatorial and the other in axial position (Figure 4a and 4b). Due to the difference in geometry the amount of strain on the bonds with Fe differs, which is reflected in a higher complexation constant for racemic o,o-FeEDDHA.

The position of the hydroxyl group on the phenolic ring affects the complexation constant of FeEDDHA components more strongly than strain: in para-position the hydroxyl group is sterically inhibited from contributing to binding Fe. As a consequence, o,o-EDDHA binds Fe more strongly than o,p-EDDHA (see Table 1.1), which in turn binds Fe more strongly than p,p-EDDHA. Rest-FeEDDHA is a very heterogeneous group, comprising of compounds that vary in molecular weight, number of functional groups, etc, and hence also in complexation constant.

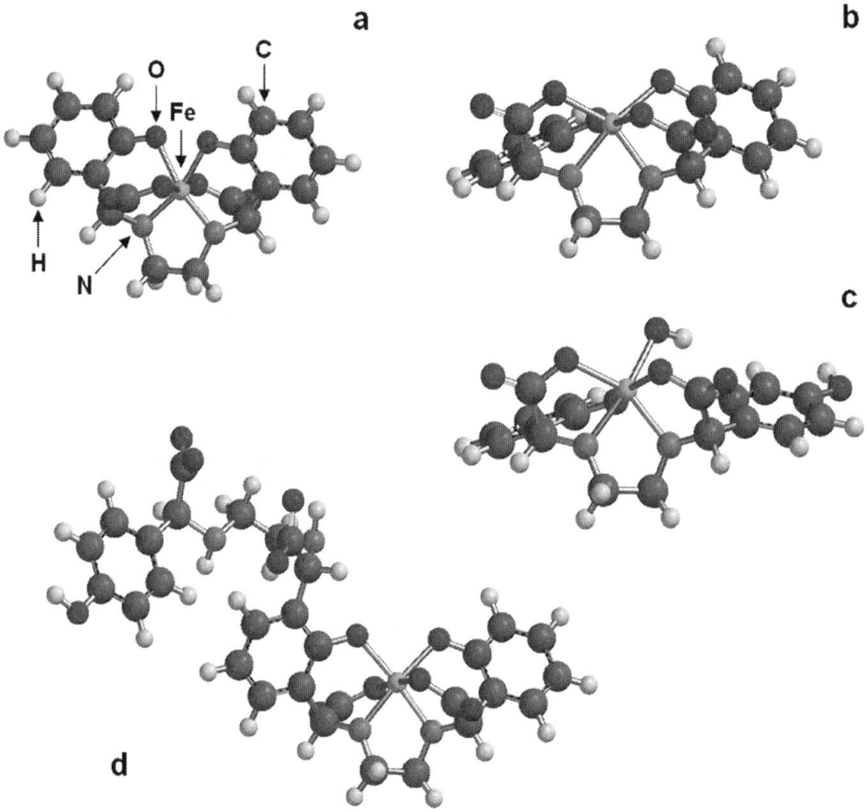

Fig. 4. Spatial structures of the FeEDDHA components **a)** racemic o,o-FeEDDHA; **b)** meso o,o-FeEDDHA; **c)** o,p-FeEDDHA with OH^- on the coordination complex; and **d)** rest-FeEDDHA (one possible polycondensate) (Schenkeveld et al., 2007).

Component	Log K (I = 0.1 M (NaCl))	
racemic o,o-FeEDDHA	35.86	Yunta et al. (2003a)
meso o,o-FeEDDHA	34.15	Yunta et al. (2003a)
o,p-FeEDDHA	28.72	Yunta et al. (2003b);

Table 1. Complexation constants of FeEDDHA components.

1.7 The market and regulation of FeEDDHA products

The market size for products based on FeEDDHA or related phenolic aminocarboxylate Fe chelates (e.g. FeEDDHMA, FeEDDHSA), is approximately 10 thousand tonnes per year, corresponding with a market value of around 60 million Euros. It is linked to areas of high soil-pH, in particular the Mediterranean area and the Middle East.

From the variability in composition of FeEDDHA products, and the difference in fertilizer value of the FeEDDHA components arose the need to ensure the quality of commercial FeEDDHA formulations. Several tests and methodologies have been developed to assess the quality of FeEDDHA products (Cantera, et al., 2002; Garcia-Marco, et al., 2003; Lucena, et al., 1992a; b). At present the quality of FeEDDHA products is guarded in the European Fertilizer Law (Regulation (EC) No. 2003/2003; amendment (EC) No. 162/2007) through the following parameters: (1) soluble Fe content of the product, (2) percentage of Fe chelated, and (3) percentages of Fe chelated by respectively o,o-EDDHA and o,p-EDDHA. Data on these parameters have to be indicated on the product label. FeEDDHA products should comprise at least 5 weight percent of water-soluble Fe, of which at least 80 percent should be chelated, and at least 50 percent should be chelated to either o,o- or o,p-EDDHA. To be included on the product label, there is a threshold value for both o,o- and o,p-EDDHA of 1 weight percent of chelated Fe.

In order to quantify the composition of FeEDDHA products, both for product information and law enforcement purposes, suitable protocols for analysis had to be developed. The method that is currently used for quantitative analysis is the high performance liquid chromatography (HPLC) method laid down by the European Committee for Standardization (CEN. EN 13368-2:2007). This method is almost identical to the ion-pair HPLC method developed by Lucena et al (1996).

1.8 The effectiveness of FeEDDHA components as Fe fertilizer

The efficacy of FeEDDHA as Fe fertilizer relies on its ability to increase the solubility of Fe, thereby enhancing its bioavailability through an increase in diffusive flux of Fe to the root. The effectiveness of individual FeEDDHA components is determined by: 1) their ability to remain in solution, 2) their susceptibility to cation competition and biodegradation, 3) their ability to transfer Fe to the plant, and 4) the ability of the corresponding EDDHA component to selectively mobilize Fe (Lucena 2003). Considerable effort has been invested to improve the understanding of these characteristics. The interaction between FeEDDHA components and soil and soil constituents has been examined by Alvarez-Fernandez, et al., 2002; Cantera, et al., 2002; Garcia-Marco, et al., 2006a; Hernandez-Apaolaza, et al., 2006; Hernandez-Apaolaza and Lucena, 2001 and Schenkeveld et al., 2007; Fe uptake from FeEDDHA components in hydroponic systems has been examined by Cerdan, et al., 2006; Garcia-Marco, et al., 2006a; Hernandez-Apaolaza, et al., 2006; Lucena and Chaney, 2006;

2007; Rojas, et al., 2008; and mobilization of Fe from Fe oxides by EDDHA ligands has been studied by Perez-Sanz and Lucena, 1995.

Still, the question of how much individual FeEDDHA components actually contribute to supplying soil-grown plants with Fe had remained unaddressed up until recently. An understanding of this issue is however particularly relevant for agricultural practice, since nowadays the composition of FeEDDHA products in terms of FeEDDHA components varies greatly. An efficient use of FeEDDHA fertilizer, implying maximizing the benefits in terms of crop yield and Fe uptake by plants, while minimizing the applied FeEDDHA dosage, is desirable both for the applier in view of cost efficiency, and from an environmental perspective to minimize the input of synthetic chemicals into the environment. In practical terms efficient FeEDDHA application translates into applying the right fertilizer (right composition) at the right moment in the right quantity. This requires a profound understanding of the effectiveness of individual FeEDDHA components in soil application. This chapter aims to inform on recent advances made in understanding the performance of FeEDDHA components in soil application (Schenkeveld et al. 2008; 2010a; 2010b). In a series of pot trial studies with soybean, FeEDDHA-facilitated Fe uptake was examined in relation to 1) the composition of the FeEDDHA treatments, 2) the soil solution concentrations of the FeEDDHA components as a function of time, and 3) the moment of FeEDDHA application.

2. Materials and methods

2.1 Soil

The calcareous soil used for the pot experiments was collected at a site located in Santomera (Murcia, Spain), from the top soil layer (0 - 20 cm). Plants grown on Santomera soil became chlorotic under field conditions. Pre-treatment of the soil consisted of air drying and sieving (1 cm). Santomera soil is a clay soil with a lutum fraction of 260 g kg^{-1} and a $CaCO_3$ content of 520 g kg^{-1}. The soil has a high pH: 8.0 (pH-$CaCl_2$) and a low soil organic carbon (SOC) content: 5 g kg^{-1}. The dissolved organic carbon (DOC) concentration amounts 30 mg l^{-1} (0.01 M $CaCl_2$), and Fe availability parameters are low: oxalate extractable Fe content: 0.30 g kg^{-1} Fe, and diethylene triamine penta acetic acid (DTPA) extractable Fe content: 3.5 mg kg^{-1} Fe. A more complete overview of soil characteristics of Santomera soil is presented in Schenkeveld et al., 2010a.

2.2 FeEDDHA solutions

Depending on the pot trial experiment, FeEDDHA solutions were prepared from EDDHA stock solutions varying in EDDHA component composition, from a solid o,o-H_4EDDHA mixture (purity: 99%), or from separated solid racemic o,o-EDDHA (purity: 100%), meso o,o-EDDHA (purity: 99.5%) and o,p-EDDHA (purity: 90%). Racemic and meso o,o-H_4EDDHA were obtained by separation of the o,o-H_4EDDHA mixture, as described in Bannochie and Martell (1989) and Bailey et al. (1981). Solid H_4EDDHA was first dissolved in sufficient 1 M NaOH solution. An $FeCl_3$ solution was added to the EDDHA solution in a 2-5% excess, based on a 1:1 stoichiometry between Fe and ethylene diamine (incorporated in the EDDHA ligands). pH was raised to 7 ± 1, and the solution was stored overnight in the dark to allow excess Fe to precipitate as Fe(hydr)oxides. Subsequently the FeEDDHA solutions were filtered over a 0.45 µm nitro cellulose filter (Schleicher & Schuell, refno: 10401114) and further diluted for application in the pot trial. The composition of FeEDDHA solutions was examined at t=0 and at the end of the experiment by ICP and HPLC analysis.

2.3 Pot trial studies

The effectiveness of FeEDDHA components in providing soybean plants from the chlorosis susceptible cultivar Mycogen 5072 with Fe was examined in three pot trial studies.

Effect of FeEDDHA treatment composition on Fe uptake– pot trial 1

In pot trial 1, soybean plants were given FeEDDHA treatments similar in Fe dose (≈ 7 mg l^{-1} Fe in the pore water; 0.13 mM), but differing in FeEDDHA component composition. Four FeEDDHA treatments (16%o,o; 34%o,o; 49%o,o and 99%o,o) and a blank treatment were included in the experiment; the composition of the treatments is presented in Table 2. The treatments are named after the combined percentage of the Fe chelated by racemic and meso o,o-EDDHA and were given at t=0. The pot trial experiment had a run time of eight weeks. A more elaborate description of the experiment is provided in Schenkeveld et al., 2008.

FeEDDHA-facilitated Fe uptake as a function of time - pot trial 2

In pot trial 2, the relation between FeEDDHA component concentrations in the pore water and Fe uptake by plants was examined as a function of time. Soybean plants were offered two different FeEDDHA treatments, (30%o,o and 100%o,o) and a blank treatment. The treatments were equal in Fe dose (≈ 4 mg l^{-1} Fe in the pore water; 0.07 mM), but differed in the percentage of Fe chelated by racemic and meso o,o-EDDHA. The composition of the treatments is presented in Table 2. FeEDDHA was applied once, at the start of the experiment. The pot trial with had a runtime of six weeks. Plants were harvested and soil solution was sampled every week. The experiment is described more elaborately in Schenkeveld et al., 2010a.

Effect of moment of application on Fe uptake from FeEDDHA components- pot trial 3

In pot trial 3, the influence of the moment of application on the effectiveness of individual FeEDDHA components in proving soybean plants with Fe was examined. The experiment involved a blank treatment and six FeEDDHA treatments: *o,p*; *meso o,o*; *racemic o,o*; *o,o-mix low*; *o,o mix-low + o,p*; and *o,o-mix high*. Two levels of FeEDDHA application were distinguished; a low level in the first four treatments, corresponding to a pore water concentration of around 0.6 mg l^{-1} Fe (i.e. 11 μM), and a high level in the latter two treatment, corresponding to a pore water concentration of around 1.8 mg l^{-1} Fe (i.e. 32 μM). The high level FeEDDHA application was included to ensure that Fe uptake had not yet reached a maximum in the low level application, which is prerequisite for comparing the effectiveness of the FeEDDHA components. The mixed treatments were included to examine potential synergetic effects. The composition of the treatments is presented in Table 2. With exception of the blank treatments, all pots received one FeEDDHA treatment, either at t=0 after transfer of the seedlings, after 3 weeks in the progressed vegetative stage, or after 6 weeks in the reproductive stage. Which FeEDDHA treatment was applied in which growth stage is indicated in Table 2. Treatments are named after the FeEDDHA treatment administered and the moment of application. The pot trial had a runtime of 8 weeks. Schenkeveld et al., 2010b describes the pot trial more elaborately.

All pot trial experiments were carried out in a greenhouse with 7 liter Mitscherlich pots containing either 6 kg (pot trial 1 and 2) or 5 kg (pot trial 3) of soil at 50% of the waterholding capacity. The experiments were done in triplicates. In pot trial 1 and 2, FeEDDHA solutions were mixed through the soil prior to filling the pots; in pot trial 3, the FeEDDHA treatments were applied through a sand column in the middle of the pot, which went up to a depth of approximately 10 cm into the soil. After FeEDDHA addition, the sand column was flushed with demineralized water.

Seeds of the Fe chlorosis susceptible soybean (*Glycine max* (L.) Merr.) cultivar Mycogen 5072 were germinated on quartz sand with demineralised water. After five days eight seedlings were transferred to each pot, which had been filled with soil one day prior to the transfer. Preparation of the pot trial, soil fertilization with macronutrients, foliar fertilization with micronutrients other than Fe, and plant care were performed as described in Schenkeveld et al. (2008). In pot trial 2, foliar fertilization was omitted. In pot trial 3 the amounts of macronutrients added to the soil were lowered, in proportion to the smaller quantity of soil used per pot.

Treatment	racemic o,o-FeEDDHA (mg l^{-1} Fe)	meso o,o-FeEDDHA (mg l^{-1} Fe)	o,p-FeEDDHA (mg l^{-1} Fe)	rest-FeEDDHA (mg l^{-1} Fe)	total Fe (mg l^{-1} Fe)	Moment of application
Pot trial 1						
blank	-	-	-	-	-	-
16%o,o	0.58 (8%)	0.61 (8%)	1.15 (16%)	5.02 (68%)	7.36	t = 0
34%o,o	1.07 (16%)	1.24 (18%)	1.26 (19%)	3.14 (47%)	6.71	t = 0
49%o,o	1.69 (22%)	2.03 (27%)	1.31 (18%)	2.50 (33%)	7.53	t = 0
99%o,o	3.44 (48%)	3.64 (51%)	-	0.10 (1%)	7.18	t = 0
Pot trial 2						
blank	-	-	-	-	-	-
30%o,o	0.60 (14%)	0.68 (16%)	0.79 (19%)	2.18 (51%)	4.25	t = 0
100%o,o	1.93 (48%)	2.00 (50%)	-	0.05 (1%)	3.98	t = 0
Pot trial 3						
blank	-	-	-	-	-	-
o,p	*	*	0.53	0.05	0,58	t = 3 and 6 weeks
meso o,o	-	0.56	-	-	0.56	t = 0, 3 and 6 weeks
racemic o,o	0.58	-	-	-	0,58	t = 0, 3 and 6 weeks
o,o-mix low	0.29	0.31	-	-	0.60	t = 3 weeks
o,o-mix low + o,p	0.29	0.31	1.06	0,10	1.76	t = 3 weeks
o,o-mix high	0.87	0.93	-	-	1.80	t = 0 and 3 weeks

Table 2. Treatment overview of the pot trials; * o,p-EDDHA standard contains traces of racemic and meso o,o-EDDHA

2.4 Sampling and measurement

SPAD measurements were done, as described in Schenkeveld et al., 2008, on the youngest leaves throughout the pot trials, to monitor chlorosis. Chlorosis was established based on a

significant difference (α = 0.05) in SPAD-indices between the blank and the treatment with the highest SPAD-index.
At harvest, the shoots were cut off directly above the soil surface. A 1 kg mixed subsample was taken from the soil. Roots were collected manually from the soil subsample, which was stored at 4 °C until further use. The shoots were washed with demineralized water and dried at 70 °C. After 48 hours, the shoots were weighed (dry weight). The mineral contents of the shoots were determined by microwave digestion with nitric acid, fluoric acid and hydrogen peroxide (Novozamsky, et al., 1996). Cu, Fe, Mn and Ni concentrations were measured by ICP-AES (Varian, Vista Pro). Fe uptake was calculated as the product of shoot dry weight yield and Fe content of the shoot. Roots were left out of consideration, due to contamination with soil material.
Pore water was collected from the soil subsample by centrifugation at 7,000 rpm for 15 minutes as described in Schenkeveld et al 2008. The pH of the pore water was measured directly after collection. Fe, Ca and Mg concentrations were measured by ICP-AES (Varian, Vista Pro); Cu, Al, Mn, Zn, Ni and Co concentrations were measured by ICP-MS (Perkin Elmer, ELAN 6000). The samples were acidified with nitric acid before ICP-analysis. FeEDDHA isomer concentrations were determined after separation by high-performance liquid chromatography (HPLC) as described in Schenkeveld et al. (2007). Preparation of experimental solutions and dilution of samples was done with analytical grade chemicals and ultra pure water.

2.5 Statistical analysis

Statistical analysis of the data was performed using SPSS 12.0. Homogeneity of variance was tested with the Levene's test (α = 0.05). A log transformation of the data was executed in case the variance proved non-homogenous. Differences among treatments were determined by applying the multivariate general linear model procedure with a Tukey post-hoc test (α = 0.05). Block effects from the tables were accounted for by including table as a random factor.

3. Results and discussion

Chlorosis

Inducing Fe deficiency chlorosis is a prerequisite for testing the effectiveness of the FeEDDHA components. In all three pot trials, the plants in the blank treatment became chlorotic, approximately a week after transfer of the seedlings to the pots. The development of chlorosis differed per pot trial; in pot trial 1 and 2, the degree of chlorosis reached a maximum after around three weeks, after which the difference in SPAD-index started to decrease. In pot trial 1, chlorosis in the youngest leaves of the blank treatment was actually entirely over-grown by the time the plants were harvested (Schenkeveld et al., 2008). Possibly, the decrease in degree of chlorosis is related to an increased root density in the pots as a result of an ongoing development of the root system. This high root density leads to increased rhizosphere effects and an enhanced ability of the plants to acquire Fe. In pot trial 3, the degree of chlorosis stabilized and remained more or less constant towards the end of the experiment.

3.1 Effect of FeEDDHA treatment composition on Fe uptake

Pore water concentrations

The Fe concentration in the pore water of the blank treatment was below detection limit, both in this and the other two pot trial experiments, indicating that FeEDDHA components were responsible for all Fe in solution in the FeEDDHA treatments. At harvest of pot trial 1, the total Fe concentration in the pore water proved linearly related to the o,o-FeEDDHA content of the FeEDDHA treatments (Figure 5). Racemic o,o-FeEDDHA accounted for approximately 80% of the Fe in solution, and meso o,o-FeEDDHA for the remaining 20%. o,p-FeEDDHA and rest-FeEDDHA had been removed from soil solution practically completely. These components have a tendency to adsorb due to a relatively high affinity for soil reactive surfaces. Moreover, upon interaction with soil, Cu may rapidly displace Fe from o,p-FeEDDHA resulting in solibilization of o,p-CuEDDHA (Garcia-Marco et al., 2006; Hernandez-Apaolaza et al., 2006; Schenkeveld et al., 2007). Hence, removal of FeEDDHA components from soil solution is to a large extent unrelated to plant processes. From the amount of Fe added with the FeEDDHA treatment only in between 4 and 20 % was retrieved at harvest. The recovery of racemic o,o-FeEDDHA and meso o,o-FeEDDHA was around 30% and 7%, respectively. The recovery of o,p-FeEDDHA and rest-FeEDDHA was below 1%.

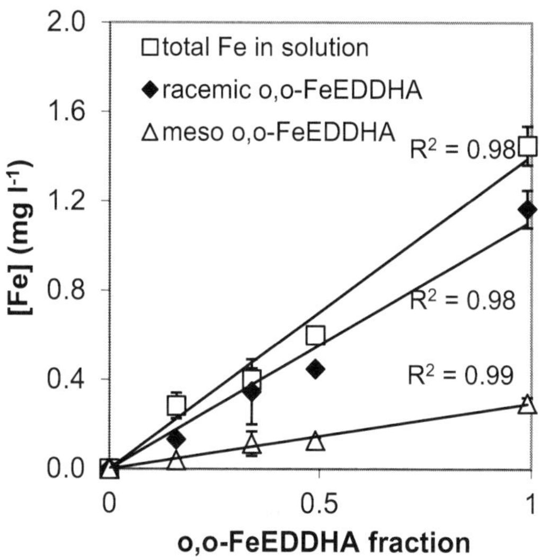

Fig. 5. Fe and FeEDDHA component concentrations in soil solution of Santomera soil at harvest as a function of the o,o-FeEDDHA content of the FeEDDHA treatment. Error bars indicate standard deviations. (based on Schenkeveld et al., 2008)

Fe uptake

Fe uptake by soybean plants increased with increasing o,o-FeEDDHA content of the FeEDDHA treatment (Figure 6a). At low o,o-FeEDDHA content, the increase in Fe uptake is relatively strong, but the slope of the curve flattens with increasing o,o-FeEDDHA content, and eventually an optimum is reached (Schenkeveld et al., 2010a).

The increase in Fe uptake with increasing o,o-FeEDDHA content suggests that Fe uptake is related to the Fe concentration in soil solution (Figure 5). This makes sense, since the limited solubility of Fe in calcareous soil is one of the primal causes for Fe chlorosis. The fact that Fe uptake, unlike Fe concentration in soil solution, is not linearly related to the o,o-FeEDDHA content suggests a saturation effect, commonly observed with micronutrient uptake in relation to bioavailability (Marschner, 1995).

As a result of the FeEDDHA treatments, Fe uptake increased from 0.70 mg pot^{-1} in the blank to 1.75 mg pot^{-1} in the 99% o,o-FeEDDHA treatment; a 150% increase. The 16%o,o FeEDDHA treatment already increased Fe uptake by approximately 75%, to 1.22 mg pot^{-1}. The additional Fe uptake in the FeEDDHA treatments in comparison to the blank only accounted for 7 to 15% of the Fe provided as FeEDDHA, and for 15 to 44% of the Fe added as o,o-FeEDDHA.

The increased Fe uptake manifested both in an increased Fe content of the shoot (Figure 6b), and in an increased dry weight yield (Figure 6c). The trends in Fe content and dry weight yield as a function of o,o-FeEDDHA content are similar as for Fe uptake. The relative effect on Fe content of the shoot: an increase from 31 to 60 mg kg(dw)$^{-1}$ (≈ 100% increase), was larger than the relative effect on dry weight yield; an increase from 22.1 to 29.0 g(dw) pot^{-1} (≈ 30% increase). Comparable results were also obtained with soybean grown on another calcareous soil (Schenkveld et al., 2008; results not shown).

An important practical implication of these results for FeEDDHA application prior to the onset of chlorosis is, that for obtaining similar results in terms of crop yield and crop quality, a smaller dosage of FeEDDHA products with a higher o,o-FeEDDHA content is required in comparison to products with a lower o,o-FeEDDHA content.

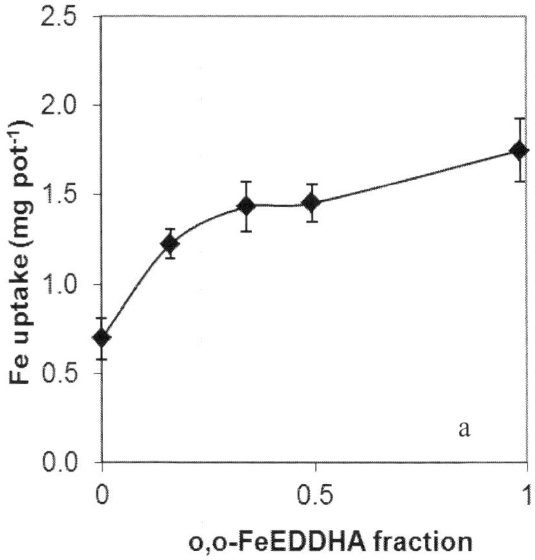

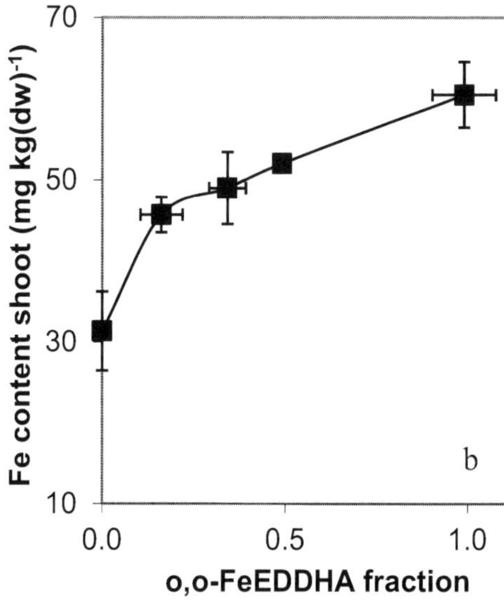

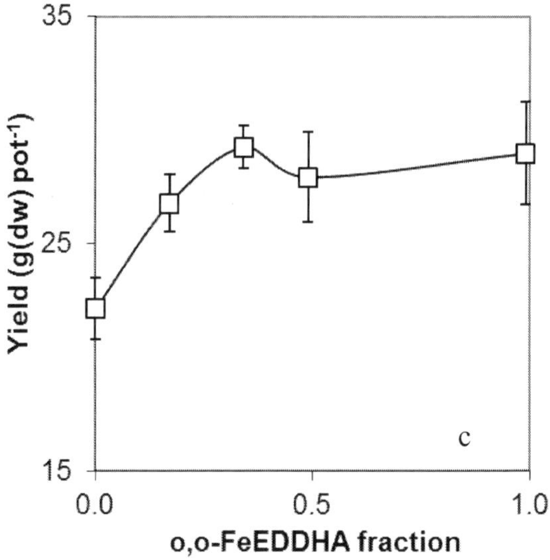

Fig. 6. a) Fe uptake; b) Fe content of the shoot; and c) dry weight yield (shoot) of soybean plants grown on Santomera soil as a function of the o,o-FeEDDHA content of the FeEDDHA treatment. Error bars indicate standard deviations. (based on Schenkeveld et al., 2008)

3.2 FeEDDHA-facilitated Fe uptake as a function of time

Pore water concentrations

In Figure 7, the Fe and FeEDDHA component concentrations are presented as a function of time for the 30%o,oL treatment from pot trial 2. Within the first week of the experiment, the Fe concentration underwent a strong drop, from 4.25 to 0.81 mg l^{-1} Fe, after which it gradually declined further (Figure 7a). This drop was largely caused by the practically complete removal of o,p-FeEDDHA and rest-FeEDDHA from soil solution. From 1 week onward, the Fe concentration was largely (> 92%) governed by racemic and meso o,o-FeEDDHA (Figure 7b). The concentration of racemic and meso o,o-FeEDDHA underwent two stages: 1) a rapid, strong decline within the first week, and 2) a gradual decline from one week onward. The initial decrease in racemic o,o-FeEDDHA concentration (≈28%) was smaller than for meso o,o-FeEDDHA (≈54%). This fast decline has been attributed to adsorption, which can be described with linear adsorption isotherms (Schenkeveld et al 2010a). The rate of the gradual decline was higher for meso o,o-FeEDDHA than for racemic o,o-FeEDDHA, resulting in a continuous increase in relative contribution of racemic o,o-FeEDDHA to the total Fe in solution. The nature of the gradual decline differed for racemic and meso o,o-FeEDDHA: for meso o,o-FeEDDHA it could be accurately described with an exponential decay function, whereas for racemic o,o-FeEDDHA no decline was observed in the second week of the experiment and from 2 weeks onward, the rate of decline was less consistent (Figure 7b). The decay constant in the exponential function describing the gradual decline in meso o,o-FeEDDHA concentration proved dependent on the applied amount of meso o,o-FeEDDHA (Schenkeveld et al., 2010a).

Fe uptake

Fe uptake as a function of time is presented in Figure 8 and was calculated by subtracting total Fe uptake of two consecutive harvesting rounds for a corresponding treatment. Fe uptake at 2 weeks actually represent the Fe taken up during the second week, and so on. During the 2nd week, in the early vegetative stage, Fe requirements were still low. Chlorosis had just developed in the soybean plants and possibly utilization of Fe which had been present in the seeds, still covered part of the Fe requirements. In the 3rd and the 4th week, during the progressed vegetative stage, Fe demand strongly increased and in the blank treatment chlorosis was most severe. In the course of the 4th and during the 5th week, the transfer from the vegetative to the reproductive stage took place; the plants flowered and started to grow pods. In the 6th week, the seed formation inside the pods progressed and Fe requirements were even larger than during the vegetative stage, in order to provide the seeds with sufficient Fe (Grusak, 1995). Throughout the experiment, the sequence in Fe uptake was: blank < 30%o,o < 100%o,o. The difference in Fe uptake among the treatments was largest in growth stages in which Fe requirements were largest. The large differences in Fe uptake during the reproductive stage did not show in an increased difference in SPAD-indices (Schenkeveld et al., 2010a).

Relation between FeEDDHA removal and Fe uptake

The amount of FeEDDHA components removed from the soil system (solid and solution phase combined) per week was calculated from the decrease in soil solution concentration (Figure 8b), assuming linear adsorption (Schenkeveld et al., 2010a), and is presented as a function of time for the 100%o,o treatment in Figure 9a. The removal of meso o,o-FeEDDHA was larger than the removal of racemic o,o-FeEDDHA throughout the experiment. Still, racemic o,o-FeEDDHA, seems to have a more pronounced influence on the shape of the total o,o-FeEDDHA removal-curve (Figure 9a).

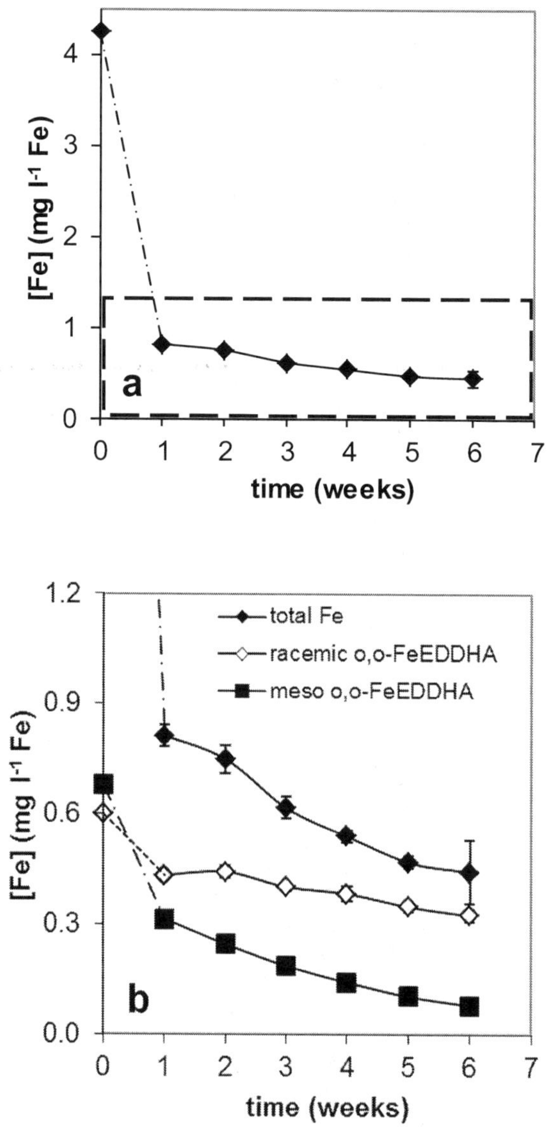

Fig. 7. Fe and FeEDDHA component concentrations in the pore water of Santomera soil as a function of time for the 30%o,o treatment. Error bars indicate standard deviations. (based on Schenkeveld et al., 2010a)

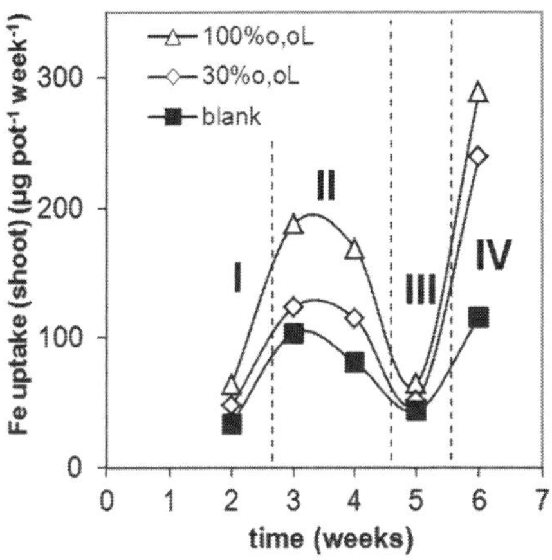

Fig. 8. Fe uptake (shoot) by soybean plants grown on Santomera soil as a function of time. Error bars have been omitted. (based on Schenkeveld et al., 2010a)

In Figure 9b, two scenarios for FeEDDHA-facilitated Fe uptake are presented as a function of time for the 100%o,o treatment. In the maximum FeEDDHA-facilitated uptake scenario, all Fe uptake by the soybean plants is assumed FeEDDHA-facilitated; in the minimum FeEDDHA-facilitated uptake scenario, only the Fe uptake in addition to Fe uptake in the blank treatment is assumed FeEDDHA-facilitated. The shape of the racemic o,o-FeEDDHA removal curve strongly resembles the shape of the FeEDDHA-facilitated Fe uptake curves (Figure 9b). This suggests that the removal of racemic o,o-FeEDDHA from the soil system is to a large extent plant-related. The fact that the gradual decline in racemic o,o-FeEDDHA concentration only started after 2 weeks, when the plants developed a strong need for Fe, further supports this reasoning. The shape of the meso o,o-FeEDDHA removal curve (Figure 9a) does not show a similar resemblance, which suggests that the removal of meso o,o-FeEDDHA from the soil system is to a large extent non-plant related. The nature of the plant-independent process causing a decline in meso o,o-FeEDDHA concentration remains unclear.

3.3 Effect of moment of application on Fe uptake from FeEDDHA components

Pore water concentrations

The FeEDDHA component concentrations in the pore water at harvest of pot trial 3 are presented in Figure 10. o,p-FeEDDHA was not detected in any of the samples and has not been included in the figure. In agreement with the results from the other two pot trails, for each of the moments of application separately, racemic o,o-FeEDDHA remained in solution to a larger extent than meso o,o-FeEDDHA. The recovered concentrations only accounted for up to 25% of the racemic o,o-FeEDDHA and up to 8% of the meso o,o-FeEDDHA applied. In particular for the treatment applied at t=6 weeks these low recoveries are remarkable; there was only 2 weeks of residence in the soil-plant system.

For corresponding treatments applied at t=0 and t=3 weeks, the recovery of the treatment applied at t=3 weeks was consequently lower than for the treatment applied at the start of the experiment. This seems counter-intuitive, because the residence time in the soil-plant

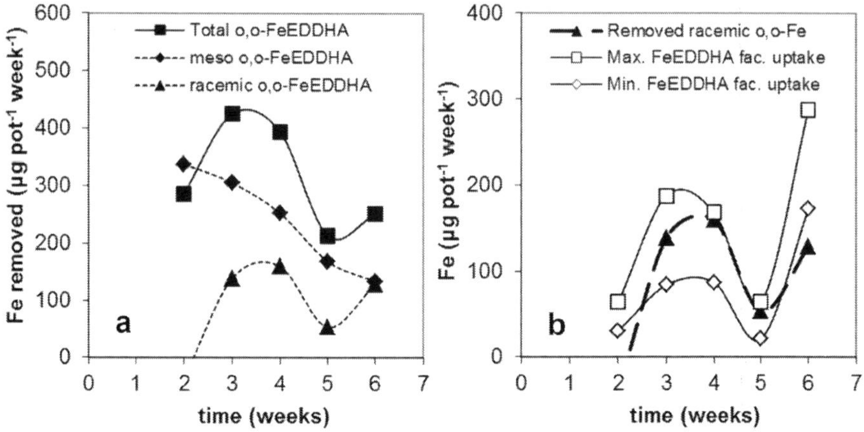

Fig. 9. a) Amount of total, racemic and meso o,o-FeEDDHA removed from the soil system per week for the 100%o,o treatment; b) Minimum and maximum FeEDDHA-facilitated Fe uptake (shoot) per week by soybean plants grown on Santomera soil as a function of time, and the amount of racemic o,o-FeEDDHA removed from the soil system per pot per week, both for the 100%o,oL treatment. Error bars have been omitted. (based on Schenkeveld et al., 2010a)

system of the treatment applied at t=0 is 3 weeks longer. An essential difference regarding the system to which the FeEDDHA treatments were applied, is that with application at t=3 weeks, the soybean plants had grown chlorotic and Fe deficiency stress mechanisms had been activated by the time the treatment was applied, whereas plants receiving FeEDDHA treatment at t=0 never grew Fe deficient to this extent in the first place. For strategy I plants like soybean, one of the stress response mechanisms involves up-regulation of the ferric chelates reductase (FCR) system at the root surface (Robinson et al., 1999; Marschner, 1995), enabling plants to more efficiently reduce and take up chelated Fe. Provided that the efficiency of the corresponding EDDHA ligand in complexing and solubilizing Fe from the soil is limited, the FeEDDHA isomer concentration in soil solution will hence decrease more swiftly and strongly in the presence of Fe deficient plants than with plants that are not Fe deficient.

Comparison of corresponding treatments applied at t=0 and t=6 weeks shows that the racemic o,o-FeEDDHA concentrations are comparable (133 and 145 $\mu g\ l^{-1}$), and the meso o,o-FeEDDHA concentrations are approximately twice as high in the t=6 weeks treatment (20 and 47 $\mu g\ l^{-1}$); still these differences in meso o,o-FeEDDHA concentration are small in comparison to the dosage applied (560 $\mu g\ l^{-1}$). The effect of stress response mechanisms on o,o-FeEDDHA concentrations equaled six weeks of residence time in the soil-plant system for racemic o,o-FeEDDHA, and over 3 weeks for meso o,o-FeEDDHA.

Fe uptake

The Fe uptake data presented in Figure 11 demonstrate that, in agreement with Rojas et al., 2008, o,p-FeEDDHA did not significantly increase Fe uptake in any of the treatments; neither applied as a single substance (*o,p 3* and *o,p 6* treatment) nor in a mixture through a

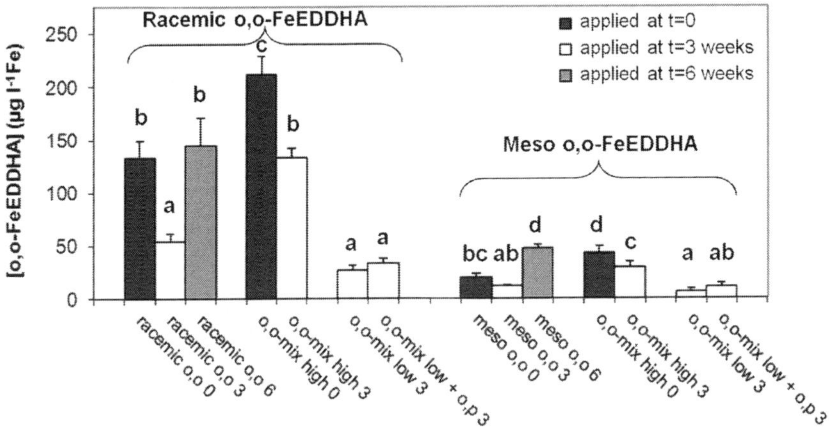

Fig. 10. Racemic and meso o,o-FeEDDHA concentration in the pore water of Santomera soil at harvest. Error bars indicate standard deviations. Letters indicate the significantly different groups as identified by the Tukey post-hoc test, for both FeEDDHA components separately. (based on Schenkeveld et al., 2010b)

synergistic effect (*o,o-mix low + o,p*). Due to its interaction with soil constituents, the residence time of o,p-FeEDDHA in soil solution can be short (Schenkeveld et al., 2007). Therefore o,p-FEDDHA had only been applied to soybean plants that were already Fe deficient at the moment of application. However, even when applied in the growth stages that Fe requirements were highest and Fe stress response mechanisms were activated, facilitating a more efficient Fe uptake, o,p-FeEDDHA still did not significantly increase the Fe uptake of soybean plants.

Both racemic o,o-FeEDDHA and meso o,o-FeEDDHA did contribute to Fe uptake (Figure 11), as shown from the fact that in all treatments with o,o-FeEDDHA, Fe uptake was significantly higher than in the blank treatment. This is in agreement with the conclusion from the study by Ryskievich and Boku (1962). For none of the moments of application, significant differences in Fe uptake were found between the *racemic o,o* and *meso o,o* treatment. Because overall Fe uptake in the *o,o-mix high* treatments was higher than in the *racemic o,o* ($p = 0.030$) and the *meso o,o* ($p = 0.012$) treatments, Fe uptake was not yet maximal in the *racemic o,o* and the *meso o,o* treatments. Therefore it can be concluded that racemic and meso o,o-FeEDDHA were approximately equally effective in facilitating Fe uptake. Lucena and Chaney (2006) reported that meso o,o-FeEDDHA was more effective in delivering Fe to hydroponically grown cucumber plants than racemic o,o-FeEDDHA, as a result of a lower stability favouring Fe reduction at the root surface. Possibly in soil, a preferential Fe uptake from meso o,o-FeEDDHA was balanced by a higher affinity for the solid phase and a faster decline in soil solution concentration.

Moreover, for both the *racemic o,o* and *meso o,o* treatments, no significant difference in Fe uptake was observed between the different moments of application. This is remarkable, because the plants receiving treatment after 6 weeks had much less time to benefit from the applied o,o-FeEDDHA. Apparently, as a result of Fe deficiency stress response mechanisms and development of the root system, the soybean plants had grown much more efficient with regard to Fe uptake. In only two weeks time, the soybean plants from the *racemic o,o 6* and *meso o,o 6* treatments took up an additional 0.36 mg of Fe per pot, which corresponds with 50% of the total Fe uptake in the blank treatment.

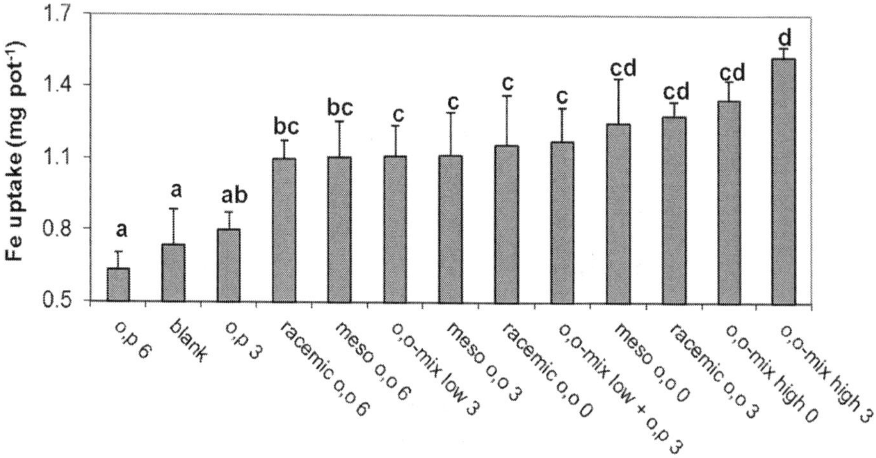

Fig. 11. Fe uptake by soybean plants grown on Santomera soil for all FeEDDHA treatments. Error bars indicate standard deviations. Letters indicate the significantly different groups as identified by the Tukey post-hoc test including all FeEDDHA treatments and all moments of application. (based on Schenkeveld et al., 2010b)

4. Conclusions, limitations and challenges for future research

In the pot trial experiments conducted, it was found that the effectiveness of FeEDDHA components in delivering Fe to soil-grown plants is largely determined by their ability to remain in solution. The residence time of o,p-FeEDDHA in soil solution proved too short to significantly contribute to facilitating Fe uptake. The residence time of both racemic and meso o,o-FeEDDHA was much longer and both isomers did contribute to Fe uptake, approximately to the same extent on the time scale considered. o,o-FeEDDHA facilitated Fe uptake increased both the Fe content and the dry weight yield of the soybean plants. Contrary to racemic o,o-FeEDDHA, the residence time of meso o,o-FeEDDHA in soil solution was substantially compromised by plant-independent processes. Due to its longer residence time, racemic o,o-FeEDDHA is likely to remain effective for a longer time-span than meso o,o-FeEDDHA. The effectiveness of rest-FeEDDHA has not been separately assessed in the pot trials. In the study examining the effect of FeEDDHA treatment composition, it was concluded that o,o-FeEDDHA governed Fe uptake; the contribution of rest-FeEDDHA was marginal, at most.

The findings from the presented pot trial studies may serve appliers of FeEDDHA fertilizer to make a better selection out of the available products and help them to optimize the dosage and frequency of application. Furthermore, they may provide producers of

FeEDDHA fertilizers with leads for optimizing the compositions of their formulations and for effectively marketing their products.

Although the processes examined in these pot trials also take place in a field situation, a translation of the results to a field situation should be treated with caution, because plant care and growth conditions differ strongly between the field and a conditioned greenhouse, not all processes affecting FeEDDHA concentration in the field have been considered, and the relative impact of the individual process may well be different in the field than in a controlled environment.

The presented studies persued insights on a level, transcending an individual soil or crop. Still, for practical reasons, only one plant species (soybean) and one soil (Santomera) have been used. This inevitably holds a risk of over-representation of soil-, species- or even cultivar specific peculiarities. Challenges for future research would therefore include carrying out comparative studies with different soils and crops, and conducting field trials to examine how the results from the pot trials relate to agricultural practice.

Another focal point for further research concerns the fate of FeEDDHA components in the soil-plant system. The results from the pot trial studies show that for most of the FeEDDHA components, the fate is determined by plant-independent processes. A better understanding of the soil processes affecting the effectiveness of FeEDDHA components, or in a more general sense, of Fe chelates applied as fertilizer, would enable a more efficient and soil specific application of Fe fertilizer products. Processes to examine more closely than reported so far would for instance include biodegradation, adsorption, cation competition and leaching.

5. Acknowledgment

The authors wish to thank the following: AkzoNobel for funding the presented research, P. Weijters and M. Bugter for initiating the research project, Prof. W.H. van Riemsdijk and Dr. A.M. Reichwein for their contributions to the articles on which this chapter is based, Prof. R.J. Goos for providing the Mycogen 5072 seeds, P. Nobels for his help with the ICP-analyses, W. Menkveld, A. Brader and P. Pellen for plant care and J. Nelemans for advice and practical support.

6. References

Alvarez-Fernandez, A.; Sierra, M.A. & Lucena, J.J. (2002). Reactivity of synthetic Fe chelates with soils and soil components. *Plant Soil*, Vol. 241, pp. 129-137, ISSN 0032-079X

Alvarez-Fernandez, A.; Garcia-Lavina, P.; Fidalgo, C.; Abadia, J. & Abadia A. (2004). Foliar fertilization to control iron chlorosis in pear (Pyrus communis L.) trees. *Plant Soil*, Vol. 263, pp. 5-15, ISSN 0032-079X

Alvarez-Fernandez, A.; Abadia, J. & Abadia, A. (2006). Iron deficiency, fruit yield and fruit quality. In: *Iron nutrition in plants and rhizospheric microorganisms*, L.L. Barton and J. Abadia (Eds.), pp. 85-101. Springer, ISBN 978-1-4020-4742-8.

Bailey, N.A.; Cummins, D.; McKenzie, E.D. & Worthington, J.M. (1981). Iron(III) compounds of phenolic ligands. The chrystal and molecular structure of the sexadentate ligand

N,N′-ethylene-bis-(o-hydroxyphenylglycine). *Inorg. Chim. Acta*, Vol 50, pp. 111-120, ISSN 0020-1693.

Bannochie, C. J. & Martell, A. E. (1989). Affinities of racemic and meso forms of N,N'-ethylenebis[2-(o-hydroxyphenyl)glycine] for divalent and trivalent metal ions. *J. Am. Chem. Soc.*, Vol. 111, pp. 4735-4742, ISSN 0002-7863.

Boxma, R. (1972). Bicarbonate as the most important soil factor in lime-induced chlorosis in the Netherlands. *Plant Soil*, Vol. 37, pp. 233-243, ISSN 0032-079X.

Cantera, R.G.; Zamarreno, A.M. & Garcia-Mina, J.M. (2002). Characterization of commercial iron chelates and their behavior in an alkaline and calcareous soil. *J. Agric. Food Chem.*, Vol.50, pp. 7609-7615, ISSN 0021-8561

Cerdan, M.; Juarez, S.A.M.; Jorda, J.D. & Bermudez, D. (2006). Fe uptake from meso and d,l-racemic Fe(o,o-EDDHA) isomers by strategy I and II plants. *J. Agric. Food Chem., Vol.* 54, pp. 1387-1391, ISSN 0021-8561

Chaney, R.L. (1984). Diagnostic practices to identify iron deficiency in higher plants. *J. Plant Nutr.*, Vol. 7, 47-67, ISSN 0190-4167, ISSN 1532-4087.

Chen, Y. & Barak, P. (1982). Iron nutrition of plants in calcareous soils. *Adv. Agron.*, Vol. 35, pp. 217-240, ISSN: 0065-2113

Cremonini, M.A.; Alvarez-Fernandez, A.; Lucena, J.J.; Rombola, A.; Marangoni, B. & Placucci, G. (2001). NMR analysis of the iron ligand ethylenediaminedi(o-hydroxyphenyl)acetic acid (EDDHA) employed in fertilizers. *J. Agric. Food Chem.*, Vol. 49, pp. 3527-3532, ISSN 0021-8561

Fairbanks, D.J. (2000). Development of genetic resistance to iron-deficiency chlorosis in soybean. *J. Plant Nutr.*, Vol. 23 pp. 1903-1913, ISSN 1532-4087.

Garcia-Marco, S.; Yunta, F.; De la Hinojosa, M.I.M; Marti, G. & Lucena, J.J. (2003). Evaluation of commercial Fe(III)-chelates using different methods. *J. Plant Nutr.*, Vol. 26, pp. 2009-2021, ISSN 1532-4087.

Garcia-Marco, S.; Martinez, N.; Yunta, F.; Hernandez-Apaolaza, L. & Lucena, J.J. (2006). Effectiveness of ethylenediamine-N(o-hydroxyphenylacetic)-N '(p-hydroxy-phenylacetic) acid (o,p-EDDHA) to supply iron to plants. *Plant Soil*, Vol. 279, pp. 31-40, ISSN 0032-079X.

Goos, R.J. & Johnson, B.E. (2000). A comparison of three methods for reducing iron-deficiency chlorosis in soybean. *Agron. J.*, Vol. 92, pp. 1135-1139, ISSN 1435-0645.

Goos, R.J.; Johnson, B.; Jackson, G. & Hargrove, G. (2004). Greenhouse evaluation of controlled-release iron fertilizers for soybean. *J. Plant Nutr.*, Vol. 27, pp. 43-55, ISSN 1532-4087.

Gruber, B. & Kosegarten, H. (2002). Depressed growth of non-chlorotic vine grown in calcareous soil is an iron deficiency symptom prior to leaf chlorosis. *J. Plant Nutr. Soil Sci.*, Vol. 165, pp. 111-117, ISSN 1532-4087.

Grusak, M.A. (1995). Whole-root iron(III)-reductase activity throughout the life-cycle of iron-grown *Pisum Sativum* L. (Fabaceae): Relevance to the iron nutrition of developing seeds. *Planta*, Vol. 197, pp. 111-117, ISSN 1432-2048.

Grusak, M.A.; Pearson, J.N. & Marentes, E. (1999). The physiology of micronutrient homeostasis in field crops. *Field Crops Res.*, Vol. 60, pp. 41-56, ISSN 0378-4290.

Hansen, N.C.; Jolley, V.D.; Naeve, S.L. & Goos, R.J. (2004). Iron deficiency of soybean in the north central US and associated soil properties. *Soil Science and Plant Nutrition,* Vol. 50, pp. 983-987, ISSN 0038-0768.

Hansen, N.C.; Hopkins, B.G.; Elsworth, J.W. & Jolley, V.D. (2006). Iron nutrition in field crops. In: *Iron nutrition in plants and rhizospheric microorganisms,* L.L. Barton and J. Abadia (Eds.), pp. 23-59. Springer, ISBN 978-1-4020-4742-8 .

Heitholt, J. J.; Sloan, J. J.; MacKown, C. T. & Cabrera, R.I. (2003). Soybean growth on calcareous soil as affected by three iron sources. *J. Plant Nutr.,* Vol. 26, pp. 935-948, ISSN 1532-4087.

Hernandez-Apaolaza, L. & Lucena, J.J. (2001). Fe(III)-EDDHA and -EDDHMA sorption on Ca-montmorillonite, ferrihydrite, and peat. *J. Agric. Food Chem.,* Vol. 49, pp. 5258-5264, ISSN 0021-8561.

Hernandez-Apaolaza, L.; Garcia-Marco, S.; Nadal, P.; Lucena, J.J.; Sierra, M.A.; Gomez-Gallego, M.; Ramirez-Lopez, P. & Escudero, R. (2006). Structure and fertilizer properties of byproducts formed in the synthesis of EDDHA. *J. Agric. Food Chem.,* Vol. 54, pp. 4355-4363, ISSN 0021-8561.

Inskeep, W.P. & Bloom, P.R. (1986). Effects of soil-moisture on soil pCO2, soil solution bicarbonate and iron chlorosis in soybeans. *Soil Sci. Soc. Am. J.* , Vol. 50, pp. 946-952, ISSN 0361-5995.

Kroll, H. (1957). The ferric chelate of ethylene-diamine di(O-hydroxyphenylacetic acid) for treatment of lime-induced chlorosis. *Soil Sci.,* Vol. 84, pp. 51-53, ISSN 0038-075X.

Kroll, H.; Knell, M.; Powers, J. & Simonian, J. (1957). A phenolic analog of ethylenediamine-tetraacetic acid. *J. Am. Chem. Soc.,* Vol. 79, pp. 2024-2025, ISSN 0002-7863..

Lindsay, W.L. (1974) Role of chelation in micronutriënt availability. In: *The plant root and its environment,* E.W. Carson (Ed.), pp. 508-524., Charlottesville University Press, VA. ISBN 9780813904115

Lindsay, W.L. (1979). *Chemical equilibria in soils.* John Wiley and Sons: N.Y. pp. 449, ISBN 9781930665118.

Lindsay, W.L. & Schwab, A.P. (1982). The chemistry of iron in soils and its availability to plants. *J. Plant Nutr.,* Vol. 5, pp. 821-840, ISSN 1532-4087.

Lucena, J.J.; Manzanares, M. & Garate, A. (1992a). Comparative study of the efficacy of commercial Fe-chelates using a new test. *J. Plant Nutr.,* Vol. 15, pp. 1995-2006, ISSN 1532-4087.

Lucena, J.J.; Manzanares, M. & Garate, A. (1992b). A test to evaluate the efficacy of commercial Fe-chelates. *J. Plant Nutr.,* Vol. 15, pp. 1553-1566, ISSN 1532-4087.

Lucena, J.J.; Barak, P.; & Hernandez-Apaolaza, L. (1996). Isocratic ion-pair high-performance liquid chromatographic method for the determination of various iron(III)chelates. *J. Chromatogr.* A, Vol. 727, pp. 253-264, ISSN 0021-9673.

Lucena, J.J. (2000). Effects of bicarbonate, nitrate and other environmental factors on iron deficiency chlorisis. A review. *J. Plant Nutr.,* Vol. 23, pp. 1591-1606, ISSN 1532-4087.

Lucena, J.J. (2003). Fe chelates for remediation of Fe chlorosis in strategy I plants. *J. Plant Nutr.*, Vol. 26, pp. 1969-1984, ISSN 1532-4087.

Lucena, J.J. (2006). Synthetic iron chelates to correct iron deficiency in plants. In: *Iron nutrition in plants and rhizospheric microorganisms,* L.L. Barton and J. Abadia (Eds.), pp. 103-128. Springer, ISBN 978-1-4020-4742-8.

Lucena, J.J. & Chaney, R.L. (2006). Synthetic iron chelates as substrates of root ferric chelate reductase in green stressed cucumber plants. *J. Plant Nutr., Vol.* 29, pp. 423-439, ISSN 1532-4087.

Lucena, J.J. & Chaney, R.L. (2007). Response of cucumber plants to low doses of different synthetic iron chelates in hydroponics. *J. Plant Nutr.,* Vol. 30, pp. 795-809, ISSN 1532-4087.

Marschner, H. (1995). *Mineral nutrition of higher plants.* Academic Press, London. pp. 889, ISBN 978-0-12-473543-9.

Mengel, K.; Breininger, M.T. & Bubl, W. (1984). Bicarbonate, the most important factor inducing iron chlorosis in vine grapes on calcareous soil. *Plant Soil,* Vol. 81, pp. 333-344, ISSN 0032-079X.

Mengel, K. (1994). Iron availability in plant tissues-iron chlorosis on calcareous soils. *Plant Soil,* Vol. 165 pp. 275-283, ISSN 0032-079X .

Mengel, K. & Kirkby, E.A. (2001). *Principles of plant nutrition.* International Potash Institute, Bern, Switzerland, ISBN 9783906535036.

Mortvedt, J.J. (1991). Correcting iron deficiencies in annual and perennial plants: Present technologies and future prospects. *Plant Soil,* Vol. 130, pp. 273-279, ISSN 0032-079X.

Nikolic, M. & Römheld, V. (2002). Does high bicarbonate supply to roots change availability of iron in the leaf apoplast? *Plant Soil,* Vol. 241, pp. 67-74, ISSN 0032-079X.

Novozamsky, I.; Van Eck, R.; Houba, V.J.G. & Van der Lee, J.J. (1996). Solubilization of plant tissue with nitric acid-hydrofluoric acid-hydrogen peroxide in a closed-system microwave digestor. *Commun. Soil Sci. Plan.,* Vol. 27, pp. 867-875, ISSN 1532-2416.

O'Conner G.A.; Lindsay, W.L. & Olsen, S.R. (1971). Diffusion of Iron and iron chelates. *Soil Sci. Soc. Am. Proc.,* Vol. 35, pp. 407-410, ISSN 0038-0776.

Perez-Sanz, A. & Lucena, J.J. (1995). Synthetic iron oxides as sources of Fe in a hydroponic culture of sunflower. In: *Iron nutrition in soils and plants.* Ed. J Abadia. pp. 241-246. Kluwer Academic Publishers, ISBN 9780792329008.

Robinson, N.J.; Procter, C.M.; Connolly, E.L. & Guerinot, M.L. (1999). A ferric-chelate reductase for iron uptake from soils. *Nature,* Vol. 397, pp. 694-697, ISSN 0028-0836.

Rojas, C.L.; Romera, F.J.; Alcantara, E.; Perez-Vicente, R.; Sariego, C.; Garcia-Alonso, J.I.; Boned, J. & Marti, G. (2008). Efficacy of Fe(o,o-EDDHA) and Fe(o,p-EDDHA) isomers in supplying Fe to Strategy I plants differs in nutrient solution and calcareous soil. *J. Agric. Food Chem.,* Vol. 56, pp. 10774-10778, ISSN 0021-8561.

Römheld, V. & Marschner, H. (1981). Rhythmic iron stress reactions in sunflower at suboptimal iron supply. *Physiol. Plantarum,* Vol. 53, pp. 347-353, ISSN 1399-3054.

Römheld, V. (2000). The chlorosis paradox: Fe inactivation as a secondary event in chlorotic leaves of grapevine. *J. Plant Nutr.*, Vol. 23, pp. 1629-1643, ISSN 1532-4087.

Ryskievich, D.P. & Boka, G. (1962). Separation and characterization of the stereoisomers of N,N′-ethylenebis-[2-(o-hydroxy-phenyl)]glycine. *Nature*, Vol. 193 pp. 472-473, ISSN 0028-0836.

Schenkeveld W.D.C; Reichwein, A.M.; Temminghoff, E.J.M. & Van Riemsdijk, W.H. (2007). The behaviour of EDDHA isomers in soils as influenced by soil properties. *Plant Soil*, Vol. 290, pp. 85-102, ISSN 0032-079X.

Schenkeveld, W.D.C; Dijcker, R.; Reichwein, A.M.; Temminghoff, E.J.M. & Van Riemsdijk, W.H. (2008). The effectiveness of soil-applied FeEDDHA treatments in preventing iron chlorosis in soybean as a function of the o,o-FeEDDHA content. *Plant Soil*, Vol. 303, pp. 161-176, ISSN 0032-079X.

Schenkeveld, W.D.C.; Temminghoff, E.J.M.; Reichwein, A.M. & Van Riemsdijk, W.H. (2010a). FeEDDHA-facilitated Fe uptake in relation to the behaviour of FeEDDHA components in the soil-plant system as a function of time and dosage. *Plant Soil*, Vol. 332, pp. 69-85, ISSN 0032-079X.

Schenkeveld, W.D.C; Reichwein, A.M.; Bugter, M.; Temminghoff, E.J.M. & Van Riemsdijk, W.H. (2010b). The performance of soil-applied FeEDDHA isomers in delivering Fe to soybean plants in relation to the moment of application *J. Agric. Food Chem.*, Vol. 58, (24), pp. 12833–12839, ISSN 0021-8561.

Vasconcelos, M. & Grusak, M.A. (2006). Status and future developments involving plant iron in animal and human nutrition. In: *Iron nutrition in plants and rhizospheric organisms*. Eds. L L Barton and J Abadia. p. 477. Springer, ISBN 978-1-4020-4742-8.

Wallace, A.; Muerller, R.; Lunt, O.R.; Ashcroft, R.T. & Shannon, L.M. (1955). Comparisons of five chelating agents in soils, in nutrient solutions, and in plant responses. *Soil Sci.*, Vol. 80, pp. 101-108, ISSN 0038-075X.

Wallace, A. (1962). A decade of synthetic chelating agents in inorganic plant nutrition, Los Angeles (California), ASIN: B000C9XS8U .

Wallace, A. (1966). Ten years of iron EDDHA use in correcting iron chlorosis in plants. In: *Current topics in plant nutrition*. Ed. A Wallace. pp 1-2. Edwards Brothers, Inc., Ann Arbor, Michigan.

Wallace, A. & Cha, J.W. (1986). Effects of bicarbonate, phosphorus, iron EDDHA, and nitrogen sources on soybeans grown in calcareous soil. *J. Plant Nutr.*, Vol. 9, pp. 251-256, ISSN 1532-4087.

Yunta, F.; Garcia-Marco, S. & Lucena, J.J. (2003a). Theoretical speciation of ethylenediamine-N-(o-hydroxyphenylacetic)-N '-(p-hydroxyphenylacetic) acid (o,p-EDDHA) in agronomic conditions. *J. Agric. Food Chem.*, Vol. 51, pp. 5391-5399, ISSN 0021-8561.

Yunta, F.; Garcia-Marco, S.; Lucena, J.J.; Gomez-Gallego, M.; Alcazar, R. & Sierra, M.A. (2003b). Chelating agents related to ethylenediamine bis(2-hydroxyphenyl)acetic acid (EDDHA): Synthesis, characterization, and equilibrium studies of the free ligands and their Mg^{2+}, Ca^{2+}, Cu^{2+}, and Fe^{3+} chelates. *Inorg. Chem.*, Vol. 42, pp. 5412-5421, ISSN 0020-1669.

Zhang, C.D.; Römheld, V. & Marschner, H. (1995). Retranslocation of iron from primary leaves of bean-plants grown under iron deficiency. *J. Plant Physiol.*, Vol. 146, pp. 268-272, ISSN 0176-1617.

6

Effects of an Agropastoral System on Soybean Production

Katsuhisa Shimoda
Japan International Research Center for Agricultural Science
Japan

1. Introduction

Increase of the cereal production in the world is desired for increase of global food demand. The savanna agro-ecosystems of South America are one of the most important potential areas for expansion of agricultural production in the world. Therefore, the opening of some 12 million ha of this area has been conducted over the last 40 years (Kerridge, 2001). For example, in the Brazilian savanna, grain crops cover 12-14 million ha and introduced pasture area is over 50 million ha (Maceado, 2001).
However, under continuous cropping, soil productivity declines due to soil loss, soil compaction, loss of organic matter and increase in pests, diseases and weeds. For example, the soybean yield in Yguazu, Alto Parana, Paraguay, increased to over 3 t/ha after the introduction of a no-tillage cultivation system in 1983. However, the yield increase stopped in the early 1990's, and recently the yield seems to be decreasing (Seki, 1999; Shimoda et al., 2010). As a means to solve these problems, Kerridge (2001) suggested that integrated agro-pastoral systems with no-tillage appeared to be the key to sustainable development. But, since the period used as pasture and the pasture manegement methods have various combination in the system, few studies have been evaluated them synthetically.
The Japan International Research Center for Agricultural Socience Japan (JIRCAS) was conducted the joint reserchs of agropastoral system with the National Beef Cattle Research Center of the Brazilian Agricultural research Corporation (EMBRAPA-CNPGC) from 1996 to 2006 and many results were obtained from several experiments (Macedo et al., 2004; Kanno et al., 2004; Miranda et al., 2004 etc.). In addition, JIRCAS was conducted with the Japan International Cooperation Agency's Paraguay Agricultural Technology Center (CETAPAR-JICA) from 2003 to 2008 (Shimoda et al., 2010; 2011). In this report, I clarify the positive effects of an agropastoral system synthetically.

2. Material and methods

2.1 Study site

Two experiments (Exp.-1 and Exp.-2) were conducted in CETAPAR-JICA for soybean productivity with agropastoral system. CETAPAR-JICA is located in Colonia Yguazu (a Japanese settlement, 35°27'S, 55°04'W) in Alto Parana, Paraguay. Soil in this area is fertile and is known as "Terras Roxas" in Brazil (Igarashi 1997). Mean annual temperature and precipitation from 1972 to 2002 were 21.6°C and 1545 mm, respectively.

2.2 Experimental design

For Exp.-1, part of a field at CETAPAR-JICA, where soybean and wheat had been continuously cropped in a no-tillage system since 1993, was converted to Guinea grass (*Panicum maximum* cv. Tanzania) pasture in 1996. Established as a permanent pasture, it was maintained without fertilizer, cutting, or renovation for 7 years after establishment, and was used as a complementary pasture. In October 2003, the pasture was converted into an agropastoral plot where soybean and wheat were cultivated. the agropastoral plot was 2.97ha. In another part of the field adjacent to the agropastoral site, where soybean and wheat had been continuously cultivated in a no-tillage system since 1993, the non-converted treatment was replicated in three plots (control plots). Each plot was 0.68 ha.

For Exp.-2, 15 plots were arranged at the study site, each plot was 0.68 ha (124 m × 55 m) where soybean and wheat had been continuously cultivated in a no-tillage system since 1993. Twelve plots were randomly converted to Guinea grass (*Panicum maximum* cv. Monbasa) pasture in November 2003. These pastures were managed as intensive grazing pastures under high grazing pressure. The strip grazing was conducted in the pasture year round, and cattle were fed supplement during four months in dry season. Fertilization was also conducted (ammonium sulfate). The stocking rate was from 4.5 to 6.0 UA/ha for 3 years. Three plots of these pastures were also reconverted to soybean-wheat fields in October 2007 as no-tillage system (agropastoral plots). The non-converted treatment was replicated in three plots (control plots). Control plots in Exp.-1 and Exp.-2 were same plots.

2.3 Chemical and physical properties of soil

To investigate the chemical properties of the soil, samples from depths of 0-10, 10-20, 20-40, and 40-60 cm were collected independently from each plot, and the concentrations of phosphate, the percentage organic matter and pH, were measured. The concentrations of phosphate were analyzed using the Mehlich-III method, and percentage organic matter was analyzed using the Walkley-Black method. The pH of soils was measured using a pH meter (Horiba Co. Ltd.).

Moreover, soil was sampled from 0–5, 5–10, 10–20, 20–40, and 40–60 cm depths for measurement of physical properties, three phases of soil, bulk density, and soil aggregates (Only Exp.-1). Three phases of soils and soil aggregates were measured using a three-phase meter and an aggregate analyzer (Daiki Rika Kogyo Co., Ltd.), respectively.

We analyzed soybean and wheat production and soil chemical and physical data between the agropastoral plots and control plots using t-test, and the annual variation of chemical data in both plots using the Tukey-Kramer method. Details of the study methods were shown in Shimoda et al (2010, 2011).

3. Results

3.1 Soybean production

Since the yield of a soybean had a large change every year, the effect of agropastoral system was evaluated by using ratio of the soybean yield in agropastoral plots to that in control plots (Table 1). As a result, the ratios of the first year when reconverted into soybean field from the pasture were 2.35 in Exp-1 and 1.02 in Exp-2, respectively. In addition, the ratios of the second year were 1.86 in Exp-1 and 1.42 in Exp-2, respectively. The effect of Exp-1 was larger than that of Exp-2.

The first, second and third year in Exp.-1 were drought years and the second year in Exp.-2 was drought year too. The ratios of drought years were larger than the ratio of normal year in both experiments.
Moreover, during the experimental period in Exp.-1, the relative yield of soybean decreased year by year from 2.31 t/ha in the first year to 1.11 t/ha in the fourth year.

	Experiment 1 (Exp-1)[1]	Experiment 2 (Exp-2)[2]
Pasture condition		
Period as a soybean field before rotation	More than 3 years	More than 10 years
Period as a pasture after rotation	7 years	4 years
Introduced grass species	*P. maxmum* cv. Tanzania	*P. maxmum* cv. Monbasa
Grazing intensity	Extensive	Intensive
Weght gain per hectare	Little(unknown)	1.34ton/ha
Soybean production (Agropastoral/continuous cropping)		
First year soybean production after rerotation	1.48ton/ha (2.35 times)*	3.71ton/ha (1.02 times)
Second year soybean production after rerotation	3.56ton/ha (1.86 times)*	1.24ton/ha (1.42 times)*
Third year soybean production after rerotation	2.84ton/ha (1.45 times)*	-
Forth year soybean production after rerotation	2.74ton/ha (1.11 times)	-

Source: [1] from Shimoda et.al. (2010) and [2] from Shimoda et al. (2011).
*:Drought year.

Table 1. Study site profile and soybean production

3.2 Chemical properties

In Exp-1, the soil organic matter content in the agropastoral plots was 1.20 times higher than that in the control plots with significant difference at the reconversion from the pasture. However, in Exp-2, the soil organic matter content in the agropastoral plots was same (1.04 times) as that in the control plots. Under intensive grazing (Exp.2), even if soybean field was converted to the pasture, the accumulation of organic matter was not promoted (Fig. 1).
On the other hand, the phosphate concentration in agropastoral plots was 0.28 times than that in the control plots in Exp.-1 and 0.19 times than that in the control plots in Exp.-2 (Tble 2). Under extensive and intensive grazing, the accumulation of phosphate at soil surface was dissolved. After conversion to the pasture, phospate concentration was reduced by half in only two years (Fig. 2). But, phosphate accumulation was promoted in control plots. In addition, pH of soil surface was also increased with dissolution of phspate accumulation, the acidic soil was improved by nutral soil (Fig. 3).

	Experiment 1 (Exp-1)[1]	Experiment 2 (Exp-2)[2]
Chemical properties	7rd year pasture/Control plot	3rd year pasture/Control plot
Organic matter	1.20 times***	1.04 times
Phosphate	0.28 times***	0.19 times**

Source: [1] from Shimoda et.al. (2010) and [2] from Shimoda et al. (2011).
**:P<0.01. **:P<0.001

Table 2. Comparison of soil chemical properties at soil surface (0-10cm in depth)

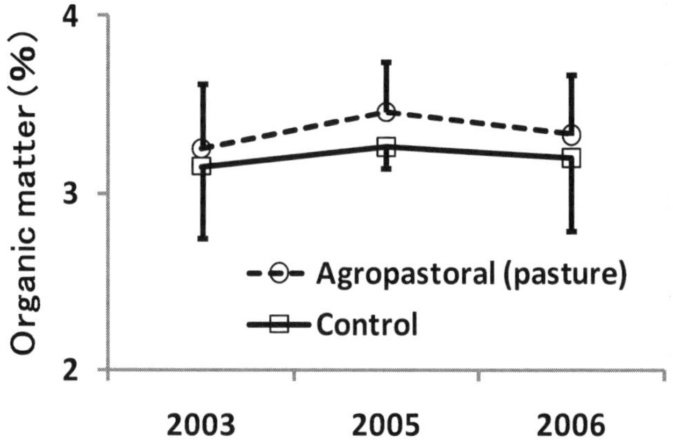

Fig. 1. Change in organic matter percentage (%) of Exp.-2 in the soil surface (0-10cm at depth).

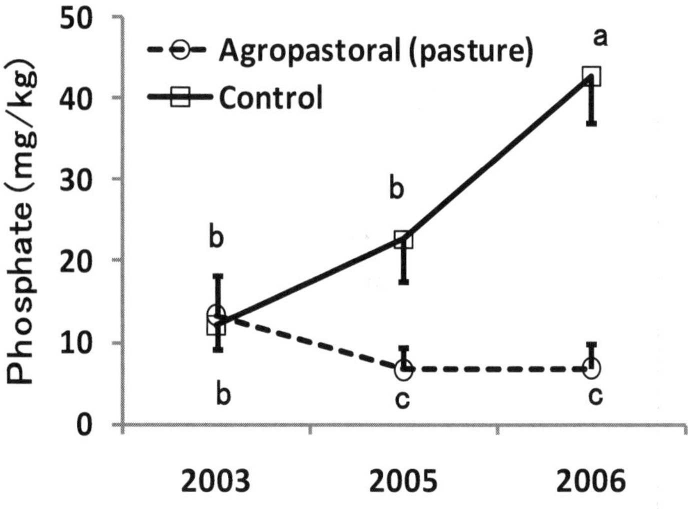

Fig. 2. Change in phosphate concentration (mg/kg) of Exp.-2 in the soil surface (0-10cm at depth). A different letter is significantly different.

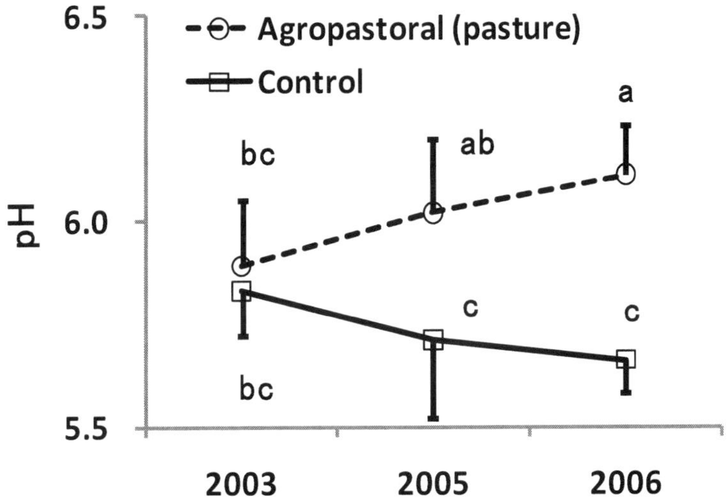

Fig. 3. Change in pH of Exp.-2 in the soil surface (0-10cm at depth).
A different letter is significantly different.

3.3 Physical properties

In soil samples from deeper than 10 cm, the gaseous phase percentage in the agropastoral plots was significantly higher than in control plots (Fig 4), but was lower at the soil surface (0–10 cm depth). The bulk density of samples from the soil surface of agropastoral plots was higher than that of control plots, and the bulk density of soil samples from deeper than 20 cm was lower than that of control plots (Fig. 4). But, they did not have significantly difference. In addition, the percentage of large aggregates in the soil of the agropastoral plots was 3 to 7% higher than that in control plots at each depth (Fig. 4). Especially, at the soil surface, it was significantly higher than that in control plots.

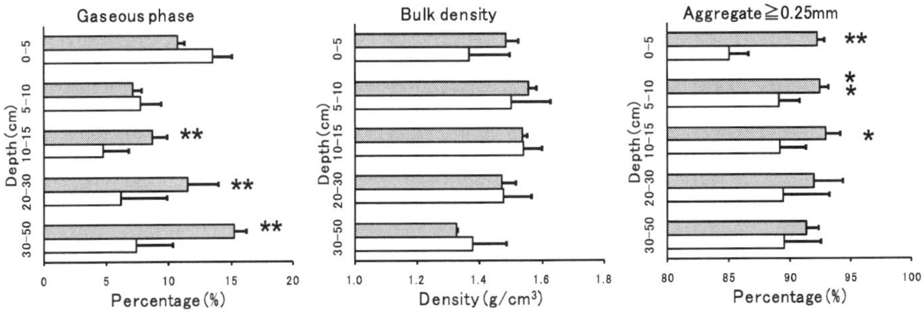

*,**:Data are significantly different at 5% and 1% level from the control plot, respectively.

Fig. 4. Physical properties of the soil in agropastoral and control plots of Exp.-1.
Gray bar is the mean value and SD in agropastoral plot and white bar is those in control plot.

4. Discussion

In the first and second year after reconversion from the pastures, the ratio of the soybean yield in agropastoral plots to that in control plots in Exp.-1 was higher than that in Exp.-2. Therefore, it was thought that the positive effects on soybean productivity of the agropastoral system with extensive grazing and long term pasture was higher than that with intensive grazing and short term pasture. Macedo et al. (2004) reported that the mean ratio of the soybean yield in the first year between all 4 years agropastoral plots and control plots was 1.12 (calculated from their table) under a conventional grazing system which the weight gain of cattle per hectare was one-third that of our intensive grazing system. But, it had higher grazing pressure than that of our extensive grazing system. Therefore, it was considered that the positive effect on soybean productivity was large as the grazing pressure was low.

However, it was thought that it would be lost in about four years even if agropastoral system with the extensive grazing and long term pasture (seven years) was conducted. So, it was important for this system to convert a field into a pasture continually.

In the drought year, the positive effect on soybean productivity was clear. It was considered as a reason that the phosphate accumulation at soil surface was dissolved. Many studies reported that the root of soybean was distributed within a shallow soil layer with no-tillage system (Iijima et al., 2007; Izumi et al., 2009). In addition, the phosphate accumulates near the soil surface, which restricts a crop root distribution within a shallow soil layer with a no-tillage system (Holanda et al., 1998; Seki et al., 2001). Plants with shallower root systems have a disadvantage for uptake and sensitive to drought (Schwinning ,1988).

In general, soybean does not grow well in acidic soil, and a pH range of 6.0 to 6.5 is best for soybean cultivation (Kokubun, 2002). In our system, soil pH in top soil improved from 5.89 to 6.11 over 3 years in the pasture, and conversely, soil pH became lower and the soil acidified in the control plots. Therefore, possibly the improvement of soil pH had same effect on the increase of the soybean yield.

Studies have reported that the accumulation of organic matter in soil is promoted by introducing agropastoral systems (Miranda et al., 2004; Salton & Lamas, 2007; Shimoda et al., 2010). In general, organic matter develops the soil aggregate structure and improves the water-holding capacity (Uwasawa, 2002). However, the accumulation of organic matter was promoted in Exp.-1 and not promoted in Exp.-2. Ogawa & Mitamura (1982) also reported that the accumulation of organic matter was not promoted by grazing. It was a reason that the root growth was inhibited by cutting the aboveground part of the grass (Davidson & Milthorpe, 1966a; 1966b). In addition, a lot of grass was grazed and much cattle meat (1.54 ton/ha) was carried out from the pasture every year under our intensive grazing. Therefore, it was thought that the positive effects on soybean productivity in Exp.-1 was larger than that in Exp.-2 by the promotion of organic matter accumulation.

In Exp.-1, at soil surface (0-10 cm at depth) in agro-pastoral plots, the percentage of gaseous was lower and bulk density was higher. Soil compaction of soil surface inhibits soybean production (Ae, 1997). However, since soil sampling was carried out immediately after killing off Guinea grass by herbicide, soil compaction at the surface would disappear rapidly by decomposition of the root of Guinea grass after that. And, since the percentage of gaseous phase of soil in agropastoral plots was higher in the soil layer from 10 cm to 50 cm at depth, it was thought that soil compaction occurred in no-tillage cultivation had improved. In addition, percentage of large aggregate of soil of agro-pastoral plots was higher than that of control plots in each depth. Higher percentage of large aggregate may promote inflow of air to underground. Inflow of air to underground promotes nitrogen fixation of soybean (Ae 1997). Therefore, it seemed that the improvement of physical properties of soil has contributed to recovery of soybean productivity.

5. Conclusion

The effects on soybean productivity of the agropastoral system with extensive grazing and long term pasture and with intensive grazing and short term pasture were positive. In addition, the positive effect on soybean productivity was clear in the drought year. It was thought the positive effects were promoted by the dissolvation of phosphate accumulates near the soil surface, accumulation of organic matter, improvement of soil compaction, and etc..

The investigation which took in intensive grazing was carried out for the income compensation of soybean farmers during the period used as a pasture. However, it was thought that productivity recovery of a soybean and the productivity of livestock had a relation of a trade-off. So, it waits for research to shorten the period used as a pasture.

6. Acknowledgment

All of the field tests described in this chapter were conducted in the joint research of JIRCAS and CETAPAR-JICA, Yguazu, Alto Parana, Paraguay. I deeply thank Mr. Toshiyuki Horita, Mr. Ken Hoshiba, and Mr. Jorge Bordon for their helpful advice and support. Special thanks are also due Dr. Kazunobu Toriyama and Dr. Ana Kojima for their helpful suggestion and encouragement.

7. References

Ae, N. (1997). Issues of fertilization in South America. *JIRCAS Working Report*, **7**: 49-53, ISSN 1341-710X

Davidson, J. L. & Milthorpe (1966) Leaf growth in *Dactylice glomerata* following defoliation. *Ann. Bot.*, **30**: 174-184, ISSN 0970-0153

Davidson, J. L. & Milthorpe (1966) The effect of defoliation on the carbon balance in *Dactylice glomerata*. *Ann. Bot.*, **30**: 185-198, ISSN 0970-0153

Holanda, F. S. R., Mengel, D. B., Paula, M. B., Carvaho, J. G. & Bertoni, J. C. (1998) Influence of crop rotations and tillage systems on phosphorus and potassium stratification and root distribution in the soil profile. *Commun. Soil Sci. Plant Anal.*, **29**: 2383-2394, ISSN 0010-3624

Igarashi, T. (1997) Characteristics and problems of soil fertility. *JIRCAS Working Report*, **7**: 15-27, ISSN 1341-710X

Iijima, M., Morita, S., Zegada-Lizarazu, W. & Izumi, Y. (2007) No-tillage enhanced the dependence on surface irrigation water in wheat and soybean. *Plant Prod. Sci.*, **10**: 182-188, ISSN 1343-943X

Izumi, Y., Yoshida, T. & Iijima, M. (2009) Effects of subsoiling to the non-tilled field of wheat-soybean rotation on the root system development, water uptake, and yield. *Plant Prod. Sci.*, **12**: 327-335, ISSN 1343-943X

Kerridge, P. C. (2001). A historical perspective of agropastoral system research in the savannas of South America. *JIRCAS Working Report*: **19**: 3-17, ISSN 1341-710X

Kokubun, M. (2002) Soybean. *In* Encyclopedia of crop science, ed. Crop Science Society of Japan, Asakura Publishing, Tokyo, Japan, 370-377, ISBN 4-254-41023-9 C3561 (in Japanese)

Macedo, M.C.M., Bono, J. A., Zimmer, A., Miranda, C. H. B., Costa, F. P., Kichel, A. N. & Kanno, T. (2001). Preliminary results of agropastoral systems in the cerrados of Mato Grosso do Sul-Brazil. *JIRCAS Working Report*, **19**: 35-42, ISSN 1341-710X

Macedo, M.C.M., Zimmer, A., Miranda, C. H. B., Costa, F. P., Kanno, T., Bono, J. A. & Fukuda, A. (2004) Results of soybean production, animal liveweight gain and soil fertility changes in agro-pastoral systems. *JIRCAS Working Report*, **36**: 15-18, ISSN 1341-710X

Miranda, C.H.B., Macedo, M.C.M., Kanda, K. & Nakamura, T (2004) Soil organic residue accumulation in agro-pastoral systems in the cerrados of Brazil. *JIRCAS Working Report*, **36**: 29-33, ISSN 1341-710X

Ogawa, Y. & Mitamura, T. (1982) Studies on cessation of grazing on permanent pasture ecosystem: 5. Effect of cessation of grazing on some chemical characteristics in the soil. *J. Japan. Grassl. Sci.*, **27**: 407-412, ISSN 0447-5933 (in Japanese with English summary)

Salton, J. C. & Lamas, F. M. (2007) Integration of farming-livestock and cultivation of the cotton plant in savannahs. *In* Cotton in the savannah of Brazil, ed. Ferire E. C., Brasilia, Brazil, 379-402, (in Portuguese)

Schwinning, S. (1988) Summer and winter drought in a cold desert ecosystem (Colorado Plateau). I. Effects on soil water and plant water uptake. *J. Arid Environ.*, 60: 547-566, ISSN 0140-1963

Seki, Y., Hoshiba, K. & Bordon, J. (2001) Root distribution of soybean plants in no-tillage fields in Yguazau district of Paraguay. *Nettai Nougyou (Jpn. J. Trop. Agr.)*, 45: 33-37, ISSN 0021-5260 (in Japanese with English summary)

Shimoda, K., Horita, T., Hoshiba, K. & Bordon, J. (2010) Evaluation of an agropastoral system introduced into soybean fields in Paraguay: Positive effects on soybean and wheat production. *JARQ*, 44: 25-31, ISSN 0021-3551

Shimoda, K., Horita, T., Hoshiba, K. & Bordon, J. (2011) Evaluation of an agropastoral system introduced into soybean fields in Paraguay: Effects on soybean and animal production under intensive grazing. *JARQ*, 45: in press, ISSN 0021-3551

Uwasawa, M. (2002) Management of organic matter and fertilizers. *In* Encyclopedia of plant nutrition and fertilizers, ed. Editorial committee of encyclopedia of plant nutrition and fertilizers, Asakura Publishing, Tokyo, Japan, 436-439, ISBN 978-4-254-43077-6 C3561 (in Japanese)

7

Application Technologies for Asian Soybean Rust Management

Carlos Gilberto Raetano, Denise Tourino Rezende and Evandro Pereira Prado
São Paulo State University "Julio de Mesquita Filho", FCABO
Brazil

1. Introduction

1.1 Occurrence

Soybean rust is a foliar disease which initially surfaced and remained for many years in Asian countries such as Taiwan, Thailand, Japan and India (Ozkan et al., 2006). After that, the disease was detected in Uganda and South Africa and more recently in South America. Asian Soybean Rust (ASR) is caused by the *Phakopsora pachyrhizi* Sydon & P. Sydon fungus and has been the worst disease in soybean culture. The disease has been present on the American continent, in Paraguay and southern regions of Brazil since 2001 (Yorinori et al., 2010). The importance of ASR disease in Brazil can be evaluated by its rapid expansion and severity and the subsequent economic losses. Over three years (2001 to 2003), ASR dispersed to all soybean producing regions of Brazil, reached the whole of the American continent and was detected in the United States of America in November, 2004 (Yorinori, 2010). ASR disease, when not controlled, can cause a total loss of production (Yorinori et al., 2004). In Brazil, crops free of disease can have an average productivity of 3,300 kg ha^{-1}. However, with the production cost included for a return, net profits of 2,436 kg ha^{-1} have been seen, thus it is recommendable to control the causal agent of the disease (Yorinori, 2005). In the 2007/08 season, ASR showed the lowest severity level since the 2002/03 season, due to farmer awareness of the necessity to obey the "period of sowing interruption", instituted by many of states of the Brazilian Federation. Another cause for improvement was the predominance in the planting of earlier varieties and the improved monitoring system of the disease (Yorinori et al., 2010).

1.2 Severity of ASR disease

The permanence of the pathogen inoculum in the fields the whole year around, due to the post-harvest sowing of the summer season and due to sowing carried out under irrigation during the winter/spring crop season, it was difficult to control the disease between the 2003 and 2005 seasons. During this control period, in the western region of Brazil, the first symptoms of rust were already visible 18 to 30 days after emergence began (V3/V4) and, therefore, some crops received up to seven applications of fungicides. In 2005, with the liberation of the genetically modified soybean culture Roundup Ready (RR), between harvests, the situation became even more serious due to the permanence of the pathogen in the fields all year round (Yorinori et al., 2010). With the "period of sowing interruption", when only the cultivation of soybeans used in research and for increasing generations provided by breeding lines is permitted under severe rust control conditions and subject to government control organs (MAPA), the severity of the disease has diminished. Nevertheless, the use of control practices of low efficiency, with inadequate fungicides, the use of reduced doses for lowering costs and inadequate number and duration of applications, unfortunately, contribute to persistence of the disease resulting in significant production losses. Continuous monitoring programs, adequate handling practices and appropriate application technology are necessary in order to guarantee the production of soybean culture. The relationship between lateness in the control of ASR and the severity of the disease is 0.25% for each day in which control is not carried out. The relationship between the return rate of the soybean and the severity of ASR is of -36 kg ha^{-1} for each severity percentage point (Calaça, 2007).

1.3 Control of the pathogenic agent

In order to define the strategies to be used for ASR control, regarding application technology, there must be an awareness of the way systemic fungicides move into plants after application and absorption has been carried out. In the present-day market, the majority of fungicides recommended for ASR control move from the base to the top of each leaf, with little chance of moving in the other direction and without the possibility of dislocation from one leaf to another (Antuniassi, 2005). Amongst the fungicides available at present for pathogen control, the triazole fungicides, when used alone, have not presented good performance, as can be seen with ciproconazole, propiconazole and meticonazole (Yorinori et al., 2010). The consistency shown in programmes for chemical control applied in a curative and preventive manner on different soybean varieties and growth stages of the crop has been evaluated by Navarini et al. (2007). The authors established that there was a tendency that higher profit rates were related to preventive applications between the R1 and R3 stages. They also established a low efficiency rate in the control of the pathogen when the fungicide propiconazole was applied in a preventive manner. A deficiency in the control of *P. pachyrhizi* was also observed 30 days after the spraying of the fungicide difeconazole on this crop, in a comparative evaluation of fungicides carried out by Soares et al. (2004). It therefore, becomes evident that triazole fungicides have some limited systemic activity (moving through the plant, especially to newly developed leaves) and are thus somewhat forgiving if the application is less than perfect. When triazole fungicides are mixed with strobilurin fungicides, they show better performance in the control of rust disease. In Brazil, it is believed that the causes of control failures may be related to technical failures in application, predominantly in a population shift which is more tolerant to triazole fungicides in some regions (*Fungicide Resistance Action Committee – FRAC*), instead of developing a tolerance or resistance to the triazole fungicides through genetic mutation of the fungus. As a precautionary measure, class representatives of the producers recommend the use of triazoles only when mixed with other groups of fungicides.

1.4 Economic impact of ASR in Brazil

An estimate of the volume of grain losses and of the economic impact of ASR in the period between 2002 and 2009 reached 34.2 million tons, a value equivalent to more than half a full soybean harvest. On the other hand, the economic impact of ASR, adding up grain losses (US$ 7.95 billions), control costs (US$ 5.76 billions) and intake losses (US$ 1.55 billions) during the same period, totalled US$ 15.25 billion (Yorinori et al., 2010). If we consider that the world demand for soybeans is strongly linked to population increase, to world riches and to bioenergy, the necessity to minimise losses in all stages of soybean processing becomes evident. These demand indicators generate worries, particularly in Brazil, where the biofuel program has demanded a mixture of 4% biodiesel in the final fuel formulation since 2009, especially since the main source of biodiesel is soybean oil (Barros & Menegatti, 2010).

2. Fungicide spray coverage

The right moment for application is determined by climate conditions, the presence and severity of the disease, plant growth stage and fungicide efficiency (Yorinori et al., 2004). These factors, together with correct calibration of the application equipment and with correct handling practices aimed at the control of *P. pachyrhizi*, have not been sufficient to impede the advance of the disease in soybean culture. The necessity for more efficient application of phyto-sanitary products has been related to various researchers such as Adam (1977), Matuo (1990) and Van De Zande et al. (1994), amongst others. It can therefore be noted that, in order to obtain the best efficiency, the study and development of new application technologies are indispensable. Phyto-sanitary products must be applied with maximum efficiency and, for this to occur, studies of spray deposition and coverage and spray drift are necessary. This last factor is responsible for losses and is also a cause of environmental contamination (Matthews, 1992).

2.1 Droplet size and spray coverage

A definition of droplet size and the volume to be applied must be a priority in the planning of an application. Further factors, such as the correct time of application, weather conditions, product recommendations and operational conditions, should be considered as a whole, looking towards maximum performance with the least losses and the least environmental impact (Antuniassi, 2010). Spray volume has the greatest impact on canopy penetration and leaf coverage. Increasing the volume improves penetration and coverage. The recommended spray volume differs for each fungicide. For aerial applications, the minimum recommended volume is 5-7 gallons per acre (47-65 L ha^{-1}). Recent research on soybean canopy coverage for ground applications at different growth stages of soybean (R1, R3 and R5) support recommendations that a spray volume of 15 gpa (140 L ha^{-1}) may provide adequate coverage of the entire canopy early in the growing season (R1 and R3) but 20 gpa (187 L ha^{-1}) is necessary later in the growing season (by R5) when the soybean canopy density and volume have increased (Brown-Rytlewski & Staton, 2010). In Brazil, the spray volume rates for conventional ground spraying of soybean have varied from 100 to 150 L ha^{-1}, but it is possible to have a reduction of 50% in spray volume using the new spray technologies and earlier varieties. In the mid-west region (Cerrado), the use of low application rates with conventional ground sprayers is limited by climatic conditions due to the high temperature (30 to 40°C) and low air humidity (12 to 30%) during the greatest part of the year. Droplet size is the second most important factor affecting canopy penetration and leaf coverage. Research has shown that fine to medium droplets, with median volume diameters (MVD) in the range of 200 to 350 μm, maximise canopy penetration and leaf coverage. Smaller droplets provide better leaf coverage but lack the momentum to penetrate the canopy. Larger droplets have the momentum to penetrate the canopy but do not provide sufficient leaf coverage. Ground speed, nozzle pressure and spray volume should be considered when selecting nozzles for the sprayer. Choose nozzles that will produce 200-350 µm droplets at 15 to 20 gallons per acre (140 to 187 L ha^{-1}) while travelling at the desired speed. In most cases, nozzles for herbicide applications should not be used for fungicide applications as they are designed to generate larger droplets at lower application rates. All nozzle manufacturers use a spray classification system (ASAE standard S-572) of six categories with corresponding colours to classify the droplet size range produced by nozzles under various operating pressures. The colour of the nozzle itself should not be confused with the colours listed in Table 1. The nozzle colour describes the flow rate for the nozzle and the colours on the table describe the nozzle's droplet size range. When using droplet size classification charts, select nozzles that produce droplets near the fine end of the medium (yellow) category.

Droplet category	Colour	Symbol	MVD (µm)
Very fine	Red	VF	<150
Fine	Orange	F	150-250
Medium	Yellow	M	250-350
Coarse	Blue	C	350-450
Very coarse	Green	VC	450-550
Extremely coarse	White	XC	>550

Michigan State University - Department of Plant Pathology, USA

Table 1. ASAE Standard S-572 Spray Quality Categories

Ground speed affects spray volume and vertical droplet velocity. Taking into consideration that in order to apply fungicides, a fine to medium category of drops are indicated, and that the maximum wind speed during spraying should not surpass 9.6 km h^{-1} (Andef, 2004), a critical new situation presents itself in the field. Auto-propelled sprayers present innovations that give greater stability to the spray booms and with this, the operational speed increases to values rearing and even above 16 km h^{-1}. The immediate consequence of this operational situation is that the relative wind between the boom in displacement and the air canopy which is present between the spray boom and the intended crop have a braking effect, contrary to the downward speed of the fine droplets generated at the tips of the sprayers. This process help with evaporation and also with the drift of the fine spray droplets and hinders its arrival on the crop canopies to be treated. A second consequence depends on middle-sized droplets that manage to maintain their falling speed in spite of the opposite effect generated by the dislocation speed of the boom. Research carried out recently on winter cereals, by the Institute DLG in Germany, shows that these droplets deposit themselves, on the whole, only on one side of the plants, with the other side ("shady side") consistently lacking in droplets (Boller & Raetano, 2011). The research also revealed that an increase in the displacement speed of the equipment implies in a greater deposit of droplets on the upper third of the plants and fewer droplets deposited on the lower leaves.

The increase in spraying pressure may partially compensate for this effect; however; one cannot emphasise too strongly that excessive working pressure is one of the most important factors that facilitate spray droplet drift. This picture deserves particular attention, due to the fact that the actual and future tendency is the increase in the displacement speed of the spray equipment by land. In the same situation, spray nozzles with flat double spray outlets show a slight increase in the quantity of droplets deposited on the side known as "the shady side". The most balanced situation was obtained when ends with flat double jets, with differentiated angles in relation to the vertical position, were utilised. The results indicate that this type of outlet may be efficient for a more even deposition of the droplets, on both sides of the plants, when the displacement speed of the boom is around 12 km h^{-1} (Boller & Raetano, 2011). There are basically two ways to increase coverage: 1) reduce droplet size and 2) increase carrier volume (application rate). Large droplets do not provide good coverage and result in chemical wastage. Increasing the application rate may be equally undesirable. It requires frequent refilling of the sprayer tank. This wastes time that may be extremely valuable when there is a short period of opportunity to spray. Ideally, we want to have as many small droplets on the target as possible. However, extremely small droplets have a tendency to drift. Research has shown that there is a rapid decrease in the drift potential of droplets whose diameters are greater than approximately 200 μm. When extremely small droplets are released from the nozzle, they quickly lose the momentum that is needed to push the droplets into the canopy. Also, these extremely small droplets do not last long after they are released from the nozzle. Most of them evaporate within a few seconds (Ozkan, 2010). The single most important factor affecting the control of ASR disease is to get a thorough coverage of soybeans with the fungicide, which is much different and more challenging than spraying for weeds and insects. The most effective coverage on soybean plants can be obtained with both the horizontal as well as vertical distribution of the fungicide on soybean leaves. Asian soybean rust usually shows its symptoms in the lower parts of the plant first and works itself up towards the top of the plant. The most effective spray equipment and methods for applying fungicides on soybean plants to control Asian soybean rust was studied by Ozkan et al. (2006). A second component of the study was to determine the effect of spray quality (fine, medium, coarse) on spray deposition and coverage using three different sizes (8002, 8004 and 8005) of the XR type of a flat fan nozzle operated at different spray pressures. The application rate was kept constant at 145 L ha^{-1} for all the treatments. The average spray coverage on the middle part of the soybean canopy (0.6 m above the ground) varied from 1.3 to 7.3% among the treatments. The Jacto sprayer provided the highest spray coverage on the middle part of the canopy, followed by Top Air sprayer and the boom sprayer with a TX-18 hollow cone nozzle that produced the lowest spray coverage on the middle part of the canopy, followed by Turbo duo, and then XR 8002 nozzles. The average spray coverage at the bottom part of the soybean canopy (0.3 m above the ground) varied from 0.5 to 3.9% among the treatments. Similarly to the coverage on the middle part of the canopy, the Jacto sprayer provided the highest spray coverage on the bottom part of the canopy, followed by the boom sprayer with the canopy opener and then the Top Air sprayer. The boom sprayer with XR 8002 nozzles produced the lowest spray coverage on the boom part of the canopy, followed by hollow cone TX-18 nozzles. XR 8002 flat fan nozzles and hollow cone nozzles had smaller MVD than other treatments with the boom sprayer. The authors observed that among the three spray qualities (fine, medium and coarse), the medium quality spray provided the highest coverage and the fine quality spray provided the lowest coverage at both middle and bottom parts of the canopy. When compared to the XR 8004 flat fan pattern nozzles with medium spray quality, Twinjet, Turbo dual pattern nozzles and hollow cone nozzles provided very low coverage on the middle and bottom parts of the canopy. Droplets from Twinjet, turbo dual pattern and hollow cone nozzles had poor penetration capabilities because these droplets had horizontal velocities. The horizontal movement of droplets consumed kinetic energy and caused droplets to easily settle on the top leaves. The influence of the size of droplets from different nozzles on soybean spray coverage was studied by Antuniassi et al. (2004). The authors verified that very fine quality spray obtained with hollow cone TX VK6 nozzle and Twinjet flat fan TJ 60 11002 nozzle, and fine quality spray with a flat fan pattern XR 11002 nozzle, provided greater coverage in middle and bottom parts of the soybean plants when compared to the extremely coarse spray quality produced by air induction flat fan nozzles. The effects of spray nozzles (flat fan pattern, pre-orifice flat fan, air induction flat fan and air induction twin flat fan) and volume rates (115 and 160 L ha^{-1}) on chemical control of rust and the deposition of tebuconazole fungicide sprayed on soybeans of the Emgopa 313 variety, were studied by Cunha et al. (2006). The results showed that, despite the fact of the volume rate of 160 L ha^{-1} and of the use of pattern flat fan nozzles, they provided larger fungicide distribution uniformity in the plant canopy. There was no influence of the nozzle type neither of the application volume in the control of the rust, as well as in the soybean yield. In part, the results described by Raetano & Merlin (2006) ratified those observations that have been made by Cunha et al. (2006). The experiments were conducted in 2004/05 and 2005/06 seasons, using soybean, IAC-19 variety, with the same sprayer equipment and near application volumes (99 and 143 L ha^{-1}; 100 and 150 L ha^{-1}). The values of spray deposition were less influenced by nozzle type (hollow cone, flat fan and twin flat fan), both with fine spray quality. It is recommended for Asian soybean rust control that droplets have a size of 200 to 300 μm (OZKAN, 2005), but droplets smaller than 100 μm can be used with drift control in spraying with air assistance delivery systems near to the sleeve boom.

3. Fungicide application techniques

Nowadays, Asian soybean rust (ASR) deserved special attention due to its severity and difficulty of control, since it develops in the aerial part of plants, damaging the physiology and contributing to a drastic reduction of grain yield. For efficient control and cost-cutting, spray techniques and spray equipment must be improved. Studies show that the use of air assistance in the sleeve boom, connected to the hydraulic system of the tractor, can reduce the drift, increase droplet penetration into the plant canopy and improve the spraying distribution (Bauer & Raetano, 2000; Cooke et al., 1990; Taylor et al., 1989; Taylor & Andersen, 1991).

3.1 Air assistance delivery system in boom sprayers

The use of air assistance in phyto-sanitary product application is very old. However, the enthusiasm in using this spray technology started in 1980, as reported by Robinson (1993). Four years later, the Degania Sprayers Company in Israel developed a sprayer, revolutionary at the time, equipped with air assistance on the spraying sleeve boom. However, only since the end of the 1980s and the beginning of the 1990s has air assistance been effectively adopted in sleeve boom sprayers. In Europe, this technology was introduced by Hardi, and in Germany, in 1996, seven manufacturers exhibited equipment with air assistance in the *Agritechnica* agricultural trade show (Koch, 1997). At that time, the Brazilian industry also incorporated this technology to tractor-driven trailing sleeve boom sprayers. The incorporation of this technology to sleeve boom sprayers was an attempt to improve spraying penetration in the target culture, reduce drift and the number of applications required, increase the time available for carrying out the spraying and enable changes to the spraying height over the culture (DEGANIA SPRAYERS Co., s.d.). For applying phyto-sanitary products on low-stem cultivation, the spraying sleeve booms equipped with air assistance appeared as the ideal tools to improve application quality (smaller droplets, in higher numbers), increase productivity (lower volume and replenishment, higher displacement speed and extended spraying times), reduce drift (the machine's wind speed is greater than the environmental wind) and exposure to these products (Sartori, 1997). After twenty years of using air assistance in sleeve boom sprayers, a great deal of information must still be clarified about the interactions between air volume and speed which are more appropriate for different cultures, the angle of the nozzles on the boom in relation to the air, spraying height and displacement speed, amongst other factors which enable wider spraying coverage and lower losses.

3.1.1 Characterisation of the technology

Tractor-driven sprayers with air assistance can be coupled to the tractor's hydraulic power take-off (third point) (those with lower capacity tanks or of the trailing type). These sprayers are equipped with one or two fans, usually axial, positioned near the centre section of the spraying sleeve boom, which distribute a very high air volume in an inflated duct assembled over the boom and nozzles (Matthews, 2000). The speed of the air generated may vary with the fan rotation (rpm), and generally it does not follow a linear relationship. Also, air speed variations could occur along the boom, at the ends, when compared to the speed achieved in its centre section (Raetano, 2002). The established standards for evaluating with accuracy the speed of the air generated by sleeve boom sprayers equipped with air assistance was necessary to standardise the measuring distance in relation to the air exit opening, as well as to specify anemometers that are able to record high air speeds (30-40 m s^{-1}). Thus, Kunz (2010) developed two methods for air speed and volume measuring in spray booms equipped with air assistance. In the first method, a wooden mould was placed at the outlet of the air curtain, in a vertical position, in the direction of the air flow, and measurements were taken with the anemometer at pre-established distances. This form of measuring became know as the "ruler method". This method makes it very difficult to determine the main vector of the air flow that comes out in a continuous manner through the rectangular opening on the lower part of the inflated sleeve, which makes it difficult to measure the air speed with precision. New air-speed readings are now taken beforehand, using a nylon thread fixed to the air outlet, to indicate the point of air flow displacement vector, which substitutes the ruler method. In this way, it makes it very much easier to identify the main air flow and increases the precision and uniformity of speed values obtained with the anemometer. In a similar manner to the ruler method, pre-defined distances are marked off on the nylon thread, so that measurements can be taken with greater ease and accuracy. This procedure is called the "thread method" (Figure 1). Due to the dynamic behaviour of the air flow, it becomes difficult to identify the vector of the air flow that comes out under high speed from the system, principally at distances of 0.25 and 0.50 m, which causes a great variation in the speed data obtained with the ruler method, as can be seen in Table 2. The air speed values obtained with the thread method present greater uniformity in relation to the ones obtained with the ruler method, especially at distances of 0.25 and 0.50 m from the air outlet. This can be observed through the variance values (%) of the data, which were smaller with the thread method (Table 2). The average values of the air speed obtained with the thread method were greater, probably due to the correct identification of the main air flow vector when measured by this method. Measuring the air speed, therefore, becomes more precise and easier, especially at longer distances in relation to the air flow at the spray boom.

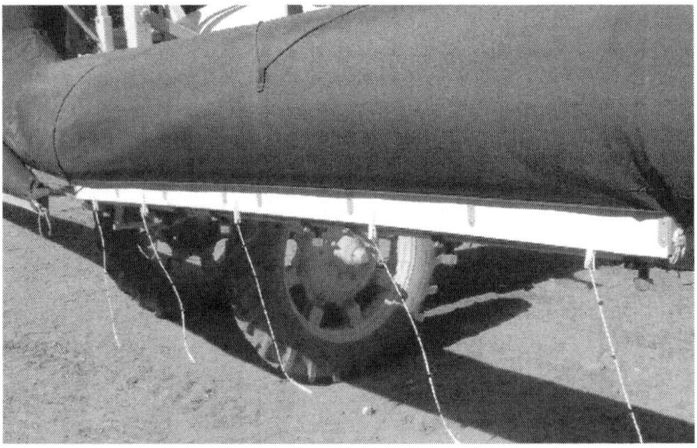

Fig. 1. "Nylon thread method" for measuring air speed in a sleeve boom sprayer.

Distance (m)	Average*	S.D.	Variance	CV %	Min*	Max*	Amplitude*
Ruler method of measurement							
0.0	70.14	10.00	100.05	14.26	53.40	97.20	43.80
0.25	41.71	5.70	32.55	13.68	31.20	54.00	22.80
0.50	29.49	8.20	67.27	27.81	20.50	51.80	31.30
Nylon thread method of measurement							
0.0	71.57	10.09	101.83	14.10	59.50	93.60	34.10
0.25	43.64	4.04	16.33	9.26	36.60	51.00	14.40
0.50	35.26	3.53	12.47	10.01	28.80	41.90	13.10

*Values expressed in km h^{-1}.

Table 2. Descriptive statistics of the air speed data obtained along the spray boom with different evaluation methods.

3.1.2 Air speed on spray deposition

Raetano & Bauer (2003) evaluated the effects from air speed variation (50%, 75% and 100% of the maximum fan rotation capacity) on the spraying sleeve boom, when depositing phyto-sanitary products on bean culture. Forty-eight days after sprouting begins, 200 g of copper oxide per 100 L of water were applied with AXI-110015 tips at 206.7 kPa and JA-1 at 1,033.5 kPa, either with air assistance or not, using a Model Falcon vortex sprayer. The broth volume was 100 L ha^{-1} in both operational conditions. The air speed variation did not influence the deposit levels in the culture, but the use of air assistance, operated at full fan capacity, resulted in better deposit levels on the abaxial surface of the leaflets positioned in the lower portion of the plants. Cereal-cultivated soil contamination can be reduced to approximately 40% when using 50% of the maximum speed of the air generated by the fan in a sprayer equipped with air assistance on the sleeve boom, when compared with conventional application (without air), as reported by Taylor & Andersen (1997). The deposit and losses of spraying broth in the cultivation of bean (*Phaseolus vulgaris*), 26 days after sprouting, and using sprayers equipped with air assistance on the sleeve boom and conventional sprayers (without air) and volumes of 60 and 100 L ha^{-1}, have been evaluated by Raetano & Bauer (2004). The higher volume resulted in greater deposits, but high losses to the soil (above 60%) have been noted, even when using air assistance with air speed corresponding to 50% of the maximum fan rotation. In part, such results have been assigned to 40% of the soil bare of vegetation at this growth stage of the culture. The air volume generated may vary from 0 to 2000 m^3 per hour per boom, depending on the number and power of the fans distributed on variable-size booms that could reach 30 m in length. The air distributed in the inflated duct is forced to pass through a continuous or intercalated opening, in a perpendicular direction to the one in which it has been generated, in a descending direction. The effects of chemical control of the rust and deposition fungicide sprayed under four speeds (zero, 9, 11 and 29 km h^{-1}) by a spray boom on soybean crop were evaluated by Christovam (2008) and Prado et al. (2010). Significant differences were obtained in the lower part of the plants for spray deposition using higher speed of air assistance. On the top part of the plants, greater levels of deposition were seen when spraying without air assistance was carried out. The rust severity was more intense in treatments without air assistance. Raetano & Bauer (2003) evaluated different velocities of air assistance near the spray boom and concluded that air assistance, with maximum air speed generated by the fan (29 km h^{-1}) and a flat fan nozzle (AXI 110015 type), provided greater spray deposition on the abaxial leaf surface, on the bottom part of the bean plants. The data of the soybean crop yield at different air speeds using an air-assisted sprayer for Asian soybean rust management was compared in the 2006/07 (Christovam et al., 2010a) and 2007/08 (Prado et al., 2010) seasons (Figure 2). Air speeds used in both studies were zero, 9, 11 and 29 km h^{-1}.

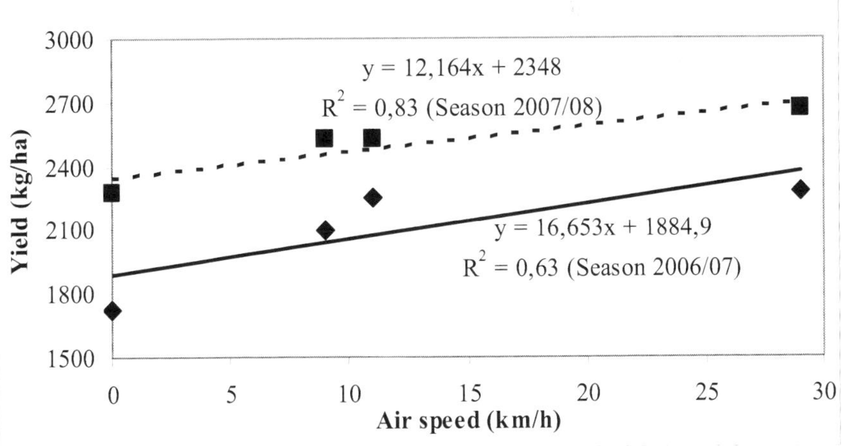

Fig. 2. Effect of the air speed on soybean crop yield, 2006/07 and 2007/08 agricultural season using a sleeve boom sprayer. Botucatu, SP, Brazil.

There is a positive correlation between air speed and soybean crop yield. When compared, the soybean yield using the maximum air speed generated by the fan (29 km h^{-1}) in conventional spraying (without air), increases of 31.9% and 17.1% can be seen in the 2006/07 and 2007/08 seasons, respectively. As can be verified, in the last agricultural season, the increase in soybean yield was lower, due the higher severity of ASR on the Conquista variety (Figure 2). The effect of different air speeds (0, 9, 11 and 29 km h^{-1}) in chemical control of pests on soybean crop, Conquista variety, was evaluated by Prado (2009), after insecticide spraying using an air-assisted sprayer. The use of air speed generated by the fan in the maximum rotation (29 km h^{-1}) provided greater control of the lower velvetbean caterpillar (*Anticarsia gemmatalis*). This effect was not observed with longer caterpillars (> 1.5 cm) because the lowest caterpillars are more easily located in the lower part of the soybean plants. Thus, the lower caterpillars received an additional amount of insecticides due the effect of air assistance with maximum air speed into the canopy. In general, there was not a statistically significant difference between air speeds on stink bug control after insecticide spraying on soybean culture.

3.1.3 Nozzle angle on spray deposition

The positioning angle of the spraying nozzle in relation to the air curtain (Figure 2), generated by the equipment (vertical, descending), as well as the nozzles and air curtain, simultaneously, in relation to the vertical position, may significantly influence the deposit levels and the spraying distribution. Nowadays, in sleeve boom sprayers equipped with air assistance, the angle variations of the nozzles and air curtain, in relation to the vertical position, pro or against the tractor-sprayer assembly displacement, are made simultaneously clockwise or counterclockwise, with the single-cylinder command. The results research carried out under controlled conditions and in the field have shown that the positioning of the nozzle at 30° forward of the displacement in conventional sprayers (without air) provides a significant increase in deposits on the leaf surface of different vegetal species: *Cyperus rotundus* (Silva, 2001), *Brachiaria plantaginea* (Tomazela, 2006) and *Glycine max* (Bauer, 2002). In England, research carried out in wind tunnels with plants cultivated on trays have confirmed that the spraying angle forward of the displacement, in the presence of air assistance, increased deposit on cereals and reduced soil contamination (Hislop et al., 1995). Nowadays, one may position the spraying nozzles and air curtain at angles of 15° and 30° in relation to the vertical position in sleeve boom sprayers equipped with air assistance, made in Brazil. The use of air angled forward of the displacement with fine droplets could substantially increase spraying deposit levels on vertical targets. These results were obtained from practical experiences published by the Hardi Int. Tech. Reports in potato culture which indicated that spraying penetration and retention are greater with air assistance positioned at an angle forward of the displacement on the leaves in the lower portion of the plants. In the upper portion, the retained broth volume was virtually not influenced by the air exit angle, pro or against the equipment displacement (Taylor & Andersen, 1997). The effect of the nozzle angle and air-jet parameters in an air-assistance sprayer on the biological effects of ASR chemical protection was studied by Christovam et al. (2010b). Four air levels (0, 9, 11 and 29 km h^{-1}) were combined at two nozzle angles 0° and 30° for the sprayings using flat fan AXI 110015 nozzles. The spraying with triazole fungicide was realised in R2 and R5.2 growth stages of soybean at 142 L ha^{-1} of volume rate. For the evaluation of spray deposition, a cupric tracer was used. At the bottom part of the plant, spraying with maximum air speed generated by the fan and nozzles angled at 30°, it was essential to promote doubled deposits on the abaxial leaf surface (Table 3). Maximum air speed (29 km h^{-1}) and nozzles angled at 30° resulted in an increase in spray deposits on adaxial surface of leaves in the bottom part of the plants (Table 3).

Air Speed (km h^{-1})	Adaxial surface				Abaxial surface			
	Angle 0°		Angle +30°		Angle 0°		Angle +30°	
	µl cm^{-2}		µl cm^{-2}		µl cm^{-2}		µl cm^{-2}	
0	1.2425	a A	0.6962	b AB	0.6585	a A	0.2444	b B
9	0.6997	a A	0.9865	a AB	0.3621	a A	0.4527	a B
11	1.147	a A	0.6395	b B	0.4651	a A	0.3292	a B
29	0.6287	b A	1.2663	a A	0.4904	b A	0.8552	a A
DMS Angle	0.46				0.24			
DMS Air speed	0.62				0.33			
CV (%)	34.17				34.16			

The same larger letters in the column, did not differ by the Tukey test ($p<0.05$).

Table 3. Average values of deposits of the copper tracer in an artificial target (filter paper) on leaf surfaces, in the bottom part of the soybean plants, Conquista variety, in relation to different spraying angles. Botucatu, SP, 2006/2007.

Nozzles angled at 30°, in the same direction of the sprayer displacement, combined with air assistance (29 km h^{-1} of air speed) positively influenced the control of disease as well as the yield of the Conquista variety crop. This fact confirms the importance of spraying performed with nozzle angles in the same direction of the sprayer movement, which can contribute significantly to Asian soybean rust control, considering the disease epidemiology. The choice of the best combination of air speed and nozzle angle in air-assisted sprayers is influenced by architecture and growth stage of the plants to obtain a desirable biological effect in soybean Asian rust chemical protection with this technology. Conventional spraying (without air) and air-assistance at 0° (vertical) and 30° (forward to displacement of the equipment) are shown in Figures 3A, 3B and 3C, respectively. The spray boom angle interference, with or without air assistance near the boom, on spray deposit levels were studied by Scudeler & Raetano (2004) in potato culture. The higher deposits were evidenced with nozzles positioned at 0° and 30°, with the presence of air assistance, both at the top and bottom part of the potato plants. The lower spray deposits were obtained with nozzles positioned at 30° in the opposite direction to the displacement of the sprayer. In addition to the volume rate, generated air speed and nozzle angle in air-assisted sprayers, other factors, such as displacement speed of the tractor-sprayer assembly, presence of vegetal coverage in the area or not, vegetal coverage type (monocotyledonous or dicotyledonous, plant density, architecture and plant cuticle characteristics), positions of insect pests and plant pathogens, agrochemical product characteristics, droplet size and environmental conditions, especially wind speed, may influence the efficacy of phyto-sanitary control. It is necessary to develop studies with variations in air speed combined at different angles of spray nozzle on spray deposition and coverage. Dynamic systems for air speed evaluation combined at different nozzle angles and the performance of these in spraying could be better studied.

Fig. 3. Air-assisted sleeve boom sprayer in the following operation modes: A - conventional spraying (without air); B - spraying with air assistance at 0° (vertical) and C - spraying with air assistance angled at 30° forward to the displacement of the tractor-sprayer assembly on soybean crop.

3.1.4 Effect of the air-assistance delivery system on soybean productivity

The influence of an air-assisted delivery system on soybean productivity, var. Conquista, 2006/07 and 2007/08 seasons are shown in Figure 4. The spraying treatments used to control *P. pachyrhizi* fungus were applied on soybean plants (R2 and R5.2 growth stages) using a triazole + strobilurin spray mixture and volume rates between 120 at 150 L ha^{-1}. The data of soybean crop productivity after two sprayings with different technologies was submitted to a variance analysis and the averages were compared by Tukey's test ($p<0.05$).

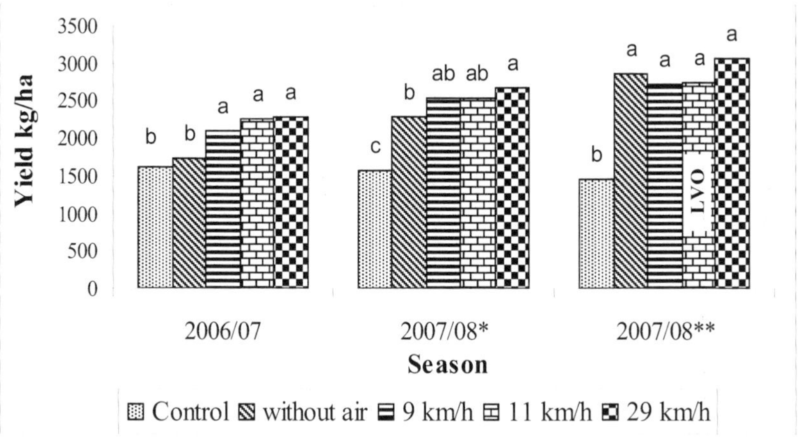

Fig. 4. Effect of treatments: control (not treated); air-assistance delivery system at zero (without air), 9, 11 and 29 km h^{-1} air speed; rotating nozzle-LVO with a third part of spray volume (40L ha^{-1}) on soybean productivity. The Conquista variety was used in the 2006/07 and 2007/08 agricultural seasons. Botucatu, SP, Brazil (2006/07 - Christovam et al. 2010b; 2007/08* - Prado et al. 2010; 2007/08** - Christovam et al. 2010a).

In general, this technology provided higher productivity, in relation to that with conventional spraying (without air) and control (not treated). A higher increase in soybean productivity was obtained with maximum air speed generated by the fan (29 km h^{-1}). The air assistance constituted an optional implement on the boom sprayers that can increase up 30% the equipment cost in relation to conventional sprayers which do not have this technology. Although the use of this technology should not be recommended on soils without vegetation or even on early stages due to the smaller foliar area. This method provides several advantages that have been mentioned before.

3.1.5 Air assistance and spray drift

Air assistance in the spraying sleeve boom significantly improves the penetration of spraying, especially in high cultures and high leaf density, such as potato, in addition to reducing drift (Koch, 1997). However, these effects were not observed when air-assisted spraying is done on bare soil or plants in the first stages of growth. Also, according to Matthews (2000), air-assisted spraying penetration is better when compared to conventional spraying on wide-leaf cultures, such as cotton. In Holland, tests with the air-assisted sprayer Twin (Hardi) have been carried out on potato cultivation. Generally, air assistance reduced drift by sedimentation by 50% and the air-carried drift by 75%. In Holland, the accepted drift percentage by sedimentation is 8-10% for a distance of 1.5 to 2.0 metres from the boom, and around 0.2% for 5.0 to 6.0 metres intervals. The recommendation for making spraying in Holland is with a wind speed less than 5.0 m sec^{-1}. In Germany, the accepted drift values by sedimentation when applying phyto-sanitary products range from 0.6 to 0.1%, respectively, for distances of 5.0 to 30.0 metres from the spraying sleeve boom (Jorgensen & Witt, 2000). Considering the drift limits accepted for spraying in Germany, the safe distance for applying near water channels (irrigation/draining) in that country is 10.0 metres for 80% of the herbicides approved for use, and 20.0 metres for other herbicides. France and Belgium comply with the drift limits accepted in Germany. Artificial targets have been also used by the Morley Research Centre for simulating venomous plants in the sugar beet. The variations in the deposit values for air-assisted spraying were lower when compared to those achieved with the conventional sprayer (Taylor & Andersen, 1997). These authors have also demonstrated the influence of air assistance on drift percentage reduction compared with conventional application (without air), by obtaining 90, 84, 83, 76, 68 and 61%, respectively, when spraying barley, bean, pea, Brussels sprouts, lettuce and leek, with fine droplets. Nowadays, studies involving computer models aim at clarifying the relationship among the air released drift risk and deposit on target. Preliminary studies have shown that the increase of the displacement speed with air-assisted sprayers may reduce drift, but provide lower evenness in the target culture treatment (Miller, 1997). However, aiming at reducing the application volume, Nordbo (1992) has demonstrated lower variation and improved deposit by using air assistance. The density, architecture, cuticle type (pilose, glabrous and waxy) and growth stage of the vegetal species in the area are factors influencing phyto-sanitary control efficiency when using air-assisted sleeve boom sprayers. Fine droplets provide larger deposits on plants, especially monocotyledons, but are very susceptible to drift. Their penetration capacity in cultures is small, and then the loss to the soil must be limited. Therefore, air assistance enables using fine droplets more efficiently, by reducing the drift and increasing the deposits on the target, in addition to providing higher penetration of these droplets in cultures with higher leaf density, and reducing losses to the soil (Jorgensen & Witt, 1997). On the other hand, coarse droplets generally provide good drift control. In dicotyledons, the deposits do not depend only on droplet size (Nordbo, 1992). Unlike the results with smaller diameter droplets, coarse droplets provide significantly lower deposits on vertical surfaces (monocotyledons), and especially in the first growth stages, by increasing the loss to soil proportionally to their size (Jorgensen & Witt, 1997). In vegetables, where droplet retention is limited by the presence of waxy layers on the cuticle, further studies are required, especially with air-assisted spraying, in order to evaluate the application quality (Koch, 1997). In the absence of vegetation (bare soil), air assistance may increase drift and deflect the air from the sprayer by the soil, unlike the effect which occurs in the presence of vegetation, with the impact of droplets on the leaf surface (Matthews, 2000).

3.2 Alternative spray technologies on soybean

Nowadays, other technologies are available to the boom sprayers enabling higher spraying droplet canopy penetration in soybean culture. The difficulty in controlling Asian soybean rust and late season diseases has favoured the development of new spraying techniques, particularly due to the difficulty in reaching the exact target to be controlled. Thus, the use of the opener, rotating system nozzles, hose drops and electrification of droplets associated with air assistance can be mentioned.

3.2.1 Opener

Conventional sprayers linked to an artefact providing the canopy opener at the same spraying way can turn out to be an economic and effective alternative to soybean growers with lower purchasing power (Zhu et al., 2008b). These authors found that spraying performed with conventional sprayers linked to a canopy opener did not results significant differences in the coverage of the spray in the middle part of soybean plants when compared to spraying carried out using air assistance. However, the canopy opener coverage and air assistance along the bar was higher compared to treatments where the spraying was conducted by the conventional system without the canopy opener. Thus, the opener and spray boom coupling can provide deposition results similar to those obtained with the use of air assistance, besides being a more economical alternative to Asian soybean rust control (Figure 5). Considering the difficulties in controlling ASR by fungicide spraying, Prado (2011) evaluated the effectiveness of the canopy opener compared to conventional sprayers and air assistance in the spray boom on spraying deposition, rust control efficiency and soybean productivity. The experiment was conducted in a randomised block

Fig. 5. Canopy opener artefact fixed to the spray boom in a soybean crop.

experimental design with six treatments: conventional spraying (T1), spraying with air assistance at maximum capacity of the fan rotation in the boom (T2), spray with a canopy opener to a depth of 0.10 m (T3), spraying with a canopy opener to a depth of 0.10 m with air assistance (T4), spraying with a canopy opener to a depth of 0.20 m (T5) spraying with a canopy opener to a depth of 0.20 m with air assistance (T6) and control treatment (without spraying) (T7) in four replicates, totalling 28 plots. The area of each plot was equivalent to 70 m^2. The depths of 0.10 and 0.20 m refer to the distance from the canopy opener in relation to the top of the soybean plant. In addition to the distance between the canopy depth opener and the top of the plants, there is also a predetermined horizontal distance of 0.15 m between the boom and the canopy opener. The function of the canopy opener is to promote the soybean plants to slope forward, opening a space in the plant canopy and thus facilitating the flow, and consequently droplet deposition, on the bottom of the soybean plants. The sprayer used in the experiment was the Advance Vortex model 2000 with an 18.5 m long boom, 37 flat fan XR 8002 nozzles spaced 0.50 m apart operating at a pressure of 295 kPa and a spray volume of 150 L ha^{-1}. The comparative effect of these different technologies on soybean productivity after three fungicide pyraclostrobin + epoxiconazole mixture sprayings at a dose of 25 + 66.5 g a.i ha^{-1} in the development stages R2, R3 and R5, as shown in Figure 6. The treatment T5 (spraying with a canopy opener at a depth of 0.20 m) had a higher productivity increase (54%) compared to control treatment. All treatments which received fungicide had significantly higher yields than the control treatment. There was no difference between the canopy opener and air assistance on soybean yield, making it an interesting and economical alternative for the control of Asian soybean rust. These results corroborate those obtained by Zhu et al. (2008).

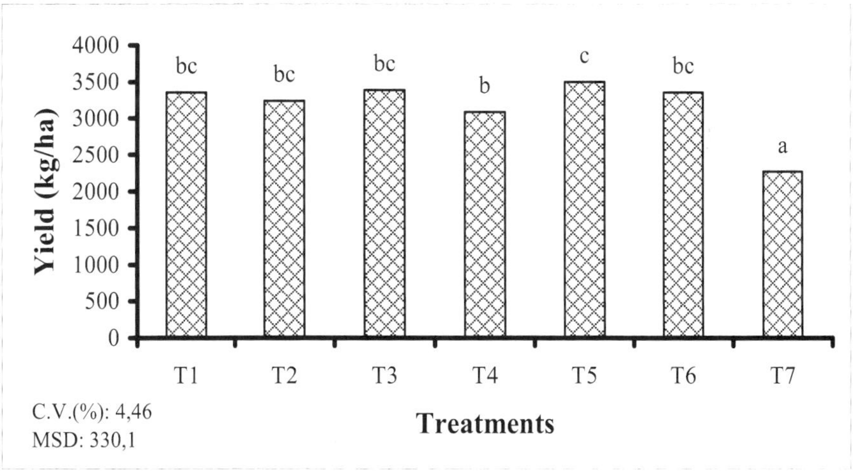

Fig. 6. Effect of different treatments: conventional spraying (T1); air-assistance (T2); *canopy opener* at a depth of 0.10 m (T3); *canopy opener* at a depth of 0.10 m with air assistance (T4); *canopy opener* at a depth of 0.20 m (T5); *canopy opener* at a depth of 0.20 m with air assistance (T6) and treatment control (without spraying) on soybean yield.

3.2.2 Rotating system nozzles

Recently, the use of centrifugal energy to produce spray droplets (rotating system nozzle) is an interesting alternative application technology to control Asian soybean rust, using oily formulations, low spraying volumes and, consequently, a greater operational performance of sprayers. In Brazil, new techniques for pesticide application using low volumes and rotating nozzles have been developed in the mid-west region (Cerrado) for soybean rust control in soybean culture. The rotating nozzle low volume oily (LVO) and different levels of air speed with an air-assisted sprayer on spray deposits were compared by Christovam et al. (2010c) on Asian soybean rust control and soybean productivity. Two experiments were carried out in the experimental area of FCA/UNESP, Botucatu, SP, Brazil, on a soybean crop of the Conquista variety, in the 2007/2008 season. In the first experiment, three air levels (0, 9 and 29 km h^{-1} air speed generated by a fan) with flat fan XR 8002 nozzles and a spray volume of 130 L ha^{-1} were compared with a rotating nozzle - using LVO at 40 L ha^{-1} of spray volume (Figure 7). The second experiment was carried out under the same conditions as the previous experiment, including a control treatment (untreated plants). The grades varied between 0.6 and 78.5% disease severity. In general, air assistance promoted the increase on deposit levels on the adaxial surface of the leaf located in the top part of the plants. Therefore, in the bottom part, there was not a significant difference in spray deposits between the spraying techniques. Also, the abaxial surface did not show differences in deposit levels, in the top or bottom part, between the spraying techniques. The use of air assistance, when compared with the rotating nozzle system, did not show significant differences in spray deposits on adaxial or abaxial surfaces of the leaves in the bottom part of the plant. Monteiro (2006) observed results very similar to those obtained in the current study. This author performed a study that aimed to evaluate the spraying efficiency of a rotating atomiser system - LVO using 25 L ha^{-1} of fungicide outflow on a soybean crop, when compared to a sprayer equipped with hydraulic nozzles at a spray volume of 150 L ha^{-1}. The treatments sprayed with the fungicidal mixture provided a weight of 1000 seeds and productivity significantly higher in comparison with untreated plants (control). The highest increase of productivity was obtained with the maximum air speed generated by the fan (29 km h^{-1}) near to the spray boom using 130 L ha^{-1} when compared with the control treatment. The spray volume

Fig. 7. Spraying with an air-assisted sprayer and rotating system nozzles fixed at the spray boom on soybean culture.

applied with the rotating system nozzle - LVO was 40 L ha^{-1}. Therefore, it did not provide the same increase in productivity compared with the treatment using air assistance at the maximum speed. The rotating system nozzle was 30% more economical than the treatment with a spray volume of 130 L ha^{-1}, with or without air assistance near the boom, using the Advance Vortex 2000 sprayer.

3.2.3 Hose drops

Another possibility to improve spraying coverage with boom sprayers is addressed by spraying with three flat fan nozzles involving the entire planting row, of which two of them are positioned near the bottom of the plants or positioned on opposite sides between the crop rows and near the bottom of the plant. The structures that support the spray nozzle from the spray boom at its lower end are called hose drops. In the USA, there are several reports on the use of flat fan nozzles placed in the hose drops ends, which move in the line between the culture, with volumes around 140 L ha^{-1} in Asian soybean rust treatment (Ozkan, 2005), although their culture is planted at greater spacing than those in Brazil with early cultivars. In Brazil, growers with difficulty will adopt hose drops in this application in order to obtain better spray coverage of the leaves on the bottom part of soybean plants. However, the differences in the growth habits, foliage degree and plant architecture of the varieties and the smaller spacing between planting rows makes the use of this technology difficult.

3.2.4 Electric charge (electrostatic) in spray droplets

Nowadays, air assistance can be combined with electrification (by induction) of the spraying droplets, aiming at reducing drift and exposure of appliers and the environment to phyto-sanitary products. An experiment was carried out in commercial areas of the soybean crop, Cidade Verde Farming, Primavera do Leste, MT, Brazil, on soybean plants of the Monsoy 8757 variety in the 2009/2010 agricultural season. Sowing was performed in 12/11/2009, leaving 0.45 m spacing between planting rows and 14 seeds per linear metre. The experimental design was in random blocks, with six treatments constituting three application techniques: conventional spraying, air-assisted spraying and air-assisted spraying combined with electrically charged droplets in two spray volumes, 50 and 100 L ha^{-1}, in four replications, totalling 24 experimental plots. The experimental plots were 24.0 x 100.0 m (width x length). The width of the plots corresponded to the boom size of the sprayer used in this research. During spraying, a self-propelling sprayer (Uniport 3000 model) was used equipped with a spray boom 24.0 m in length with hollow conical nozzles spaced every 0.35 m. The spray hollow conical nozzles used were of the JA-1 and JA-2 type, operated at working pressures of 690 and 828 kPa respectively. The spray displacement speed was 15 km h^{-1}, usually practiced by farmers in the Brazilian mid-western region (Cerrado). This sprayer operated with or without air assistance on the spray boom (conventional) combined with electric charge transference to the spray droplets in turn-on or turn-off mode. For air supply into the sleeve boom, two axial fans were positioned on the central point of the boom and operated at the maximum rotation speed. For the quantification of spray deposits, a tracer dye (Brilliant Blue) was used at a concentration of 0.3%, according to qualitative and quantitative evaluation studies of spray deposits validated by Palladini et al. (2005). The spraying of the tracer dye was performed in the R5.1 growth stage, 80 days after sowing. The average values of height and foliar area were respectively 0.92 m and 0.158 m^2 at this growth stage. The mean values of spray deposits with different application technologies on soybean plants of the Monsoy 8757 variety are shown in Tables 4 and 5. Greater spray deposits were obtained with air assistance and this spray technology combined with electrically charged droplets in relation to conventional spraying at 100 L ha^{-1} on leaflets positioned at the top part of soybean plants (Table 4). At a low volume rate, there was no observed difference in deposit levels with the different spray technologies. With the higher spray volume using air assistance combined with electric charge transference to the droplets, it was possible obtain greater spray deposits when compared to low spray volume. The best spray deposits on leaflets at the bottom position of soybean plants was obtained with air assistance combined with electric charge transference technology at a volume of 100 L ha^{-1} (Table 5). With this spray volume, air assistance technology combined with electrically charged droplets was better when compared the other two spraying technologies. There was no significant difference between the spraying techniques on tracer deposits using the volume of 50 L ha^{-1} (Table 5). These results obtained with a new spray technology, employing air assistance combined with electric charge transference to droplets, is very promising in disease management in this culture, especially for Asian soybean rust management.

Spray technique	**Volume (L ha^{-1})**		**LSD values ($p< 0.05$)**
	100	**50**	
Air-assistance + electric charge	0.744 Aa	0.352 Ba	
Air-assistance	0.684 Aa	0.480 Aa	0.350
Conventional	0.254 Ab	0.313 Aa	
LSD values ($p< 0.05$)	0.427		
CV (%)		11.27	

Original means and data transformed in root square of x + 0.5 for analysis.
Means followed by the same letter, smaller in the column and bigger in the line, did not differ by Tukey's test at the 5% significance level.

Table 4. Mean values of Brilliant Blue tracer deposits (µL cm^{-2}) on leaflets at the top position of soybean plants after spraying with different techniques. Botucatu, SP, Brazil, 2009/10.

Spray technique	**Volume (L ha^{-1})**		**LSD values ($p< 0.05$)**
	100	**50**	
Air-assistance + electric charge	1.166 Aa	0.016 Ba	
Air-assistance	0.215 Ab	0.029 Aa	0.622
Conventional	0.033 Ab	0.115 Aa	
LSD values ($p< 0.05$)	0.758		
CV (%)		21.11	

Original means and data transformed in root square of x + 0.5 for analysis.
Means followed by the same letter, smaller in the column and bigger in the line, did not differ by Tukey's test at the 5% significance level.

Table 5. Mean values of Brilliant Blue tracer deposits (µL cm^{-2}) on leaflets at the bottom position of soybean plants after spraying with different techniques. Botucatu, SP, Brazil, 2009/10.

4. Considerations of ASR management

Despite new techniques and equipment available for the application of fungicides targeting Asian soybean rust management, other factors such as climate conditions, varieties, disease severity, plant architecture, fungicide characteristics, sowing in the same season and application time are important in ensuring culture productivity. Associated with chemical control, plant disease resistance and the adoption of the period of sowing interruption (inter-season) in most Brazilian states have contributed to decreasing the severity of Asian soybean rust. Disease monitoring time of application and choice of fungicide are important factors for the success of Asian soybean rust control. Beforehand sowing and choosing an early variety can also contribute to the control of this disease. Nowadays, multidisciplinary research development is necessary to achieve suitable management of Asian soybean rust. Only knowledge of pesticide application techniques is not sufficient to improve control of the *P. pachyrhizi* pathogen.

5. References

Adam, A.V. (1977). Importance of pesticide application equipment and related field practices in developing countries. In: *Pesticide management and insect resistance*, D.L. Watson & A.W. Brown, (Ed.), 217-225, Academic Press, New York

ANDEF - Associação Nacional de Defesa Vegetal. (2004) *Manual de tecnologia de aplicação de produtos fitossanitários*. Línea Creativa, Campinas, 50p

Antuniassi, U.R. (2010). Tecnologia de aplicação de defensivos: Conceitos básicos de tecnologia de aplicação de defensivos, In: *Boletim de Pesquisa de Soja 2010*, Vol.1, No.14, Fundação Mato Grosso, (Ed.), pp. 347- 372, Mato Grosso, Brazil, ISSN 1807-7676

Antuniassi, U.R. (2005). Tecnologia de aplicação para o controle da ferrugem da soja, *Proceedings of 1rst Workshop brasileiro sobre a ferrugem asiática*, pp. 193-219, UDUFU, Uberlândia, Minas Gerais, Brazil

Antuniassi, U.R.; Camargo, T.V.; Bonelli, M.A.P.O. & Romagnole, E.W.C. (2004). Avaliação da cobertura de folhas de soja em aplicações terrestres com diferentes tipos de pontas, *Proceedings of 3rd Simpósio internacional de tecnologia de aplicação de agrotóxicos*, pp. 48-51, FEPAF, Botucatu, São Paulo, Brazil

Bauer, F.C. (2002). *Distribuição e deposição da pulverização sob diferentes condições operacionais na cultura da soja [Glycine max (L.) Merrill]*. 130p. Tese (Doutorado em Agronomia / Proteção de Plantas) - Faculdade de Ciências Agronômicas, Universidade Estadual Paulista, Botucatu, São Paulo, Brazil

Bauer, F.C. & Raetano, C.G. (2000). Assistência de ar na deposição e perdas de produtos fitossanitários em pulverizações na cultura da soja. *Scientia Agricola*, Vol.57, No.2, pp. 271-276, ISSN 0103-9016

Barros, A. L.M.B. & Menegatti, A.L. (2010). Tendências de médio prazo para o mercado de soja no Brasil e no mundo, In: *Boletim de Pesquisa de Soja 2010*, Vol.1, No.14, Fundação Mato Grosso, (Ed.), pp. 25-27, Mato Grosso, Brazil, ISSN 1807-7676

Boller, W. & Raetano, C.G. (2011). *Bicos e pontas de pulverização de energia hidráulica, regulagens e calibração de pulverizadores de barras*, 31p. (Documento no prelo)

Brown-Rytlewski, D. & Staton, M. (2010). *Fungicide application technology for soybean rust - 2006: Field crop advisory alert*, 2p, Michigan State University Extension, Department of Plant Pathology, Michigan, USA

Calaça, H.A. (2007). *Ferrugem Asiática da soja: relações entre atraso do controle químico, rendimento, severidade e área foliar sadia de soja (Glycine max L. Merrill)*. 80p.

Dissertação (Mestrado) - Escola Superior de Agricultura Luiz de Queiroz, Piracicaba, São Paulo, Brazil

Christovam, R.S. (2008). *Assistência de ar e aplicação em volume baixo no controle da ferrugem asiática da soja. Botucatu.* 68p. Dissertação (Mestrado em agronomia/ Proteção de Plantas) - Faculdade de Ciências Agronômicas, Universidade Estadual Paulista, Botucatu, São Paulo, Brazil

Christovam, R. S.; Raetano, C.G., Aguiar-Júnior, H.O., Dal Pogetto, M.H.F.A., Prado, E.P., Junior, H.O.A., Gimenes, M.J. & Kunz, V.L. (2010a). Assistência de ar em barra de pulverização no controle da ferrugem asiática da soja. *Bragantia,* Vol.69, No.1, pp. 231-238, ISSN 0006-87065

Christovam, R. S.; Raetano, C.G., Dal Pogetto, M.H.F.A., Prado, E.P., Aguiar-Junior, H.O., Gimenes, M.J. & Serra, M.E. (2010b). Effect of nozzle angle and air-jet parameters in air assisted sprayer on biological effect of soybean Asian rust chemical protection. *Journal of Plant Protection Research,* Vol.50, No.3, pp. 347-353, ISSN 1427-4345

Christovam, R. S.; Raetano, C.G., Prado, E.P., Dal Pogetto, M.H.F.A., Aguiar-Junior, H.O., Gimenes, M.J. & Serra, M.E. (2010c). Air-assistance and low volume application to control of Asian rust on soybean crop. *Journal of Plant Protection Research,* Vol.50, No.3, pp. 354-359, ISSN 1427-4345

Cooke, B.K. et al. (1990). Air-assisted spraying of arable crops in relation to deposition, drift and pesticide performance. *Crop Protection,* Vol.9, No.4, pp. 303-311, ISSN 0261-2194

Cunha, J.P.A.R.; Reis, E.F. & Santos, R.O. (2006). Controle químico da ferrugem asiática da soja em função de ponta de pulverização e de volume de calda. *Ciência Rural,* Vol.36, No.5, pp. 1360-1366, ISSN 0103-8478

Hislop, E.C.; Western, N.M. & Butler, R. (1995). Experimental air-assisted spraying of a maturing cereal crop under controlled conditions. *Crop Protection,* Vol. 14, No. 1, pp. 19-26, ISSN 0261-2194

Jorgensen, M.K. & Witt, K.L. (1997). Spraying technique in relation to approval and use of pesticides in Northern Europe, *Proceedings of 14th Danish Plant Protection Conference: Pesticides and Environment, SP-report* No.7, 1997

Jorgensen, M.K. & Witt, K.L. (2000). Spraying and the impact on the environment: Spraying technique in relation to approval and use of pesticides in Northern Europe, *Proceedings of Hardi international application technology course,* Vol.1, Chap.1, pp. 4-16, Taastrup, September, 2000

Koch, H. (1997). The evolution of application techniques in Europe, *Proceedings of 1rst Simpósio Internacional de Tecnologia de Aplicação de Produtos Fitossanitários,* pp. 30-38, Águas de Lindóia, São Paulo, Brazil, March 26-29, 1996

Kunz, V.L. (2010). *Dinâmica do ar em barra pulverizadora, com saída única e dupla, deposição da calda e controle da ferrugem asiática da soja.* 48p. Tese (Doutorado em agronomia/ Proteção de plantas) - Faculdade de Ciências Agronômicas, Universidade Estadual Paulista, Botucatu, São Paulo, Brazil

Matthews, G.A. (1992). *Pesticide application methods* (2nd. ed.), Longman, ISBN 0-470-21818-5, London, England

Matthews, G.A. (2000). *Pesticide application methods,* (3rd. ed.), Blackwell Science, ISBN 9780632054732 , Oxford, UK

Matuo, T. (1990). *Técnicas de aplicação de defensivos agrícolas,* FUNEP, Jaboticabal, 139p.

Miller, P. (1997). Engineering research and development related to ground-based crop sprayers, *Proceedings of 1rst Simpósio Internacional de Tecnologia de Aplicação de Produtos Fitossanitários,* pp. 102-109, Águas de Lindóia, São Paulo, Brazil, March 26-29, 1996

Monteiro, M.V.M. (2006). BVO Terrestre, In: *Manual de Operação Para Aplicações Terrestres em BVO*, 9 p., Centro Brasileiro de Bioaeronautica, Sorocaba, São Paulo, Brazil

Navarini, L.; Dallagnol, L.J.; Balardin, R.S.; Moreira, M.T.; Meneghetti, R.C. & Madalosso, M.G. (2007). Controle químico da ferrugem asiática (*Phakopsora pachyrhizi* Sidow) na cultura da soja. *Summa Phytopathologica*, [online], Vol.33, No.2, pp. 182-186, ISSN 0100-5405

Nordbo, E. (1992). Effects of nozzle size, travel speed and air assistance on artificial vertical and horizontal targets in laboratory experiments. *Crop Protection*, Vol.11, N o.3, pp. 272-277, ISSN 0261-2194

Ozkan H. E. (2005). Best spraying strategies to fight against Soybean Rust. Available from: http://www.jacto.com/soybean_rust. html.

Ozkan, H.E. (2010). Spraying Recommendations for Soybean Rust. Available at Ohio State University Extension's web site "Ohionline" (http://ohionline.usu.edu).

Ozkan, H.E.; Zhu, H.; Derksen, R.C.; Guler, H. & Krause, C. (2006). Evaluation of various spraying equipment for effective application of fungicides to control Asian soybean rust. *Aspects of Applied Biology*, Vol.77, pp. 1-8

Palladini, L.A.; Raetano, C.G. & Velini, E.D. (2005). Choice of tracers for the evaluation of spray deposits. *Scientia Agricola*, Vol.62, No.5, pp. 440-445, ISSN 0103-9016

Prado, E.P. 2009. *Assistência de ar em pulverização no manejo fitossanitário na cultura da soja [Glycine max (L.) Merrill.]*. 94p. Dissertação (Mestrado em agronomia/ Proteção de Plantas) - Faculdade de Ciências Agronômicas, Universidade Estadual Paulista. Botucatu, São Paulo, Brazil

Prado, E. P.; Raetano, C.G.; Aguiar-Junior, H.O.; Dal Pogetto, M.H.F.A.; Christovam, R.S.; Gimenes, M.J. & Araújo, D. (2010). Velocidade do ar em barra de pulverização na deposição da calda fungicida, severidade da ferrugem asiática e produtividade da soja. *Summa Phytopathologica*, Vol.36, No.1, pp. 45-50, ISSN 0100-5405

Prado, E.P. (2011). Different fungicides application technologies to control of Asian rust in the soybean culture, 120p. (Not published)

Raetano, C.G. (2002). Assistência de ar em pulverizadores de barra. *O Biológico*, Vol.64, No.2, pp. 221-225, ISSN 0366-0567

Raetano, C.G. & Bauer, F.C. (2003). Efeito da velocidade do ar em barra de pulverização na deposição de produtos fitossanitários em feijoeiro. *Bragantia*, Vol.62, No.2, pp. 329-334, ISSN 0006-8705

Raetano, C.G. & Bauer, F.C. (2004). Deposição e perdas da calda em feijoeiro em aplicação com assistência de ar na barra pulverizadora. *Bragantia*, Vol.63, No.2, pp.309-315, ISSN 0006-8705

Raetano C.G. & Merlin A. (2006). Avanços tecnológicos no controle da ferrugem da soja, In: *Ferrugem Asiática da Soja*, L. Zambolim, (Ed.), 115-138, UFV, ISBN 85-60027-14-9, Viçosa, Minas Gerais, Brazil

Robinson, T.H. (1993). Large-scale ground-based application techniques, In: *Application technology for crop protection*, G.A. Matthews & E.C. Hislop, (Eds.), 163-186, CAB International, ISBN 0-85198-834-2, Wallingford, Oxon, UK

Sartori, S. (1997). Equipamentos tratorizados para culturas de baixo fuste: situação no Cone-Sul, *Proceedings of 1rst Simpósio Internacional de Tecnologia de Aplicação de Produtos Fitossanitários*, pp. 110-112, Águas de Lindóia, São Paulo, Brazil, March 26-29, 1996

SCudeler, F. & Raetano, C.G. (2004). Assistência de ar e angulação da barra pulverizadora na deposição e perdas da pulverização na cultura da batata. Depto. de Produção Vegetal, Faculdade de Ciências Agronômicas, Universidade Estadual Paulista, Botucatu, São Paulo, Brazil, 35p. (Relatório Científico)

Silva, M. A. S. (2001). *Depósitos da calda de pulverização no solo e em plantas de tiririca (Cyperus rotundus L.) em diferentes condições de aplicação.* 53p. Tese (Doutorado em Agronomia / Agricultura) - Faculdade de Ciências Agronômicas, Universidade Estadual Paulista, Botucatu, São Paulo, Brazil

Soares, R.M.; Rubin, S.A.L.; Wielewicki, A.P. & Ozelame, J.G. (2004). Fungicidas no controle da ferrugem asiática (*Phakopsora pachyrhizi*) e produtividade da soja. *Ciência Rural,* Vol.34, No.4, pp.1245-1247, ISSN 0103-8478

Taylor, W.A. & Andersen, P.G. (1991). Enhancing conventional hydraulic nozzle use with the Twin Spray System. *British Crop Protection Council Monograph,* No.46, pp. 125-136, UK

Taylor, W.A. & Andersen, P.G. (1997). A review of benefits of air assisted spraying trials in arable crops. *Aspects of Applied Biology,* Vol.48, (January 1997), pp. 163-174, ISSN 0265-1491, Wellesbourne, Warwick, UK

Taylor, W.A.; Andersen, P.G. & Cooper, S. (1989). The use of air assistance in a field crop sprayer to reduce drift and modify drop trajectories, *Proceedings of 3rd Brighton Crop Protection Conference Weeds,* p.631, ISBN 0948404361, British Crop Protection Council, Brighton, Farnham, UK, November 20-23, 1989

Tomazela, M.S.; Martins, D.; Marchi, S.R. & Negrisoli, E. (2006). Avaliação da deposição da calda de pulverização em função da densidade populacional de *Brachiaria plantaginea,* do volume e do ângulo de aplicação. *Planta Daninha,* Vol.24, No.1, pp.183-189, ISSN 1806-9681

Van De Zande; J.C.; Meier, R. & Van Ijzendoorn, M.T. (1994). Air-assisted spraying in winter wheat-results of deposition measurements and the biological effect of fungicides against leaf and ear diseases. *Proceedings of British Crop Protection Conference - pests and diseases,* pp.313-318, BCPC, Brighton, UK, 1994

Zhu, H.; Brazee, R.D.; Fox, R.D.; Derksen, R.C. & Ozkan, H.E. (2008a). Development of a canopy opener to improve spray deposition and coverage inside soybean canopies: Part 1. Mathematical models to assist opener development. *Transactions of the ASABE,* Vol.51, No.6, pp.1905-1912, ISSN 2151-0032

Zhu, H.; Derksen, R.C.; Ozkan, H.E.; Reding, M.E. & Krause, C.R. (2008b). Development of a canopy opener to improve spray deposition and coverage inside soybean canopies: Part 2. Opener design with field experiments. *Transactions of the ASABE,* Vol.51, No.6, pp.1913-1921, ISSN 2151-0032

Yorinori, J.T.; Júnior, J.N. & Lazzarotto, J.J. (2004). Ferrugem asiática da soja no Brasil: Evolução, importância econômica e controle, 36p, EMBRAPA Soja, Londrina, Paraná, Brazil (Documentos, 247)

Yorinori J.T. (2005). A ferrugem asiática da soja no continente americano: evolução, importância econômica e estratégias de controle. *Proceedings of Workshop Brasileiro sobre a Ferrugem Asiática,* pp.21-37, EDUFU, Uberlândia, Minas Gerais, Brazil, 2005

Yorinori, J.T.; Yuyama, M.M. & Siqueri, F.V. (2010). Doenças da soja, In: *Boletim de Pesquisa de Soja 2010,* Vol.1, No.14, Fundação Mato Grosso, (Ed.), pp. 218-274, Mato Grosso, Brazil, ISSN 1807-7676

Yorinori, J.T. (2010). Controle da ferrugem "Asiática" da soja na safra 2006/2007, 3p, EMBRAPA soja, Londrina, Paraná, Brazil

8

Competition for Nodulation

Julieta Pérez-Giménez, Juan Ignacio Quelas and Aníbal Roberto Lodeiro
IBBM-Facultad de Ciencias Exactas, Universidad Nacional de La Plata-CONICET
Argentina

1. Introduction

Nitrogen (N) is the nutrient that most often becomes limiting for plant growth. Soybean may obtain this nutrient from the air, thanks to its ability to perform a symbiosis with bacteria of the genera *Bradyrhizobium* (*B. japonicum, B. elkanii, and B. liaoningense*), *Sinorhizobium* (*S. fredii* and *S. xinjiangense*) and *Mesorhizobium* (*M. tianshanense*). These bacterial species are collectively known as soybean-nodulating rhizobia, but only *B. japonicum, B. elkanii*, and *S. fredii* were used as commercial inoculants for soybean crops, with *B. japonicum* being the most widely employed. In this symbiosis the rhizobial partner reduces the atmospheric N_2 to NH_3 in a reaction catalyzed by the nitrogenase enzymatic complex, while the plant partner supplies the C sources that provide the energy required for the N_2 reduction reaction. Since atmospheric N_2 is an unlimited source of N, the process of N_2 fixation is of great potential for sustainable agriculture, and in the special case of legumes, the symbiosis is so efficient that in hydroponic culture the plant may satisfy all its N needs without resorting to any other N source. In addition, this symbiosis is a biological process that does not require fossil energy consumption, and does not leak any contaminant byproduct to the biosphere. Therefore, the inoculation of legume crops with selected rhizobial strains of high N_2 fixation performance is an extended practice in agriculture since decades ago. In parallel, the industry of inoculants is very active, commercializing a variety of formulations with different strains and combinations with other plant-promoting rhizobacterial species such as *Azospirillum brasilense* or *Pseudomonas fluorescens*. For the farmers, inoculating a legume crop with active rhizobia is a simple procedure, and its economic cost is much lower than applying chemical fertilizers. All these advantages are, however, obscured by the fact that in field crops the symbiotic N_2 fixation seldom provides the expected results, and the plants may consume the N from the soil.

Several factors account for this low performance of N_2 fixation in field crops. In energetic terms, N_2 fixation is more costly for the plant than soil N uptake and therefore soil N is preferred when this source is not limiting, or when N_2 fixation is inefficient (Salon et al., 2009). This may be appreciated if one takes into account that the symbiosis only occurs in a specialized organ known as root nodule. It is there where the rhizobia differentiate into the state able to reduce N_2 −the bacteroid− and where the O_2 concentration is lowered at levels compatible with nitrogenase activity (Patriarca et al., 2004). Therefore, rhizobia must infect the roots and trigger the development of nodules, which finally will be occupied by the rhizobia. During the earliest steps of nodule development and root infection (Ferguson et al., 2010), the plant-rhizobia relationship is more similar to a pathogenesis than to a mutualistic symbiosis: rhizobia invade plant tissues, consume plant energy resources, induce a tumor-like development, and proliferate inside plant cells, without any benefit for the plant until nodules are completely developed and N_2 fixation starts. Indeed, rhizobial strains with low N_2-fixing efficiency or unable to fix N_2 trigger typical plant defense reactions and lead to a weakening of the plant. In such a scenario, N removal from soil may be quite significant. It has been estimated that in soils with low N content, a good N_2-fixing symbiosis may provide 70 % of N that plant needs (i.e. 70 % of all plant N coming from the air) while an inefficient symbiosis only provides 20-30 % of N (Unkovich & Pate, 2000). Therefore, the difference between a good and a bad symbiosis would be around 40 % plant N being obtained from the air or the soil, respectively. N contents of soybean grains are around 2.5 % w/w, depending on the cultivar, all this N being removed from the ecosystem at grain harvest. Hence, considering a mean yield of 2,500 kg ha^{-1} the difference between a good and a bad N_2-fixing symbiosis equals 25 kg N ha^{-1} that are respectively conserved or removed from the soil each year.

In soybean-producing countries like Argentina, these crops are grown in soils with low N content, either because the soils were previously cropped with species of high soil N demand, such as wheat or corn, or because they are in marginal areas. Therefore, N_2 fixation becomes a key input of sustainable soybean cropping, because this species has also a high N demand, and thus, the N that cannot be provided by N_2 fixation must be supplied as chemical fertilizer, which involves many environmental problems and has a higher cost.

The efficiency of the N_2-fixing symbiosis depends on many factors. Of primary importance are the total number of nodules formed in each plant, and the N_2-fixing activity of each nodule. As mentioned before, nodulation has an energetic cost to the plant and therefore the number of nodules cannot be too high. Instead, an optimal number of nodules able to provide the necessary amount of fixed N_2, with a reasonable energetic cost involved in its maintenance, is regulated by the plant. In this way, once a given number of active nodules are established, the plant progressively inhibits the formation of new nodules (see below). This indicates that both the plant genotype (by its ability to assimilate fixed N_2 and its ability to control the number of nodules) and the rhizobial genotype (by its N_2-fixing activity) are key determinants of the symbiosis performance. However, being a biological process, the environment also plays a fundamental role, not only by its influence on the activity of each partner, but also by its interaction with both genotypes.

Competition for nodulation between inoculated rhizobia and different rhizobial strains resident in the soil is a striking example of this complexity. Normally, the soils are populated with rhizobia either from the indigenous bacterial population or introduced by the inoculants used in previous crop seasons. Since rhizobia are soil bacteria, they readily adapt to a new soil and exchange genetic material with the local soil microbiota. However, in the soil there is not a high selective pressure for high N_2 fixation performance and therefore, genetic drift leads to dispersion of the N_2-fixing potential among different genotypes with diverse efficiencies. Therefore, the soil rhizobial population is often of high efficiency to nodulate the plants, but of medium to low efficiency to fix N_2. Hence, the competition of this population for plant nodulation may prevent the newly inoculated strains to occupy a significant proportion of the nodules, leading to lack of N_2 fixation response to inoculation (Toro, 1996). To get an approximation to the problem, we can consider that each nodule contains a clone of bacteria derived from a single precursor bacterium that initiated the root infection. Occasionally, two bacterial clones may share a nodule, but it is extremelly unfrequent to find nodules with more than two clones. Given that a single soybean plant might possess, at most, in the order of 10^2 nodules at maturity, this is the order of magnitude of the bacterial individuals that can survive the root penetration. However, in soils with several years of soybean cropping there may be in the order of 10^5 to 10^7 soybean-nodulating rhizobia colonizing the proximity of the root –the rhizosphere– and therefore, for each bacterial cell that succeeds in penetrating the root, there are 100 that remain outside. Considering that nodules are protected environments for rhizobia, a harsh competition is established among rhizospheric rhizobia to gain access to the root nodules. In this process, the bacterial genotypes will have a prominent role in defining their competitiveness, but also the plant genotype will dictate how and when the bacteria will be allowed to penetrate the roots, the interaction between the bacterial and plant genotypes will determine which strains will be favored, and the interaction of the environment with both genotypes will determine the relative advantages for each bacterial strain.

Soybean seeds are inoculated with around 10^5-10^7 rhizobia seed^{-1}, but more than 80 % of the rhizobia die within the first 2 h after inoculation (Streeter, 2003). From the survivors, only a small percentage reaches the rhizosphere after sowing (López-García et al., 2002). Thus, given the above figures, obtaining around 10 % of all the nodules occupied by the inoculated strain, as is currently accomplished in soybean crops, may be considered quite successful, but still it is completely insufficient to get a significant increase in plant N_2 fixation above the levels obtainable without inoculation. Therefore, to get a significant proportion of plant N coming from the air it is imperative to improve inoculant's competitiveness to obtain significantly higher percentages of nodules occupation, and active research about this problem is being done since decades. In this chapter we will provide a look on the methods employed to study the problem of competition for nodulation, the bacterial and plant traits that may influence the competition, the ecological aspects that modulate the plant and rhizobia interaction, and how these factors may be managed in order to profit the symbiosis to increase soybean crop yields.

2. Methods employed to study competition for nodulation

In any method, at least one of the competing strains has to be selectively labelled in order to discriminate the nodules occupied by it. Labels employed are not different than those used for strain selection or tracking, including antibiotic resistances, fluorescent proteins, antibodies, DNA probes or reporter proteins such as β-galactosidase or GUS. Special properties of certain rhizobial strains, like melanin production, were also employed (Castro et al., 2000). However, there are some caveats to be kept in mind for labelling. In the first place, it has to be demonstrated to which extent the label may alter the nodulating ability or the competitiveness of the strain. Ideally, labelling should not alter any of them, but provided the effect is accurately measured in comparison to the unlabeled wild type, strains altered after labelling may be confidently used (Thomas-Oates et al., 2003). The introduction of labels in replicative plasmids has the advantages of avoiding undesired modifications in the rhizobial genomic background, and the possibility of introducing the same plasmid in different host strains. However, the plasmid replication as well as the expression of genes carried by it may consume energy from the bacterial cell, and therefore may affect competitiveness. This may explain why strains labelled in this way tend to have diminished their competitiveness (Mongiardini et al., 2009). By difference, the introduction of labels in the genomic DNA of the bacterial cell may cause undesired loss or alteration of a given function. Several rhizobial genomes were completely sequenced (http://genome.kazusa.or.jp/rhizobase) and therefore now it is possible to choose the exact location where a label may be introduced without affecting any coding region (Pistorio et al., 2002). However, both replicative plasmids and reporter genes recombined in the genomic DNA render gene-manipulated strains that cannot be safely released into the environment. An alternative may be the choice of a natural selection, such as for antibiotic resistance; however, enrichment in a strain tolerant to high antibiotic doses is also not desirable for the environment, because this trait may be dispersed by horizontal gene transfer, and in addition, several reports also indicate that selection for resistance to high antibiotic concentration may yield strains with diminished competitiveness (Lochner et al., 1989; Spriggs & Dakora, 2009; Thomas-Oates et al., 2003). Therefore, if this method is chosen, a careful assessment of symbiotic and competitive abilities of the antibiotic-resistant derivatives must be carried out before employing them in competition studies (López-García et al., 2002; Spriggs & Dakora, 2009). Intrinsic antibiotic resistance to low doses (without enrichment by selection) may be used but this method is not so efficient to distinguish an indicator strain in a population (Spriggs & Dakora, 2009). Other methods, which do not involve any manipulation of the strains, are antibody or DNA labelling. Antibody labelling of different strains recovered from nodules may be efficiently carried out by ELISA (Spriggs & Dakora, 2009); however, antibodies cannot distinguish among genotypically related strains. DNA labelling may be carried out either with probes or by PCR. In this last methodology, DNA fingerprints may be obtained, which have better discriminative potential than antibodies. The disadvantage of these methods is that they involve more manipulation and equipment, and are more expensive than the other methods mentioned.

Once the labeled indicator strain is available, the next step is plant culture and inoculation. Here there also exist a wide variety of methods, which can be divided in field assays and laboratory assemblies. In the last case, the rooting substrate may be sterile or nonsterile soil, or an artificial substrate such as sand, vermiculite, perlite or a mixture of them. Even more artificial root environments were employed, including liquid plant nutrient solution, agarized media, and plastic growth pouches. Depending on how close to the natural environment are the experimental conditions used, more accurate predictions on rhizobial competitiveness in field crops may be obtained, but more variables need to be considered. In many cases, an unnecessarily high number of variables may obscure the conclusions about a given factor under study, such as the effect of a bacterial single mutation. The influence of some important environmental variables will be considered in another section.

Field trials are normally done in plots arranged in complete random blocks. Factorial analysis also is employed and is useful when more than one variable is under study (López-García et al., 2009). As mentioned before, field trials are the most close to the real crop, but at the same time are subjected to the whole ensemble of natural variables. For this reason, field trials should not be restricted to a single location or crop season, and an analysis of weather, soil, and previous land use should accompany the nodulation experiment. Generalizations require repeating the experiment in more than one season and ideally, at several locations. Otherwise, it must be clearly stated that the conclusions are restricted to the site and season where the experience took place.

The most popular laboratory assemblies used to contain the rooting substrate are the pots and the Leonard jars (Vincent, 1970). They differ in the watering method, which in turn lead to differences in the water content of the substrate. Pots are watered from the top at given intervals and therefore, it is possible to maintain the field capacity within reasonable limits, provided that drainage is incorporated to the pots (for instance, by means of holes at the bottom). The disadvantage of this method is that, since periodic watering is required, pots are quite exposed to cross-contamination. To avoid it, layers of inert material are commonly placed on top to prevent penetration of undesired rhizobia to the pots. In addition, it is always necessary to distribute uninoculated pots among the test pots to serve as negative controls and for assessing the absence of cross-contamination. By contrast to pots, Leonard jars are not irrigated from top. They are assemblies consisting in an upper container filled with the desired rooting substrate and a bottom reservoir that contains the irrigation solution. The rooting substrate is moistened by a wick running the length of the upper container and extending out into the solution reservoir. In this way, the irrigating solution rises by capillary from the reservoir avoiding the need of irrigation by hand (the only requirement is to maintain the liquid in the reservoir). Hence, Leonard jars are less prompt to cross-contamination than pots but provide higher moisture around roots. Later we will see how this variable may affect competitiveness.

When soil is used as rooting substrate it may be sterile or nonsterile, depending on whether the influence of biotic factors is to be analyzed. Since this experimental technique is commonly performed in a greenhouse, variables associated to climatic factors are eliminated. However, it is often overlooked that if several soil samples are pooled and sieved, soil variables like soil structure and patching are also eliminted. Various methods were also used for soil sterilization. When available, gamma radiation is preferable to heat (dry or wet), since that method is less likely to provoke changes in organic matter. The other rooting substrates mentioned above are chemically inert and possesses water retention capacity, normally in the order vermiculite > perlite > sand. Although these substrates do not substitute for soil, they share some common properties like water retention and porosity. Thus, they may be used in combination with defined plant nutrient solutions to provide a controlled environment where some of the most important soil variables (e.g. moisture, porosity, pH, and nutrients level) may be manipulated. Moreover, plant nutrient solutions are used in many experiments as liquid or agarized media, or to wet plant stands, like a paper towel, without employing any other rooting substrate. In these cases the influence of both water potential and porosity is lost, which could lead to misinterpretations when phenomena like bacterial motility or chemical diffusion may play a role on the results.

The essence of measuring competition for nodulation rests on the count of the proportion of nodules occupied by a given strain on a plant. Therefore, these approaches need also an appropriate statistical analysis. The methods more widely employed are the analysis of variance (ANOVA) and the χ^2 test. Whatever the method applied, two important aspects are the number of replicas of each treatment and the proper data input in the statistical analysis. The number of replicas depends on the variability of the experimental material but it is not recommended to pool all the nodules from different plants. Instead, nodules from each plant should be treated separately in order to express the results in a per plant basis, thus considering each plant as an experimental unit. Regarding data input, it has to be kept in mind that the above mentioned statistical methods suppose that certain conditions are obeyed by the data. One important condition is the homogeneity of the variance, which means that all the experimental groups have the same variance. Although the number of nodules in different samples obeys this condition, the proportion of nodules occupied by a given strain does not. In this case, the variance is maximal for a proportion of 0.5, and tends to zero as the proportion approaches 0.0 or 1.0 (Lison, 1968). To obtain a dataset obeying the homogeneity of variance, these proportions must be transformed to the arc sin root square before applying a test such as

ANOVA to them. Then, the whole analysis is carried out with the transformed data, but if averages are to be compared (for instance, with the Tukey test) the data need to be used with the original values, i.e. are not used with the transformed values. A rather frequent error in the literature is the use of the proportions (or percentages) of nodule occupation without transformation in ANOVA tests, which in this case lose sensitivity.

A special method was developed in the 80s by Amarger and Lobreau (1982). Since its proposal, the method was widely employed and allows the determination of strains competitiveness quite accurately. It is based on the use of two competitor strains at a range of concentration ratios. For instance, strain A is competed against strain B at 100:1; 10:1; 1:1; 1:10, and 1:100 A:B initial concentrations ratios in the inocula, which are termed $I_A : I_B$. Then, nodules hare harvested and the proportions of nodules occupied by strain A and strain B, N_A and N_B respectively, are registered for each $I_A : I_B$ ratio. With these data a plot of log (N_A/N_B) against log (I_A/I_B) is constructed, which gives a straight line that cuts the ordinate log (N_A/N_B) axis when I_A/I_B equals 1.0, thus giving the $C_{A:B}$ value that represents the competitiveness of strain A on strain B when both are inoculated at exactly the same concentration. Although this method is laborious because several inoculum rates must be tested for each pair of competing strains, the result of $C_{A:B}$ is more exact than the obtained in a single competition at approximately equal concentrations. This accuracy is especially valuable when differences in competitiveness are not wide (as may be the case when soil isolates are evaluated against a collection strain).

3. Bacterial and plant traits that affect competition for nodulation

Both the rhizobia and the plant genotypes influence the competition for nodulation. In particular, genes determinant for symbiosis, like those for production, transduction and binding of symbiotic soluble signals, development of infections and nodules, and N_2 fixation, will have obvious effects on competitiveness, since they affect rhizobial and plant activities that are prerequisite for competition. Therefore, these genes will not be reviewed here and the reader is forwarded to excellent recent reviews on the subject of plant infection and nodulation (Ferguson et al., 2010; Patriarca et al., 2004). Furthermore, rhizosphere colonization is a previous and necessary step for nodulation and therefore, any gene or set of genes that favor the adaptation of rhizobia to the environment may have a positive effect on competition for nodulation, although as we will see later, the relation between efficiency in rhizosphere colonization and competitiveness for nodulation is not so straightforward. Anyway, tolerance against environmental stress, metabolic efficiency in the use of nutrients from the rhizosphere, growth speed and persistence in the environment, and resistance against predation are also related with competitiveness. After rhizosphere colonization, adhesion to root surfaces is also required for nodulation and therefore, surface components of both symbionts have a role in competitiveness. However, other genes unrelated to these activities may also influence competitiveness in unexpected ways. In the following, some important traits for which specific effects on competitiveness for nodulation were observed, are reviewed.

3.1 Bacterial traits

Some approaches were developed to identify genes associated to competitiveness, which required solving two problems: 1) testing a high number of candidate genes in plant assays and 2) screen the competitiveness of all these candidates. These problems were addressed in two ways, although none of them in soybean-nodulating rhizobia.

The earliest attempt was to employ a highly competitive strain of *Rhizobium etli* to extract from it large DNA fragments (in the order of 25 kb) that were introduced into cosmids, which were used to transform the less competitive reference strain. These transconjugants were inoculated in mass on common bean plants, the nodules were obtained, and rhizobia occupying these nodules were extracted and used in a second round of inoculation, nodules extraction and rhizobia recovering (Beattie & Handelsman, 1993). According to the authors, if competitiveness conferred by the cosmids was as high as that of the donor strain, these two rounds of enrichment by nodules passage should allow a high chance of recovery of clones with enhanced competitiveness. The authors recovered nine such clones, but surprisingly, when these clones were cured from the cosmids their high competitiveness remained intact. In addition, the introduction of the cosmids in the reference strain did not render this strain more competitive, whereby the enhanced competitiveness of these nine clones seems to be the consequence of the enrichment procedure and not a trait conferred by the DNA fragments obtained from the more competitive strain. Unfortunately this strategy was not continued and the causes for the higher competitiveness obtained are unknown. More recent studies indicated that bacterial strains cultured continually under laboratory conditions tend to lost some traits related with their adaptation to the natural environment (Marks et al., 2010). Therefore, the enrichment procedure by nodules passage might have reversed a laboratory adaptation of the reference strain, which could include heritable genetic changes. If this is the case, such an enrichment procedure may be considered for strains improvement without genetic manipulation.

The second approach was to employ signature-tagged mutants (Pobigaylo et al., 2008). Briefly, different short DNA fragments are introduced in transposons in such a way that a collection of transposons is obtained, where each transposon can be identifyed by the DNA sequence tag that it carries. Then, these transposons are used to mutagenize a given bacterial strain and as a consequence, a set of mutants where each member can be distinguished from the others is obtained. By combining this technique with microarray screening, each tag can be mapped into a corresponding insertion sequence. Pobigaylo et al. (2008) employed this technique to inoculate two sets of 378 tagged *S. meliloti* mutants on alfalfa plants and were able to recover 67 mutants attenuated in symbiosis among which 23 were altered in genes that affected competitiveness but are not obviously related to nodulation or N_2 fixation. Many of these mutants were affected in metabolic or transport functions and two encode hypothetical proteins. The question as to whether modification of the expression of any of these genes could improve rhizobia competitiveness remains to be elucidated.

The above-mentioned work was done in aeroponic plant cultures, which allowed inoculation and homogeneous distribution of a high number of mutant strains as well as the recovery of many nodules for screening. However, as mentioned earlier, such an artificial assembly does not allow evaluation of other important features, like rhizosphere colonization. Rhizosphere is a nutrient enriched zone in comparison to the rest of the soil, due to the many compounds released in plant root exudates, among which various sugars, organic acids, aminoacids and vitamins serve as sources for microbial growth, and flavonoids and related compounds may act as signal molecules. Therefore, rhizobia colonize this soil compartment and an important question is if this colonization involves an active movilization of the rhizobia towards rhizosphere. Rhizobia are motile bacteria, expressing active flagella and able to move by swimming and swarming (Bahlawane et al., 2008; Braeken et al., 2007; Daniels et al., 2006; Nogales et al., 2010; Soto et al., 2002; Tambalo et al., 2010). For many years, various studies informed that rhizobia can move from soil to legume

roots to initiate root colonization and infection (Brencic & Winans, 2005; Fujishige et al., 2006; González & Marketon, 2003; Yost et al., 2003). Evidence includes measurements of rhizobial dispersal in soil (Lowther & Patrick, 1993), the in vivo observation of *S. meliloti* motility towards infection sites on legume roots (Gulash et al., 1984), the characterization of rhizobial attraction by root exuded molecules and specific flavonoids (collectively referred to as chemoattractants) in *R. leguminosarum*, *S. meliloti*, and *B. japonicum* (Barbour et al., 1991; Caetano-Anollés et al., 1988a; Chuiko et al., 2002; Gaworzewska & Carlile, 1982; Pandya et al., 1999), and the observation of diminished root adsorption, colonization, and nodulation rates in motility defective mutants of *S. meliloti* and *R. leguminosarum* (Ames & Bergman, 1981; Caetano-Anollés et al., 1988b; Hunter & Fahring, 1980; Mellor et al., 1987; Parco et al., 1994).
This notion of rhizobial movement in soil towards root exudates and surfaces underlies also the agronomic practice of seed inoculation for soybean crops. Accordingly, it is expected that rhizobia will move in some way from the inoculation site on the seed surface to the infectable root cells that lie near the root tip to produce a nodule, even when these infection sites continuously migrate away from the inoculation site as the roots grow (Bhuvaneswari et al., 1980). However, most of the above evidence was obtained in laboratory experiments performed in saturated aqueous media. When porous media more similar to soil were employed, some conflicting results were obtained. It was reported long ago that vertical motility of rhizobia in the soil profile is restricted to a few millimeters unless other factors, like percolating water, earthworms, or tillage, aid in moving rhizobia to a greater depth (Madsen & Alexander, 1982). This correlates with a poor root apex colonization of seed-inoculated *B. japonicum* when these root apical regions −the infectable zone− penetrate a few centimeters into the rooting substrate, and with the observation that *B. japonicum* inoculated on soybean seeds sowed in vermiculite where a rhizobial population, isogenic with the inoculant, was previously established, occupied less than 20 % of the nodules regardless of its intrinsic infectivity (López-García et al., 2002). More recently, the motility of *S. meliloti* towards root exudates in a peat substrate was again observed as being very restricted, unless nematodes able to be attracted by specific volatile compounds produced by *Medicago truncatula* are also present. In these experiments, rhizobia were observed to be transported both on the nematodes surface and into the nematodes gut (Horiuchi et al., 2005). In agreement with these results, Liu et al. (1989) found that lack of motility barely affected nodulation competitiveness of *B. japonicum* in unsaturated, non-sterile soil, and suggested that the encounter between roots and rhizobia depends not on rhizobial movement, but on soil exploration by growing roots. In agreement with these observations, *B. japonicum* non-motile flagella-defective mutants were similarly competitive as the wild type in vermiculite at field capacity but were totally displaced from nodules occupation by the wild type when the vermiculite was flooded, indicating that bacterial swimming may be a factor of competition for nodulation only in this last situation (Althabegoiti et al., 2011). Therefore, in soils at field capacity rhizobial motility may be retarded by many factors that are absent both in flooded rooting substrates or liquid media (Horiuchi et al., 2005; Liu et al., 1989; López-García et al., 2002; Madsen and M. Alexander, 1982; McDermott & Graham, 1989). Among these factors we can mention chemoattractant diffusion, which at field capacity is slower due to the lower water potential, paths impairement due to the tortuosity and size of the soil pores, and retardation of bacterial displacement due to attachment/detachment to and from soil particles (Tufenkji, 2007; Watt et al., 2006). Hence, it was suggested that the limited motility of rhizobia in soils at field capacity might be a primary factor in the problem of competition for nodulation (López-García et al., 2002). To solve this problem with the

inoculants, two measures were proposed: the use of in-furrow inoculation instead of seed inoculation, and the selection of inoculant strains with higher motility (Althabegoiti et al., 2008; Bogino et al., 2008; López-García et al., 2009). Both techniques yielded promising results in soybean field assays (López-García et al., 2009), but the in-furrow inoculation method still needs technical improvement for its application to soybean crops at a similar cost-benefit relationship as seed inoculation.

Since plants roots often release protons, rhizosphere is usually an acidic compartment (Hinsinger et al., 2005). In addition, the interior of root hairs where rhizobia will penetrate is also acidic. Therefore, acid tolerance was raised as a trait related with efficiency of rhizosphere colonization and root infection. However, few studies were carried out in *B. japonicum* since it was early recognized that slow-growing *Bradyrhizobium* species are more tolerant to acid stress than fast-growing rhizobial species, being able to tolerate pH below 5.0 (Graham, 1992). Moreover, regarding acid stress, soybean plants seem more sensitive than *B. japonicum* and hence, most of the efforts directed towards improvement of acid tolerance were directed to the plant partner of this symbiosis. Despite this, acid tolerance may be involved in competitiveness even in acid tolerant species. Studies in *R tropici* indicated that the substitution of Ala for Ser in a domain of AtvA, a protein homologous to the virulence protein AcvB from *Agrobacterium tumefaciens,* caused a significant drop in competitiveness. The authors observed that mutation of the membrane protein LpiA, whose gene lies in the same operon as *atvA,* caused a similar effect; however, these competitiveness defects took place also under non-acidic conditions, and none of these changes altered nodulation or N_2 fixation (Vinuesa et al., 2003). Therefore, the requirements of these genes seems to authentically affect competitiveness, and might be related to coping with the acidic environment of the root surfaces or the interior of plant cells rather than a general environmental adaptation.

Rhizosphere may be also a dry environment because of the continuous root water suction activity. Therefore, drought tolerance is also a trait that might be relevant for rhizosphere colonization and competitiveness for nodulation. Among the strategies employed by microbia to cope with desiccation, the accumulation of solutes such as trehalose seems widespread. *B japonicum* cannot use trehalose as C source, and therefore, trehalose incorporation to growth media leads to its accumulation in the cytoplasm. This treatment yielded *B. japonicum* cells more tolerant to desiccation and led to increased survival of rhizobia inoculated on soybean seeds (Streeter, 2003). Moreover, trehalose spontaneously accumulates in *B. japonicum* during desiccation. Three pathways of trehalose biosynthesis were found in this bacterial species: trehalose synthase, trehalose-6-phosphate synthetase, and maltooligosyltrehalose synthase. A transcriptomics study of *B. japonicum* during desiccation showed that the expression of the genes encoding trehalose-6-phosphate synthetase (*otsA*), trehalose-6-phosphate phosphatase (*otsB*), and trehalose synthase (*treS*) was significantly induced and in parallel, the activity of trehalose-6-phosphate synthetase and the trehalose intracellular concentration were increased thus indicating that trehalose accumulation is a regulatory response of desiccation tolerance in *B. japonicum* (Cytryn et al., 2007). However, no studies are available about the role of trehalose accumulation on competition for nodulation in soybean. In *R. leguminosarum* bv *trifolii* a mutant unable to accumulate trehalose was less competitive for nodulation than the parental strain but capable of nodulation and N_2 fixation (McIntyre et al., 2007). In agreement with this result, an *S. meliloti* triple mutant in *treS, treY,* and *otsA* was impaired in competitiveness although its intrinsic nodulation and N_2-fixation abilities were not altered (Domínguez-Ferreras et al.,

2009). By difference, an *otsA* mutant of *R. etli*, although still able to accumulate trehalose, was defective in nodulation and nitrogenase activity in common bean (Suárez et al., 2008). Therefore, the requirements of these genes for nodulation in the absence of competitors might be more stringent for determinate nodules formation. An interesting result was obtained by Ampomah et al. (2008), who observed that both *S. meliloti* and *S. medicae thuB* mutants, unable to catabolyze trehalose (and therefore, accumulating this solute in the absence of stress conditions) were improved in competition for nodulation in two cultivars of *M. sativa* and one of *M. truncatula* but were equally competitive as the wild type for rhizosphere colonization. Moreover, the authors observed that *thuB* expression is induced during *S. meliloti* penetration of root hairs and suggested that osmotic stress and threhalose accumulation occur in this environment. Therefore, desiccation tolerance appears to be necessary for root infection, independently of the drought conditions in the rhizosphere.

In addition to physicochemical factors such as barriers to motility, acidity, and dryness, the ability of rhizobia to use rhizospheric nutrients for growth is also an important factor for rhizosphere colonization, root infection and competitiveness (Toro, 1996). Moreover, several rhizobial species may induce the production of specific nutrients by the plant, in an analogous manner as *A. tumefaciens* induces the production of opines, which only this bacterial species may catabolize. Therefore, these substances, derived from *myo*-inositol, were termed rhizopines (Murphy et al., 1987). Rhizopines are produced by bacteroids into the nodules, exported to the rhizosphere, and consumed there by free-living rhizobia. Only a limited range of strains of *S. meliloti* and *R. leguminosarum* were found to produce and consume rhizopines, so that this ability is considered a selective advantage for rhizosphere colonization and competition for nodulation. In both species, rhizopine catabolism requires a functional *myo*-inositol catabolic pathway (Bahar et al., 1998; Galbraith et al., 1998). Although rhizopine production/consumption was not observed in soybean-nodulating rhizobia, *myo*-inositol catabolism was found as related with nodulation competitiveness in this symbiosis. An *S. fredii* mutant in *idhA*, which encodes *myo*-inositol dehydrogenase, nodulates normally but is severely impaired in competition for nodulation. In addition, this mutant is defective in N_2 fixation and bacteroid morphology (Jiang et al., 2001). In an *R. leguminosarum* bv *viceae* strain that does not produce rhizopines, catabolism of *myo*-inositol was also found as required for competition for nodulation, although this requirement was not observed for rhizosphere colonization, nodulation or N_2 fixation, whereby the authors concluded that *myo*-inositol catabolism is required during early plant root infection (Fry et al., 2001). Other genes related with catabolism of specific rhizosphere substances were found as determinant for competition for nodulation. Rosenblueth et al. (1998) searched for genes induced by bean root exudates in a library of *R. tropici* and found the *teu* genes, which are induced only by *Phaseolus vulgaris* and *Macroptilium atropurpureum* root exudates (both plants are symbionts of *R. tropici*). However, the compound responsible for *teu* operon induction was not identified. To search whether a similar pathway existed in other rhizobial species, the authors incubated the bean root exudate with bacteria from different rhizobial species to sequester the inducer and found that only *R. etli*, *R. leguminosarum* bv *phaseoli* and *R. giardinii*, all symbionts of *P. vulgaris*, had this capacity. Therefore, the system of *teu* induction by root exudates seems a specific trait of this group of rhizobia and plant species. An *R. tropici* CIAT 899 mutant in *teuB* was not affected in nodulation when inoculated alone, but was less competitive than the wild type for nodulation at various inoculum rates.

In addition to the ability of metabolizing specific substrates, the general metabolic activity of rhizobia also influences their competitiveness. Normally, soybean is cultivated in N-limited soils, and in addition, N limitation is a prerequisite for nodulation and N_2 fixation. It is well known that legumes are able to inhibit nodulation in the presence of abundant soil N-sources, but research about the influence of N-limitation on the rhizobial side is scarcer. López-García et al. (2001) found that *B. japonicum* can grow with minute amounts of NH_4^+ in the culture medium, and N-limitation leads to derepression of glutamine synthetase I and II, change in C-sinks with accumulation of exopolysaccharides (EPS) at the expense of polyhydroxybutirate (PHB), and higher sensitivity to genistein for *nodC* expression. Correlating these physiological changes, the rhizobia are more infective and competitive for nodulation. These changes can be exploited in the formulation of improved inoculants. Likewise, *Burkholderia mimosarum* competitiveness to nodulate three *Mimosa* species was increased in low-N incubation media (Elliott et al., 2009). Nitrogen assimilation is regulated by the *ntr* system, which includes *ntrB-ntrC-ntrY-ntrX*. Two genes downstream this system is encoded the small RNA binding protein Hfq. Transcriptomic and proteomic analyses of *S. meliloti* mutants in *hfq* indicated that alteration of this gene leads to imbalance in C and N metabolism, suggesting that the mutant strain tends to use aminoacids instead of primary C substrates as energy sources (Torres-Quesada et al., 2010). Furthermore, the authors found that *hfq* mutation did not affect nodulation but severely diminished N_2-fixation and when the mutants were coinoculated with the wild type in 1:1 relationship on alfalfa plants, no nodules were found occupied only by the mutant. However, this diminished competitiveness might be due to the repression of *iolC, iolD, iolE* and *iolB*, encoding *myo*-inositol catabolic activities, which as we saw, are required for competition at an early step of nodulation (Fry et al., 2001).

The adaptations mentioned before, i.e. acid tolerance, drought/osmotic tolerance, and ability to metabolize specific organic nutrients from root exudates, are important examples of improvement in cell fitness to the root environment. However, competence may be exerted not only by doing things better than competitors, but also by precluding competitor's activity. Soybean-nodulating rhizobia are known to have natural resistance to several antibiotics, among them chloramphenicol, neomycin, and penicillin (Cole et al., 1979). In addition, some strains of other rhizobial species are able to produce an anti-rhizobial substance known as trifolitoxin. This is a ribosomally synthesized, posttranslationally modified peptide, which was found in *R. leguminosarum* bv *trifolii* T24, against which various α-proteobacterial species are sensitive (Triplett, 1994). Among the soybean-nodulating rhizobia, *S fredii* strains are sensitive, while *B. japonicum* seems resistant. Trifolitoxin production and resistance was considered as an interesting trait to enhance competitiveness for nodulation. Therefore, the ability to produce trifolitoxin was introduced with a replicative plasmid in *R. etli*. Then, the competitiveness of trifolitoxin-producer or non-producer *R. etli* strains against a sensitive strain for nodulation of common beans was assayed in field trials, in soils with a low bean-nodulating rhizobial population (in the order of 10^2 rhizobia g of soil^{-1}). As a result, the trifolitoxin-producer strain occupied significantly more nodules than the sensitive strain, while the non-producer strain did not differ from the sensitive strain. In turn, grain yield was not modified in the inoculated or in uninoculated beans, indicating that the indigenous soil population was proficient for N_2 fixation (Robleto et al., 1998). However, the release of high numbers of rhizobia genetically modified to produce antimicrobial substances involves a number of serious environmental concerns, whereby this technology needs more studies before its commercial implementation.

After rhizosphere colonization, rhizobial adhesion to root surfaces is also a key aspect in competitiveness. Therefore, cell-surface characteristics conferred by surface polysaccharides are important for competitiveness (Bhagwat et al., 1991; Parniske et al., 1993; Quelas et al., 2010; Zdor et al., 1991). Rhizobial adhesion depends on several factors such as the medium composition where rhizobia and plants are put in contact, the composition of the culture medium where rhizobia were grown previously, and rhizobial growth state at the moment of their contact with the roots (Vesper & Bauer 1985; Smit et al., 1992). In addition, Vesper & Bauer (1985) observed that in a batch culture of *B. japonicum* only a subpopulation of bacterial cells is proficient for adhesion. This is consistent with more recent findings indicating that bacterial culture populations are not homogeneous even under controlled growth conditions (Ito et al., 2009). Moreover, bacterial cells seem to recognize specific adhesion sites in the plant roots, as mediated by plant and rhizobial agglutinins (Laus et al., 2006; Lodeiro & Favelukes, 1999; Loh et al., 1993; Mongiardini et al., 2008). These agglutinin-mediated modes of rhizobial adhesion are related to infectivity and competitiveness. In *B. japonicum* a lectin called BJ 38 was described, which mediates both polar adhesion among rhizobial cells that form special structures known as stars or rosettes, and polar adhesion of rhizobial cells to the soybean cells (Ho et al., 1990; 1994) A *B. japonicum* mutant defective in BJ38 activity was less infective on soybean plants (Ho et al., 1994). This bacterial lectin is located at one cell pole of the rhizobia (Loh et al., 1993) but it is unknown whether it is part of a larger cell appendage. Vesper & Bauer (1986) found that *B. japonicum* pili are required for adhesion to soybean roots, but it is uncertain whether BJ 38 is part of these pili. Similarly to BJ38, a bacterial agglutinin called RapA1 was found in the cell poles of *R. leguminosarum* and *R. etli* (Ausmees et al., 2001). Overproduction of this agglutinin in *R. leguminosarum* bv *trifolii* led to higher rhizobial adhesion to different plant roots, but had no effect on the speed of nodulation in clover (Mongiardini et al., 2008). However, the overproducing rhizobia were more competitive than a control strain for clover nodulation (Mongiardini et al., 2009). In addition to bacterial agglutinins, the plant agglutinins also exert an influence on rhizobial adhesion and competitiveness. Pretreatment of *B. japonicum* in low concentrations of soybean seed lectin before plants inoculation improves rhizobial adhesiveness, infectivity, and competition for nodulation (Halverson & Stacey, 1986; Lodeiro et al., 2000). This lectin is bound by the bacterial EPS at the opposite cell pole of BJ38, in a growth state-dependent manner (Bhuvaneswari et al., 1977). It was found that the sugar receptor in the EPS is galactose (Bhuvaneswari et al., 1977), and mutants unable to incorporate galactose in their EPS are severely impaired in lectin binding, adhesion to soybean roots, infectivity, and N_2 fixation (Pérez-Giménez et al., 2009; Quelas et al., 2006; 2010). This sugar moiety is modified according to the physiological state of the bacteria: rhizobia in exponential growth have in their EPS acetylated galactose, while rhizobia in stationary phase have methylated galactose. Likewise, acetylated galactose has higher affinity for lectin than methylated galactose, and rhizobia in exponential phase bind more lectin, adhere better to the roots and are more infective and competitive than rhizobia in stationary phase (Bhuvaneswari et al., 1977; Lodeiro & Favelukes, 1999; López-García et al., 2001; Vesper & Bauer, 1983). Soybean lectin binding to rhizobia may be enhanced by culture conditions that increase the amount of EPS: as we mentioned before, *B. japonicum* cultured under N-limiting conditions produces more EPS at the expense of PHB, and this EPS overproduction leads to higher soybean lectin binding activity of the bacterial cells, which become more infective and competitive to nodulate soybean against isogenic bacteria grown in normal N media (López-García et al., 2001). Likewise, in *R. leguminosarum* bv *trifolii* and *R. etli* the overexpression of the regulatory genes *pssA* and *rosR* leaded to increased

EPS production and competitiveness for nodulation (Bittinger et al., 1997; Janczarek et al., 2009). Therefore, the possibilities of manipulating the expression of agglutinins in the rhizobia (Ho et al., 1994; Mongiardini et al., 2009) or increasing rhizobial sensitivity to plant agglutinins by culture conditions (López-García et al., 2001) or gene manipulation are interesting ways to increase rhizobial competitiveness.

Adhesion of bacteria to diverse surfaces leads to development of biofilms, which are complex structures, where bacteria differentiate from the single-cell planktonic state (Stoodley et al., 2002). Therefore, many determinants of bacterial adhesion to plant roots also play a role in biofilm formation on inert surfaces (Danhorn & Fuqua, 2007). However, it is controversial whether biofilm formation is related to nodulation. Although biofilms may be formed on legume roots, the time required for the development of a mature biofilm is larger than the time required for root infection and nodule initiation. In addition, factors affecting both processes like soybean seed lectin or the basic core of lipochitooligosacaride Nod factors seem to be required in different manners for legume root infection or biofilm formation (Fujishige et al., 2008; Pérez-Giménez et al., 2009). Hence, biofilm formation and nodulation were regarded as alternative strategies for rhizobial survival in the soil rather than sequential steps of the symbiosis (Pérez-Giménez et al., 2009). In the soil, free-living rhizobia may tend to form biofilms on biotic or abiotic surfaces (Seneviratne & Jayasinghearachchi, 2003) and this may also explain their low motility at field capacity (see above).

3.2 Plant traits

The importance of competition for nodulation is also highlighted by the fact that plants exert a control on the number of nodules formed in the roots. The earliest nodules are often the most active in N_2 fixation. Therefore, the occupation of the earliest nodules by the inoculated strain is of prime importance to determine the global N_2 fixing activity of the plant. Control of nodulation involves a systemic signaling mechanism that was described thanks to the availability of hypernodulating mutants, which loss the autoinhibition of the formation of new nodules once sufficient nodules were formed. This autoregulation is systemic and the signals responsible for this pathway were not yet found. The mutations are related to the ability to nodulate in the presence of high concentrations of combined N compounds, and the insensivity to ethylene or light (Oka-Kira & Kawaguchi, 2006). In soybean, the process of autoregulation of nodulation involves the production of a cue signal in the nodulated root, called "Q", which travels to the shoot where it is perceived by a LRR RLK with a a serine/threonine kinase domain called *GmNARK*. As a response to the perception of "Q" in the leaves, a shoot-derived inhibitor (SDI) is produced and released to the roots, where it inhibits further nodulation (Ferguson et al., 2010). Current work is in progress to elucidate the chemical nature of "Q" and SDI. The "Q" signals may be CLAVATA3/ESR-related (CLE) peptides, higly modified by proline hydroxylation and glycosilation, and recent work identifyed three such CLE peptides in soybean, two of which may be related to nodulation inhibition by *B. japonicum* and the other, by nitrogen (Reid et al., 2011). In turn, the SDI seems a low molecular weight (< 1,000 Da) molecule, which is heat-stable and seems not RNA or protein (Lin et al., 2010).

The plant growth regulator ethylene is also an inhibitor of nodulation, although it is not clear whether ethylene takes part in the autoregulation response (Ding et al., 2009; Oka-Kira & Kawaguchi, 2006). As mentioned before, nodulation is an energy-consuming process, and the role of ethylene might be related to prevent this process if environmental conditions are not adequate (or alternative N-sources are available in the soil) in order to save photosynthates in stressful situations. Nevertheless, strains able to locally counteract the nodulation inhibition by ethylene seem more competitive for nodulation. Some strains of *B. elkanii* are able to produce an ethylene synthesis inhibitor called rhizobitoxine [2-amino-4-(2-amino-3-hydropropoxy)-transbut-3-enoic acid]. Rhizobitoxine-producing strains were more competitive for nodulation of *M. atropurpureum* (Siratro) than non-producing strains (Okazaki et al., 2003). Likewise, the expression of 1-cyclopropane-1-carboxylate (ACC) deaminase in free-living cells of *M. loti* enhanced their competitiveness to nodulate *Lotus japonicus* and *L. tenuis.* This enzyme degrades the ethylene precursor ACC and therefore, lowering ACC levels in the rhizosphere or during initial infection might have been avoided inhibition of nodulation in a local manner (Conforte et al., 2010). However, this strategy may be not useful in the soybean symbiosis, since nodulation of this species seems not sensitive to ethylene (Schmidt et al., 1999).

A very interesting phenomenon that seems not related with autoregulation is restriction of nodulation of certain soybean genotypes by some serogroups of *B. japonicum* and *S. fredii* (Cregan & Keyser, 1986; 1988). Despite the initial interpretation of this phenomenon, discrimination of serogroups is not absolute, since restriction is not necessarily involving all the members of a given serogroup (Scott et al., 1995). Genetic studies lead to the identification of plant loci controlling restriction of nodulation, in particular, the dominant genes Rj2, Rj3, and Rj4, which restrict nodulation by strains in the USDA 122 and c1 serogroups, and the recessive genes Rj1, Rj5, and Rj6 that restrict nodulation by virtually all bradyrhizobia strains (Pracht et al., 1993; Weiser et al., 1990; Williams & Lynch, 1954). Therefore, the use of such genes in cultivated soybeans may help in selecting a set of soybean-nodulating rhizobial strains for nodulation in detriment of an undesired soil rhizobial population (Devine & Kuykendall, 1996). However, discrimination in host-controlled restriction of nodulation seems not so specific and more studies are required before this strategy can be transferred to farmers. Some advances were done in the molecular characterization of the soybean genes involved in host restriction of nodulation. The Rj1 gene was mapped to the soybean molecular linkage group D1b of chromosome 2 and was identified as the Nod factor receptor *Gm NFR1a.* Its recessive allele has a 1-bp deletion that introduces a premature stop codon eliminating the protein kinase domain of *Gm NFR1a* (Lee et al., 2011). The dominant Rj2 gene as well as its allele Rfg1, which restricts *S. fredii* serogroups related with USDA 257, were mapped to the soybean linkage group J in chromosome 16 and encodes Toll interleukin receptor/nucleotide-binding site/leucine-rich repeat (TIR-NBS-LRR) class of plant resistance (R) proteins (Yang et al., 2010). Therefore, this genetic system of host restriction of nodulation has similarity with the gene-for-gene resistance system against plant pathogens. Counterpart genes in the bacterial side were also identified. The PI (plant introduction) 417566 genotype restricts the USDA 110 serogroup (Lohrke et al., 1995). Interestingly, host-restriction of nodulation requires high inoculum doses of *B. japonicum* USDA 110, while low-inocula are not restricted (Lohrke et al., 2000). It was found that USDA 110 mutants in *nodD2* are not restricted at high cell-densities, and are more competitive for nodulation than the non-restricted USDA 123 serogroup (Jitacksorn & Sadowsky, 2008). *nodD2* is part of the complex circuit of nodulation genes expression in *B. japonicum* (Loh & Stacey, 2003). This gene is activated by *nolA,* which in turn is activated by a special quorum signal of *B. japonicum* called bradyoxetin. Furthermore, *nodD2* is an inhibitor of *nodD1,* the activator of nodulation genes in *B. japonicum.* Thus, at high cell densities, *nodD2* indirectly responds to quorum sensing and inhibits the expression of the other *nod* genes. It still remains to be elucidated whether *nodD2* fits also to the gene-for-gene model.

Similarly as the case of soybean, in other legumes such as pea and *M. truncatula*, the plant cultivar also exerts some selection on the rhizobial genotypes that will have preference for nodulation, although a genetic system of host-restriction of nodulation was not described with the same detail as in soybean (Depret & Laguerre, 2008; Rangin et al., 2008). Of particular interest is the case of pea, where Depret & Laguerre (2008) observed that not only the plant genotype exerts a strain selection for nodulation, but also the set of strains that preferentially occupy the nodules changes across the different phenological stages of the plants. The authors employed three pea cultivars, two of which (Austin and Athos) share a common ancestor, and the other (Frisson) is more ancient. In addition, two hypernodulating mutants of Frisson were included. The *R. leguminosarum* bv *viceae* strains came from two different soils and were genotypically classified according to the neutral 16S intergenic space and the symbiotically functional *nodD* gene into 68 genotypes, 5 of which predominated in one of the soils and 6 in the other. In each soil, the cv. Austin and Athos tended to be nodulated by a similar rhizobial population, which had a different structure in the nodules from cv. Frisson. In addition, there was a difference in nodule population structure between cv. Frisson and its hypernodulating mutant P118, indicating that the single gene change in the latter was enough to induce a change in the nodule bacterial population. Moreover, the *nodD* genotypes were diverse in the nodules produced before the beginning of flowering, but tended to be dominated by a single genotype in all three cultivars in the nodules produced between the beginning of flowering and the beginning of seed filling. The authors attributed these differences at least in part to the differences in rhizosphere composition at the different phenological states, which might lead to differences in the rhizobial population structuring. In addition, the metabolic state and structure of the roots may also influence the plant preference for certain rhizobial genotypes.

4. Ecological aspects of the plant-rhizobia interaction

Rhizobia are world-wide distributed bacteria and therefore traits for adaptation to almost every environment where agriculture is carried out may be found. In addition, local populations of rhizobia are not restricted to nodulate the indigenous legume species. Horizontal gene transfer of rhizobial symbiotic genes was documented not only for those rhizobial species that carry this information in transmisible symbiotic plasmids (Torres-Tejerizo et al., 2011) but also for those species that carry this information in the bacterial chromosome (Gomes-Barcellos et al., 2007; Sullivan & Ronson, 1998), and therefore the ability to nodulate a newly introduced legume species is rapidly acquired by the local population. A mathematical model was developed to simulate the propagation of horizontally transferred symbiotic genes to a local non-symbiotic population and the prediction is that such genes can be fixed in the local population in a few generations (Provorov & Vorobyov, 2000). Nevertheless, the major richness in rhizobial genotypes for a given legume species is often found in the centers of origin of that species. Soybean is originated in Asia and therefore, it may be presumed that most of the soybean-nodulating rhizobia are originated in this same region. Therefore, searches for new soybean-nodulating strains with special adaptations were conducted there. A recent survey was performed in soils from four different regions in China (Thomas-Oates et al., 2003). Many fast-growing rhizobia were isolated and classified according to various physiological, biochemical and genetic characteristics, to build a catalogue of fast-growing soybean-nodulating strains with different adaptations that can be used according to local requirements.

Competitiveness of inoculant strains for nodulation of certain plant cultivars in special areas may benefit from such collections. Nevertheless, it is important to understand how the peculiarities of a region may influence competitiveness. It is not a simple task to predict the most limiting factors in a given environment. Sometimes these factors are climate and soil characteristics that can be readily noted, but in many instances there are environmental influences that are hard to identify. These influences may take place at the onset of root infection or during rhizosphere colonization. Rhizosphere is a complex and dynamic habitat (Hinsinger et al., 2005; Watt et al., 2006) that requires a particular approach for its analysis. New cell labeling and microscopy tools are expanding our knowledge on rhizosphere events, since now it is possible to follow a single living cell in real time in the rhizosphere. The knowledge that we are gaining thanks to these methodologies will bring new ideas and applications in the near future.

Although rhizosphere colonization is of prime importance in competition for nodulation, there are reports indicating that the relationship is not so straightforward. In two pea fields (named sites I and II) inoculated with *R. leguminosarum* bv *viceae* it was found that, although the rhizobial indigenous populations were similar, the inoculant resulted more competitive at site I but less competitive at site II, even when at this site the inoculum was applied at very high concentration; however, in laboratory tests of competition of the inoculant against the dominant strain from site II, both resulted equally competitive (Meade et al., 1985). Strains of *R. leguminosarum* bv *trifolii* that were the most abundant nodule occupiers in field-grown subclover were not the most competitive in laboratory experiments carried out in Leonard jars with perlite-vermiculite as substrate (Leung et al., 1994a). In particular, there were four isolates, classified according to their electrophoretic types (ET), which were the most abundant in the field crops. Two of them, named ET 2 and 3, were studied in more detail. Although the ET 3 isolate was in general more competitive than the others, the authors recovered some isolates that, although were rarely present in the field nodules, were more competitive than ET 3 in the laboratory. Regarding the isolate ET 2, although it was found in a large proportion of field nodules, it had a low competitiveness in laboratory. The authors attributed these behaviors to a series of factors, among them the obvious possibility that the environment exerted a decisive influence in the field, but also considered more subtle alternatives, such as a possible non-random distribution in the soil of the more competitive, yet rare isolates. Similarly, a dominant serotype of *R. leguminosarum* bv *trifolii* was found in field nodules of subclover, although there were 13 distinct serotypes in that site. However, the rhizosphere effect, i.e. the ratio between rhizosphere and non rhizosphere population densities increased much more among the rare serotypes than in the dominant one in spring, while this effect was similar in other seasons (Leung et al., 1994b). Moreover, in a study carried out by Laguerre et al. (2003) in France, a difference between fave bean and pea was observed for the relationship of dominance in soil or nodules in the *R. leguminosarum* bv *viceae* population. While the success in nodule occupancy of rhizobial genotypes in fava bean was mainly determined by the rhizobial symbiotic genotype independently of the soil conditions, in pea there was a stronger influence of the rhizosphere colonization ability (linked to the genomic background but not necessarily to the symbiotic phenotype) on the competition for nodulation. These results underscore the complexity of the environmental effect on genotypic expression, which is differentially exerted on a given rhizobial population according to the plant genotype, the season, and the soil structure.

It has been argued that rhizobia from soil or rhizosphere are more competitive than rhizobia from rich broths due to a physiological state induced by nutrients limitation, which predispose the rhizobia to seek for plant root infection. Such physiological conditioning was observed for processes that may be related with rhizosphere colonization and nodulation, such as motility and roots infectivity (Lodeiro et al., 2000). However, nutrients limitation achieved by suspending in poor media rhizobia previously grown in rich broths is different than nutrients limitation achieved by rhizobia themselves at stationary growth phase, when nutrients from the broth are exhausted. In the last case, infectivity is diminished (López-García et al., 2001). Several reports indicated that strains isolated from highly competitive soil rhizobial populations frequently lack their superior competitiveness in laboratory tests or when reintroduced into soil containing an established population. In turn, in several instances an apparently poor competitor for nodulation in laboratory experiments resulted very competitive in soil (Lochner et al., 1989 and references therein). Hence, López-García et al. (2002) tested directly this concept on the basis of competition between two nearly isogenic *B. japonicum* strains (Fig. 1). They established a population of strain LP 3004 (USDA 110 Sm-resistant) in vermiculite pots by inoculating the pots with the bacteria in N-free plant nutrient solution and leaving the pots without plants for at least 1 month in the greenhouse. After an initial period of cell divisions, the rhizobial population stabilized in around 10^6 rhizobia ml^{-1} without decaying along this period. After this incubation, when the nutrient-limited physiological state seemed to be acquired by this rhizobial population,

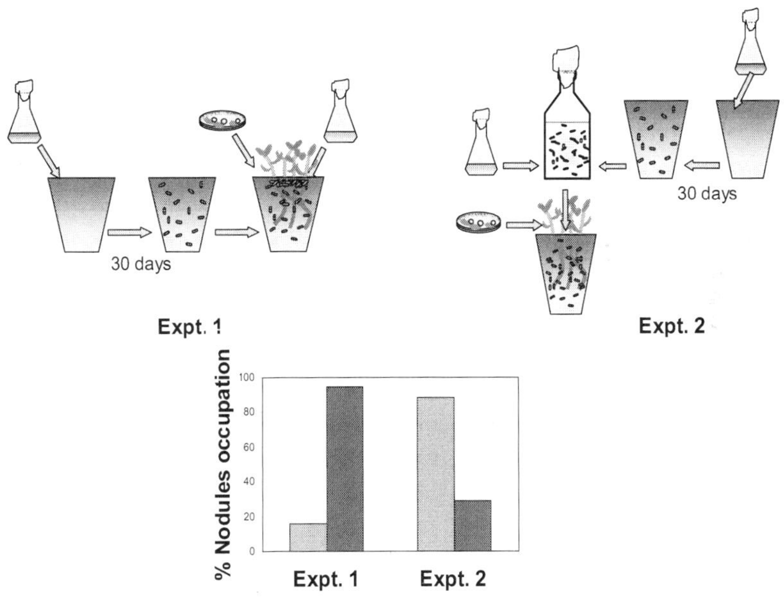

Fig. 1. Demonstration that the rhizobial position in the rooting substrate is determinant for competition for nodulation. Two nearly isogenic strains (indicated in red and green) were inoculated either on the seeds and the rooting substrate or mixed homogeneously before added to the pots. As result, the strain inoculated on the seeds occupied few nodules, despite being intrinsically more competitive. For futher details see text.

the pots were divided in two groups. To one of them soybean plantlets were planted and inoculated on the seedlings with around 10^8 rhizobia $plant^{-1}$ of the strain LP 3001 (USDA 110 Sp-resistant) freshly obtained from a rich broth. From the other half of the pots the rhizobia were removed, suspended in N-free plant nutrient solution, and mixed there with another aliquot of rich broth-grown LP 3001 in 1:1 relationship (10^6 rhizobia ml^{-1} of each strain). This mixture was added to fresh vermiculite pots, and soybean plantlets were planted. After 20 days in the greenhouse the nodules were recovered and their contents were identified according to their antibiotic resistances. As a result, it was observed that in the pots where the rich broth rhizobia were inoculated on the seedlings, these rhizobia formed around 20 % of the nodules, but in the pots where both strains were homogeneously mixed before pouring them into the vermiculite, the rhizobia from the rich broth formed more than 80 % of the nodules. This experiment demonstrated that the superior competitiveness of the established population is not caused by a nutrient-limited physiological state, but simply by the better position that they had in the vermiculite with respect to the growing roots: the authors also observed that at field capacity the movement of the rhizobia in the vermiculite is very scarce, which was recently corroborated with non-flagellated mutants (Althabegoiti et al., 2011), and supported the idea that the initial cells that colonize the rhizosphere do not arrive swimming but are "scavenged" by the displacement of the growing roots.
Life in the soil is nevertheless very important for the rhizobia. They may persist even in the absence of legume crops, and it was observed that *Bradyrhizobium sp. (Lotus)* retains its nodulation, competition, and N_2 fixing characteristics even after 10 years in the soil (Lochner et al., 1989). If we consider a soybean crop season, and take into account that nodules start to senesce at grain filling, we can estimate that the rhizobia are into the nodules for less than 40 % of a year; the other 60 % of the time they have to survive in the free-living state in the soil. During this period the rhizobia have to face diverse threats, including UV irradiation, temperature changes, predation, drought, flooding, etc. Since these microorganisms are unable to sporulate, a preferred state of endurance is the biofilm (Danhorn & Fuqua, 2007). To this end, the cell surface components play a major role, but flagella are lost and biofilm rhizobia are not motile, which may in part explain the lack of effects of motility on competition for nodulation in non-flooded soils. In addition, plant lectins may help in developing the biofilms. It was observed that soybean lectin enhances the biofilm formation by *B. japonicum* in a way that is dependent on the presence of the receptor EPS molecule in the bacteria (Pérez-Giménez et al., 2009). Since this process seems not related with plant infection, it was argued that lectin-assisted biofilm formation may favor *B. japonicum* biofilms in the vicinity of decaying soybean roots or even on dead roots, where soybean lectin may have been released. This is supported by the observation that soybean lectin is remarkable stable, being unaltered even after a week of incubation at 70 $^{\circ}C$. Thus, this enhancement of biofilms formation where soybeans were recently cultivated may keep a localized high rhizobial population for the next nodulation cycle, thus explaining the heterogeneities of rhizobial distribution in the soil previously postulated by Leung et al. (1994).

5. Conclusion

Competition for nodulation still remains a very complex and largely unknown phenomenon, yet a very important issue for N_2 fixation technology. Nevertheless, understanding of this phenomenon has advanced in the last years, and several measures to improve competitiveness of rhizobial inoculated strains may be proposed. Among that are the manipulation of host-controlled restriction of nodulation, the genetic manipulation of the plant and bacterial partners, selection of superior strains, improvement of inoculant formulations by manipulating the culture media and the physiological and metabolic state of the bacteria, and the improvement of inoculant application technologies, particularly with in-furrow inoculation. These methods, as well as the new developments that are in progress, are necessary for the sustainable agriculture of the future.

6. Acknowledgment

The work of the authors is funded by Agencia Nacional de Promoción de la Investigación Científica y Tecnológica (ANPCyT), Argentina. JPG and JIQ are fellows of Consejo Nacional de Investigaciones Científicas y Técnicas (CONICET), Argentina. ARL is member of the scientific career of CONICET.

7. References

Althabegoiti, M.J., López-García, S.L., Piccinetti, C., Mongiardini, E.J., Pérez-Giménez, J., Quelas, J.I., Perticari, A. & Lodeiro, A.R. (2008) Strain selection for improvement of *Bradyrhizobium japonicum* competitiveness for nodulation of soybean. *FEMS Microbiology Letters* Vol. 282 pp. 115-123.

Althabegoiti M.J., Covelli, J.M., Pérez-Giménez, J., Quelas, J.I., Mongiardini, E.J., López, M.F., López-García, S.L. & Lodeiro, A.R. (2011) Analysis of the role of the two flagella of *Bradyrhizobium japonicum* in competition for nodulation of soybean. *FEMS Microbiology Letters*. In press.

Amarger, N. & Lobreau, J.P. (1982) Quantitative study of nodulation competitiveness in *Rhizobium* strains. *Applied and Environmental Microbiology* Vol. 44 pp. 583-588.

Ames, P. & Bergman, K. (1981) Competitive advantage provided by bacterial motility in the formation of nodules by *Rhizobium meliloti*. *Journal of Bacteriology* Vol. 148 pp. 728-729.

Ampomah, O.Y., Beck-Jensen, J. & Bhuvaneswari, T.V. (2008) Lack of trehalose catabolism in *Sinorhizobium* species increases their nodulation competitiveness on certain host genotypes. *New Phytologist* Vol. 179 pp. 495–504.

Ausmees, N., Jacobsson, K. & Lindberg, M. (2001) A unipolarly located, cell-surfaceassociated agglutinin RapA belongs to a family of *Rhizobium*-adhering proteins (Rap) in *Rhizobium leguminosarum* bv. *trifolii*. *Microbiology* Vol. 147 pp. 549–559.

Bahar, M., De Majnik, J., Wexler, M., Fry, J., Poole, P. S. & Murphy, P. J (1998) A model for the catabolism of rhizopine in *Rhizobium leguminosarum* involves a ferredoxin oxygenase complex and the inositol degradative pathway. *Molecular Plant-Microbe Interactions* Vol. 11 pp. 1057-1068.

Bahlawane, C., McIntosh, M., Krol, E. & Becker, A. (2008) *Sinorhizobium meliloti* regulator MucR couples exopolysaccharide synthesis and motility. *Molecular Plant-Microbe Interactions* Vol. 21 pp. 1498-1509.

Barbour, W.M., Hattermann, D.R. & Stacey, G. (1991) Chemotaxis of *Bradyrhizobium japonicum* to soybean exudates. *Applied and Environmental Microbiology* Vol. 57 pp. 2635-2639.

Beattie, G.A. & Handelsman, J. (1993) Evaluation of a strategy for identifying nodulation competitiveness genes in *Rhizobium leguminosavum* biovar *phaseoli*. *Journal of General Microbiology* Vol. 139 pp. 529-538.

Bhagwat, A.A., Tully, R.E. & Keister, D.L. (1991) Isolation and characterization of a competition-defective *Bradyrhlizobium japonicum* mutant. *Applied and Environmental Microbiology* Vol. 57 pp. 3496-3501.

Bhuvaneswari, T.V., Pueppke, S.G. & Bauer, W.D. (1977) Role of lectins in plant microorganism interactions. I. Binding of soybean lectin to rhizobia. *Plant Physiology*. Vol. 60 pp. 486-491.

Bhuvaneswari, T.V., Turgeon, B.G. & Bauer, W.D. (1980) Early events in the infection of soybean (*Glycine max* L. Merr) by *Rhizobium japonicum*. I. Localization of infectible root cells. *Plant Physiology* Vol. 66 pp. 1027-1031.

Bittinger, M.A Milner, J.L., Saville, B.J. & Handelsman, J. (1997) *rosR*, a determinant of nodulation competitiveness in *Rhizobium etli*. *Molecular Plant-Microbe Interactions*. Vol. 10, pp. 180–186.

Bogino, P., Banchio, E., Bonfiglio, C. & Giordano, W. (2008) Competitiveness of a *Bradyrhizobium* sp. strain in soils containing indigenous rhizobia. *Current Microbiology* Vol. 56 pp. 66–72

Braeken, K., Daniels, R., Vos, K., Fauvart, M., Bachaspatimayum, D., Vanderleyden, J. & Michiels, J. (2007) Genetic determinants of swarming in *Rhizobium etli*. *Microbial Ecology* DOI: 10.1007/s00248-007-9250-1.

Brencic, A. & Winans, S.C. (2005) Detection of and response to signals involved in host-microbe interactions by plant-associated bacteria. *Microbiology and Molecular Biology Reviews* Vol. 69 pp. 155-194.

Caetano-Anollés, G., Crist-Estes D.K. & Bauer, W.D. (1988a) Chemotaxis of *Rhizobium meliloti* to the plant flavone luteolin requires functional nodulation genes. *Journal of Bacteriology* Vol. 170 pp. 3164-3169.

Caetano-Anollés, G., Wall, L.G., De Micheli, A.T., Macchi, E M, Bauer, W.D & Favelukes, G. (1988b) Role of motility and chemotaxis in efficiency of nodulation by *Rhizobium meliloti*. *Plant Physiology* Vol. 86 pp. 1228-1235.

Castro, S., Carrera, I. & Martínez-Drets G. (2000) Methods to evaluate nodulation competitiveness between *Sinorhizobium meliloti* strains using melanin production as a marker. *Journal of Microbiological Methods* Vol. 41 pp. 173–177.

Chuiko, N.V., Antonyuk T.S. & Kurdish I.K. (2002) The chemotactic response of *Bradyrhizobium japonicum* to various organic componunds. *Microbiology (Russia)* Vol. 71 pp. 391-396.

Cole, M.A. & Elkan, G.H. (1979) Multiple antibiotic resistance in *Rhizobium japonicum*. *Applied and Environmental Microbiology* Vol. 37 pp. 867-870.

Conforte, V.P., Echeverría, M., Sánchez, C., Ugalde, R.A., Menéndez, A.B. & Lepek, V.C. (2010) Engineered ACC deaminase-expressing free-living cells of *Mesorhizobium loti* show increased nodulation efficiency and competitiveness on *Lotus* spp. *Journal of General and Applied Microbiology* Vol. 56 pp. 331-338.

Cregan, P.B. & Keyser, H.H. (1986) Host restriction of nodulation by *Bradyrhizobium japonicum* strain USDA 123 in soybean. *Crop Science* Vol. 26 pp. 911-916.

Cregan, P.B. & Keyser, H.H. (1988) Influence of *Glycine* spp. on competitiveness of *Bradyrhizobium japonicum* and *Rhizobium fredii*. *Applied and Environmental Microbiology* Vol. 54 pp. 803-808.

Cytryn, E.J., Sangurdekar, D.P., Streeter, J.G., Franck, W.L., Chang, W-S., Stacey, G., Emerich, D.W., Joshi, T., Xu, D. & Sadowsky, M.J. (2007) Transcriptional and physiological responses of *Bradyrhizobium japonicum* to desiccation-induced stress. *Journal of Bacteriology*, Vol. 189 pp. 6751–6762.

Danhorn, T. & Fuqua, C. (2007) Biofilm formation by plant-associated bacteria. *Annual Review of Microbiology* Vol. 61 pp. 401-22.

Daniels, R., Reynaert, S., Hoekstra, H., Verreth, C., Janssens, J., Braeken, K., Fauvart, M., Beullens, S., Heusdens, C., Lambrichts, I., De Vos, D.E., Vanderleyden, J., Vermant, J. & Michiels, J. (2006) Quorum signal molecules as biosurfactants affecting swarming in *Rhizobium etli*. *Proceedings of the National Academy of Sciences USA* Vol. 103 pp. 14965-14970.

Depret, G. & Laguerre, G. (2008) Plant phenology and genetic variability in root and nodule development strongly influence genetic structuring of *Rhizobium leguminosarum* biovar *viciae* populations nodulating pea. *New Phytologist* Vol. 179 pp. 224–235.

Devine, T.E. & Kuykendall, L.D. (1996) Host genetic control of symbiosis in soybean (*Glycine max* L.). *Plant and Soil* Vol. 186 pp. 173–187.

Ding, Y. & Oldroyd, G.E.D. (2009) Positioning the nodule, the hormone dictum. *Plant Signaling & Behavior* Vol. 4 pp. 89-93.

Domínguez-Ferreras, A., Soto, M.J., Pérez-Arnedo, R., Olivares, J. & Sanjuán J. (2009) Importance of trehalose biosynthesis for *Sinorhizobium meliloti* osmotolerance and nodulation of alfalfa roots. *Journal of Bacteriology* Vol. 191 pp. 7490–7499.

Elliott, G.N., Chou, J-H., Chen, W-M., Bloemberg, G.V., Bontemps, C., Martínez-Romero, E., Velázquez, E., J. Young, P.W., Sprent, J.I. & James, E.K. (2009) *Burkholderia* spp. are the most competitive symbionts of *Mimosa*, particularly under N-limited conditions. *Environmental Microbiology* Vol. 11 pp. 762–778.

Ferguson, B.J., Indrasumunar, A., Hayashi, S., Lin, M-H., Lin, Y-H., Reid, D.E. & Gresshoff, P.M. (2010) Molecular analysis of legume nodule development and autoregulation. *Journal of Integrative Plant Biology* Vol. 52 pp. 61–76.

Fry, J., Wood, M. & Poole, P.S. (2001) Investigation of *myo*-inositol catabolism in *Rhizobium leguminosarum* bv. *viciae* and its effect on nodulation competitiveness. *Molecular Plant-Microbe Interactions* Vol. 14 pp. 1016–1025.

Fujishige, N.A., Kapadia, N.N., De Hoff, P.L. & Hirsch A.M. (2006) Investigations of *Rhizobium* biofilm formation. *FEMS Microbiology Ecology* Vol. 56 pp. 195-206.

Fujishige, N.A., Lum, M.R., De Hoff, P.L., Whitelegge, J.P., Faull, K.F. & Hirsch, A.M. (2008) *Rhizobium* common *nod* genes are required for biofilm formation. *Molecular Microbiology* Vol. 67 pp. 504–515.

Galbraith, M.P., Feng, S. F., Borneman, J., Triplett, E. W., De Bruijn, F. J. & Rossbach, S. (1998) A functional *myo*-inositol catabolism pathway is essential for rhizopine utilization by *Sinorhizobium meliloti*. *Microbiology (U.K.)* Vol. 144 pp. 2915-2924.

Gaworzewska, E.T. & Carlile, M.J. (1982) Positive chemotaxis of *Rhizobium leguminosarum* and other bacteria towards root exudates from legumes and other plants. *Journal of General Microbiology* Vol. 128 pp. 1179-1188.

Gomes-Barcellos, F., Menna, P., da Silva Batista J.S. & Hungria, M. (2007) Evidence of horizontal transfer of symbiotic genes from a *Bradyrhizobium japonicum* inoculant strain to indigenous diazotrophs *Sinorhizobium* (*Ensifer*) *fredii* and *Bradyrhizobium elkanii* in a Brazilian savannah soil. *Applied and Environmental Microbiology* Vol. 73 pp. 2635–2643.

González, J.E. & Marketon, M.M. (2003) Quorum sensing in nitrogen-fixing rhizobia. *Microbiology and Molecular Biology Reviews* Vol. 67 pp. 574-592.

Graham, P.H. (1992) Stress tolerance in *Rhizobium* and *Bradyrhizobium*, and nodulation under adverse soil conditions *Canadian Journal of Microbiology* Vol. 38 pp. 475-484.

Gulash, M., Ames, P., Larosiliere, R.C. & Bergman, K. (1984) Rhizobia are attracted to localized sites on legume roots. *Applied and Environmental Microbiology* Vol. 48 pp. 149-152.

Halverson, L.J. & Stacey, G. (1986) Effect of lectin on nodulation by wild-type *Bradyrhizobium japonicum* and a nodulation–defective mutant. *Applied and Environmental Microbiology* Vol. 51 pp. 753-760.

Hinsinger, P, Gobran, G.R., Gregory, P.J. & Wenzel, W.W. (2005) Rhizosphere geometry and heterogeneity arising from rootmediated physical and chemical processes. *New Phytologist* Vol. 168 pp. 293–303.

Ho, S-C., Wang, J.L. & Schindler, M. (1990) Carbohydrate binding activities of *Bradyrhizobium japonicum*. I. Saccharide-specifc inhibition of homotypic and heterotypic adhesion. *Journal of Cellular Biology* Vol. 111 pp. 1631-1638.

Ho, S-C., Wang, J.L., Schindler, M. & Loh, J.T. (1994) Carbohydrate binding activities of *Bradyrhizobium japonicum*. III. Lectin expression, bacterial binding, and nodulation efficiency *Plant Journal*. Vol. 5 pp. 873-884.

Horiuchi, J., Prithiviraj, B., Bais, H.P., Kimball, B.A. & Vivanco, J.M. (2005) Soil nematodes mediate positive interactions between legume plants and rhizobium bacteria. *Planta* Vol. 222 pp. 848-857.

Hunter, W.J. & Fahring, C.J. (1980) Movement by *Rhizobium*, and nodulation of legumes. *Soil Biology and Biochemistry* Vol. 12 pp. 537-542.

Ito, Y., Toyota, T., Kaneko, K., & Yomo, T. (2009) How selection affects phenotypic fluctuation. *Molecular Systems Biology* Vol. 5; Article number 264.

Janczarek, M., Jaroszuk-Scisel, J. & Skorupska, A. (2009) Multiple copies of *rosR* and *pssA* genes enhance exopolysaccharide production, symbiotic competitiveness and clover nodulation in *Rhizobium leguminosarum* bv. *trifolii*. *Antonie van Leeuwenhoek* Vol. 96 pp. 471–486.

Jiang, G., Krishnan, A.H., Kim, Y-W., Wacek, T.J., & Krishnan1 H.B. (2001) Functional *myo*-inositol dehydrogenase gene is required for efficient nitrogen fixation and competitiveness of *Sinorhizobium fredii* USDA191 to nodulate soybean (*Glycine max* [L.] Merr.) *Journal of Bacteriology* Vol. 183 pp. 2595–2604.

Jitacksorn, S. & Sadowsky, M.J. (2008) Nodulation gene regulation and quorum sensing control density-dependent suppression and restriction of nodulation in the *Bradyrhizobium japonicum*-soybean symbiosis *Applied and Environmental Microbiology* Vol. 74 pp. 3749–3756.

Laguerre, G., Louvrier, P., Allard, M-R. & Amarger, N. (2003) Compatibility of rhizobial genotypes within natural populations of *Rhizobium leguminosarum* biovar *viciae* for nodulation of host legumes. *Applied and Environmental Microbiology* Vol. 69 pp. 2276–2283.

Laus, M.C., Logman, T.J., Lamers, G.E., Van Brussel, A.A.N., Carlson R.W. & Kijne, J.W. (2006) A novel polar surface polysaccharide from *Rhizobium leguminosarum* binds host plant lectin. *Molecular Microbiology* Vol. 59 pp. 1704-1713.

Lee, W.K., Jeòng, N., Indrasumunar, A., Gresshoff, P.M. & Jeong, S-C. (2011) *Glycine max* non-nodulation locus *rj1*: a recombinogenic region encompassing a SNP in a lysine motif receptor-like kinase (*GmNFR1a*). *Theoretical and Applied Genetetics* Vol. 122 pp. 875–884.

Leung, K., Wanjage, F.N. & Bottomley, P.J. (1994a) Symbiotic characteristics of *Rhizobium leguminosarum* bv. *trifolii* isolates which represent major and minor nodule-occupying chromosomal types of field-grown subclover (*Trifolium subterraneum* L.). *Applied and Environmental Microbiology* Vol. 60 pp. 427-433.

Leung, K., Yap, K., Dashti, N. & Boytomley, P.J. (1994b) Serological and ecological characteristics of a nodule-dominant serotype from an indigenous soil population of *Rhizobium leguminosarum* bv. *trifolii*. *Applied and Environmental Microbiology* Vol. 60 pp. 408-415.

Lin Y.H., Ferguson, B.J., Kereszt, A. & Gresshoff, P.M. (2010) Suppression of hypernodulation in soybean by a leaf-extracted, NARK and Nod factor-dependent low molecular mass fraction. *New Phytologist* Vol. 185 pp. 1074-1086

Lison, L. (1968) *Statistique appliquée a la biologie expérimentale*. Gauthier-Villars, Paris, France.

Liu, R., Tran, V.M. & Schmidt, E.L. (1989) Nodulating competitiveness of a nonmotile Tn7 mutant of *Bradyrhizobium japonicum* in nonsterile soil. *Applied and Environmental Microbiology* Vol. 55 pp. 1895-1900.

Lochner, H.H., Strijdom, B.J. & Law, I.J. (1989) Unaltered nodulation competitiveness of a strain of *Bradyrhizobium* sp. (Lotus) after a decade in soil. *Applied and Environmental Microbiology* Vol. 55 pp. 3000-3008.

Lodeiro, A.R. & Favelukes, G. (1999) Early interactions of *Bradyrhizobium japonicum* and soybean roots: specificity in the process of adsorption. *Soil Biology and Biochemistry* Vol. 31 pp. 1405-1411.

Lodeiro, A.R., López-García, S.L., Vázquez, T.E.E. & Favelukes, G. (2000) Stimulation of adhesivenes, infectivity, and competitivenesss for nodulation of *Bradyrhizobium japonicum* by its pretreatment with soybean seed lectin. *FEMS Microbiology Letters* Vol. 188 pp. 177-184.

Loh, J. & Stacey, G. (2003) Nodulation gene regulation in *Bradyrhizobium japonicum*: a unique integration of global regulatory circuits. *Applied and Environmental Microbiology* Vol. 169 pp. 10–17.

Loh, J.T., Ho, S-C., De Feijtert, A.W., Wang, J.L. & Schindler, M. (1993) Carbohydrate binding activities of *Bradyrhizobium japonicum*: Unipolar localization of the lectin BJ38 on the bacterial cell surface *Proceedings of the National Academy of Sciences USA* Vol. 90, pp. 3033-3037.

Lohrke, S.M., Orf, J.H. Martínez-Romero, E. & Sadowsky, M.J (1995) Host-controlled restriction of nodulation by *Bradyrhizobium japonicum* strains in serogroup 110. *Applied Environmental Microbiology* Vol. 61 pp. 2378–2383.

Lohrke, S.M., Madrzak, C.J., Hur, H.G., Judd, A.K., Orf, J.H & Sadowsky, M.J. (2000) Inoculum density-dependent restriction of nodulation in the soybean-*Bradyrhizobium japonicum* symbiosis. *Symbiosis* Vol. 29 pp. 59–70.

López-García S.L., Vázquez, T.E.E., Favelukes, G. & Lodeiro, A.R. (2001) Improved soybean root association of N-starved *Bradyrhizobium japonicum. Journal of Bacteriology* Vol. 183 pp. 7241-7252.

López-García, S.L., Vázquez, T.E.E., Favelukes, G. & Lodeiro, A.R. (2002) Rhizobial position as a main determinant in the problem of competition for nodulation in soybean. *Environmental Microbiology* Vol. 4 pp. 216-224.

López-García, S.L., Perticari, A., Piccinetti, C., Ventimiglia, L., Arias, N., De Battista, J.J., Althabegoiti, M.J., Mongiardini, E.J., Pérez-Giménez, J., Quelas, J.I. & Lodeiro, A.R. (2009) In-furrow inoculation and selection for higher motility enhances the effi cacy of *Bradyrhizobium japonicum* nodulation. *Agronomy Journal* Vol. 101 pp. 357–363.

Lowther, W.L. & Patrick, H.N. (1993) Spread of *Rhizobium* and *Bradyrhizobium* in soil. *Soil Biology and Biochemistry* Vol. 25 pp. 607-612.

Madsen, E.L. & Alexander, M. (1982) Transport of *Rhizobium* and *Pseudomonas* through soil. *Soil Science Society of America Journal* Vol. 46 pp. 557-560.

Marks, M.E., Castro-Rojas, C.M., Teiling, C., Du, L., Kapatral, V., Walunas, T. L. & Crosson, S. (2010) The genetic basis of laboratory adaptation in *Caulobacter crescentus Journal of Bacteriology* Vol. 192 pp. 3678-3688.

McIntyre, H. J., Davies, H., Hore, D. T., Miller, S. H., Dufour, J.-P. & Ronson, C. W. (2007) Trehalose biosynthesis in *Rhizobium leguminosarum* bv. *trifolii* and its role in desiccation tolerance. *Applied and Environmental Microbiology* Vol. 73 pp. 3984– 3992.

Meade, J., Higgins, P. & O'Gara, F. (1985) Studies on the inoculation and competitiveness of a *Rhizobium leguminosarum* strain in soils containing indigenous rhizobia. *Applied and Environmental Microbiology* Vol. 49 pp. 899-903

Mellor, H.Y., Glenn, A.R., Arwas, R. & Dilworth, M.J. (1987) Symbiotic and competitive properties of motility mutants of *Rhizobimn trifoii* TAl. *Archives of Microbiology* Vol. 148 pp. 34-39.

Mongiardini, E.J., Ausmees, N., Pérez-Giménez, J., Althabegoiti, M.J., Quelas, J.I., López-García, S.L. & Lodeiro, A.R. (2008) The Rhizobial Adhesion Protein RapA1 is involved in adsorption of rhizobia to plant roots but not in nodulation. *FEMS Microbiology Ecology* Vol. 65 pp. 279–288.

Mongiardini, E.J., Pérez-Giménez, J., Althabegoiti, M.J., Covelli, J., Quelas, J.I., López-García, S.L. & Aníbal R. Lodeiro (2009) Overproduction of the rhizobial adhesin RapA1 increases competitiveness for nodulation. *Soil Biology & Biochemistry* Vol. 41 pp. 2017–2020.

Murphy, P. J., Heycke, N., Banfalvi, Z., Tate, M. E., Debruijn, F., Kondorosi, A., Tempe, J. & Schell, J. (1987) Genes for the catabolism and synthesis of an opine-like compound in *Rhizobium meliloti* are closely linked and on the Sym plasmid. *Proceedings of the National Academy of Sciences USA* Vol. 84 pp. 493-497.

Nogales, J., Domínguez-Ferreras, A., Amaya-Gómez, C.V., van Dillewijn, P., Cuéllar, V., Sanjuán, J., Olivares, J. & Soto, M.J. (2010) Transcriptome profiling of a *Sinorhizobium meliloti fadD* mutant reveals the role of rhizobactin 1021 biosynthesis and regulation genes in the control of swarming. *BMC Genomics* Vol. 11: 157.

Oka-Kira, E. & Kawaguchi, M. (2006) Long-distance signaling to control root nodule number. *Current Opinion in Plant Biology* Vol. 9 pp. 496–502.

Okazaki, S., Yuhashi, K-I. & Minamisawa, K. (2003) Quantitative and time-course evaluation of nodulation competitiveness of rhizobitoxine-producing *Bradyrhizobium elkanii*. *FEMS Microbiology Ecology* Vol. 45 pp. 155-160.

Pandya, S., Iyer, P., Gaitonde, V., Parekh, T. & Desai, A. (1999) Chemotaxis of *Rhizobium* sp.S2 towards *Cajanus cajan* root exudate and its major components. *Current Microbiology* Vol. 4 pp. 205-209.

Parco, S.Z., Dilworth, M.J. & Glenn, A.R. (1994) Motility and the distribution of introduced root nodule bacteria on the root system of legumes. *Soil Biology and Biochemistry* Vol. 26 pp. 297-300.

Parniske, M., Kosch, K., Werner, D. & Muller, P. (1993) ExoB mutants of *Bradyrhizobium japonicum* with reduced competitiveness for nodulation of *Glycine max*. *Molecular Plant-Microbe Interactions* Vol 6: pp. 99-106.

Patriarca, E.J., Tatè, R., Ferraioli, S. & Iaccarino, M. (2004) Organogenesis of legume root nodules. *International Reviews of Cytology* Vol. 234 pp. 201-262.

Pérez-Giménez, J., Althabegoiti, M.J., Covelli, J., Mongiardini, E.J., Quelas, J.I., López-García S.L. & Lodeiro, A.R. (2009) Soybean lectin enhances biofilm formation by *Bradyrhizobium japonicum* in the absence of plants. *International Journal of Microbiology* Vol. 2009:719367.

Pistorio, M., Balagué, L.J., Del Papa, M.F., Pich-Otero, A., Lodeiro, A., Hozbor, D.F., Lagares A. (2002) Construction of a *Sinorhizobium meliloti* strain carrying a stable and non-transmissible chromosomal single copy of the green fuorescent protein GFP-P64L/S65T. *FEMS Microbiology Letters* Vol. 214 pp. 165-170.

Pobigaylo, N., Szymczak, S., Nattkemper, T.W. & Becker, A. (2008) Identification of genes relevant to symbiosis and competitiveness in *Sinorhizobium meliloti* using signature-tagged mutants. *Molecular Plant-Microbe Interactions* Vol. 21 pp. 219–231.

Pracht, J.E., Nickell, C.D. & Harper, J.E. (1993) Genes controlling nodulation in soybean: Rj5 and Rj6. *Crop Science* Vol. 33 pp. 711–713.

Provorov, N.A. & Vorobyov, N.I. (2000) Population genetics of rhizobia: Construction and analysis of an "infection and release" model. *Journal of Theoretical Biology* Vol. 205 pp. 105-119.

Quelas, J.I., López-García, S.L., Casabuono, A., Althabegoiti, M.J., Mongiardini, E.J., Pérez-Giménez, J., Couto, A. & Lodeiro, A.R. (2006) Effects of N-starvation and C-source on exopolysaccharide production and composition in *Bradyrhizobium japonicum*. *Archives of Microbiology* Vol. 186 pp. 119-128.

Quelas, J.I., Mongiardini, E.J., Casabuono, A., López-García, S.L., Althabegoiti, M.J., Covelli, J.M., Pérez-Giménez, J., Couto, A. & Lodeiro, A.R. (2010) Lack of galactose or galacturonic acid in *Bradyrhizobium japonicum* USDA 110 exopolysaccharide leads to different symbiotic responses in soybean. *Molecular Plant-Microbe Interactions* Vol. 23 pp. 1592–1604.

Ramey, B.E., Koutsoudis, M., von Bodman, S.B. & Fuqua, C. (2004) Biofilm formation in plant–microbe associations. *Current Opinion in Microbiology* Vol. 7 pp. 602–609.

Rangin, C., Brunel, B., Cleyet-Marel, J-C., Perrineau, M.M. & Béna, G. (2008) Effects of *Medicago truncatula* genetic diversity, rhizobial competition, and strain effectiveness on the diversity of a natural *Sinorhizobium* species community. *Applied Environmental Microbiology* Vol. 74 pp. 5653–5661.

Reid D.E., Ferguson, B.J. & Gresshoff, P.M. (2011) Inoculation- and nitrate-induced CLE peptides of soybean control NARK-dependent nodule formation. *Molecular Plant-Microbe Interactions* In press. DOI: 10.1094/MPMI-09-10-0207.

Robleto, E.A., Kmiecik, K., Oplinger, E.S., Nienhuis, J. & Triplett1, E.W. (1998) Trifolitoxin production increases nodulation competitiveness of *Rhizobium etli* CE3 under agricultural conditions. *Applied and Environmental Microbiology* Vol. 64 pp. 2630–2633.

Rosenblueth, M., Hynes, M.F. & Martínez-Romero, E. (1998) *Rhizobium tropici teu* genes involved in specifc uptake of *Phaseolus vulgaris* bean-exudate compounds. *Molecular and General Genetics* Vol. 258: pp. 587-598.

Salon, C., Lepetit, M., Gamas, P., Jeudy, C., Moreau, S., Moreau, D., Voisin, A-S., Duc, G., Bourion, V. & Munier-Jolain, N (2009) Analysis and modeling of the integrative response of *Medicago truncatula* to nitrogen constraints. *Comptes Rrendus Biologies* Vol. 332 pp. 1022–1033

Schmidt, J.S., Harper, J.E., Hoffman, T.K. & Bent, A.F. (1999) Regulation of soybean nodulation independent of ethylene signaling. *Plant Physiology* Vol. 119 pp. 951–959.

Seneviratne, G. & Jayasinghearachchi, H.S. (2003) Mycelial colonization by bradyrhizobia and azorhizobia. *Journal of Biosciences* Vol. 28 pp. 243–247.

Smit, G., Swart, S., Lugtenberg, B.J.J. & Kijne, J.W. (1992) Molecular mechanisms of attachment of *Rhizobium* bacteria to plant roots. *Molecular Microbiology* Vol. 6 pp. 2897-2903.

Soto, M.J., Fernández-Pascual, M., Sanjuan, J. & Olivares, J. (2002) A *fadD* mutant of *Sinorhizobium meliloti* shows multicellular swarming migration and is impaired in nodulation efficiency on alfalfa roots. *Molecular Microbiology* Vol. 43 pp. 371-382.

Spriggs, A.C. & Dakora, F.D. (2009) Assessing the suitability of antibiotic resistance markers and the indirect ELISA technique for studying the competitive ability of selected *Cyclopia* Vent. rhizobia under glasshouse and field conditions in South Africa. *BMC Microbiology* Vol. 9: 142 doi:10.1186/1471-2180-9-142.

Stoodley, P., Sauer, K., Davies, D.G. & Costerton J.W. (2002) Biofilms as complex differentiated communities. *Annual Review of Microbiology* Vol. 56, pp. 187–209.

Streeter J.G. (2003) Effect of trehalose on survival of *Bradyrhizobium japonicum* during desiccation. *Journal of Applied Microbiology* Vol. 95 pp. 484–491.

Suárez, R., Wong, A., Ramírez, M., Barraza, A., Orozco, M.C., Cevallos, M. A., Lara, M., Hernández, G. & Iturriaga G. (2008) Improvement of drought tolerance and grain yield in common bean by overexpressing trehalose-6-phosphate synthase in rhizobia. *Molecular Plant-Microbe Interactions* Vol. 21 pp. 958–966.

Sullivan, J.T. & Ronson, C.W. (1998) Evolution of rhizobia by acquisition of a 500-Kb symbiosis island that integrates into a *phe-tRNA* gene. *Proceedings of the National Academy of Science USA* Vol. 95 pp. 5145–5149.

Tambalo, D.D., Yost, C.K. & Hynes, M.F. (2010) Characterization of swarming motility in *Rhizobium leguminosarum* bv. *viciae*. *FEMS Microbiology Letters* Vol. 307 pp. 165-174.

Thomas-Oates, J., Bereszczak, J., Edwards, E., Gill, A., Noreen, S., Zhou, J. C., Chen, M.Z., Miao, L.H., Xie, F.L., Yang J.K., Zhou, Q, Yang, S.S., Li, X.H., Wang, L., Spaink,

H.P., Schlaman, H.R.M., Harteveld, M., Díaz, C.L., van Brussel, A.A.N., Camacho, M. Rodríguez-Navarro, D.N., Santamaría, C., Temprano, F., Acebes, J.M., Bellogín, R.A., Buendía-Clavería, A.M., Cubo, M.T., Espuny, M.T., Gil, A.M., Gutiérrez, R., Hidalgo, A., López-Baena, F.J., Madinabeitia, N., Medina, C., Ollero, F.J., Vinardell, F.J. & Ruiz-Sainz J.E. (2003) A Catalogue of molecular, physiological and symbiotic properties of soybean-nodulating rhizobial strains from different soybean cropping areas of China. *Systematics and Applied Microbiology* Vol. 26 pp. 453–465.

Toro N. (1996) Nodulation competitiveness in the *Rhizobium*-legume symbiosis. *World Journal of Microbiology and Biotechnology* Vol. 12 pp. 157-162.

Torres Tejerizo, G., Giusti, M.A., Del Papa M.F., Lozano, M.J., Draghi, W.O., Jofre, E., Lagares, A. & Pistorio, M. (2011) Horizontal gene transfer, generalities and lessons from rhizobia, In: *Bacterial Populations: Basic and Applied Aspects of Their Structure and Evolution*. Anibal Lodeiro (Ed.) pp. 37-59. Transworld Research Network. Kerala, India.

Torres-Quesada, O., Oruezabal R.I., Peregrina, A., Jofré, E., Lloret, J., Rivilla, R., Toro, N. & Jiménez-Zurdo, J.I. (2010) The *Sinorhizobium meliloti* RNA chaperone Hfq influences central carbon metabolism and the symbiotic interaction with alfalfa. *BMC Microbiology* Vol. 10:71.

Triplett, E.W., Breil, B.T. & Splitter, G.A. (1994) Expression of *tfx* and sensitivity to the rhizobial peptide antibiotic trifolitoxin in a taxonomically distinct group of α-proteobacteria including the animal pathogen *Brucella abortus*. *Applied and Environmental Microbiology* Vol. 60 pp. 4163–4166.

Tufenkji, N. (2007) Modeling microbial transport in porous media: Traditional approaches and recent developments. *Advances in Water Research* Vol. 30 pp. 1455-1469.

Unkovich, M.J. & Pate, J.S. (2000) An appraisal of recent field measurements of symbiotic N_2 fixation by annual legumes. *Field Crops Research* Vol. 65 pp. 211-228.

Vesper, S.J. & Bauer, W.D. (1985) Characterization of *Rhizobium* attachment to soybean roots. *Symbiosis* Vol. 1 pp. 139-162.

Vesper, S.J. & Bauer, W.D. (1986) Role of pili (fimbriae) in attachment of *Bradyrhizobium japonicum* to soybean roots. *Applied and Environmental Microbiology* Vol. 52 pp. 134-141.

Vincent, J.M. (1970) *A manual for the practical study of the root nodule bacteria. IBP Handbook No. 15*. Blackwell Scientific Publ., Oxford, UK.

Vinuesa, P., Neumann-Silkow, F., Pacios-Bras, C., Spaink, H.P., Martínez-Romero, E. & Werner D. (2003) Genetic analysis of a pH-regulated operon from *Rhizobium tropici* CIAT899 involved in acid tolerance and nodulation competitiveness. *Molecular Plant-Microbe Interactions* Vol. 16 pp. 159–168.

Watt, M., Silk, W.K. & Passioura, J.B. (2006) Rates of root and organism growth, soil conditions, and temporal and spatial development of the rhizosphere. *Annals of Botany* Vol. 97 pp. 839-855.

Weiser, G.V., Skipper H.D. & Wollum A.G. (1990) Exclusion of inefficient *Bradyrhizobium japonicum* serogroups by soybean genotypes. *Plant and Soil* Vol. 121 pp. 99–105.

Williams, L.F. & Lynch, D.L. (1954) Inheritance of a nonnodulating character in the soybean. *Agronomy Journal* Vol. 46 pp. 28–29.

Yang, S., Tang, F., Gao, M., Krishnan, H.B. & Zhu, H. (2010) R gene-controlled host specificity in the legume–rhizobia symbiosis. *Proceedings of the National Academy of Sciences USA* Vol. 107 pp. 18735–18740.

Yost, C.K., Clark, K.T., Del Bel, K.L. & Hynes, M.F. (2003) Characterization of the nodulation plasmid encoded chemoreceptor gene *mcpG* from *Rhizobium leguminosarum*. *BMC Microbiology* Vol. 3: doi: 10.1186/1471-2180-3-1.

Zdor, R.E. & Pueppke, S.G. (1991) Nodulation competitiveness of Tn5-induced mutants of *Rhizobium fredii* USDA208 that are altered in motility and extracellular polysaccharide production. *Canadian Journal of Microbiology* Vol. 37 pp. 52-58.

9

Symbiotic Nitrogen Fixation in Soybean

Ali Coskan and Kemal Dogan
Süleyman Demirel University, Mustafa Kemal University
Turkey

1. Introduction

Nitrogen is the key component of vegetable protein for human and animal consumption. Although 78% of the atmosphere by volume is consisted of molecular nitrogen, this huge amount is not available for plants, animals or human. Only the bacteria that have nitrogenase enzyme can reduce atmospheric nitrogen and thus they called as a "nitrogen fixers". Nitrogen fixers reduce molecular nitrogen to amino acids and protein through ammonia (Fritsche, 1990; Lindemann and Glower, 2003). Nitrogen fixation process realizes either by free-living, associative or symbiotic nitrogen fixers. In symbiotic relation microorganism infects the plant root through infection thread and lives in the nodule forming structure. Afterwards plant supply component of nitrogenase and organic compounds to microorganism whereas microorganism supply reduced nitrogen to plant. Associative microorganisms are not infecting to plant however they colonize in rhizosphere and use of root exudates to successful nitrogen fixation. Free living fixers are independent and they need neither infect to plant nor rhizosphere exudates for fixation. Although the fixation rate vary depends on the nitrogen fixer type, the most effective fixation occurs in symbiotic relation with legumes. Soybean itself represents 77% of the N fixed by the crop legumes by fixing 16.4 Tg N annually, fixation by soybean in the U.S., Brazil and Argentina is calculated at 5.7, 4.6 and 3.4 Tg, respectively (Herridge et al. 2008).

Plant and microorganism are particular for each other, thus only certain microorganism can infect particular plant whereas the appropriate rhizobium of soybean is called as *Bradyrhizobium japonicum*. The shape and size of the nodules are also particular for legumes, the soybean nodules are round and in same cases big as pea. Effective nodules are large in size and reddish in the inside colour.

Legumes have an important role for both human nutrition and animal feeding, however, soybeans are unique in legumes with contents of 40% protein and 21% oil as well as isoflavones. Thus, soybean is the most widely grown protein/oilseed crop in the world, with both North and South America producing large portions of the world's supply of this remarkable crop.

In case of legume introduce to soil for the first time, appropriate rhizobium strain has to be inoculated for successful nitrogen fixation. In many cases some rhizobium bacterium might be existed in the soils, nevertheless, due to the insufficient number and activity (Gok and Onac, 1995), inoculation should be repeated. No successful nodulation as well as nitrogen fixation should be expected without inoculation with appropriate and healthy rhizobium strain by convenient inoculation method. A number of methods available to used in inoculation of soybean by *Bradyrhizobium japonicum*, however, inoculation by irrigation water and seed bad inoculation methods are more effective according to nitrogen fixation parameters (Isler and Coskan, 2009). On the other hand organic compound such as fulvic and humic acids have stimulatory effect on soybean-rhizobium symbiosis (Coskan et al., 2010). Moreover, biological nitrogen fixation of soybean influenced by the number of factor such as pH, salinity, partial oxygen pressure, soil water content, ambient temperature as well as soil mineral N content.

2. Cultivation of soybean for a first time: In scope of inoculation view

Cultural plants need considerable amount of macro and micro nutrients in mineral form to produce high quality of yield and these nutrients should be provided to correct yield-limiting factors. Mineral and organic fertilizations are the pathways to enhance soil mineral nutrient budget. Due to the plant can only use mineral forms of nutrients, mineral fertilizers are readily available for the plants. Nutrient in organic fertilizers are in organic form that not readily available for the plants, thus the organic fertilizers have to be convert mineral form via the process called "mineralization".

Considerable amount of nitrogen is removed from soil when protein-rich grain or hay harvested, thus nitrogen is the most commonly deficient nutrient among macro and micro nutrients. Due to the nitrogen is the key component of healthy growing, all plants other than legumes should be fertilized by nitrogenous fertilizer. Legume plants are unique for their ability to fix nitrogen from atmosphere by symbiotic relationship with rhizobium bacteria. Rhizobia require a plant host therefore they cannot independently fix nitrogen. These bacteria located around root hair and fixing atmospheric nitrogen using particular enzyme called "nitrogenase". When this mutualistic symbiosis established, rhizobia use plant resources for their own reproduction whereas fix atmospheric nitrogen to meet nitrogen requirement of both itself and the host plants. Supply of nitrogen through biological nitrogen fixation has ecological and economical benefits. Farmers are not taking advantage of rhizobial inoculation to a number of reasons, thus they are passed up the potential of biological nitrogen fixation.

In many cases *Rhizobium spp.* might be existed in the soils, nevertheless, due to the insufficient number and activity (Gok and Onac, 1995), inoculation should be repeated. A number of studies indicate that no nodule formation appeared in the soybean roots if inoculated soybean isn't grown previously. Biren (2002) carried out the experiment to evaluate the effects of rhizobium inoculation in Turkish Republic of Northern Cyprus where soybean is not cultivated previously. He reported that there was no nodule formation in the non-inoculated control plants. Similarly, in Isparta where the soybean is not cultivated regularly there was no nodule occurrence (Coskan et al., 2009). In some circumstances it is possible to observe very limited number of infection even in first cultivation at non inoculated condition. Isler and Coskan (2009) reported that in the first cultivation in non-inoculated condition there was a very few nodule formation in very light weight.

In scope of inoculation view, rhizobium inoculation should be realized with appropriate and healthy rhizobium strain in first cultivation of soybean plant and inoculation should be repeated every 2 to 3 year to sustain successful symbiotic nitrogen fixation. Depends on the rhizobium variety used, amount of nitrogen fixation greatly changed (Gok et al., 2001; Coskan et al., 2003). Thus, results obtained from local research should be considered in designating the effective strain.

3. Effects of inoculation methods on fixation

A number of inoculation methods are available, however, wetting the seeds by sugar, water and strain mixture or inoculation with peat culture are the most common methods in practise. Due to the rhizobium strain sensitive to the sunshine, inoculation and drying should be realized in indoor environment and seeds should also be protected from direct sunlight at sowing. Inoculation by wetting the seed methods has a number of disadvantages as follows: (1) Seeds are clinging to each other or to any surface of sowing equipment. Thus, farmers are abstained from inoculation to prevent time loss and extra workload. (2) The use of excessive water damages to the shell of the seed during inoculation therefore seeds become vulnerable to external conditions. Deaker et al. (2004) reported that seed inoculation method causes reduction of viable cell number when seed passes through machinery or lifting the seed coat out of the ground during germination. (3) Less amount of strain can be introduced to the seed especially in smaller seeds. Therefore, the higher amount of rhizobium bacteria per seed can be used in soil inoculation method compared to seed inoculation, especially for small seeded legumes (Brockwell, 1977).

Isler and Coskan (2009) tested the five different inoculation practises in pot experiment to evaluate the most effective method. They use the methods as follows: seed inoculation with sugar as an adhesive (SI), top inoculation with first irrigation (TI), two times top inoculation, one with first irrigation and one after germination (TTI), seed bad inoculation (SBI) and inoculation with peat culture-rhizobium suspension IWP). Result revealed that all practices other than control increased both number of nodule and nodule weight (Fig. 1). SI which commonly used inoculation technique was not effective as the other techniques tested. Observed nodule formation in TI proved that inoculants may reach rhizosphere area without any difficulties. Therefore inoculation with irrigation water may be used as an alternative inoculation technique considering the salt contents of the irrigation water. TTI application realized to compare with TI, however there wasn't statistical differences between TI and TTI.According to yield and the weight of seeds, SBI was the most effective inoculation techniques. Moreover SBI is the method that can easily adapt to sowing machinery with small changes.

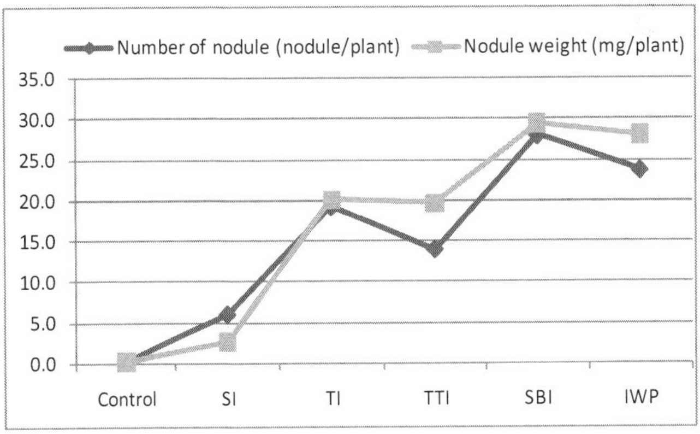

Fig. 1. Effects of inoculation methods on nodule formation and nodule weight (SI: seed inoculation, TI: top inoculation, TTI: two times top inoculation, SBI: seed bad inoculation, IWP: inoculation with peat culture)

In this method seed is not directly contacted to the inoculants material, instead, seed located nearby or above the soil which rhizobia applied. Thus, the difficulties reported by Deaker et al. (2004) and Brockwell (1997) surmounted. Due to the inoculation material mostly stored in peat culture, inoculation with peat culture is another common inoculation method. But, if the peat culture dries out after inoculation, peat removed from seed and accumulate bottom of sowing machine (Gault 1978). Besides, when dry peat is wetted, great heat occurred which may reduce the number of viable rhizobia (Deaker 2004). Thus, in IWP application, water added to peat before adding the rhizobia to prevent high temperature occurrence, then this suspension applied to seed bed. Nodule count and nodule weight results revealed that the problem mentioned above is not realized and effective infection occurred in IWP. Results strongly indicate that, in the case of inoculants were not contact with seeds directly, the success of symbiotic relation increased. In general, SBI were the most effective methods among all methods tested. This method is also ripe for development of automated sowing machines.

4. Effects of different tillage system

Biological N_2 fixation (BNF) was effected by different tillage system including agricultural practices, pesticides applications, addition of organic material, residue chopping. The ways in which these operations are implemented affect the physical and chemical properties of the soil, which in turn affect soil microorganisms as BNF bacteria.
The amount of nitrogen actually fixed by a legume depends not only on the genetics of the bacteria and host plant but also on the environment and agricultural practices. Among the common agricultural practices, fertilization with P and N has important effects in nitrogen fixation. It is a well-established fact that, when legumes are grown in soils high in available nitrogen, the nitrogen fixation rate is reduced.
According to different research, by definition, biological N_2 fixation (BNF) is synonymous with sustainability. Advances in agricultural sustainability will require an increase in the utilization of BNF as a major source of nitrogen for plants. The process of BNF offers an economically attractive and ecologically sound means of reducing external nitrogen input and improving the quality and quantity of internal resources.
Soil tillage methods have complex effects on physical, chemical and biological properties of soil. Because of the changing physical and chemical properties of soil by soil tillage methods, the biological properties of soil may also change. Actually these changes are indirect results of tillage. Changed physical and chemical soil properties by soil tillage methods effect the parameters directly related with soil microbial activities such as organic matter, soil humidity, temperature and ventilation as well as the degrees of interaction between soil mineral and organic matter. As a result of these effects, significant differences can be observed in the population of microbial activities in soil (Kladivko, 2001; Lavelle, 2000; Wardle 1995; Saggar et al. 2001).
Plant and microorganism interactions in rhizosfer region are very important for plant growth. In the rhizosphere region, rhizobial activities occur as reciprocal and compulsory interactions (symbiosis) of plant-microorganism (Altieri, 2000; Garcia and Altieri, 2005). One of the important activities related to soil qualities is beneficial microorganism activities. The most important of these activities is a root nodule bacterium which provides to biological N_2-fixation (Ferreira et al., 2000).
Microorganisms, that are important parts of the nature, are considerably affected by the environmental conditions. These organisms which rapidly reproduce and function in proper

environmental conditions, also struggle to continue their functions under poor conditions (Doğan et al., 2007).
As a result of symbiotic N_2-fixation, legumes supply nitrogen to the soil not only with their nodules, but also by decomposition of their roots and shoots. Nitrogen might have formed by mixing the separated dead nodule tissues into the soil. This situation can be accelerated by cutting of the plant's shoots (Werner, 1987; Goormachting et al., 2004).
In a study of Dogan et al. (2011), the effects of six different soil tillage methods (Table 1) on some parameters related with nitrogen fixation have been investigated. According to the findings of the research in the No-Tillage with Direct Seeding (NTDS) plots, root weights (6.9 g/plant), number of nodules (96 number/plant), weight of nodules (0.318 g/plant) and root nitrogen content (% 0.71) are found to be statistically higher than with the other tillage applications. In the Reduced tillage with rotary tiller (RTR) plots, the values of up-root dry weight (51.3 g/plant), mean nodule weight (3.91 mg/nodule), root N content (2.38%), are found higher on the lands than in NTDS plots.

Soil Tillage Methods	Soil Tillage for winter wheat cultivation	Soi Tillage for second crop soybean plant
Conventional Tillage with Residue (CTR)	Chopping the residues Plowing (30-33 cm) Disk horrow (13-15 cm) (2 times) Packing (2 times) Wheat planting with a universal planter (4 cm)	Chopping the residues Heavy disk horrow (18-20 cm) Disk horrow (2 times) (13-15 cm) Packing (2 times) Soybean planting with Pneumatic-precision seeding machine (8 cm)
Conventional Tillage with Burnt Residue (CTBR)	Burning the residues Plowing (30-33 cm) Disk horrow (13-15 cm) (2 times) Packing (2 times) Wheat planting with a universal planter (4 cm)	Burning the residues Chiselling (35-38 cm) Disk horrow (13-15 cm) (2times) Packing (2 times) Soybean planting with Pneumatic-precision seeding machine (8 cm)
Reduced Tillage with Heavy Disking (RTHD)	Chopping the residues Heavy disking (18-20 cm) (2 times) Packing (2 times) Wheat planting with a universal planter (4 cm)	Chopping the residues Rotary tilling (13-15 cm) Packing (2 times) Pneumatic-precision seeding machine (8 cm)
Reduced tillage with rotary tiller (RTR)	Chopping the residues Rotary tilling (13-15 cm) Packing (2 times) Wheat planting with a universal planter (4 cm)	Chopping the residues Rotary tilling (13-15 cm) Packing (2 times) Soybean planting with Pneumatic-precision seeding machine (8 cm)
No-Tillage with Heavy Disking (NTHD)	Chopping the residues Heavy disking (18-20 cm) Doting (2 times) Wheat planting with a universal planter (4 cm)	Chopping the residues Herbicide application Soybean planting with Pneumatic-precision seeding machine (8 cm)
No-Tillage with Direct Seeding (NTDS)	Chopping the residues Herbicide application Wheat seeding with direct seeder (4 cm)	Chopping the residues Herbicide application Soybean planting with Pneumatic-precision seeding machine (8 cm)

Table 1. Soil tillage methods in the major and secondary crop (soybean) production (Dogan et al., 2011)

Among the applications, in the plots of Reduced Tillage (RTHD and RTR) rhizobial nitrogen fixation parameters have been found considerably higher compared with the other applications (Fig. 2). However, some soil tillage methods used in this study negatively affected some soil parameters. For the Reduced Tillage with Rotary tiller (RTR) plots the

dry root weight (4,8 g/plant), up-root weight (35,7 g/ plant) and root N content (% 0,68) values and for the Conventional Tillage with Burnt Residue (CTBR) plots, number of nodules and weight of nodule values were found to be lower than in the other tillage applications. The values of dry nodule weights, like in Conventional Tillage with Burnt Residue (CTBR) were low in the plots of Conventional Tillage with Residue (CTR) and Reduced Tillage with Heavy Disking (RTHD) with the values 0,071 and 0,088 g/plant, respectively. Besides, the lowest mean nodule weights (2,06 mg/nodule) have been observed in Conventional Tillage with Residue (CTR) plots and the lowest up root N content (%1,98) have been observed in Reduced Tillage with Heavy Disking (RTHD) plots.
The results of the study have been showed that, parameters of nitrogen rhizobial fixation has been affected negatively by the conventional tillage methods in which 3-5 tillage operations are applied and soil is disturbed . There were differences among the tillage methods and these differences were found to be statistically significant. In general, the best results related with rhizobial activity have been obtained with No-Tillage with Direct Seeding (NTDS) and No-Tillage with Heavy Disking (NTHD). However, other soil tillage methods decreased the nitrogen fixation (Dogan et al., 2011). Similar studies have also showed that zero and reduced soil tillage methods have increased the soil microbial activity and population (Ferreria, 2000; Alvarez et al., 1995; Gassen and Gassen 1996).

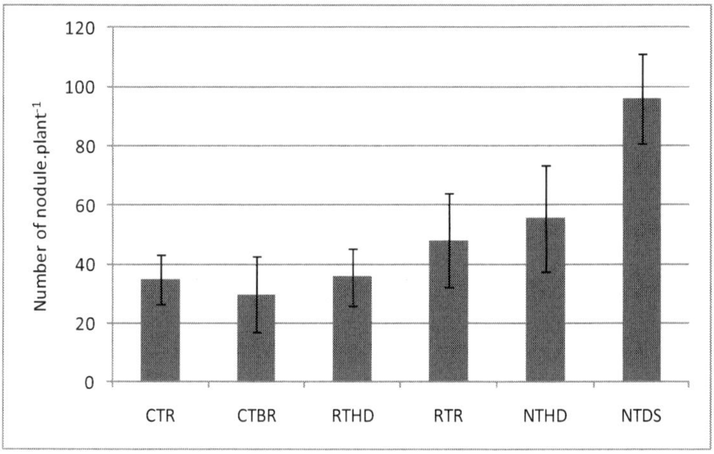

Fig. 2. The effects of different soil tillage methods on number of nodule in secondary crop soybean plant (CTR: Conventional Tillage with Residue, CTBR: Conventional Tillage with Burnt Residue, RTHD: Reduced Tillage with Heavy Disking, RTR: Reduced tillage with rotary tiller, NTHD: No-Tillage with Heavy Disking, NTDS No-Tillage with Direct Seeding)

Generally, soil microbial activity is affected negatively by soil tillage (Jinbo et al., 2007; Kladivko, 2001; Hussain et al., 1999; Saggar et al., 2001). Therefore rhizobial activity is also be affected negatively by soil tillage (Hassen et al., 2007; Ferriera et al., 2000). Soil organic matter decreased by soil tillage operations is also important for the vital activities of soil microorganisms. The decrease of organic matter in the soil can also cause decreases in soil microbial activity (Saggar et al., 2001; Eliot el al., 1984). As it can be seen in the similar studies, the effects of soil tillage methods may differ depending on climate, regional, and

environmental factors. These factors must be taken into consideration before applying tillage methods. Otherwise, biological activity, fertility and sustainability of soil will be destroyed.
On high-input farms, microorganisms are generally thought to play a minor role in soil fertility because most nutrients in inorganic fertilizers are readily available for the plants and do not require degradation or mineralization (Smith et al., 2001).
Many studies have concluded that herbicides affect nitrogen fixation largely via indirect effects on plant growth and consequent availability of photosynthate to the root nodules (Wally et al., 2006; Abd-Alla et al. 2000); there is evidence that some pesticides might impair the ability of the rhizobia to recognize appropriate host plants. As a consequence, early nodulation events can be disrupted. However, according to their research, not all pesticides had a negative impact on nodulation and the degree to which nodulation was inhibited was dependant on pesticide concentrations. In some instances, results from various studies have been contradictory. For example, when examining the effects of chlorsulfuron under laboratory conditions, Anderson et al. (2004) observed that even at rates equivalent to two times field rates, chlorsulfuron did not influence rhizobial growth. However, although rhizobial growth was not influenced, the subsequent ability of these rhizobia to form nodules was reduced. Thus, they reported that when rhizobia were exposed to relatively high levels of chlorsulfuron, subsequent nodule size and total nitrogen fixation was reduced. In contrast, Martensson (1992) reported that nodulation ability was unaffected by previous exposure to chlorsulfuron. These contrasting results suggest that the impact of various herbicides on specific nodulation events may be highly dependent on specific environmental conditions, including different soil characteristics (i.e., pH, organic matter, moisture, etc.) and weather conditions. Martensson (1992) examined the impact of various herbicides on root hair formation. Rhizobia infect plant roots through root hairs and thus it was hypothesized that herbicides affecting root hair development might interfere with nodulation. Author reported that some herbicides, including glyphosate, caused root hair deformations that apparently resulted in fewer nodules being formed. It is important to note, however, that this was a laboratory study and consequently the herbicide rates used in these experiments were not necessarily similar to rates that would be encountered in soils under field conditions. Thus, although the research demonstrates the possibility for herbicides to affect nodulation via root hair deformations, it is not known if this phenomenon occurs under field conditions (Walley et al. 2006).
Saggar et al. (2001) studied the effect of cultivation on soil organic C, functional chemical composition of SOM, and soil structure in soils of contrasting mineralogy. They found that soil susceptibility to structural degradation increased with years of cultivation, and from light textured to heavier textured soils. Because cultivation causes profound changes in the soil physical and chemical properties, and populations of microfauna and macrofauna, it is relevant to quantify its effects on soil microbial and microfaunal populations and on SOM dynamics.

5. Mycorhiza-rhizobium interaction in light limited condition

Vesicular Arbuscular Mycorrhiza (VAM) is symbiotically living organism with many crops and they enhance plant P uptake along with other micronutrients especially Zinc. Phosphorus efficiency in highly limy soil (in high pH) is considerably low whereas mycorrhiza assists plant to receive that immobile phosphorus by exudates and/or enhancing soil contact area. As mentioned previously, rhizobium is a microorganism, capable of fixing aerial nitrogen (N_2) to soil/plant via symbiotic relations with legumes. Both organisms utilize the photosynthesis products that assimilated by host plants to survive. In non-limiting conditions those organism supports the plants for the most important macro and micro nutrients as N, P, Fe and Zn. On the other hand unsuitable soil or climatic conditions in growth season may result negative Rhizobium x Mycorrhiza interactions. Although both microorganisms use organic compounds formed by plants, they use trace amount of that compound compared to the plant biomass formation. Coskan et al. (2003) carried out the pot experiment to evaluate cross interactions of rhizobium and mycorrhiza in light limited condition. Results revealed that no nodule formation appeared in non-rhizobia-inoculated control variant whereas rhizobial inoculation increased number of nodule (Fig. 3) while decreased biomass weight (Fig. 4).

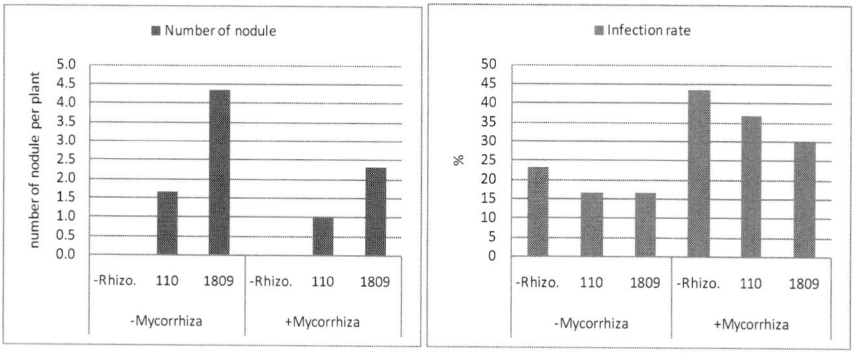

Fig. 3. Nodule formation (left) and mycorrhizal infection rate (right) in the light limited condition

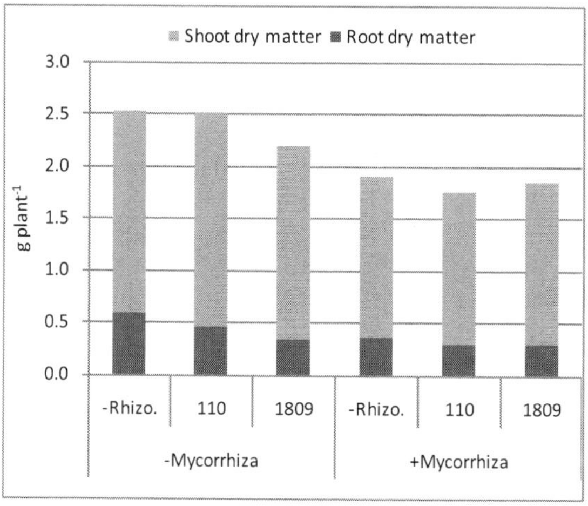

Fig. 4. Effect of dual inoculation on root and shoot dry weight of soybean in light limited condition

Although both of the rhizobium strains give rise to nodule formation, *B. japonicum* 1809 strain cause considerably higher nodule number. On the other hand, mycorrhiza inoculation increased the infection rate (Fig 3). Rhizobial inoculation decreased mycorrhizal infection rate in both with mycorrhiza and without mycorrhiza applications. Bacterial inoculation has no significant effect on plant growth except nodule formation. It is clearly seen that both rhizpbium and mycorrhiza applications reduced total plant dry weight. However, plant dry weight and phonological observations revealed that plant development is adversely effected due to mycorrhizal inoculation in light-limited growing session.

6. Effects of humic+fulvic acid on symbiotic nitrogen fixation

Organic matter is one of the most important issues of agriculture and it contains three very important components: humic acids, fulvic acids and humin. Plants and microorganisms in soil benefit from applications of humic acid in several ways. Humic acid stimulate root growth, increase carbohydrate production, have a hormone-like affect within the plant, and increase soil microorganisms (Lawn Care Academy, 2010). The incorporation of humic acid fractions in media designed for the enumeration of soil micro-organisms belonging to specific physiological groups was found to result for some groups in appreciably higher counts. It is suggested that by influencing the enzyme systems of certain micro-organisms, humic compounds may affect the range of substrates which they can utilize. The effect could have implications on the activity of organisms in environments in which humic substances are normally present, such as soils and natural waters (Visser, 1984).

Coskan et al. (2010) carried out a pot experiment to represent effects of humic + fulvic acid (HFA) applications on biological nitrogen fixation under soybean vegetation. Humic + fulvic acid application realized by either incorporate to soil or admixing by irrigation water. Seeds are inoculated by appropriate Bradyrhizobium japonicum strain, before sowing. In flowering stage, roots are removed from soil and the number of nodule determined (Fig 5).

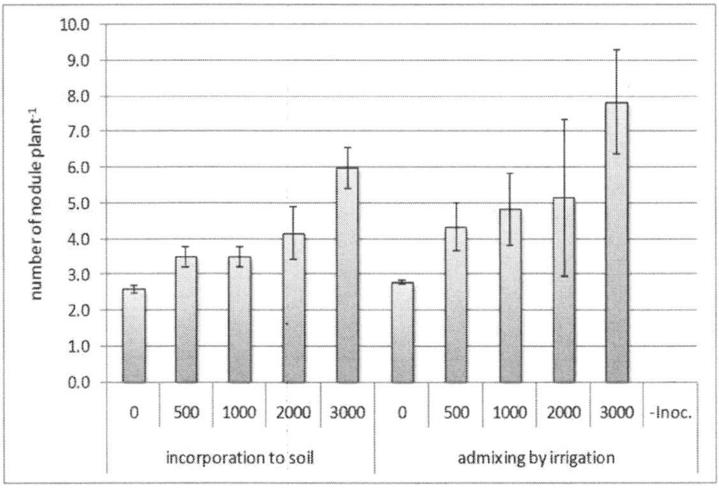

Fig. 5. Effects of humic + fulvic acid application on nodule number of soybean (-Inoc: non-inoculated)

Due to the fields where soil was taken is not previously introduced with the rhizobia that match with soybean; no nodule occurrence was observed in the pot which non-inoculated control variant. A few nodules observed in non-HFA applied pots however HFA application increased nodule occurrence considerably. Both "incorporation to soil" and "admixing by irrigation" applications were effective on formed nodule number; however because of the dilution effect in incorporation to soil application, admixing by irrigation application seems to be more effective than incorporation to soil. (dilution effect). Increasing doses of HFA increased the number of nodule, thus findings expressing the considerable positive effects of HFA on biological nitrogen fixation.

7. Effects of organic matter application

Soil organic matter is important for biological nitrogen fixation (BNF) because of its influence on soil physical, chemical and biological properties and processes. It helps to create a favourable medium for physical processes, chemical reactions, and biological activity. The multi-faceted role of soil organic matter (SOM) must therefore be taken into consideration in any assessment of `soil quality' and `sustainable land management. Low concentration of organic matter can have a deleterious effect on offsite environment because it is often associated with decreased soil fertility, water holding capacity and water infiltration, and increased erosion. Further, SOM turnover controls the fluxes of nutrients. Microbial biomass measurements combined with soil respiration have frequently been used as an index of soil development or degradation (Insam and Domsch, 1988) and to assess the quality of organic matter input (Anderson and Domsch, 1990). Interactions between soil microorganisms and soil microfauna, particularly Protozoa and Nematoda, largely control SOM turnover (Elliot et al., 1984).

Harvested crop residues and rotation has a major impact on the soil organic matter content. However, some features, such as the type and degree of decomposition of organic matter, affected BNF in different ways. Many field trials which were applied organic material by Gok et al. (2001) show that with the organic material application, organic matter content of soil increased in the short term but at the end of the trial, soil organic matter content decreased due to high rate of mineralization. To gain long-term soil organic matter, all kind of organic substrates should be regularly added to the soils which under the effect of semi-arid climate condition. Moreover, degradation resistant substrates such as lignin and cellulose should be preferred to dump mineralization. With this way nitrogen is temporarily immobilized in biomass that preventing (Asmus and Hubner 1985; Gok 1987). In a research, Limon-Ortega et al. (2002) studied to evaluate the effects of burning and natural wheat or maize stubble on some properties of soil. Results indicated that the positive effect of that substrates appeared after 2 or 3 year continuously stubble applications. The result obtained at 5th - 6th years were more expressive than those obtained in the 1st to 3rd years. When the stubble is burned almost all nutrients in organic substrates converted to available form for plants in seconds. Therefore, compared with burned or natural stubble applied plots, in the beginning years burned stubble seems to be more efficient, but in following years the effects of stubble become much more effective on the soil parameters.

A two year field experiment at soybean cultivation was undertaken for determining the effects of stubble burning, a widely performed practice in Cukurova Region, along with admixing 0, 5000 and 10000 kg ha^{-1} tobacco wastes on symbiotic nitrogen fixation, grain yield and biomass production (Coskan et all, 2009). Results revealed that applications were significantly effective on nitrogen fixation, yield and biomass production. According to overall averages, the highest biomass production of root and shoot were observed at wheat burned and 10000 kg ha^{-1} tobacco waste applied plot as 830 and 4730 kg ha^{-1}. The highest nitrogen contents at harvest stage were determined in the plot wheat and 5000 kg ha^{-1} tobacco waste applied (root, 0.87%; shoot, 0.95%). At the end of experiment determined grain yield amounts in first year were higher in the stubble burned plots. No statistical difference was determined between burned and non-burned stubble in the second year. When the variants of tobacco waste applications were compared according to their tobacco rates, the productivity was increased at plots of waste application in both years. The determined highest yield 4520 and 5280 kg ha^{-1} at stubble burned and non-burned plots in which 10000 kg ha^{-1} waste was applied in the first and second years, respectively

8. Factors that effective on symbiotic nitrogen fixation in soybean

Nitrogen fixation is one of the important soil microbial activity which was affected by all ongoing processes in soil as well as other soil microorganisms. The biological nitrogen fixing process depends on the occurrence and survival of Rhizobium in soils and also on their efficiency (Adamovich, Klasens, 2001).

The rate of the nitrogen fixation was affected by many different physiological and environmental factors in soil, such as temperature, water holding capacity, water stress, salinity, nitrogen level, pH and other nutrients. Many of these factors, including temperature, affect many aspects of nitrogen fixation and assimilation, as well as factors such as respiratory activity, gaseous diffusion and the solubility of dissolved gasses, which ultimately affect plant growth (Dogan et al 2010; Keerio et al., 2001).

High amount of mineral nitrogen in soil has negative effect on nodulation. Wide or narrow C:N ratio decreases nodule formation, therefore nitrogen fixation. If the C:N ratio is in expected ratio (15-30) nodulation and N_2-fixation regularly realizes. Inhibitory effect of nitrate causes the reduction of capillary roots development as well as preventing particular infection's strands. This effect is very similar to herbicides' effect. Many researches have shown that adequate nitrate, nitrite, ammonium and urea concentrations in soil causes to decrease the number of infections, to delay to the first formation of nodules, to decrease to the nodule number and weight. Temperature is the main factor affecting N_2-fixation; however, optimum temperature for N_2-fixation is depending on various soil properties. Optimum N_2-fixation temperature value is between 20-40 °C. Nodulation and nitrogen fixation in soybean is composed of between 20-30 °C. High soil temperature diminishes root growth as well as nodule formation. Furthermore, temperature changes affects to the competitive ability of Rhizobium/Bradyrhizobium species. Low temperatures decreased to nodule formation and N_2-fixation. However, N_2-fixation in natural legumes is not influenced extreme cold conditions (Bordeleau and Prevost, 1994).

Soil reaction (pH) is one of the most important factor influencing legume and Rhizobium symbiosis. A higher concentration of H^+ ions increases the solubility of Al, Mn and Fe, and higher amount of these elements may become toxic for rhizobium. *Sinorhizobium meliloti* and *Rhizobium galegae* are highly sensitive to acid pH and soluble Al when the critical soil pH is 4.8–5.0 (Bordeleau ve Prevost, 1994). *Rhizobium leguminosarum bv. trifolii* and *Rhizobium leguminosarum bv. Viciae* in comparison with alfalfa rhizobia are more tolerant to soil acidity. However, pH less than 4.6 inhibits their activity. Legumes and Rhizobium have form an efficient symbiosis and fix high amounts of biological nitrogen when soil pH is no less than 5.6-6.1. Soil acidification inhibited the root-hair infection process and nodulation. Optimum soil pH for nodulation and yield for soybean is between 6.2 and 6.8 (Lapinskas, 1998).

The results of a study indicate that *Rhizobium leguminosarum* bv. trifolii is widely distributed in slightly acid soils with pH_{KCl} 5.6–6.0. The average content of rhizobia was 540.0 • 10^3 cfu g^{-1} of soil. Less *Rhizobium leguminosarum* bv. viciae and significantly less *Sinorhizobium meliloti* and *Rhizobium galegae* were found. Rhizobium significantly declined in acid soils (pH_{KCl} 4.1–5.0). Most of biological nitrogen was fixed at soil pH_{KCl} 6.1–7.0. In this case, *Rhizobium galegae* accumulated 196 to 289 kg N ha^{-1} of nitrogen, whereas rhizobia of alfalfa and clover were less, and it depended on strain efficiency and soil pH. Soil liming had a positive effect on nitogenase activity in red clover. The soil liming ($CaCO_3$ rate 6.2 t ha^{-1}) in combination with inoculation have increased biological nitrogen fixation by red clover at 106 kg N ha^{-1}. Associative diazotrophes in non-legume rhizoplane have fixing the biological nitrogen too. The effective strains of *Rhizobium spp., Agrobacter radiobacter* and *Arthrobacter mycorens* have made up an active association with barley, timothy and spring rape and accumulated 11.0 to 20.4 kg N ha^{-1} of biological nitrogen (Lapinskas, 2008).

Soil moisture can affect to nitrogen fixation both directly and indirectly. In low moisture condition in soil, nodule respiration decreases and nitrogen in nodule moves out slowly. This case is direct effect of low soil moisture. However in the same condition, nitrogen fixation decreased due to deterioration of generating photosynthesis units assimilate and in this case, N_2-fixation was affected indirectly.

Iron (Fe) and molybdenum (Mo) are located in structure of the Nitrogenase enzyme which is working with legumes for symbiotic nitrogen fixation (Fig. 6). Therefore, the amount of these nutrients in the soil and plant uptake affects the symbiotic N_2-fixation of legumes directly (Werner, 1987; Durrant, 2001).

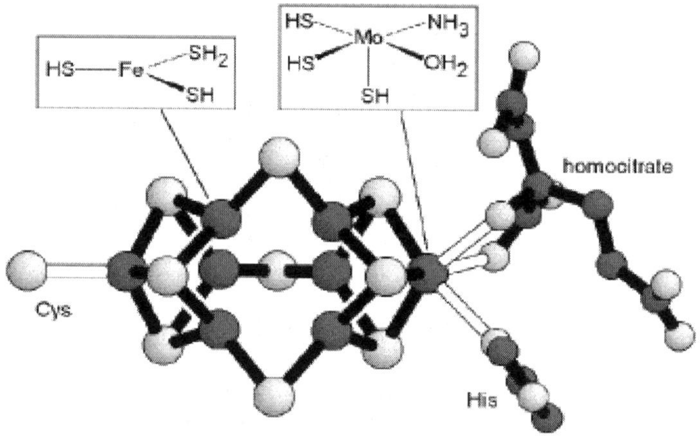

Fig. 6. Structure of nitrogenase enzyme (Durrant, 2001)

Nitrogen fixation in soybean is negatively affected by increasing salt contents of the soil. N_2-fixation of Rhizobium bacteria and their activities decreased in accordance with increasing soluble salt contents. Thus, increasing salt concentration in irrigation water was found to reduce a significant amount of grain and nodule weight in soybean (FAO, 1982).
According to many research it was determined to development of soybean was decreased in soil condition of 0.08% $CaCl_2$ and 1.5% $ZnSO_4$ (Anonymous, 1982). According to the results of many similar studies show that salt tolerance of rhizobium bacteria, optimum pH, antibiotic resistance and so on has revealed important differences (Gok and Martin, 1993).

9. References

Abd-Alla, MH, Omar, SA and Karanzha, S. (2000). The impact of pesticides on arbuscular mycorrhizal and nitrogen-fixing symbioses in legumes. *Appl Soil Ecol.* 14:191-200.

Adamovich, A, Klasens, V. (2001). Symbiotically fixed nitrogen in forage legume-grass mixture. *Grassland Science in Europe.* Pp:12.

Altieri, MA. (2000). The ecological impacts of transgenic crops on agroecosystem health. *Ecosystem Health,* 6:13-23.

Alvarez, R, Dõaz RA, Barbero N, Santanatoglia OJ, Blotta L (1995). Soil organic C, microbial biomass & CO_2-C production from three tillage systems. *Soil Tillage Res.*33:17-28.

Anderson, A, Baldock, JA, Rogers, SL, Bellotti, W and Gill, G. (2004). Influence of chlorsulfuron on rhizobial growth, nodule formation, and nitrogen fixation with chickpea. *Aust. J. of Agric. Res.* 55:1059-1070.

Anderson, TH, Domsch, KH. (1990). Application of ecophysiological quotients (qCO_2 and qD) on microbial biomass from soils of different cropping histories. *Soil Biol. Biochem.* 22:251-255.

Asmus, F, Hubner, C. (1985). Untersuchungen zur N-Immobilizierung nach Strohdüngung. *Arch. Acker-Pflanzenb. Bodenkd.* 29:39-45.

Biren, S. (2002). The Effects of Bacterial (*Bradyrhizobium japonicum*) inoculation on nodulation and yield in soybean (*Glycine Max L.*) at Turkish Republic of Northern Cyprus. MSc Thesis, University of Cukurova. Available at http://tez2.yok.gov.tr

Bordeleau, LM, Prevosi, D. (1994). Nodulation and nitrogen fixation in extreme environments. *Plant and Soil* 161:115-125.

Brockwell, J, (1977). Application of legume seed inoculants. In: *A Treatise on Denitrogen Fixation.* Hardy, RWF, Gibson, AH. Wiley, Sydney, pp:277-309.

Coskan, A, Gok, M, Erol, H, Dogan, K. (2010). Humic + fulvic acid as a bio-stimulator on biological nitrogen fixation. *9. Symposiums des Verband deutsch-türkischer Agrar- und Naturwissenschaftler (VDTAN),* 22-27 Marz, Mustafa Kemal Univ., Hatay, Turkey.

Coskan, A, Gok, M, Onac, I, Ortas, I. (2003). The effects of rhizobium and mycorrhiza interactions on N_2-fixation, biomass and P uptake. *Journal of Cukurova University Faculty of Agriculture.* 18 (1):35-44.

Coskan, A, Isler, E, Kucukyumuk, Z, Erdal, I. (2009). Effects of bacterial inoculation of soybean on Nodulation and yield of soybean grown in Isparta conditions. *Journal of Suleyman Demirel University Faculty of Agriculture.* 4(2):17-27.

Deaker, R, Roughley, RJ and Kennedy, IR. (2004). Legume Seed Inoculation Technology - a review. *Soil Biology & Biochemistry.* (36):1275-1288.

Dogan, K, Celik, I, Gok, M. and Coskan, A. (2011). Effect of different soil tillage methods on rhizobial nodulation, biyomas and nitrogen content of second crop soybean. *African Journal of Biology* (In press)

Dogan, K, Gok, M, Coskan, A. (2007). Effects of bacteria inoculation and iron application on nodulation and N-fixation in peanut as a 2nd crop. *Journal of Suleyman Demirel University Faculty of Agriculture.* 22(3):43-52.

Dogan, K, Gok, M, Coskan, A. (2010). Effects of bacteria inoculation and iron application on biomass, yield and nitrogen contents in Cukurova region. *5th National Plant Nutrition and Fertilization Congress.* 15-17 September, Izmir.

Durrant, MC, (2001). Controlled protonation of iron-molybdenum cofactor by nitrogenase: a structural and theoreticel analysis. *Biochem. J.* 355:569-576.

Elliot ET, Coleman DC, Ingham RE, Trofymow JA (1984). Carbon and energy low through microflora and microfauna in the soil subsystem of terrestrial ecosystems. In: *Perspectives in Microbial Ecology,* Klug, MJ, Reddy, CA (Eds.), Current American Society of Microbiology, Washington, DC.

FAO, 1982. Anonymous, 1982. Application of Nitrogen Fixing Systems in soil Management. Technical Papers, www.fao.org, Rome.

Ferreira MC, Andrade DS, Maria L, Chueire O, Takemura SM, Hungria M (2000). Tillage method and crop rotation efects on the population sizes and diversity of bradyrhizobia nodulating soybean. *Soil Biology & Biochemistry* 32:627-637.

Fritsche, W., 1990. Mikrobiologie. Gustav Fischer Verlag. Jena. DOI: 10.1002/food.19920360516

Garcia AG, Altieri MA (2005). Transgenic Crops: Implications for Biodiversity and Sustainable Agriculture. *Bulletin of Science, Technology & Society,* 25, 4: 335-353

Gassen D, Gassen F (1996). *Plantio Direto, o Caminho do Futuro. Aldeia do Sul, Passo Fundo,* Brazil (207 pp.).

Gault, R. R., 1978. A Study of Developments and Trends in New Zealand, The USA and Canada in the Technology Associated with the Exploitation of the Nitrogen-Fixing Legume Root Nodule Bacteria, Rhizobium spp. For Use in Legume Crops New to Australian Agriculture, Winston Churchil Memorial Trust. Canberra.

Gok, M and Onac, I. (1995). Some microbiological properties of wide spread soil series of the plains Hilvan and Baziki in Turkey. In: *Ilhan Akalan Soil and Environment Symposium,* Volume II, pp 158-167.

Gok, M, (1987). Einfluss energiereicher Subsrate (Cellulose oder Stroh) und O_2-Partialdruck auf Quantitaet und Qualitaet der Denitrifikation eines sandigen Lehms. PhD thesis, University of Hohenheim, Stuttgart, Germany.

Gok, M, Martin, P. (1993). Effect of different rhizobia inoculation symbiotic N-fixation on soybean, clover and vetch. *Doga-Tr. J. of Agricultural and Forestry.* 17:753-761.

Gok, M, Saglamtimur, T, Coskan, A, Inal, I., Onac, I, Tansı, V. (2001). Effects of organic and mineral fertilizers on denitrification loss and nitrogen turnover in soil. *Final Report of TARP-1785,* Tubitak, Ankara-Turkey

Goormachting, S, Capoen, W, Holsters, M. (2004). Rhizobium infection: lessons from the versatile nodulation behavior of water-tolerant legumes. *Trends Plant Sci.* 9:518-522.

Hassen, AA, Xu, J, Yang, J (2007). Growth conditions of associative nitrogen-fixing bacteria enterobacte cloace in rice plants. *Agricultural journal* 2(6):672-675.

Herridge, DF, Peoples, M B and Boddey, RM (2008). Global inputs of biological nitrogen fixation in agricultural systems. *Plant Soil.* 311:1-18. DOI 10.1007/s11104-008-9668-3

Hussain, I, Olson, KR, Ebelhar, SA (1999). Long-Term Tillage Effects on Soil Chemical Properties and Organic Matter Fractions. *Soil Sci. Soc. Am. J.* 63:1335–1341

Insam, H, Domsch, KH, (1988). Relationship between soil organic carbon and microbial biomass on chronosequences of reclamation sites. *Microb. Ecol.* 15, 177-188.

Işler, E. and Coskan, A., 2009. Effects of Different Bacterium (*Bradyrhizobium japonicum*) Inoculation Techniques on Biological Nitrogen Fixation and Yield of Soybean. *Tarim Bilimleri Dergisi,* 15 (4) 324-331.

Jinbo, Z, Changchun, S, Wenyan, Y. (2007). Effects of cultivation on soil microbiological properties in a freshwater marsh soil in Northeast China. *Soil Tillage Res.* 93:231–235

Keerio, MI, 2001. Nitrogenase activity of soybean root nodules inhibited after heat stress. *Online J. Biol. Sci.* 1:297–300.

Kladivko, JK. (2001). Tillage systems and soil ecology. *Soil Tillage Res.* 61(1-2):61-76.

Lapinskas, E. (1998) Biologinio azoto fiksavimas ir nitraginas. *Akademija,* 218 p.

Lapinskas, E., 2008. Biological nitrogen fixation in acid soils of Lithuania. *Žemės Ūkio Mokslai.* 15(3)67–72.

Lavelle, P. (2000). Ecological challenges for soil science. *Soil Sci.* 165(1):73-86.

Lawn Care Academy (2010) Research and Benefits of Humic Acid [online] <http://www.lawn-care-academy.com/humic-acid.html> Access Date and Time: 20.04.2010, 15:00

Limon-Ortega, A, Sayre, KD, Drijber, RA. (2002). Soil attributes in furrow-irrigated bed planting system in Northwest Mexico. *Soil & Tillage Research,* 63:123-132.

Lindemann, WC and Glover, CR. (2003). Nitrogen fixation by legumes. New Mexico State Uni. Electronic Distribution May 2003, http://aces.nmsu.edu/pubs/_a/a-129.pdf

Martensson, AM, (1992). Effects of agrochemicals & heavy metals on fast-growing rhizobia and their symbiosis with small-seeded legumes. *Soil Biol. Biochem.* 24:435-445

Saggar, S, Yeates, GW, Shepherd, TG. (2001). Cultivation effects on soil biological properties, microfauna and organic matter dynamics in Eutric Gleysol and Gleyic Luvisol soils in New Zealand. *Soil & Tillage Research.* 58, 55-68.

Smith, E., Leeflang, P., Gommans, S., Broek, J, Mil, S. and Wernars, K. (2001). Diversity and seasonal fluctuations of the dominant members of the bacterial soil community in a wheat field as determined by cultivation and molecular methods. *Applied and Environmental Microbiology.* 67(5):2284–2291.

Visser, SA. (1984) Effect of humic acids on numbers and activities of micro-organisms within physiological groups. *Organic Geochemistry.* 8(1):81-85

Walley, F, Taylor, A and Lupwayi, N. (2006). Herbicide effects on pulse crop nodulation and nitrogen fixation. *Farm Tech 2006 Proceedings.* pp. 121-123.

Wardle, DA. (1995). Impacts of disturbance on detritus food webs on agro-ecosystems of contrasting tillage and weed management practices. In: *Advances in Ecological Research,* Begon, M, Fitter, AH. Vol: 26. Academic Press. New York p. 105-185.

Werner, D. (1987). Pflanzliche und Mikrobielle Symbiosen. Georg Thieme Verlag Stuttgart-New York. ISBN 3-13-698301-7

10

Intensity of Powdery Mildew in Soybean Under Changes of Temperature and Leaf Wetness

Marcelo de C. Alves[1] et al.*
[1]*Federal University of Mato Grosso*
Brazil

1. Introduction

Soybean *Glycine max* L. Merr. is cultivated in several tropical and subtropical regions of the world. United States (USA) and Brazil are the world's largest producers and exporters of oilseed (Agrianual, 2008; Miyasaka & Medina, 1981).

Despite the high production and export of Brazilian soybeans, many factors have affected the quality or quantity of production of that crop, causing reduction in financial returns per unit area, such as disease epidemics. Among the diseases, powdery mildew, whose etiologic agent is *Microsphaera diffusa* Cke. & Pk., suddenly began to cause significant damage in soybean, despite having a broad host range and have been reported in Brazil, Canada, Republic of China, India, Puerto Rico, South Africa, United States (Sinclair, 1999), Germany, Argentina and Bolivia (Sartorato & Yorinori, 2001) .

According to Yorinori & Hiromoto (1998), crops widely affected by the disease, had estimated reductions between 30 and 40% of yield, in the same order of magnitude as those reported abroad by Dunleavy (1978) and Philips (1984). The susceptibility of cultivars and the influence of climate favored epidemics with high rates of disease progress, in successive years in Brazil. Considering the lack of resistance of most of the cultivars, chemical control is required, especially in the south and the high plains of the savannah biome (Sartorato & Yorinori, 2001). In 1996/97, epidemics of powdery mildew in soybean in a great extent of Brazil, from the Central West region to the Rio Grande do Sul state, resulted in average losses of 15 and 20% in susceptible cultivars, with extremes ranging from 50 to 60% (Yorinori & Hiromoto, 1998 ; Seganfredo & Silva, 1999).

M. diffusa is distinguished from *M. polygoni* by presenting cleistothecium with appendages forked at its end (Sartorato & Yorinori, 2001; Grau, 1975). The fungus is an obligate parasite that develops throughout the soybean shoot, including leaves, stems, petioles and pods. Symptoms can range from chlorosis, green islands, rusty spots, defoliation or severe combination of these symptoms, depending on the reaction of cultivars. Chlorotic spots and necrosis on the leaf veins indicate a hypersensitivity reaction. However, the most obvious is the very structure and powdery white fungus on the surface of infected parts (Yorinori, 1982; Yorinori, 1986, Tanaka et al., 1993; Yorinori et al., 1993; Sinclair, 1999; Sartorato &

*Edson A. Pozza[2], João de C. do B. Costa[3], Josimar B. Ferreira[4], Dejânia V. de Araújo[5], Luiz Gonsaga de Carvalho[2], Fábio Moreira da Silva[2] and Luciana Sanches[1]
[1]*Federal University of Mato Grosso, Brazil,* [2]*Federal University of Lavras,* [3]*CEPEC/CEPLAC l,* [4]*Federal University of Acre,* [5]*State University of Mato Grosso, Brazil*

Yorinori , 2001). In general, the lower leaves of young plants are more susceptible than the upper leaves (Mignucci & Lim, 1980).

In relation to physiological changes in the host, Mignucci & Boyer (1979) studied the inhibition of photosynthesis and transpiration of soybean infected with powdery mildew and found lower photosynthesis and transpiration with increased infection. With 82% of leaf area infected, more than half of the leaf photosynthetic activity had been lost and transpiration dropped to 36% compared to control, considering the direct result of the change in metabolic activity induced by the pathogen. Because infection occurs primarily in the lower leaves and poorly lit, it is unlikely that the reduction in rates of photosynthesis and leaf transpiration resulted in great reduction in soybean yield, however, favorable climatic conditions may enabled the infection of upper leaves leading to high losses (Mignucci & Boyer, 1979; Sartorato & Yorinori, 2001).

Susceptibility of cultivars and influence of the climate has caused outbreaks of powdery mildew in successive years in Brazil. The lack of resistance in most cultivars have required chemical control mainly in the south and the high plateaus of the savannahs. In the U.S.A., powdery mildew caused economic damage reached in the 70's and early 80's. Since then, the use of resistant cultivars has dispensed chemical control (Sartorato & Yorinori, 2001).

Reactions of different soybean varieties to powdery mildew and the effect of environmental variables in the progress of the disease have been reported (Arny et al. 1975; Buzzell et al. 1975; Degree & Laurence, 1975, Johnson & Phillips, 1961; Mignucci 1977; Mignucci & Boyer, 1979; Mignucci & Lim, 1980; Lohnes & Bernard, 1992; Lohnes & Nickell, 1994). According to Bedendo (1995), in Brazil, powdery mildew may occur in the humid and cold climates, but are favored by hot dry conditions (20-25 °C). According to the author, conidia do not germinate when is present a film of water on the leaf surface, however, relative humidity near 95% is required for germination.

Mignucci et al. (1977) reported temperatures of 18 °C as favorable to the development of powdery mildew on susceptible cultivars and at temperatures of 30 °C disease progress was inhibited. Degree & Laurence (1975) also observed lower disease severity at 30 °C. According to Sartorato & Yorinori (2001) the information about the effects of relative humidity, leaf wetness, rainfall, solar radiation or other environmental factors in the progress of powdery mildew in soybeans was not precise.

Therefore, the intensity of powdery mildew of soybean under different temperatures and periods of leaf wetness on the cultivars conquista and suprema was evaluated.

2. Material and methods

Seeds of soybean cultivar conquista (MG/BR 46) and suprema were sown in pots containing 5 kg of soil mixture, sand and organic matter (manure) in the proportion 2:1:0.5 in a green house. Thinning was performed 15 days after planting, leaving two plants per pot, forming the experimental unit. The plants were kept in a green house until the V3 stage, according to the soybean phenological scale proposed by Ritchie et al. (1982). During the same period, inoculation of *M. diffusa* was done stirring soybean plants on healthy plants which were then randomly placed next to diseased plants (Demski & Phillips, 1974). According to Grau (1975), because of the ease with which conidia are disseminated, it becomes hard to test inoculation of *M. diffusa* with different isolates without contamination.

The plants were transferred to growth chambers and arranged in randomized blocks, factorial 4 x 5 with three replicates, considering four air temperatures (15, 20, 25 and 30 degrees C) and five leaf wetness periods (0 , 6, 12, 18 and 24 hours). For the different periods of leaf wetness, recently sprayed plants were kept in a moist chamber with transparent plastic bags, during the period used for each treatment. In the treatment of 0 h of wetness, the plants were taken without a moist chamber for the growth chambers. Irrigation was performed by spraying water directly on the stem of the plants.

There were four incidence and severity assessments every five days after the beginning of the experiment. The severity was assessed on all central leaflet of each plant with trifoliate leaves at 9, 11, 13 and 15 days after inoculation, using the grading scale published Sartorato & Yorinori (2001), adopting grade 1 = 1% of affected leaf area; grade 2 = 5%, grade 3 = 10%, grade 4 = 25%, grade 5 = 50%, grade 6 = 100%.

The intensity data were integrated using the area under incidence progress curve over time, according to Campbell & Madden (1990).

$$AUDPCS = \sum_{i}^{n-1} \left(\frac{ys_i + ys_{i+1}}{2}\right)(t_{i+1} - t_i) \qquad (1)$$

$$AUDPCI = \sum_{i}^{n-1} \left(\frac{yi_i + yi_{i+1}}{2}\right)(t_{i+1} - t_i) \qquad (2)$$

Where:

AUDPCI was the area under the progress curve of powdery mildew incidence; AUDPCS was the area under the progress curve of powdery mildew severity; ys and yi were the disease severity and incidence over time i and i+1, respectively; t was the time in days and n was the number of evaluations along the time.

Plants with chlorosis, green islands, rusty stains, and combination of these symptoms were considered infected by powdery mildew (Figure 1).

Fig. 1. Evaluated signals of powdery mildew in soybean plants.

The significant variables in the F test by the variance analysis of AUDPCS were subjected to regression analysis, linear and nonlinear adjustment models (Leite & Amorim, 2002, Reis et al., 2004). In the case of significant interaction, the combined effect of temperature and leaf wetness duration in disease intensity was modeled (Reis et al., 2004).

3. Results and discussion

Symptoms of powdery mildew evaluated at 9, 11, 13 and 15 days after inoculation, were characterized by chlorosis, green islands, rusty stains, and combination of these symptoms in the cultivars suprema and conquista. However, the most striking characteristic evaluated was the presence of the fungus, and powdery white structure on the surface of infected leaves. This symptomatology was consistent with those described by Sartorato & Yorinori (2001), Tanaka et al. (1993), Yorinori et al. (1993) and Yorinori & Hiromoto (1998). According to Sartorato & Yorinori (2001) and Yorinori et al. (1998), may also be variations in the symptoms of powdery mildew due to climatic variations, genetic variability between populations of *M. diffusa*, genetic resistance of cultivars, stage of plant development and adopted agronomic practices. Tanaka et al. (1993) studying the occurrence of powdery mildew (*M. diffusa*) in a collection of 27 soybean genotypes, in a green house, observed differences of severity symptoms presented by By Hampton cultivars (more susceptible), followed by IAC-Foscarin 31 and IAC-Santa Maria 702, respectively. According to Lohnes & Bernard (1992), Lohnes & Nickell (1994) and Mignucci & Lim (1980), the differing responses of powdery mildew in soybeans are consequences of three alleles at locus Rmd: Rmd-c (resistant), Rmd (resistance in adult plants) and rmd (susceptibility) and according to Dunleavy (1978) and Phillips (1984), these differences may be evidenced by a 35% loss of productivity in soybean cultivars susceptible to powdery mildew in the field.

A significant interaction in the F test between temperature and leaf wetness was observed for the AUDPCS in conquista cultivar (P = 0.0242) and the isolated effect of temperature in conquista (P <0.0001) and suprema (P <0.0001) . Thus, models were adjusted using non-linear regression to describe the monocyclic process of the epidemic based on the dependent variables. With regard to temperature, a greater amount of disease was observed at temperatures around 23 °C for the conquista and 24 °C for the suprema cultivar. Temperatures above 15 °C and 30 °C were not favorable to the development of powdery mildew in both cultivars (Figures 2, 3 and 4).

Likewise Leath & Carroll (1982) in a study on powdery mildew of soybean cultivars, evaluating 38 cultivars, observed greater susceptibility of cultivars Ware, Falcon, AP350, V76-438, Emerald, AgDSR232, AgDSR532, Md71-583 as well as smaller and larger disease progress (*M. diffusa*) in Georgetown, under temperature of 29.6 °C and 23.2 °C, respectively. However, Mignucci et al. (1977) found at temperatures of 18 °C in a green house, the greater progress of powdery mildew in Flambeau, Norchief, Chippewa 64, Corsoy, Harosoy 63, Wells cultivars,, grown in the USA and Puerto Rico. Seedlings were subjected to temperatures of 18, 24 and 30 °C per 14 hours, with alternating 10 hour temperature of 20 °C, to simulate day and night temperatures. In further studies, the same authors, after inoculation of *M. diffusa*, in cultivar Harosoy, in a green house, diseased plants were kept in a growth chamber under daytime temperature of 26 ± 2 °C and night 21 ± 2 °C (Mignucci & Chamberlain, 1978) and at 25 ± 0.25 °C (Mignucci & Boyer, 1979). However, both Mignucci et al. (1977), Mignucci (1989) and Leath & Carroll (1982) agreed that temperatures around 30 °C were not favorable to disease progress, similar to that observed in the cultivars evaluated in this study. In another pathosystem, powdery mildew of grape (*Uncinula necator* (Schw.) Burr.), The optimum temperature for growth of the fungus was 25 °C, while in the temperature between 21 and 30 °C there was germination of spores and increased sporulation. At temperatures above 33 °C occurred death of spores and colonies (Thomas et al., 1994; Reis, 2004).

$$AUDOCS = (54.5327)\exp(-(2.86394)(((T-(23.1927))/(9.03343))^2 +((M-(8.24455))/(14.5439))^2)) \quad R2 = 0.85*$$

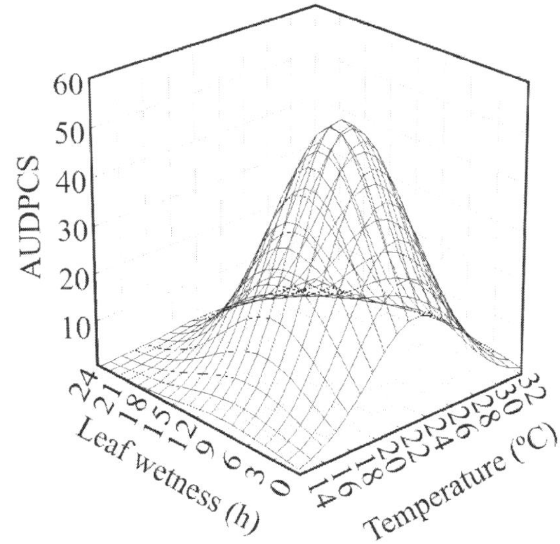

Fig. 2. Nonlinear regression of the progress curve of disease severity (AUDPCS) of powdery mildew of soybean in conquista cultivar according to the interaction between temperature and leaf wetness.

With regard to leaf wetness in conquista cultivar, there were signs of the disease from 0 to 8 hours of leaf wetness, with growth up values of AUDPCS until 8 hours, with the maximum of temperature of 23 °C, indicating the need of water for germination of spores and fungus infection. From that point, higher values of leaf wetness reduced the AUDPCS (Figure 2). With respect to the isolated effects of leaf wetness periods in suprema cultivar, the maximum point of leaf wetness in the AUDPCI occurred in the period of 12.9 hours, with significant reduction near 0 and 24 hours (Figure 5). There is little information in the literature about the effects of leaf wetness on powdery mildew in soybeans (Sartorato & Yorinori, 2001), however, according to Bedendo (1995), this disease can occur in humid regions, but is favored by dry environments. Mignucci (1989) reported that the low relative humidity is highly favorable precisely described the development of powdery mildew in soybeans, though not presented values to describe precisely. Similarly, Brodie & Neufeld (1942) studying the development of conidial structures of *Erysiphe polygoni* DC., found germination in relative humidity ranging from 0-100%, while Mattiazzi (2003) studying the effect of mildew on the soybean production, observed greater progress at a relative humidity of 80%. Thus, the relative humidity of the growth chamber, with an average of 50%, may have given the conidial germination, even in treatments with no leaf wetness.

$$AUDPCS = (0.132503)((T-(12.0232))^{(1.51746)})(((30.41)-T)^{(0.98158)})$$
$$R^2=0.99^{**}$$

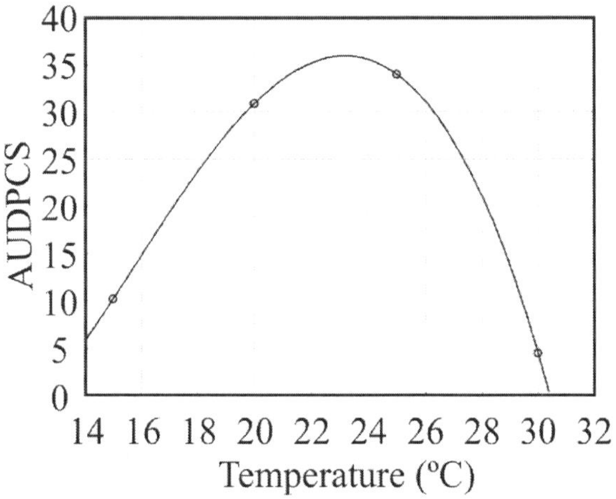

Fig. 3. Nonlinear regression of the progress curve of disease severity (AUDPCS) of powdery mildew of soybean in conquista cultivar according to the isolated effect of temperature.

$$AUDPCS = (0.103698)((T-(11.6493))^{(1.60307)})(((30.2061)-T)^{(0.900076)})$$
$$R^2=0.99^{**}$$

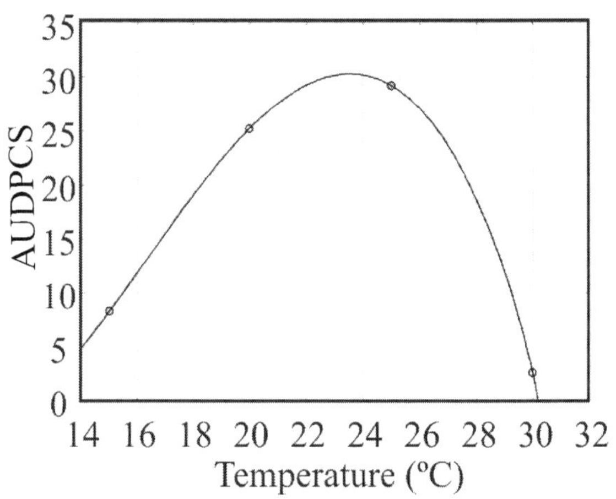

Fig. 4. Nonlinear regression of the progress curve of disease severity (AUDPCS) of powdery mildew of soybean in suprema cultivar according to the isolated effect of temperature.

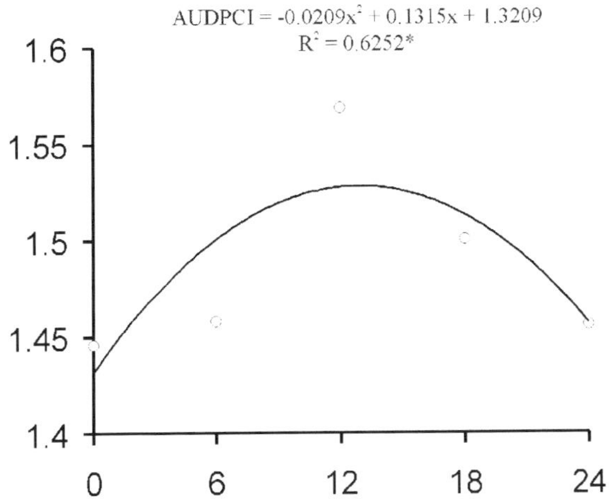

Fig. 5. Linear regression with polynomial quadratic fit of the progress curve of disease incidence (AUDPCI) of powdery mildew of soybean in suprema cultivar (Y axis) according to the isolated effect of leaf wetness (X axis).

Therefore, the results on the effect of leaf wetness in the progress of powdery mildew (*M. diffusa*) are contradictory and studies on the interaction effect between temperature and duration of leaf wetness on disease progression had not yet been assessed. The results presented potential use in prospective studies on the effects of weather on the progress of powdery mildew of soybean cultivars in Brazil.

4. Conclusion

The progress of the severity of powdery mildew (AUDPCS) in suprema and conquista cultivars was favored by air temperatures around 23 °C and 24 °C, respectively.
Leaf wetness of 8h and air temperature of 23 °C provide the maximum progress of disease severity in conquista cultivar.
Temperatures above 30 °C and 15 °C reduced the intensity of the disease.

5. Summary

In this study the effects of temperature and leaf wetness period on the intensity progress of powdery mildew in soybean conquista and suprema cultivars were evaluated. Plants at the V3 stage were inoculated in greenhouse. Subsequently, the plants were conditioned in growth chambers at temperatures of 15, 20, 25 and 30°C and leaf wetness periods of 0, 6, 12, 18 and 24 hours. Severity data was integrated in time by the disease progress curve for incidence (AUDPCI) and severity (AUDPCS). Non-linear regression models were adjusted for the disease severity and a polynomial fit was adjusted for disease incidence data. Temperatures near 23 °C and 24 °C favored the powdery mildew intensity progress

(AUDPCS) in Conquista and Suprema cultivars, respectively. Leaf wetness period of 8 h allowed the maximum progress of the disease in conquista at temperatures of 23 °C. Temperatures near 30 °C and 15 °C reduced powdery mildew intensity. The maximum point of leaf wetness in the AUDPCI occurred in the period of 12.9 hours, with significant reduction near 0 and 24 hours

6. References

Agrianual (2008). *Anuário da Agricultura Brasileira.* 2007. São Paulo: FNP Consultoria & Agroinformativos. 504 p.

Arny, D.C.; Hanson, E.W.; Worf, G.L.; Oplinger, E.S.; Hughes, W.H. (1975). Powdery mildew on soybean in Wisconsin. *Plant Disease Reporter*, Beltsville, v. 59, p. 288-290.

Bedendo, I.P. Oídios. (1995). In: Bergamin Filho, A.; Kimati, H.; Amorim, L. *Manual de fitopatologia*: princípios e conceitos. 3. ed. São Paulo: Agronômica Ceres. p. 866-871.

Brodie, H.J.; Neufeld, C.C. (1942). The development and structure of the conidia of *Erysiphe polygoni* DC and their germination at low humidity. *Canadian Journal Research*, Ontario, v. 20, p. 41-61.

Buzzell, R.I.; Haas, J.H. (1975). Powdery mildew of soybeans. *Soybean Genetics Newsletter*, Ames, v. 2, p. 7-9.

Campbell, C.L.; Madden, L.V. (1990). *Introduction to plant disease epidemiology*. New York: Academic Press. 532 p.

Dunleavy, M.M. (1978). Soybean seed yield losses caused by powdery mildew. *Crop Science*, Madison, v. 18, p. 337-339.

Demski, J.W.; Phillips, D.V. (1974). Reactions of soybean cultivars to powdery mildew. *Plant Disease Reporter*, Beltsville, v. 58, p. 723-726.

Embrapa. (1976). Centro Nacional de Pesquisa de Arroz e Feijão. *Manual de métodos de pesquisa em feijão*. EMBRAPA-CNPAF, Goiânia. 80 p.

Grau, C.R.; Laurence, J.A. (1975). Observations on resistance and heritability of resistance to powdery mildew of soybean. *Plant Disease Reporter*, Beltsville, v. 59, p. 458-460.

Johnson, H.W.; Phillips, D.V. (1961). Other legumes prove susceptible to a powdery mildew of *Psoralea tenax*. *Plant Disease Reporter*, Beltsville, v. 45, p. 542-543.

Leath, S.; Carroll, R.B. (1982). Powdery Mildew on Soybean in Delaware. *Plant Disease*, St. Paul, v. 66, p. 70-71.

Leite, R.M.V.B.C.; Amorim, L. (2002). Influência da temperatura e do molhamento foliar no monociclo da mancha de alternaria em girassol. *Fitopatologia brasileira*, v. 27, p. 193-200.

Lohnes, D.G.; Bernard, R.L. (1992). Inheritance of resistance to powdery mildew in soybeans. *Plant Disease*, St. Paul, v. 76, p. 964-965.

Lohnes, D.G.; Nickell, C.D. (1994). Effects of Powdery Mildew Alleles *Rmd-c, Rmd,* and *rmd* on Yield and Other Characteristics in Soybean. *Plant Disease*, St. Paul, v. 78, p. 299-301.

Mattiazzi, P. (2003). *Efeito do oídio (Microsphaera diffusa Cooke e Peck) na produção e duração da área foliar sadia da soja.* Tese (Doutorado em Fitotecnia). Escola Superior de Agricultura Luiz de Queiroz, Piracicaba.

Mignucci, J.S. (1989). Powdery mildew. In: Sinclair, J.B.; Backman, P.A. (3. Ed.) *Compendium of soybean diseases.* Saint Paul: APS Press. p. 21-23.

Mignucci, J.S.; Boyer, J.S. (1979). Inhibition of photosynthesis and transpiration in soybean infected by *Microsphaera diffusa* Powdery mildew. *Phytopathology*, St. Paul, v. 69, p. 227-230.

Mignucci, J.S.; Chamberlain, D.W. (1978). Interactions of *Microsphaera diffusa* with Soybeans and Other Legumes. *Phytopathology*, v. 68, p. 169-173.

Mignucci, J.S.; Lim, S.M. (1980). Powdery mildew development on soybeans with adult plant resistance. *Phytopathology*, St. Paul, v. 70, p. 919-921.

Mignucci, J.S.; Lim, S.N.; Hepperly, P.R. (1977). Effects of temperature on reactions of soybean seedlings to powdery mildew. *Plant Disease Reporter*, Beltsville, v. 61, p. 122-124.

Miyasaka, S.; Medina, J.C. (1981). *A soja no Brasil.* Campinas: ITAL. 1026 p.

Phillips, D.V. (1984). Stability of *Microsphaera diffusa* and the effect of powdery mildew on yield of soybeans. *Plant Disease*, St. Paul, v. 68, p. 953-956.

Reis, E.M. (2004). *Previsão de doenças de plantas.* Passo Fundo, UPF. 316 p.

Reis, E.M.; Sartori, A.F.; Câmara, R.K. (2004). Modelo climático para previsão da ferrugem da soja. *Summa Phytopathologica*, Botucatu, v. 30, p. 290-292.

Ritchie, S.; Hanway, J.J.; Thompson, H.E. (1982). *How a soybean plant develops.* Ames: Iowa State University of Science and Technology, Cooperative Extension Service. 20 p.(Special Report, 53).

Sartorato, A.; Yorinori, J.T. (2001). Oídios de leguminosas: feijoeiro e soja. In: Standnick, M. J.; Rivera, M. C. *Oídios.* Jaguariúna: Embrapa Meio Ambiente. p. 255-284.

Silva, O.C.; Seganfredo, R. (1999). Quantificação de danos ocasionados por doenças de final de ciclo e Oídio, em dois cultivares de soja. In: *Anais do Congresso Brasileiro de Soja*, Londrina, 1999. Brasil, p. 460. (resumo 299).

Sinclair, J.B. (1999). Powdery mildew. In: Hartman, G.L.; Sinclair, J.B.; Rupe, J.C. *Compedium of Soybean Diseases.* 4 ed. APS Press, St. Paul, 100p.

Tanaka, M.A.S.; Ito, M.F.; Mascarenhas, H.A.A.; Dudienas, C.; Miranda, M.A.C. (1993). Desenvolvimento do oídio da soja em casa-de-vegetação. *Summa Phytopathologica*, Botucatu, v. 19. p. 125-126.

Thomas, C.S.; Gubler, W.D.; Leavitt, G. (1994). Field testing of a powdery mildew disease forecast model on grapes in California. *Phytopathology*, v. 84, p. 1070.

Yorinori, J.T. (1982). Doenças na soja no Brasil. In: A soja no Brasil Central. 2 ed. Campinas: Fundação Cargill, p. 301-364.

Yorinori, J.T. (1998). Controle integrado das principais doenças da soja. In: Câmara, G. M. S. *Soja*: tecnologia da produção. Piracicaba. p. 139-192.

Yorinori, J.T. (1986). Doenças da soja no Brasil. In: Fundação Cargill (3. Ed.) *A soja no Brasil Central.* Campinas. p. 301-363.

Yorinori, J.T.; Charchar, M.; D´Avila, J.; Nasser, L.C.B.; Henning, A.A. (1993). Doenças da soja e seu controle. In: Arantes, N. E.; Souza, P. I. M. *Cultura da Soja nos Cerrados.* Piracicaba: POTAFOS. p. 333-397.

Yorinori, J.T.; Hiromoto, D.M. (1998). *Determinação de perdas em soja causadas por doenças fúngicas*. In: EMBRAPA - Centro Nacional de Pesquisa de Soja. Embrapa- CNPSo, Londrina. p. 112-114 (Embrapa-CNPSo. Documentos, 118).

11

Future Biological Control for Soybean Cyst Nematode

Masanori Koike[1], Ryoji Shinya[2], Daigo Aiuchi[3], Manami Mori[1], Rui Ogino[1], Hiroto Shinomiya[1], Masayuki Tani[1] and Mark Goettel[4]

[1]*Department of Agro-environmental Science*
Obihrio University of Agriculture & Veterinary Medicine
[2]*Graduate School of Agriculture, Kyoto University*
[3]*National Research Center of Protozoan Disease*
Obihrio University of Agriculture & Veterinary Medicine
[4]*Lethbridge Research Centre, Agriculture and Agri-Food Canada, Lethbridge*
[1,2,3]*Japan*
[4]*Canada*

1. Introduction

The soybean cyst nematode (SCN) *Heterodera glycines* Ichinohe, is widely distributed in soybean-producing countries. The losses in total yield caused by SCN are greater than those for any other pest of soybean (Wrather et al., 2001). These nematodes have generally been controlled by rotating soybeans with nonhost crops, planting of resistant cultivars, application of effective nematocides and organic materials, and physical control techniques such as solarisation. The combination of biological control with above methods will enhance the effectiveness of nematode control. Recently, numerous studies have been conducted on the fungal antagonist of SCNs (Chen and Dickson, 1996; Kim and Riggs, 1991, 1995; Liu and Chen, 2001; Meyer and Huettel, 1996; Meyer and Meyer, 1996; Timper et al., 1999); however, few biological control agents have been commercialized to date.

Lecanicillium spp. (formally, *Verticillium lecanii*) have been studied as potential biological control agents for SCN. Entomopathogenic *Lecanicillium* spp. are ubiquitously distributed in soils, although these fungi are mainly isolated from insects. Numerous strains have been commercialized worldwide as biopesticides namely of aphids, thrips and mites (Faria and Wraight, 2007; Kabaluk et al, 2010) . In addition, it is known that *Lecanicillium* spp. have a broad host range, *e.g.*, insects, phytopathogenic fungi, and plant-parasitic nematodes (Hall, 1981; Meyer et al., 1990; Goettel et al., 2008) providing the possibility that strains could be found that could be developed for simultaneous control of multiple pest problems. For instance, a strain of *L. longisporum* was found to effectively control both cucumber powdery mildew and aphids (Kim et al, 2007, 2008, 2010).

One strain of *Lecanicillium* sp was found to exhibit high virulence to SCNs, although it was found to be a poor colonizer of the soybean rhizosphere (Meyer and Wergin, 1998). However, it is quite likely that other strains are more aggressive rhizosphere colonizers because *Lecanicillium* spp. (*V. lecanii*) possess varied abilities among different strains

(Sugimoto et al., 2003). The objective of this chapter is to review the development of entomopathogenic *Lecanicillium* hybrid strains with effects on the SCN, and discuss the future prospects for its use in the biological control of the SCN.

2. Genus *Lecanicillium*, as pathogen of plant parasitic nematodes

Until recently, the form genus *Verticillium* contained a wide variety of species with diverse host ranges including arthropods, nematodes, plants and fungi (Zare and Gams, 2001). The genus has been recently redefined using rDNA sequencing, placing all insect pathogens into the new genus *Lecanicillium* (Zare et al., 2000; Gams and Zare, 2001; Zare and Gams, 2001). These include *L. attenuatum, L. lecanii, L. longisporum, L. muscarium* and *L. nodulosum*, which were all formerly classified as *V. lecanii*. These recent reclassifications bring forth the possibility that several different species were actually involved in previous studies. There is also evidence that in recent literature, some authors have simply replaced the genus name *Verticillium* with *Lecanicillium* without conducting the necessary rDNA sequencing, adding to the confusion (Sugimoto et al., 2003; Koike et al., 2007a). In this review, we refer to the former name, *Verticillium lecanii*, as *Lecanicillium* spp. unless it is specifically known that the species in question was verified using the new nomenclature.

Species of *Lecanicillium* are well known and important nematophagous fungi with potential for development as biopesticides against plant-parasitic nematodes. For instance, *L. psalliotae, L. antillanum*, and other *Lecanicillium* spp. infect the eggs of the root-knot nematode *Meloidogyne incognita* (Gan et al., 2007; Nguyen et al., 2007). *Lecanicillium* spp. infect females, cysts and eggs of *Heterodera glycines*, the soybean cyst nematode (SCN), reducing nematode populations in laboratory and greenhouse studies (Meyer et al., 1997). Mutant strains of an SCN active strain were induced through UV radiation which resulted in increased efficacy against this nematode (Meyer and Meyer, 1996).

Some reports indicated that immature eggs are more susceptible to fungal attack than the mature eggs containing second stage juveniles (J2) (Chen and Chen, 2003; Irving and Kerry, 1986; Kim and Riggs, 1991). Furthermore, Meyer et al. (1990) demonstrated that one strain of *Lecanicillium* sp. (as *V. lecanii*) decreased the number of viable SCN eggs from yellow females, whereas the viability of eggs from cysts was not affected. This strain also reduced the viability of SCN eggs without colonization of the egg; however, no such effect was observed in other strains. This suggested that *V. lecanii* produced a natural substance that could affect egg viability and there was a remarkable variation in the ability for producing such a substance among strains.

3. Genetic improvement of entomopathogenic *Lecanicillium* spp. using protoplast fusion

Mycotal® (*L. muscarium*) and Vertalec® (*L. longisporum*) are strains commercialized by Koppert, The Netherlands, for insect control. Strain B-2 of *L. muscarium*, which was isolated from the peach aphid (*Myzus persicae*) in Japan, has high epiphytic ability on cucumber leaves (Koike et al., 2004). Protoplast fusion was performed using three strains of *Lecanicillium* spp. (as *V. lecanii*) to obtain new strains possessing useful characteristics as biological control agents (Aiuchi et al. 2004, 2008). From the combination of Vertalec-Mycotal, B-2 -Mycotal, and B-2-Vertalec, many hybrid strains were detected. Nit (nitrate non-utilizing) mutants (Correll et al., 1987) were used for visually selecting protoplasts (Fig.1).

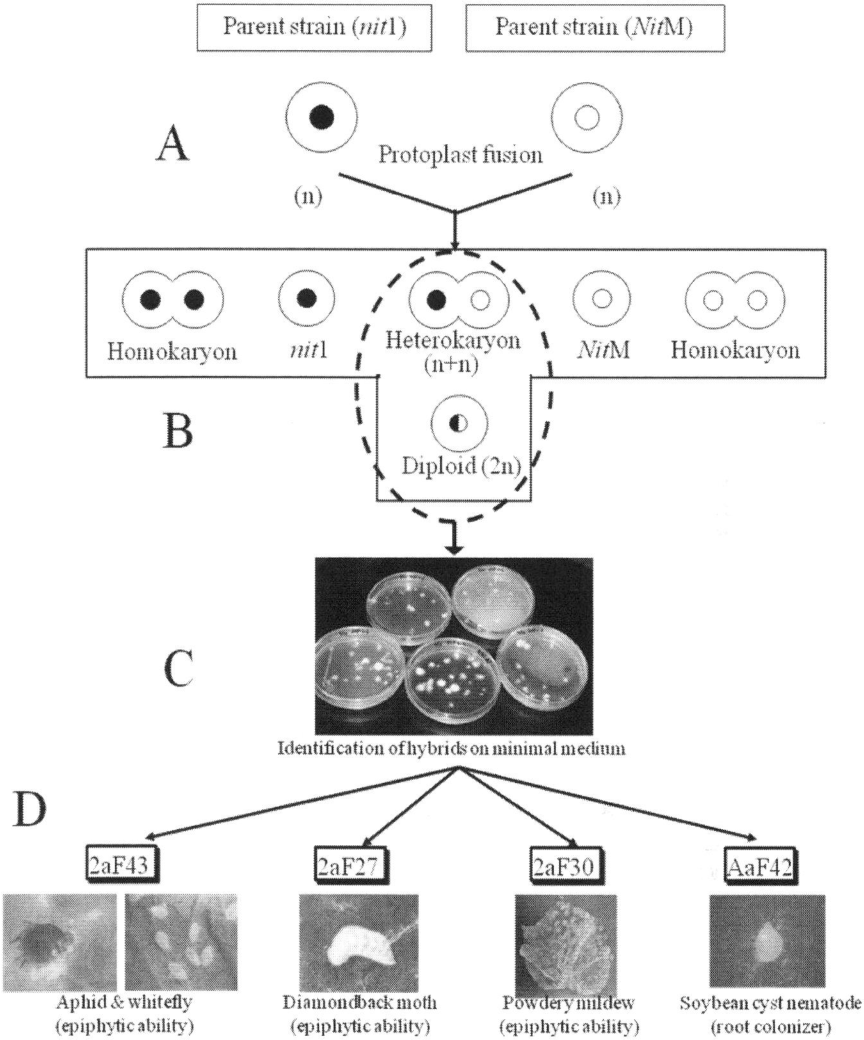

Fig. 1. Model for identification of hybrids on protoplast fusion procedure and selection sequence for hybrid strains of *Lecanicillium* spp. A) Protoplast fusion was conducted on complemental combination of *nit* mutants. B) Protoplast suspension after fusion treatment contain heterokaryon, diploid, homokaryon (self-fusing) and nit mutant (non-fusing). C) Only heterokaryon and diploid could develop the colony as prototrophic growth on minimal medium. D) Screening procedure based on various parameters and candidates of hybrid strains as BCAs.

The morphological characteristics of the hybrid strains differed from those of their parental nit mutants. Furthermore, genomic analyses were done to ascertain the success of protoplast fusion. These confirmed protoplast fusions were in genomic DNA but not in mitochondrial DNA (mtDNA). In both analyses, they observed a uniform biased tendency of the banding pattern, depending on the combination of the parental strains. Some of these genomic analyses confirmed successful fusion and/or genetic recombination. These results demonstrated the usefulness of conducting genomic analyses such as polymerase chain reaction-restriction fragment length polymorphism, arbitrarily primed-PCR and genome profiling for discovering nucleotides that exhibit high polymorphism in order to ascertain success of protoplast fusion (Aiuchi et al., 2008, Kaibara et al., 2010).

Further studies were conducted to screen desirable *Lecanicillium* hybrid strains that have a wide host range or increased efficacy (Aiuchi et al., 2007). Initially, 43 hybrid strains were used in bioassays against the cotton aphid, *Aphis gossypii*. Of these, 30 strains induced mortality equal to or higher than Vertalec (42%). Secondly, 50 hybrid strains were used in bioassays against the greenhouse whitefly, *Trialeurodes vaporariorum*. Of these, 37 strains exhibited an equal or higher infection rate as compared to that of Mycotal (36.2%). Finally, 50 hybrid strains were applied to cucumber leaves in order to test strain viability under low humidity conditions (ca.13% RH). Two weeks after application, 17 hybrid strains exhibited viabilities equal to or higher than B-2 (1.5×10^3 cfu/cm^2). These results identified hybrid strains whose parental characteristics had not only recombined but also whose pathogenicity or viability had improved, with a hybrid isolate even producing conidia on a leaf hair. Finally, 13 candidate hybrid strains were selected that exhibited improved qualities, and these hybrid strains can be expected to be highly effective as biological control agents (Fig.1).

3.1 Selection of *Lecanicillium* hybrid strains against the SCN

Shinya et al. (2008a) investigated whether the protoplast fusion technique was an effective tool for development of more efficient nematode control agents. Three parental strains (Vertalec, Mycotal, and B-2) and their 162 hybrid strains were screened in greenhouse pot tests against the soybean cyst nematode *H. glycines*. Some of these hybrid strains reduced the density of SCN in the soil and suppressed damage to soybean plants. In particular, one hybrid strain, AaF42 (Vertalec: *L. longisporum* ×Mycotal: *L. muscarium*), reduced nematode egg density by 93% as compared with the control providing excellent protection to soybean plants. Furthermore, this strain significantly reduced cyst and egg densities compared to the parental strains (Fig.2, Table 1).

Fig. 2. *Lecanicillium* hybrid strain AaF42 (Vertalec × Mycotal) protected soybean plants from soybean cyst nematode (*Heterodera glycine*) 4, 6 and 8 weeks after treatment in SCN infested soil (Shinya et al. 2008a) .

Strains [1]	Cysts/50 g soil	Eggs/g soil	Eggs/cyst	Fresh root weights (g)
AaF17	12.3 ± 4.4 [de]	18.7 ± 7.5 [d]	73.3 ± 17.0 [e]	1.10 ± 0.09 [bc]
AaF23	26.7 ± 9.2 [cde]	39.3 ± 15.6 [d]	78.2 ± 16.6 [e]	0.93 ± 0.08 [ab]
AaF42	11.9 ± 3.6 [e]	16.8 ± 8.3 [d]	70.9 ± 29.4 [e]	1.52 ± 0.21 [c]
AaF80	27.8 ± 7.3 [cd]	65.0 ± 22.3 [cd]	114.9 ± 13.5 [cd]	0.85 ± 0.22 [ab]
AaF103	13.0 ± 4.6 [de]	27.0 ± 11.6 [d]	100.9 ± 19.0 [de]	0.83 ± 0.14 [ab]
Mycotal	39.4 ± 9.4 [bc]	106.0 ± 29.3 [bc]	133.3 ± 12.3 [bc]	0.62 ± 0.08 [ab]
Vertalec	47.9 ± 11.6 [b]	157.4 ± 54.5 [b]	161.1 ± 20.6 [ab]	0.72 ± 0.03 [ab]
Control 1 (without fungus) [2]	69.1 ± 17.2 [a]	248.6 ± 75.7 [a]	179.2 ± 25.5 [a]	0.60 ± 0.09 [ab]
Control 2 (untreated) [3]	ND [4]	ND	ND	1.00 ± 0.10 [ab]

The values are the means ± standard deviation of three replicates. The different letters in the columns indicate significant differences ($P < 0.01$, Tukey's HSD test).

1 The hybrid strains, AaF were derived from protoplast fusion of Vertalec × Mycotal.

2 Control 1: SCN was inoculated but fungus was not.

3 Control 2: Neither SCN nor fungus was inoculated.

4 ND: not detected.

Table 1. The effects of selected strains of *Verticillium lecanii* on the density of *Heterodera glycine* cysts and eggs, and the growth of soybean roots in pots (Shinya et al., 2008a).

3.2 Effects of culture filtrates of the *Lecanicillium* hybrid strains to SCN

Shinya et al. (2008b) also evaluated the effects of fungal culture filtrates of the *Lecanicillium* hybrid strains on mature eggs, embryonated eggs (eggs fertilized but without development of juveniles), and J2 of SCN and compared these effects to those of their parental strains. The fungal culture filtrates of some hybrid strains inhibited egg hatch of mature eggs. Furthermore, the fungal culture filtrates of two hybrid strains, AaF23 and AaF42 (Vertalec: *L. longisporum*× Mycotal: *L. muscarium*), exhibited high toxicity against embryonated eggs. However, most of the fungal culture filtrates did not inactivate J2.

These results suggested that the enzymes or other active compounds in the fungal culture filtrates exhibit activity against specific stages in the SCN life cycle. In addition, based on a visual assessment of the morphological changes in eggs caused by filtrates of each strain, there were differences between the hybrid strains and their respective parental strains with regard to the active substances produced by *Lecanicillium* spp. against the embryonated eggs (Fig. 3). It is known that some entomopathogenic fungi produced nematicidal and insecticidal metabolites, for example entomopathogenic *Verticillium* sp. FKI-1033 (*Lecanicillium* sp.) produced Verticilide (Shiomi et al., 2006). As a result of promoting recombination of whole genomes via protoplast fusion, several hybrid strains may have enhanced production of active substances that are different from those produced by their parental strains. It was concluded that natural substances produced by *Lecanicillium* hybrids are important factors involved in the suppression of SCN damage.

3.3 Parasitism of the *Lecanicillium* hybrid strains to SCN

Shinya et al., (2008c) also investigated the pathogenicity and mode of action of the *Lecanicillium* hybrid strains to the sedentary stages of SCN. Three different sedentary stages (pale yellow female, yellow brown cyst, and dark brown cyst) of SCN were treated and incubated on water agar. After 3 weeks incubation, eggs were investigated for the following:

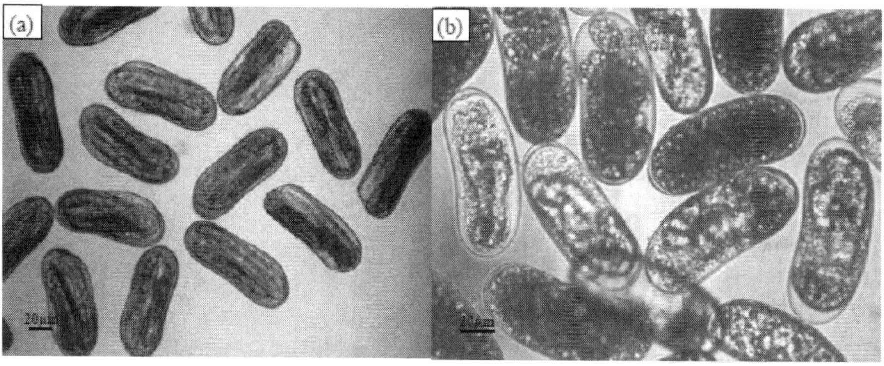

Fig. 3. Effect of fungal culture filtrates on the development of embryonated eggs. (a) Mature eggs containing J2 in the control well after 10 days incubation. (b) Abnormal eggs treated with fungal culture filtrates of *Lecanicillium* hybrid strain AaF23 after 10 days incubation (Shinya et al., 2008b).

(i) the infection frequencies of eggs, (ii) the number of eggs laid, and (iii) the number of mature and healthy eggs. Subsequently, the fecundity of SCN treated with the *Lecanicillium* hybrids was investigated in greater detail.

Most *Lecanicillium* hybrid strains examined appeared to have higher infection rates of pale yellow female (PYF) eggs than those of yellow brown cysts (YBCs) and dark brown cysts (DBCs). Meyer and Wergin (1998) reported that cysts tended to be more rapidly colonized by *V. lecanii* (*Lecanicillium* sp.) than females and also described that the cyst wall apparently was not a barrier to *V. lecanii*, so it is possible that these results show differences in egg development. PYFs contained more immature eggs than cysts. It is thought that *Lecanicillium* hybrid strains infected more eggs that had not completed their embryonic development than mature eggs containing J2 individuals.

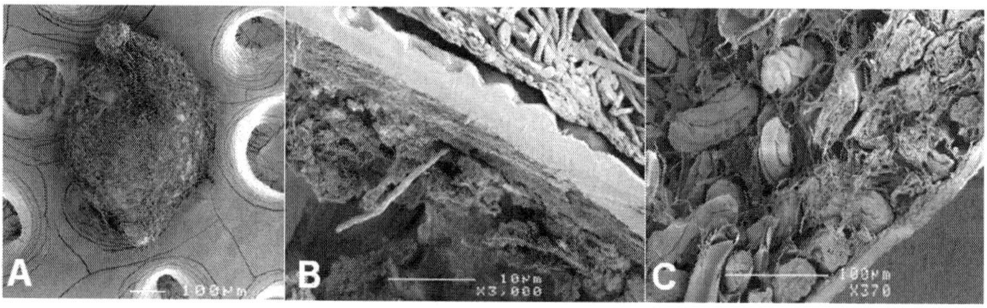

Fig. 4. Scanning electron micrographs of *Lecanicillium* hybrid strain AaF42 infected soybean cyst nematodes. A: Colonized mature female, B: Penetration of cyst wall, C: Infected eggs of SCN.

Moreover, infection with some *Lecanicillium* hybrid strains reduced the number of eggs of PYFs. Egg laying by females treated with AaF42 terminated approximately 3 d after incubation. The body wall of these females rapidly tanned and the individuals subsequently encysted. A cyst can be considered a dead female (Niblack, 2005); therefore, the formation of cysts indicated that females treated with AaF42 died before the completion of egg laying. Meyer and Wergin (1998) observed that some females colonized by *V. lecanii* contained few eggs and hypothesized that *V. lecanii* infected and killed some females before a full complement of eggs was produced. Our results also support this hypothesis. In addition, Kerry (1990) indicated that *V. chlamydosporium* (*Lecanicillium chladosporia*) reduced the fecundity of *Heterodera schachtii* infected individuals forming small cysts containing few healthy eggs. In this study, four *Lecanicillium* hybrid strains (AaF42, AaF17, AaF103, and AaF23) that suppressed SCN populations and damage to soybean plants in a preliminary greenhouse test tended to reduce the number of eggs and also the number of mature eggs in PYFs; however, no significant difference was observed in the effect on YBCs among individual strains in YBCs, and AaF42, which caused remarkable suppression of SCN populations in a greenhouse test, did not exhibit a high percentage of egg infection in cysts (Shinya et al., 2008c). This suggests that *Lecanicillium* hybrid strains may have colonized and rapidly weakened or killed SCN females before the completion of egg laying and reduced the number of mature and healthy eggs in soil.

Since the evaluation method using estimates of the number of mature and healthy eggs is largely accurate over several modes of action, it appears that this method is an appropriate and simple *in vitro* test to evaluate the pathogenicity of *Lecanicillium* hybrid strains to nematode eggs. However, testing the efficacy of these fungi in soil is essential, since fungi that perform well in laboratory tests may not be effective under field conditions (Kerry, 2001).

Based on the results of this study, we conclude that *Lecanicillium* hybrid strains are more effective against female SCN than against cysts, and the following could be its modes of action: (i) the colonization of females and the reduction of their fecundity, (ii) the prevention of embryonic development or the killing of immature eggs, and (iii) the infection of immature or dead eggs (Fig.4). From this viewpoint, the ability to attack females and the ability to colonize soybean root surfaces, from which females emerge, may be important to control SCN by *Lecanicillium* hybrid strains, and at least these two abilities should be high in potentially useful strains. It is quite likely that AaF42 which exhibited a high reduction of fecundity has high potential as a biological control agent against SCN.

3.4 *Lecanicillium* hybrid strain AaF42 as rhizosphere colonizer and endophyte

There has been little unequivocal evidence of true rhizosphere competence (growth of the fungus within the root zone utilizing plant carbon) in entomopathogenic fungi. The mechanisms of interaction between fungus and plant root needs to be elucidated (Vega et al., 2009). Gaining an understanding of the population structure of rhizosphere colonizers and how they change throughout the season is imperative for development of strategies for controlling plant parasitic nematodes, root diseases and improving root health. The current soil treatment with methyl bromide: chloropicrin can improve plant growth and yield even in the absence of known soilborne pathogens (Martin, 2003).

The ecology of fungal entomopathogens in the rhizosphere is an understudied area of insect pathology. The rhizosphere is the region of soil in which the release of root exudates influences the soil microbiota, and may provide a favorable environment for fungal entomopathogens (Bruck, 2010). We performed studies to determine the persistence of *Lecanicillium* hybrid strain AaF42 as soybean root colonizer. It was found that AaF42 was a better root colonizer compared with parental strains (Vertalec & Mycotal, Fig. 5).

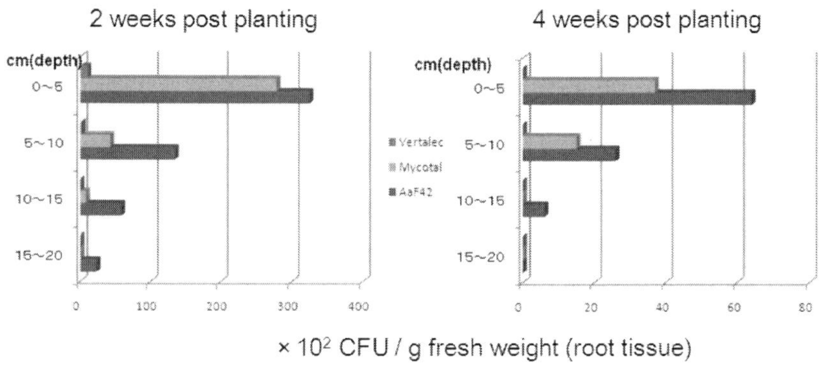

Fig. 5. Fungal populations on soybean roots (cfu g-1 fresh weight of root tissue) of *Lecanicillium* hybrid strain AaF42 and the parental strains (Vertalec & Mycotal) using a cylinder pot (height 50mm, ϕ 85mm). Dilution plate method was done on the surface of soybean roots thus avoiding propagules within the root tissues

Two weeks after planting soybeans into pots pretreated with the fungi, Mycotal and AaF42 could be detected ca. 3 X 10^4 cfu per g root fresh weight and there were no significant differences at the soil depth 0~5cm. However, as the soil depth increased, more AaF42 was detected than Mycotal. At four weeks after planting, there was one order difference in detection between AaF42 and Mycotal. In contrast, the detections of Vertalec were nil or very low. Bruck (2010) described the role of fungal entomopathogens in the rhizosphere for controlling root-feeding insects. Currently, data on the pest management potential of rhizosphere competent fungal entomopathogens are scant. However, the prospective ramifications of this relationship are tremendous. A simple calculation of the economic benefits that can be realized by utilizing rhizosphere competent fungal entomopathogens yields savings significant enough to warrant further investigation (Bruck, 2010). It can be said that *Lecanicillium* hybrid strain AaF42 with high culture filtrate toxicity, pathogenicity and parasitisim to SCN, and a good root-colonization ability, shows considerable promise for development as a biological control agent for SCN.

Recently, molecular and micro-ecological trials with *Lecanicillium* hybrid strain AaF42 were designed to do elucidate the tritrophic interactions among the fungi, SCN and soybean root (unpublished data). This was accomplished by employing a gfp gene driven by a constitutive promoter which strongly labeled the fungus with no impact on fungal growth or pathogenicity (Fig. 6). Preliminary results indicated that AaF42 might act as an endophyte, however, further studies are required before firm conclusions can be made.

3.5 Stage specificity of *Lecanicillium* against SCN and its importance in the control of SCN

As described above, the *stage* in the SCN life cycle attacked by *Lecanicillium* hybrid strain AaF42 has a profound effect on the viability of SCN and damage to soybean crops. This is a very significant point in the control of plant parasites, especially cyst nematodes. The cyst nematodes generally have a high reproductive potential, producing approximately 200-500 eggs per cyst (female), and they can survive for several years at least in the soil without a host plant. Therefore, several thousand nematodes appear in the next generation even if

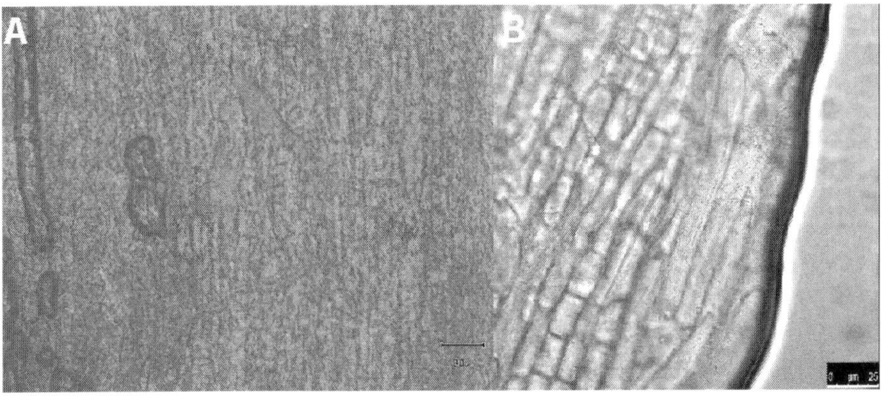

Fig. 6. *Lecanicillium* hybrid strain AaF42 as possible endophyte (A: AaF42 mycelium with normal Lactophenol Cotton Blue Stain, B: Recombinant AaF42 with GFP gene within soybean root tissue)

only several J2 nematodes successfully invade the root of a host plant. The *J2* of cyst nematodes which emerge from the eggs can quickly invade roots near the root tip of a host plant. Thus, the sedentary stage, especially immature female or immature cyst, would be the most appropriate target stage in the biological control using nematophagous fungi. The nematode-trapping fungi, e.g., *Arthrobotrys spp.,* are the well *known* group of nematophagous fungi, probably owing to their remarkable morphological adaptations and their dramatic infection of nematodes. However, these fungi are known as a poor colonizers of eggs and sedentary stages of cyst nematodes (Chen et al., 1996). From this view point, the nematode-trapping fungi seemed to be unsuitable as biological control agents against cyst nematode. We demonstrated that *Lecanicillium* hybrid strain AaF42 has a distinguished infectivity against sedentary stages of SCN, especially immature females and eggs, and a high ability as root colonizer and endophyte. It would be inferred from these exceptional talents that hybrid strain AaF42 has high potential as a biological agent against SCN.

4. Future prospect (potential of biological control agents for SCN and other complex diseases)

Fungi traditionally known for their entomopathogenic characteristics, such as *Beauveria bassiana* and *Lecanicillium* spp., have recently been shown to engage in plant-fungus interactions (Vega, 2008; Vega et al., 2008), and both have been reported to effectively suppress plant disease (Goettel et al. 2008; Ownley et al., 2004, 2008). Biological control of plant pathogens usually refers to the use of microorganisms that reduce the disease causing activity or survival of plant pathogens. Several different biological control mechanisms against plant pathogens have been identified. The biocontrol organism is directly involved in some mechanisms such as antibiosis, competition, and parasitism. With other modes of biological control, such as induced systemic resistance and increased growth response, endophytic colonization by the biocontrol organism triggers responses in the plant that reduce or alleviate plant disease (Ownley et al., 2010).

Lecanicillium spp. have activity against numerous phytopathogenic fungi including powdery mildews (Verhaar et al., 1997, 1998; Askary et al., 1997, 1998, 1999; Dik et al., 1998; Miller et al., 2004), rusts (Spencer and Atkey, 1981; Leinhos and Buchenauer, 1992) green molds (Benhamou and Brodeur, 2000) and *Pythium* (Benhamou and Brodeur, 2001). Fungi that may control phytopathogenic fungi can act through antibiosis and mycoparasitism (Kiss, 2003). Some *Lecanicillium* isolates act as mycoparasites, attaching to powdery mildew mycelia and conidia, producing enzymes such as chitinase, that allow penetration of the mildew spores and hyphae, killing the pathogen (Askary et al., 1997). Leinhos and Buchenauer (1992) demonstrated that several *Lecanicillium* spp. were able to penetrate and colonize uredial sori of *Puccinia coronata*. In *Penicillium digitatum*, the mode of action was attributed to changes in host cells prior to contact by the *Lecanicillium* spp. (Benhamou and Brodeur, 2000) while in *P. ultimatum*, in addition to mycoparasitism of the plant pathogen, the mode of action was linked to colonization of host plant tissues, triggering a plant defense reaction (Benhamou and Brodeur, 2001). Hirano et al. (2008) found that applying *L. muscarium* blastospores to cucumber roots induced systemic resistance. *L. muscarium* pre-inoculated plants suffered significantly fewer lesions and reduced disease severity compared with non-inoculated plants. Kusunoki et al. (2006) and Koike et al. (2007b) found that root treatment with *L. muscarium* reduced disease incidence and wilting score in other soil-borne disease combinations such as tomato—*Verticillium dahliae*, Japanese radish—*V. dahliae*, and melon—*Fusarium oxysporum* f.sp. *melonis*.

In the case of soilborne pathogens, further opportunities exist for interactions with other microorganisms occupying the same ecological niche. The significant role of nematodes in the development of diseases caused by soilborne pathogens has been demonstrated in many crops throughout the world. In many cases, such nematode-fungus disease complexes involve root-knot nematodes (*Meloidogyne* spp.), although several other endoparasitic (*Globodera* spp., *Heterodera* spp., *Rotylenchulus* spp., *Pratylenchus* spp.) and ectoparasitic (*Xiphinema* spp., *Longidorus* spp.) nematodes have been associated with diseases caused by soilborne fungal pathogens (Back et al., 2002). In the case of SCN, Sudden Death Syndrome (SDS) caused by *F. solani* is a major disease of soybean which, among other symptoms, induces root rot, crown necrosis, interveinal chlorosis, defoliation and abortion of pods (Rupe, 1989; Nakajima *et al.*, 1996). Recent research on SDS has focused on identifying genes for dual resistance against both nematode and fungus (Chang *et al.*, 1997; Meksem *et al.*, 1999; Prabhu *et al.*, 1999).

It is known that entomopathogenic *Lecanicillium* spp. have antagonistic effects to soil-borne fungi such as *Fusarium oxysporum, F. solany, Pythium* spp. and *Verticillium dahlia* (Koike et al., 2006, Goettel et al., 2008). Therefore, it might be possible to develop *Lecanicillium* hybrid strains with potential for biological control of a complex of plant diseases, plant parasitic nematodes and insect pests.

5. Conclusion

Much research is still needed to fully understand the role that rhizosphere competent fungal entomopathogenic *Lecanicillium* hybrid strains play in regulating SCN populations and how we can use this knowledge to design and implement more effective SCN biological control programs. Questions of particular importance to consider are highlighted by Vega et al. (2009) and include the following: (1) Do plants benefit from a rhizosphere association with fungal entomopathogens? (2) Is the 'bodyguard' concept relevant in soil? If so, what is the signaling mechanism between trophic levels? (3) Do different phylogenetic groups of fungal entomopathogens display different strategies in their association with plants? (4) How do soil-borne fungal entomopathogens interact between above and below ground ecosystems? (5) What is the mechanism of yield increases in biological control target plant? (6) Does plant diversity impact fungal entompathogen diversity at the landscape or local level, and what is its impact on natural pest control? In addition to the basic scientific questions posed above, there are a number of questions that require further investigation as well: (1) What is the most effective approach for inoculating roots with rhizosphere competent isolates? Approaches will need to be identified for plants propagated via seed treatment, because there are a lot of problems in the direct treatment of soil such as costs & labor requirements. (2) How long do rhizosphere competent isolates persist on the root system of soybean or other host plants of plant parasitic nematodes? (3)Will the use of rhizosphere competent isolates provide consistent and acceptable levels of pest including plant parasitic nematode control?

At present there has been only limited success with field applications of biological controls against SCN. Chen (2004) pointed out factors involved in their biological control, 1) stage of nematode infected, 2) ability to colonize soil, roots, cysts and gelatinous matrices, 3) competition with other fungi, 4) cropping systems and tillage, and 5) edaphic and environmental factors. In our research, all experiments were done *in vitro* and in glasshouses. Although there is still much to be learned at the field level, it has been demonstrated that *Lecanicillium* hybrid strains have multiple effects (toxic and parasitism) for SCN and soybean plant roots (as root colonizer and endophyte) as well as on plant pathogens and insect pests, making these strains promising for development as broad spectrum biopesticides that include SCN.

6. References

Aiuchi, D.; Koike, M.; Tani, M.; Kuramochi, K.; Sugimoto, M. & Nagao, H. (2004). Protoplast fusion, using nitrate non-utilizing (nit) mutants in the entomopathogenic fungus *Verticillium lecanii* (*Lecanicillium* spp.). IOBC/WPRS Bulletin 27 (8), 127-130.

Aiuchi, D.; Baba, Y.; Inami, K.; Shinya, R.; Tani, M.; Kuramochi, K.; Horie, S. & Koike, M. (2007). Screening of *Verticillium lecanii* (*Lecanicillium* spp.) hybrid strains based on evaluation of pathogenicity against cotton aphid and greenhouse whitefly, and viability on the leaf surface. Japanese Journal of Applied Entomology & Zoology 51, 205-212.

Aiuchi, D.; Inami, K.; Sugimoto, M.; Shinya, R.; Tani, M.; Kuramochi, K. & Koike, M. (2008). A new method for producing hybrid strains of the entomopathogenic fungus *Verticillium lecanii* (*Lecanicillium* spp.) through protoplast fusion by using nitrate non-utilizing (nit) mutants. Micologia Aplicada International 20, 1-16.

Askary, H., Benhamou, N., Brodeur, J. ((1997)). Ultrastructural and cytochemical investigations of the antagonistic effect of *Verticillium lecanii* on cucumber powdery mildew. Phytopathology 87, 359-368.

Askary, H., Benhamou, N., Brodeur, J., (1999). Ultrastructural and cytochemical characterization of aphid invasion by hyphomycete *Verticillium lecanii*. Journal of Invertebrate Pathology 74, 1-13.

Askary, H., Carriere, Y., Belanger, R.R., Brodeur, J., (1998). Pathogenicity of the fungus *Verticillium lecanii* to aphids and powdery mildew. Biocontrol Science & Technology 8, 23–32.

Benhamou, N., Brodeur, J. (2000). Evidence for antibiosis and induced host defense reactions in the interaction between *Verticillium lecanii* and *Penecillium digitatum*, the causal agent of green mold. Phytopathology 90, 932–943.

Benhamou, N., Brodeur, J. (2001). Pre-inoculation of Ri T-DNA transformed cucumber roots with the mycoparasite, *Verticillium lecanii*, induces host defense reactions against *Pythium ultimum* infection. Physiological & Molecular Plant Pathology 58, 133–146.

Bruck, D. (2009) Fungal entomopathogens in the rhizosphere. BioControl 55:103-112.

Chen, S.Y., Chen, F.J. (2003). Fungal parasitism of *Heterodera glycines* eggs as influenced by egg age and pre-colonization of cysts by other fungi. Journal of Nematology 35, 271–277.

Chen, S.Y., Dickson, D.W. (1996). Pathogenicity of fungi to eggs of *Heterodera glycines*. Journal of Nematology 28, 148-158.

Claydon, N., Grove, J.F. (1982). Insecticidal secondary metabolic products from the entomogenous fungus *Verticillium lecanii*. Journal of Invertebrate Pathology 40, 413–418.

Correll, J.C., Klittich, C.J.R., Leslie, J.F. (1987). Nitrate nonutilizing mutants of *Fusarium oxysporum* and their use in vegetative compatibility tests. Phytopathology 77, 1640–1646.

Dik, A.J., Verhaar, M.A., Belanger, R.R., 1998. Comparison of three biological control agents against cucumber powdery mildew (*Sphaerotheca fuliginea*) in semicommercial-scale glasshouse trials. European Journal of Plant Pathology 104, 413–423.

Faria, M.R., Wraight, S.P., 2007. Mycoinsecticides and mycoacaricides: a comprehensive list with worldwide coverage and international classification of formulation types. Biological Control 43, 237–256.

Gams, W., Zare, R., 2001. A revision of *Verticillium* Sect. Prostrata. III. Genetic classification. Nova Hedwigia 72, 329–337.

Gan, Z., Yang, J., Tao, N., Liang, L., Mi, Q., Li, J., Zhang, K.-Q., 2007. Cloning of the gene *Lecanicillium psalliotae* chitinase Lpchi1 and identification of its potential role in the biocontrol of root-knot nematode *Meloidogyne incognita*. Applied Microbiology and Biotechnology 76, 1309–1317.

Gindin, G., Barash, I., Harari, N., Raccah, B., 1994. Effect of endotoxic compounds isolated from *Verticillium lecanii* on the sweet potato whitefly, *Bemisia tabaci*. Phytoparasitica 22, 189–196.

Goettel, M.S., Eilenberg, J., Glare, T.R., 2005. Entomopathogenic fungi and their role in regulation of insect populations. In: Gilbert, L.I., Iatrou, K., Gill, S. (Eds.), Comprehensive Molecular Insect Science, vol. 6. Elsevier, Oxford, pp. 361–406.

Goettel, M.S., M. Koike, J.J. Kim, D. Aiuchi, R. Shinya, and J. Brodeur. 2008. Potential of *Lecanicillium* spp. for management of insects, nematodes and plant diseases. Journal of Invertebrate Pathology 98: 256-261.

Hall, R.A. (1981) The fungus *Verticillium lecanii* as a microbial insecticide against aphids and scales. In: Burges, H.D. (ed), Microbial control of pests and plant disease 1970-1980, Academic Press, London. pp.115-146.

Hirano, E., Aiuchi, D., Tani, M., Kuramochi, K., Koike, M. (2008). Pre-inoculation of cucumber roots with *Verticillium lecanii* (*Lecanicillium muscarium*) induces resistance to powdery mildew. Research Bulletin Obihiro University of Agriculture & Veterinary Medicine 29, 82-94.

Irving, F., Kerry, B.R., 1986. Variation between strains of the nematophagous fungus, *Verticillium chlamydosporium* Goddard. II. Factors affecting parasitism of cyst nematode eggs. Nematologica 32, 474–485.

Kabaluk, J.T., Svircev, A.M., Goettel, M.S., & Woo, S.G. (eds) (2010), The Use and Regulation of Microbial Pesticides in Representative Jurisdictions Worldwide., IOBC Global, 99 pages available online http://www.iobc-global.org/publications.html#microbial_regulation_book

Kaibara, F., Koike, M., Aiuchi, D., Oda, H., Hatakeyama,Y., Iwano, H. (2009) DNA polymorphisms in hybrid strains of entomopathogenic fungi *Lecanicillium* spp. *IOBC/wprs Bulletin* 45 : 279-282 (2009)

Kang, C.S., Goo, B.Y., Gyu, L.D., Heon, K.Y., 1996. Antifungal activities of *Metarhizium anisopliae* against *Fusarium oxysporum*, *Botrytis cinerea* and *Alternaria solani*. Korean Journal of Mycology 24, 49–55.

Kavkova, M., Čurn, V., 2005. *Paecilomyces fumosoroseus* (Deuteromycotina: Hyphomycetes) as a potential mycoparasite on *Sphaerotheca fuliginea* (Ascomycotina: Erysiphales). Mycopathologia 159, 53–63.

Kerry, B. R. (2001) Exploitation of the nematophagous fungal *Verticillium chlamydosporium* Goddard for the biological control of root-knot nematodes (*Meloidogyne* spp.). In: Butt, T. M., Jackson, C., Magan, N. (eds), Fungi as biocontrol agents: progress, problems and potential, CAB International, Wallingford. pp. 155-167.

Kim, D.G., Riggs, R.D., 1991. Characteristics and efficacy of a sterile hyphomycete (ARF18), a new biocontrol agent for *Heterodera glycines* and other nematodes. Journal of Nematology 23, 275–282.

Kim, D.G., Riggs, R.D., 1995. Efficacy of the nematophagous fungus ARF18 in alginate-clay pellet formulation against *Heterodera glycines*. Journal of Nematology 23:275-282.

Kim, J.J., Goettel, M.S., Gillespie, D.R., 2007. Potential of *Lecanicillium* species for dual microbial control of aphids and the cucumber powdery mildew fungus, *Sphaerotheca fuliginea*. Biological Control 40, 327–332.

Kim, J.J., Goettel, M.S., Gillespie, D.R., 2008. Evaluation of *Lecanicillium longisporum*, Vertalec for simultaneous suppression of cotton aphid, Aphis gossypii, and cucumber powdery mildew, *Sphaerotheca fuliginea*, on potted cucumbers. Biol. Control., 45: 404-409

Kim, J.J., M.S. Goettel and D.R. Gillespie. 2010 Evaluation of *Lecanicillium longisporum*, Vertalec against the cotton aphid, *Aphis gossypii* , and cucumber powdery mildew, *Sphaerotheca fuliginea* in a greenhouse environment. Crop Protection 29, 540-544.

Kiss, L., 2003. A review of fungal antagonists of powdery mildews and their potential as biocontrol agents. Pest Management Science 59, 475–483.

Koike, M., Higashio, T., Komori, A., Akiyama, K., Kishimoto, N., Masuda, E., Sasaki, M., Yoshida, S., Tani, M., Kuramochi, K., Sugimoto, M., Nagao, H., 2004. *Verticillium lecanii* (*Lecanicillium* spp.) as epiphyte and their application to biological control of pest and disease in a glasshouse and a field. IOBC/WPRS Bulletin 27 (8), 41–44.

Koike, M., Sugimoto, M., Aiuchi, D., Nagao, H., Shinya, R., Tani, M., Kuramochi, K., 2007a. Reclassification of Japanese isolate of *Verticillium lecanii* to *Lecanicillium* spp. Japanese Journal of Applied Entomology and Zoology 51, 234–237 (in Japanese with English summary).

Koike, M., Yoshida, S., Abe, N., Asano, K., 2007b. Microbial pesticide inhibiting the outbreak of plant disease damage. US National Phase Appl. No. 11/568,369, 371(c).

Kusunoki, K., Kawai, A., Aiuchi, D., Koike, M., Tani, M., Kuramochi, K., 2006. Biological control of Verticillium black-spot of Japanese radish by entomopathogenic *Verticillium lecanii* (*Lecanicillium* spp.). Research Bulletin of Obihiro University 27, 99–107 (in Japanese with English summary).

Leinhos, G.M.E., Buchenauer, H., (1992) Hyperparasitism of selected fungi on rust fungi of cereal. Z. Pflanzenkr. Pflanzenschutz 99, 482–498.

Liu, X.,Z., Chen, S.,Y., (2001) Screening isolates of *Hirsutella* species for biocontrol of *Heterodera glycines*. Biocontrol Science and Technology 11:151-160.

Martin, F.N. (2003) Development of alternative strategies for management of soilborne pathogens currently controlled with Methyl bromide. Annual Review of Phytopathology 41:325–50

Meyer, S.L.F., Huettel, R.N., Sayre, R.M. (1990) Isolation of fungi from *Heterodera glycines* and *in vitro* bioassays for their antagonism to eggs. Journal of Nematology 22, 532–537.

Meyer, S.L.F., Huettel, R.N. (1996) Application of a sex pheromone, pheromone analogs, and *Verticillium lecanii* for management of *Heterodera glycines*. Journal of Nematology 28:36-42.

Meyer, S.L.F., Meyer, R.J. (1996.) Greenhouse studies comparing strains of the fungus *Verticillium lecanii* for activity against the nematode *Heterodera glycines*. Fundamentals of Applied Nematology 19, 305–308.

Meyer, S.L.F., Johnson, G., Dimock, M., Fahey, J.W., Huettel, R.N., 1997. Field efficacy of *Verticillium lecanii*, sex pheromone, and pheromone analogs as potential management agents for soybean cyst nematode. Journal of Nematology 29, 282–288.

Meyer,S.L.F., Wergin, W.P. (1998) Colonization of soybean cyst nematode females, cyst, and gelatinous matrices by the fungus *Verticillium lecanii*. Journal of Nematology 34:1-8

Miller, T.C., Gubler, W.D., Laemmlen, F.F., Geng, S., Rizzo, D.M., 2004. Potential for using *Lecanicillium lecanii* for suppression of strawberry powdery mildew. Biocontrol Science and Technology 14, 215–220.

Nguyen, N.V., Kim, Y.-J., Oh, K.-T., Jung, W.-J., Park, R.-D., 2007. The role of chitinase from *Lecanicillium antillanum* B-3 in parasitism to root-knot nematode *Meloidogyne incognita* eggs. Biocontrol Science and Technology 17, 1047–1058.

Ownley, B.H., Griffin, M.R., Klingeman, W.E., Gwinn, K.D., Moulton, J.K., Pereira, R.M., 2008. *Beauveria bassiana:* endophytic colonization and plant disease control. Journal of Invertebrate Pathology 98, 267–270.

Ownley, B. H., Gwinn, K. D., Vega, F.E. (2010) Endophytic fungal entomopathogens with activity against plant pathogens:ecology and evolution. BioControl 55:113-128.

Ownley, B.H., Pereira, R.M., Klingeman, W.E., Quigley, N.B., Leckie, B.M., 2004. *Beauveria bassiana*, a dual purpose biocontrol organism, with activity against insect pests and plant pathogens. In: Lartey, R.T., Caesar, A.J. (Eds.), Emerging Concepts in Plant Health Management. Research Signpost, India, pp. 255–269.

Shinya, R., Aiuchi, D., Kushida, A.,Tani, M., Kuramochi, K., Koike, M. (2008c) Pathogenicity and its mode of action in different sedentary stages of *Heterodera glycines* (Tylenchida: Heteroderidae) by *Verticillium lecanii* hybrid strains. Applied Entomology & Zoology 43(2):227-233.

Shinya, R., Aiuchi, D., Kushida, A., Tani, M., Kuramochi, K., Kushida, A., Koike, M. (2008b) Effects of fungal culture filtrates of *Verticillium lecanii* (*Lecanicillium* spp.) hybrid strains on *Heterodera glycines* egg and juveniles. Journal of Invertebrate Pathology 97: 291–297.

Shinya, R., Watanabe, A., Aiuchi, D., Tani, M., Kuramochi, K., Kushida, A., Koike, M. (2008a) Potential of *Verticillium lecanii* (*Lecanicillium* spp.) hybrid strains as biological control agents for soybean cyst nematode: Is protoplast fusion an effective tool for development of plant-parasitic nematode control agents? Japanese Journal of Nematology 38(1):9-18.

Shiomi, K., Matsui, R., Kakei, A., Masuma, R., Arai, N., Isozaki, M., Monma, S., Sunazuka, T., Kobayashi, S., Tanaka, H., Turberg, A., Omura, S. (2006) A Novel Ryanodine Binding Inhibitor, Verticilide, Produced by *Verticilium* sp. FKI-1033. In International Symposium on the Chemistry of Natural Products 2006, 23.07.2006, Available from http://ci.nii.ac.jp/els/110006802293.pdf?id=ART0008750947&type=pdf&lang=jp&host=cini&order_no=&ppv_type=0&lang_sw=&no=1301800760&cp=

Spencer, D.M., Atkey, P.T., 1981. Parasitic effects of *Verticillium lecanii* on two rust fungi. Trans. Br. Mycol. Soc. 77, 535–542.

Sugimoto, M., Koike, M., Hiyama, N., Nagao, H., 2003. Genetic, morphological, and virulence characterization of the entomopathogenic fungus *Verticillium lecanii*. Journal of Invertebrate Pathology 82, 176–187.

Timper, P., Riggs, R.D., Crippen D.L. (1999) Parasitism of sedentary stages of *Heterodera glycines* by isolates of a sterile nematophagous fungus. Phytopathology 89:1193-1199.

Vega, F.E. (2008) Insect pathology and fungal endophytes. Journal of Invertebrate Pathology 98:277–279.

Vega, F.E., Goettel, M.S., Blackwell, M., Chandler, D., Jackson, M.A., Keller, S., Koike, M., Maniania, N.K., Monzo´n ,A., Ownley, B.H., Pell, J.K., Rangel, D.E.N., Roy, H.E. (2009) Fungal entomopathogens: new insights on their ecology. Fungal Ecology 2:149–159

Verhaar, M.A., Hijwegen, T., Zadoks, J.C., 1998. Selection of *Verticillium lecanii* isolates with high potential for biocontrol of cucumber powdery mildew by means of components analysis at different humidity regimes. Biocontrol Science and Technology 8, 465–477.

Verhaar, M.A., Ostergaard, K.K., Hijwegen, T., Zadoks, J.C., 1997. Preventive and curative applications of *Verticillium lecanii* for biological control of cucumber powdery mildew. Biocontrol Science and Technology 7, 543–551.

Zare, R., Gams, W., 2001. A revision of *Verticillium* section Prostrata. IV. The genera *Lecanicillium* and *Simplicillium*. Nova Hedwigia 73, 1–50.

Zare, R., Gams, W., Culham, A., 2000. A revision of *Verticillium* sect. Prostrata. I.Phylogenetic studies using ITS sequences. Nova Hedwigia 71, 465–480.

12

Soybean Utilization and Fortification of Indegenous Foods in Times of Climate Changes

J. C. Anuonye
Food Science and Nutrition, Federal University of Technology Minna Niger State Nigeria

1. Introduction

Food security remain an unfulfilled dream for more than 800 million people (Combes *et al.*, 1996) who are unable to lead healthy and active lives because they lack assess to safe and nutritious food. More than 840 million people lack access to enough food to meet their daily basic needs, while more than one third of the world's children are stunted due to diets inadequate in quantity and quality (WHO, 2001). Widespread nutritional problems are steadily reported in less developed countries (LDCS). This is manifested in protein energy malnutrition indicated within vulnerable groups such as infants, children, the elderly, and pregnant and lactating mothers, who often have high nutrient needs.

Anon (2003), reports that the World Health Organization (WHO) called protein energy malnutrition, (PEM), the silent emergency. According to this report, it declared that PEM is an accomplice in at least half of 10.4 million child deaths each year. WHO (2001) reports that malnutrition cast long shadows, affecting close to 800 million people with 20% of all such people in the LDC. Reports of these wide growing nutritional problems have been steadily mentioned even in Nigeria (Smith and Oluwoye, 1988). Majority of this class is found in the rural areas and urban slums where common heritage of poverty, ignorance, poor sanitation and other conditions contribute to the problems of malnutrition, interfere with its solution, and thus perpetuate a vicious cycle.

Most malnourished people live in Asia and Africa; and the staple of most people in Asia and Africa are starchy pastes. These pastes are made from cereals (sorghum, rice, maize, wheat, millet, acha*)* roots and tubers (cassava, yam, sweet potato and plantain). These crops do not only provide marginal nutrition (especially for children) but also require high inputs of time, labour and fuel to prepare. In most cases they are consumed as combinations in the home because the blends provides complementary balance of amino acids (proteins) in the diet (FAO, 1985). That Africa and especially sub Saharan Africa is in danger of food shortages is no longer news. What is news however is the inability of this region to rise to the great danger facing this region in terms of provision of adequate food.

It was in response to this bleak future that the Bill Gate foundations (2007-2009) sponsored recent research on the possibility of development of drought resistant legumes including soybeans for the areas prone to drought. This was in the realization that these legumes would not only provide needed protein there by improving the nutritional status of the farming populations it would also enhance the socio- economic status of the populace through value chain addition.

The broad objective of the soybean component of the study was to enhance promiscuous multipurpose soybean productivity and production in drought-prone areas in sub-Saharan Africa. The specific objectives included: To increase production of soybean by 15%, through increasing on farm yields in drought years by 20%, on 60% of target area planted, and also by increasing value chain marketing by 20%, income by 30% and house-hold consumption by 25%. These objectives were to be achieved through; Testing of promiscuous multipurpose existing soybean breeding lines for drought and low P tolerance. At least 20 promiscuous elite soybean lines with resistance to bacterial pustule, frog-eye spot, rust and shattering were to be evaluated for adaptation to drought and low P tolerance, and for promiscuous nodulation.

Part of the overall goal of the work was to develop soybean value chains to increase income and improve nutrition of smallholder farm families and other rural entrepreneurs. At least ten thousand households across the target countries were to be informed about profitable and environmental friendly value- addition technologies. At least 25% increase in consumption of soybean products and 25% increase in household income for at least 25% of the population in the target areas was envisaged. At least six training courses for soybean processing and utilization was to be organised in each project site in each country . At least 5 pilot sites for community-based soybean value addition operational in each of the target countries was to be established. At least 10 best-bet technologies for value addition in soybean used by at least 10% of households in the target countries was to be developed.

1.1 Effect of drought on the nutritional status of soybean

A study to evaluate the nutritional qualities of soybean grown under limited rainfall (drought) was carried out. The results showed that drought grown soybeans grains were smaller and ranged between 6.81-7.88 mm in length and 4.42-5.20 mm in width compared to 7.47-8.22 mm in length and 5.15-5.72 mm in width of the rainfed soybeans grains.

The functional properties of drought and rain fed samples(Table 1) showed that packed and loose bulk density of the milled flours were not significant ($P \geq 0.05$). Water absorption capacity had been reported by Oyelade *et al* (2002) to denote the maximum amount of water that a food material can take up and retain under formulation conditions. It is known to be related to the degree of dryness and porosity. The high water absorption capacities and index showed that incorporation of the drought sample flour to other food supplement would yield similar results as use of rainfed soybean flours. Foaming capacity of the drought samples were significantly different ($P \leq 0.05$) from the rainfed soybean sample. Though there were no significant differences ($P \geq 0.05$) in the foaming stability of the drought and rainfed samples, however, drought grown soybean samples had higher foaming capacities but lower foaming stability. Soy protein is used in food formulations for its foaming properties (Iwe, 2003). The results showed that drought materials have potential capacities to be used as foaming agents. Thus it could be a success in replacement or partial replacement of traditional ingredients for foaming such as egg white (INTSOY, 1998).

Emulsion capacity showed no significant differences ($P \leq 0.05$) between the drought and rainfed samples. However the drought soybeans showed higher emulsion capacities than the rainfed samples indicating that there were no loss of critical functional property.

The least gelation capacity were also not significantly different ($P \geq 0.05$). Gelation is an important functional property that soy protein can impart to comminuted sausage products. The results showed that even at 1%, the drought soybeans like the rainfed, soybean produced a gel.

Similarly there were significant ($P \leq 0.05$) differences in all the proximate parameters measured (Table 2). The results followed already established principles that drought materials are low in moisture but higher in protein percentage. The reduction in moisture content and the hardness of the grain kernel makes oil extraction difficult. But results gotten from this work showed that even fat content of the drought material was significantly ($P \leq 0.05$) higher than the rain fed samples. The vitamin and mineral contents of the materials showed similar trends in their profile. The amino acid profile further elucidated the nutritional superiority of the drought samples over the rainfed samples. Compared to the FAO recommendations for infants and adults the results showed that drought materials exceeded the recommendations for infant nutrition in all the amino acids.
The implication of this is that while rainfed material may record higher yields due to bigger seed size ,drought materials would be better nutritionally.

Functional Properties	Drought Soybean	Rainfed Soybean
Packed bulk density (g/ml)	0.04	0.42
Loose bulk density (g/ml)	0.55	0.60
Water absorption capacity (%)	114.67	108.3
Water absorption index (%)	2.15	2.09
Oil absorption capacity (%)	151.00	131.00
Foaming capacity (%)	4.00	2.17
Foaming stability (%)	1.75	2.25
Emulsion capacity (%)	37.29	32.08
Emulsion stability (%)	28.16	30.14
Least gelation capacity (%)	49.74	51.87

Table 1. The Functional Properties of the Drought and Rainfed Soybean

Proximate	Rainfed Soybean	Drought Soybean
Protein (%)	34.07	38.25
Fat (%)	15.85	18.21
Crude Fibre (%)	4.37	4.88[a]
Ash (%)	4.91	5.14
Dry matter (%)	92.50	95.88
Moisture (%)	7.49	4.13
Carbohydrate (%)	37.67	34.27
Energy (Kcal)	429.60	454.20

Table 2. The Proximate Composition of Drought and Rainfed Soybean.

Vitamins	Drought Soybeans	Rainfed Soybeans
Retinal (Vit A) (μg/100g)	241.75	293.05
Tocopherol (Vit E) (μg/100g)	60.63	82.13
Riboflavin (Vit B2) (Mg/100g)	0.20	0 .30
Niacin (Vit.B3) (Mg/100g)	1.39	1.81
Thiamine (Vit.B1) (Mg/100g)	0.92	1.07

Table 3. Vitamin Content of Drought and Rainfed Soybeans

Minerals (Mg/100g)	Drought Soybean	Rainfed Soybean
Phosphorus	0.58	04.8
Potassium	0.98	0.98
Sodium	0.51	0.51
Calcium	0.15	0.13
Magnesium	0.62	0.53
Manganese	27.53	70.00
Iron	75.23	60.87

Table 4. Minerals Content of Drought and Rainfed Soybeans

Sensory Attributes					
Sample	Appearance	Aroma	Taste	Texture	Overall Acceptability
Drought	87.5	6.6	6.7	6.6	7.0
Rainfed	8.9	6.6	6.2	6.4	6.9
Commercial	8.0	6.6	7.2	7.0	8.0

Table 5. Sensory Attributes of Soymilk Produced from Drought and Rainfed

Amino acids (g/100g protein)	Drought Soybean	Rainfed Soybean	FAO Amino Acid Ref. Pattern	
			Children	Adult
Lysine	7.22	7.46	5.50	2.40
Histidine	3.00	2.32	1.40	-
Arginine	6.21	5.87	-	-
Aspartic acid	8.59	8.96	-	-
Threonine	4.00	3.11	4.00	1.40
Serine	3.37	3.02	-	-
Glutamic acid	16.07	15.1	-	-
Proline	2.97	2.24	-	-
Glycine	3.99	3.70	-	-
Alanine	4.32	3.78	-	-
Cystine	1.32	1.19	-	-
Valine	5.00	4.36	5.00	2.00
Methionine	1.13	0.94	-	-
Isoleucine	3.48	3.70	4.00	2.00
Leucine	8.15	7.52	7.00	2.80
Tyrosine	3.22	3.54	-	-
Phenylalanine	4.90	4.23	-	-

Table 6. The Amino Acid Profile of Drought and Rainfed Soybean Compared to FAO Reference Pattern

Organoleptic properties (Table5) showed that drought soybean samples had higher mean scores for taste (6.7), consistency (6.6) and overall acceptability (7.0) compared to 6.2, 6,4 and 6.9 respectively for rainfed soybeans samples. The results indicated that the only issue of serious consideration is the smaller size of drought soybean seeds which translates to lower yield. Growing soybean under limited rains may reduce its physical size, but have no reduction in its chemical, functional, nutritional or organoleptic properties. It is therefore necessary for increase in yield and to facilitate adoption and incorporation of soybean into both the farming systems and recipes of those in areas prone to drought that drought resistant soybean varieties be developed.

2. Fortification of indegenous meals with soybean for effective food security in changing climates. Introduction

According to FAO(2001) across the African continent , protein energy malnutrition affect 40% of children under three years. This situation may not be unconnected with the weaning culture. A semisolid cereal starch reconstituted to a gruel is the major weaning food. The fermented cereal starch is stored for a few days by leaving in fresh water that must be changed every other day. The high moisture content(78-80%) of the extracted starch paste predisposes it to quick microbial and other physico-chemical degradation resulting in low shelf-life. Due to low shelf life and low nutrient density there is great imperative in complementation of cereal weaning foods with legumes in developing nations in the complementation of available weaning foods in developing nations.

While Development of drought resistant soybean varieties is imperative in view of the challenges of changing climate, however the greatest challenge to soybean utilization remained the significant changes in the colour taste and texture of foods complemented with soybean flour. Flours from tuber crops like cassava yam etc and pulp fruits such as plantain and banana flours loss of firmness and moudability of reconstituted dumplings remained a major challenge in the utilization of soybeans and its products. According to Anuonye (2001) development of weaning foods of cereal /soybean blends is greatly impeded by the instability of soybean products at ambient temperatures, thus posing serious storage problems. This is made worse by unstable electric power supply, ruling out refrigeration and other cold preservation considerations at the house hold and small scale industrial levels. Complementing cereal flours with roasted soybean flours would have been an alternative but the coarseness of the end product and inherent raw soybean after taste(beany flavor) limits the acceptability of the end products.

Fortification of weaning foods of cereal origin with soybean and development of new weaning foods with soybean incorporation and having extended shelf life would be one sure way of combating the weaning food crisis in several developing nations. This section presents the process technologies for producing multi purpose soybean flour and cereal starch flours by ambient drying to give a whiter flour end product with reduced changes in colour perception. It also presents the process technologies of fortifying tubers and fruits with the multi purpose soybean flour for enhanced nutrition. The functional, nutritional, pasting and other organoleptic properties of such fortified products are also reported.

3. Process technologies for preparation of multi-purpose soybean flours and cereal starches by ambient temperature drying

The technology for preparation of multipurpose soybean flour and cereal starch flour dried at ambient temperature is shown in figs 1 and 2. Soybean flour was added to cereal starch extracted and dried at ambient temperatures at 25% levels of substitution.

The addition of soybean flour at 25% levels of substitution increased the protein and fat contents significantly ($P \leq 0.05$) as expected (Table7). Conversely there was a drastic reduction in the carbohydrate content of the blend. In connection with the pasting properties the peak viscosity decreased significantly ($P \leq 0.05$) in all fortified meals due majorly to reduced bulk density of the fortified samples following the modification of the fiber content of the samples.

Soybeans
↓
Sorting
↓
Tempering(Soak in clean tap water and dry for 3-4hrs)
↓
Crack(Break into grits using attrition mill)
↓
Winnowing (manually separate the chaff and the grits)
↓
Soaking (soak overnight 12-17hrs in cold water)
↓
Washing
↓
Boiling(For 5mins at 100°C)
↓
Drying(ambient temp25-28°C)
↓
Milling
↓
Soybean flour with enhanced white colour appeal

Fig. 1. Process flow diagram for preparation of soybean flour with greater white colour appeal.

Cereal Grain(maize, Sorghum or millet)

Sorting
↓
king(2days with change of soak water)
↓
Wet Milling
↓
Filtration
↓
Sedimentation
↓
Dewatering
↓
Drying(Thinly spread starch paste is dried at ambient temp 25-28°C)

Fig. 2. Process flow diagram for preparation of cereal starch flour with minimal colour change.

Samples	Moisture(%)	Fat(%)	Protein(%)	Ash(%)	Cho(%)
Mi	10.30	0.20	9.50	0.3	80.50
Mii	9.70	0.30	27.20	1.20	78.50
Gc	11.30	0.10	8.50	0.60	79.50
Gcii	10.00	0.30	20.00	0.90	68.80
Ma	9.50	0.20	9.50	0.30	80.50
Maii	9.00	0.40	22.00	1.20	67.40

Mi=Millet; Mii=Millet flour fortified with 25% soybean flour Gc=Guinea corn flour; Gci=Guinea corn flour fortified with 25% soybean flour; Ma=maize flour; Mai=Maize flour fortified with 25% soybean flour.

Table 7. Proximate Composition of Cereal Starch/Soybean flour

	SAMPLES					
Rheological Properties	**Mi**	**Mii**	**Gc**	**Gci**	**Ma**	**Mai**
Pasting Temp(°C)	76	79	75	76	70	80
Gel Time(mins)	31	31	29	27	25	27
Tvp(°C)	91	89	87	90	87	89
Vp(BU)	770	30	470	360	780	150
Mn(mins)	38	36	38	34	38	31
Vis at95°C(BU)	680	340	410	340	640	140
Cooking Time (mins)	9	5	9	8	13	4

KEY: Gel Time=Gelatinization Time; Tvp=Temperature at peak viscosity ; Vp=Peak viscosity during heating ; Mn=Time to reach peak viscosity; Visat 95=Viscosity at 95°C; Mi=Millet; Mii=Millet flour fortified with 25% soybean flour Gc=Guinea corn flour; Gci=Guinea corn flour fortified with 25% soybean flour; Ma=maize flour; Mai=Maize flour fortified with 25% soybean flour.

Table 8. Amylograph Pasting Viscosity of Fortified and unfortified Cereal Meal

Igbian (2004) reported that peak viscosity is an indication of the maximum increase in that value for the starch-water solution upon heating. Therefore lower values of peak viscosities indicated that a greater amount of gelatinization had occurred in the initial samples or there had been fortification of flours with oilseeds. Peak viscosity also indicates the water binding capacity of starch or mixtures, and also provides indication of the viscous load likely to be encountered by a mixing cooker. The lower peak viscosities showed that there fortified samples will imbibe more water and subsequently swell more. This also would translate to serious reduction in cooking time as evidenced by the reduced cooking time of the fortified samples. Despande *etal* (1988) Maria *etal* (1983) and Igbian (2004) have all reported decreased cooking times occasioned by addition of legumes to cereals. These properties showed that such cereal/soybean paste would remain fluid with higher nutrient density and lowered bulkiness. Reduced bulkiness is an indication that infants would take in more than the would have taken the unfortified meals.

The reduced peak time also showed that less energy would be required to cooking the paste and the problem of retrogradation or hardening might not arise.

The extraction of the cereal starch is to solve the problem of coarseness of the roasted cereal flour that would lead to textural and consistency problems of the reconstituted cereal gruel. The ambient temperature drying and subsequent reduction of moisture content to as low as 9-10% is to ensure long term storage. This solves the problem of unhygienic keeping of the watery paste at ambient temperatures by rural women which results in recontamination and infection at the rural and sub urban levels.

4. Developing new weaning foods to meet the challenges of changing climates and nutrional needs of the most vunerable

One of the greatest challenges of changing climate patterns is the decreased productivity of the familiar food crops that could mitigate hunger and infant malnutrition. There is the overhanging fear that infants and nursing mothers may be more affected nutritionally when there is less food available. The situation is made worse in Sub Saharan Africa where animal sources of protein continue to be out of the reach of the average family. Low wages combined with increased joblessness and difficulty in assessing credit have nearly wiped out the middle class creating a new social order of the rich and the poor. This situation is aggravated by the extended family systems which entails that the average working class person will cater for his or her extended family. This lead to a vicious circle of poverty. Children are therefore born into this unfortunate web hence weaning children presents peculiar challenges.

Weaning food is a meal given to infant prior to withdrawal of breast milk. It begins when parent gradually introduce semi-solid food, other than breast milk in to their baby's diet. This specifically done because young children have high nutritional requirement, and in part because they are growing fast (Aldermal *et al.,* 2004)

Traditionally, most weaning foods of Africa are based on starchy staples food such as cereals including corn (zeamays) Sorghum (Sorghum Bicolour), legume such as soybeans (Glycine max) Cowpea (vigna Unguiculata) and oil seeds such as peanut (Arachis hypogea) (Mosha and Vincent, 2005) It is therefore necessary to evolve combination of locally available foods to complement each other in such a way that new pattern of nutrients can be created.

The Food Agricultural Organization and World Health Organization (1970) reported that most of the infant foods formulated and consumed in communities of developing nations are deficient in essential nutrients. Osundahunsi, (2006) also reported that most weaning foods prepared traditionally in African countries are inadequate in energy and protein, which has been a major cause of protein energy malnutrition (PEM) in preschool children in Nigeria. The first few years of life is usually the vulnerable period for developing under-nutrition, which usually coincides with the introduction of weaning foods. Protein-energy malnutrition(PEM) and micronutrients deficiency therefore become serious problems during the weaning period, as most weaning foods given to the infant do not supply adequate amount of nutrients needed to support optimal growth (Mosha and Vincent, 2005)

Effort have been made to improve the nutritional quality of the weaning foods, including fortifying the locally produced food with specific nutrients or blending then with other nutrient rich foods to form nutritious composite mixtures (Ngoddy *et al.,* 1994 ;Anuonye, *etal* 2001;Obatolu 2003.) There are however several fruit-like staples including plantain, banana etc that their nutrient composition and functionality recommends them as foods for fighting hunger and infant malnutrition in the coming years. Innovative processing and development of complementary foods high in protein will go along way in mitigating infant malnutrition and hunger.

According to Manihot and Lancaster (1983) when plantain is cooked, the fruit is extremely low in fat(0.20-0.30%), high in fiber(6-7%) and carbohydrate(35%) while protein is about, (1.2%)and ash (0.8%). It is also a good source of potassium, magnesium, phosphorous, calcium and iron

as well as vitamin A and vitamin C. According to Ferson and Sharrock, (1998) banana and plantain represent more than 25 percent of the food energy requirements of Africa .
The starch of plantain flour is very low in cholesterol and salt. An average sized plantain fruit (50 to 80gms) will yield on cooking 2 -3gms of protein, 4 - 6gms of fibre and about 0.01 to 0.3gms of fat. It's very rich in potassium, and is commonly prescribed by doctors for people having low level of potassium in their blood (At well, 1999). The potassium in plantain is very good for the heart and helps to prevent hypertension and heart attack. Cooked unripe plantain is very good for diabetics as it contain complex carbohydrate that is slowly released overtime. A diet of green plantain is filling and can be a good inclusion in a weight loss diet plan.

4.1 Processing and utilization of unripe plantain

Unripe plantain is traditionally processed into flour in Nigeria and in other West Africa and Central African countries (Ukhum *etal;* 1991). This traditional technology is equally present in Amazonian, Bolivia. The preparatory method consist of peeling the fruit with hands, cutting the pulp into small round pieces and sun drying them for few days. The dried pulp is then ground in wooden mortar or a corn grinder. The flour produced is mixed with boiling water to prepare an elastic dumpling (amala in Nigeria and fufu in Cameroon) which is eaten with sauces. Some improvement of this traditional method by blanching the plantain pulp at 80°C for some few minutes and cutting them into round pieces (or by soaking for about 3minutes in sodium metabisulfite solution) followed by draining and drying in an oven at 65°C for 48hours or in the sun for some days resulted in the production of a more improved flour that can be reconstituted into staple foods and eaten with soups or break fast meal or a gruel for weaning purposes.
Combining plantain flour (good for diabetics as it contain complex carbohydrate that is slowly released overtime) and soybean flour (a versatile pulse with the richest, cheapest and best source of vegetable protein available to mankind, containing high protein, high polyunsaturated fat with absence of cholesterol and lactose, an excellent source of the essential amino acids vital for body growth, maintenance and reproduction) will give weaning diets having the recommended nutrient density and functionality.
The proximate composition of the blend(table 9) showed that the moisture content of the blends ranged between 4.30-8.53%. The low moisture content of the products indicated the longer storage potentials of the blend compared to conventional pastery weaning foods.
The proximate values (Table 9) indicated that unripe plantain contained low amounts of protein (6.33%) which significantly ($p \leq 0.05$) increased as soybean substitution increased. This was expected and agreed with earlier reports(Osho and Adenekan 1995, Iwe,2003 and Obatolu,2003).
Similarly the carbohydrate content and the bulkiness of the samples reduced indicating a modification in the product structure mainly due to the breakdown of the strong amylase and amylopectin bond by the sulphurdral linkages. The mineral composition of the blends followed similar patterns of increases as the soybean substitution levels increased.
The sensory evaluation (Table11) showed that non of the formulations were rejected. Each had over 50% acceptance. The mean scores of the fortified blends for aroma and overall acceptability were higher than the unripe plantain flour. However the 62.5:37.5% formulation was preferred to other samples. Overall acceptability increased with increased levels of soybean flour substitution showing that the process formulation of the soybean

flour was adequate in eliminating the offensive and objectionable after taste. Both the proximate composition and sensory evaluation results indicated that soybean flour could be added beyond the 50% levels without noticeable objectionable flavor.

The amino acids profile (AAP) of unripe plantain/soybean flour(Table12) showed that blending significantly improved the amino acid profile. Compared with FAO (1970) reference pattern for children and adult nutrition, the results showed that the blend was only deficient in its isoluecine content (1.02 compared to 4.00 recommended). However the blend exceeded the recommendations for adult nutrition in all the amino acids showing that it would be a wise nutritional choice for adult nutritional management.

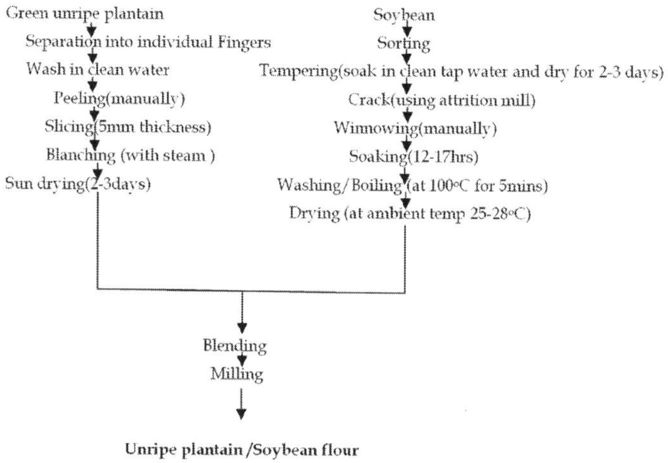

Fig. 3. Flow Process for production of Unripe Plantain and Soybean flour for weaning and Break fast and other diabetic Preparations.

Samples	Parameters Evaluated					
	Moisture (%)	Fat (%)	Protein (%)	Crude Fiber (%)	Ash(%)	Cho (%)
A	5.50	1.50	6.33	1.36	3.13	82.18
B	6.03	1.70	9.13	1.03	4.03	78.08
C	7.13	1.75	15.26	1.13	4.03	70.7
D	7.53	2.00	17.48	1.36	4.03	67.60
E	8.53	2.00	16.97	1.03	4.03	67.44

KEY
A=100:0 Unripe plantain flour to Soybean flour
B=87.50:12.50 Unripe plantain flour to Soybean flour
C=75:25 Unripe plantain flour to Soybean flour
D=62.5:27.5 Unripe plantain flour to Soybean flour
E=50:50 Unripe plantain flour to soybean flour

Table 9. Proximate composition of unripe plantain /Soybean flour mixtures.

Samples	Ca(mg/100g)	Mg(mg/100g)	K(mg/100g)	Na(mg/100g)
A	0.76	0.28	1.13	0.50
B	0.96	0.42	1.13	0.80
C	1.12	0.49	0.98	0.80
D	1.21	0.21	0.83	0.40
E	1.40	0.70	0.83	0.40

KEY
A=100:0 Unripe plantain flour to Soybean flour
B=87.50:12.50 Unripe plantain flour to Soybean flour
C=75:25 Unripe plantain flour to Soybean flour
D=62.5:27.5 Unripe plantain flour to Soybean flour
E=50:50 Unripe plantain flour to soybean flour

Table 10. Mineral composition of Unripe plantain dflour/Soybean Flour mixtures

Samples	Taste	Appearance	Arroma	Texture	Overall Acceptability
A	5.33	6.33	5.93	6.33	5.87
B	5.53	5.87	5.60	5.60	5.67
C	5.13	5.73	5.86	6.00	6.13
D	6.67	7.07	6.93	6.67	7.60
E	6.53	6.40	7.20	6.67	6.67

A=100:0 Unripe plantain flour to Soybean flour
B=87.50:12.50 Unripe plantain flour to Soybean flour
C=75:25 Unripe plantain flour to Soybean flour
D=62.5:27.5 Unripe plantain flour to Soybean flour
E=50:50 Unripe plantain flour to soybean flour

Table 11. Acceptability of Reconstituted Unripe Plantain/Soybean Flour

Amino Acids (g/100g) Protein	SAMPLES			FAO Recommended Pattern	
	Unripe plantain	Soybeans	Blend of Unripe plantain/soybean	Children	Adults
Lysine	2.31	6.24	4.00	5.50	2.40
Histidine	0.88	2.38	1.10	1.40	2.00
Arginine	2.30	7.49	3.91		
Aspartic Acid	3.00	9.33	4.61		
Threonine	1.00	3.77	3.00	4.00	1.40
Serine	2.05	3.02	2.59		
Glutamic Acid	4.10	14.26	3.40		
Proline	3.08	3.19	2.97		
Glycine	3.06	4.55	3.51		
Alanine	2.08	3.94	2.49		
Cystine	0.40	1.59	0.79		
Valine	3.49	5.08	4.00	5.00	2.00
Methionine	0.39	1.23	0.70		
Isoleucine	0.78	4.64	1.02	4.00	2.00
Leucine	1.02	7.91	6.20	7.00	2.80
Tyrosine	2.42	3.54	3.06		
Phenylalanine	0.76	5.41	3.13		

Table 12. Amino Acid Profile of Unripe Plantain/Soybean Flour Blends Compared to FAO Reference Pattern(1970)

5. Fortification of traditional delicacy (pounded yam) meal with soybean flour

Yam, a member of the genus "*Dioscorea*" is an important staple in Nigeria and other West African countries (Cliff *et al.*, 2007). Yam is the perennial herbaceous vine cultivated for the consumption of their starchy tubers in Africa, Asia, latin America and oceanic. Due to their abundance and consequently, their importance to survival, yam was highly regarded in Nigeria ceremonial culture and even worshipped

Before the introduction of cereals and grains in West Africa, yam was the major source of carbohydrate. Ukpabi (1992), reports that yam is considered a man's crop and has ritual and socio-cultural significance. Today, yams are grown widely throughout the tropics. In 2005 48.7million tones of yam were produce world wide. Besides their importance as food source, yam also play a significant role in the socio-culture of some producing regions like the celebrated New Yam festivals in West Africa

The greater part of the worlds yam is kept and eventually consumed in the fresh state. Nevertheless, as a result of the combination of high degree of perishability, bulkiness, distance from production area to the consuming centre and the seasonal nature of production, attention has therefore been drawn to the processing of tubers into flour which depend on some vital functional properties of yam varieties.

Holford (1998) reported that, yams are high in vitamin C, dietary fiber, vitamin B6, potassium and manganese, while being low in saturated fat and sodium. Further more, yam products are high in potassium - sodium balance in the human body and so protect against osteoporosis and heart disease.

Yam products generally have a lower glycemic index than potato products which means that they will provide a more sustained form of energy and give better protection against obesity and diabetes (Schlitz, 1993).According to Rickard (1978) and Igbeka, (1985) harvested tubers are frequently attacked by several viruses, bacteria, fungi and insects. Also rodent feed on some of the harvested tubers stored in the barns, therefore there is need for processing

5.1 Processing and utilization of yam flour

Processing will greatly increase the utilization of root crops, the flour can be use as a component of multi mix baby foods and in composite flour for making bread .

The Food and Agricultural Organization (1987) have reported that, processing of yam involves peeling the root then cutting into slices, blanching, and dried. Peeling can be effected by immersion in 10% lye solution or by steaming at high temperature (150°C) for short period. Dried product require less storage space and have a longer shelf life. They can be quickly reconstituted into pounded yam and prepared for eating.

According to Bourdoux *et al.*, (1983), composite flour incorporating yam has been used in extruded products such has noodles and macaroni, similar processes could be used in production of flour products from other root crops.

Raw yam flour has also found increasing uses in bakery as dough conditioners in bread making and as stabilizers in ice-cream and as thickener in soups . Pregelatinized flour is also used for making instant pounded yam which brings succor to pounded yam lovers as the drudgery of pounding is eliminated (Adeyemi and Oke, 1991).

Production of Yam flour and subsequent reconstitution leads to a dumping lacking in firmness, texture and rigidity of the conventional pounded yam. This witling down of the conventional pounded yam consistency makes many not to accept reconstituted yam flour meal as pounded yam.

This meal which reduces drudgery of pounding, faces limited local, ethnic and regional acceptance. It becomes necessary therefore to fortify yam flour with locally available firming agents to reconstitute a yam flour meal close to the conventional pounded yam. Addition of soybean to such fortified yam flour would increase the nutritional status and also its functionality. This was accomplished by firming-up yam flour with cassava starch.

Cassava *"Manihot escullenta"* is a staple food consumed in both rural and urban areas of Nigeria.

Starch is one of the most important plant product to man (Landry and Moreax, 1982). It is an essential component of food providing a large proportion of the daily colorific in take (Scott *et al.,* 2000). Cassava starch is recommended for use in extruded snacks for improved expansion (Senthiikumar and Subburam, 2001). It is also used as a thickener in foods that are not subjected to rigorous processing conditions (Okezie and Kosikowki, 1982).

Cassava starch, which is very bland in flavour is used in processed baby foods as a filler materials and bonding agent in confectionary and biscuit industries (Fregene *et al.,* 2003) Cassava starch can perform most of the function where maize, rice and wheat starch are currently used.

A technology of adding cassava starch to yam flour (Figure4) with 25% levels of cassava starch was developed(Fig5).The yam flour (Figure 5) strengthened with cassava starch was then fortified with soybean flour up to 30% levels of substitution(Figure 6).

Yam Tuber
↓
Washing
↓
Peeling
↓
Slicing
↓
Drying
↓
Milling
↓
Yam Flour

Fig. 4. Flow diagram for Traditional Yam flour Preparation

Yam Tuber
↓
Washing
↓
Peeling
↓
Slicing
↓
Sulphating(NaH_2SO_3 in 1L distilled water)
↓
Boiling (10mins at 100°C)
↓
Drying(5-6hrs at 50-60°C)
↓
Milling
↓
Improved yam flour

Fig. 5. Flow process for Improved Yam flour Preparation (FIIRO,2003)

Pre enrichment of yam tuber flours with native cassava starch up to 25% produced very firm gels close to the traditionally pounded yam meals(Table13). Sensory evaluation (Table 14) showed that yam flour fortified with cassava starch as gelling agent was generally more acceptable in appearance colour, taste, consistency and overall acceptability than those fortified with corn starch(Table14). Addition of 10% soybean flour enhanced the protein content of the meal as well as had no noticeable rheological problems on the firmness or moudability. Addition of 10% soybean flour brought the rheological characteristics of the sample to nearly the same with conventional pounded yam. This improvement is as a result of the increased stability of the yam starch due to added cassava starch. This increased stability is reflected in the high sensorial scores of the fortified meals(Table15). With the high sensorial rating obtained for samples at 10% levels of substitution it is concluded that firming yam flour with 25% cassava starch and fortifying with soybean will produce a dumpling in the mould of conventional pounded yam. Adoption of this technology would lead to greater utilization of cassava produced maximally in this part of the world

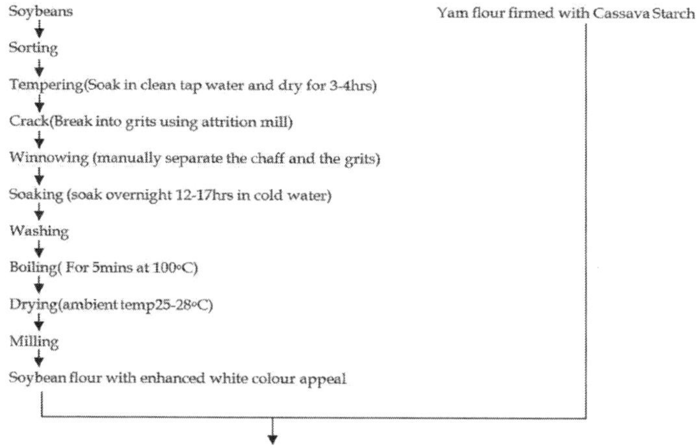

Fig. 6. Process Flow Diagram for production of Firmed Yam / Soybean flour for pounded Yam Preparation

Samples	**Pasting Parameters**						
	PV (RVU	TR (RVU)	BDU)	FV (RVU)	SB (RVU)	PT (mins)	PT (°C)
A	90.83	80.16	10.66	140.96	60.79	6.96	61.85
B	126.00	114.0	12.00	158.08	44.08	6.70	61.65
C	120.4	105.75	14.79	192.04	86.29	6.96	61.92
D	100.08	88.33	11.75	153.46	65.12	7.00	61.87

PV=Pasting Viscosity TR=Trough BD=Break down FV=Final Viscosity SB=Set Back Viscosity PT= Time at Peak Viscosity PT=Temperature at Peak Viscosity
A=Conventional Pounded Yam B=Pounded Yam with 25% cassava starch C= Pounded Yam with 25% corn starch D= Pounded Yam with 5% cassava starch

Table 13. Pasting Characteristics of Yam / Cassava or corn Starch flour for Pounded Yam Preparation

Samples	Appearance	Colour	Taste	Consistency	Overall Acceptability
A	5.40	5.60	6.70	4.70	7.10
B	7.20	6.80	6.60	6.50	6.20
C	3.50	3.00	3.00	3.20	2.90
D	8.80	8.30	8.90	8.90	8.60

A=Conventional Pounded Yam
B=Pounded Yam with 25% cassava starch
C= Pounded Yam with 25% corn starch
D= Pounded Yam with 5% cassava starch

Table 14. Acceptability of Reconstituted Yam Flour/ Firmed Cassava / Corn Starch

PARAMETER	Samples A	B	C	D	E
Color	8.19	7.88	6.81	6.93	6.69
Smell	7.81	7.87	7.18	7.43	6.68
Texture	8.438	7.56	6.87	7.25	7.06
Taste	8.00	7.75	7.06	7.50	7.1
Moudability	8.43	7.68	7.06	7.68	7.31
Overall acceptability	8.813	8.62	7.37	7.56	7.43

A=Conventional Pounded Yam
B=Reconstituted Fortified Pounded Yam Flour with 10% Soybean flour
C= Reconstituted Fortified Pounded Yam Flour with 15% Soybean flour
D= Reconstituted Pounded Yam Flour with 20% Soybean flour
E= Reconstituted Pounded Yam Flour with 30% Soybean flour

Table 15. Sensory Evaluation of Reconstituted Yam/Soybean Flour enriched with Cassava Starch

Samples	**Pasting Parameters**						
	PV (RVU	TR (RVU)	BD (RVU)	FV (RVU)	SB (RVU)	PT (mins)	PT (°C)
A	90.83	80.16	10.66	140.96	60.79	6.99	61.85
B	90.21	85.84	2.25	142.13	56.29	6.44	61.65
C	77.55	74.50	3.05	129.63	55.13	6.44	62.05
D	74.80	70.46	3.54	126.29	55.84	6.52	61.70
E	60.92	58.57	4.38	116.42	57.50	6.36	61.25

PV=Pasting Viscosity TR=Trough BD=Break down FV=Final Viscosity SB=Set Back Viscosity PT= Time at Peak Viscosity PT=Temperature at Peak Viscosity
A=Conventional Pounded Yam
B=Fortified Pounded Yam Flour with 10% Soybean flour
C=Fortified Pounded Yam Flour with 15% Soybean flour
D= Pounded Yam Flour with 20% Soybean flour
E= Pounded Yam Flour with 30% Soybean flour

Table 16. Pasting Characteristics of Yam flour for Pounded Yam Preparation

6. Blending soybeans with lesser known cereals

Blending legumes and cereals hold the key to food security for the greater number of the world population. Indigenous foods especially those identified for their health benefits and those that can by innovative processes be enriched calorie-wise need be exploited in order to halt the devastating effects of hunger. Such cereals include acha. Blending acha and soybean therefore would provide a wide range of both high calorie and high protein food if properly processed. As already stated, most malnourished people live in Asia and Africa; and the staple of most people in Asia and Africa are starchy pastes. These pastes are made from cereals such as sorghum, maize, millet, acha etc; roots and tubers such as cassava, yam, sweet potato etc. These crops do not only provide marginal nutrition (especially for children) but also require high inputs of time, labour and fuel to prepare. In most cases they are customarily consumed as combinations in the home because the blends provides complementary balance of amino acids (proteins) in the diet

'Acha' occupies about 300,000 hectares in West Africa and provides foods for about 4 million people (kwon-ndung and Misari, 2000). It is not known to grow outside of West Africa and is also not known to grow in a wild state. Is said to be the oldest West Africa cereal whose cultivation dates back to about 5000 BC (Pulse glove, 1975). It remains a very important crop from areas scattered from Cape Verde to Lake Chad even though many have not heard of it. In Nigeria, acha is popularly grown in five states (Bauchi, Kaduna, Kebbi, Plateau Niger) and the Federal Capital Territory. In some of these areas, the crop forms the staple where the very small grains are processed into different menu.

Acha is one of the world's best tasting cereals. In recent times, comparison of dishes of acha and rice showed that majority preferred acha dish. The protein content of acha grains is rich in methionine, cysteine (above the recommended levels). These levels are unusual for cereals. Acha is also used in dietary preparations for diabetic patients (Victor and James, 1991). Traditionally, acha is used in preparation of unfermented porridge food. It is also made into "gwette" and *acha-jollof*. With the exception of methionine, the essential amino acid content of acha is lower than in maize, rice sorghum, millet, wheat, barley and oats. While acha is a cheap source of carbohydrate for man, and livestock, particularly in dry infertile areas, in the tropics, Victor and James (1991) advocates its complementation with protein rich foods to make a balance diet. Another reason why acha is not popular is that its food uses are not yet established, except for the limited ones already mentioned (Jideani and Akingbala, 1993).

The low protein intake in most Africa countries including Nigeria is attributed to t he increasingly high cost of animal sources such as beef, mutton, fish and game (bush meat and also to inadequate utilization of most plant protein source. Soybean is an inexpensive source of protein used in supplementation of various cereals, legumes root and tuber based diets. Soybeans have also been used in several novel food products such as soy-ogi as well as other cereal and tuber products to complement their amino acid profiles (Iwe and Onuh 1992).

Acha like sorghum and millet has been cultivated in West Africa since ancient times. Acha grows with reasonable yields in areas of low rainfall and poor sandy or ironstone soils. Though grass- like acha reaches heights of 30-80 cm and can resist periods of droughts and heavy rains (Jean Francis, 2004).

Acha (D *exilis)* is a semi erect /straggling annual plant which is hairless, having a height ranging from 102-123cm and rooting sometimes at the lower nodes. The stem, known as culms is sparingly branched from below with 5-8 nodes. A single grain of the crop can produce a multiple of stems on a single stand. The leaf sheaths are usually held tight to the stem while the leaf blade is approximately 13 to 15cm long depending on accession (Dachi, 2000).

According to Dunsmore *etal* (1976) acha matures around early September before the main harvest period for other staple crops when food and money are traditionally in short supply. Varieties with very short cycle (70-85 days) allow farmers to harvest early and enable them

Acha and ibura can completely substitute for rice in different rice dishes such as cooking in water. Jideani (1999) reported that dehulled acha and ibura cook soft in boiling water within 3-8 min compared to 20 – 30 min for some rice varieties. According to him, this beneficial property of acha would mean less use of energy in preparation that needs to be exploited for developing quick cooking non-conventional food products including weaning foods and break fast cereals. Again, whole acha grains could be made into products similar to 'quarker oats'. Unlike most other cereals grains, porridge made from products containing whole grains provides the necessary fiber component. Further more, the small size and location of constituents in these grains give them the advantage of minimal processing. (Jideani ,1999; Irving and Jideani, 1997).

Acha and ibura can be used for weaning foods of low dietary bulk and high caloric density. Anuonye(2006) have established that extrusion of acha/soybeans presents an interesting case of food complementality. However the findings of that study cannot be implemented immediately due to dearth of extruders.

A technology (Fig7) of enriching acha flour with soybeans was developed to produce break fast cereal/soybean meal having adequate nutrient balance.

The results showed that adding soybeans flour at 37.5% produced acceptable breakfast meal. Recent studies (Anuonye 2006) showed that soybean could be added to cereals up to 37.5%. Complementary weaning foods developed from this process technology showed that there were significant ($p \leq 0.05$) increases in protein form 7% in acha flour to 22% in blends of % samples. Similarly the fat increased from 4% in sole acha to 17% in blended samples. 50:50 ratio. Addition of soybean to acha flour also led to increased water absorption index form 3.6 in acha flour to 5.6 in soybean flour fortified samples. There was also and a corresponding decrease in bulk density form 8.5 in acha to 7.0 g/m^3 in the blended samples. The pasting viscosity showed that peak viscosity, peak time, peak temperature etc were all significantly ($p \leq 0.05$) lowered by addition of soybean at 37.5%. Sensory analysis showed that panelist preferred sample blends of 37.5% soybean to other samples.

The amino acid profile of the blended samples showed that blending with soybean increased all amino acids levels compared to the acha flour index. Compared to the FAO reference pattern the results showed that the blend of 62.7:37.5 meet the recommendation for infant nutrition while it surpassed all the recommendation for adult nutritional management. The meeting of the nutritional recommendation by the blend may not be unconnected to the processing of the soybean flour. Anuonye (2006) have noted that raw soybean flour addition to acha flour may be affected by lypoxygenase enzyme activity reducing the values in analytical tests. The present results lend credence to this observation. Animal feeding trails showed that protein digestibility of the blend was over 90% while protein efficiency ratio was 0.05g/g with feed conversion ratio at 0.2g/g. Serum profile showed that all parameters evaluated were within the recommended normal range.

	Proximate Parameters Evaluated						
Samples	Moisture (%)	Fat (%)	Protein (%)	Ash (%)	Crude Fiber(%)	Cho (%)	Energy (Kcal/100g)
A	4.10	3.93	6.99	4.23	2.16	80.75	394.02
B	4.39	7.09	8.87	4.08	2.35	75.56	408.08
C	5.01	11.03	11.36	3.11	2.53	69.51	432.01
D	5.35	13.06	15.41	2.51	2.67	63.66	445.40
E	5.50	17.07	22.04	2.50	2.74	52.93	464.04

KEY : A=100:00 Acha flour to Soybean flour
B=87.50:12.50 Acha flour to Soybean flour
C=75:25 Acha flour to Soybean flour
D=62.50:37.50 Acha flour to Soybean flourE=50:50 Acha flour to Soybean flour

Table 17. Proximate Composition of Acha/Soybean Blends

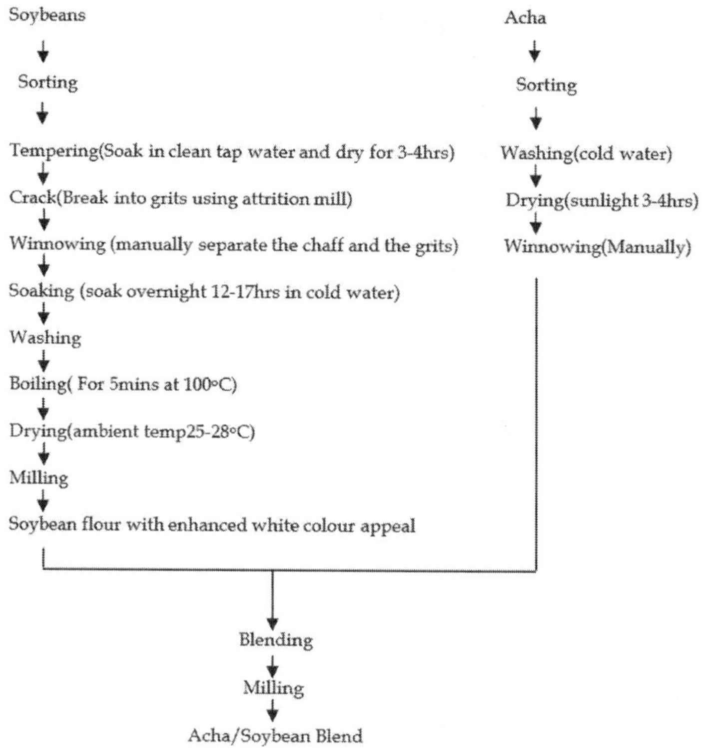

Fig. 7. Process Flow Diagram for the Production of Acha/Soybean flour For break fast and Dietetic Applications

	Sensory Parameters					
Samples	Appearance	Aroma	Taste	Texture	Mouth feel	Overall Acceptability
A	7.40	5.50	6.55	7.60	7.45	6.40
B	7.15	6.75	6.45	7.25	7.20	6.65
C	7.40	6.95	7.20	7.25	7.60	7.35
D	7.35	7.70	7.20	7.55	7.55	7.55
E	7.00	7.30	7.05	7.15	7.15	6.65

KEY

A=100:00 Acha flour to Soybean flour
B=87.50:12.50 Acha flour to Soybean flour
C=75:25 Acha flour to Soybean flour
D=62.50:37.50 Acha flour to Soybean flour
E=50:50 Acha flour to Soybean flour

Table 18. Acceptability of Acha/Soybean Blends

Amino Acids(g/100g) Protein	Acha/Soybean Blends		FAO Recommended Pattern	
	Acha/soybean flour (62.50:37.50)*	Acha / Raw Soybean Flour (62.50:37.50)*	Children	Adults
Lysine	4.17	3.51	5.50	2.40
Histidine	2.08	2.55	1.40	2.00
Arginine	2.98	4.25		
Aspartic Acid	4.30	5.27		
Threonine	4.00	3.41	4.00	1.40
Serine	3.05	3.51		
Glutamic Acid	6.20	9.67		
Proline	3.08	1.02		
Glycine	3.45	4.16		
Alanine	2.86	3.71		
Cystine	1.82	1.71		
Valine	5.05	5.31	5.00	2.00
Methionine	2.51	2.20		
Isoleucine	4.24	3.81	4.00	2.00
Leucine	8.04	8.01	7.00	2.80
Tyrosine	3.05	3.19		
Phenylalanine	5.16	4.72		

*Anuonye,(2010)

Reference Pattern(1970)

Table 19. Amino Acid Profile of Acha/Soybean Blend Compared to FAO

7. Conclusions

Development of drought varieties of soybean have been highlighted as very necessary for adoption of soybean in the drought prone areas. However adoptable processing technologies for house hold and small-scale industrial concerns remain the basic issue in adoption of the multiplied use soybean. The development of the multipurpose soybean flour provides answers to several challenges of soybean utilization. The use to which the multi purpose soybean flour can be put appear limitless. Adaptability of its production to the rural and sub urban settings and conditions makes it a novel approach to soybean processing and utilization. Its low moisture content and low water activity assures of longer keeping time solving the problem of shelf instability of many soybean products. It also solves the sanitary problems of many rural and sub urban dwellers.

Innovative processing of diverse crops and subsequent fortification with soybean flour and its allied products is one sure way of contending with the nutritional challenges posed by changing climate. Balance in the amino acid profile of such fortified meals and improvement in rheological functional organoleptic and keeping qualities as evidenced from the works reported herein show that there is much that could be accomplished through product complementation. While the battle to feed the teeming world populations go on we advocate product complementation as a means of addressing part of the global food crisis.

8. References

Adeniji, A. L., Ega, M. Akoroda, A. Adeniyi, B.U. and Balogun, A. (1997): Cassava Development Department of Agriculture Federal Ministry of Agriculture and Natural Resources, Lagos Nigeria.

Adeyemi, A. N., Oke, O.P., (1991). Yam production and its future prospects, outlook Agriculture, 16:105-110.

Alderman H. and Hoddinott, J. (2004): "Hunger and malnutrition" (International Food Policy Research InstituteWashington, D.C (2004 draft); L. Haddad, H. Alderman, S.ppleton L. song and y.yohannes.

Anon (2003). Malnutrition the silent emergency. Watch Tower Bible Tract Society N.Y. p3-11

Anuonye, J. C. (2006) Effect of extrusion process variables on physiochemical sensory microbial and storage properties of product from Acha digeterial in esilis and soybeans (Glycine Max (L.)Merr). PHD, Thesis University of Agriculture Makurdi.

Bourdoux, P., Seghers, P., Mafuta, M., Vender pas, J., Varinder pasrivera, M., Delange, F. and Ermans, M. A. (1983). Traditional Cassava detoxification process and nutrition education in Zaire. In delange, R. eds. Cassava loxicity and thyroid. Research and public health issues P. 134 - 137. Ottawa, IDRC (IDRC 207e)

Clif, K. R. Sarafadeen, A. A. Andrew, O. W. and Hallen, N. A. (2007): 'Surface properties of yam (Dioscorea Sp) starch powders and potential for use as bunders and disintegrates in drugs formulation" journal of science food Agric, 20: 165 -171

Combes G.F. Welch, R.M., Duxbury, J.M., Uphoff, N.T and Neshelm, M.C (1996). Food-based approaches to preventing micronutrient malnutrition; an International Research Agenda. Cornnell University, Ithaca, N. York pp68.

Dachi, S.N. (2000). The effects of different rates of NPK and organo-mineral fertilizers on the growth and yield of Acha (digitaria exilis). Kippis stap in South western Nigeria MSc thesis. Department of Agronomy. University of Ibadan. 82pp.

Deshpande, P.D. Rangnekar, P.D, Sathe, S. K. and Salunkhe, D. K. (1983). Functional properties of wheat -bean composite flours. J. Food Sci. 48:1659-1662.

Dunsmore, J.R, Ram, A.B, Lowe, G.D.N, Moffat, D.JL: Anderson, I.P, Williams, J.B. (1976).The agricultural development of the Gambia: An agricultural and socio economic analysis, Land Resources study 22, ministry of overseas Development, Survey, U.K.

FAO/WHO, (1970) Amino Acids content of foods and Biological Data on protein; pp 59 - 62. nutritional studies No. 24.

FAO (1973). Amino acid content of foods and biological data on proteins. (FAO Nutritional Studies No. 24) FAO. Rome, Italy.

Fregene, M. A., Suarez, M. Mkumbir, A. J., Kulembeka, H., Ndeya, E., Kulaya, A., Mittcheal, S., Gullberg, U., (2003) Handbook of methods of Anlysis C.A.B. international.

Holford, P. (1998). The optimum Nutrition Bible ISBN No. 7499-1855-1

Igbeka, J. C. (1985). Storage practice for yam in Nigeria, Agricultural mechanization in Asia Africa and Latin America. 16,55-58.

Ingbian, E. K. (2004). Characterization of physical, chemical and microbiological properties of mamu, a roasted maize meal. Ph.D thesis. Department of Food Science and Technology University of Ibadan. pp112-193.

Integrated Crop Management, (2008) . Maturation of Protein and Sugars in Desiccation: Tolerance of developing Soybean Seeds. Plant Physiol. 100 (1): 225-230.

INTSOY, (1998). International Soybean Program on Soybean Processing for food uses. Pg. 48-52, 73-83.

Irving, D. W., and Jideani, I. A. (1997). Microstructure and composition of Digitaria exilis stapf (*acha*): A potential crop. Cereal Chem. 74:224-228.

Iwe, M. O., (2003) The Science and Technology of Soybeans. Rejoint Communication Services LTD 65 Adelabu St. Nwani Enugu, Nigeria. First Edition Pg6--28

Iwe, M.O. and Onuh J.O. (1992). Functional and sensory properties of soybeans and sweet potato flour blends. Lebensm - wiss. U. - Technol. 25:569-573.

Jean -Francois, C. (2004). Fonio: a small grain with potential. Leisa magazine, march; 2004. 20: 1pp16-17.

Jideani, A.I (1999). Acha (Digiteni erilis) the Neglected Cereal Agricultural International, may Agric. International 42: 5 -7.

Jideani, I. A. (1999). Traditional and possible technological uses of Digitaria exils (*acha*) and Digitaria iburica (*iburu*). Plant Foods for Human Nutrition 00:1-13

Jideani, I. A. and Akingbala, J. O. (1993). Some physicochemical properties of *acha* (*Digitaria exilis stapf*) and Ibura (*Digitaria Ibura stapf*) grains. J. Sci Food and Agric, 63:369 - 374.

Jill, J.H.(2006) Cassava a basic energy source in the tropics. Science vol. 218 no. 4514. p. 755-762

Kwon - Ndung, E.H and Misari, S.M. (2001). Over view of research and development of *acha* digitaria exilis kippis stapf and prospects of genetic improvement in Nigeria. Genetics and food security in Nigeria, pp71 - 76.

Manihot J. and Lancaster, P.A. (1983) Bananas and Plantain hand book of Tropical Foods Harvey T.C J.R (Edition), pg 35 - 145 Marcel Dekker Inc.

Maria, L. T. Manual, S and Maria, G. S (1983). Physical, chemical, nutritional and sensory properties of corn-based fortified food products. J. Food Sci. 48:1637-1643.

Mosha A.C and U. Suvanberg, (1990). The acceptance and mistake of bulk reduced weaning food, the language village study. Food Nutr. Bull 12: 156 - 162.

Obatolu, V. A. (2002). Nutrient and sensory qualities of extruded malted or unmalted millet/soybean mixture. Food Chem. 76:129-133.

Okezie, B. O. and kosikowski, F. V. (1982) Cassava as a food. Critical Review of Food Science and Nutrition, Vol. 17, no. 3, P. 259-275.

Osho, S. M. and Adenekan, I. G. (1995). Production and Nutritional evaluation of soybean fortified malted sorghum meal extrudate. In: Processing and Industrial Utilization of Sorghum and Related Cereals in Africa. Eds. Menyinga, J. M; Bezuznch, I., Nwasike, C.C, Sedojo, P. M. and Tenkouano, A. OAU/STRC -SAFGRAD Publ. pp 109-118.

Oyelade, O. J. Suny-Igweji, E. O., Otunola, E. T. and Ayorinde, A. (2002) Effect of Tempeh Addition on selected Physico-chemical and sensory Attributes of Reconstituted Cassava flour-Research Communications in Food Science (in-press).

Park, S. (2004). Root crops in Fiji, part 2: Development and future food production strategy, Fiji Agriculture Journal, (42): 11-17

Pulseglove,J.W. (1975)Tropical Crops: Monocotyledons vol 1 and 2 combined. English Longman pp142-143.

Rickard, M. I. (1978). The use of Cassava starch in the formulation of gelation capsules. Journal de Pharmade de belgique vol. 48, No. 5, P. 325-334.

Scott, G.J., Rosergrant, M.W. and Ringeer, C. (2000) . Root and tubers for the 21st century: Trends, projections, and policy options, P. I-71.

Senthiikumar, P. and Subburam, V. (2001) Carbon from cassava peel, an agricultural waste, as an adsorbent in the removal of dyes and metal ions from aqeuos solution Bio-resource Technology 3 : 233-235.

Smith, I.F. and Oluwoye. O.R (1988). Energy, protein and selected intakes of low, middle and high income Nigerians. Nutri. 8:249 -254.

Ukhum, M.E, Ukpebor, I.E (1991) Production of instant plantain flour, sensory evaluation and physiochemical changes during storage food chemistry 42: 287 - 299.

Ukpabi, K.J. (1992). Socio cultural significance of yam. Nigeria Journal of food crop, 7:34

Victor, J. T. and James, D. B. (1991). Proximate Chemical Composition of Acha (Digitaria exilis) grain. J. Sci Food And Agric, 56 :561-563.

13

Enhancement of Soybean Seed Vigour as Affected by Thiamethoxam Under Stress Conditions

Ana Catarina Cataneo[1], João Carlos Nunes[2], Leonardo Cesar Ferreira[1], Natália Corniani[1], José Claudionir Carvalho[2] and Marina Seiffert Sanine[1]
[1]*Department of Chemistry and Biochemistry; Institute of Biosciences UNESP - São Paulo State University, Botucatu*
[2]*Syngenta Crop Protection, São Paulo*
[1,2]*São Paulo State*
Brazil

1. Introduction

Cruiser ® (thiamethoxam), developed and registered by Syngenta, is a chloronicotinic insecticide, belonging to the class of neonicotinoids for seed treatment and has long residual control for a wide range of chewing and sucking insects present in seeds, soil and leaves (Maienfisch et al., 2001).

Thiamethoxan acts by contact and ingestion and the insect stops eating within 24 h after contact with the insecticide. The primary mode of action involves interference with, or by binding to nicotinic acetylcholine receptors (Maienfisch et al., 2001).

Surprisingly, it has been noticed that the treatment of soybean seeds with Cruiser results in a "stand" more uniform, vigorous and more productive, thus acting on germination.

However, seed germination and seedling development of crops are negatively affected by adverse conditions, such as drought (Davidson & Chevalier, 1987; Passioura, 1988, Soltani et al., 2004), salinity (Hampson & Simpson, 1990; Ramoliya & Pandey, 2003, Soltani et al., 2004, Luo et al., 2005; Athar et al., 2008) and high concentrations of soluble forms of aluminum (Matsumoto, 2000; Echart & Cavalli-Molina, 2001, Rout et al., 2001).

A common characteristic of various stress types is the increased production of reactive oxygen species (ROS), which are generally considered harmful to plant cells (Alscher et al. 1997; Smirnoff, 1993, Richards et al., 1998). The ROS include superoxide radical ($O2^{\bullet -}$) and hydroxyl (•OH), hydrogen peroxide (H_2O_2) and singlet oxygen ($^1O_2$). There are evidences that increased production of ROS under environmental adversities may induce oxidative stress in plants. It has been reported the induction of oxidative stress under conditions of water stress (Smirnoff, 1993; Alscher et al., 1997), salinity (Rio-Gonzalez et al. 2002; Bor et al., 2003; Athar et al., 2008) and excessive concentrations of aluminum in soils (Tamás et al., 2004).

For protection against ROS, plant cells contain an antioxidant system, including various enzymes, among wich, superoxide dismutase (SOD) and peroxidase (POD) (Fridovich, 1978, Bowler et al., 1992, Foyer et al., 1994; Cataneo et al., 2005; Ferreira et al., 2010). SOD and

POD are metalloenzymes acting in the elimination of, respectively, $O2^{\bullet -}$ radical and H_2O_2 produced in stress conditions. Peroxidases are active in many physiological and development processes and are involved both in consumption, as in the production of H_2O_2 and other ROS (Silva et al. 1994; McQueen-Mason & Cosgrove, 1994; McQueen-Mason, 1995, Bacon et al. 1997; Amaya et al. 1999; Passardi et al., 2004).
Thus, the aim of this study was to evaluate the effect of Cruiser on the enzymes involved in protection against oxidative stress (SOD and POD) caused by drought, salinity and presence of high concentrations of aluminum during soybean germination.

2. Methods

2.1 Plant material and conduction of experiments

In this study were used seeds from two different cultivars of soybean (*Glycine max* L.): Pintado, representative of the Brazilian Midwest region, characterized by the predominance of the Brazilian savanna (cerrado) features and BRS 133, representative of the South region, with features adapted to the soil and climate of this geography.
Three experiments were carried out in the Xenobiotic Lab from Department of Chemistry and Biochemistry, Institute of Biosciences, UNESP, Botucatu, in a germination chamber at 25°C in the dark.
Seeds were germinated on filter paper rolls moistened with distilled water or with different solutions. The volume of such solutions used in the treatments was 2.5 mL X g filter paper weight. The germination rolls were placed into plastic containers, each with a perforated lid. In the germination evaluations, seeds presenting root length equal to or greater than to 2 mm were considered germinated (Duran & Tortosa, 1985).
In the three experiments were adopted the experimental design completely randomized, with four replicates and twenty-five seeds per plot. The results were subjected to analysis of variance. The treatments were compared by Tukey test at 1% probability. The experiments were conducted in three phases.

2.2 First experiment

Seeds of two soybean cultivars were treated with the recommended level of Cruiser 350 FS - **D1** - (100 mL f.p./100Kg seed), with twice the recommended level of Cruiser 350 FS - **D2** - (200 mL f.p./100Kg seeds) and the control seeds were treated only with distilled water - **D0**. The counting of germinated seeds of the three treatments was performed at 24, 36, 48, 60 and 72 h of imbibition.

2.3 Second experiment

Seeds of two soybean cultivars were treated with the recommended level of Cruiser 350 FS - **D1** - (100 mL f.p./100Kg seed) and the control seeds were treated only with distilled water - **D0**.

2.3.1 Presence of heavy metal – aluminum

Followed by treatment with the levels D0 and D1 of Cruiser, germination paper leaves were moistened with solutions of aluminum sulphate at concentrations of 0; 5; 10 and 15 mmol L^{-1}. Germination evaluations were performed at 24, 36, 48, 60 and 72 h of imbibition in the solutions of different concentrations of aluminum sulfate. At the end of the experiment (72 h) the embryo axis were removed and weighed.

2.3.2 Salinity – NaCl

Followed by treatment with the levels D0 and D1 of Cruiser, germination paper leaves were moistened with solutions of sodium clhoride at concentrations of 0; 25; 50; 100 and 150 mmol L^{-1}. Germination evaluations were performed at 24, 36, 48, 60, 72 and 84 h of imbibition in the solutions of different concentrations of NaCl. At the end of the experiment (84 h) the embryo axis were removed and weighed.

2.3.3 Water deficit

Treated seeds with levels D0 and D1 of Cruiser were germinated on filter paper rolls moistened with solutions of polyethylene glycol 6000 (PEG) that simulate different situations of water deficit. PEG solutions at the water potentials -0.1; -0.2 and -0.3 MPa were prepared according to Michel & Kaufmann (1973). Distilled water was used in the control. Germination evaluations were performed at 24, 36, 48, 60, 72 and 84 h of imbibition in the solutions of different concentrations of PEG. At the end of the experiment (84 h) the embryo axis were removed and weighed.

2.4 Third experiment

To develop the third experiment, were chosen for each cultivar, the concentrations of the solutions of aluminum sulfate, NaCl, PEG and the period of imbibition that provided the biggest differences between the treatment with Cruiser and control, from the second study. Seeds of two soybean cultivars were treated with the recommended level of Cruiser 350 FS - **D1** - (100 mL f.p./100Kg seed) and the control seeds were treated only with distilled water - **D0**. The concentrations of the solutions and the periods of imbibition used in the different treatments are shown in the Table 1.

Seed treatment	Concentration of solutions (* chosen from second experiment)		Periods of Imbibition (h)	
	cv. BRS 133	**cv. Pintado**	**cv. BRS 133**	**cv.Pintado**
H_2O (D0)	Distilled H_2O	Distilled H_2O	24 and 36	24 and 36
Cruiser (D1)	Distilled H_2O	Distilled H_2O	24 and 36	24 and 36
H_2O (D0)	Al sulfate 10 mmol.L^{-1}	Al sulfate 10 mmol.L^{-1}	24 and 36	36 and 48
Cruiser (D1)	Al sulfate 10 mmol.L^{-1}	Al sulfate 10 mmol.L^{-1}	24 and 36	36 and 48
H_2O (D0)	NaCl 50 mmol.L^{-1}	NaCl 100 mmol.L^{-1}	24 and 36	36 and 48
Cruiser (D1)	NaCl 50 mmol.L^{-1}	NaCl 100 mmol.L^{-1}	24 and 36	36 and 48
H_2O (D0)	PEG -0,3 MPa	PEG -0,3 MPa	60 and 72	72 and 84
Cruiser (D1)	PEG -0,3 MPa	PEG -0,3 MPa	60 and 72	72 and 84

Table 1. Concentration of solutions (*) used in third experiment – aluminum (Al sulfate), salinity (NaCl) and water deficit (PEG) and periods of imbibition in which were collected the samples of Embryo Axis of soybean cv. BRS 133 and Pintado.

For each treatment and imbibition period described in Table 1, were collected samples of embryo axis in two imbibition periods to determine activity of the antioxidant enzymes, peroxidase (POD) and superoxide dismutase (SOD).

Enzymatic extracts used for determination of SOD and POD activities were obtained according to the method described by Ekler et al. 1993. POD and SOD activities were assayed according to the method described by Teisseire & Guy (2000) e Bor et al. (2003), respectively.

3. Results

3.1 First experiment: Action of cruiser on the germination of soybean seeds

In the cultivar BRS 133 the treatment with Cruiser used in the recommended level (D1) and at twice the recommended level (D2) accelerated the germination in the first 24 h of imbibition (Figure 1). The increase in germination was higher at D2 treatment.

In the cultivar Pintado (Figure 2) Cruiser caused acceleration of germination until 36 h of imbibition, being observed that at 24 h of imbibition the increase in germination was higher at the twice-recommended level of Cruiser and at 36 hours of imbibition, germination did not differ statistically between the two levels of Cruiser. Germination in both cultivars did not differ significantly between the control seeds (D0) and seeds treated with two levels of Cruiser (D1 and D2) between 48 and 72 h of imbibition.

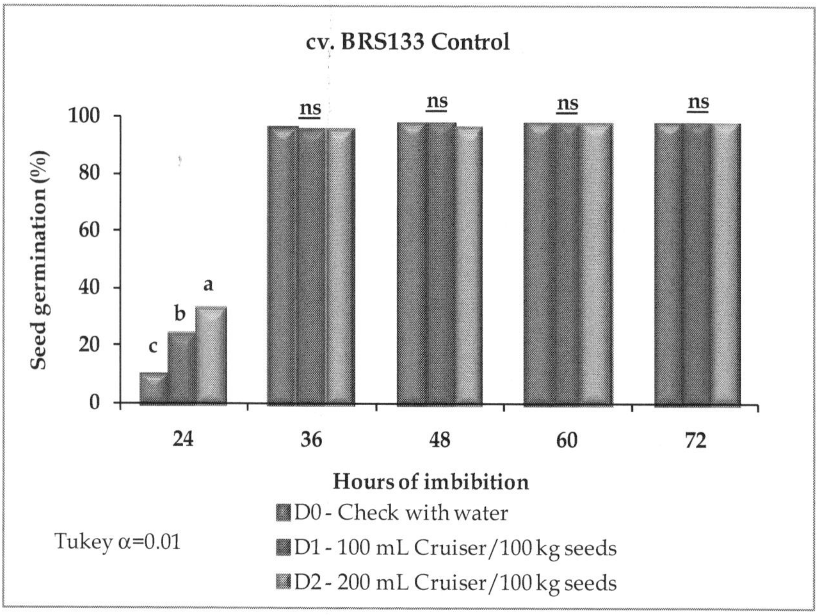

Fig. 1. Soybean germination percentage cv. BRS 133 treated at recommended dose of Cruiser (D1), double of recommended dose (D2) and check (D0). Average followed by the same letter did not differ significantly for each imbibition period. **ns**: not differ significantly for each imbibition period.

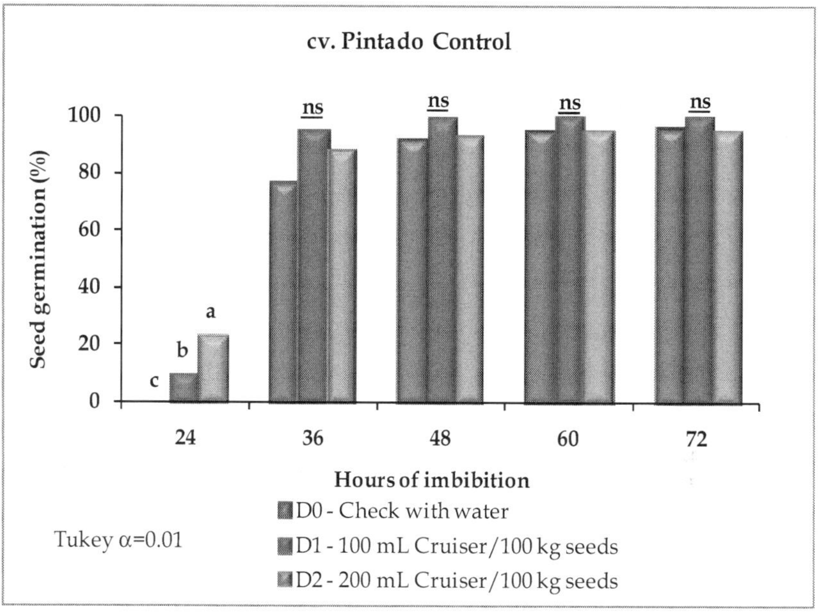

Fig. 2. Soybean germination percentage cv. Pintado treated at recommended dose of Cruiser (D1), double of recommended dose (D2) and check (D0). Average followed by the same letter did not differ significantly for each imbibition period. **ns**: not differ significantly for each imbibition period.

3.2 Second Experiment: Cruiser action on the germination of soybean seeds subjected to stress conditions induced by heavy metal (aluminum), salinity (NaCl) and water deficit

In the presence of aluminum in different concentrations (Figures 3 to 5), the treatment of soybean seeds of cultivar BRS 133 with the recommended level of Cruiser (D1) caused acceleration of germination, when compared with the control (D0), up to 36 h of imbibition. In the soybean seeds of cultivar Pintado, the same pattern of cultivar BRS 133 was observed within 36 h of imbibition in aluminum concentration of 5 mmol L^{-1} (Figure 3) and up to 48 h of soaking in aluminum concentrations of 10 and 15 mmol L^{-1} (Figures 4 and 5, respectively). In the Figure 6 is shown comparisons of the effect of Cruiser on the germination of cultivar BRS 133 in the different concentrations of aluminum, at 24 and 36 h of imbibition. In the cultivar Pintado comparisons were performed at 36 and 48 h of imbibition (Figure 7).

Analyzing the results can be considered that: a) aluminum delays germination in both cultivars studied; b) in the two soybean cultivars, the increase in aluminum concentration caused a decrease in germination; c) Cruiser increase germination in aluminum stress conditions and d) on cultivar BRS 133 at 36 h of imbibition (Figure 6) and on cultivar Pintado at 48 h of imbibition (Figure 7) greater the stress by the presence of aluminum, greater was the effect of Cruiser. Therefore, the results of soybean germination in response to treatment of seeds with Cruiser, under stressful aluminum conditions, indicate that the insecticide acts by reducing the toxic effect of aluminum on germination.

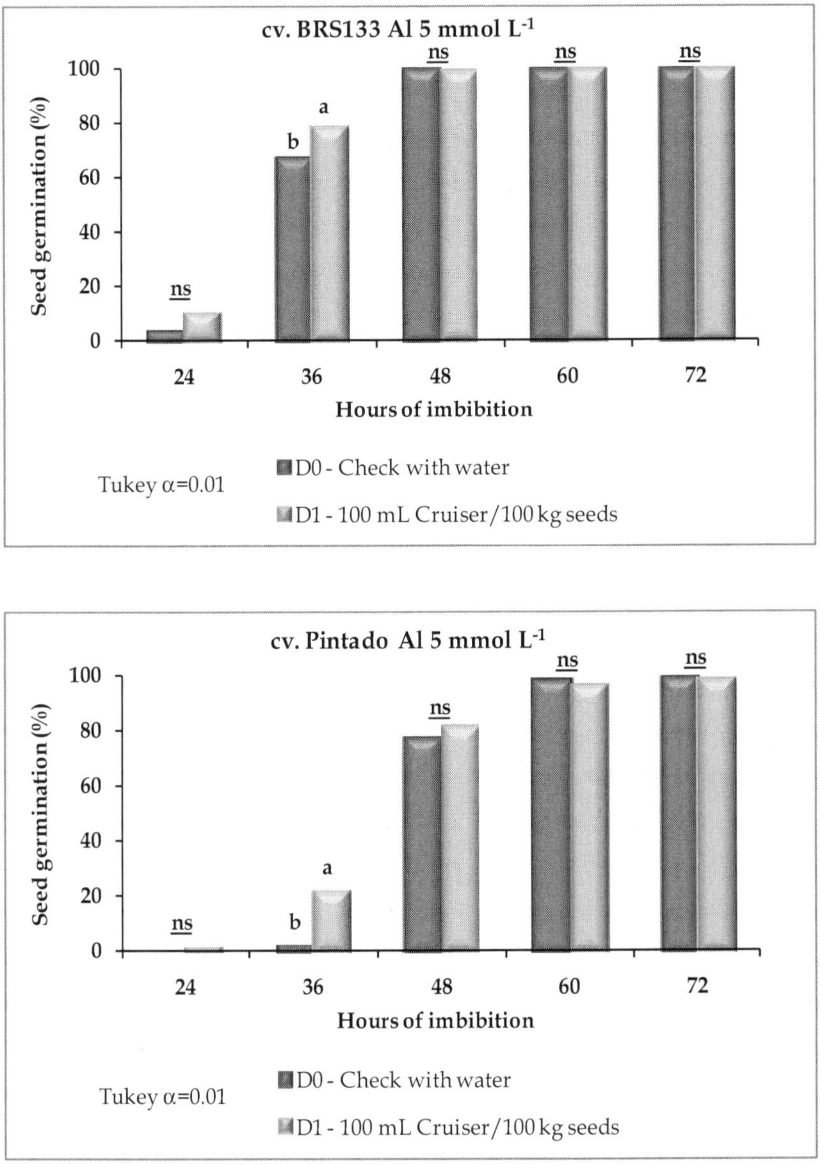

Fig. 3. Soybean germination percentage cv. BRS 133 and cv. Pintado treated at recommended dose of Cruiser (D1) and check (D0), under aluminum sulfate 5 mmol L^{-1}. Average followed by the same letter did not differ significantly for each imbibition period. **ns**: not differ significantly for each imbibition period.

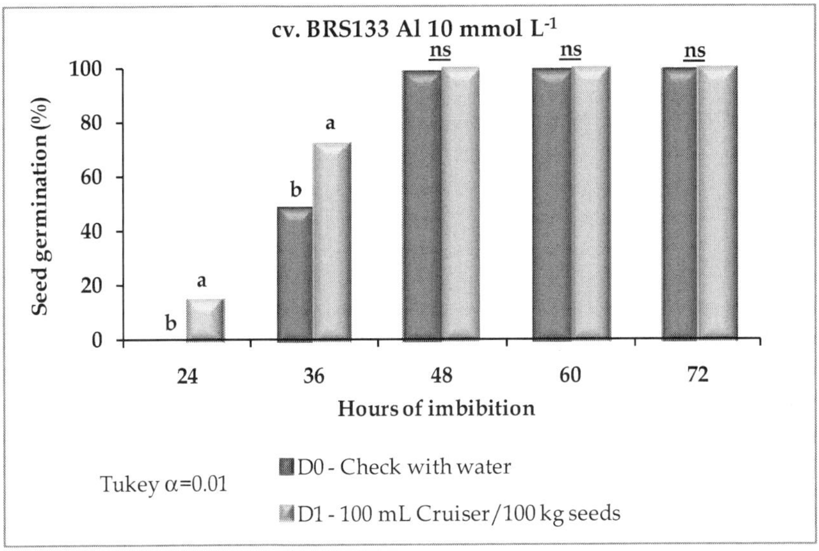

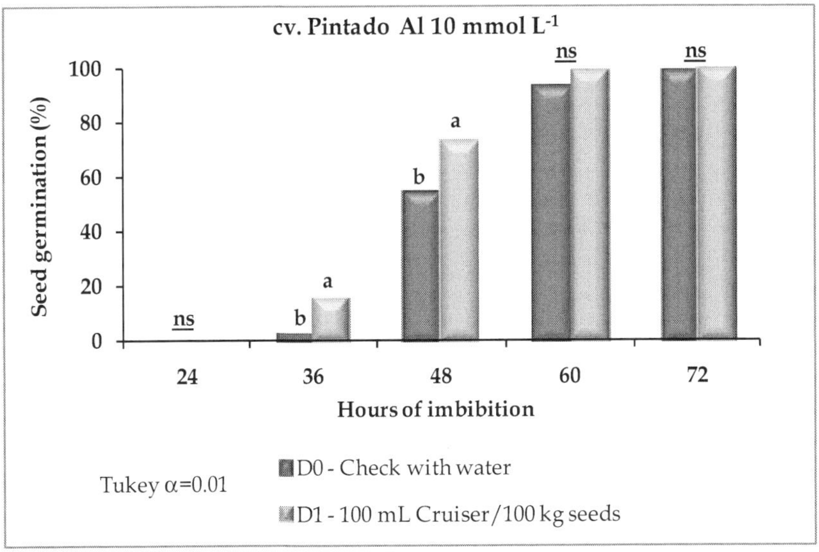

Fig. 4. Soybean germination percentage cv. BRS 133 and cv. Pintado treated at recommended dose of Cruiser (D1) and check (D0), under aluminum sulfate 10 mmol L^{-1}. Average followed by the same letter did not differ significantly for each imbibition period. **ns**: not differ significantly for each imbibition period.

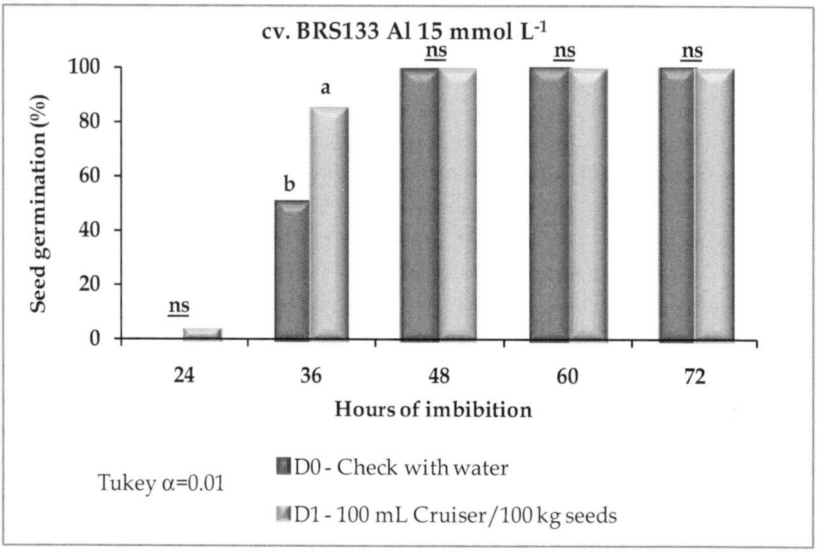

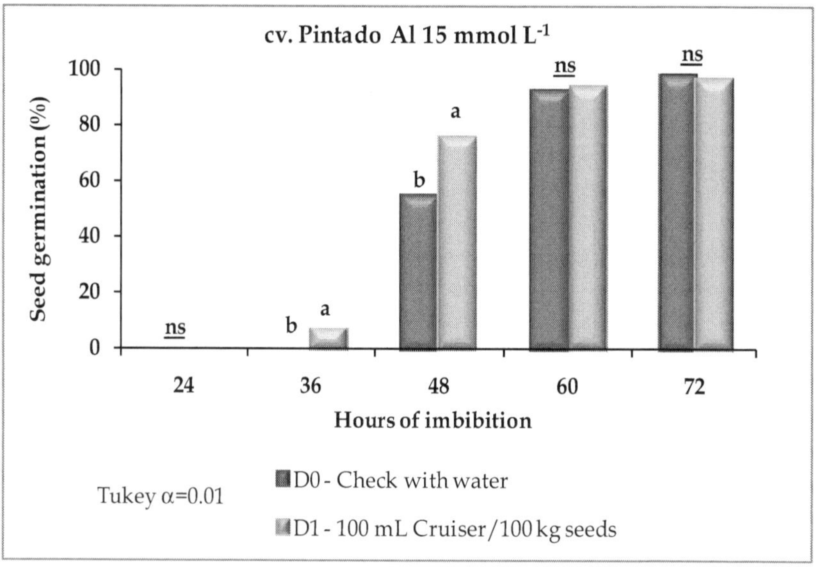

Fig. 5. Soybean germination percentage cv. BRS 133 and cv. Pintado treated at recommended dose of Cruiser (D1) and check (D0), under aluminum sulfate 15 mmol L^{-1}. Average followed by the same letter did not differ significantly for each imbibition period. **ns**: not differ significantly for each imbibition period.

Under salinity conditions in the presence of NaCl (Figures 8 to 11), the treatment of soybean seeds of cultivar BRS 133 with Cruiser caused acceleration of germination in the first periods of imbibition evaluated. It was observed that higher the concentration of NaCl, the effect mentioned was observed in the later periods of imbibition, reaching up to 48 h in NaCl concentration of 150 mmol L^{-1} (Figure 11). It was observed that Cruiser had no effect on germination of cultivar Pintado at concentrations of NaCl 25 (Figure 8) and 100 mmol L^{-1} (Figure 10). In NaCl concentration of 50 mmol L^{-1} (Figure 9) Cruiser decreased germination, but in the concentration of 150 mmol L^{-1} (Figure 11) it increased germination at 48 h of imbibition.

Comparing the results of Cruiser effect on germination of cultivar BRS 133 in the different concentrations of NaCl in the imbibition periods of 24, 36 and 48 h (Figure 12), can be made several considerations: a) NaCl causes decrease in germination, the effect being more pronounced greater the salinity stress; b) at 24 h of imbibition, Cruiser had effect until the NaCl concentration of 100 mmol.L^{-1}; c) at 36 h of imbibition, Cruiser eliminated the effect of salt stress up to the salt concentration of 50 mmol.L^{-1} and in higher salinity stress, greater was the effect of Cruiser; d) at 48 h of imbibition, Cruiser eliminated any effect of salt stress.

Analyzing the comparisons of the results in Figure13 can be considered that in cultivar Pintado Cruiser had no effect on germination under salt stress, during imbibition of 24, 36 and 48 h.

The effect of Cruiser on germination of cultivar BRS 133 under water deficit induced by PEG solutions of different water potentials are shown in Figures 14 to 16. At the water potentials of -0.1 and -0.2 MPa, Cruiser had no effect on germination, but at the water potential of -0.3 MPa, Cruiser has caused a significant increase in germination at 72 h of imbibition. In respect of germination of cultivar Pintado under water deficit conditions induced by PEG solutions of different water potentials, it was observed that in the water potential of -0.1 MPa, Cruiser caused increase on germination at 48 and 60 h of imbibition. In water potential of -0.2 MPa the increase on germination by Cruiser effect were observed from 60 to 84 h of imbibition and in the potential of -0.3 MPa only at 72 and 84 h of imbibition.

Comparing the results of the effect of Cruiser on germination of cultivar BRS 133 at different imbibition periods (Figure 17) in the different water potentials, can be made some considerations: a) the decrease of water potential delays germination; b) there is consistency of Cruiser effect in increasing the germination for the three water potentials; c) at 72 h of imbibition, the largest increase in germination under Cruiser effect occurred where the water deficit was higher.

Comparing the effects of Cruiser on germination of cultivar Pintado, at the water potentials used (Figure 18), can be made some considerations: a) water deficiency causes delayed germination; b) Cruiser has effect in combating water stress for all the three tested water potentials; c) at 72 and 84 h of imbibition Cruiser has a greater effect on germination in the largest water deficit.

In Figure 19 is represented, the effect of Cruiser on the weights of embryo axis of soybean cultivars BRS 133 and Pintado under conditions of aluminum presence. Can be inferred that in all concentrations of aluminum used Cruiser has caused increased growth of the embryo axis but, this increase was significantly higher in the absence of aluminum, in the concentration of 10 mmol L^{-1} for BRS 133 and in the absence of aluminum (0 mmol L^{-1}) to cultivar Pintado. The effect of Cruiser on development of embryo axis occurred in the absence of aluminum in both cultivars. The weight of the embryo axis tended to be equal between the treated and untreated seeds with Cruiser, with the increase of aluminum stress (Figures 57 and 58).

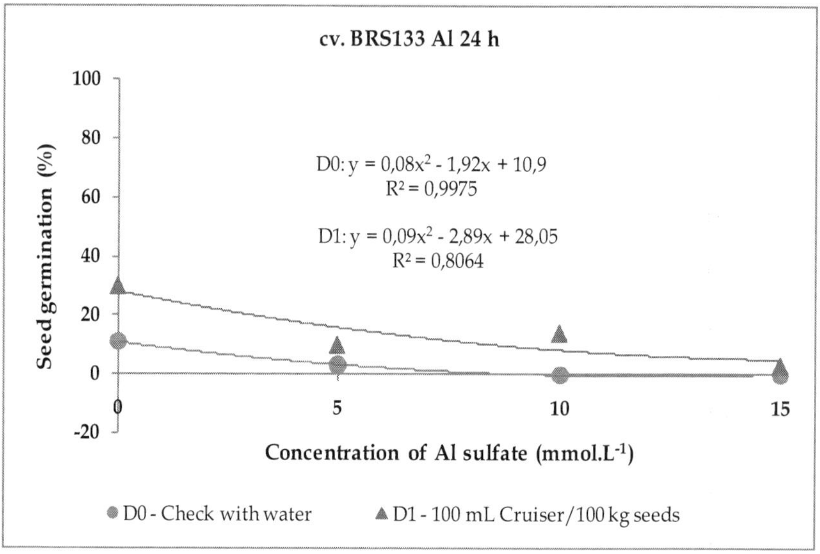

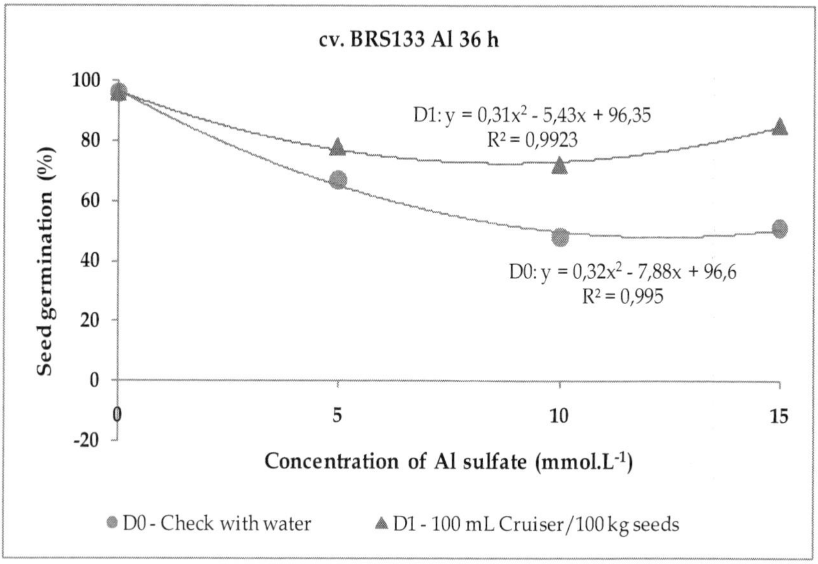

Fig. 6. Comparison of soybean germination percentage cv. BRS 133 treated at recommended dose of Cruiser (D1) and check (D0), under different aluminum sulfate concentrations (5, 10 and 15 mmol L^{-1}) at 24 and 36 h of imbibition.

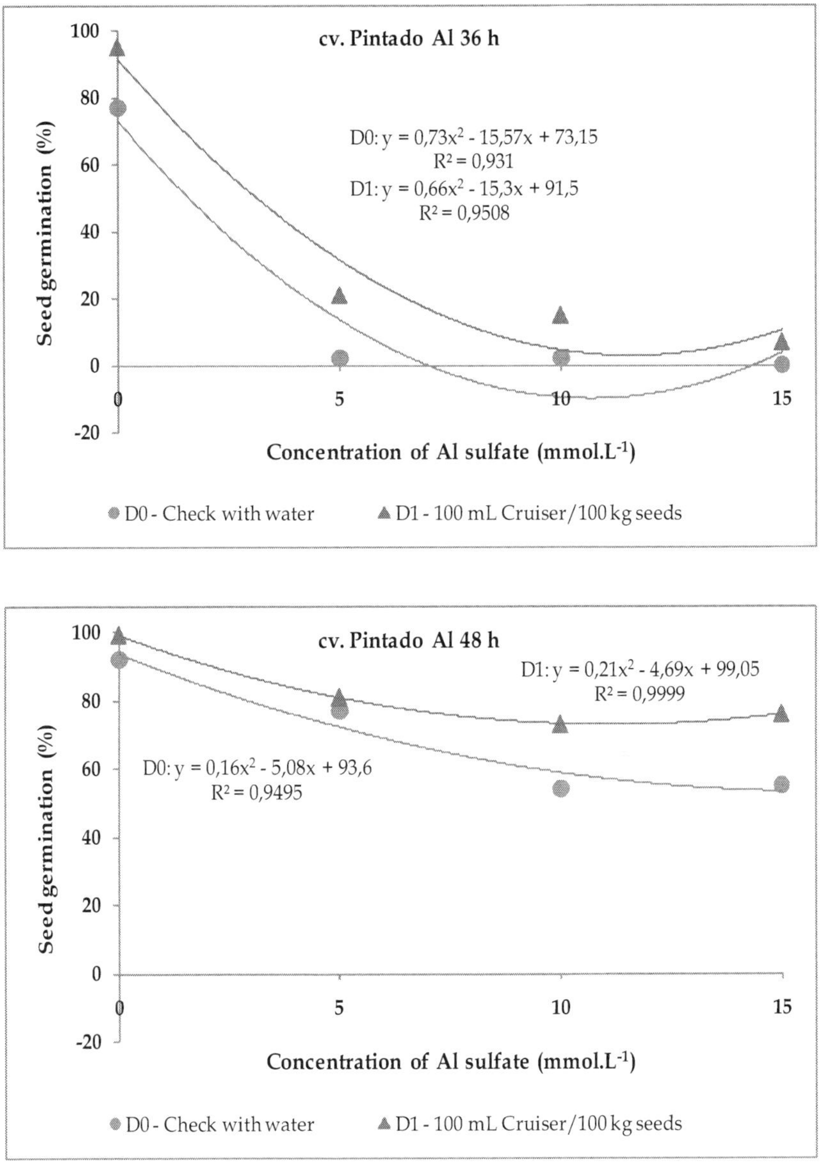

Fig. 7. Comparison of soybean germination percentage cv. Pintado treated at recommended dose of Cruiser (D1) and check (D0), under different aluminum sulfate concentrations (5, 10 and 15 mmol L^{-1}) at 24 and 36 h of imbibition.

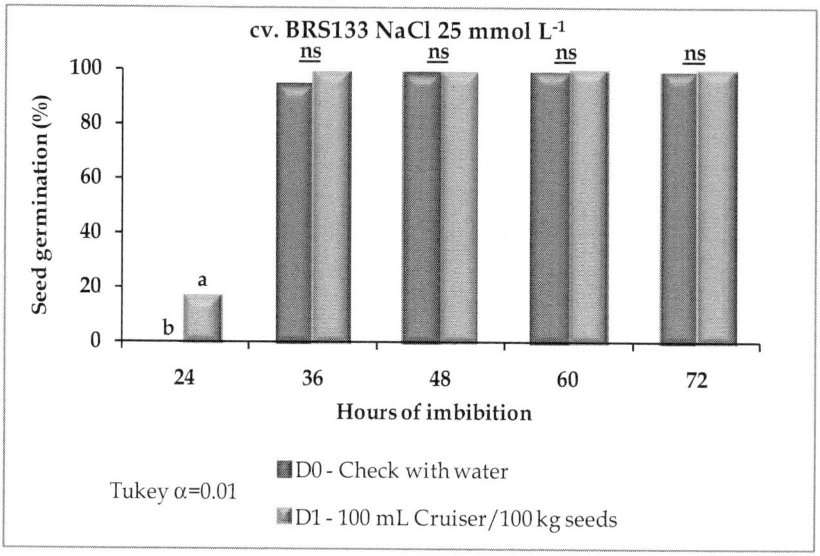

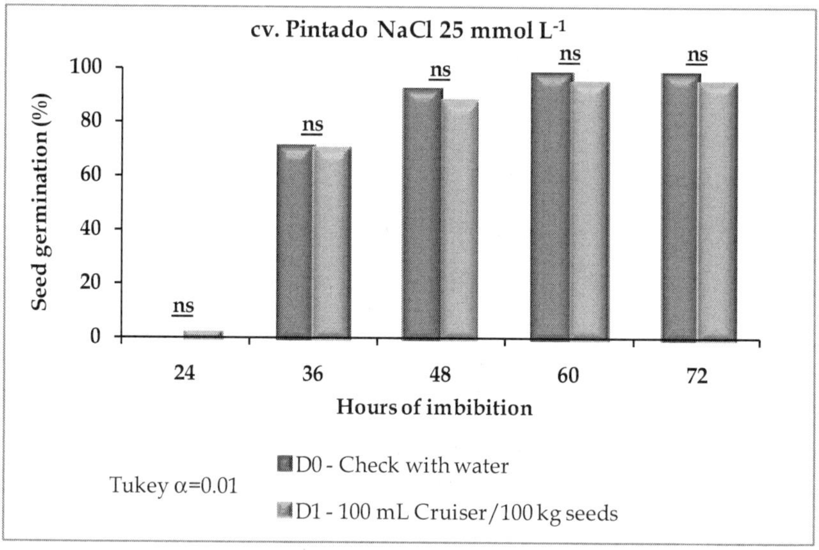

Fig. 8. Soybean germination percentage cv. BRS 133 and cv. Pintado treated at recommended dose of Cruiser (D1) and check (D0), under NaCl 25 mmol L^{-1}. Average followed by the same letter did not differ significantly for each imbibition period. **ns**: not differ significantly for each imbibition period.

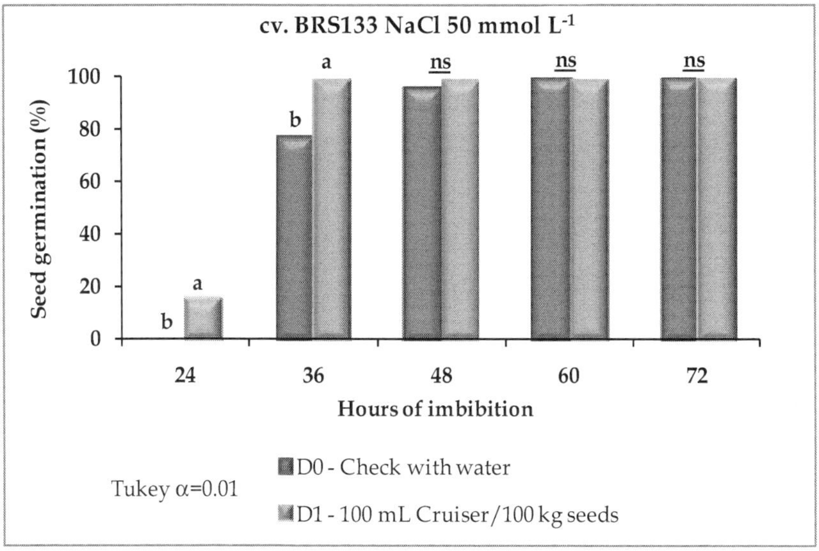

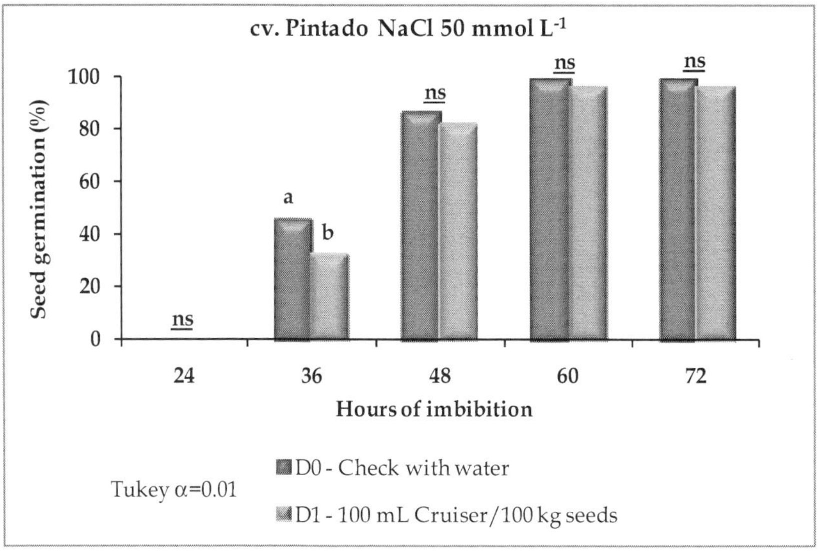

Fig. 9. Soybean germination percentage cv. BRS 133 and cv. Pintado treated at recommended dose of Cruiser (D1) and check (D0), under NaCl 50 mmol L^{-1}. Average followed by the same letter did not differ significantly for each imbibition period. **ns**: not differ significantly for each imbibition period.

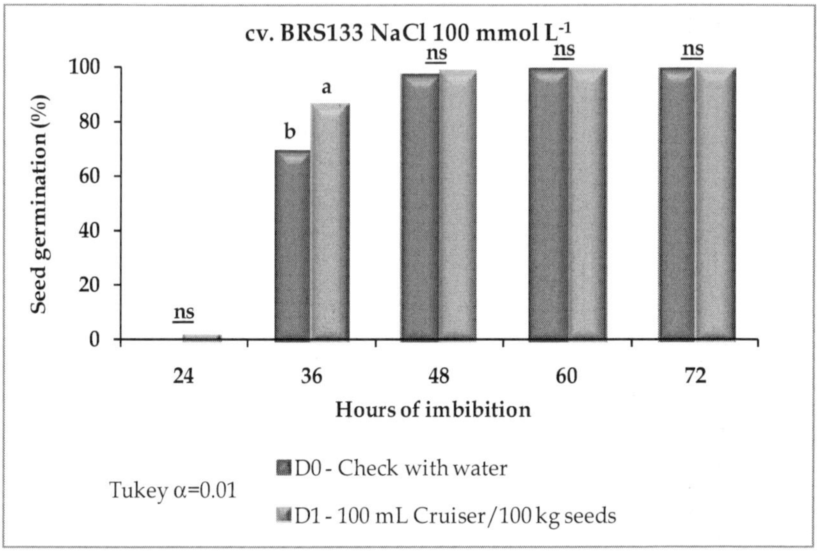

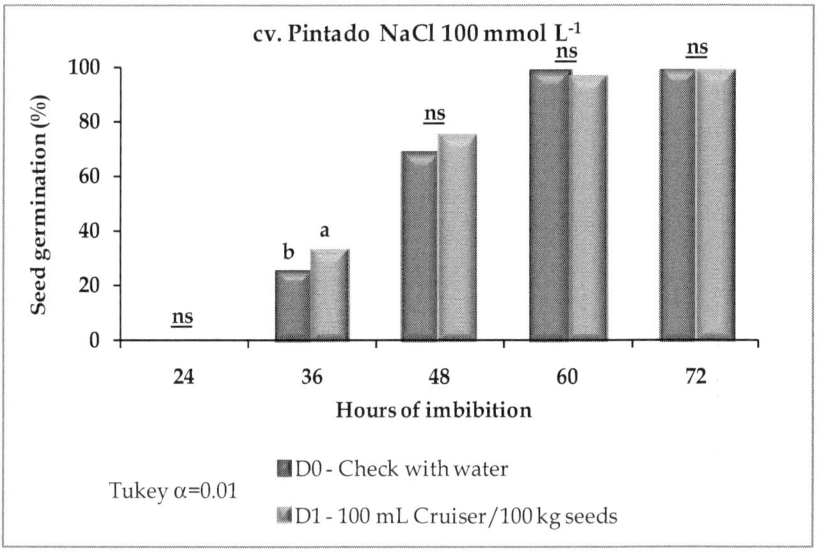

Fig. 10. Soybean germination percentage cv. BRS 133 and cv. Pintado treated at recommended dose of Cruiser (D1) and check (D0), under NaCl 100 mmol L^{-1}. Average followed by the same letter did not differ significantly for each imbibition period. **ns**: not differ significantly for each imbibition period.

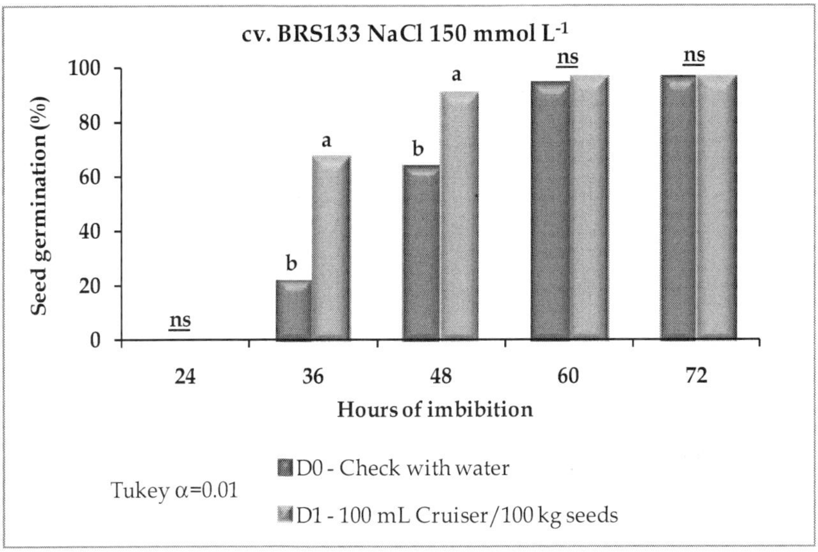

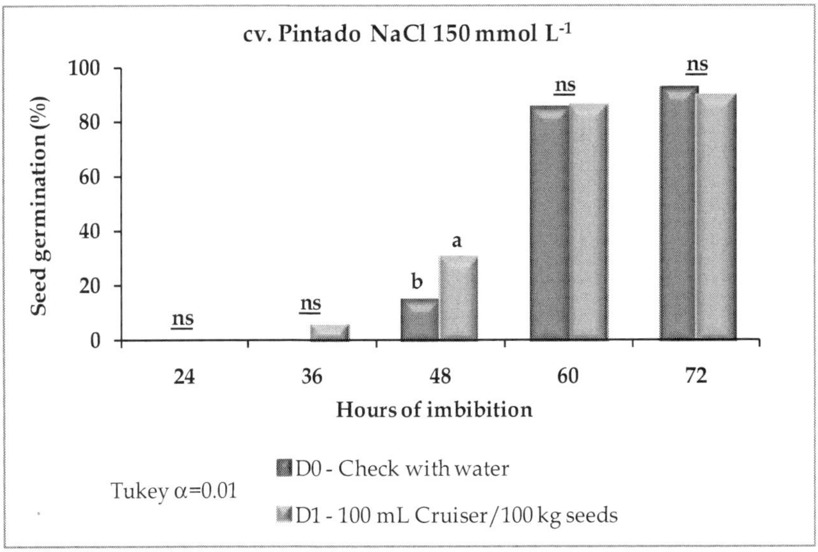

Fig. 11. Soybean germination percentage cv. BRS 133 and cv. Pintado treated at recommended dose of Cruiser (D1) and check (D0), under NaCl 150 mmol L^{-1}. Average followed by the same letter did not differ significantly for each imbibition period. **ns**: not differ significantly for each imbibition period.

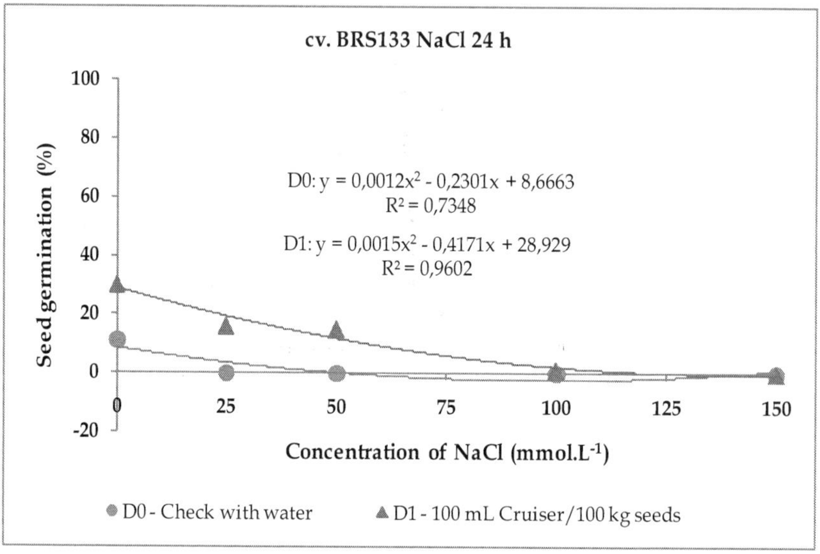
cv. BRS133 NaCl 24 h
100
80
60
40
20
0
-20
Seed germination (%)
D0: y = 0,0012x² - 0,2301x + 8,6663
R² = 0,7348
D1: y = 0,0015x² - 0,4171x + 28,929
R² = 0,9602
0
25
50
75
100
125
150
Concentration of NaCl (mmol.L⁻¹)
D0 - Check with water
D1 - 100 mL Cruiser/100 kg seeds

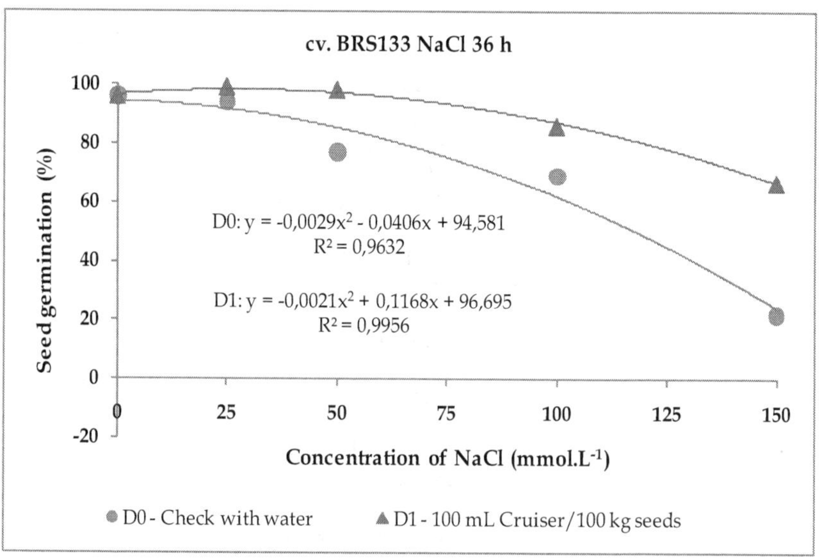
cv. BRS133 NaCl 36 h
100
80
60
40
20
0
-20
Seed germination (%)
D0: y = -0,0029x² - 0,0406x + 94,581
R² = 0,9632
D1: y = -0,0021x² + 0,1168x + 96,695
R² = 0,9956
0
25
50
75
100
125
150
Concentration of NaCl (mmol.L⁻¹)
D0 - Check with water
D1 - 100 mL Cruiser/100 kg seeds

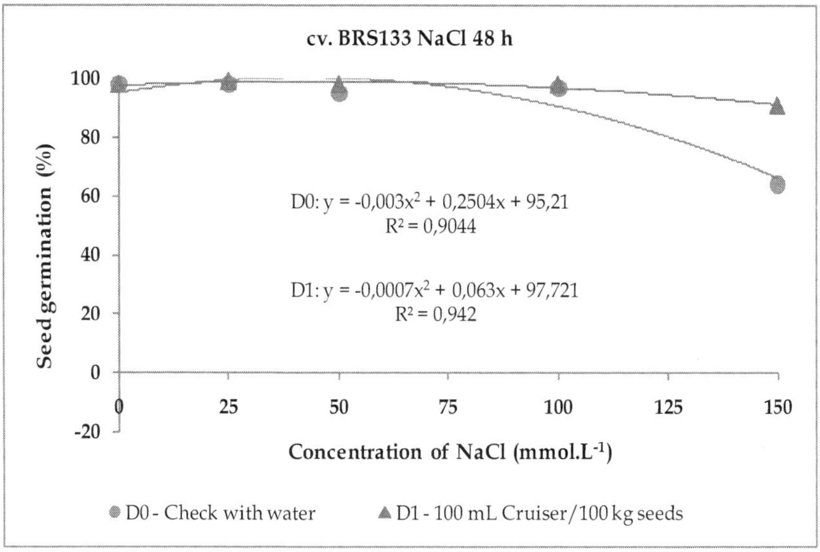

Fig. 12. Comparison of soybean germination percentage cv. BRS133 treated at recommended dose of Cruiser (D1) and check (D0), under different NaCl concentrations (25, 50, 100 and 150 mmol L^{-1}) at 24, 36 and 48 h of imbibition.

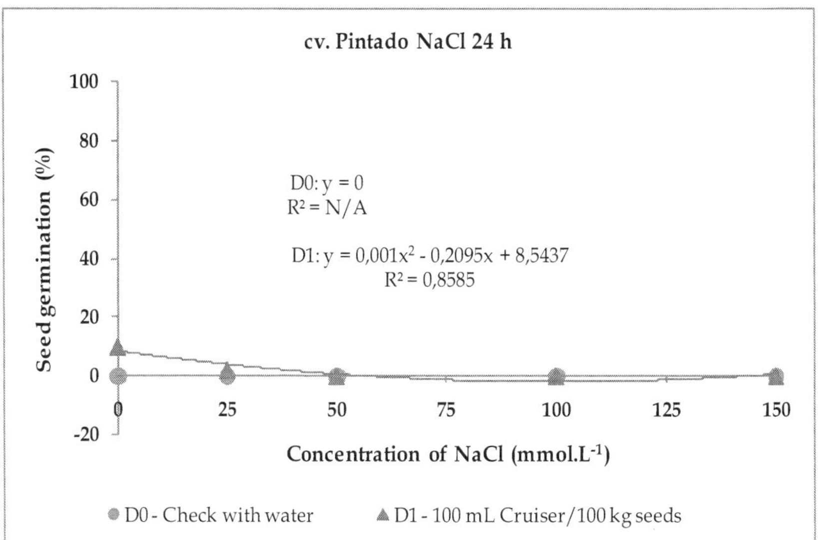

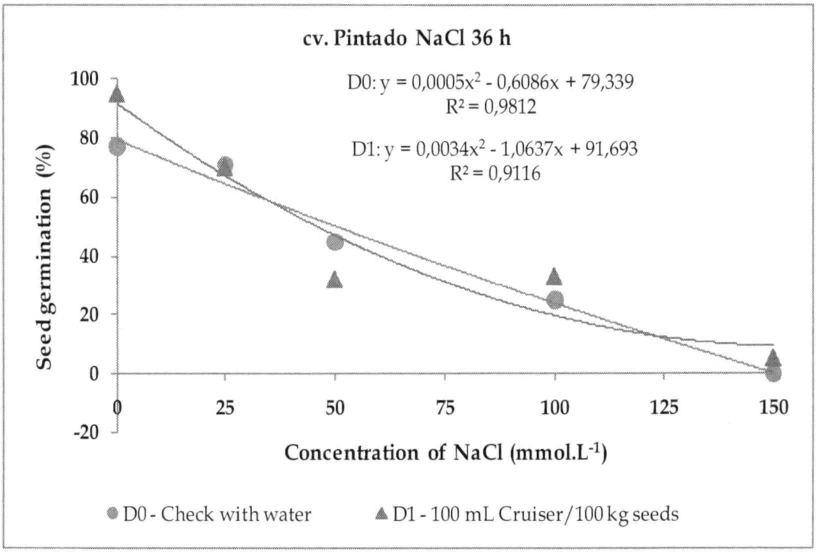

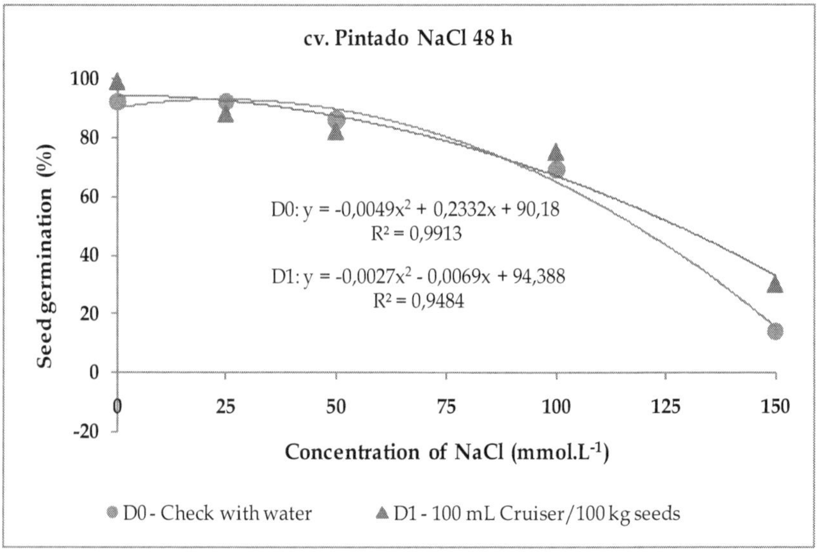

Fig. 13. Comparison of soybean germination percentage cv. Pintado treated at recommended dose of Cruiser (D1) and check (D0), under different NaCl concentrations (25, 50, 100 and 150 mmol L^{-1}) at 24, 36 and 48 h of imbibition.

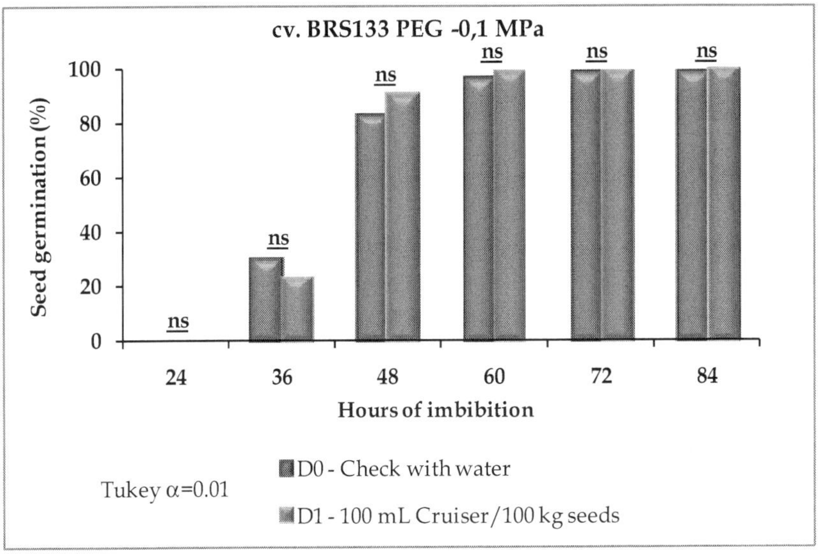

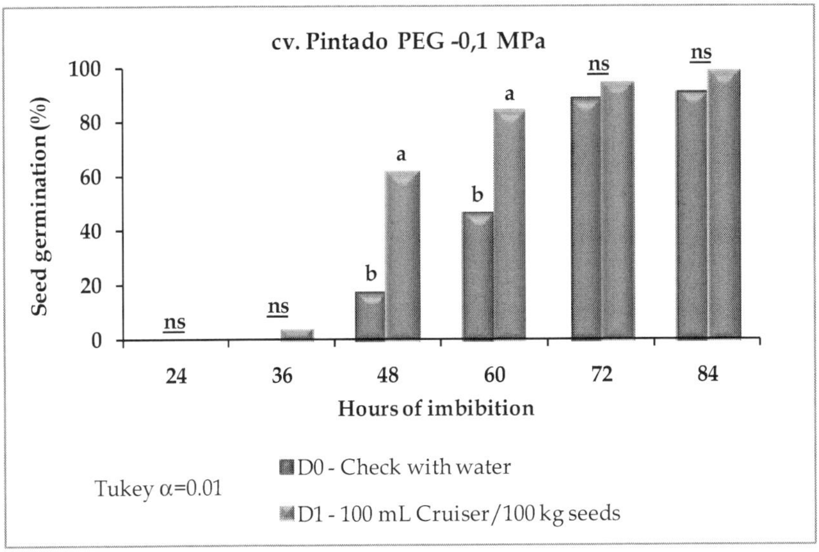

Fig. 14. Soybean germination percentage cv. BRS 133 and cv. Pintado treated at recommended dose of Cruiser (D1) and check (D0), under PEG potential -0,1 MPa. Average followed by the same letter did not differ significantly for each imbibition period. **ns**: not differ significantly for each imbibition period.

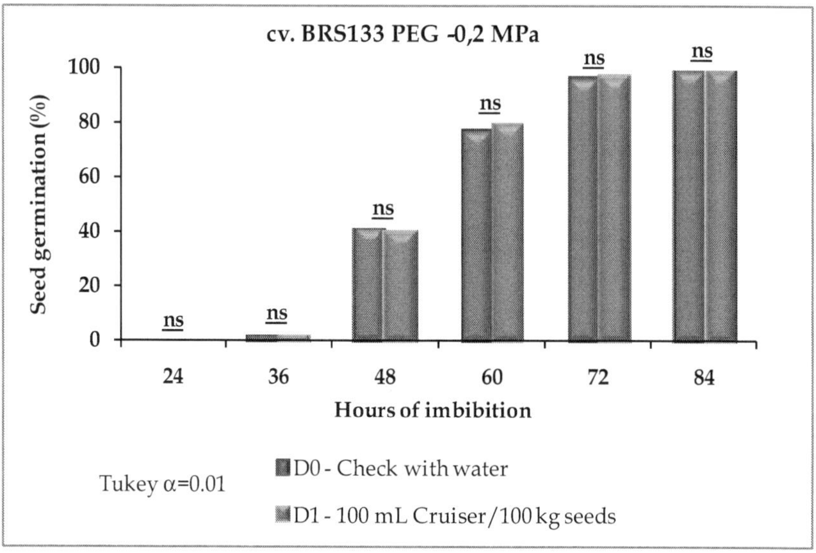

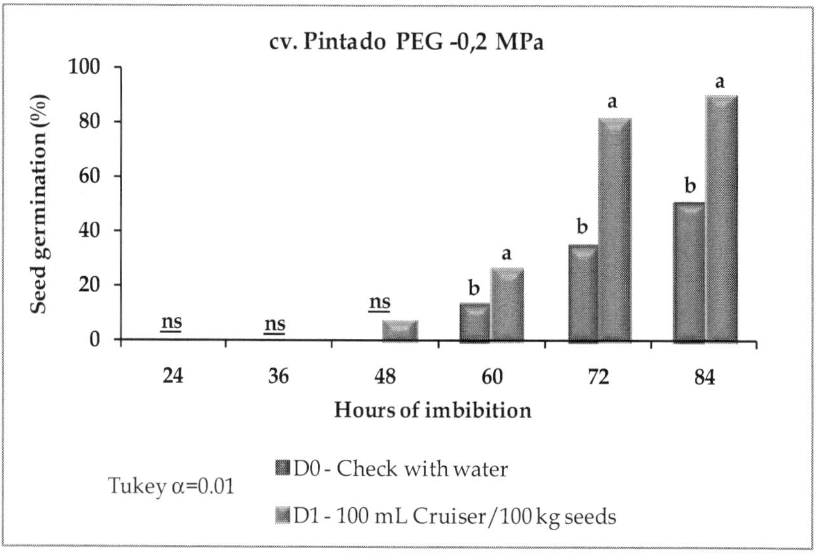

Fig. 15. Soybean germination percentage cv. BRS 133 and cv. Pintado treated at recommended dose of Cruiser (D1) and check (D0), under PEG potential -0,2 MPa. Average followed by the same letter did not differ significantly for each imbibition period. **ns**: not differ significantly for each imbibition period.

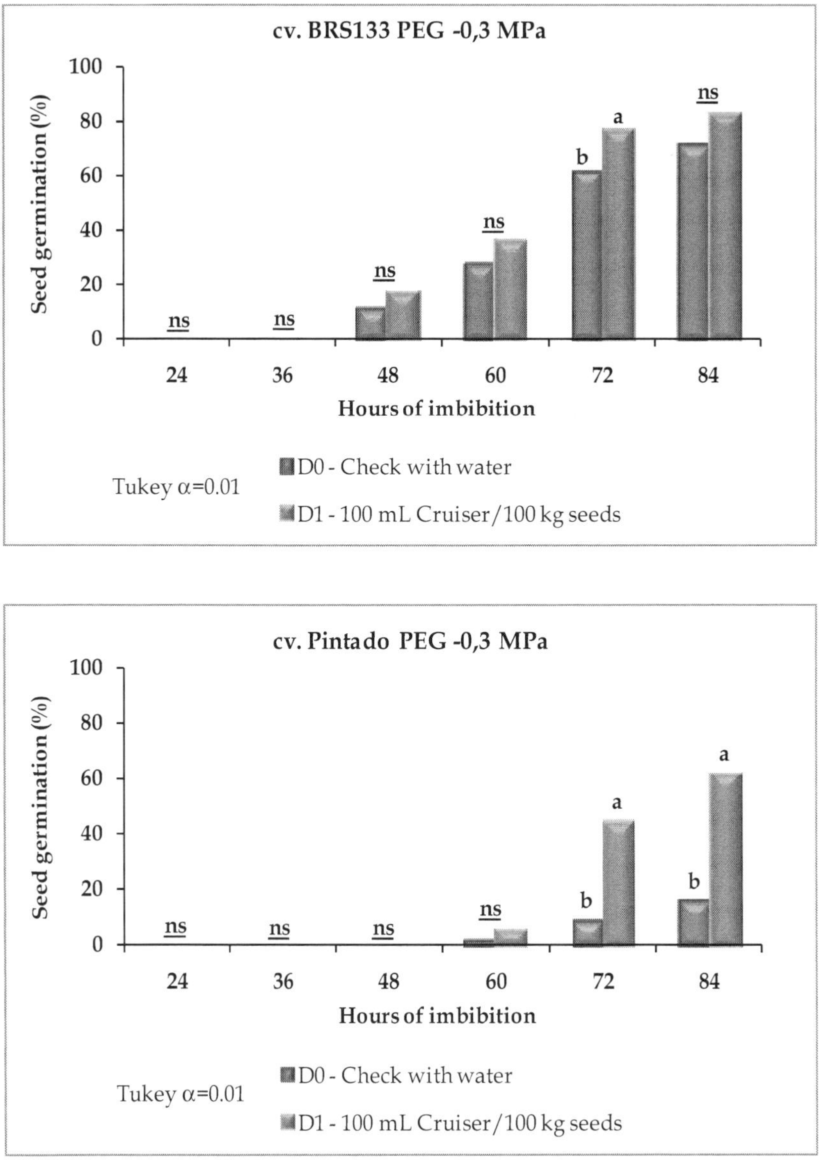

Fig. 16. Soybean germination percentage cv. BRS 133 and cv. Pintado treated at recommended dose of Cruiser (D1) and check (D0), under PEG potential -0.3 MPa. Average followed by the same letter did not differ significantly for each imbibition period. **ns**: not differ significantly for each imbibition period.

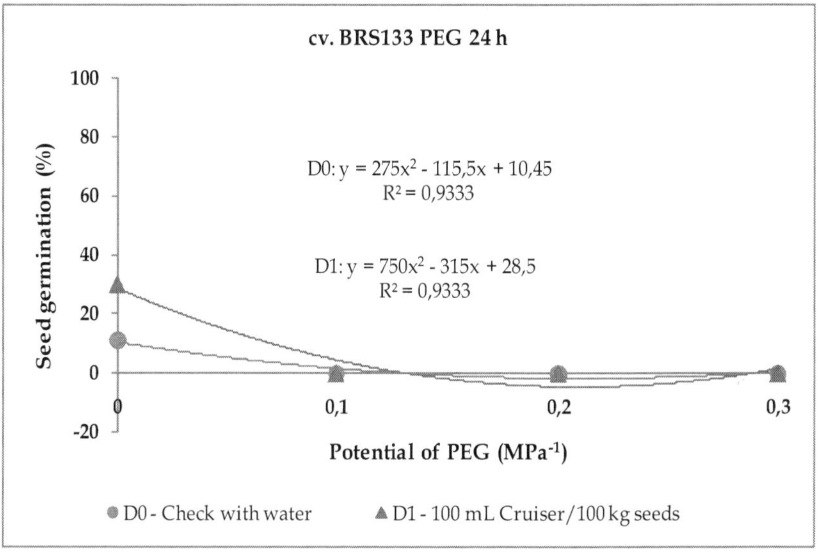
cv. BRS133 PEG 24 h
100
80
60
40
20
0
-20
Seed germination (%)
D0: y = 275x² - 115,5x + 10,45
R² = 0,9333
D1: y = 750x² - 315x + 28,5
R² = 0,9333
0
0,1
0,2
0,3
Potential of PEG (MPa⁻¹)
D0 - Check with water
D1 - 100 mL Cruiser/100 kg seeds

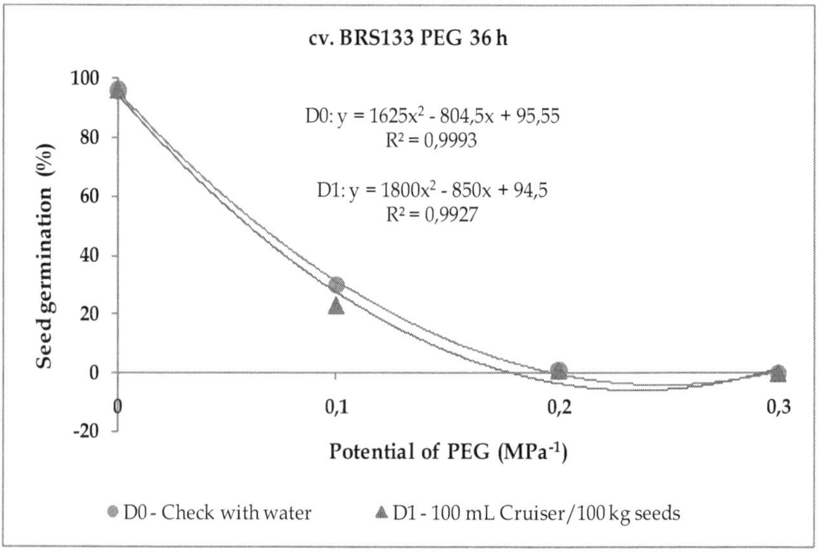
cv. BRS133 PEG 36 h
100
80
60
40
20
0
-20
Seed germination (%)
D0: y = 1625x² - 804,5x + 95,55
R² = 0,9993
D1: y = 1800x² - 850x + 94,5
R² = 0,9927
0
0,1
0,2
0,3
Potential of PEG (MPa⁻¹)
D0 - Check with water
D1 - 100 mL Cruiser/100 kg seeds

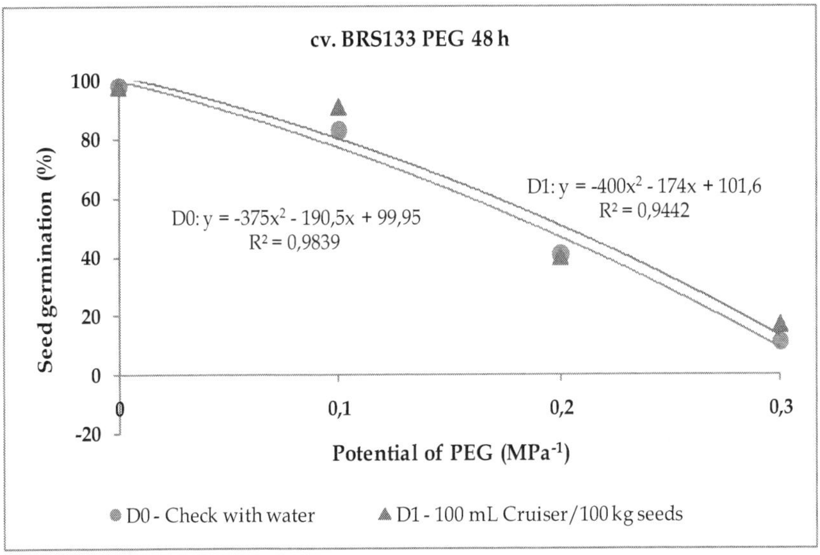
cv. BRS133 PEG 48 h
Seed germination (%)
100
80
60
40
20
0
-20
0
0,1
0,2
0,3
D1: y = -400x² - 174x + 101,6
R² = 0,9442
D0: y = -375x² - 190,5x + 99,95
R² = 0,9839
Potential of PEG (MPa-1)
D0 - Check with water
D1 - 100 mL Cruiser/100 kg seeds

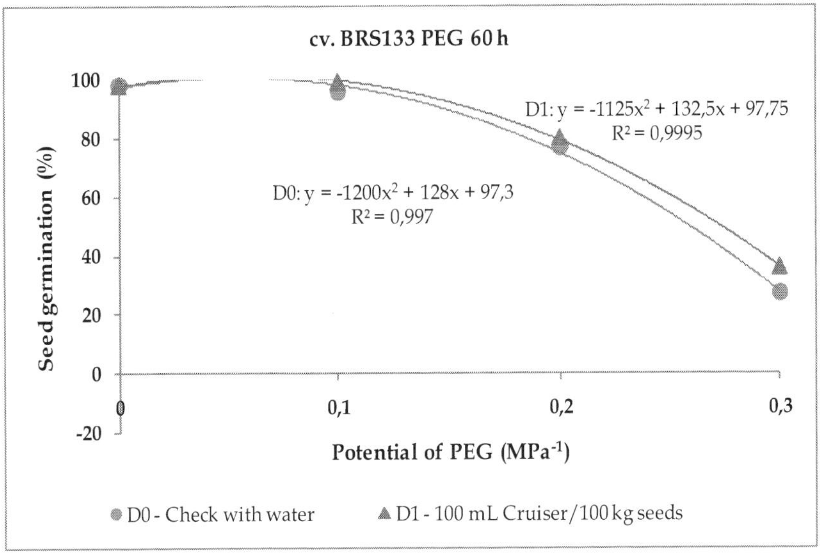
cv. BRS133 PEG 60 h
Seed germination (%)
100
80
60
40
20
0
-20
0
0,1
0,2
0,3
D1: y = -1125x² + 132,5x + 97,75
R² = 0,9995
D0: y = -1200x² + 128x + 97,3
R² = 0,997
Potential of PEG (MPa-1)
D0 - Check with water
D1 - 100 mL Cruiser/100 kg seeds

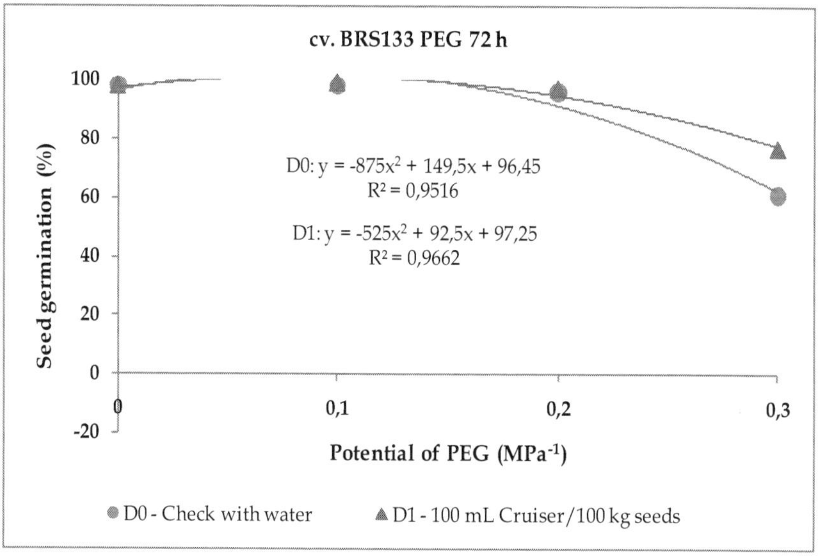

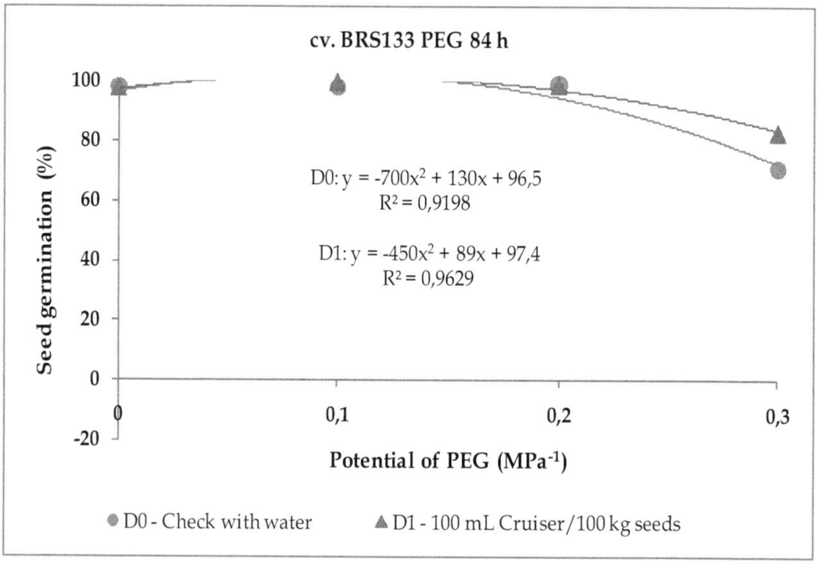

Fig. 17. Comparison of soybean germination percentage cv. BRS133 treated at recommended dose of Cruiser (D1) and check (D0), under different PEG potentials (-0,1; -0,2 and -0,3 MPa) at 24, 36, 48, 60, 72 and 84 h of imbibition.

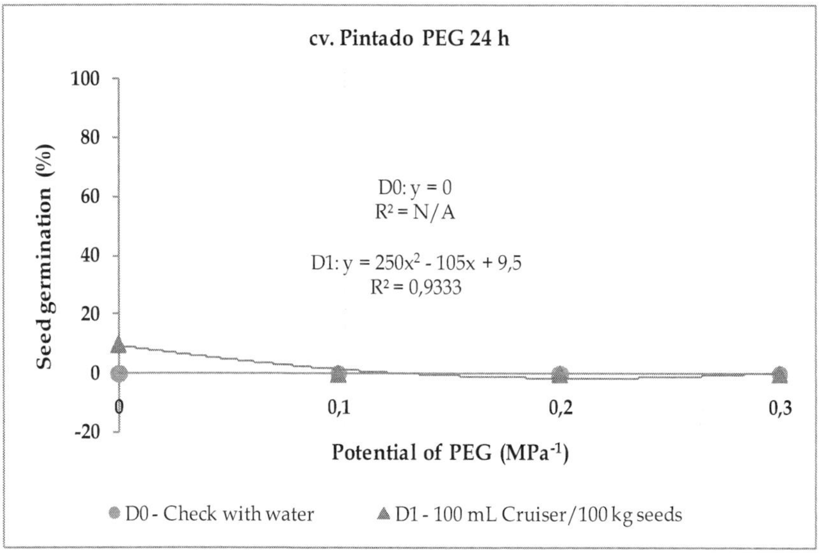
cv. Pintado PEG 24 h
100
80
60
40
20
0
-20
Seed germination (%)
D0: y = 0
R² = N/A
D1: y = 250x² - 105x + 9,5
R² = 0,9333
0
0,1
0,2
0,3
Potential of PEG (MPa-1)
D0 - Check with water
D1 - 100 mL Cruiser/100 kg seeds

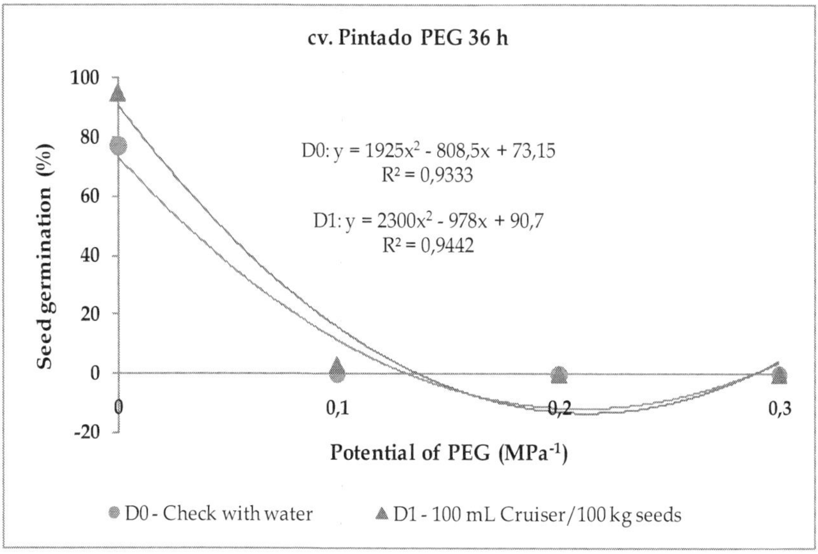
cv. Pintado PEG 36 h
100
80
60
40
20
0
-20
Seed germination (%)
D0: y = 1925x² - 808,5x + 73,15
R² = 0,9333
D1: y = 2300x² - 978x + 90,7
R² = 0,9442
0
0,1
0,2
0,3
Potential of PEG (MPa-1)
D0 - Check with water
D1 - 100 mL Cruiser/100 kg seeds

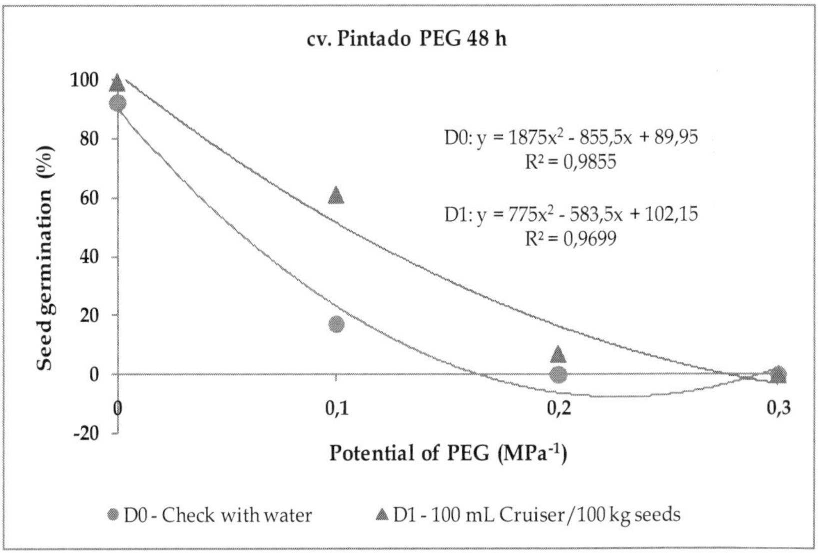
cv. Pintado PEG 48 h
Seed germination (%)
100
80
60
40
20
0
-20
0
0,1
0,2
0,3
Potential of PEG (MPa-1)
D0: y = 1875x2 - 855,5x + 89,95
R2 = 0,9855
D1: y = 775x2 - 583,5x + 102,15
R2 = 0,9699
D0 - Check with water
D1 - 100 mL Cruiser/100 kg seeds

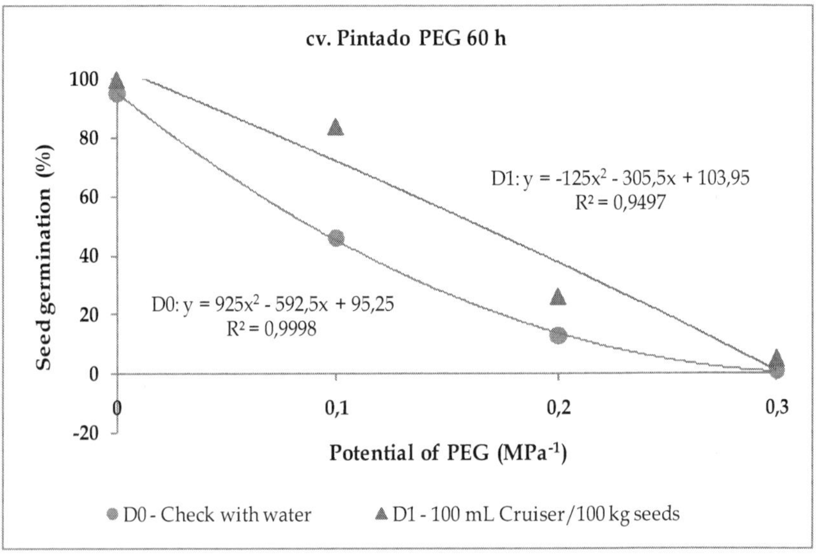
cv. Pintado PEG 60 h
Seed germination (%)
100
80
60
40
20
0
-20
0
0,1
0,2
0,3
Potential of PEG (MPa-1)
D1: y = -125x2 - 305,5x + 103,95
R2 = 0,9497
D0: y = 925x2 - 592,5x + 95,25
R2 = 0,9998
D0 - Check with water
D1 - 100 mL Cruiser/100 kg seeds

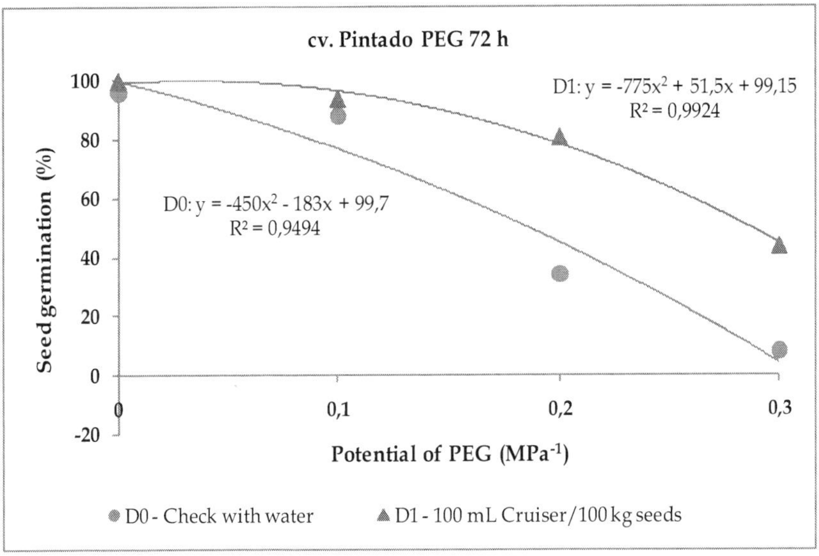

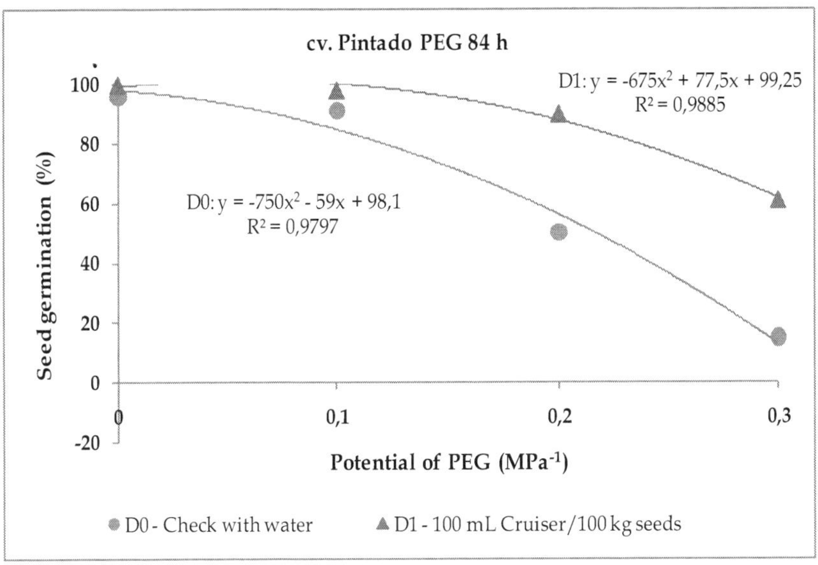

Fig. 18. Comparison of soybean germination percentage cv. Pintado treated at recommended dose of Cruiser (D1) and check (D0), under different PEG potentials (-0,1; -0,2 and -0,3 MPa) at 24, 36, 48, 60, 72 and 84 h of imbibition.

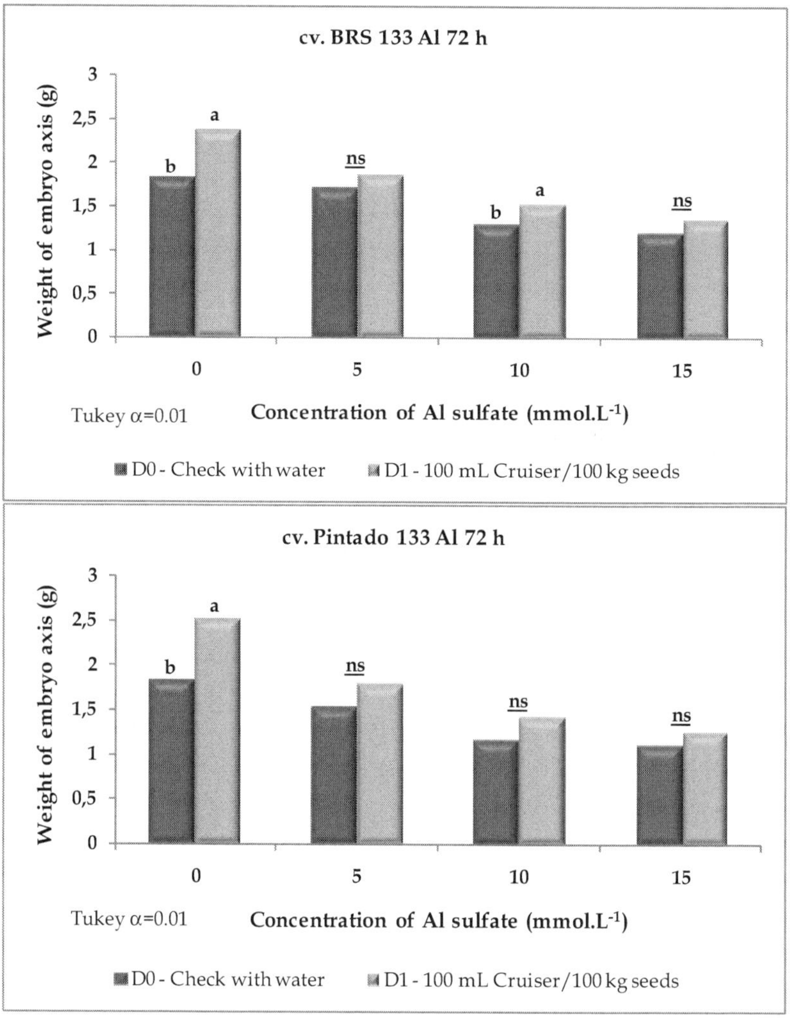

Fig. 19. Weight (g) of embryo axis of soybean seeds cv. BRS 133 and cv. Pintado treated at recommended dose of Cruiser (D1) and check (D0), under different Al concentrations at 72 h of imbibition. Average followed by the same letter did not differ significantly for each concentration or each potential. **ns**: not differ significantly for each imbibition period.

The weights of embryo axis of soybean cultivars BRS 133 and Pintado under salinity are shown, respectively, in Figure 20. In the cultivar BRS 133 Cruiser, generally, caused an increase in the weight of embryo axis at all concentrations of NaCl used except at a concentration of 100 mmol L^{-1} where there was no significant difference between seeds treated and untreated. In cultivar Pintado was observed a significant increase in the weight of embryo axis in the absence of NaCl and at concentration of 25 mmol L^{-1}; however, at the

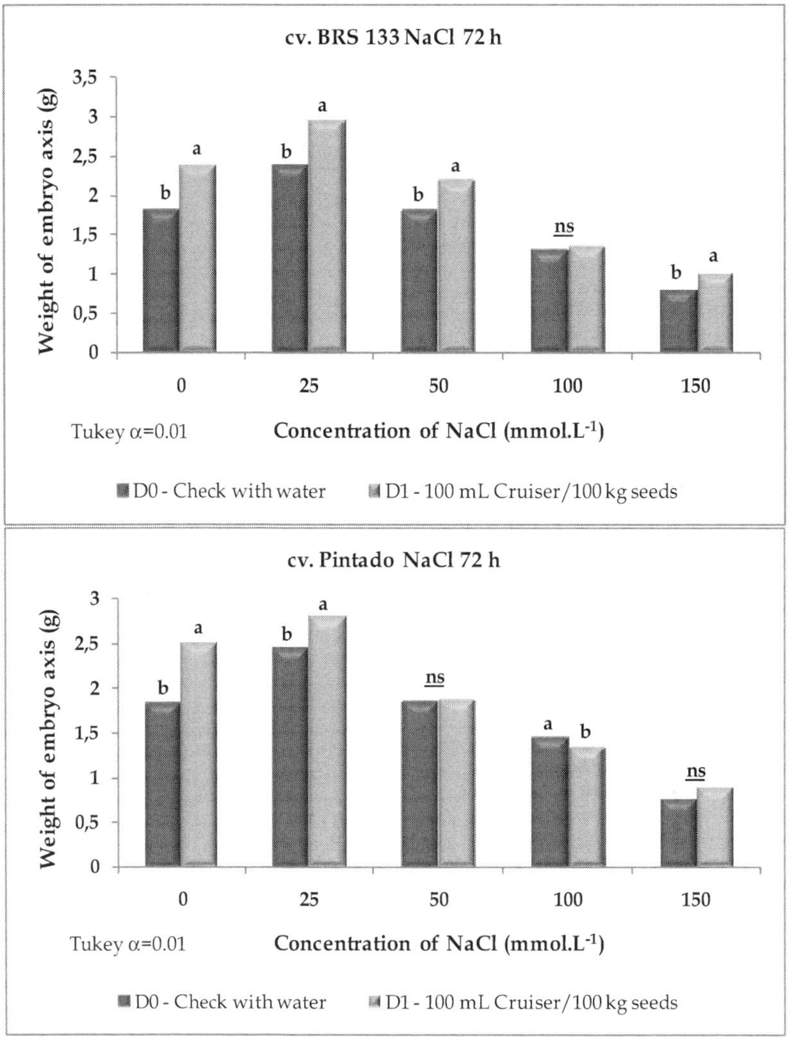

Fig. 20. Weight (g) of embryo axis of soybean seeds cv. BRS 133 and cv. Pintado treated at recommended dose of Cruiser (D1) and check (D0), under different and NaCl concentrations at 72 h of imbibition. Average followed by the same letter did not differ significantly for each concentration or each potential. **ns**: not differ significantly for each imbibition period.

concentration of 100 mmol L^{-1} Cruiser caused a decrease in axis weight. In saline conditions, Cruiser's effect on the development of the axis in cultivar BRS 133 is smaller with the increase of salt stress and in cultivar Pintado Cruiser has no effect under these conditions. The effect of Cruiser on weight of the embryo axis of cultivars BRS 133 and Pintado under water stress conditions are represented in Figure 21. It was observed that in both cultivars,

Cruiser increased the development of the embryo axis in the water potentials of 0 and -0.1 MPa. In situations of greater water deficit (-0.2 and -0.3 MPa) there was no significant difference between treated and untreated seeds. The effect of Cruiser on the development of embryo axis in conditions of water stress is smaller with increasing of water deficit.

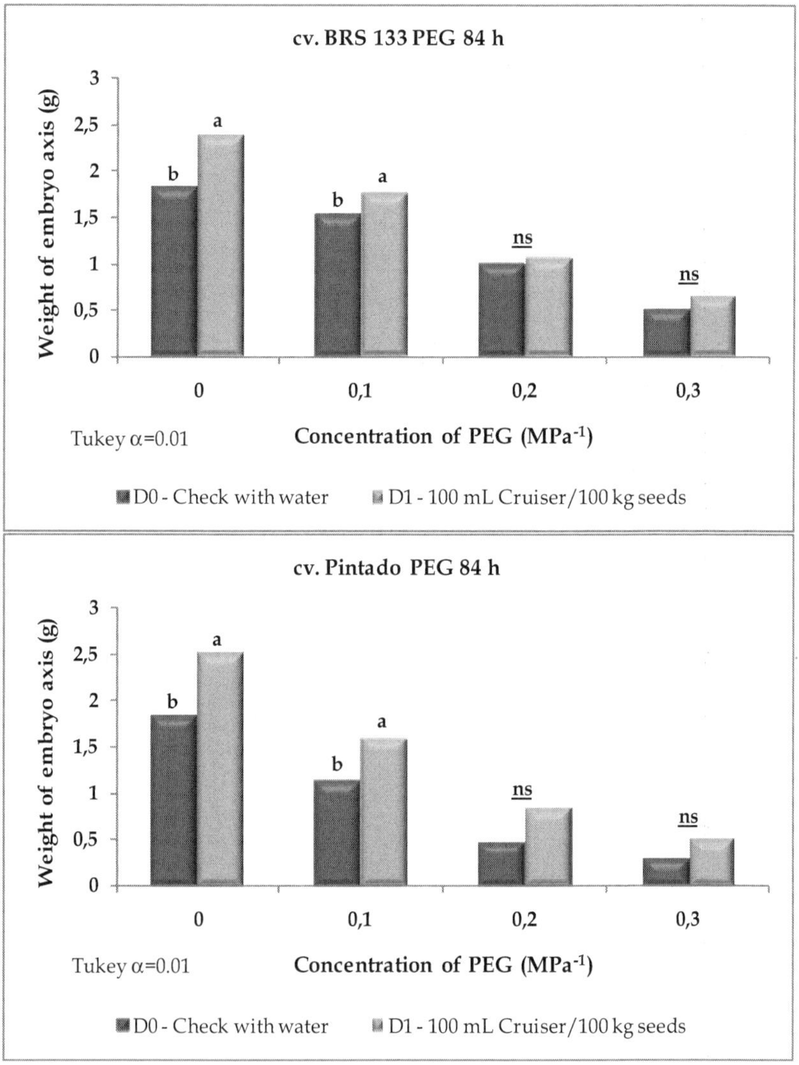

Fig. 21. Weight (g) of embryo axis of soybean seeds cv. BRS 133 and cv. Pintado treated at recommended dose of Cruiser (D1) and check (D0), under different PEG potentials at 84 h of imbibition. Average followed by the same letter did not differ significantly for each concentration or each potential. **ns**: not differ significantly for each imbibition period.

3.3 Third experiment: Cruiser's action on the enzymes involved in the response to oxidative stress induced by aluminum presence, salinity and water deficit

Cruiser has caused significant increase in peroxidase activity (POD) in BRS 133 and Pintado cultivars at 24 and 36 h of imbibition (Figure 22) when seeds were placed to germinate in distilled water (control).

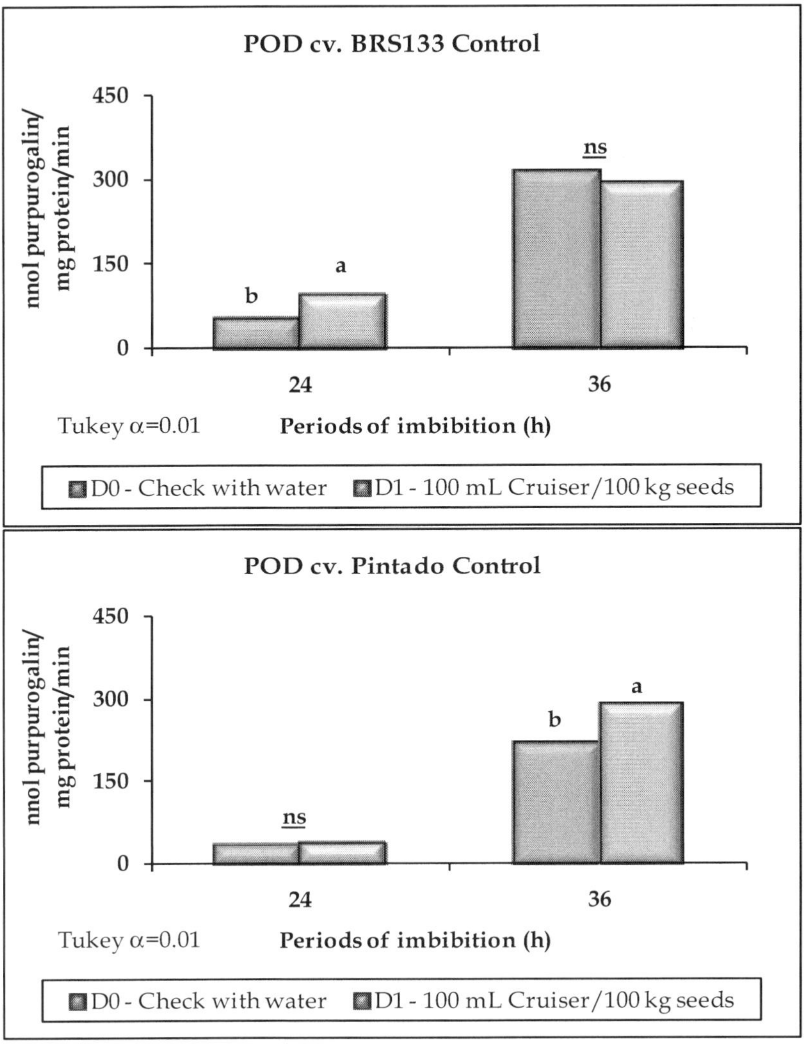

Fig. 22. Peroxidase activity (nmol purpurogalin mg protein^{-1} min^{-1}) in soybean seeds cv. BRS133 and Pintado treated at recommended dose of Cruiser (D1) and check (D0), under distilled water (control). Average followed by the same letter did not differ significantly for each imbibition period. **ns**: not differ significantly for each imbibition period.

In the cultivar BRS 133 Cruiser has caused increase in POD activity at the aluminum concentration of 10 mmol L^{-1} (Figure 23) in the two imbibition periods analyzed, at 24 and 36 h. In the cultivar Pintado, Cruiser caused a decrease in POD activity at 36 h of imbibition and increased at 48 h.

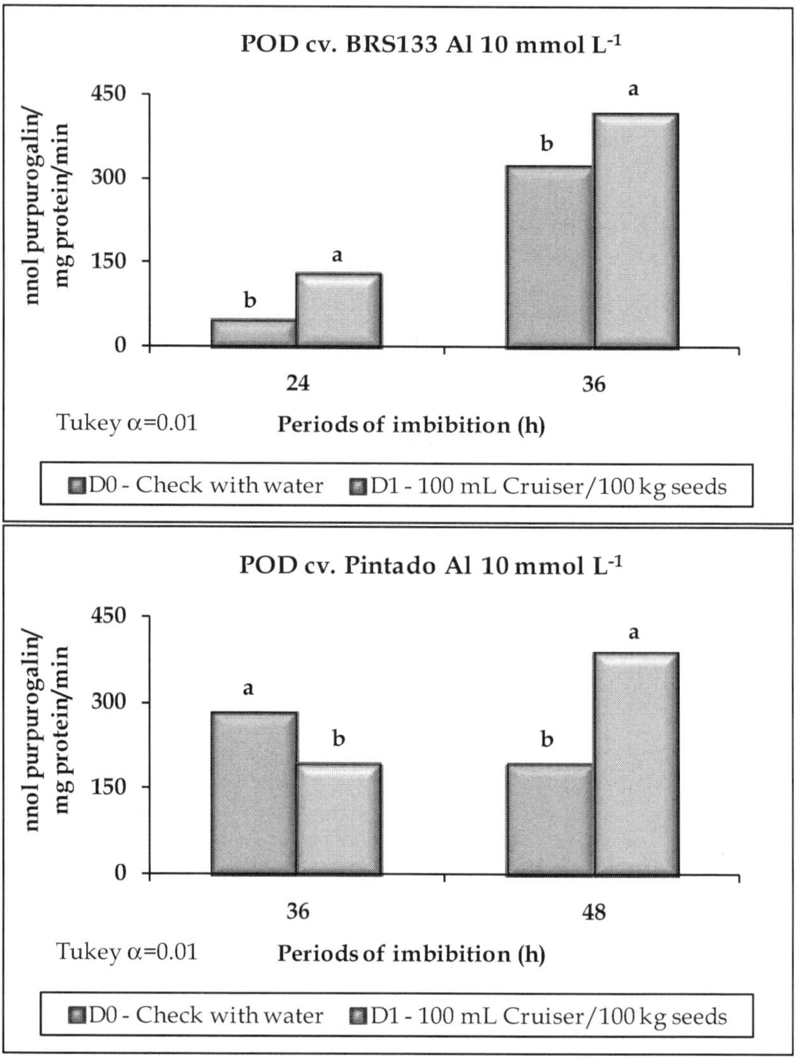

Fig. 23. Peroxidase activity (nmol purpurogalin mg $protein^{-1}$ min^{-1}) in soybean seeds cv. BRS133 and Pintado treated at recommended dose of Cruiser (D1) and check (D0), under aluminum. Average followed by the same letter did not differ significantly for each imbibition period.

In NaCl concentration of 50 mmol L^{-1}, Cruiser caused increase of POD at 36 hours of imbibition in the cultivar BRS 133 (Figure 24). Cruiser used under conditions of NaCl concentration of 100 mmol L^{-1} in the cultivar Pintado caused a increase in POD activity at 36 h of imbibition and decreased enzyme activity at 48 h of imbibition.

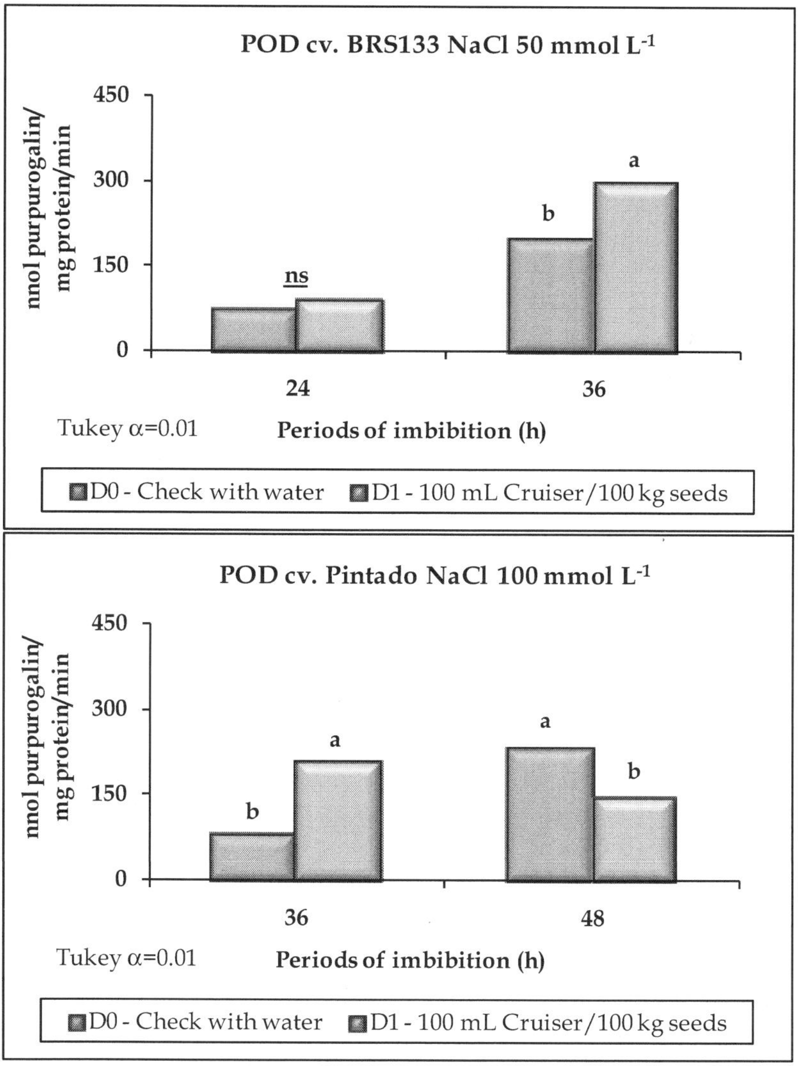

Fig. 24. Peroxidase activity (nmol purpurogalin mg $protein^{-1}$ min^{-1}) in soybean seeds cv. BRS133 and Pintado treated at recommended dose of Cruiser (D1) and check (D0), under NaCl. Average followed by the same letter did not differ significantly for each imbibition period. **ns**: not differ significantly for each imbibition period.

Under water deficit conditions of -0.3 MPa, Cruiser increased activity of POD at 60 and 72 h of imbibition in cultivar BRS 133 (Figure 25), however, in the cultivar Pintado decreased it at 72 h of imbibition and did not alter the enzyme activity at 84 h of imbibition.

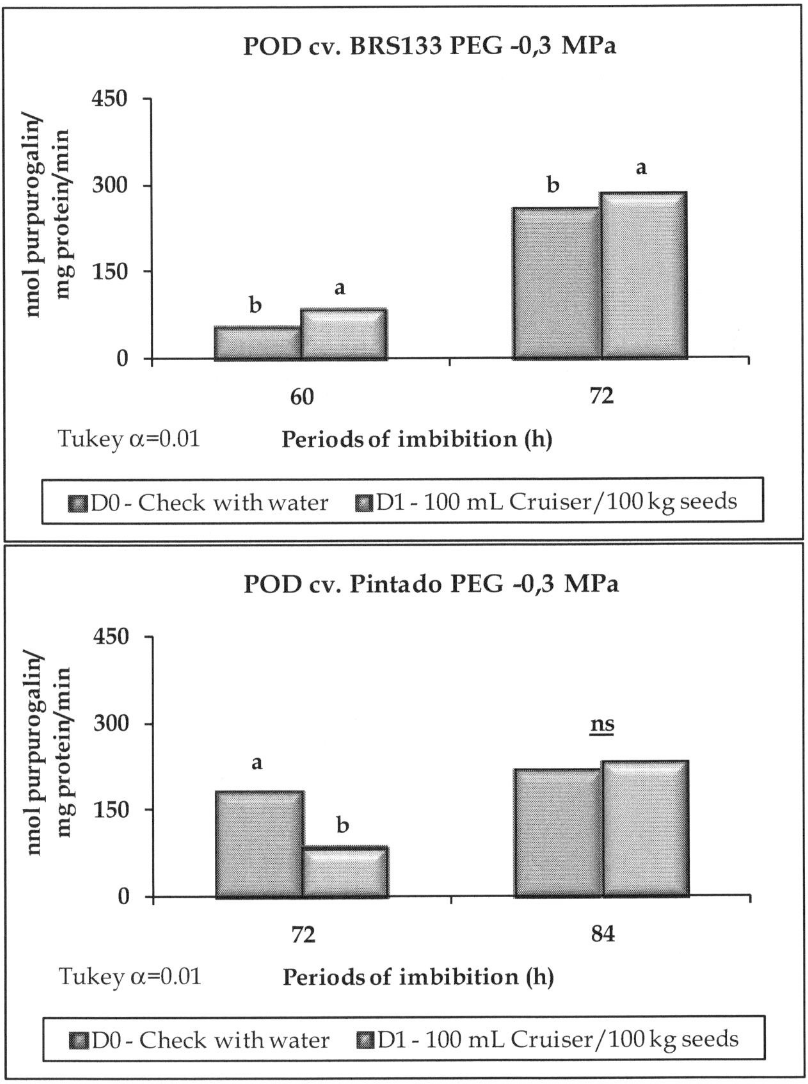

Fig. 25. Peroxidase activity (nmol purpurogalin mg protein^{-1} min^{-1}) in soybean seeds cv. BRS133 and Pintado treated at recommended dose of Cruiser (D1) and check (D0), under PEG. Average followed by the same letter did not differ significantly for each imbibition period. **ns**: not differ significantly for each imbibition period.

Regarding the activity of superoxide dismutase (SOD) (Figures 26 to 29), this did not change as effect of Cruiser when the soybean seeds of both cultivars were germinated under the same conditions of stress, the same imbibition periods analyzed to determine the POD.

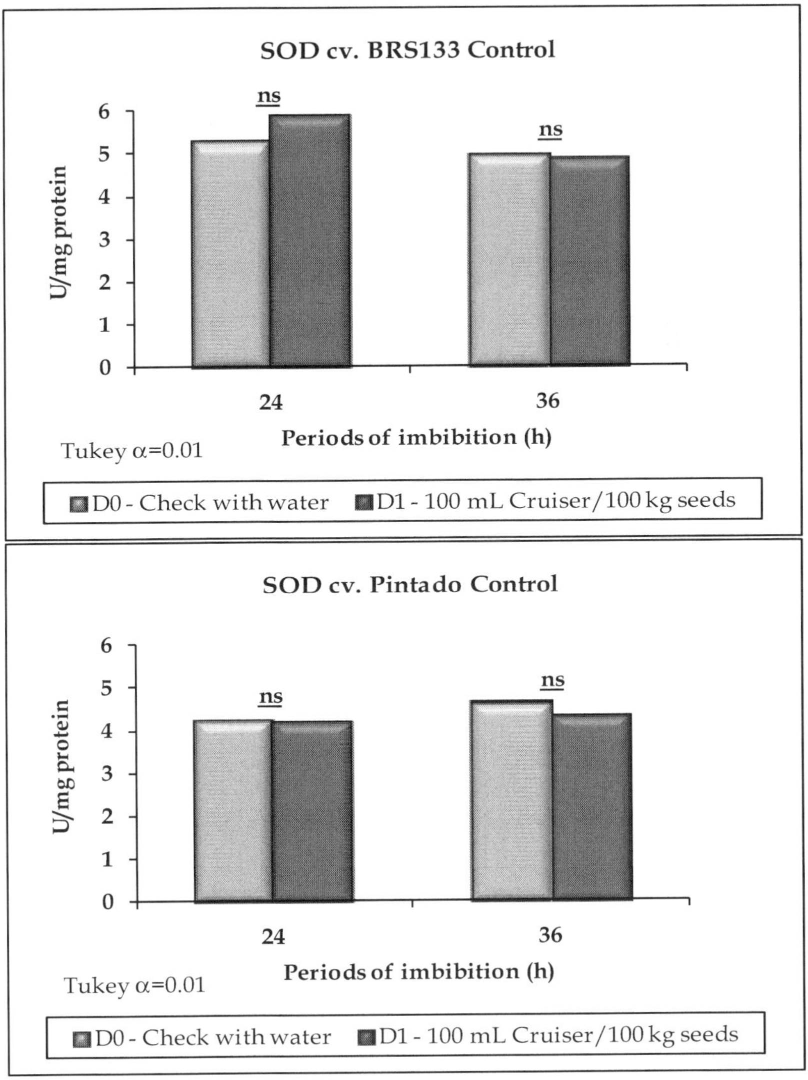

Fig. 26. Superoxide dismutase activity (U mg proteína^{-1}) in soybean seeds cv. BRS133 and Pintado treated at recommended dose of Cruiser (D1) and check (D0), under distilled water (control). **ns**: not differ significantly for each imbibition period.

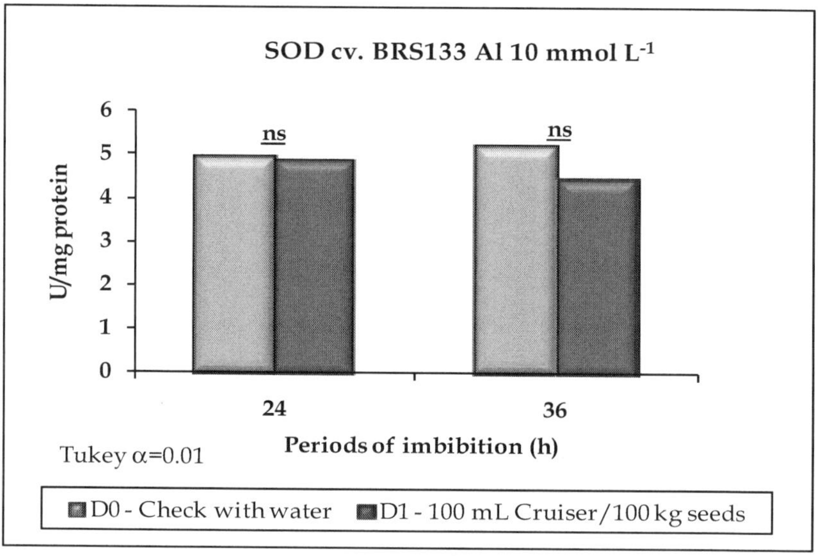

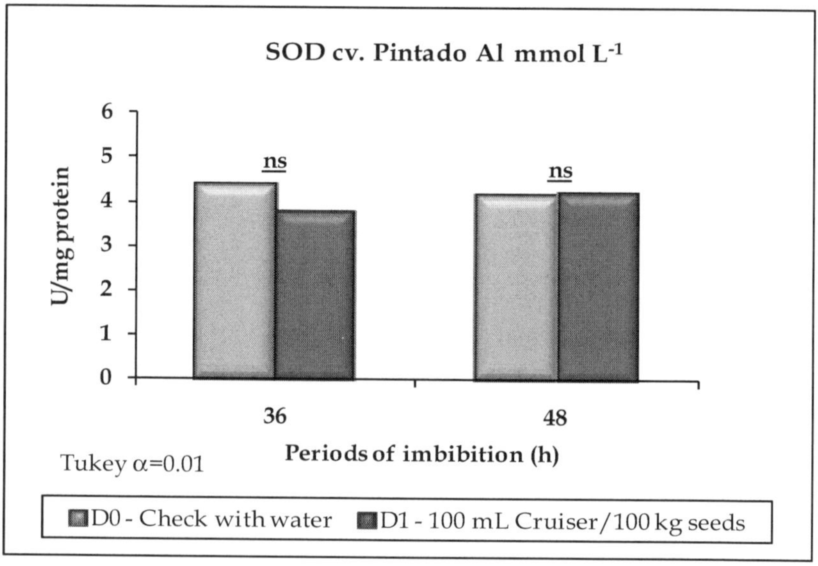

Fig. 27. Superoxide dismutase activity (U mg proteína^{-1}) in soybean seeds cv. BRS133 and Pintado treated at recommended dose of Cruiser (D1) and check (D0), under aluminum. **ns**: not differ significantly for each imbibition period.

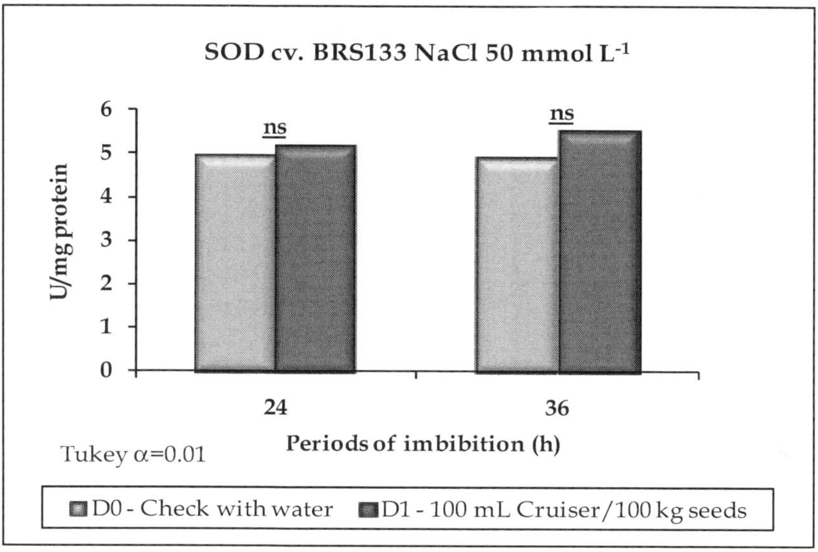

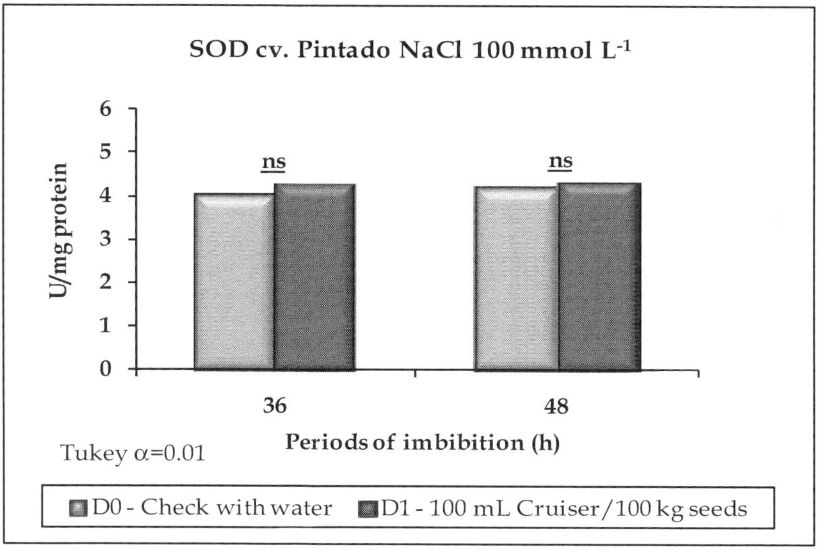

Fig. 28. Superoxide dismutase activity (U mg proteína^{-1}) in soybean seeds cv. BRS133 and Pintado treated at recommended dose of Cruiser (D1) and check (D0), under NaCl. **ns**: not differ significantly for each imbibition period.

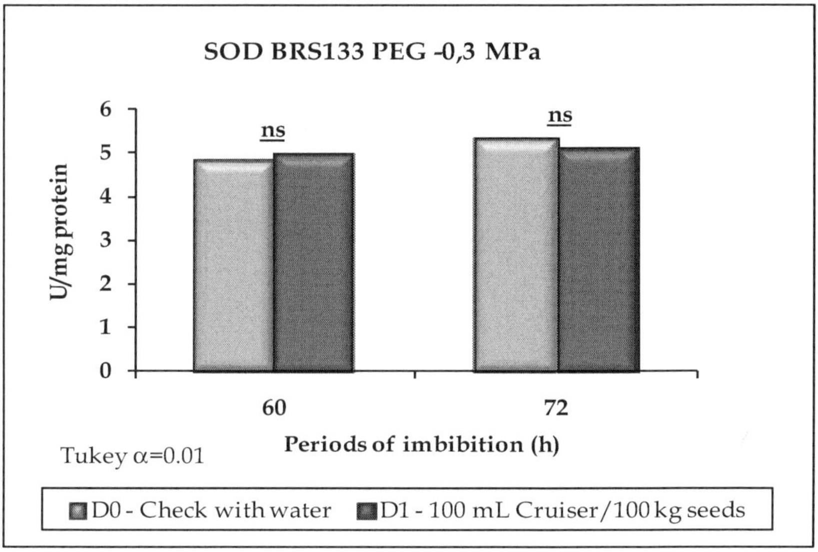

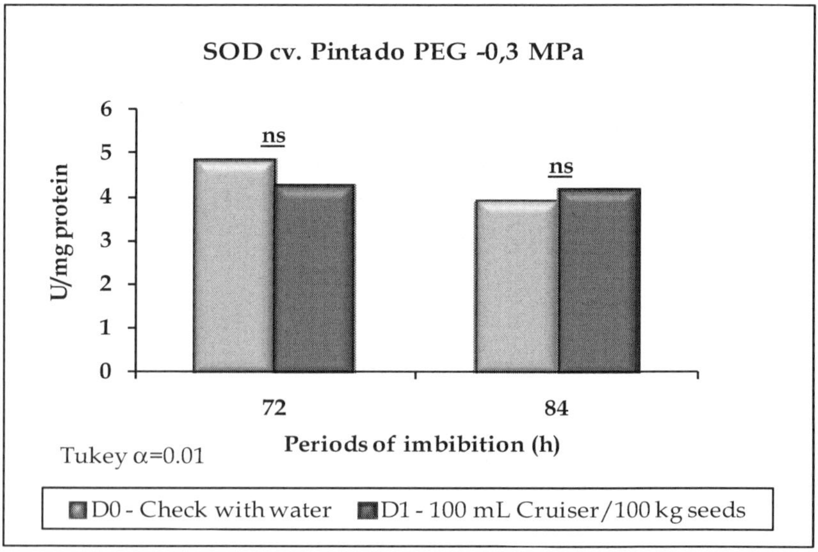

Fig. 29. Superoxide dismutase activity (U mg proteína^{-1}) in soybean seeds cv. BRS133 and Pintado treated at recommended dose of Cruiser (D1) and check (D0), under PEG. **ns**: not differ significantly for each imbibition period.

4. Discussion

Cruiser used as treatment for soybean seeds cultivars BRS 133 and Pintado, accelerated germination, the effect being more pronounced at twice the recommended level. Therefore, the Cruiser's action on the germination reduces the time for crop establishment in the field, reducing the negative effects of competition with weeds or essential nutrients in the soil.

Have been reported that seed germination and seedling development are delayed by high concentrations of aluminum (Matsumoto, 2000; Echart & Cavalli-Molina, 2001, Rout et al., 2001), salinity (Ashraf & McNeily, 1988; Hampson & Simpson, 1990; Ramoliya & Pandey, 2003, Soltani et al., 2004, Luo et al., 2005) and drought (Davidson & Chevalier, 1987; Passioura, 1988, Soltani et al., 2004).

According to Kochian (1995), Matsumoto (2000) and Rout et al. (2001) high aluminum concentrations inhibit root elongation, being proposed that the effect is due to inhibition of cell division, disjunction of cell wall, inhibition of ions flow, loss of membrane integrity and increased production of reactive oxygen species (ROS).

Aluminum causes a delay in germination of the two soybean cultivars in the control treatment and least in treatment with Cruiser, being more pronounced at higher concentrations of this heavy metal.

Salinity causes growth inhibition, being related to a decrease in extensibility of cell walls in the regions of root expansion (Neumann et al. 1994; Chazen et al., 1995), decreases the hydration of the seed (Allen et al. 1986), affects the physiological activities of the embryo due the toxicity of the absorbed ions (Khan et al., 1989), change the metabolism of carbohydrates (Corchete & Guerra, 1986), proteins (Ramagopal, 1990; Dell'Áquila & Spada, 1993) and nucleic acids (Gomes Filho et al., 1983). These changes make difficult to mobilize seed reserves, delaying the emergence of embryonic tissues, or even become non-viable seed (Rogers et al. 1995; Khan & Ungar, 1997).

NaCl causes a delay in germination but Cruiser reduces the negative effect of salinity on germination of soybean cultivar BRS 133, being more evident higher is the concentration of NaCl. To cultivar Pintado no answer was observed.

Cruiser has no effect on germination of soybean cultivar BRS 133 in conditions of drought, but in the cultivar Pintado, Cruiser accelerates germination being the effect more clear in situations of severe water stress.

The reduction on percentage of seeds germination in water stress conditions is attributed to lower diffusion of water through the integument. Water stress causes a prolongation of the stationary phase of the imbibition due to reduced enzyme activity and, consequently, a smaller meristematic development and delay on radicle protrusion (Falleri, 1994).

Seed germination and seedling development of various cultures decrease, influenced by conditions of low water availability, as reported by Owen (1972); Kiem & Krostad (1981), Davidson & Chevalier (1987); Passioura (1988); Soltani et al. (2004).

According to Soltani & Galeshi (2002) the decrease in germination and seedling development, as effect of environmental adversities, with consequent deficiency on crop establishment can cause: a) decreasing the competitiveness of the crop with weeds; b) less protection of soil surface and subsequently greater loss of soil water through evaporation and therefore, less available water for crop; c) lower light interception and yield potential.

It can also be considered that the loss in germination in situations of water stress might result in lower seedling development in the morning period, when the vapor pressure deficit is low and as result decreases CO_2 fixation (Tanner & Sinclair, 1983; Condon et al., 1993).

It was detected in the two soybean cultivars used on this study that Cruiser induced more development of the embryonic axis in presence of aluminum, salinity and water deficit, the effect being less evident with increasing of stress intensity.

The present results suggest it can be considered that Cruiser reduces the negative effects of stressful situations studied on germination of soybean seeds.

ROS generation during germination and root growth is generally accepted as an active physiological process, controlled in plant development (Chen & Schopfer, 1999; Schopfer et al., 2001), whose basal production is increased during conditions of biotic and abiotic stresses.

POD activity results indicate that Cruiser promotes this enzyme activity under stressful conditions, but has no effect on SOD activity during soybean germination under the same conditions.

According to Passardi et al. (2004), the peroxidases can be considered as bifunctional enzymes that can oxidize many substrates in H_2O_2 presence, but also produce ROS. They can promote cell elongation by ROS generation, or are involved in regulating H_2O_2 concentration, whose reactions cause restriction of growth.

Lin & Kao (2001) suggested that elevated production of H_2O_2 in rice roots during osmotic stress is probably involved in cell wall stiffening catalyzed by peroxidase, as explanation for the reduction of root growth. It was also suggested that the increase of peroxidase activity in situations of salinity and water stress induced inhibition of growth (Bacon et al. 1997; Lin & Kao, 2001).

The peroxidases can also participate in the lignification of new xylem elements in the embryo, hypocotyl, radicle and the hydroxyl radical (•OH) produced by its action could help on the break of seed tegument and subsequent cell elongation (Passardi et al., 2004). Amaya et al. (1999), related that the increase on expression of peroxidase associated with cell wall caused higher rates of germination on tobacco seeds, for providing water retention under conditions of osmotic stress induced by NaCl.

Looking at the results of Cruiser's action on the induction of POD activity and compare it with the results of germination determined in the same periods of imbibition and stressful situations, can be generally considered that the increases in germination are related to increased activity of POD, which had one of two consequences:

a) consumption of ROS originated in stressful situations, thereby preventing the damage caused by these molecules on the cell components and their metabolism or

b) increased production of ROS, arising in situations of stress and for Cruiser's action, which would cause the stimulation of cell elongation, promoting greater radicle development.

As Cruiser had no effect on SOD activity, future work should be focused on investigating the action of the insecticide on other enzymes such as catalase, ascorbate peroxidase, glutathione peroxidase and lipoxygenase, participants of the enzymatic complex involved in protection against the oxidative stress triggered by the presence of aluminum, salinity and water deficit. It would also be of interest to investigate the action of Cruiser on activity of peroxidase associated with the cell wall, whereas in this study was determined only the total peroxidase.

5. Conclusions

Cruiser used in the treatment of soybean seeds cultivars BRS 133 and Pintado:

- accelerates the germination during the process of imbibition, and the effect is more pronounced at twice recommended level.

- induces further development of the embryonic axis, minimizing the negative effects in situations as presence of aluminum, salinity and water deficit.
- accelerates germination during the imbibition process in the presence of aluminum, being more evident in situations of greater concentration of this heavy metal.
- reduces the negative effect of salinity on germination during the imbibition process for cultivar BRS 133 and has no answer for the cultivar Pintado.
- accelerates germination of the cultivar Pintado under water deficit conditions, the effect being more pronounced with increased stress conditions and has no answer for cultivar BRS 133.
- accelerates germination, stimulates the activity of peroxidase, which can act both in consumption of ROS, preventing oxidative stress, as in the production of ROS, stimulating cell elongation.

6. References

Allen, S.G.; Dobrenz, A.K.; Bartels, P.G. Physiological responses of SALT tolerant and non-tolerant alfalfa to salinity during germination. *Crop Sci.*, v.26, p.1004-8, 1986.

Alscher, R.G.; Donahoe, J.L.; Cramer, C.L. Reactive oxygen species and antioxidants; relationships in green cells. *Physiol. Plant.*, v.100, p.224-33, 1997.

Amaya, I.; Botella, M.A.; La Calle, M.; Medina, M.I.; Heredia, A.; Bressam, R.A.; Hasegawa, P.M.; Quesada, M.A.; Valpuesta, V. Improved germination under osmotic stress of tobacco plants overexpressing a cell wall preoxidase. *FEBS Letters*, v.457, p.80-4, 1999.

Ashraf, M.; Mcneily, T. Variability in salt tolerance of nine spring wheat cultivars. *J. Agron. Crop. Sci.*, v.160, p.14-21, 1988.

Athar, H.; Khan, A.; Ashraf, M. Exogenously applied ascorbic acid alleviates salt-induced oxidative stress in wheat. *Environ. Exp. Bot.*, v.63, p.224-31, 2008.

Bacon, M.A.; Thompson, D.S.; Davis, W.J. Can cell wall peroxidase activity explain the leaf growth response of *Lolium temulentum* L. during drought? *J. Exp. Bot.*, v.48, p.2075-85, 1997.

Bailey, C.J.; Boulter, D. Urease, a typical seed protein of Leguminosae. In: Chemotaxonomy of the Leguminosae, eds Harborne J.B., Boulter, D. & Turner, B.L. Academic Press, New York, p.485-502, 1971.

Bor, M.; Özdemir, F; Türkan, I. The effect of salt stress on lipid peroxidation and antioxidants in leaves of sugar beet *Beta vulgaris* L. and wild beet *Beta maritima* L. *Plant Sci.*, v.164, p. 77-84, 2003.

Bowler, C.; Van Montagu, M.; Inzé, D. Superoxide dismutase and stress tolerance. *Annu. Rev. Plant Physiol. Plant Mol. Biol.*, v.43, p.83-116, 1992.

Cataneo, A.C.; Chamma, K.L.; Ferreira, L.C.; Déstro, G.F.G.; Sousa, D.C.F. Atividade de superóxido dismutase em plantas de soja (*Glycine max* L.) cultivadas sob estresse oxidativo causado por herbicida. *Revista Brasileira de Herbicidas,* v. 4, p.23-31, 2005.

Chazen, O.; Hartung, W.; Neumann, P.M. The different effects of PEG 6000 and NaCl on leaf development are associated with differential inhibition of root water transport. Plant Cell Environ., v.18, p.727-35, 1995.

Chen, S.; Schopfer, P. Hydroxyl-radical production in physiological reactions. A novel function of peroxidase. *Eur. J. Biochem.*, v.260, p.726-35, 1999.

Condon, A.G.; Richards, R.A.; Farguhar, G.D. Relationships between carbon isotope discrimination, water use efficiency and transpiration efficiency for dryland wheat. *Aust. J. Agric. Res.*, v.44, p.1693-711, 1993.

Corchete, H.; Guerra, H. Effect of NaCl and polyethylene glycol on soluble content and glycosidase activities during germination of lentil seeds. *Plant Cell Environ.*, v.9, p.589-93, 1986.

Davidson, D.J.; Chevalier, P.M. Influence of polyethyleneglycol induced water deficits on tiller production in spring wheat. *Crop Sci.*, v.27, p.1185-7, 1987.

Dell'aquila, A.; Spada, P. The effect of salinity stress upon protein synthesis of germinating wheat embryos. *Ann. Bot.*, v.72, p.97-101, 1993.

Duran, J.M.; Tortosa, M.E. The effect of mechanical and chemical scarification on germination of charlock *S. arvensis*. *Seed Science and Technology*, Zurich, v.13, p.155-63, 1985.

Echart, C.L.; Cavalli-MolinA, S. Fitotoxicidade do alumínio: efeitos, mecanismo de tolerância e seu controle genético. *Ciência Rural,* Santa Maria, v.31, n.3, p.531-41, 2001.

Ekler, Z.; Dutka, F.; Stephenson, G.R. Safener effects on acetochlor toxicity, uptake, metabolism and glutathione S-transferase activity in maize, *Weed Res.*, v.33, p.311-8, 1993.

Falleri, E. Effect of water stress on germination in six provenances of *Pinus pinaster* Ait. *Seed Science and Technology*, Zurich, v.22, p.591-9, 1994.

Ferreira, L.C.; Cataneo, A.C.; Remaeh, L.M.R.; Corniani, N.; Fumis, T.F.; Souza, Y.A.; Scavroni, J.; Soares, B.J.A. Nitric oxid reduces oxidative stress generated by lactofen in soybean plants. *Pest. Biochem. Physiol.*, v.97, p.47-54, 2010.

Foyer, C.H.; Descourviéres, P.; Kunert, K.J. Protection against oxygen radicals: an important defence mechanism studied in transgenic plants. *Plant Cell Environ.*, v.17, p.507-23, 1994.

Fridovich, I. The biology of oxygen radicals. Science, v.201, p.875-80, 1978.

Gomes-Filho, E.; Prisco, J.T.; Campos, F.A.P.; Filho, J.E. Effect of NaCl salinity in vivo and in vitro on ribonuclease activity of *Vigna unguiculata* cotyledons during germination. *Physiol. Plant.*, v.59, p.183-8, 1983.

Hampson, C.R.; Simpson, G.M. Effects of temperature, salt and osmotic pressure on early growth of wheat (*Triticum aestivum*). 1. Germination. *Can. J. Bot.;* v.68, p.524-8, 1990.

Khan, A.H.; Azmi, A.R.; Ashraf, M.Y. Influence of NaCl on some aspects of sorghum varieties. *Pak. J. Bot.*, v.21, p.74-80, 1989.

Khan, M.A.; Ungar, I.A. Effect of thermoperiod on recovery of seed germination of halophyte from saline conditions. *Am. J. Bot.*, v.84, p.279-83, 1997.

Kiem, D.L.; Krostad, W.E. Drought response of winter wheat cultivars grown under field stress conditions. *Crop Sci.*, v.21, p.11-5, 1981.

Kochian, L.V. Cellular mechanisms of aluminum toxicity and resistance in plants. *Annu. Rev. Plant Physiol. Plant Mol. Biol.*, v.46, p.237-60, 1995.

Lin, C.C.; Kao, C.H. Cell wall peroxidase against ferulic acid, lignin, and NaCl-reduced root growth of rice seedlings. *J. Plant Physiol.*, v.158, p.667-71, 2001.

Luo, Q.; Yu, B.; Liu, Y. Differential sensitivity to chloride and sodium ions in seedlings of *Glycine max* and *G. soja* under NaCl stress. *J. Plant Physiol.*, v.162, p.1003-12, 2005.

Maienfisch, P.; Angst, M.; Brandl, F.; Fischer, W.; Hofer, D.; Kayser, H.; Kobel, W.; Rindlisbacher, A.; Senn, R.; Steinemann, A.; Widmer, H. Chemistry and biology of thiamethoxam: a second generation neonicotinoid. *Pest Manage. Sci.*, v.57, p.906-13, 2001.

Matsumoto, H. Cell biology of aluminum toxicity and tolerance in higher plants. *Int. Rev. Cytol.*, v.200, p.1-46, 2000.

McQueen-MASON, S.J. Expansins and cell wall expansion. *J. Exp. Bot.*, v.46, p.1639-50, 1995.

McQueen-mason, s.j.; cosgrove, d.j. Expansin mode of action on cell walls: analysis of wall hydrolysis, stress relaxation and binding. *Plant Physiol.*, v.107, p.87-100, 1994.

Michel, B.E. & Kaufmann, M.R. The osmotic potential of polyethylene glycol 6000. *Plant Physiology*, v.51, p. 914-6, 1973.

Neumann, P.M.; Azaizeh, H.; Leon, D Hardening of root cell walls: a growth inhibitory response to salinity stress. *Plant Cell Environ.*, v.17, p.303-9, 1994.

Owen, P.C.J. The relation of germination of wheat to water potential. *J. Exp. Bot.*, v.3, p.188-92, 1972.

Passardi, F.; Penel, C.; Dunand, C. Performing the paradoxical: how plant peroxidases modify the cell wall. *TRENDS in Plant Science*, V.9, p.534-40, 2004.

Passioura, J.B. Root signals control leaf expansion in wheat seedlings growing in drying soil. *Aust. J. Plant Physiol.*, v.15, p.687-93, 1988.

Ramagopal, S. Inhibition of seed germination by salt and its subsequent effect on embryo protein synthesis in barley. *J. Plant Physiol.*, v.136, p.621-5, 1990.

Ramoliya, P.J.; Pandey, A.N. Effect of salinization of soil on emergence, growth and survival of seedlings of *Cordia rothii*. *Forest Ecology and Management*, v.176, p.185-94, 2003.

Richards, K.D.; Schott, E.J.; Sharma, Y.K.; Davis, K.R.; Gardner, R.C. Aluminum induces oxidative stress genes in *Arabidopsis thaliana*. *Plant Physiol.*, v.116, p.409-18. 1998.

Rios-Gonzalez, K.; Erdei, L.; Lips, H. The activity of antioxidant enzymes in maize and sunflower seedlings as affected by salinity and different nitrogen sources. *Plant Sci.*, v.162, p.923-30, 2002.

Rogers, M.E.; Noble, C.L.; Halloran, G.M.; Nicolas, M.E. The effect of NaCl on the germination and early seedling growth of white clover (*Trifolium repens* L.) populations selected for high and low salinity tolerance. *Seed Sci. Technol.*, v.23, p.277-87, 1995.

Rout, G.R.; Samantaray, S.; Das, P. Aluminium toxicity in plants: a review. *Agronomie*, v.21, P.3-21, 2001.

Schopfer, P.; Plachy, C.; Frahry, G. Release of active oxygen intermediates (superoxide radicals, hydrogen peroxide, and hydroxyl radicals) and peroxidase in germinating radish seeds controlled by light, gibberillin, and abscisic acid. *Plant Physiol.*, v.125, p.1591-602, 2001.

Silva, J.; Arrowsmith, D.; Hellyer, A.; Whiteman, S.; Robinson, S. Xyloglucan endotransglycosylase and plant growth. *J. Exp. Bot.*, v.45, p.1693-701, 1994.

Smirnoff, N. The role of active oxygen in the response of plants to water deficit and desiccation. *New Phytol.*, v.125, p.27-58, 1993.

Soltani, A.; Galeshi, S. Importance of rapid canopy closure for wheat production in a temperate sub-humid environment: experimentation and simulation. *Field Crops Res.*, v.77, p.17-30, 2002.

Soltani, A.; Holipoor, M.; Zeinali, E. Seed reserve utilization and seedling growth of wheat as affected by drought and salinity. *Environ. Exp. Bot.*, in press, 2004.

Tamás, L.; Simonovicová, M.; Huttová, J.; Mistrík, I. Aluminium stimulated hydrogen peroxide production of germinating barley seeds. *Environ. Exp. Bot.*, v.51, p.281-8, 2004.

Tanner, C.B.; Sinclair, T.R. Efficient water use in crop production: research or re-search. In: Taylor, H.M.; Taylor, W.R.; Sinclair, T.R. (Eds.). *Limitations to Efficient Water Use in Crop Production.* ASA/CSSA/SSSA, Madison, WI, p.1-27, 1983.

Teisseire, H.; Guy, V. Copper-induced changes in antioxidant enzymes activities in fronds of duckweed (*Lemna minor*). *Plant Sci.*, v.153, p.65-72, 2000.

14

Effect of Row-Spacing and Planting Density on Podding and Yield Performance of Early Soybean Cultivar 'Enrei' with Reference to Raceme Order

Kuniyuki Saitoh
Okayama University
Japan

1. Introduction

In Japan, the genetically modified herbicide-tolerant soybean cultivar cannot be grown in the commercial field without permission due to the public concern about the effects on the ecosystem and human health. Recently, interest for no-tilling, narrow row-spacing and dense cultivation in soybean has been increasing as a labour-saving technique. The no-tilling cultivation has an advantage in saving labor and drainage of soil, but the merit of narrow row and dense planting has not been clarified. The dense planting increases the competition among plants from the early stage and the risk of excessive growth which results in lodging. On condition that the planting density is equal, narrow row-spacing decrease the competition with plants during the earlier growth stage than wide row-spacing, and result in rapid leaf area expansion, higher crop growth rate and higher seed yield due to the development of branches, increase in the node number and pod number per node (Cooper 1977, Costa et al. 1980, Duncan 1986, Miura and Gemma 1986, Miura et al. 1987, Board et al. 1990a, 1990b, Bullock et al. 1998, Ikeda 2000). However, narrow row-spacing did not increase the yield (Beatty et al. 1982, Nakano 1989) and has been reported to even decrease the yield (Cooper and Nave 1974).

In this chapter, the factors affecting the increase in yield of narrow row and dense planting in soybean and yield determining process was clarified with reference to pod position (main stem/branches, raceme order). In order to analyze the advantages and disadvantages of narrow row and dense planting, we examined the effects of planting pattern and density on solar radiation utilization, dry-matter production and emergence of weeds.

2. Materials and methods

2.1 Plant cultivation and experimental plots

The field experiment was conducted at the Field Science Centre of Okayama University (34°41′ N, 133°55′ E, Japan) in 2001 and 2002. The texture of the soil was sandy clay and preceding crop was pumpkin. Indeterminate soybean (*Glycine max* (L.) Merr.) cv. 'Enrei' (maturity group III) was used. Two seeds were sown on 13 and 14 June in 2001 and 2002, respectively, with an 80cm (wide) and 30cm (narrow) row-spacing, and sparse (11.1 plants

m^{-2}, 11.25 and 30cm plant spacing in wide and narrow row-spacing, respectively) and dense (22.2 plants m^{-2}, 5.6cm and 15cm in wide and narrow row-spacing, respectively) planting density. Each plots size was 57.6 m^2 (3.2×18.0m) with no replication. A basal fertilizar was applied at the rate of 2.1g N, 4.4g P and 10.0g K. Herbicide was applied to the soil surface to avoid weed emergence. The plants were thinned to a plant per hill when primary leaves were fully expanded. In wide row plots, soil molding was conducted by a rotary cultivator. The crop was irrigated with a water-spraying vinyl hose placed on every other row. Recommended pesticides were applied for the control of insects and diseases.

2.2 Growth and yield observation

Thirty plants were harvested from each plots, and ten standard plants were selected to examine the node number, main stem length, stem diameter, stem weight, and seed/stem weight ratio. Pods were distinguished on the position, main stem/branches and raceme order (Fig. 1.), and seeds were depodded manually, then weighed to record the data on yield and yield components.

The raceme orders were defined as follows (Torigoe et al. 1982). The terminal racemes appeared at the top of the stems, and first order racemes differentiate from the axil just above the petiole on the stem. The secondary racemes differentiate from both sides of the first order raceme and tertiary racemes differentiate from the sides of the secondary racemes. Racemes differentiating from both sides of the branch were classified as secondary racemes. The terminal and first order racemes, and those over secondary raceme will be collectively called basal raceme and lateral raceme, respectively. Some lateral racemes had compound leaves. The lodging score was recorded every week by measuring the angle of the main stem, and ranked 0 (erect), 1 (inclined 15 degrees), 2 (inclined 45 degrees), 3 (inclined 75 degrees) and 4 (inclined horizontally), then the average score was obtained.

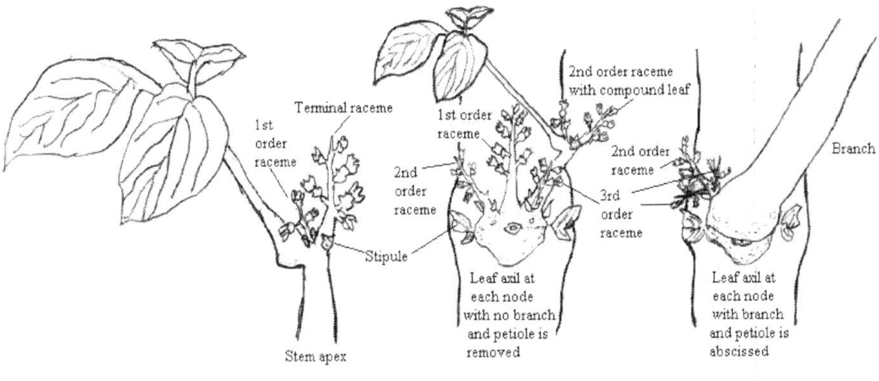

Fig. 1. Classification of raceme order in determinate type of soybean.

2.3 Dry matter production and canopy structure

Five plants (three replication for each plots) were sampled and three (nine plants for each plots) were separated into leaves, petioles, stems and pods on each main stem and branch, then measured the leaf area of a standard plant (AAM-8, Hayashidenko). Samples were air-dried at 80 degrees C for 48 hours and weighed. At the beginning of flowering and full seed growth stage relative PAR (photosynthetically active radiation) at each height of the canopy were measured with a long PAR sensor (LI-191S, LI-COR) in the evening under diffuse light condition. Then, canopy structures were surveyed by the stratified clip method (Monsi und Saeki 1953). From the logarithmic relationships between cumulative LAI of the canopy top and relative PAR, the canopy light extinction coefficient (k) was obtained. In addition, the relative PAR at the height of 0, 60 and 120cm above the ground was measured every 2.5 hours from 7 a.m. to 17 p.m., and diurnal change in light extinction coefficient under direct light condition was obtained.

2.4 Cumulative solar radiation within canopy

Integrated solarimeter films (R-2D, Taisei E&L) were used for the measurement of cumulative solar radiation. Film was cut in 1cm width and 2cm length, then placed at 10cm intervals on the square bars, 1cm width and 100cm length, which were installed horizontally every 15cm height from the soil surface. The dye percentages were measured every six hours by a spectro-photometer (UV-1200, Shimadzu). The dye percentages had been calibrated with the cumulated solar radiation measured by radiation sensor (LI-200SA, LI-COR). Accordingly, the distribution of solar radiation within a canopy was calculated.

2.5 Weed emergence

Three quadrats (80cm*60cm) were randomly arranged within each plots. At the beginning of flowering stage, all weeds were sampled and the number and dry-weight of each weed species were recorded.

3. Results

3.1 Growth characters

In 2001, the precipitation was 14% lower, the average mean temperature was 0.8 degree higher, and the sunshine hours was 13% longer than the normal year, and it was characterized by low rainfall, high temperature and much sunshine. In 2002, the precipitation was 56% lower, the average mean temperature was 0.9 degree higher, and the sunshine hours was 7% longer than the normal year, and it was characterized by drought, high temperature and much sunshine though lower than in 2001. The field was hit by a typhoon on Aug. 21 in 2001. There was no typhoon damage in 2002.

In both years, the number (per square meter) of nodes on the main stem, racemes with compound leaves and in total was higher, but in the number of branches was lower than in sparse plots (Table 1). The node number on the branches and in total was larger in wide plots than in narrow plots except that in sparse plots in 2001, and also that of racemes with compound leaf in 2001. The main stem length in dense plots was 2-12 cm longer than in sparse plots, and that in narrow plots was 7-16 cm shorter than in wide plots. The weight, diameter and section area of stem were larger than in sparse and narrow plots than dense and wide plots, respectively. The seed/stem weight ratio in dense plots was smaller than in sparse plots among the narrow plots, but not among the wide plots. The ratio in narrow plots was larger than in wide plots among the sparse plots, but not among the dense plots.

Year / Plot	Node number (m^{-2}) Main stem	Node number (m^{-2}) Bran-ch	Node number (m^{-2}) Rac. with leaf	Node number (m^{-2}) Total	Main stem length (cm)	Stem weight (g)	Bran-ch no. (m^{-2})	Stem dia-meter (mm)	Stem section area (mm^2)	Seed / stem weight ratio
2001										
Wide/Sparse	150	316	137	602	63.4	18.3	66	9.4	53.0	2.20
Wide/ Dense	290	192	183	665	69.6	8.7	74	6.9	30.4	2.51
Narrow/Sparse	141	239	211	591	47.4	18.1	60	9.2	56.4	2.58
Narrow/Dense	296	342	307	944	55.7	12.2	102	7.8	37.1	2.49
$LSD_{(0.05)}$	9	ns	33	54	3.3	1.4	9	0.4	4.4	ns
2002										
Wide/Sparse	159	272	71	502	61.2	12.5	60	8.5	43.5	1.98
Wide/ Dense	301	248	121	670	63.5	7.7	89	7.0	27.8	2.06
Narrow/Sparse	162	324	89	576	53.6	13.6	70	9.1	49.4	3.39
Narrow/Dense	318	347	122	787	65.3	10.2	111	7.8	38.2	2.25
$LSD_{(0.05)}$	7	53	24	67	2.5	1.3	14	0.1	3.4	0.50

Values are means of twelve plants. 'ns' means no siginificant difference at 5% level.

Table 1. Growth characteristics (2001, 2002).

3.2 Seed yield and yield components

In both years, seed yields in dense plots and narrow plots were larger than sparse plots and wide plots, respectively, and those in 2001 were higher than in 2002 because of the much sunshine hours (Fig. 2, Table 2). The highest yield, 668 g m^{-2}, was obtained in narrow/dense plots in 2001. A close correlation (r=0.934, P<0.01) was observed between seed yield and pod

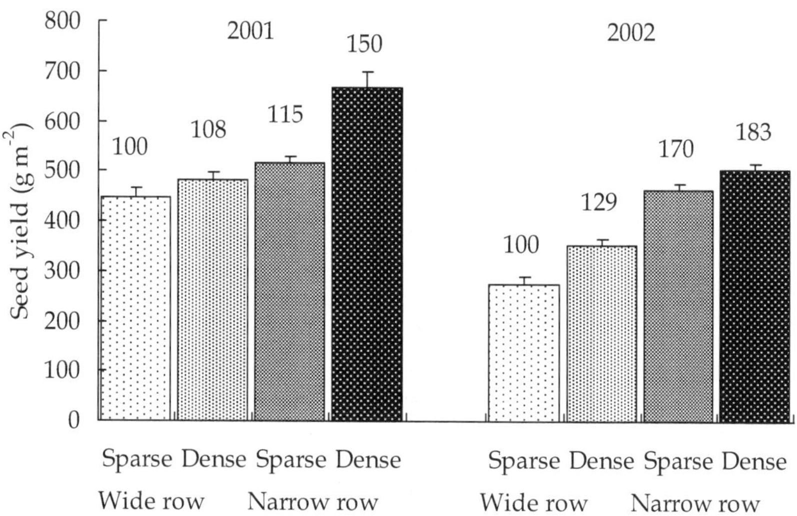

Fig. 2. Effects of planting pattern on seed yield of soybean.

number, indicating that seed yield was determined by the pod number. Seed number per pod and seed setting ratio were not significantly different among plots, and 100 seeds weight in narrow plots tended to be slightly heavier than in wide plots, but the difference was not significant.

The pod number on the main stem relative to the total was higher in dense plots than in sparse plots, and that on the branches was higher in narrow plots than in wide plots (Table 3). The percentage share of basal raceme was higher in 2002 than in 2001. The percentage share of racemes with compound leaves was higher in dense plots than in sparse plots, and was also higher in narrow plots than in wide plots, especially in 2001.

Year / plot	Seed yield (g m^{-2})	Pod number (m^{-2})	Seed number per pod	100 seeds weight (g)	Seed setting ratio (%)
2001					
Wide/Sparse	446	894	2.03	30.2	95.0
Wide/ Dense	483	904	1.99	31.1	96.5
Narrow/Sparse	515	1011	1.96	32.3	95.5
Narrow/Dense	668	1256	1.99	31.5	97.4
$LSD_{(0.05)}$	45	84	ns	ns	ns
2002					
Wide/Sparse	274	766	2.04	26.5	83.7
Wide/ Dense	354	893	2.01	27.5	91.3
Narrow/Sparse	464	910	2.02	33.6	87.5
Narrow/Dense	503	993	2.04	31.8	92.2
$LSD_{(0.05)}$	26	83	ns	ns	4.5

Values are means of twelve plants.
'ns' means no siginificant difference at 5% level.

Table 2. Seed yield and yield components (2001, 2002).

Year / plot	Main stem	Branch	Basal raceme	Raceme with leaf	Upper raceme	Total
2001						
Wide/Sparse	377 (42)	518 (58)	367 (41)	262 (29)	266 (30)	894 (100)
Wide/ Dense	685 (76)	219 (24)	359 (40)	321 (36)	223 (24)	904 (100)
Narrow/Sparse	384 (38)	627 (62)	333 (33)	416 (41)	262 (22)	1011 (100)
Narrow/Dense	702 (56)	553 (44)	456 (36)	524 (42)	276 (22)	1256 (100)
$LSD_{(0.05)}$	61	93	ns	66	ns	85
2002						
Wide/Sparse	337 (44)	429 (56)	446 (58)	119 (16)	201 (26)	766 (100)
Wide/ Dense	567 (63)	326 (37)	464 (52)	205 (23)	223 (25)	893 (100)
Narrow/Sparse	292 (32)	618 (68)	480 (53)	154 (17)	276 (30)	910 (100)
Narrow/Dense	607 (61)	387 (39)	536 (54)	244 (25)	213 (21)	993 (100)
$LSD_{(0.05)}$	48	72	ns	43	ns	83

Values are means of twelve plants. 'ns' means no siginificant difference at 5% level.
Values in parentheses are relative to total (100).

Table 3. Pod number on main stem or branch and raceme order (2001, 2002).

3.3 Dry weight and leaf area index

At each growth stage, the dry-weight tended to be heavier in dense plots than in sparse plots, but the difference was not significant (Fig. 3). The dry-weight tended to be heavier in narrow plots than in wide plots except that in sparse plots at 44 days after sowing (DAS) and in dense plots at 65 DAS. At 107 DAS, the dry-weight was heaviest in narrow/dense plots and became lighter in the order of wide/dense plots > narrow/sparse plots > wide/sparse plots.

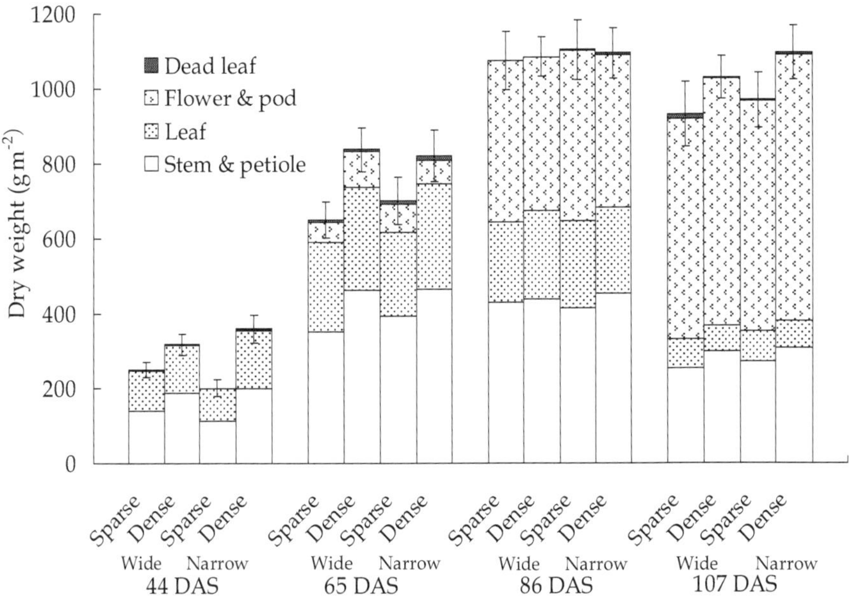

Fig. 3. Changes in cumulative dry-weight of different plant parts during growth (2001).

The leaf area index (LAI) tended to be larger in dense plots than in sparse plots, and in narrow plots than in wide plots especially at 65 DAS, when LAI in dense plots exceeded 8 (Fig. 4).

3.4 Canopy structure

At the flowering stage, the higher the canopy layer, the larger the leaf area from 20 to 100 cm above the ground in wide/dense plots, and the larger leaf area was distributed at a 40-100 cm height in narrow/sparse plots (Fig. 5). In dense plots, leaf area was concentrated in the 80-100 cm layer above the ground especially in narrow plots. The total dry-weight of non-assimilative organ was heavier in narrow plots than in wide plots. The light extinction coefficients (k), the lower value indicates that the canopy has a good light-intercepting characteristic, was in the order of narrow/dense (0.60) < wide/dense (0.68) < narrow/sparse (0.73) < wide/sparse (0.81). It was clear that the light penetrated into a deeper layer of the canopy when planted dense and narrow row-spacing. The order of k at the seed growth stage coincided with that at the flowering stage (data not shown).

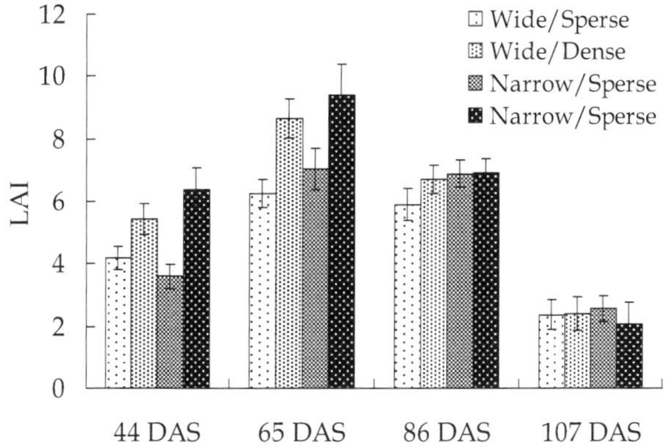

Fig. 4. Changes in LAI during growth (2001).

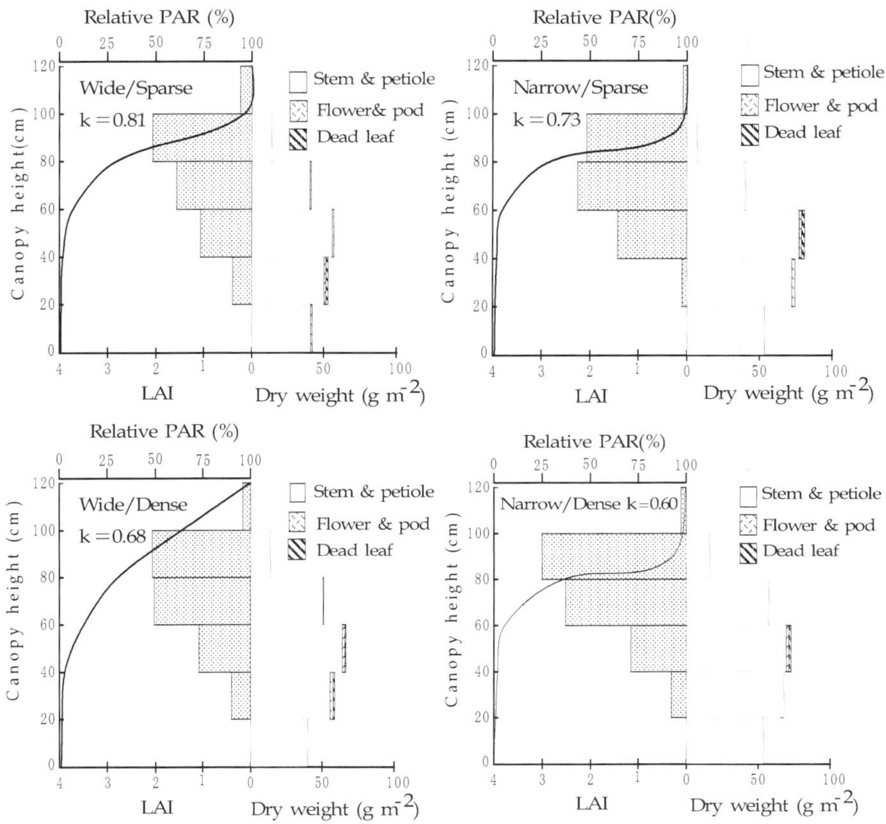

Fig. 5. Canopy structures at the full-flowering stage (2001).

3.5 Diurnal change in canopy light extinction coefficient (k)

The k-value measured under direct sunlight was higher in the morning and evening, and decreased during the daytime (Fig. 6). The k-values in the morning and evening were similar to those measured under diffuse light (Fig. 5), which were lower in dense and narrow row plots. At midday, k showed the lowest value in wide plots, which suggested that the direct sunlight reached the furrow surface in the non-closed canopy in wide row plots. The extent of variation during the daytime was small in narrow plots due to the closed canopy.

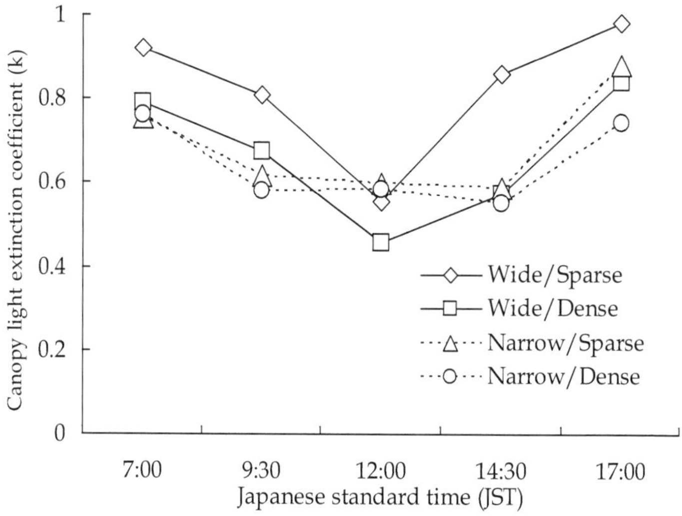

Fig. 6. Diurnal change in canopy light extinction coefficient at the beginning of the flower stage (2001).

3.6 Distribution of cumulative solar radiation at each height within canopy

The cumulative solar radiation at every height was lower in dense plots than in sparse plots, and was lower near the row (plant) and higher at the furrow in a direction perpendicular to the row (Fig. 7). In narrow row plots, the cumulative solar radiation was lower in dense plots than in sparse plots, and the difference between that on the row and furrow was small.

3.7 Changes in lodging score

In 2002, lodging did not occur in any plot. In 2001, the lodging score increased in narrow/sparse plots at 34 DAS due to a rainstorm, followed by the gradual increase in wide/sparse plots, and was larger in narrow row plots than in wide row plots (Fig. 8). At 71 DAS, when a typhoon hit, the lodging score increased markedly in dense plots, and was slightly larger in narrow/dense plots than in wide/dense plots. After lodging, plants could not recover during the later growth period.

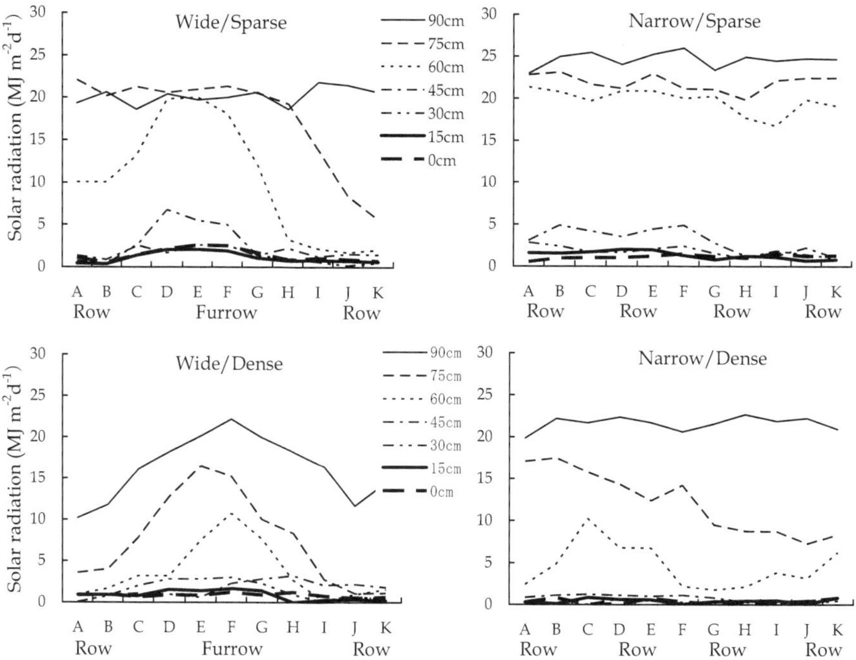

Fig. 7. Distribution of cumulative solar radiation at each height within canopy in a direction perpendicular to the row at the beginning flower stage (2001).

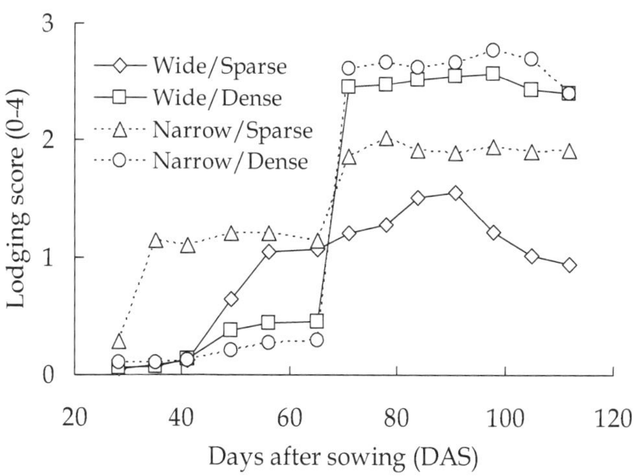

Fig. 8. Changes in lodging score (2001).

3.8 Weed emergence

More weed plants appeared in 2002 than in 2001. *Portulaca* and *Cyperus* species were dominant in 2001, and *Digitaria* and *Galinsoga* in 2002. In both years, there were fewer emerged weeds in narrow plots than in wide plots.

Year/Plot	*Amaranthus viridis*	*Portulaca oleracea*	*Digitaria ciliaris*	*Cyperus*	*Rorippa indica*	*Galinsoga ciliata*	*Setaria viridi*	*Chenopodium album*	*Euphorbia supina*	*Mollugo pentaphylla*	Total
2001											
Wide/Sparse	7.6	9.7	-	8.3	2.1	2.1	2.1	2.1	-	-	34.0
Wide/ Dense	5.2	16.0	-	7.6	2.1	2.1	2.1	9.0	-	3.1	47.2
Narrow/Sparse	-	2.1	-	-	-	-	-	-	-	-	2.1
Narrow/Dense	-	14.6	-	-	-	2.1	2.1	2.1	-	-	20.8
2002											
Wide/Sparse	-	52.8	11.1	-	-	60.2	-	-	28.7	-	124.1
Wide/ Dense	-	-	23.1	-	-	94.4	-	-	-	-	117.6
Narrow/Sparse	-	63.9	9.3	-	-	11.1	3.7	-	25.0	-	78.7
Narrow/Dense	-	-	18.5	-	-	45.4	4.6	-	-	-	68.5

Values indicate the number of weed plants. Average of three quadrats (80cm * 60cm) .

Table 4. Emergence of weeds at the beginning of flowering of soybean.

4. Discussion

In soybean, dense planting has been reported to increase the node number, pod number and therefore seed yield without the consideration of lodging (Nakaseko and Goto 1975, Costa et al. 1980, Miura et al. 1987, Saitoh et al. 1998a). The square- or triangular-shape planting increased the space occupied by plants than rectangular-shape planting, and promoted the development of branches, thus increasing the seed yield (Cooper 1977, Costa et al. 1980, Duncan 1986, Miura and Gemma 1986, Miura et al. 1987, Board et al. 1990b, Ikeda 2000). Nakano et al. (2001) also reported that planting pattern affected the light environment within the canopy, which determined the branch node number, pod number and seed yield. In the present study, the seed yield was in the order of narrow/dense > narrow/sparse > wide/dense > wide/sparse (Table 2, Fig. 2), and the yield increase in narrow row planting was due to the yield increase on the branches especially on the raceme with compound leaves (Table 3).

The raceme with compound leaves is morphologically the same as a branch. The branch differentiates on the leaf axil just above the petiole on the main stem, and the raceme with compound leaves differentiates on the left and right axils of the basal raceme in the upper node of the main stem and branches, and develops a stem with one to four leaves. In a previous study, the differentiated racemes developed compound leaves when assimilates were supplied to the raceme (Saitoh et al. 2001). In the present two- year study, seed yield was positively correlated with total pod number ($r=0.934$, $P<0.001$) and pod number on racemes with compound leaves ($r=0.864$, $P<0.01$). Thus the increase in the pod number on the raceme with compound leaves contributed to the increase in seed yield.

The longer sunshine hours accelerated the source activity and increased assimilates were supplied to the axil of each node. Our three-year planting density experiment showed that the number of floral buds on racemes with compound leaves increased markedly in the year with longer sunshine hours (Saitoh et al. 1998a), and the pod number on racemes with compound leaves increased especially when the twelfth node was isolated by pruning the top above the twelfth node and removing all of the leaves, petioles and floral organs except those on the twelfth node at the flowering stage. Under such conditions, assimilates were concentrated to the twelfth node (Saitoh et al. 1998b), and the number of racemes with compound leaves on the main stem and branches increased when the leaves on branches and main stem were removed, respectively (Saitoh et al. 2001).

The present study revealed that the increase in pod number by narrow row planting was due to the increase in that on the racemes with compound leaves suggesting that the microclimate within canopy affected the development of racemes with compound leaves in narrow row-spacing. The narrow row-spacing canopy had a lower light extinction coefficient, i.e., better light-intercepting characteristics (Fig. 5).

In wide row-spacing, solar radiation was distributed non-uniformly, penetrated a deeper layer of the canopy due to fewer leaves distributed within the furrow, and decreased markedly above the row space (Fig. 7). In narrow row-spacing, solar radiation was distributed uniformly, the difference between the row and furrow was small, so that many racemes developed compound leaves due to the surplus assimilative supply to the raceme from the upper layer of canopy. The raceme with compound leaves is not only a sink organ, but also a source organ.

The canopy light extinction coefficients (k) measured under direct sunlight decreased during the daytime (Fig. 6). The decrease in k-value means that the sunlight penetrated uniformly into a deeper layer of the closed canopy with a higher LAI, however, sunlight reached a deeper layer directly and leaves received the excess light in non-closed canopy with lower LAI like wide row-spacing. This suggests that the k-value during the daytime can not evaluate the light intercepting characteristics in non-uniformly foliage distributed canopy.

The comparison of dry matter production in the plants with different planting patterns revealed that dry-weight was heavier and LAI was larger in dense plots than in sparse plots along as shown by others (Shibles and Weber 1965, Sugiyama et al. 1967, Asanuma et al. 1977, and also in narrow row-spacing than wide row-spacing (Fig. 3, 4) in accordance with the previous studies (Bullokck et al. 1998, Duncan 1986,Shibles and Weber 1965, 1966). In narrow row-spacing, the distance between plants was longer than in wide row-spacing, so that the canopy had a better light-intercepting environment, which accelerated the development of branches and racemes with compound leaves and the expansion of leaf area during the earlier stage, though, LAI in dense planting at 65 DAS exceeded 8, which means over luxuriant growth (Sugiyama et al. 1967).

Next, we should consider the effects of lodging. The lodging score was larger in narrow plots than in wide plots, (Fig. 8). This is because the distance between plants was longer in narrow row-spacing, and there was less mutual support with the neighboring plants. After the full flowering stage, a large amount of foliage was distributed in the upper layer of the canopy in the narrow/sparse plots (Fig. 5), and the higher the center of gravity, the higher the susceptibility to lodging. In narrow row-spacing, the main stem length was 15cm shorter and 0.9mm thicker than in wide row-spacing in 2001 (Table 1) because the competition between plants for elongation growth decreased due to the longer distance between plants. Despite this, the lodging score was larger in narrow row-spacing, meaning that the lodging of soybean was influenced by the above ground weight and center of gravity than the main stem length and stem thickness. Further study is needed to analyze the factors affecting the lodging tolerance in soybean.

Finally, let me consider about the weed management. In narrow row-spacing, we should eradicate weeds by hand if early weed control fails. It is impossible to kill weeds by cultivator after sowing. It was already demonstrated that the narrow row cultivation decreased weeds emergence and the alternative application of herbicide to soil or foliage (Gramineae weeds) could control weeds with labour saving and stability (Ohdan et al. 2005). Present results also showed that the less number of weeds were appeared in narrow row plots than in wide row plots (Table 4), in both plots herbicide was applied to the soil surface after sowing and the soil molding was conducted with a rotary cultivator in wide row plots. The dry-weight of weeds per square-meter was about 2g, which was extremely less than that of soybean, 300-400 g m^{-2}, i.e., weeds could be controlled sufficiently. We considered that weeds could be controlled by one application of herbicide to the soil surface after sowing. If we failed to kill weeds by the soil applied herbicide, the additional application of bentazone, newly registered foliar applied herbicide in Japan, can be used after sowing.

5. Conclusion

The narrow row-spacing (wide distance between plants) and dense planting in soybean increase seed yield than in the wide row-spacing (narrow distance between plants), which was caused by the decrease in competition among plants for elongation growth, the promotion of branch development, the development of racemes with compound leaves, and the increase in pod number due to the uniform light environment within the upper layer of canopy. The improvement of lodging tolerance and perfect weed control will be needed in the narrow row and dense planting of soybean were considered to be needed.

6. References

Asanuma, K., Naka, J. & Kogure, K. (1977). On the relation between dry matter production and plant density in autumn type soybeans. *Technical Bulletin Faculty of Agriculture Kagawa University*, Vol.28, pp. 11-18, ISSN 0368-5128.

Beatty, K.D., Eldridge, I.L. & Simpson, A.M.Jr. (1982). Soybean response to different planting patterns and dates, *Agronm Journal*, Vol.74, pp. 859-862, ISSN 0002-1962.

Bullokck, D., Khan, S. & Rayburn, A. (1998) Soybean yield response to narrow rows is largely due to enhanced early growth, *Crop Science*, Vol.38, pp. 1011-1016, ISSN 0011-183X.

Board, J.E., Harville, B.G. & Saxton A.M. (1990a). Narrow-row seed-yield enhancement in determinate soybean. *Agronm Journal*, Vol.82, pp. 64-68, ISSN 0002-1962.

Board, J.E., Harville, B.G. & Saxton, A.M. (1990b). Branch dry weight in relation to yield increases in narrow-row soybean, *Agronm Journal*, Vol.82, pp. 540-544, ISSN 0002-1962.

Cooper, R.L. (1971). Influence of soybean production practices on lodging and seed yield in highly productive environments, *Agronm Journal*, Vol.63, pp. 490-493, ISSN 0002-1962.

Cooper, R.L. & Nave, W.R. (1974). Effect of plant population and row width on soybean yield and harvesting loss, Transactions of the ASAE, Vol.17, pp. 801-805. ISSN 0001-2351

Cooper, R.L. (1977). Response of soybean cultivars to narrow rows and planting rates under weed-free conditions, *Agronm Journal*, Vol.69, pp. 89-92, ISSN 0002-1962.

Costa, J.A., Oplinger, E.S. & Pendleton, J.W. (1980). Response of soybean cultivars to planting patterns, *Agronm Journal*, Vol.72, pp. 153-156, ISSN 0002-1962.

Duncan, W.G. (1986). Planting patterns and soybean yield, *Crop Science*, Vol.26, pp. 584-588, ISSN 0011-183X.

Ikeda, T. (2000). Some factors related with net production of soybean population, *Japanese Journal of Crop Science*, Vol.69, pp. 12-19, ISSN 0011-1848.

Miura, H. & Gemma, T. (1986). Effect of square planting on yield and its components of soybean under different levels of planting density, *Japanese Journal of Crop Science*, Vol.55, pp. 483-488, ISSN 0011-1848.

Miura, H., Wijeyathungam , K. & Gemma T. (1987). Variation in seed yield of soybean as affected by planting patterns, *Japanese Journal of Crop Science*, Vol.56, pp. 652-656, ISSN 0011-1848.

Monsi, M. & Saeki, T. (1953). Über den Lichtfactor in den Pflanzengesell schaften und seine Bedeutung für die Stoffproduktion, *Japanese Journal of Botany,*. Vol.14, pp. 22-52.

Nakano, H. (1989). Effect of narrow spacing on grain yield in soybean, *Japanese Journal of Crop Science*, Vol.58, pp. 133-134, ISSN 0011-1848.

Nakano, H., Komoto, K. & Ishida, K. (2001). Effect of planting pattern on development and growth of the branch from each node on the main stem in soybean, *Japanese Journal of Crop Science*, Vol.70, pp. 40-46, ISSN 0011-1848.

Nakaseko, K. & Goto, K. (1975). Comparative studies on dry matter production, plant type and productivity in leguminus crop. III. Dry matter production of soybean plant at various population densities, *Japanese Journal of Crop Science*, Vol.44 (Extra issue 2), pp. 71-72, ISSN 0011-1848.

Ohdan, H., Sumiyoshi, T. & Koarai, A. 2005. Efficient weed control using herbicides and narrow row-dense planting for cultivation with no-intertillage of soybean cv. 'Sachiyutaka'. *Report Kyushu Branch Crop Science Society of Japan*, Vol.71, 30-32, ISSN 0285-3507.

Saitoh, K., Isobe, S. & Kuroda T. (1998a). Significance of flower differentiation and development in the process of determining soybean yield : Relation between the number of pods and flowers, *Japanese Journal of Crop Science*, Vol.67, pp. 70-78, ISSN 0011-1848.

Saitoh, K., Isobe, S. & Kuroda T. (1998b). Pod elongation and seed growth as influenced by nodal position on stem and raceme order in a determinate type of soybean, *Japanese Journal of Crop Science*, Vol.67, pp. 85-90, ISSN 0011-1848.

Saitoh, K., Isobe, S., Seguchi, Y. & Kuroda T. (2001). Effects of source and/or sink restriction on the number of flowers, yield and dry-matter production in field-grown soybean, *Japanese Journal of Crop Science*, Vol.70, pp. 365-372, ISSN 0011-1848.

Shibles, R.M., & Weber. C.R. (1965). Leaf area, solar radiation and dry matter production by soybeans, *Crop Science*, Vol.5, pp. 575-578, ISSN 0011-183X.

Shibles, R.M., & Weber. C.R. (1966). Interception of solar radiation and dry matter production by various soybean planting patterns *Crop Science*, Vol.6, pp. 55-59, ISSN 0011-183X.

Sugiyama, S., Matsuzawa, H., Horiuch, T. &. Kawashima, R. (1967) Effect of various planting density and pattern on growth and seed yield of soybean, Bulletin of the Nagano Agricultural Experiment Station, Vol.32, pp.34-35, ISSN 0388838X.

Torigoe Y., Shinji H. & Kurihara H. (1982). Studies on developmental morphology and yield determining process of soybeans. II. Developmental regularity of flower clusters and flowering habit from a view point of gross morphology, *Japanese Journal of Crop Science*, Vol.51, pp. 89-96, ISSN 0011-1848.

15

Physiological Quality of Conventional and RR Soybean Seeds Associated with Lignin Content

Cristiane Fortes Gris[1] and Edila Vilela de Resende Von Pinho[2]

[1]*Federal Institute of Southern Mines*

[2]*Federal University of Lavras*

Brazil

1. Introduction

The sale of genetically modified soybean seed resistant to the Roundup Ready (RR) herbicide has revolutionized the worldwide soybean market in recent years. According to data from the International Service for the Acquisition of Agri-Biotech Applications-ISAAA (2009), in 2009, for the first time, more than three-quarters (77%) of the 90 million hectares of soybeans grown globally were biotech; followed by cotton, with almost half (49%) of the 33 million hectares being biotech; by maize, with over a quarter (26%) of the 158 million hectares grown globally being biotech; and finally by canola, with 21% of the 31 million hectares being biotech. These numbers indicate not only increases in hectares, but also a strong and growing adherence of farmers around the world to this technology.

Considering the area planted to RR soybeans in the 2009/10 growing season throughout the world, from these 69.3 million hectares, a demand of approximately 4.2 million tons of RR soybean seeds may be estimated, which makes the international soybean seed market ever more expressive and competitive. In Brazil alone, up to November 2010, nearly 35% of the total soybean cultivars registered in the Ministry of Agriculture were RR genetically modified, this number having increased more than 443% in the last four growing seasons, a result of the increase in the number of breeding programs for obtaining RR cultivars.

It is known that the physiological quality of soybean seeds is controlled in large part by the genotype or cultivar, features of the plant, and more specifically those of the pod and the seed itself, determining a differential response of each cultivar and its levels of tolerance to seed deterioration, to adverse field conditions and even to mechanized harvesting. Among seed characteristics, the seed coat is one of the principal conditioning factors for germination vigor and longevity of seeds, with its characteristics being associated with susceptibility to mechanical damage, longevity and potential for seed deterioration, which may be influenced by the lignin content and the degree of seed coat permeability. Understanding of the structure and properties of the seed coat has contributed to explaining and altering seed behavior under certain environmental conditions.

In the case of soybeans, differences in the lignin content among seed coat have been observed by various authors (Tavares et al., 1987; Carbonell et al., 1992; Alvarez, 1994; Carbonell & Krzyzanowski, 1995; Panobianco, 1997; Menezes, 2008). In addition, a great deal of speculation has been generated in relation to the lignin content in the plant between RR genetically modified soybean cultivars and conventional cultivars (Coghlan, 1999; Gertz

Junior et al., 1999; Kuiper et al., 2001; Edmisten et al., 2006; Nodari & Destro, 2006), indicating overproduction of this substance of up to 20% more in RR cultivars. Such variation may occur not only in the vegetative parts of plants, but also in reproductive parts, such as pods and seeds.

The term lignin is used to designate a group of substances with similar chemical units indicated as polymers derived from "p-coumaryl", "conyferyl" e "sinapyl" alcohols (Lewis & Yamamoto, 1990). Impermeable to water, lignin is also very resistant to pressure and not very elastic and it is the most abundant plant polymer after cellulose, being found in greater quantity in the cell wall, around 60% to 90% (Egg-Mendonça, 2001), and its deposition occurs during the formation of the cell wall.

According to the authors, overproduction of lignin observed in the RR soybean plant in the US, and more recently in Brazil, is leading to deep stem fissures, with a significant number of plants in the field presenting bent or broken stems, and this effect possibly arises in the presence of water deficit and high temperatures.

Although the exact cause of the lignin behavior in this mechanism is still unknown, the hypothesis of overproduction of lignin in RR soybean plants is based on the fact of the precursors of the lignin molecule being formed in the same metabolic pathway, the pathway of shikimic acid, inhibited by the glyphosate herbicide. The inhibition of EPSPS enzymes by glyphosate present in this pathway leads to a deficiency in the production of amino acids and consequent death of the plants. That way, the sequence CP4 EPSPS, introduced in the genome of commercial soybean cultivars responsible for production of the protein CP4 enolpyruvylshikimate-3-phosphate synthase (EPSPS), an enzyme that participates in the biosynthesis of aromatic amino acids in plants and microorganisms, may be presenting the pleiotropic effect, thus modifying the lignin content in the plant.

Nevertheless, research in this area is still quite limited and the few results published do not compare conventional cultivars with their respective RR genetically modified versions, but refer to comparison between diverse genotypes and therefore do not isolate the effect of the inserted transgene. In this context, it is relevant to discuss the results of more recent research dealing with this issue in this chapter, principally looking at comparisons between conventional materials and their RR versions, which are essentially derivatives.

For that reason, in this chapter we will discuss results of research dealing with the physiological quality and the lignin content in RR and conventional soybean seeds submitted to different harvest times and spraying with glyphosate herbicide, produced in two different time periods and submitted to direct imbibition in water.

2. The lignification process and RR soybeans

The term lignin is used to designate a group of substances with similar chemical units. According to Panobianco (1997), the chemical structure of lignin is very complex and still not very well defined. Butler & Bailey (1973), cited by Silva (1981), refer to lignin as a polymer, 3-methoxy-phenyl-propanol and 3-5-dimethoxy-phenyl-propanol, bonded in varied proportions and in random sequence, leading to a great variety of products, which makes exact definition difficult. According to Esau (1976), lignin consists of an organic substance or mixture of organic substances with high carbon content, but different from carbohydrates, and which is found associated with cellulose on the walls of numerous cells.

The term lignin is used to designate a group of substances with similar chemical units reported as polymers derived from "p-coumaryl", "conyferyl" e "sinapyl" alcohols (Lewis &

Yamamoto, 1990). Impermeable to water, lignin is also very resistant to pressure and not very elastic and it is the most abundant plant polymer after cellulose, being found in greater quantity in the cell wall, around 60% to 90% (Egg-Mendonça, 2001), and its deposition occurs during the formation of the cell wall.

The growth and development of the cell wall may be divided into two phases: growth of the primary wall, a phase in which the cell increases in size, and growth of the secondary wall, a phase in which deposition of lignin polymers occurs to the extent that the cell wall becomes progressively thicker as of the internal edge of the primary wall, in the direction of the center of the cell. The inclusion of lignin on the cell wall originates in the middle lamella, going in the direction of the interior of the secondary wall. According to Jung & Alen (1995), the effect of this lignin deposition pattern makes the middle lamella/primary cell wall region more intensely lignified.

This lignin deposition is important not only to lend rigidity and resistance to plant tissue, such as stem and leaves, but especially for the seed coat of soybean seeds, it has been correlated with resistance to mechanical damage (Alvarez, 1994; Panobianco, 1999), providing mechanical resistance to the tissue and protection against infestations by microorganisms to the cell wall (Rijo & Vasconcelos, 1983, cited by Tavares et al., 1987).

2.1 Lignification and the soybean seed coat

The seed coat is one of the main factors which determine germination capacity, vigor and longevity of seeds. It has a protective function during imbibition, avoiding cell rupture and loss of intracellular substances (Duke & Kakefuda, 1981), and also protects the embryonic axis (Carvalho & Nakagawa, 2000). It is derived from the integuments of the ovule where the primine gives rise to the testa and the secundine gives rise to the tegmen.

By means of a cross section of the testa of a soybean seed, three layers may be distinguished, the epidermis, the hypodermis and the inner parenchyma (Swanson et al., 1985). This last layer, composed of the spongy parenchyma, is present in the entire testa of the seed, except for the hilar region. It has from 6 to 8 cell layers, tangential to the surface of the testa, formed by thin walls and absent protoplasm, with the outermost part of this parenchyma being formed by large, elongated cells, while the innermost part by smaller and significantly branched cells (Esau, 1977).

The intermediate layer of the testa, the hypodermis, is formed of cells in hourglass form, or pillar cells, or even osteosclereid cells. It consists of a uniform cell layer through the entire testa, except for the hilar region. The cell wall of its sclerenchyma cells is not uniform, with the presence of large intercellular spaces (Corner, 1951).

The epidermis, outside of the testa, remains uniseriate and gives rise to the palisade layer, characteristic of leguminosae seeds. This layer consists of macrosclereids (Malpighi cells) with wall of unequal thickness, having a cuticle present over their outermost wall. It cells are elongated and arranged perpendicular to the surface of the testa, with thick cell walls (Esau, 1976).

In soybean seeds, the thickness of the four testa layers altogether, including the cuticle, starting from the surface, may vary from 70 to 100 micrometers, there being variation among cultivars. Nevertheless, this characteristic is a constant with each cultivar and is controlled genetically (Caviness & Simpson, 1974). The presence or lack of pores and their quantity, shape and size on the surface of the testa is also controlled genetically. The pores seem to be related to water absorption, such that in hard seeds they are either absent or they exist in small quantity (Calero, 1981).

Morphological characteristics associated with the thickness and structure of the seed coat has also been related to the quality of soybean seeds. With the aid of a Scanning Electron Microscopy (SEM), it is possible to obtain a direct image of the atoms on the surface of a material, formed by secondary electrons and emitted from the surface of the irradiated specimen by the beam of primary electrons or by those scattered, which, in spite of generating poorer quality images, may indicate differences in the elementary composition of the sample. Designed basically for surface examination of samples, SEM allows the observation of internal surfaces if fractured and exposed, using principally secondary electrons (Alves, 2006).

Silva (2003), by means of scanning micrography of transversal sections of the testa of soybean seeds of the cultivars M-Soy 8400 and M-Soy 8411 observed three visible cell layers: palisade cell layers, an hourglass cell layer, and spongy parenchyma cells. The author evaluated the behavior of these cell layers that compose the testa of soybean seeds when they were exposed to five periods of accelerated aging (0, 24, 48, 72 and 96 hours) at 42° C and approximately 100% relative air humidity. For the cultivars evaluated, reduction in the thickness of the testa of the soybean seed was verified, which suggests collapse of the cells that compose such layers, which may be related to reduction of germination potential.

Menezes et al. (2009) evaluating the thickness and structure of the soybean seed coat (Figures 1 and 2) and the association of these characteristics with the physiological quality of the seeds, concluded that traits used for evaluation of physiological quality may be correlated with the lignin content of the seed coat. Nevertheless, according to the author, it was not possible to establish a relationship between the physiological quality of the soybean seeds and the anatomical aspects of the seed coat evaluated by SEM, emphasizing the need for refining the methodologies available for this purpose due to the difficulties of establishing the work area of common structures on the seeds, and of having observed that cell structures vary in different positions on the seed coat, which makes comparison of these structures among seeds of different genotypes difficult. In spite of that, in a general way, it was possible to observe that the lignin thickness on the palisade cell layers was greater when compared to the hourglass cell layers.

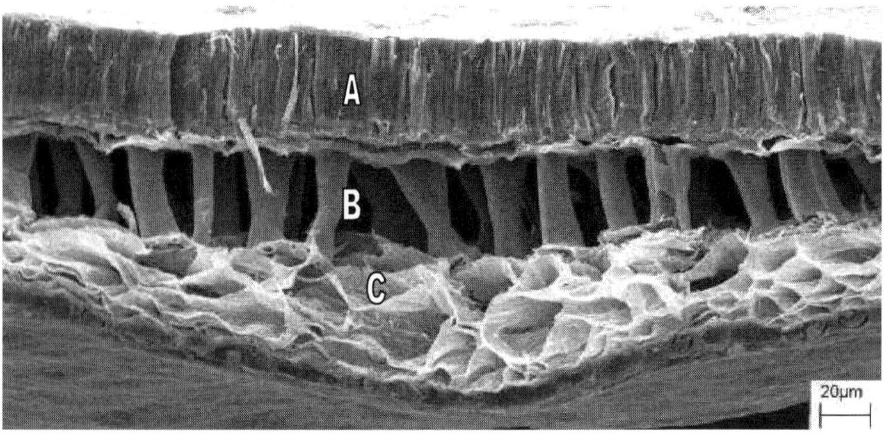

Fig. 1. Scanning micrography of the testa of the cultivar CD 201; A: palisade cell layer; B: hourglass cell layer and C: spongy parenchyma. Source: Menezes et al. (2009).

As is common in leguminosae, there is a particularly impermeable region on the walls of the upper part of the macrosclereids, which reflects light more intensely than the rest of the wall (Esau, 1965). What is called the conspicuous light line is visible in many wild soybean species, but is less prominent in cropped species (Alexandrova & Alexandrova, 1935, cited by Carlson & Lersten, 1987). This palisade layer drew the interest of researchers through the fact of its structure, and in certain hard seeds of leguminosae, being the cause of the high degree of impermeability of the seed coat, consequently affecting germination capacity (Esau, 1976).

Hard or impermeable seeds, according to Woodstock (1988), may be the result of compacted organization of cellulose microfibriles on the cell wall. This, for its part, may be impregnated with waterproof substances, such as lignin, waxes, suberins or tannin. They are abundantly composed of cellulose and hemicellulose polysaccharides, and of phenylpropanoid polymers such as lignin (McDougall et al., 1996).

In accordance with McDougall et al. (1996), the impermeability of the seed coat provided by lignin, exercises a significant effect on the speed and capacity of water absorption through it, thus interfering in the quantity of leached materials released to the outside during the imbibition phase of the seed germination process. Crocker (1948) already mentioned the need for better understanding of this mechanism since it was considered to be the best example of efficiency against water penetration and should therefore be better utilized by breeders in adjusting this characteristic to their needs. As general characteristics of soybean cultivars with a less permeable seed coat, one may cite better conservation potential, lower levels of infection by pathogens, greater vigor and viability, as well as resistance to reabsorption of moisture after maturation (Panobianco, 1999).

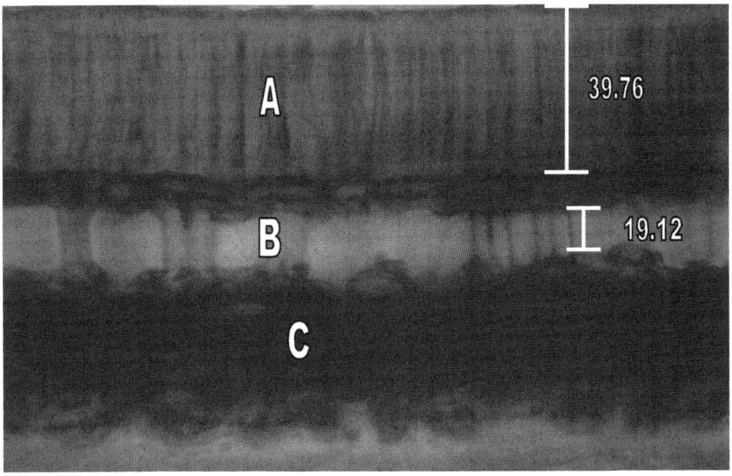

Fig. 2. Comparson of the thickness (μm) of the palisade cell and hourglass cell layers obtained by SEM from the cultivar CD 206. A: palisade cell layer; B: hourglass cell layer and C: spongy parenchyma. Source: Menezes et al. (2009).

Tavares et al. (1987), studying structural characteristics of the seed coat of seeds of soybean lines, concluded that the total fiber content is not connected with impermeability; however, in regard to the type of fiber, an accentuated increase in the lignin values was observed in the lines with impermeable seed coats (4.69% to 7.70%), differentiated from the values 1.80% to 3.18% found in lines with permeable seed coats. According to Brauns & Brauns (1960), cited by Tavares et al. (1987), the hydrophobic trait of lignin affects the hydrophilic bonds of the middle lamella and the removal of lignin interferes in the biological resistance of hydration in around 10.5% to 17% of the original tissue.

The occurrence of hard seeds in leguminosae has been attributed to both genetic and environmental factors (Donnelly, 1970). The percentage of hard seed exhibits considerable variability depending on the species or cultivar, the degree of maturity, the maturation conditions and the storage time. Thus, low air humidity during maturation results in a considerable increase in seed hardness (Baciu-Miclaus, 1970; Martins, 1989).

In soybeans, differences in the lignin content of the seed coat has been observed by various authors (Tavares et al., 1987; Carbonell et al., 1992; Alvarez, 1994; Carbonell & Krzyzanowski, 1995; Panobianco, 1999; Menezes et al., 2009; Gris et al., 2010;), and, in addition, differences have been reported in regard to the lignin content in the plant between genetically modified RR and conventional cultivars.

2.2 Lignin biosynthesis and RR soybeans

The advent of genetically modified soybeans, tolerant to the Roundup Ready© herbicide (RR), revolutionized the world soybean market. With the introduction of the CP4 EPSPS sequence in the genome of commercial soybean cultivars, which confers tolerance to the active ingredient glyphosate, the protein CP4 enolpyruvylshikimate-3-phosphate-synthase (EPSPS) is produced, an enzyme that participates in the biosynthesis of aromatic amino acids in plants and microorganisms. In the case of conventional cultivars, the inhibition of these enzymes by glyphosate, present in the shikimic acid pathway, leads to a deficiency in production of essential amino acids and consequent death of the plants, which does not occur in RR cultivars.

A great deal of speculation has been generated in relation to the lignin contents in the plant between genetically modified RR cultivars and conventional cultivars (Coghlan, 1999; Gertz Junior et al., 1999; Kuiper et al., 2001; Edmisten et al., 2006; Nodari & Destro, 2006).

In the late 1990s, some farmers in Georgia complained about the poor performance of their RR soybeans in years with a spring with drought and heat conditions. Scientists then carried out a comparative laboratory study of genetically modified and conventional soybeans (Gertz Junior et al. 1999). They found that the genetically modified plants were shorter, had a lower fresh weight, had less chlorophyll content, and, at high soil temperature of 40 °C to 50°C, suffered from stem splitting. According to Coghlan (1999), the elevated levels of lignin deposited in the stem of soybean plants would be leading to this splitting due to the stiffening of the plants under high temperatures (45ºC), a problem also detected in genetically modified RR soybean crops in the USA, and which was to have led to considerable losses through falling of plants in hotter years (Nodari & Destro, 2006) as a consequence of overproduction of lignin in RR cultivars (Kuiper et al., 2001).

According to these authors, under stress conditions, losses in RR soybeans can arrive at 40% in comparison with conventional soybeans, brought about by greater production of lignin, up to 20% greater (Coghlan, 1999; Gertz et al., 1999). Nodari & Destro (2006), in a study undertaken in nine soybean crops in the state of Rio Grande do Sul (Brazil), observed that in the presence of drought and high temperatures, the RR soybean crops suffered more losses than conventional soybeans. The authors observed a large number of plants with deep stem splitting and a significant quantity of these plants had bent or broken stems, around 50% to 70% of the plants, according to the authors, possibly due to overproduction of lignin in the RR material (Figure 3).

Fig. 3. Plants of the "Maradona" variety with broken stem (left), split (middle) and intact stem (right). Source: Nodari & Destro (2006).

The plants are responsible for the production of secondary metabolites that perform innumerable functions, among which the terpenes, the phenolic compounds and the alkaloids are considered as the most important. The secondary compounds are biosynthesized through three basic metabolic pathways, the acetate-mevalonate, the acetate-malonate and the acetate-shikimate (Érsek & Kiraly, 1986), also denominated simply as mevalonic acid pathway, malonic acid pathway and shikimic acid pathway, respectively (Taiz & Zeiger, 1998).

In superior plants, the shikimic acid pathway occurs in plastids, there also being evidence that it is present in the cytosol (Hrazdina & Jensen, 1992). This important metabolic pathway begins with phosphoenolpyruvate (PEP), derived from glycolysis, and the erythrose 4-P coming from the monophosphate pentose pathway and the Calvin cycle, resulting in the biosynthesis of the phenylalanine amino acids, tyrosine and tryptophan (Salisbury & Ross, 1992) (Figure 4).

According to Resende et al. (2003), the enzymes that participate in the initial and intermediary steps of the lignin biosynthesis pathway are common to the phenylpropanoid pathway (Figure 5). The metabolism of the phenylpropanoids includes a complex series of biochemical pathways that provide the plants with thousands of combinations. Many of these, according to Boatright et al. (2004), are intermediate in the synthesis of structural substances of the cells, such as lignin, if formed from shikimic acid, which forms the basic units of the cinnamic and p-coumaric acids (Simões & Spitzer, 2004).

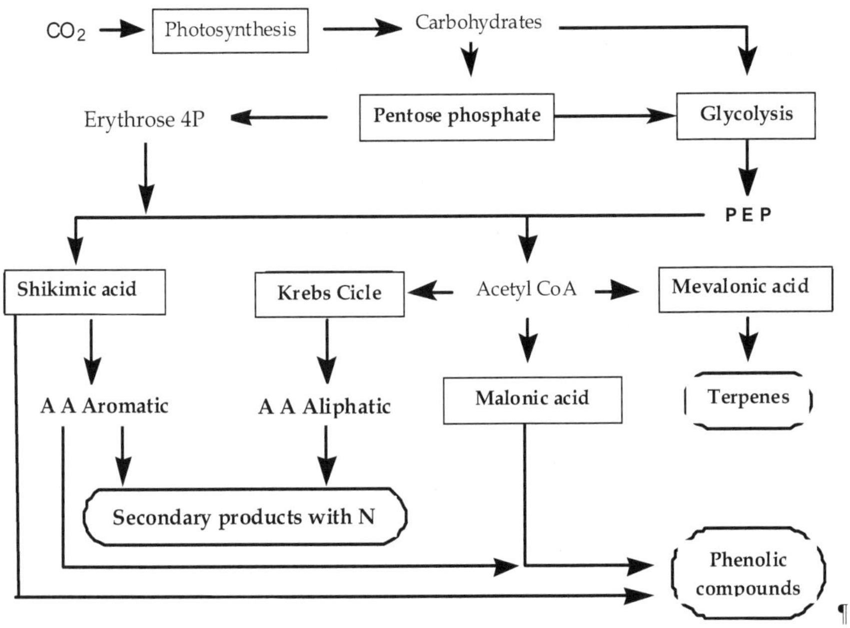

Source: Adapted from Taiz & Zeiger (1998).

Fig. 4. Schematic representation of the shikimic, malonic and mevalonic acid pathways. ¶

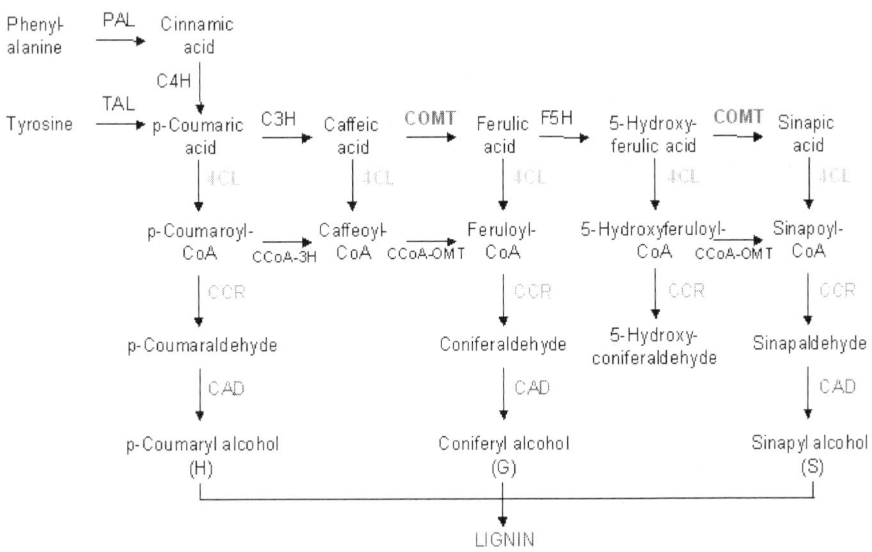

Fig. 5. Lignin biosynthesis pathway. Source: Baldoni (2010).

Lignin synthesis involves various enzymes and knowledge of them is important in studies in which the quality of soybean seeds and the lignin content is related (Baldoni, 2010). The complexity of the lignin biosynthesis pathways is attributed to various multifunctional enzymes, which also correspond to different gene families (Xu et al., 2009).

A considerable quantity of genes is attributed as participant in lignin synthesis, such as genes which regulate the activity of the enzymes phenylalanine ammonia-lyase (PAL), Cinnamate 4-Hydroxylase (C4H), 4-cumarate-CoA ligase (4CL), 4 Hydroxycinnamate 3-Hydroxylase (C3H), 5-Adenosyl-Methionine: Caffeate/5-Hydroxy (OMT), Ferulate-5-Hydroxylase (F5H), Hydroxycinnamoyl COA Reductase (CCR), cinnamyl alcohol dehydrogenase (CAD) (Boudet, 2000; Boudet, 2003; Darley et al., 2001).

Although the exact cause of lignin behavior under stress conditions in RR soybean cultivars is still unknown (Coghlan, 1999), possibly the alterations in the content of this biopolymer in the plant is due to the fact of the precursors of the lignin molecule being formed in the shikimic acid pathway, which is inhibited by the glysophate herbicide in conventional plants. The inhibition of EPSPS enzymes, present in this pathway by the glyphosate, lead to a deficiency in the production of amino acids and consequent death of the plants. That way, the CP4 EPSPS sequence introduced in the genome of the commercial soybean cultivars denominated RR, responsible for the production of the protein CP4 enolpyruvylshikimate-3-phosphate synthase (EPSPS), an enzyme that participates in the biosynthesis of aromatic amino acids in plants and microorganisms, may present the pleiotropic effect, thus modifying the lignin content in the plant.

In spite of all those studies suggesting the pleiotropic effect of the transgene under high stress conditions in laboratory tests in the USA, some authors suggest that it might not be detected until specific environmental conditions are observed, which usually does not occur in field conditions. In this sense, the quantification of lignin in the plant, and consequently in pods and the seed coat of soybeans, become necessary in field conditions, principally with a view toward comparisons between conventional materials and their RR versions, which are essentially derivatives, since the previous reports refer to diverse genotypes, thus not isolating the effect of the inserted transgene. It is worth highlighting that scientific studies that truly prove the pleiotropic effect of the RR transgene under any characteristics are rare in the literature, with most of them being based only on observations and not on scientific results.

Therefore, we will further discuss some results of research obtained in Brazil in which the relation lignin versus RR and conventional soybean cultivars under diverse aspects was evaluated, emphasizing contents of this polymer in the plant, pod and seed coat.

3. Conventional and RR genetically modified soybeans: Some results in Brazil

3.1 Physiological quality and lignin content in the seed coats submitted to different harvest times

The viability period of the soybean seed is extremely variable, depending both on genetic characteristics and environmental effects during the phases of development, harvest, processing and storage. Once unfavorable conditions occur in some of these phases, physiological damages may result in losses to seed quality, with the intensity of these damages varying with the genetic factors intrinsic to each cultivar. Various researchers have emphasized the possibility of use of the seed with seed coat with a certain degree of impermeability to water as an alternative for avoiding loss of quality in the field (Gilioli & França Neto, 1982; Peske & Pereira, 1983; Hartwig & Potts, 1987), with delay in harvest and determination of the lignin content in the seed coat being methodologies suggested for genetic breeding programs for evaluation of the quality of soybean seeds (França Neto & Krzyzanowski, 2003).

Within this context, the work presented below (Gris et al., 2010) was conducted with the purpose of evaluating the physiological quality and lignin content in the seed coat of the conventional and RR soybean seeds collected at three different times in Lavras (MG), Brazil. Thus, the seeds of ten cultivars collected at stages R7, R8 and 20 days of harvest delay (R8+20) were submitted to tests for evaluation of physiological quality and lignin content. Harvest stages were determined according to Fehr & Caviness (1977).

We observed differences in the physiological quality of seeds among the different harvest times for the cultivars BRS 134, BRS 247 RR, Conquista, Jataí and Silvânia RR, with reduction in viability with harvest delay (R8 + 20). In a similar way, when submitted to accelerated aging, the seeds of the cultivars BRS 245 RR, BRS 134, BRS Jataí and Silvânia RR also underwent a reduction in vigor with harvest delay (Table 1). Braccini et al. (2003), studying the response of 15 genotypes of soybeans to harvest delay, also observed a significant reduction in germination percentage and vigor of seeds when they were submitted to harvest 30 days after the R8 stage of development.

Cultivars	Germination			Accelerated Aging			Electrical Conductivity		
	R7	R8	R8 + 20	R7	R8	R8 + 20	R7	R8	R8 + 20
Celeste	94.75a	96.50a	95.50a	94.75a	97.50a	91.50a	77.01a	82.42a	94.76a
Baliza RR	94.25a	93.00a	91.00a	91.50a	88.50a	84.00a	83.47b	90.86b	118.01a
BRS 133	91.25a	93.00a	88.00a	91.25a	96.50a	87.50a	93.66b	82.61b	107.15a
BRS 245 RR	91.75a	96.50a	90.50a	97.75a	99.50a	87.50b	94.79a	99.11a	97.37a
BRS 134	91.75a	90.50a	79.00b	94.00a	95.00a	75.50b	86.81a	87.02a	97.65a
BRS 247 RR	96.50a	98.50a	87.50b	94.00a	96.00a	87.00a	76.38a	85.18b	102.44b
Conquista	85.75a	90.00a	78.00b	89.75a	88.50a	84.00a	93.87b	85.23b	118.25a
Valiosa RR	89.75a	83.00a	84.50a	87.50a	92.00a	87.50a	98.15b	90.01b	112.56a
Jataí	91.50a	89.00a	76.50b	93.25a	87.00a	64.00b	83.42b	88.43b	152.70a
Silvânia RR	93.00a	91.50a	82.00b	92.50a	92.00a	71.00b	92.92b	89.61b	143.74a

Means followed by the same letter in the line for each determination do not differ among themselves by the Scott-Knott test at the 5% significance level.

Table 1. Means of the germination and accelerated aging test (% of normal seedlings) from seeds of soybean cultivars and their respective RR genetically modified forms, 2007/08 harvest. UFLA, Lavras, MG, Brazil.

We observed that the greatest decreases in seed vigor by the accelerated aging test (Table 1), when the harvest delay and the mean of stages R7 and R8 are contrasted, occurred for the cultivars Jataí and Silvânia RR, which presented, on average, losses in vigor of 40.82% and 29.93% respectively, indicating that not always cultivars that have high seed quality when collected near physiological maturity have greater tolerance to deterioration with delay of harvest. And, moreover, the greatest values of electrical conductivity were observed for the majority of seeds of the cultivars collected 20 days after the R8 stage, with exception of the cultivar BRS 247 RR, in which reduction in seed vigor was observed as of the R8 stage, and of the cultivars Celeste, BRS 245 RR and BRS 134 that did not undergo any alterations with the time of harvest.

As degradation of the cellular membranes is constituted hypothetically in the first event of the deterioration process (Delouche & Baskin, 1973), tests that evaluate membrane integrity, such as the electrical conductivity test, would theoretically be the most sensitive for estimating seed vigor, which is in agreement with the results obtained in this study, in which said test stood out in detecting differences of viability between the harvest times in seven of the ten cultivars evaluated. We emphasize that the electrical conductivity values observed in this study were situated from 77.01 µS cm^{-1} g^{-1} to 98.15 µS cm^{-1} g^{-1} for the R7 harvest time, 82.42 µS.cm^{-1}.g^{-1} to 99.11 µS.cm^{-1}.g^{-1} for the R8 harvest time and 94.76 µS.cm^{-1}.g^{-1} and 152.70 µS.cm^{-1}.g^{-1} for the 20 days after R8, values which demonstrate the growing trend of leachates released by the seeds with delay in harvest.

When we analyze the percentage of mechanical damage in seeds (Table 2), we observe the greatest values with delay of harvest for the cultivars Conquista (12.5%), Jataí (16.0%) and Silvânia RR (15.0%), which was not observed for the other cultivars studied. In addition, we also observed that by the germination test of seeds submitted to the water immersion test, three of the ten cultivars evaluated were differentiated in regard to the percentage of normal seedlings, however, with distinct responses. The lowest germination values when collected in R8 were observed in seeds of the cultivar BRS 245 RR; in those of the cultivar BRS 247 RR there was a reduction in germination when collected in R8 and R8 + 20; and finally in those of the cultivar Silvânia RR the lowest germinative power was verified when collected in R7 and R8. Various authors emphasize that soybean cultivars and lines behave differently in regard to degree of tolerance to delay of harvest (Lin & Severo, 1982; Rocha, 1982; Boldt, 1984), indicating that this trait may influence maintenance of the physiological quality of the seeds.

For the lignin content in the soybean seed coat, we can observe greater lignin content in the seed coat of seeds collected in the R7 and R8 + 20 stages, as well as for the cultivar Silvânia RR, when contrasted with its conventional version Jataí (Table 3).

When we observe the data of percentage of deformed abnormal seedlings, characterized by root curling, typical of damage by rapid imbibition, we observe a smaller number of abnormal seedlings due to the greater number of dead seeds with harvest delay. Giurizatto et al. (2003) affirm that the deteriorated seeds imbibe more rapidly and are therefore more prone to greater damage through imbibition, which is in agreement with the results obtained in this study.

According to Alpert & Oliver (2002) the cellular membranes have two main states, one more fluid or "crystalline liquid" and another less fluid or "gel", remaining, when organized, in the crystalline phase. In a dry seed, the membranes are found in the gel phase and therefore do not constitute an efficient barrier to contain the release of solutes. When the seeds are exposed to rapid imbibitions, the water penetrates before the membrane can be reverted to the crystalline liquid phase, with damage occurring to the cells; thus, the transition between these two phases in the configuration of the membrane constitutes the fundamental cause of possible injuries during imbibition of seeds, which makes the study of the role of lignin in the seed coat even more important.

Cultivars	Mechanical Damage			Germination after Immersion Normal seedlings			Germination after Immersion Abnormal curled seedlings		
	R7	R8	R8 + 20	R7	R8	R8 + 20	R7	R8	R8 + 20
Celeste	3.50a	2.50a	3.00a	62.50a	70.50a	62.00a	20.00a	14.50a	11.00a
Baliza RR	3.00a	3.00a	6.00a	50.00a	46.50a	44.50a	17.50a	24.00a	12.00a
BRS 133	3.00a	1.00a	2.00a	55.00a	49.50a	43.50a	14.50b	26.00a	17.00b
BRS 245 RR	2.50a	2.50a	5.00a	46.00a	22.50b	43.50a	19.00a	17.00b	32.00b
BRS 134	1.50a	1.50a	1.00a	51.00a	47.50a	36.00a	26.00a	26.50a	23.00a
BRS 247 RR	1.50a	1.00a	3.50a	63.00a	50.50b	41.00b	15.50b	32.00a	23.00b
Conquista	6.00b	4.50b	12.50a	38.00a	33.00a	35.00a	10.50a	6.00a	1.00a
Valiosa RR	5.50a	4.50a	5.50a	35.50a	25.50a	36.00a	9.00a	4.50a	3.50a
Jataí	2.50b	3.50b	16.00a	20.50a	29.50a	26.00a	32.00a	41.50a	2.50b
Silvânia RR	4.50b	5.00b	15.00a	21.50b	28.50b	40.50a	30.00a	26.00a	1.50b

Means followed by the same letter in the line for each determination do not differ among themselves by the Scott-Knott test at the 5% significance level.

Table 2. Means obtained for mechanical damage (%) and germination after water immersion (% of normal seedlings and abnormal curled) of soybean cultivar seeds and their genetically modified RR forms, 2007/08 harvest. UFLA, Lavras, MG, Brazil.

These differences observed for the lignin content among the harvest times are not biologically explainable, having possibly been detected due to the low coefficient of variation (CV) obtained for this variable. When we analyze the sole significant contrast, for its part, the genetically modified cultivar Silvânia RR presented greater lignin content in the seed coat than its respective conventional cultivar Jataí. Nevertheless, as an isolated fact, among the five RR combinations versus the conventional versions tested, in our view it does not justify a greater inference regarding pleiotropy of the RR transgene.

Lignin Content				
Harvest Stages			Cultivars	
R7	R8	R8 + 20	Jataí	Silvânia RR
0.2685a	0.2385b	0.2615a	0.3008b	0.4167a

Means followed by the same letter in the column do not differ among themselves by the Scott-Knott test at the 5% significance level.

Table 3. Means with a significant difference obtained for lignin content in the soybean seed coat (%), 2007/08 harvest, UFLA, Lavras, MG, Brazil.

In general, we can conclude that in spite of there being behavioral differences in regard to tolerance to harvest delay among the different cultivars evaluated, we did not observe consistent results in regard to a comparison of the RR versus conventional cultivar, not indicating, for the conditions of this test, any sign of pleiotropy.

3.2 Physiological quality and lignin content in the plants submitted to spraying with glyphosate

Glyphosate (N-phosphonomethyl glycine) is one of the most used herbicides in weed control throughout the world, making up nearly 12% of global herbicide sales and presenting more than 150 commercial brands (Kruse et al., 2000). The emergence of RR genetically modified soybeans increased the use of this molecule in soybeans crops in a considerable way and, along with this, also the environmental concern due to exclusive and indiscriminate use of this herbicide.

According to Sanino et al. (1999), although pesticides (especially glyphosate) may have a beneficial effect on agricultural productivity, the potential risk of these chemical compounds in the environment must be considered, which makes greater studies regarding the behavior of glyphosate under tropical conditions relevant. Within this context we aimed to evaluate the physiological quality of genetically modified RR soybean seeds and the lignin contents of plants submitted to spraying with glyphosate herbicide (Gris, 2009).

In Tables 4 and 5 we present the mean results for the variables analyzed when the soybean plants were submitted to spraying with glyphosate herbicide and water (greenhouse test) and spraying with glyphosate herbicide or manual weeding (field test) respectively.

Cultivars	Germination		Accelerated Aging		Seed Coat Lignin	
	Water	Herbicide	Water	Herbicide	Water	Herbicide
Valiosa RR	91.0a	89.50a	98.75a	97.50a	0.33a	0.26a
BRS 245 RR	94.25a	93.75a	98.50a	96.75a	0.21a	0.22a

Means followed by the same letter in the line for each determination do not differ among themselves by the Scott-Knott test at the 5% significance level.

Table 4. Means of germination and Accelerated aging (% of normal seedlings) and lignin content in the seed coat (%) of genetically modified RR soybean seeds submitted to spraying with water and glyphosate herbicide, 2007/08 harvest, Lavras, MG, Brazil, greenhouse test.

We observe that application of the glyphosate herbicide did not alter the physiological quality of the soybean seeds nor the lignin contents in the seed coat and in the plant for the two tests evaluated. These results are not in agreement with those obtained by Sanino et al. (1999), who studying the effect of application of glyphosate herbicide in soybeans observed, in a general way, reduction in the physiological quality of the RR seeds, as well as considerable reduction in activity of the enzyme α-amylase in terms of time. It is worth emphasizing that such a study was carried out comparing only 2 soybean cultivars, one conventional and one genetically modified RR variety, and that the two did not represent the same genotype, since they originated from different parentages.

In this study (Gris, 2009) we obtained a significant response only for the interaction cultivar versus treatments, when the values of electrical conductivity of the seeds produced in the field test were evaluated (Table 5), in which we observed that seeds of the cultivars Baliza RR and BRS 247 RR had their values reduced and increased respectively when the same spraying was performed. Such a differential response may possibly be explained by the different capacity of the genes inserted in the RR cultivars in expressing tolerance to the glyphosate herbicide, which according to Lacerda & Matallo (2008) may or may not occur in a homogeneous manner among cultivars and even within the same cultivar, as well as other factors inherent to the genetics of each cultivar.

Cultivars	Germination		Accelerated Aging		Mechanical Damage		ESI	
	Weeding	Herbicide	Weeding	Herbicide	Weeding	Herbicide	Weeding	Herbicide
Baliza RR	93.50a	96.50a	88.00a	90.50a	0.75a	0.75a	7.07a	7.14a
BRS 245 RR	89.00a	93.50a	97.00a	95.50a	1.00a	0.00a	7.24a	7.12a
BRS 247 RR	96.50a	97.75a	94.50a	91.50a	0.75a	1.75a	7.19a	7.07a
Silvânia RR	93.00a	93.00a	87.00a	88.00a	3.00a	2.00a	7.44a	7.00a
Valiosa RR	86.00a	89.00a	84.50a	83.50a	2.00a	2.25a	7.42a	7.39a
Cultivars	Electrical Conduct.		Seed Coat Lignin		Pod Lignin		Stem Lignin	
	Weeding	Herbicide	Weeding	Herbicide	Weeding	Herbicide	Weeding	Herbicide
Baliza RR	61.0a	48.0b	0.24a	0.23a	8.66a	8.09a	12.71a	13.61a
BRS 245 RR	69.0a	70.0a	0.19a	0.19a	8.41a	7.81a	12.20a	13.07a
BRS 247 RR	52.0b	70.0a	0.20a	0.20a	9.26a	8.57a	13.43a	12.91a
Silvânia RR	69.0a	69.0a	0.29a	0.30a	9.61a	9.03a	20.23a	18.50a
Valiosa RR	46.0a	40.0a	0.27a	0.30a	7.87a	7.70a	14.62a	13.35a

Means followed by the same letter in the line for each determination do not differ among themselves by the Scott-Knott test at the 5% significance level.

Table 5. Means of germination and Accelerated aging (% of normal seedlings), Mechanical damage (%), Emergence speed index - ESI (days), Electrical conductivity ($\mu S.cm^{-1}.g^{-1}$), Lignin content in the seed coat, pod and stem (%) of genetically modified RR soybean cultivars submitted to manual weeding and spraying with glyphosate herbicide, 2007/08 harvest, Lavras, MG, Brazil, field test.

It is worth emphasizing that since degradation of the cellular membranes is constituted hypothetically in the first event of the deterioration process (Delouche & Baskin, 1973), tests such as electrical conductivity that evaluate membrane integrity are theoretically most sensitive for estimating seed vigor, which possibly, allied with the affirmations of Lacerda & Matallo (2008), would explain the alterations only in the conductivity values.

The absence of a significant response for treatments with weeding and spraying with the glyphosate herbicide indicate that in a general way they did not influence the physiological quality of the seeds, nor the lignin content in the soybean plants. According to Cole & Cerdeira (1982) the blocking of the shikimate pathway due to the action of the glyphosate leads to the accumulation of shikimic acid with many physiological and ecological implications, which, according to Duke & Hoagland (1985) and Becerril et al. (1989), may result in synthesis of indol acetic acid of other plant hormones, chlorophyll synthesis, phytoalexin and lignin synthesis and protein synthesis, and affect photosynthesis, respiration, transpiration, permeability of membranes and other factors.

In addition, other studies have shown that applications of glyphosate in crops interfere in nutrient absorption, increase pests and diseases, reducing crop vigor and yield (Antoniou et al., 2010). According to compilation of data made by these authors, glyphosate reduces nutrient absorption by plants, immobilizing trace elements such as iron and manganese in the soil, as well as avoiding their transport from the roots to the above ground part

(Strautman, 2007). As a result, RR soybean plants treated with glyphosate have lower levels of manganese and other nutrients and reduction in growth of budding and roots (Zobiole et al., 2010). It is worth emphasizing that the seeds produced in the two tests described in this secondary heading are being tested in regard to variation in chemical composition, data which should soon be published.

Both in the field test and in the greenhouse test, it was not possible to relate physiological quality of the seeds and lignin content in their seed coat. We observed significant differences only among the cultivars evaluated, which presented different responses when submitted to the different vigor tests, as well as lignin content, which was already expected, in terms of the great genetic variability among them.

We conclude from these tests that there is a differential response for the electrical conductivity values of the seeds when the plants of different soybean cultivars are submitted to spraying with the glyphosate herbicide; nevertheless, we did not observe a difference in the lignin contents in the stem, in the pod and in the seed coat of the soybean seeds in the cultivars evaluated when submitted to spraying with the glyphosate herbicide.

3.3 Agronomic characteristics and quality of soybean seeds produced at different times

It is known that different planting times, influenced by different environmental conditions, may be determining factors for the development of seed deterioration tolerance mechanisms and therefore for the quality of soybean seeds. Considered as a seed deterioration tolerance mechanism, the impermeability of the seed coat, characterized principally by seeds with greater lignin content, hinders water penetration in the seed coat. In a similar way to alterations in the germination process and in manifestation of vigor, in terms of the climate in the seed production phase, environmental conditions may also in some way affect the metabolism and chemical constitution of the seeds.

As we have already seen in this chapter, according to some authors, overproduction of lignin in RR soybean plants may be associated with the presence of water deficit and high temperatures during cropping, indicating that the environmental conditions found in the field during crop development may affect lignin production in the plant in an expressive way.

With this objective, we compared agronomic traits of the plant, physiological quality and seed health and lignin content in the seed coat of RR and conventional seeds produced in different time periods, summer and winter (Gris, 2009), with the determinations: plant height, height of insertion of the first pod and number of pods per plant, weight of 1000 seeds (Brasil, 1992), lignin content in the seed coat (Capeleti et al., 2005), incidence of mechanical damage (Marcos Filho et al., 1987), germination and dry matter of normal seedling from germination (Brasil, 1992), emergence speed index and germination speed index (Edmond & Drapala, 1958), final stand in the seed bed (counting at 24 days after seeding), accelerated aging at 42°C for 72h (Marcos Filho, 1999), electrical conductivity (Vieira, 1994), water immersion test of seeds and seed health, evaluating the infestation percentage (Machado, 2000) and intensity of the inoculums. The data of inoculum density were weighted by the McKinney formula (1923):

$$\text{II}(\%)=\frac{\Sigma(\text{F}\times n)\times 100}{(N\times M)}$$

In which: II = inoculum intensity, F = number of seeds with a determined score, n = score observed, N = total number of seeds evaluated and M = maximum score of the scale.

In Table 6, we present a summary of the mean results for the variables in which the contrasts (RR cultivar versus conventional cultivar) presented a significant difference, for both harvests, in which among all the characteristics evaluated, significant results for the contrasts evaluated were few.

For the electrical conductivity test, we observed a greater value for the conventional cultivar Jataí (76.54 $\mu S.cm^{-1}.g^{-1}$) when compared to the cultivar Silvânia RR (100.25 $\mu S.cm^{-1}.g^{-1}$). According to Vieira & Krzyzanowski (1999) for lots of high vigor soybean seeds, the standard conductivity values should be situated at most up to 70-80 $\mu S.cm^{-1}.g^{-1}$, however with a strong trend to present medium vigor. Nevertheless, in spite of the high value of electrical conductivity observed in seeds of the cultivar Silvânia RR, we did not observe differences between the two cultivars in the germination and vigor tests, which, according to José et al. (2004), may indicate that there are cultivars with greater efficiency in membrane reorganization, not resulting in damages, strictly speaking.

Variables	Means - Summer 2006/07 harvest				
Plant height (m)	Jataí	1.56 a	vs	Silvânia RR	1.41 b
Number Pods/plant	Jataí	110.00 a	vs	Silvânia RR	57.50 b
Germination (%)	BRS 133	95.50 a	vs	BRS 245 RR	87.25 b
Weight of 1000 seeds (g)	BRS 134	155.50 a	vs	BRS 247 RR	142.70 b
Emergence Speed Index	BRS 134	7.16 b	vs	BRS 247 RR	7.55 a
Lignin Seed Coat (%)	Celeste	0.20 b	vs	Baliza RR	0.26 a
Variables	**Means - Winter 2007 harvest**				
Electrical conductivity ($\mu S.cm^{-1}.g^{-1}$)	Jataí	76.54 b	vs	Silvânia RR	100.25 a

Capital letters followed by the same letter in the line do not differ among themselves by the Scheffe Test, at the 5% significance level.

Table 6. Mean values for some variables in which the contrasts between the conventional soybean cultivar and its genetically modified RR version presented significance, summer and winter harvest, Lavras, MG, Brazil.

Panobianco (1997) upon reporting variation in electrical conductivity of soybean seeds and the lignin content in their seed coat affirms that the genotype may alter the electrical conductivity for seeds with the same standard of physiological quality. Nevertheless, we did not observe significant differences between the cultivars Jataí and Silvânia RR in regard to lignin content in the seed coat, indicating that, in this case, it may not have been responsible for the variation in electrical conductivity observed. In the same way, it was not possible to relate the difference in the lignin contents in the seed coat, observed between the cultivars Celeste (0.20%) and Baliza RR (0.26%), and the results of physiological quality of the two, produced in the summer harvest, since they differed only for this characteristic. It is worth highlighting that in spite of the differences found for these two cultivars, it was not possible through the incidence of mechanical damage to detect any differences between the cultivars studied.

Upon observing the contrasts established between the RR and conventional cultivars, we can infer that the cultivars Jataí and Silvânia RR presented the greatest number of significant differences among the variable studied (Table 6), not only in relation to the physiological quality of the seeds, but also in regard to agronomic traits, such as plant height and number of pods per plant.

When we analyze the mean values of plant height and number of pods per plant, we verify once more that the conventional cultivar Jataí showed superiority to the cultivar Silvânia RR, such that for number of pods/plant, these values were up to 91.3% greater. Nevertheless, it is worth emphasizing that for these two cultivars in field conditions, we observed the greatest variations in regard to the phenological cycle, with greater uniformity in maturation and a shorter cycle, around 10 days, of the conventional cultivar Jataí in relation to the genetically modified RR cultivar. It is fitting to highlight that in spite of the RR cultivars tested in this study being essentially derivatives of the respective conventional cultivars, by means of backcrossings, the genotype of the recurrent genitor is not always recovered, due to number fewer recurrence cycles which may consequently result in variations between both materials. Nevertheless, for these cultivars, there is no information on the number of backcrossing cycles used.

When we evaluate the physiological quality of the seeds by means of the germination test in the summer harvest and of the germination speed index (IGV) in the winter harvest, we do not observe a relationship between the significant results for these variables, with the contrasts BRS 133 versus BRS 245 RR and Conquista versus Valiosa RR being differentiated respectively. For both results, the conventional cultivars showed superiority to the genetically modified RR cultivars, with the conventional cultivar BRS 133, with 95% of normal seedlings, overcoming the cultivar BRS 245 RR, with 87%, by approximately 9.5%, when they were produced in the summer harvest. Nevertheless, by the results in reference to the Emergence Speed Index, we observe a lower value for the genetically modified cultivar BRS 247 RR (7.55 days) in comparison with the conventional cultivar BRS 134 (7.16), which once more shows the inconsistency of data that justify a pleiotropic effect of the RR gene on lignin production.

It is worth emphasizing that in spite of the results found in this study, with exception of the variables Emergence Speed Index and lignin in the seed coat, the RR cultivars stood out in relation to the conventional cultivars; most of the significant contrasts, were seen to be isolated, in only one of the harvests or one of the tests in the midst of various comparisons among physiological quality of the seeds, therefore not indicating substantial differences of quality between the RR and conventional materials.

According to Menezes (2008) the physiological quality of soybean seeds is influenced by the maternal or extra-chromosome effect, just as is the cytoplasmatic inheritance, with the physical characteristics of the seed coat, of maternal origin, not being sole determinants of the physiological quality of the seeds. According to this author, the study of genetic control for seed quality indicates the effect of the general and specific combination capacity, which suggests the presence of additive and non-additive gene effects for physiological quality of soybean seeds. Therefore, the quality of seeds may not be attributed only to their seed coat and consequently to their lignin contents, but also to genes present in the nucleus.

When we analyze the results obtained in the seed health test (Figure 6), we observe that the cultivars BRS 133, BRS 245 RR, BRS 134 and BRS 247 RR presented the lowest percentages of infection and infection indexes (severity), when produced in the summer, indicating that the environmental conditions during the seed maturation period were responsible for seed health quality. In these cultivars a shorter phenological cycle and semi-early maturity was observed, which provided for the maturation period outside of the rainy period.

According to Delouche (1975), the alternating of dry and wet days during the maturation phase until harvest, which occurs with greater facility in the summer, can increase the incidence of diseases in a differentiated way at the end of the cycle of the seeds produced. Within this context, the seed becomes not only an easy target for the action of microorganisms, which considerably reduce its viability, but they also come to be efficient vehicles for dissemination of pathogens (Machado, 2000). This situation may be visualized principally for the cultivars Jataí and Silvânia RR, which remained for a greater period in the field, and presented the greatest percentages of infection, 39% and 38% (Figure 6A), and also the greatest indexes of infection by the pathogen Phomopsis, 35% and 26% (Figure 6B), respectively. It is worth emphasizing that when produced in winter conditions, under a controlled irrigation system, without rains in the seed maturation period, the presence of pathogens was not observed for any seeds.

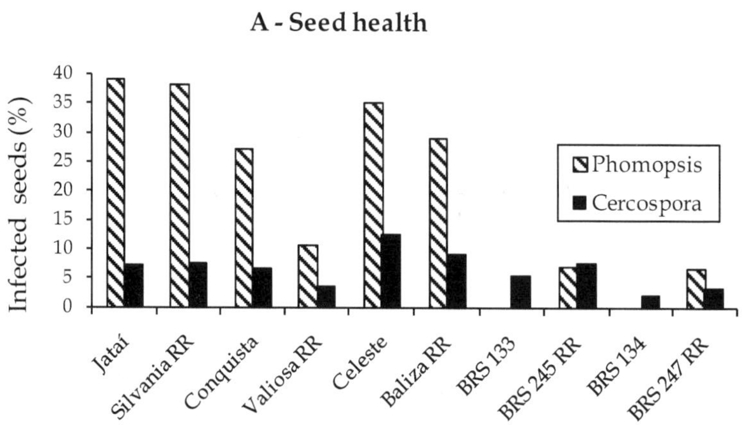

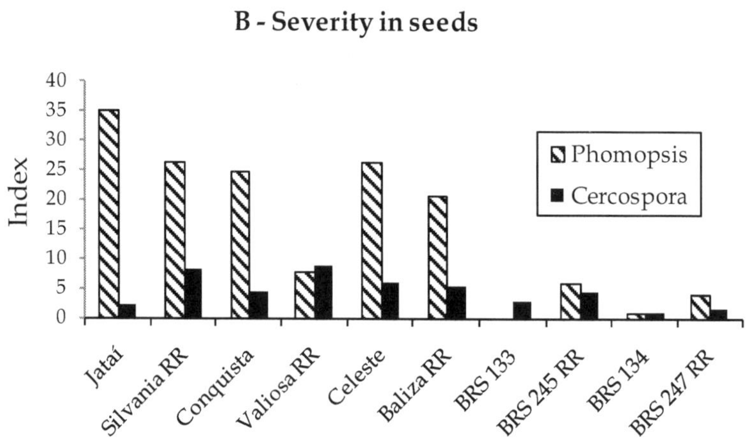

Fig. 6. Average values for infection percentage (A) and infection indexes - severity (B) in the seed health test of conventional soybeans and the genetically modified RR versions, summer harvest.

In relation to the RR versus conventional contrasts, we observe that in spite of the cultivars tested in this study having their origin in the same genotype by successive backcrossings, when observed in the field, we verified that some presented perceptible cycle variations, maintaining the cultivars Conquista and Celeste for more days in the field in relation to the cultivars Valiosa RR and Baliza RR, respectively; enough so that the first, subjected to rains and high temperatures, presented slightly greater values in the seed health and severity test. In this case, we cannot attribute the differences of RR versus conventional contrast, observed in Figure 6A and 6B, to the effect of the RR transgene, but rather to environmental conditions associated with difference of cycle.

In view of the above, in spite of some authors suggesting the pleiotropic effect of the transgene CP4 EPSPS on lignin overproduction in the plant, it was not possible for us to identify the pleiotropic effect in the cultivars studied in this and in the other studies described here, which indicates that the alterations of lignin content in the plant, observed by those authors under normal climatic conditions, are not due to the fact of the lignin molecule precursors being formed in the shikimic acid pathway. Thus, the sequence CP4 EPSPS, introduced in the genome of commercial soybean cultivars, responsible for the production of the protein CP4 enolpyruvylshikimate-3-phosphate-synthase (EPSPS), an enzyme that participates in the biosynthesis of aromatic amino acids in plants and microorganisms, seems not to be associated with lignin contents in the plant and in the soybean seed coat, and it seems that there are no substantial differences in regard to the agronomic traits and physiological quality of seeds between conventional and genetically modified RR cultivars.

4. References

Alpert, P.; Oliver, M.J. (2002). *Drying without dying*. In: BLACK, M.; PRITCHARD, H.W. (Ed.). Desiccation and survival in plants: drying without dying. Wallingford: CABI, 2002. p.4-43.

Alvarez, P. J. C. (1994). *Relação entre o conteúdo de lignina no tegumento da semente de soja e sua relação ao dano mecânico*. 43 p. Dissertação (Mestrado em Genética e Melhoramento) - Faculdade Estadual de Londrina, Londrina.

Alves, E. (2006). *Apostila do curso introdutório à microscopia eletrônica de varredura*. Lavras: UFLA, 43 p.

Antoniou, M.; Brack, P.; Carrasco, A.; Fagan, J.; Habib, M.; Kageyama, P.; Leifert, C.; Nodari, R.O.; Pengue, W. (2010). *Soja Transgênica:* Sustentável? Responsável? 2010. Available HTTP: http://www.gmwatch.org/files/GMsoy_Sust_Respons_FULL_POR_v2.pdf (11/20/2010).

Baciu-Miclaus, D. (1970). Contribuition to the study of hard seed and coat structure properties of soybean. *Proccedings of the International Seed Testing Association*, Vollebekk, v.35, n.2, p.599-617.

Baldoni, A. (2010). *Análises fisiológicas, ultraestruturais e expressão gênica de lignina em sementes de soja*. 64p. Thesis (Mestrado em Produção e Tecnologia de sementes) - Universidade Federal de Lavras, Lavras.

Becerril, J.M.; Duke, S.O.; Lydon, J. (1989). Glyphosate Effects on Shikimate Pathway Products in Leaves and Flowers of Velvetleaf. *Phytochemistry*, v.28, p.695-99.

Bervald, C.M.P. (2006). *Desempenho fisiológico e metabolismo de sementes de soja convencional e transgênica submetidas ao glifosato.* Pelotas. Dissertação (Mestrado). Universidade Federal de Pelotas.

Boatright, J. et al. (July 2004). Understanding in vivo benzenoid metabolism in petunia petal tissue. *Plant Physiology*, Bethesda, v. 135, n. 4, p. 1993-2011.

Boldt, A.F. (1984). *Relação entre os caracteres de qualidade da vagem e da semente de soja (Glycine max* (L.) Merrill). 70f. Dissertation (Mestrado em Fitotecnia)-Universidade Federal de Viçosa, Viçosa, MG.

Boudet, A. M. (Feb. 2000). Lignins and lignification: selected issues. *Plant Physiology and Biochemistry*, New Delhi, v. 38, n. 1/2, p. 81-96.

Boudet, A.M. (Aug. 2003). Lignins and lignocellulosics: a better control of synthesis for new and improved uses. *Trends in Plant Science*, Oxford, v. 12, n. 8, p. 576-581.

Braccini, A.deL. e; Albrecht, L.P.; Ávila, M.R.; Scapim, C.A.; Bio, F.E.I.; Pelegrinello, S.R. (Mar./abr. 2003). Qualidade fisiológica e sanitária das sementes de quinze cultivares de soja (Glycine max (L.) Merrill) colhidas na época normal e após o retardamento de colheita. *Acta Scientiarum Agronomy*, Maringá, v.25, n.2, p.449-457.

Brasil. (1992). Ministério da Agricultura. *Regras para análise de sementes.* Brasília: MA/SNDA/DNDV/CLV. 365p.

Calero, E.; West, S. H.; Hinson, K.(Nov./Dec. 1981). Water absorption of soybean seeds and associated causal factors. *Crop Science*, Madison, v. 21, n. 6, p. 926-933.

Capeleti, I.; Ferrarese, M.L.L.; Krzyzanowski, F.C.; Ferrarese Filho, O. (July 2005). A new procedure for quantification of lignin in soybean (*Glycine max* (L.) Merrill) seed coat and their relationship with the resistance to mechanical damage. *Seed Science and Technology*, Zurich, v.33, n.2, p.511-515.

Carbonell, S.A.M.; Krzyzanowski, F.C. (1995). The pendulum test for screening soybean genotypes for seeds resistance to mechanical damage. *Seed Science and Tecnology*, Zurich, v.23, n.2, p.331-339.

Carbonell, S.A.M.; Krzyzanowski, F.C.; Kaster, M. (Mar./abr. 1992). Avaliação do "teste de queda" para seleção de genótipos de soja com semente resistente ao dano mecânico. *Revista Brasileira de Sementes*, Brasília, v.14, n.2, p.215-219.

Carlson, J.B.; Lersten, N.R. (1987). *Reproductive morphology.* In: Wilcox, J.R. Soybeans: improvement, production and uses. Madison: ASA/CSSA/SSSA. p.95-134.

Carvalho, N. M.; Nakagawa, J. (2000). *Sementes:* ciência, tecnologia e produção. 4. ed. Jaboticabal: FUNEP, 588 p.

Caviness, C.E.; Simpsom, A.M.JR. (1974). Influence of variety and location on seed coat thickness of mature soybean seed. *Proceendings Association of seed Analanalysis*, Wellington, v. 64, p. 102-108.

Coghlan, A. (1999). *Splitting headache:* Monsanto's modified soya beans are cracking up in the heat. Saint Louis: Monsanto. Available at: <http://www.mindfully.org/GE/Monsanto-RR-Soy-Cracking.htm>. Accessed on: Mar. 10, 2009.

Cole, A.W.; Cerdeira, A.L. (1985). Southernpea response to glyphosate desiccation. *HortScience*, Alexandria, v.17, n.2, p.244-246, 1982.

Corner, E.J. (1951). The leguminous seeds. *Phytomorphology*, New Delhi, v.1, p.117-150.

Crocker, W. (1948). *Growth of plants.* New York: Reinohold. 459p.

Darley, C. P. et al. (Sept. 2001). The molecular basis of plant cell wall extension. *Plant Molecular Biology*, Dordrecht, v. 47, n. 1/2, p. 179-195.

Delouche, J.C.; Baskin, C.C. (1973). Accelerated aging techniques for predicting the relative storability of seed lots. *Seed Science and Technology*, Zurich, v.1, n.2, p.427-452.

Donelly, E.D. (1970). Persistance of hard seed in Vicia lines derived from interspecific hybridization. *Crop Science*, Madison, v.10, n.6, p.661-662.

Duke, S. H.; Kakefuda, G. (Mar. 1981). Role of the testa in preventing cellular rupture during imbibition of the legume seeds. *Plant Physiology*, Bethesda, v. 67, n. 2, p. 449-456.

Duke, S.O.; Hoagland, R.E. (1985). *Effects of glyphosate on metabolism of phenolic compounds*. Washington: CAB. Available at: <http://www.cababstractsplus.org/abstracts/Abstract.aspx?AcNo=19850776767>. Accessed on: Mar. 11, 2009.

Edmisten, K.L.; Wells, R.; Wilcut, J.W. (2000). *Investigation of the cavitation and large boll shed in roundup ready cotton*. Available at: <http://www.cottoninc.com/projectsummaries/2000ProjectSummaries/detail.asp?projectID=119>. Accessed on: Mar. 12, 2006.

Edmond, J.B.; Drapala, W.S. (June 1958). The effects of temperature, sand and acetone on germination of okra seed. *Proceedings of the American Society for Horticultural Science*, New York, v.71, p.428-434.

Egg-Mendonça, C.V. do C. (2001). *Caracterização química e enzimática de famílias de feijões obtidas do cruzamento das linhagens Amarelinho e CI - 107*. 48p. Dissertation (Mestrado em Agrobioquímica)-Universidade Federal de Lavras, Lavras.

Érzek, T. & Kiraly, Z. (1986). Phytoalexins and questions that remain unresolved forty-five years after their discovery. *Acta Phytopathologica et Entomolologica Hungarica*, 21: 5-14.

Esau, K. (1965). *Plant anatomy*. 2.ed. New York: J.Wiley. 767p.

Esau, K. (1976). *Anatomia das plantas com sementes*. São Paulo: E.Blucher. 293p.

Esau, K. (1977). *Anatomy of seeds plants*. New York: J.Wiley. 550p.

Fehr, W.R.; Caviness, C.E. (1977). *Stage of soybean development*. Ames: Iowa State University. 11p.

França Neto, J. de B.; Krzyzanowski, F.C. (2003). Estratégias do melhoramento para produção de sementes de soja no Brasil. In: SIMPÓSIO SOBRE ATUALIZAÇÃO EM GENÉTICA E MELHORAMENTO DE PLANTAS: Melhoramento de plantas e produção de sementes no Brasil, 7., 2003, Lavras. *Anais*... Lavras: UFLA, 2003. Available at: <http://www.nucleoestudo.ufla.br/gen/eventos/simposios/7simpo/resumos/20036.pdf>. Accessed on: Apr. 22, 2006.

Gertz Junior, J.M.; Vencill, W.K.; Hill, N.S. (1999). Tolerance of transgenic soybean (Glycine mar) to heat stress. In: brighton crop protection conference: weeds, 3., Brighton. *Proceedings*... Brighton: BCP, 1999. p.835-840.

Gilioli, J.L.; França Neto, J.B. (1981). Efeito da escarificação mecânica e do retardamento de colheita sobre a emergência de sementes de soja com tegumento impermeável. In: SEMINÁRIO NACIONAL DE PESQUISA DE SOJA, 2., Brasília. *Anais*... Londrina: EMBRAPA-CNPSo, 1982. v.1, p.601-609. (EMBRAPA-CNPSo. Documentos, 1).

Giurizatto, M.I.K.; Ouza, L.C.F.; Robaina, A.D.; Gonçalves, M.C. (Jul./ago. 2003). Efeito da época de colheita e da espessura do tegumento sobre a viabilidade e o vigor de sementes de soja. *Ciência e Agrotecnologia*, Lavras, v.27, n.4, p.771-779.

Gris, C. F. (2009). Qualidade fisiológica de sementes de soja convencional e RR associada ao conteúdo de lignina. 134p. Thesis (Doutorado em Produção e Tecnologia de sementes) - Universidade Federal de Lavras, Lavras.

Gris, C. F.; Von Pinho, E.V. de R.; Andrade, T.; Baldoni, A.; Carvalho, M.L. de M. (Mar./Abr. 2010). Qualidade fisiológica e teor de lignina no tegumento de sementes de soja convencional e transgênica RR submetidas a diferentes épocas de colheita. *Ciência e Agrotecnologia*, Lavras, v. 34, n. 2, p. 374-381.

Hartwig, E.E.; Potts, H.C. (May/June 1987). Development and evaluation of impermeable seed coats for preserving soybean seed quality. *Crop Science*, Madison, v.27, n.3, p.506-508.

Hrazdina, G. & Jensen, R. A. (1992). Spatial organization of enzymes in plant metabolic pathways. *Annual Review of Plant Physiology and Plant Molecular Biology*, 43: 241-267.

International Service For The Acquisition Of Agri-Biotech Applications - ISAAA. (2009). *Brief 38-2009:* Global Status of Commercialized Biotech/GM Crops: 2009. Available at HTTP:
http://www.isaaa.org/resources/publications/briefs/38/executivesummary/default.html

José, S.C.B.R.; Pinho, É.V.R. von; Pinho, R.G. von; Silveira, C.M. da. (Set./out. 2004). Tolerância de sementes de linhagens de milho a alta temperatura de secagem. *Ciência e Agrotecnologia*, Lavras, v.28, n.5, p.1107-1114.

Jung, H.G., Allen, M.S. (1995). Characteristics of plants cell walls affecting intake and digestibility of forages by ruminants. *Journal Animal Science*, Champaign, v.73, n.9, p.2774-2790.

Kruse, N.D.; Trezzi, M.M.; Vidal, R.R. (2000). Herbicidas inibidores da EPSPS: revisão de literatura. *Revista Brasileira de Herbicidas*, Brasília, v.1, n.2, p.139- 146.

Kuiper, H.A.; Kleter, G.A.; Noteborn, H.P.J.M.; Kok, E.J. (Dec. 2001). Assessment of the food safety issues related to genetically modified foods. *The Plant Journal*, Oxford, v.27, n.6, p.503-528.

Lacerda, A.L. de S.; Matallo, M.B. (2008). Verificação do ácido chiquímico em soja geneticamente modificada. In: Reunião Anual da SOCIEDADE BRASILEIRA PARA O PRPGRESSO DA CIÊNCIA, 60, Campinas. *Anais*... Campinas: UNICAMP, 2008. Available at: http://www.sbpcnet.org.br/livro/60ra/resumos/resumos/R2708-1.html>. Accessed on: Mar. 11, 2009.

Lewis, N.G., Yamamoto, E. (1990). Lignin: occurrence, biogenesis and biodegradation. *Ann. Rev. Plant Physiol. Plant Mol. Biol.*, Palo Alto, v.41, p.455-496.

Lin, S.S.; Severo, J.L. (1982). Efeito e atraso da colheita sobre a qualidade da semente e rendimento de soja (Glycine max (L.) Merrill). *Agronomia Sulriograndense*, Porto Alegre, v.18, n.1, p.37-46.

Machado, J.C.(2000). *Tratamento de sementes no controle de doenças*. Lavras: LAPS/UFLA/FAEPE. 138p.

Marcos Filho, J.(1999). Teste de envelhecimento acelerado. In: KRZYZANOWSKI, F.C.; VIEIRA, R.D.; FRANÇA NETO, J.B. (Ed.). *Vigor de sementes:* conceitos e testes. Londrina: ABRATES, 1999. cap.3, p.1-24.

Marcos Filho, J.; Cicero, S.M.; Silva, W.R. da. (1987). *Avaliação da qualidade da semente.* Piracicaba: FEALQ. 230p.

Martins, L.A.M. (1989). *Avaliação da área habilidade genética do caráter semente dura de linhagens melhoradas de soja (Glycine max* (L.) Merrill). 119f. Dissertation (Mestrado em Fitotecnia)-Universidade Federal de Viçosa, Viçosa, MG.

Mcdougall, G. J., Morrison, I. M., Stewart, D., Hillman J. R. (Feb. 1996). Plant cell walls dietary fibre: range, structure, processing and function. *Journal Science and Food Agriculture,* Londres, v. 70, n. 2, p. 133-150.

Mckinney, R.H. (1923). Influence of soil temperature and moisture on infection of wheat seedlings by *Helminthosporium sativum. Journal of Agricultural Research,* Washington, v.26, n.3, p.195-218.

Menezes, M. de. (2008). *Aspectos genéticos associados à qualidade fisiológica de sementes de soja.* 112p. Thesis (Doutorado em Fitotecnia)-Universidade Federal de Lavras, Lavras.

Menezes, M.; Von Pinho, E.V. de R.; Roveri José, S.C.B.; Baldoni, A.; Mendes, F.F. (December 2009). Aspectos químicos e estruturais da qualidade fisiológica de sementes de soja. *Pesquisa Agropecuária Brasileira,* Brasília, v. 44, n. 12, p. 1716-1723.

Nodari, R.O.; Destro, D. (2006). *Relatório sobre a situação de lavouras de soja da região de Palmeira das Missões, RS, safra 2001/2002, cultivadas com cultivares convencionais e com cultivares transgênicas:* notícias no AgirAzul. Available at: <http://www.agirazul.com.br/123/noticias/000000a3.htm>. Accessed on: Apr. 22, 2006.

Panobianco, M. (1997). *Variação na condutividade elétrica de sementes de diferentes genótipos de soja e relação com o conteúdo de lignina no tegumento.* 59p. Dissertation (Mestrado em Agronomia)-Universidade Estadual Paulista, Jaboticabal.

Panobianco, M.; Vieira, R. D.; Krzyzanowski, F. C.; França Neto, J.B. (1999). Electrical conductivity of soybean seed and correlation with seed coat lignin content. *Seed Science Technology,* Zurich, v.27. n.3, p.945-949.

Peske, S.T.; Pereira, L.A.G. (Jun. 1983). Tegumento da semente de soja. *Tecnologia de Sementes,* Pelotas, v.6, n.1/2, p.23-34.

Resende, M. L. V.; Salgado, S. M.; Chaves, Z. M. (Mar./abr. 2003). Espécies ativas de oxigênio na resposta de defesa de plantas a patógenos. *Fitopatologia Brasileira,* Brasília, v. 28, n. 2, p. 123-130.

Rocha, V.S. (1982). *Avaliação da qualidade fisiológica de sementes de genótipos de soja (Glycine max (L.) Merrill), em três épocas de colheita.* 109f. Dissertation (Mestrado em Fitotecnia)-Universidade Federal de Viçosa, Viçosa, MG.

Salisbury, F. K. & Ross, C. W. (1992). *Plant Physiology.* Wadsworth, Belmont.

Salisbury, F. K. & Ross, C. W. (1992). *Plant Physiology.* Wadsworth, Belmont.

Sanino, F.; Filazzola, M. T.; Violante, A. (1999). Fate of herbicides influenced by biotic and abiotic interactions. *Chemosphere,* [S.l.], v. 39, n. 2, p. 333-341.

Silva, D.J. (1981). *Análise de alimentos:* métodos químicos e biológicos. Viçosa, MG: UFV. 166p.

Silva, M. A. D. (2003). *Morfologia da testa e potencial fisiológico de sementes de soja.* 84p. Thesis (Doutorado em Produção e Tecnologia de sementes) - Universidade Federal de Lavras, Lavras.

Simões, C.M. O.; Spitzer, V. (2004). *Óleos voláteis*. In: Simões, C.M. O. et al. (Org.). Farmacognosia da planta ao medicamento. 5. ed. Porto Alegre: UFRGS; Florianópolis: UFSC, cap. 18, p. 467-496.

Strautman, B. (2007). Manganese affected by glyphosate. *Western Producer*. http://www.gefreebc.org/gefree_tmpl.php?content=manganese_glyphosate

Swanson, B. G.; Hughes, J. S.; Rasmussen, H. (1985). Seed microstructure: review of water imbibition in legumes. *Food Microstructure*, Chicago, v. 4, p. 115-124.

Taiz, L. & Zeiger, E. (1998). *Plant Physiology*. Sinauer Associates, Sunderland. 792p.

Tavares, D.Q.; Miranda, M.A.C.; Umino, C.Y.; Dias, G.M. (Jan./mar. 1987). Características estruturais do tegumento de sementes permeáveis e impermeáveis de linhagens de soja, *Glycine max* (L.) Merrill. *Revista Brasileira de Botânica*, São Paulo, v.10, n.1, p.147-153.

Vieira, R.D.; Krzyzanowski, F.C. (1999). *Teste de condutividade elétrica*. In: KRZYANOWSKI, F.C.; VIEIRA, R.D.; FRANÇA NETO, J.B. (Ed.). Vigor de sementes: conceitos e testes. Londrina: ABRATES. cap.4, p.1-26.

Woodstock, L.W. (Feb. 1988). Seed imbibition: a critical period for successful germination. *Journal Seed Technology*, Springfield, v.12, n.1, p.1-15.

Xu, Z. et al. (Oct. 2009). Comparative genome analysis of lignin biosynthesis gene families across the plant kingdom. *BMC Bioinformatics*, New York, v. 10, n. 11, p. 1-15. Supplement.

Zobiole, L.H.S., Oliveira, R.S., Visentainer, J.V., Kremer, R.J., Bellaloui, N., Yamada, T. (2010). Glyphosate affects seed composition in glyphosate-resistant soybean. *J. Agric. Food Chem*. 58, 4517–4522.

16

A Comparative Study of the Chelating Effect Between Textured Soya Aqueous Extract and EDTA on Fe^{3+}, Pb^{2+}, Hg^{2+}, Cd^{2+} and Ni^{2+} Ions

Guajardo Jesús, Morales Elpidio, López Francisco, Quintero Cristina, Compean Martha, Noriega María-Eugenia, González Jesús and Ruiz Facundo
Universidad Autónoma de San Luis Potosí
Mexico

1. Introduction

Metal pollution of soils, water, foods, and the environment is a grave problem. Various in-situ and ex-situ remediation techniques have been employed, e.g., solidification, stabilization, flotation, soil ashing, electroremediation, bioleaching, and phytoremediation (Mulligan, 2001). One remediation technique is ex-situ soil washing using chelating agents. The soil is removed from the site, treated in a closed reactor with the chelating agent, and returned to the site after separation of the extraction solution that now contains the extracted heavy metals (Peters & Hazard,1999). The problem is that the used chelating agent is not a natural compound. For that reason we propose the use of the textured soya extract, which is environmentally friendly, as a natural chelating agent.

EDTA (ethylenediamine tetraacetic acid) and its salts are substituted diamines. HEDTA (hydroxyethyl ethylenediamine triacetic acid) and its trisodium salt are substituted amines. These ingredients function as chelating agents in cosmetic formulations. The typical concentration of use of EDTA is less than 2%, with the other salts in current use at even lower concentrations. The lowest dose reported to cause a toxic effect in animals was 750 mg/kg/day.

These chelating agents are cytotoxic and weakly genotoxic, but not carcinogenic. Oral exposures to EDTA produced adverse reproductive and developmental effects in animals. Clinical tests reported no absorption of an EDTA salt through the skin. These ingredients are likely, however, to affect the passage of other chemicals into the skin because they will chelate calcium. Exposure to EDTA in most cosmetic formulations, therefore, would produce systemic exposure levels well below those seen to be toxic in oral dosing studies. Exposure to EDTA in cosmetic formulations that may be inhaled, however, was a concern. An exposure assessment done using conservative assumptions predicted that the maximum EDTA dose via inhalation of an aerosolized cosmetic formulation is below that shown to produce reproductive or developmental toxicity. Because of the potential to increase the penetration of other chemicals, formulators should continue to be aware of this when combining these ingredients with ingredients that previously have been determined to be safe, primarily because they were not significantly absorbed. Based on the available data, the Cosmetic Ingredient Review Expert Panel found that these ingredients are safe as used in cosmetic formulations.

Ethylenediaminetetraacetic acid (EDTA) is a very effective chelating agent but has the disadvantage that is quite persistent in the environment owing to its low biodegradability. For that reason different chelating agents were investigated, such as [S,S,]-ethylenediaminedisuccinic acid, iminodisuccinic acid, methylglycine diacetic acid, etc. but the problem is the dependence of the pH on the extraction efficiency. (Tandy et al., 2004)

Major industrial processes involve the sequestration of metal ions in an aqueous solution. In the textile industry, this prevents metal ion impurities from modifying colors of dyed products. In the pulp and paper industry, EDTA inhibits the ability of metal ions, especially Mn^{2+}, to catalyze disproportionate amounts of hydrogen peroxide, which is used in "chlorine-free bleaching." Similarly, EDTA is added to some foods as a preservative or stabilizer to prevent a catalytic oxidative discoloration which is catalyzed by metal ions.

Oral exposures have been noted to cause reproductive and developmental effects (Elliot & Brown, 1989). The same study by Lanigan also found that both dermal exposure to EDTA in most cosmetic formulations and inhalation exposure to EDTA in aerosolized cosmetic formulations would produce systemic effects below those seen to be toxic in oral dosing studies (Lanigan & Yamarik, 2002).

A crucial factor to be considered in comparing studies on chelating agent is the pH of the extraction solution. While extraction was investigated at various pH values in some studies (Elliot & Brown, 1989 ,; Pichtel, 1998; Pichtel, 1997; Kim, 2003;. Ghestem, 1998), some only stated the pH of the solution (Reed, 1996; Cline, 1995; Van Benschoten, 1997), while others did not consider pH at all (Pichtel, 2001;). In general, the lower the pH of the chelating agent solution, the greater is the extraction efficiency of the toxic metals.

The history and chemistry of the industrial use of natural products and their derivatives have a rich technological tradition. Many modern products, such as plastics, fuels, chemical intermediates and fibers, find their origins in natural products derived from plants and animals. Given the recent social emphasis on the environment and resource renewability, utilizing natural materials as potential resources for industrial products receives a ready welcome. Among the most versatile of raw materials is the soybean. (Liu, 1997)

Together, the oil and protein contents of dry soybeans account for about 60% of the weight; protein being 40% and oil 20%. The remainder consists of 35% carbohydrate and about 5% ash. Most soy protein is a relatively heat-stable storage protein. This heat stability enables the manufacture of soy food products requiring high temperature cooking, such as tofu, soy milk and textured vegetable protein (soy flour).

This article focuses on the application of natural "green" textured soya extract as a substitute for EDTA in its role as a metals-sequestering agent in foods.

2. Antecedents

2.1 What is a chelating agent?

The word chelation is derived from Greek, meaning "claw." The ligands lie around the central atom like the claws of a lobster.

The IUPAC definition of chelation is the formation or presence of two or more separate bindings between a polydentate (multiple bonded) ligand and a single central atom. Usually these ligands are organic compounds and are called chelants, chelators, chelating agents, or sequestering agents. (IUPAC)

The ligand forms a chelate complex with the substrate. Chelate complexes are contrasted with coordination complexes composed of monodentate ligands, which form only one bond with the central atom. (Morgan & Drew, 1920)

The terms bidentate (or didentate), tridentate, tetradentate,... multidentate are used to indicate the number of potential binding sites of the ligand, at least two of which must be used by the ligand in forming a "chelate". For example, the bidentate ethylenediamine forms a chelate with **Cu (I)** in which both nitrogen atoms of ethylenediamine are bonded to copper. (The use of the term is often restricted to metallic central atoms). (Kramer, Cotter-Howells, Charnock, Baker & Smith. 1996)

Chelants, according to ASTM-A-380, are "chemicals that form soluble, complex molecules with certain metal ions, inactivating the ions so that they cannot normally react with other elements or ions to produce precipitates or scale".

2.2 The chelate effect

The increased stability of complexes containing chelating ligands over those containing comparable monodentate ligands can be envisaged as having the following physical basis. Suppose we have a metal ion in solution, and we attach to it a monodentate ligand, followed by a second monodentate ligand, figure 1. These two processes are completely independent of each other. But suppose we have a metal ion and we attach to it one end of a chelating ligand (it is reasonable to assume that when we put a chelate ligand onto a metal, it happens in a stepwise fashion, i.e. one end attaches first and then the other end). The point is that the attachment of the second end of the chelate is now no longer an independent process: once one end is attached, the other end, rather than floating around freely in solution, is anchored by the linking group in reasonably close proximity to the metal ion, and is therefore more likely to join onto it than a comparable monodentate ligand would be.

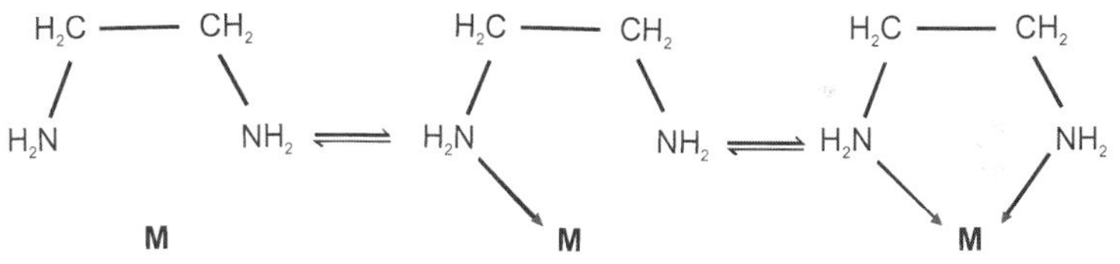

Fig. 1. Complexes formation.

The figure 2 shows the EDTA ligand binding to a central copper ion.

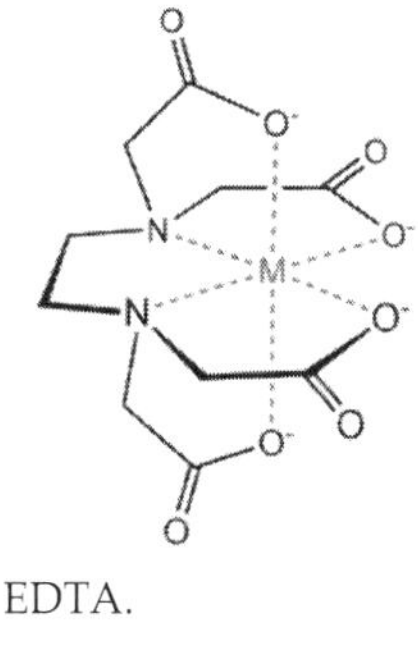

Fig. 2. Copper ion complexes with EDTA.

Amino acids are classified into different ways base don polarity, structure, nutricional requirement, metabolic fate, etc.

Generally used classification is based on polarity. Based on polarity amino acids are classified into four groups.

- *Non-polar amino acids.- They have equal number of amino and carboxyl groups and are neutral. These amino acids are hydrophobic and have no charge on the 'R' group. The amino acids in this group are alanine, valine, leucine, isoleucine, phenyl alanine, glycine, tryptophan, methionine and proline.*
- *Polar amino acids with no charge.- These amino acids do not have any charge on the 'R' group. These amino acids participate in hydrogen bonding of protein structure. The amino acids in this group are - serine, threonine, tyrosine, cysteine, glutamine and aspargine.*
- *Polar amino acids with positive charge.- Polar amino acids with positive charge have more amino groups as compared to carboxyl groups making it basic. The amino acids, which have positive charge on the 'R' group, are placed in this category. They are lysine, arginine and histidine.*
- *Polar amino acids with negative charge.- Polar amino acids with negative charge have more carboxyl groups than amino groups making them acidic. The amino acids, which have negative charge on the 'R' group are placed in this category. They are called as dicarboxylic mono-amino acids. They are aspartic acid and glutamic acid.*

Chelates of glycine with cations such as iron, zinc and copper have been fully studied. The chelates usually contain two moles of ligand (glycine) and one mol of metal as demonstrated in the figure 3.

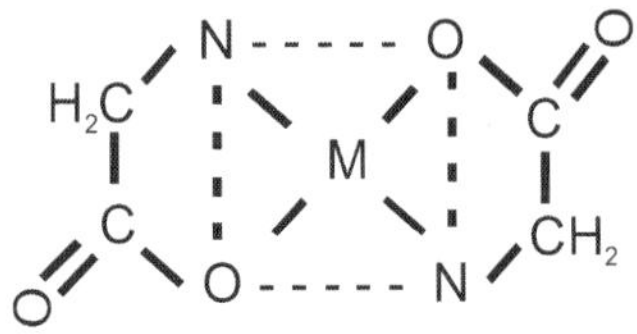

Fig. 3. Chelate of glycine with some metal M.

Consider the two equilibriums, in an aqueous solution, between the copper (II) ion, Cu^{2+} and ethylenediamine (en) on the one hand and methylamine, $MeNH_2$ on the other.

$$Cu^{2+} + en \rightleftharpoons [Cu(en)]^{2+} \quad (1)$$

$$Cu^{2+} + 2\ MeNH_2 \rightleftharpoons [Cu(MeNH_2)_2]^{2+} \quad (2)$$

In (1) the bidenate ligand ethylene diamine forms a chelate complex with the copper ion. Chelation results in the formation of a five-member ring. In (2) the bidentate ligand is replaced by two monodentate methylamine ligands of approximately the same donor power, meaning that the enthalpy of formation of Cu—N bonds is approximately the same in the two reactions. Under conditions of equal copper concentrations and when the concentration of methylamine is twice the concentration of ethylenediamine, the concentration of the complex (1) will be greater than the concentration of the complex (2). The effect increases with the number of chelate rings so the concentration of the EDTA complex, which has six chelate rings, is much higher than a corresponding complex with two monodentate nitrogen donor ligands and four monodentate carboxylate ligands. Thus, the phenomena of the chelate effect are a firmly established empirical fact.

The thermodynamic approach to explaining the chelate effect considers the equilibrium constant for the reaction: the larger the equilibrium constant, the higher the concentration of the complex.

The formation of a chelant compound is an equilibrium reaction as shown in the reaction (3)

$$aM^{n+} \quad + \quad b\,L \quad <==> \quad c\,ML \tag{3}$$

Metal Ligand Metal-chelate

The reaction rates of the forward and reverse reactions are generally not zero but, being equal; there are no net changes in any of the reactant or product concentrations. Since forward and backward rates are equal:

$$k_1\,[M^{n+}]^a\,[L]^b = k_2\,[ML]^c \tag{4}$$

and the ratio of the rate constants is also a constant, now known as an equilibrium constant.

$$K = \frac{[ML]^c}{[M^{n+}]^a[L]^b} \tag{5}$$

The concentration of ligand does not change during the reaction. For that reason the equilibrium constant can be expressed only in function of metal ion and metal-complex, as showing in the equation 6.

$$K = \frac{[ML]^c}{[M^{n+}]^a} \tag{6}$$

2.3 Common chelating agents

There are many chelating agents used in the industry as Na, Ca-ethylenediaminetetraacetic (EDTA), diethylenetriaminepentaacetic acid (DTPA), nitriloacetic acid, ethylene glycol-bis8aminoethyl)tetraacetic acid (EGTA), D,L-mercaptosuccinic acid (MSA), meso-2-3-dimercaptopropanesuccinic acid (DMSA), D,L-2,3-dimercaptopropane-1-sulfonic acid (DMPS), penicillamine (PA), N-acetylpenicillamine (NAPA), vitamins as: thiamine (B1), pyridoxine (B6), cobalim (B12) and ascorbic acid, and many more. The most common is EDTA.

2.4 Naturals chelating agents

Virtually all biochemicals exhibit the ability to dissolve certain metals cations. Thus, proteins, polysaccharides, and polynucleic acids are excellent polydentate ligands for many metal ions. In addition to these adventitious chelators, several biomolecules are produced to specifically bind certain metals. Histidine, malate and phytochelatin are typical chelators used by plants. (U Kramer, 1996; Jurandir, 2006 & Suk-Bomg Há, 1999)

Virtually all metalloenzymes feature metals that are chelated, usually to peptides or cofactors and prosthetic groups (Lippard & Berg, 1994). Such chelating agents include the porphyrin in hemoglobin and chlorophyll. Many microbial species produce water-soluble pigments that serve as chelating agents, termed siderophо. For example, species of *Pseudomonas* are known to secrete pycocyanin and pyoverdin that bind iron. Enterobactin, produced by E. coli, is the strongest chelating agent known.

In earth science, chemical weathering is attributed to organic chelating agents, *e.g.* peptides and sugars that extract metal ions from minerals and rocks. (Michael) Most metal complexes in the environment and in nature are bound in some form of chelate ring, *e.g.* with a humic acid or a protein. Thus, metal chelates are relevant to the mobilization of metals in the soil, the uptake and the accumulation of metals into plants and micro-organisms. Selective chelating of heavy metals is relevant to bioremediation *e.g.* removal of ^{137}Cs from radioactive waste. (Prasad, 2001)

2.5 Applications

Chelators are used in chemical analysis as water softeners, and are ingredients in many commercial products such as shampoos and food preservatives. Citric acid is used to soften water in soaps and laundry detergents. A common synthetic chelator is EDTA. Phosphona are also well known chelating agents. Chelators are used in water treatment programs and specifically in steam engineering, e.g., boiler water treatment system.

Chelation therapy is the use of chelating agents to detoxify poisonous metal agents such as mercury, arsenic, and lead by converting them to a chemically inert form that can be excreted without further interaction with the body, and was approved by the U.S. Food and Drug Administration in 1991. In alternative medicine, chelation is used as a treatment for autism, though this practice is controversial due to an absence of scientific plausibility, lack of FDA approval, and its potentially deadly side-effects. (Doja & Can, 2006).

Though they can be beneficial in cases of heavy metal poisoning, chelating agents can also be dangerous. The U.S. CDC reports that use of disodium EDTA instead of calcium EDTA has resulted in fatalities due to hypocalcemia. (U. S. Center for Disease Control)

Homogeneous catalysts are often chelated complexes. A typical example is the ruthenium (II) chloride chelated with BINAP (a bidentate phosphine) used in e.g. Noyori asymmetric hydrogenation and asymmetric isomerization. The latter has the practical use of manufacture of synthetic mentol.

Products such as Evapo-Rust are chelating agents sold for the removal of rust from iron and steel.

2.6 Chemical composition of the soybean seed

Together, oil and protein content account for about 60% of dry soybeans by weight; protein at 40% and oil at 20%. The remainder consists of 35% carbohydrate and about 5% ash. Soybean cultivars comprise approximately 8% seed coat or hull, 90% cotyledons and 2% hypocotyl axis or germ.

Most soy protein is a relatively heat-stable storage protein. This heat stability enables soy food products requiring high temperature cooking, such as tofu, soy milk and textured vegetable protein (soy flour) to be made.

The principal soluble carbohydrates of mature soybeans are the disaccharide sucrose (range 2.5–8.2%), the trisaccharide raffinose (0.1–1.0%) composed of one sucrose molecule connected to one molecule of galactose, and the tetrasaccharide stachyose (1.4 to 4.1%) composed of one sucrose connected to two molecules of galactose. While the oligosaccharides raffinose and stachyose protect the viability of the soy bean seed from desiccation (see above section on physical characteristics) they are not digestible sugars and therefore contribute to flatulence and abdominal discomfort in humans and other monogastric animals; compare to the disaccharide trehalose.

Since soluble soy carbohydrates are found in the whey and are broken down during fermentation, soy concentrate, soy protein isolates, tofu, soy sauce, and sprouted soy beans are without flatus activity. On the other hand, there may be some beneficial effects to ingesting oligosaccharides such as raffinose and stachyose, namely, encouraging indigenous bifidobacteria in the colon against putrefactive bacteria.

The insoluble carbohydrates in soybeans consist of the complex polysaccharides cellulose, hemicellulose and pectin. The majority of soybean carbohydrates can be classed as belonging to dietary fiber.

The following Table 1 shows the composition of mature, raw soybean seeds.

Energy	1,866 kJ (446 kcal)
Carbohydrates	30.16 g
Sugars	7.33 g
Dietary fiber	9.3 g
Fat	19.94 g
Saturated	2.884 g
monounsaturated	4.404 g
polyunsaturated	11.255 g
Protein	36.49 g
Tryptophan	0.591 g
Threonine	1.766 g
Isoleucine	1.971 g
Leucie	3.309 g
Lysine	2.706 g
Methionine	0.547 g
Phenylalanine	2.122 g
Tyrosine	1.539 g
Valine	2.029 g
Arginine	3.153 g
Histidine	1.097 g
Alanine	1.915 g
Aspartic acid	5.112 g
Glutamic acid	7.874 g
Glycine	1.880 g
Proline	2.379 g
Serine	2.357 g
Water	8.54 g
Vitamin A equiv	1 μg
Vitamin B_6	0.377 mg
Vitamin B_{12}	0 μg
Vitamin C	6.0 mg
Vitamin K	47 μg
Calcium	277 mg
Iron	15.70 mg
Magnesium	280 mg
Phosphorus	704 mg
Potassium	1797 mg
Sodium	2 mg
Zinc	4.89 mg

Source: USDA Nutrient database.

Table 1. Composition of soybean, mature, rawNutritional value per 100 g (3.5 oz)

2.7 How soybeans are used

When the farmer sells soybeans to a grain dealer, the beans may then go to a number of ultimate destinations. When processed, a 60-pound bushel will yield about 11 pounds of crude soybean oil and 47 pounds of soybean meal. Soybeans are about 18% oil and 38% protein. Because soybeans are high in protein, they are a major ingredient in livestock feed. Most soybeans are processed for their oil and protein for the animal feed industry. A smaller percentage is processed for human consumption and made into products including soy milk, soy flour, soy protein, tofu and many retail food products. Soybeans are also used in many non-food (industrial) products.

Fuel for diesel engines can be produced from soybean oil with simple processing. Soy biodiesel is cleaner burning than petroleum-based diesel oil. Its use reduces particulate emissions, and it is non-toxic, renewable and environmentally friendly. Soy crayons made by the Dixon Ticonderoga Company replace the petroleum used in regular crayons with soy oil making them non-toxic and safer for children. Candles made with soybean oil burn longer but with less smoke and soot.

Soy oil produces an environmentally friendly solvent that safely and rapidly removes oil from creeks, streams and shorelines without harming people, animals, and the environment. Soy is an ingredient in many industrial lubricants, solvents, cleaners, and paints. Soy ink is superior to petroleum-based inks because soy ink is not toxic, is renewable and also environmentally friendly. Furthermore, it cleans up easily. Soy-based lubricants are as good as petroleum-based lubricants, but can withstand higher heat. More importantly, they are non-toxic, renewable and environmentally friendly. Soy-based hydraulic fluid and rail flange lubricants are among the more recent products developed with check-off funds.

Soy-based foams are currently being developed for use in coolers, refrigerators, automotive interiors and even footwear. Beginning in October 2007, Ford Mustangs rolled off the production line with soy flexible foam in the seats. (2009 Annual Report of the North Carolina Soybean Producers Association)

2.8 Textured soya

Textured or textures vegetable protein (TVP), also known as textured soya protein (TSP), soy meat, or soya meat is a meat analogue or nutritious meat extender made from defatted soy flour, a by-product of extracting soybean oil. It is quick to cook, with a protein content equal to that of the meat, and contains no fat. (Riaz, 2006)

TVP is made from a mixture of proteins extracted primarily from soybeans, but also cotton seeds, wheat, and oats. It is extruded into various shapes (chunks, flakes, nuggets, grains, and strips) and sizes, exiting the nozzle while still hot and expanding as it does so. (Foote, 1996)

TVP can be made from soy flour or concentrate, containing 50% and 70% soy protein respectively, and is relatively flavorless. Both require rehydration before use, sometimes with flavoring added in the same step. TVP is extruded, causing a change in the structure of the soy protein which results in a fibrous spongy matrix that is similar in texture to meat. In its dehydrated form TVP has a shelf life of longer than a year, but will spoil within several days after being hydrated. In its flaked form TVP can be used similarly to ground meat. (Hoogenkamp & Wallingford, Oxon, 2005; Endres, 2001)

3. Materials and synthesis

3.1 Reagents

$Fe(NO_3)_3$, $NiCl_2$, $CuSO_4$, $HgCl_2$, $CdSO_4$, $Pb(NO_3)_2$ and EDTA (Etilenediaminetetraacetic acid), purchased from Sigma-Aldrich, were used without any further purification. Textured soya was purchased. Mili-Q water (18.2 Ω) was used throughout the experiment.

3.2 Characterization

The amount of metallic ions present in the solutions was determined by using a conductivity meter.

3.3 Determination of the chelating agent in textured soya extract

The Biuret test is a chemical test used for detecting the presence of peptide bonds. In the presence of peptides, a copper (II) ion forms a violet-colored complex in an alkaline solution. Several variants on the test have been develop

In order to find the chelating component in the textured soya extract, first a textured soya extract was prepared by heating to boiling point 1000 ml of deionizer water with 30 grams of textured soy for 20 minutes. 10 ml of textured soya extract is treated with an equal volume of 1% strong base (sodium or potassium hydroxide most often) followed by a few drops of aqueous copper (II) sulfate. The solution turns violet, for that reason we can affirm that proteins are present in the textured soya extract and these are the chelating agents.

A Fehling test was made too, and the Fehling reaction was positive, in which the green color characteristic of mono-saccharides was obtained. For that reason we can affirm that mono-saccharides present in the textured soya extract are present but the amount is not significant (< 3%).

3.4 Experimental method

First, it was necessary to find a concentration of textured soya equivalent to an EDTA solution $5x10^{-4}$ (the maximum concentration permitted in foods). For that reason we prepared six solutions of $CuSO_4$ with concentrations 0.05M, 0.1M, 0.15M, 0.2M, 0.25M and 0.3M. The conductivity of each one was then measured. Next, we mixed 1 ml of each $CuSO_4$ solution with 10 ml of an EDTA solution $5x10^{-4}$ M and we measured the conductivity of each one. Several textured soya aqueous solutions were prepared by dissolving 1 grams, 2 grams, 3 grams, and 5 grams of textured soya, each one in 100 ml of water, and heating them to boiling point for 10 minutes. The fiber was then separated by filtration. Afterwards, we mixed 1 ml of each solution of $CuSO_4$ with 10 ml of each prepared textured soya extract solution and measured the conductivity of each sample.

With the aim of studying the comparative chelating effect between textured soya extract and EDTA on some metals, we prepared five different aqueous solutions of each metal ion, Fe^{3+}, Pb^{2+}, Hg^{2+}, Cd^{2+}, Ni^{2+} and Cu^{2+} with different concentrations, with 0.01, 0.03, 0.05, 0.07 and 0.1 grams of each salt dissolved in 10 ml of deionizer water. Then we measured their conductivity. An EDTA aqueous solution of $5x10^{-4}$M was prepared. A solution of 15 grams of textured soya in 500 ml of deionizer water was heated to boiling point for 10 minutes. Afterwards, we measured the conductivity and ppm (parts of million) of each ion solution, chelant solution of textured soya extract and EDTA. In order to determine the chelating capacity of EDTA and textured soya extract, we added 0.01, 0.03, 0.05, 0.07 and 0.1 grams of each salt in 10 ml of EDTA solution, and then in the same form in 10 ml of textured soya extract solution and we measured the conductivity of each one, using a conductivity meter. All measurements were made at room temperature and at average room pressure, and pH 7.

4. Results and discussion

Table 2 shows the conductivity and parts per million of EDTA solution ($5x10^{-4}$ M), measured with the conductivity meter, and the four different solutions of textured soya extract prepared.

	1 gram textured soya	2 grams textured soya	3 grams textured soya	5 grams textured soya	EDTA ($5x10^{-4}$ M)
Conductivity (μs)	459.1	454.7	836.6	1191	63.6
ppm	308.5	303	564.6	820.3	40.51

Table 2. Conductivity y ppm of EDTA and textured soya extract solutions

Table 3 shows the resulting conductivity after mixing each of the $CuSO_4$ solutions with the EDTA solution and the four textured soy extract solutions. These results are the differences between the measurement of the mixture of the $CuSO_4$ solution with the soy chelating solution and the pure extraction solution.

Concentration $CuSO_4$ (M)	1 gram textured soya (μs)	2 grams textured soya (μs)	3 grams textured soya (μs)	5 grams textured soya (μs)	EDTA (μs)
0.025	481.8	402	401.4	443	820
0.05	958.9	645.3	853.4	835	1350.4
0.1	2157.9	1857.3	1619.4	1685	2172.4
0.15	2569.9	2969.3	2479.4	2614	3057.4
0.2	3195.9	3503.3	2876.4	3340	3682.4
0.25	3407.9	4310.3	3968.4	3952	4200.4
0.3	3780.9	4836.3	4590.4	4631	4919.4

Table 3. Conductivity of EDTA solution and textured soya extract solutions with the prepared $CuSO_4$ solutions

Figure 4 shows graphically the results of the Table 3.
From these results, we can conclude that is necessary to prepare the textured soya extract solution by using 2 or 3 grams of textured soy in 100 ml of water, heating it to boiling point for 10 minutes and separating out the fiber by filtration.
Table 4 contains the conductivity and ppm of the aqueous EDTA solution and the aqueous textured soya extract using 3 grams of textured soya in 100 ml of deionizer water we prepared.

	Textured soya extract	EDTA ($5x10^{-4}$M)
Conductivity (μs)	1414	150.3
ppm	981	67.06

Table 4. conductivity and ppm of EDTA and textured soy extract solutions.

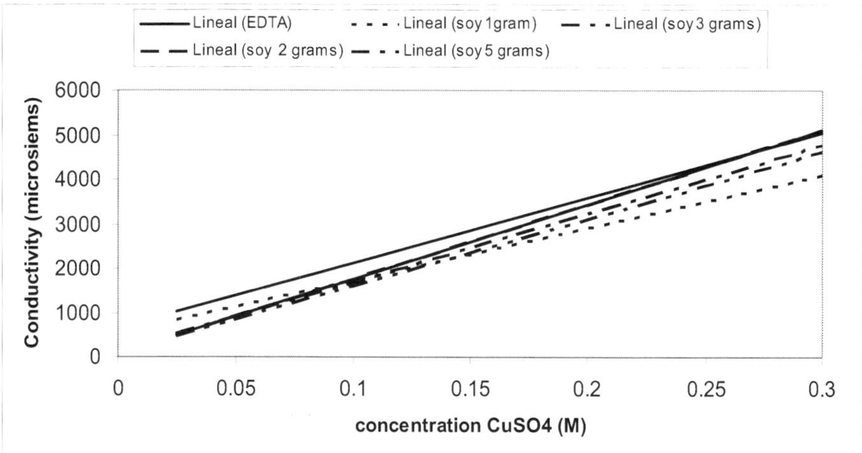

Fig. 4. Conductivity of the mixture of different textured soya extracts solutions with EDTA solution with the $CuSO_4$ solutions .

Table 5 shows the results of the conductivity and ppm of five aqueous Pb^{2+} solution prepared dissolved 0.01, 0.03, 0.05, 0.07 and 0.1 grams of $Pb(NO_3)_2$ in 10 ml of deionizer water, each one. In a similar process were prepared five aqueous solutions EDTA-Fe^{3+} dissolved 0.01, 0.03, 0.05, 0.07 and 0.1 grams of $Pb(NO_3)_2$ in 10 ml of aqueous EDTA solution and in the same way were prepared five aqueous solutions of textured soya extract-Pb^{2+} dissolved 0.01, 0.03, 0.05, 0.07 and 0.1 grams of $Pb(NO_3)_2$ in 10 ml of textured soya extract.

Grams $Pb(NO_3)_2$	Pb^{2+} aqueous solutions ppm (μs)		Textured soya extract-Pb^{2+} Ppm (μs)		EDTA- Pb^{2+} (5x10^{-4}M) ppm (μs)	
0.01	472.45	704.54	116	160	629	907.7
0.03	1193.35	1703.34	572	762	933	1328.7
0.05	2020.35	2764.34	1332	1720	2363	2323.7
0.07	2759.35	3687.34	2016	2569	2519	3354.7
0.1	4031.35	5203.34	3961	3815	3896	5013.7

Table 5. Conductivity and ppm of aqueous solutions: Pb^{2+} , EDTA- Pb^{2+} and textured soy extract-Pb^{2+}.

The conductivity and ppm of the mixture of EDTA and textured soya extract with aqueous Pb^{2+} solution, shown in the Table 5, are a result of subtracting the conductivity or ppm of the mixtures and conductivity and ppm from the EDTA and texture soya extract with $Pb(NO_3)_2$.

Figure 5 is a graphic representation of Table 5 results.

Table 6 shows the results of the conductivity and ppm of five aqueous Fe^{3+} solution prepared dissolved 0.01, 0.03, 0.05, 0.07 and 0.1 grams of $Fe(NO_3)_3$ in 10 ml of deionizer water, each one. In a similar process were prepared five aqueous solutions EDTA-Fe^{3+} dissolved 0.01, 0.03, 0.05, 0.07 and 0.1 grams of $Fe(NO_3)_3$ in 10 ml of aqueous EDTA solution and in the same way were prepared five aqueous solutions of textured soya extract-Fe^{3+} dissolved 0.01, 0.03, 0.05, 0.07 and 0.1 grams of $Fe(NO_3)_3$ in 10 ml of textured soya extract.

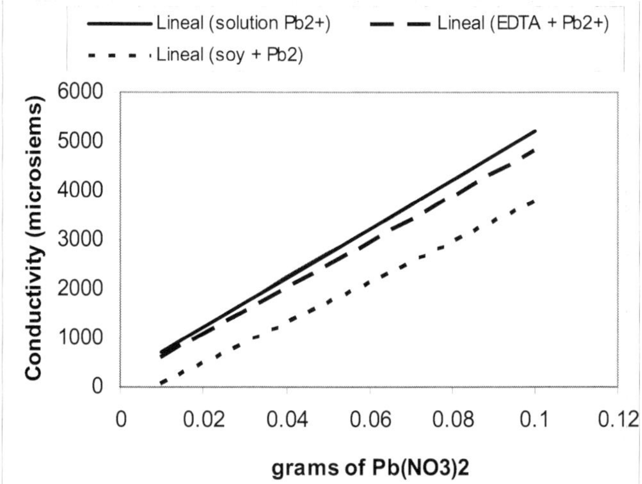

Fig. 5. Conductivity of aqueous solutions: Pb^{2+}, EDTA- Pb^{2+} and textured soya extract-Pb^{2+}.

Grams $Fe(NO_3)_3$	Fe^{3+} aqueous solutions ppm (μs)		Textured soya extract-Fe^{3+} Ppm (μs)		EDTA- Fe^{3+} ($5x10^{-4}M$) ppm (μs)	
0.01	932.64	1351.34	138	177	902.1	1286.7
0.03	2335.34	3159.34	995	1261	2097	2826.7
0.05	3910.34	5072.34	2785	3421	3573	4620.7
0.07	5078.34	6446.34	4431	5336	5008	6302
0.1	7329.34	8964.34	6065	7141	6900	8442

Table 6. Conductivity and ppm of aqueous solutions: Fe^{3+}, EDTA- Fe^{3+} and textured soya extract-Fe^{3+}.

The conductivity and ppm of the mixture of EDTA and textured soya extract with aqueous Fe^{3+} solution, shown in the Table 6, are a result of subtracting the conductivity or ppm of the mixtures and conductivity and ppm from the EDTA and texture soya extract with $Fe(NO_3)_3$. Figure 6 is a graphic representation of the Table 6 results.
Table 7 shows the results of the conductivity and ppm of five aqueous Cd^{2+} solution prepared dissolved 0.01, 0.03, 0.05, 0.07 and 0.1 grams of $CdSO_4$ in 10 ml of deionizer water, each one. In a similar process were prepared five aqueous solutions EDTA-Cd^{2+} dissolved 0.01, 0.03, 0.05, 0.07 and 0.1 grams of $CdSO_4$ in 10 ml of aqueous EDTA solution and in the same way were prepared five aqueous solutions of textured soya extract-Cd^{2+} dissolved 0.01, 0.03, 0.05, 0.07 and 0.1 grams of $CdSO_4$ in 10 ml of textured soya extract.

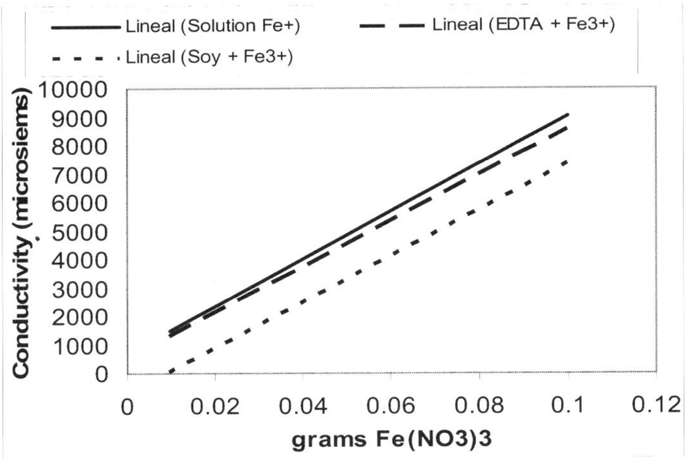

Fig. 6. Conductivity of the aqueous solution of: Fe^{3+}, EDTA-Fe^{3+} and textured soya extract-Fe^{3+}.

Grams $CdSO_4$	Cd^{2+} aqueous solutions ppm (μs)		Textured soya extract-Cd^{2+} Ppm (μs)		EDTA-Cd^{2+} ($5x10^{-4}$M) Ppm (μs)	
0.01	393.75	591.54	82	110	439	643.4
0.03	948.55	1373.34	528	704	521.4	757.2
0.05	1346.35	1906.34	861	1135	1458	1947.7
0.07	1760.35	2442.34	1326	1301	2078	2807.7
0.1	2189.35	2979.34	1760	2256	2704	3584.7

Table 7. Conductivity and ppm of aqueous solutions: Cd^{2+} , EDTA- Cd^{2+} and textured soya extract-Cd^{2+}.

The conductivity and ppm of the mixture of EDTA and textured soya extract with aqueous Cd^{2+} solution, shown in the Table 7, are a result of subtracting the conductivity or ppm of the mixtures and conductivity and ppm from the EDTA and textured soya extract solutions with $CdSO_4$.

Figure 7 is a graphic representation of Table 7 results. Table 8 shows the results of the conductivity and ppm of five aqueous Hg_2^{2+} solution prepared dissolved 0.01, 0.03, 0.05, 0.07 and 0.1 grams of $HgCl_2$ in 10 ml of deionizer water, each one. In a similar process were prepared five aqueous solutions EDTA-Hg_2^{2+} dissolved 0.01, 0.03, 0.05, 0.07 and 0.1 grams of $HgCl_2$ in 10 ml of aqueous EDTA solution and in the same way were prepared five textured soya extract-Hg_2^{2+} dissolved 0.01, 0.03, 0.05, 0.07 and 0.1 grams of $HgCl_2$ in 10 ml of textured soya extract.

The conductivity and ppm of the mixture of EDTA and extract of soybeans with aqueous Hg_2^{2+} solution, shown in Table VII, is a result of subtracting the conductivity or ppm of the mixtures from the EDTA and textured soya extract with $HgCl_2$.

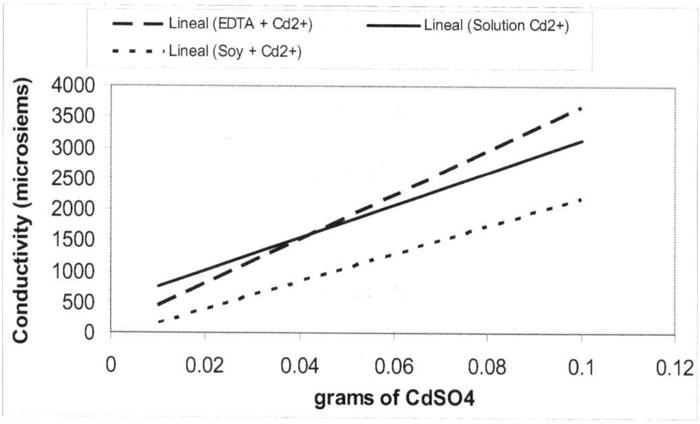

Fig. 7. Conductivity of the aqueous solution of: Cd^{2+}, EDTA-Cd^{2+} and textured soya extract-Cd^{2+}.

Grams $HgCl_2$	Hg_2^{2+} aqueous Solutions ppm (μs)		Textured soya extract-Hg_2^{2+} Ppm (μs)		EDTA-Hg_2^{2+} ($5x10^{-4}M$) Ppm (μs)	
0.01	25	39.41	46	5	113	171
0.03	26.92	42.28	58	74	107.2	163.7
0.05	33.82	53.24	78	94	100	151.7
0.07	37.56	59.09	79	100	107	160.7
0.1	48.58	76.74	100	134	113.7	173

Table 8. Conductivity and ppm of aqueous solutions: Hg_2^{2+} , EDTA- Hg_2^{2+} and textured soya extract-Hg_2^{2+}.

Figure 8 is a graphic representation of Table 8 results.
Table 9 shows the results of the conductivity and ppm of five aqueous Ni^{2+} solution prepared dissolved 0.01, 0.03, 0.05, 0.07 and 0.1 grams of $NiCl_2$ in 10 ml of deionizer water, each one. In a similar process were prepared five aqueous solutions EDTA-Ni^{2+} dissolved 0.01, 0.03, 0.05, 0.07 and 0.1 grams of $NiCl_2$ in 10 ml of aqueous EDTA solution and in the same way were prepared five aqueous textured soya extract-Ni^{2+} dissolved 0.01, 0.03, 0.05, 0.07 and 0.1 grams of $NiCl_2$ in 10 ml of textured soya extract.

Grams $NiCl_2$	Ni^{2+} aqueous solution ppm (μs)		Textured soya extract-Ni^{2+}solution Ppm (μs)		EDTA-Ni^{2+} ($5x10^{-4}M$) ppm (μs)	
0.01	720.25	1052.34	210	273	450.5	716.7
0.03	1849.35	2553.34	1108	1404	997	1413.7
0.05	2677.35	3585.34	1845	2306	2653	3519.7
0.07	4093.35	5287.34	3225	3940	4088	5194.7
0.1	6069.35	7575.34	4658	5599	5607	7006.7

Table 9. Conductivity and ppm of aqueous solutions: Ni^{2+} , EDTA- Ni^{2+} and textured soya extract-Ni^{2+}.

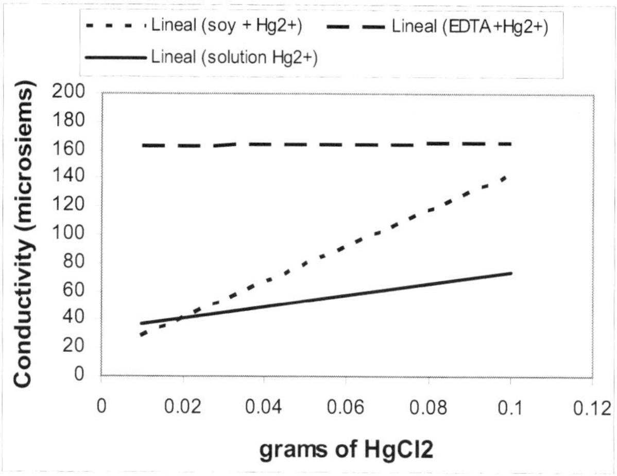

Fig. 8. Conductivity of the aqueous solution of: Hg_2^{2+}, EDTA-Hg^{2+} and textured soya extract-Hg^{2+}.

The conductivity and ppm of the mixture of EDTA and textured soya extract with aqueous Ni^{2+} solution, shown in the Table VIII, is a result of subtracting the conductivity or ppm of the mixtures from the EDTA and textured soya extract with $NiCl_2$.
Figure 9 is a graphic representation of Table 9 results.

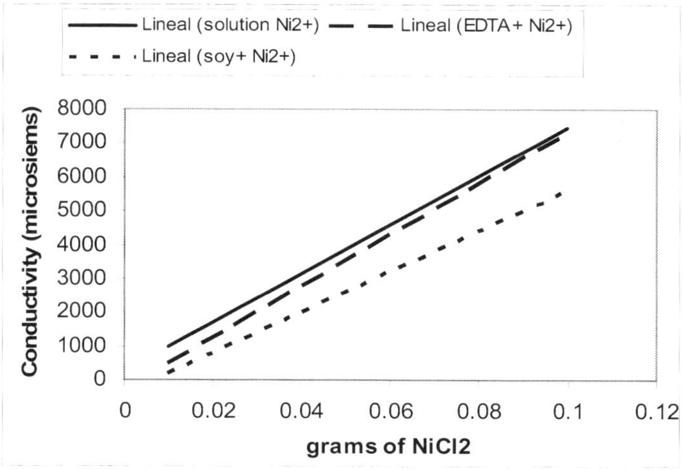

Fig. 9. Conductivity of the aqueous solution of: Ni^{2+}, EDTA-Ni^{2+} and textured soya extract-Ni^{2+}.

In Table 10 we can see the amount of metal ion sequestering for the textured soya extract in five different amounts of each salt: 0.01, 0.03, 0.05, 0.07 and 0.1 grams.

	0.01 grams	0.03 grams	0.05 grams	0.07 grams	0.1 grams
Pb^{2+}	0.0076	0.008	0.019	0.018	0.027
Fe^{3+}	0.0087	0.017	0.016	0.0089	0.021
Cd^{2+}	0.0082	0.013	0.020	0.017	0.025
Hg^{2+}	-----	-----	-----	-----	-----
Ni^{2+}	0.0075	0.012	0.018	0.014	0.027

Table 10. Amount of metal ion sequestering using textured soya extract

Figure 10 is a graphic representation of Table 10 results.

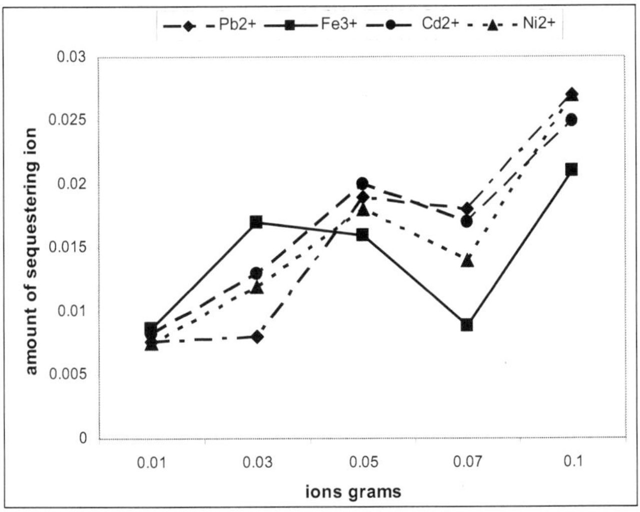

Fig. 10. Amount of metal ion sequestering using textured soya extract.

In Table 11 we can see the amount of metal ion sequestering for the EDTA in five different amounts of each salt: 0.01, 0.03, 0.05, 0.07 and 0.1 grams.

	0.01 grams	0.03 grams	0.05 grams	0.07 grams	0.1 grams
Pb^{2+}	0.0025	0.013	0.008	0.006	0.004
Fe^{3+}	0.0005	0.003	0.045	0.001	0.006
Cd^{2+}	0.0026	0.013	----	----	----
Hg^{2+}	----	----	----	----	----
Ni^{2+}	0.0032	0.013	0.001	0.0085	0.008

Table 11. Amount of metal ion sequestering using EDTA

Figure 11 is a graphic representation of Table 11 results
From the results obtained in Table 10 and Table 11, we can say that the amount of metal ion chelating increases with the increase of the concentration but the amount of salt chelated with textured soya extract is considerable major in comparison to the EDTA. In the case of Hg_2^{2+} ions, the textured soya extract and EDTA is not effective as a chelating agent. Another

difference is that EDTA is effective as chelating agent of Cd^{2+} only with low concentrations (less to 0.04 g of $CdSO_4$ in 10 ml of EDTA solution $5x10^{-4}M$) but without exception the textured soya extract is a good chelating agent.

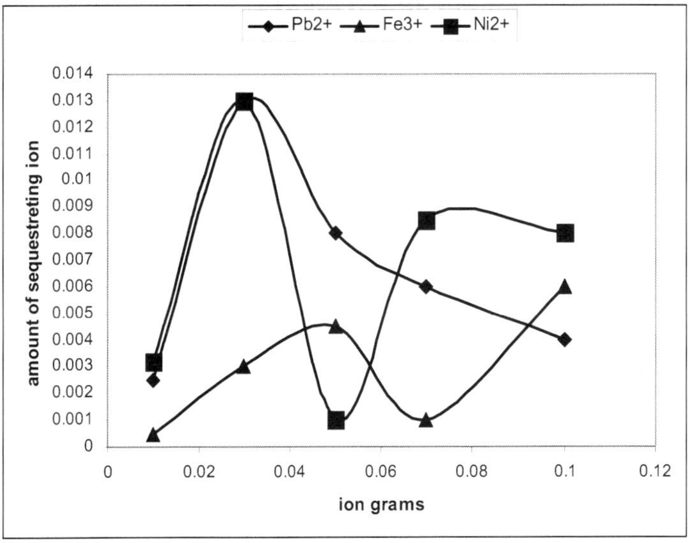

Fig. 11. Amount of metal ion sequestering using EDTA

In the case of $HgCl_2$ it is a weak electrolyte, the ionization is partial, as showing in the reaction 7.

$$HgCl_2\,(s) = HgCl^+\,(aq) + Cl^-\,(aq) \quad (7)$$

Maybe for this reason the EDTA and textured soya extract are not effective as chelating agents. But the textured soya extract is effective as chelating agent at low concentrations; nevertheless, the EDTA is not an effective chelating agent even in low concentrations.
From the results of Table 11, we are able to calculate the equilibrium constant for the textured soya extract, using the equation 6, and the results being shown in Tables 12.

	0.01 grams	0.03 grams	0.05 grams	0.07 grams	0.1 grams
	Textured soya extract	Textured soya extract	Textured soya extract	Textured Soya extract	Textured soya extract
Pb^{2+}	0.76	0.11	0.38	0.25	0.27
Fe^{3+}	0.87	0.56	0.32	0.12	0.21
Cd^{2+}	0.82	0.43	0.4	0.24	0.25
Hg_2^{2+}	------	----	------	----	------
Ni^{2+}	0.75	0.4	0.36	0.2	0.27

Table 12. Equilibrium constant of metal ion sequestering

From the results of Table 11, we are able to calculate the equilibrium constant for the EDTA, using the equation 6, and the results being shown in Tables 13.

	0.01 grams	0.03 grams	0.05 grams	0.07 grams	0.1 grams
	EDTA	EDTA	EDTA	EDTA	EDTA
Pb^{2+}	0.25	0.2	0.16	0.08	0.04
Fe^{3+}	0.05	0.1	0.09	0.014	0.06
Cd^{2+}	0.26	0.43	------	----	-----
Hg_2^{2+}	------	----	------	----	-----
Ni^{2+}	0.32	0.43	0.02	0.12	0.08

Table 13. Equilibrium constant of metal ion sequestering

5. Conclusions

In the case of the ion Pb^{2+}, it can be seen that the solution of the complex EDTA with the ion P^{2+} gives a line which is very close to the reference line of the ionic solution Pb^{2+}. This indicates that the amount of Pb^{2+} ion chelated is small in comparison to the solution of the chelate formed from the textured soya extract and the Pb^{2+} ion which has a line that is way below the reference line and the EDTA.

A similar conclusion for the study with the Fe^{3+} ion can be given.

With respect to the Cd^{2+} ion, the EDTA only acts as a chelate in concentrations lower than 0.04 grams of $CdSO_4$ per 10ml of deionized water. On the other hand, the textured soya extract is a good chelate in a wider concentration range (between 0.01 and 0.1 grams of $CdSO_4$ per 10ml of deionized water).

The chelate solution of EDTA for the Hg_2^{2+} ion does not have any effect on the Hg_2^{2+} ion in the test range from 0.01 up to 0.1 grams of $HgCl_2$ per 10ml of water. However, the textured soya extracts act as a chelate only in concentrations lower than 0.15 grams of $HgCl_2$ per 10ml of deionized water. The problem presented by this salt rests on the fact that it is a weak electrolyte and when it is placed in the water, it decomposes into two ions. Since the conductivity of the solution is measured in this study, the formation of two ions has a negative effect on the measurements obtained.

Just as in the case of Pb^{2+}, Fe^{3+} and Ni^{2+} ions, the textured soya extract is a much better chelate than EDTA.

There is normally a low concentration (parts per million) of heavy ions in food. Thus our proposal of using the textured soya extract as the chelate for heavy ions instead of EDTA. In addition, as a result of the low concentration of ions, a solution with a low concentration of the textured soya extract will be used in order not to change the color, scent or taste of the food.

A problem to be considered in this application is that food is prepared with water which has salts that are ionized, and the textured soya extract will chelate some of these ions also. It will be necessary to perform tests on the food sample to determine if the application will be practical or not.

Another application possible is the extraction of heavy metals in water, cosmetics, and soils employing textured soya extract as the chelate for heavy ions instead of EDTA.

6. Acknowledgments

This work was supported by CONACYT. Morales Elpidio, Guajardo Jesús, Lopez Francisco, Quintero Cristina, Compean Martha, Noriega María-Eugenia, would like to thank CONACYT for the grant of scholarships.

7. References

[1] Mulligan, C.N.; Young, R.N.; Gibbs, B. *Eng. Geol.* 2001, *60*, 193-207.
[2] Peters, R.W. J. Hazard. *Mater.* 1999, *66*, 151-210.
[3] Tandy Susan, Bossart Karin, Mueller Roland, Ritschel Jens, Hauser Lukas, Sculin Rainer, and Nowack Bernd, *Environ. Sci. Technol.* 2004, *38*, 937-944.
[4] Elliot, H. A.; Brown, G.A. *Water, Air, Soil Pollut.* 1989, *45*, 361-369.
[5] Lanigan RS and Yamarik TA (2002). "Final report on the safety assessment of EDTA, calcium disodium EDTA, diammonium EDTA, dipotassium EDTA, disodium EDTA, TEA-EDTA, tetrasodium EDTA, tripotassium EDTA, trisodium EDTA, HEDTA, and trisodium HEDTA". Int J Toxicol. 21 Suppl 2: 95–142.
[6] Steel, M. C; Pichtel, J. J. *Environ.Eng.* 1998, *124*, 639-645.]
[7] Pichtel, J. ; Vine, B.; Kuula-Vaisanen, P. ; Niskanen, P. *Environ. Eng. Sci.* 2001, *18*, 91-98
[8] Pichtel, J.; Pichtel, T.M. *Environ. Eng. Sci.*1997, *14*, 97-104.
[19] Kim, C.; Lee, Y.; Ong, S.K. *Chemosphere* 2003, *51*, 845-85.
[10] Ghestem, J.; Bermond, A. *Environ. Thecnol.* 1998, *19*, 409-416.
[11] Reed, B.E.; Carriere,P.C.; Moore, R. *J. Environ. Eng.* 1996, *122*, 48-50.
[12] Cline, S.R.; Reed,B.E. *J. Environ. Eng.* 1995, *121*, 700-705
[13] Van Benschoten, J.E.;Matsumoto,M.R.; Young,W.H.*J.Environ. Eng.* 1997, *127*, 217-224.
[14] Pichtel,J.; Vine,B.; Kuula-Vaisanen, P.; Niskanen, P. *Environ. Eng. Sci.* 2001, *18*, 91-98.
[15] Papassiopi, N.; Tambouris, S. Kontopoulos, A. *Water, Air, Soil, Pollut.* 1999, *109*, 1-15
[16] Xie, T.; Marshall, W.D. *J. Environ. Monit.* 2001, *3*, 411-416.
[17] Liu, Ke Shun (1997-05-01), Soybeans: *Chemistry technology and Utilization. Springer.* 532
[18] IUPAC definition of chelation.
[19] Morgan, Gilbert T.; Drew, Harry D. K. (1920).
[20] U. Kramer, JD Cotter-Howells, JM Charnock, A HJM Baker, J A C Smith. *Nature*,1996, 379: 635-638.
[21] Jurandir Vieira Magalhaes, *Proc. Natl. Acad. Sci. USA*, 2006, 103 (26):9749.
[22] Suk-Bomg Há, Aaron P. Smith, Ross Howden, Wendy M. Dietrich, Sarah Bugg, Matthew J. O'Connell, Peter B. Goldsbrough, and Chistopher S. Cobbe, *Plant Cell*, 1999, 11 (6): 1153-1164.
[23] S. J. Lippard, J. M. Berg, "Principles of Bioinorganic Chemistry" University Science Books: Mill Valley, CA; 1994 ISBN 0-935702-73-3.
[24] Michael Pidwirny, physical geography.net/fundamentals/10r
[25] Prasad (ed). Metals in the Environment. University of Hyderabad. Dekker, New York, 2001.
[26] Doja A, Roberts W., Can J. Neurol. Sci., 2006, 33 (4): 341-346.
[27] U. S. Center for Disease Control, "Death Associated with Hypocalcemia from Chelation Therapy" ,2006
[28] The 2009 Annual Report of the North Carolina Soybean Producers Association, realeased Jannary 14, 2010.

[29] Riaz MN, *Soy applications in food.* 2006. CRC Press. Pp 155-84, ISBN 0-8493-2981-7.

[30] Foote R, *Food preparation an cooking,* 1996, Cheltenham. Stanley Thornes, pp. 393. ISBN 0-7487-2566-0.

[31] Hoogenkamp, Henk W., 2005, Wallingford, Oxon, UK: CABI Pub. ISBN 0-85199-864-X.

[32] Joseph G. Endres, Soy Protein Products, 2001, AOCS Publishing. ISBN 1-893997-27-8.

17

Effect of Nitrate on Nodulation and Nitrogen Fixation of Soybean

Takuji Ohyama[1,2] et al.*
[1]*Faculty of Agriculture, Niigata University*
[2]*Quantum Beam Science Directorate, Japan Atomic Energy Agency*
Japan

1. Introduction

1.1 Biological nitrogen fixation and nitrogen nutrition in soybean plants

Biological nitrogen fixation is one of the most important processes for ecosystem to access available N for all living organisms. Although N_2 consists 78% of atmosphere, but the triple bond between two N atoms is very stable, and only a few group of prokaryotes can fix N_2 to ammonia by the enzyme nitrogenase. Annual rate of natural nitrogen fixation is estimated about 232 x 10^6 t, and the 97% depends on biological nitrogen fixation (Bloom, 2011). This exceeds the rate of chemical nitrogen fertilizer uses about 100 x 10^6 t in 2009. Soybean can use N_2, though symbiosis with nitrogen fixing soil bacteria, rhizobia, to make root nodules for harboring them.

Soybean (*Glycine max* [L.] Merr.) is a major grain legume crop for feeding humans and livestock. It serves as an important oil and protein source for large population residing in Asia and the American continents. The current global soybean production was 231 x 10^6 t in 2008 (FAOSTAT). It is a crop predominantly cultivated in U.S.A., Brazil, Argentina and China, which together contribute nearly 87 percent of the total world produce in 2008. Soybean has become the raw materials for diversity of agricultural and industrial uses.

Soybean seeds contain a high proportion of protein, about 40% based on dry weight, therefore, they require a large amount of nitrogen to get a high yield. About 8 kg N is required for 100 kg of soybean seed production. Soybean can use atmospheric dinitrogen (N_2) by nitrogen fixation of root nodules associated with soil bacteria, rhizobia. Soybean plants can absorb combined nitrogen such as nitrate for their nutrition either from soil mineralized N or fertilizer N.

It is well known that heavy supply of nitrogen fertilizer often causes the inhibition of nodulation and nitrogen fixation. Therefore, only a little or no nitrogen fertilizer is

* Hiroyuki Fujikake[1], Hiroyuki Yashima[1], Sayuri Tanabata[3], Shinji Ishikawa[1], Takashi Sato[4], Toshikazu Nishiwaki[5], Norikuni Ohtake[1], Kuni Sueyoshi[1], Satomi Ishii[2] and Shu Fujimaki[2]
[1] *Faculty of Agriculture, Niigata University,*
[2] *Quantum Beam Science Directorate, Japan Atomic Energy Agency,* [3]*Agricultural Research Institute, Ibaraki Agricultural Center,* [4]*Faculty of Bioresource Sciences, Akita Prefectural University,* [5] *Food Research Center, Niigata Agricultural Research Institute, Japan*

practically applied for soybean production. However, soybean plants only depend on the nitrogen fixation shows poor growth and low seed yield, because of the early decline in photosynthesis by decreased supply of nitrogen during the pod filling stage. Harper (1974) reported that both soil N and symbiotic N are required for the optimum soybean production.

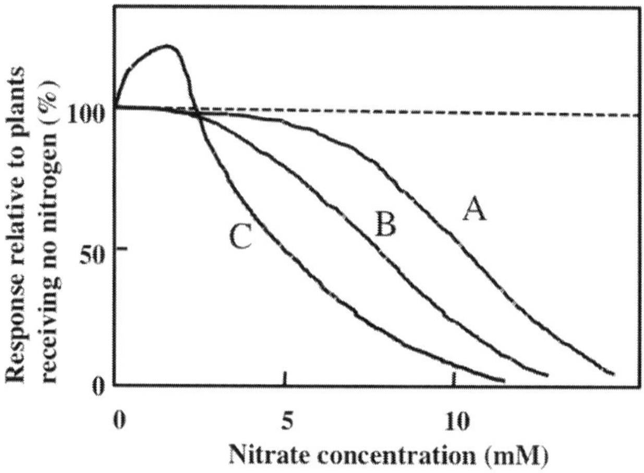

A: nodule number per a plant, B: Nitrogen fixation activity per g dry weight of nodules, C: Nodule mass per a plant.

Fig. 1. Response of legume nodules to nitrate proposed by Streeter.

The inhibitory effect of nitate on nodulaion was early reported by Fred and Graul (1916) as cited in Streeter (1988), however, the precise mechanism for the inhibition of nodulation and nitrogen fixation has not been fully understood. In the review article for inhibition of legume nodule formation and N_2 fixation by nitrate written by Streeter (1988), he proposed the responses to nitrate illustrated in Fig. 1. Curve A represents nodule number per a plant, which appears a relatively high nitrate concentration. Curve B is on nitrogen fixation activity per unit mass (g dry weight) of nodules. Curve C shows the growth response (nodule mass per a plant), this response is most sensitive to nitrate concentration, although a low concentration of nitrate as low as about 1-2 mM nitrate promotes nodule growth.

1.2 Nodule structure and function of soybean

Soybean nodule appears about 10 days after sowing when inoculated with compatible strain of rhizobia, and it grows about 3mm until about 20 days after planting (Fig. 2. A.) . The nodules start to fix nitrogen (Sato et al., 2001, Ito et al., 2006). The maximum size of nodule reaches maximum about 6-7 mm diameter, and then they eventually senesce and degrade. Soybean nodule is classified to a determinate type nodule, which has a spherical form, and nodule growth is mainly due to cell expansion after initial cell proliferation and development.

Fig. 2.B shows the structural model of a soybean nodule attached to the root. The soybean nodule has the symbiotic region (or infected region in synonym) in the center, which

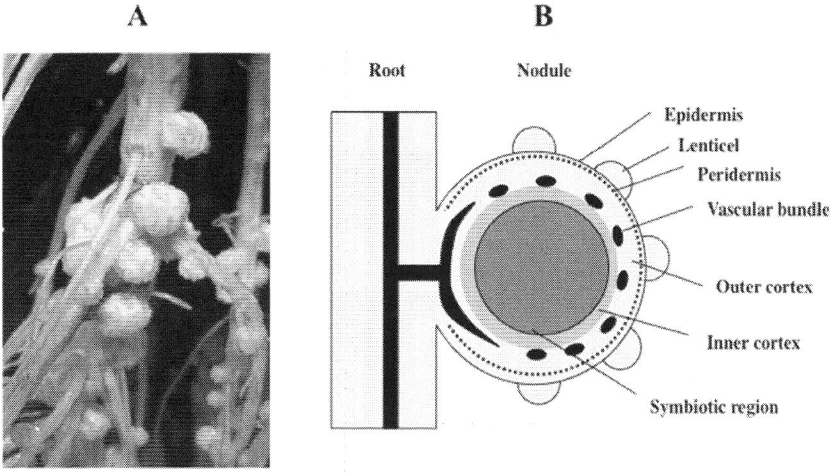

A: Photograph of soybean nodules formed in the roots of a plant cultivated by hydroponics. B: Structural model of soybean nodule.

Fig. 2. Soybean root nodules.

consists the mosaic of large infected cells and small uninfected cells. The infected cells are filled with bacteroids (the symbiotic forms of rhizobia) and they are easily recognized by the red color with nodule specific protein, leghemoglobin (Lb). The nitrogenase, an enzyme to fix N_2 in bacteroid, is very susceptible to free O_2 and irreversibly destroyed by O_2, therefore, free O_2 concentration should be kept very low in symbiotic region of nodules. There are four major components of Lb, Lba, Lbc1, Lbc2, and Lbc3 (Sato et al., 1998, 1999a). The Lb in legume nodules solves the dilemma to keep free O_2 concentration low and sufficient supply of O_2 for bacteroid respiration to support nitrogen fixation and the assimilation. Lb is a most abundant protein in nodules (about 20% of total protein) and it can bind with O_2 to form LbO_2 to decrease free O_2 concentration in the infected cells. On the other hand, nitrogen fixation and assimilation processes require a large amount of energy and reductant produced by O_2 respiration, therefore, nodule respiration is about four times higher than that of roots based on dry weight. To support active respiration, abundant supply of O_2 is essential.

Symbiotic region is surrounded by nodule cortex where the network of vascular bundles surrounding the symbiotic region to supply photoassimilate to bacteroids and to receive N_2 fixation products from them. Nodule cortex consists of inner cortex with small cells and outer cortex with large loosely packed cells. The sclerenchyma cells, which have thick cell wall were located in the outer cortex. O_2 concentration decreases sharply through the inner cortex, and the O_2 permeability is flexibly controlled by this layer (Witty and Minchin 1990, Hunt and Layzell 1993). It is hypothesized that a reversible exchange of intercellular water by the inner cortical cells plays a role in the regulation of nodule conductance to O_2 diffusion (Serraj et al., 1995, 1998, Fleurat-Lessard et al., 2005). There are lenticels outside of nodules and one layer of epidermis. Under the epidermis, there is a peridermis, a tightly packed one layer of cells, which may restrict free diffusion of solutes between inside the nodule and medium solution.

The group of positron-emitting tracer imaging system (PETIS) for plant analysis in Quantum Beam Science Directorate, Japan Atomic Energy Agency, developed a novel method of non-invasive observation and quantification of nitrogen fixation in intact soybean plants (cv. Williams) with nodules using ^{13}N-labeled nitrogen gas ($[^{13}N]N_2$) tracer and a PETIS (Ishii et al., 2009, Fujimaki et al., 2010). CO_2 gas was irradiated with a proton beam delivered from a cyclotron (Takasaki Advanced Radiation Research Institute, Japan Atomic Energy Agency) to produce ^{13}N nuclei by the ^{16}O (p, α) ^{13}N nuclear reaction. $[^{13}N]N_2$ was isolated from the resulting gas using gas chromatography and then mixed with appropriate composition of oxygen and (non-radioactive) nitrogen gases for the following feeding experiment. The total time required for the purification procedures was approximately 15 min, which is about 1.5 times the half-life of ^{13}N (only 9.97 min) and short enough to yield sufficient radioactivity of the tracer.

PETIS is one of the most advanced imaging methods today, which provides serial images of movement of positron-emitting radiotracers inside living plant bodies, like a video camera. The root of an intact test plant with nodules was immersed in a hydroponic culture solution in an acrylic box sealed with plastic clay to prevent leakage of the fed gas. This set-up was placed at the midpoint between the opposing detector heads of the PETIS apparatus so that the underground part in the acrylic box was in the field of view (Fig. 3.). The tracer gas was introduced into the box and the solution level was lowered simultaneously, then this was kept for 10 min for exposure of the nodules to the tracer gas. Finally, the tracer gas was flushed out by flowing the ambient air into the box. The two-dimensional distribution of ^{13}N in the field of view was continuously monitored by PETIS for 1 h.

As a result, obvious signals of ^{13}N were observed at the positions of the nodules (Fig. 4.). Moreover, the rates of nitrogen fixation in the whole nodules were quantitatively estimated from the PETIS data. The nitrogen fixation rate of the whole nodules was estimated at 7 μg N_2 h^{-1} in this case. The largest advantage of this method is that it is non-invasive. The instant response of fixation activities to nitrate application will be examined in a future study.

Soybean nodule is highly organized complex organ as shown by the distribution of minerals examined by EPMA (Electron Probe X-ray Microanalysis) (Mizukoshi et al., 1995). Fig. 5. shows the distribution of minerals in nodulated roots. The concentrations of N and P were higher but those of K and Cl were lower in the symbiotic region compared with nodule cortex. Ca was locally distributed in the surface layer, sclerenchyma cells and inner cortex, but the content was low in the central symbiotic region. Mg specifically accumulated in the inner and outer cortex inside sclerenchyma cells but not out side them (Mizukoshi et al., 1995).

Fig. 6. shows the outline of N metabolism in soybean nodule (Ohyama et al., 2009). Ammonia is known to be the initial product of nitrogenase. After discovering a new enzyme glutamate synthase (GOGAT) in *Aerobacter aerogenes*, it is confirmed that ammonia can be assimilated via alternative of GDH, via glutamine synthetase (GS) and GOGAT pathway (Ohyama et al. 2009). From the result obtained by the $^{15}N_2$ pulse chase experiment, the ammonia fixed by nitrogenase in bacteroids is rapidly incorporated into of amido-N of glutamine, followed by glutamate, and amino-N of glutamine in this sequence was in accordance with the initial assimilatory pathway be GS/GOGAT pathway rather than GDH. This was supported by the evidence that the rapid decline of ^{15}N in glutamine but not glutamate immediately after changing to the chase period. A major part of fixed N was used for purine synthesis in infected cells then uric acid is transported to the uninfected cells, then degraded into allantoin and allantoate. All the species in Phaseoleae (soybean, common bean, cowpea etc.) and some species in Robinieae, Indigoforeae and Desmodieae transport ureides (Atkins, 1991). Reviews on ureide biosynthesis in legume nodules were published (Schubert, 1986, Tajima et al., 2004). We compared the labeling patterns of ureides and amino acids from $^{15}N_2$ and $^{15}NO_3^-$ (Ohyama & Kumazawa, 1979), and the labeling pattern indicated that most of ureides derived from fixed N rather than absorbed N.

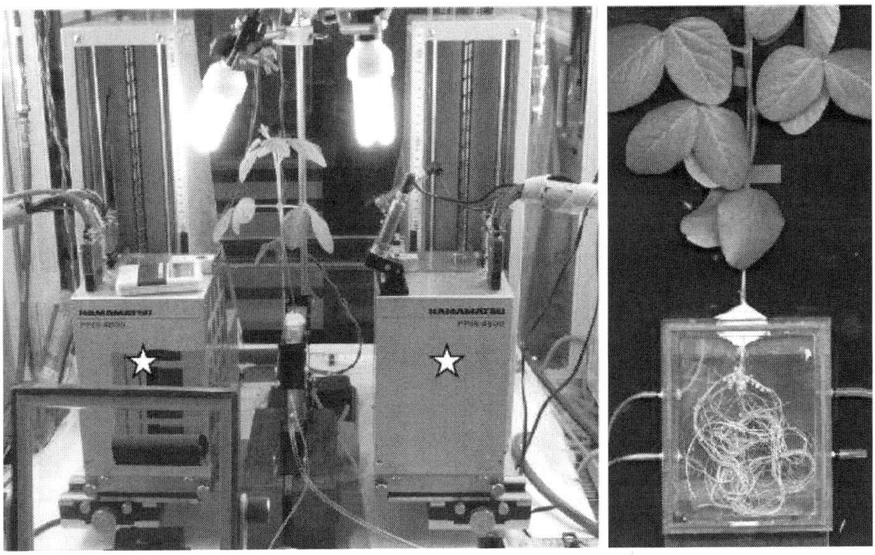

A B

Fig. 3. Set-up for the PETIS experiment (A) and a test plant (B). Star signs indicate the opposing detector heads of the PETIS apparatus.

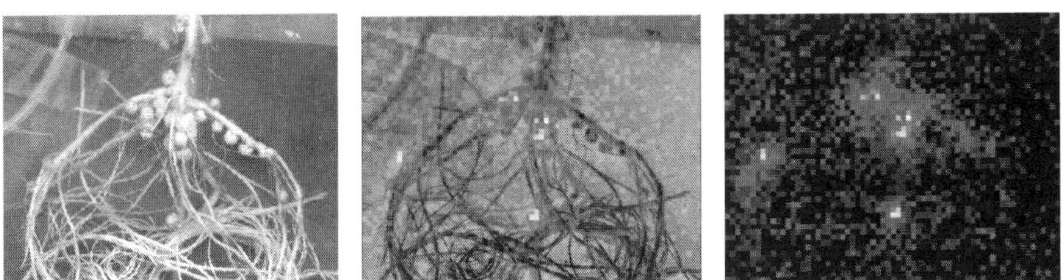

A B C

A: Nodulated roots of a test plant.
B: The merged image of nodulated root and radioactivity in the same view.
C: PETIS image of radioactivity.

Fig. 4. Image of radioactivity after exposure the nodulated soybean roots to $^{13}N_2$.

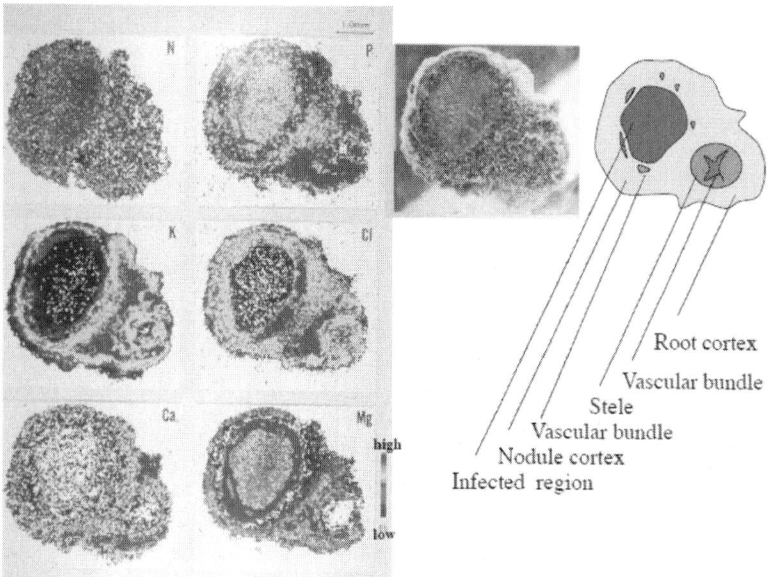

Fig. 5. Distribution of N, P, K, Cl, Ca and Mg in a nodule and root. The concentration is higher in red, orange, yellow, green, blue and white in this sequence.

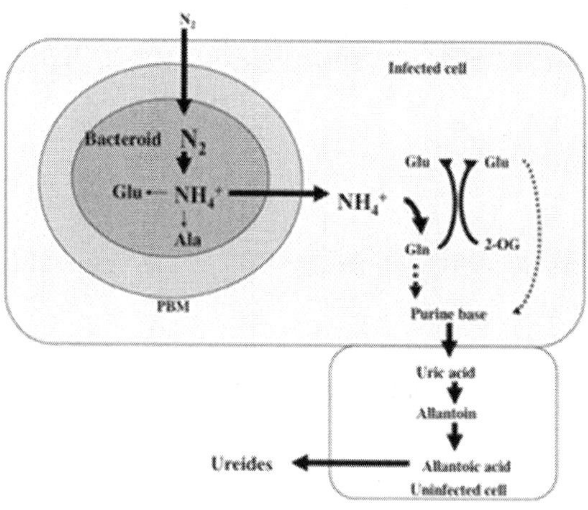

Gln: glutamine, Glu: glutamate, 2-OG: 2-oxoglutarate, PBM: peribacteroid membrane.

Fig. 6. A model of the N flow of fixed N_2 in soybean nodules.

Fig. 7. shows the model of nitrogen assimilation and transport of N derived from N_2 fixation and NO_3^- absorption in soybean plants (Ohyama &Kumazawa, 1978, 1979, 1980abc, 1981ab, 1983, 1984, Ohyama et al., 2009) . The N fixed in noudle is exported to the host plant as in the form of allantoin and allantoate about 80-90% of total N. On the other hand, some part of the NO_3^- absorbed in the roots are reduced in the roots to NO_2^- by nitrate reductase, then the NO_2^- is further reduced to NH_4^+ by plastidic nitrite reductase, then the NH_4^+ is assimilated by GS/GOGAT pathway in the roots, and mainly metabolized to asparagine then transported to shoot via xylem. Some part of NO_3^- is directly transported through xylem to the shoots and reduced in leaves. Ohtake et al. (1995) reported the seasonal changes in amino acid composition in xylem sap of soybean and they confirmed that asparagine was the principal amino acids in xylem sap collected from basal cut end of the stem at any stages.

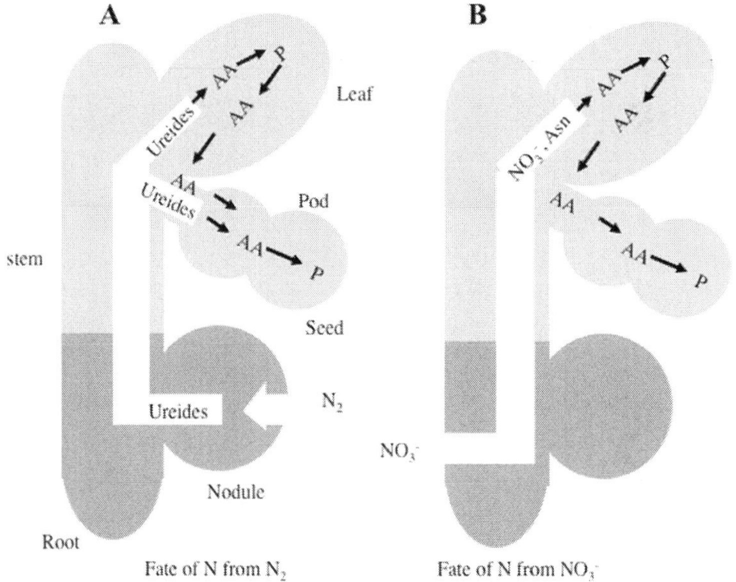

A: Flow of N from N_2 fixed in root nodules.
B: Flow of N from NO_3^- absorbed from roots.
AA: amino acids, P: protein.

Fig. 7. A model of N flow in soybean plant.

1.3 Nitrate inhibition of nodule growth and nitrogen fixation

The inhibitory effects of externally supplied N especially NO_3^- have been reviewed (Streeter, 1988, Harper, 1987). The nitrate inhibition is complex and it cannot be explained by a single mechanism. It has been suggested that there are multiple effects of nitrate inhibition, such as the decrease in nodule number, nodule mass, and N_2 fixation activity, as well as the acceleration of nodule senescence or disintegration (Streeter, 1988, Harter, 1987). In addition, nitrate inhibition of nodules is complex, because the effects of nitrate on nodule formation and growth are influenced by nitrate concentration, placement and treatment period as well as legume species (Harper & Gibson, 1984, Gibson & Harper, 1985, Davidson & Robson, 1986).

Nitrate inhibition is primarily host plant dependent and it is independent of nitrate metabolism of rhizobia (Gibson & Harper, 1984, Carrol & Mathews, 1990). Many hypothesis are proposed for the cause of nitrate inhibition of nodulation and N_2 fixation, i.e. carbohydrate deprivation in nodules (Streeter, 1988, Vessy & Waterer, 1992), feedback inhibition by a product of nitrate metabolism such as glutamine (Neo & Layzell, 1997), asparagine (Bacanambo & Harper, 1996, 1997), and decreased O_2 diffusion into nodules which restricts the respiration of bacteroids (Schuller et al., 1988, Vessey et al., 1988, Gordon et al., 2002). Kanayama and Yamamoto proposed that NO formed from NO_3^- binds to Lb to make nitrosylleghemoglobin and defect the O_2 binding activity (Kanayama & Yamamoto, 1990). On the other hand, Giannakis et al. (1988) suggested that nitrate metabolism does not occur in symbiotic region of soybean nodule, even when a dissimilatory NR is expressed, because of restricted access of nitrate.

Leghemoglobin (Lb) plays a crucial role in N_2 fixation of leguminous nodules by facilitating O_2 supply to the bacteroids. There are four major components of Lb in soybean nodules, Lba, Lbc1, Lbc2, and Lbc3, and different roles are suggested among components (Fuchsman et al., 1976), because Lba has higher affinity for O_2 than has Lbc. The concentrations of Lba and Lbc were separated by Native PAGE (Nishiwaki and Ohyama, 1995). All the four components Lba, Lbc1, Lbc2, and Lbc3 were separately determined by capillary electrophoresis (Sato et al., 1997). The concentration and component ratios in the hypernodulation mutant NOD1-3, NOD2-4, and NOD3-7 from Williams parent, and in En6500 from Enrei parent were compared in relation to their nodulation characteristics. Three mutants (NOD1-3, NOD3-7 and En6500) were controlled by a single recessive allele rj_7, but NOD2-4 was non-allelic mutant to them (Vuong et al. 1996). Plants were hydroponically cultivated in N free solution, and the nodules were separated by size. Concentration and composition of Lb components in the same size nodules were analyzed by gel-electrophoresis and capillary electrophoresis. In all NOD mutants Lb concentration was about 70% of that in the parent Williams, irrespective of nodule size and growth stages. In the hypernodulation mutant En6500, the total Lb concentration was only 25% of that in the parent Enrei, irrespective of nodule size. In Enrei, relative compositions of Lba, Lbc1, Lbc2 and Lbc3 were 36, 26, 18 and 17%, respectively, and very stable irrespective of nodule size. En6500 had relatively equal amounts of each component in which the relative compositions of Lba, Lbc1, Lbc2 and Lbc3 were 30, 22, 22 and 26%. The concentration of Lbc forms in nodules was decreased by addition of nitrate to Enrei plants, but not to En6500. When the nodule morphology was compared among hypernodulation mutant lines and parent lines, we noticed that mutant line had thick cortical regions relative to the comparable parent nodules. The relative volume of symbiotic regions was about 50-60% of total nodule volume in Williams, but it accounted for only 40-50% in NOD mutants.

Sato et al. (2001) investigated the changes in four leghemoglobin components in nodules of NOD1-3 and its parent in the early nodule developmental stage. The hydroponically grown NOD1-3 and Williams were periodically sampled. All the visible nodules were collected from the roots and then the four Lb components in the largest nodules were analyzed by capillary electrophoresis. In NOD1-3 nodule development was faster than those of Williams. Acetylene reduction activity was detected at 19 days after planting in NOD1-3 and at 22 days after planting in Williams. In addition the Lbs were initially detected at 19 days after planting in NOD1-3, a few days earlier than in Williams at 22 days after planting. The Lbcs (Lbc1, Lbc2 and Lbc3) were the main components at the earliest nodule growth stage, and the relative proportion of Lba increased with nodule growth in both NOD 1-3 and Williams.

The hypernodulation soybean mutant lines (NOD1-3, NOD2-4, NOD3-7) and the parent Williams and mutant line En6500 and the parent Enrei were cultivated in a sandy dune field in Niigata, and the nodules and root bleeding xylem sap were analyzed at 50, 70, 90 and 120 days after planting (Sato et al., 1998). The number of nodules of the hypernodulation mutant lines was about two to three times higher than that of the parent lines irrespective of sampling date. The concentration of Lb components was measured by capillary electrophoresis. The concentration of Lb components in the hypernodulation mutant lines tended to be lower than in the parents, but the component ratios were not different between mutants and the parents.

It is well recognized that plant growth is affected by various environmental factors, such as temperature, moisture, photoperiod, light intensity and quality, as well as physical, chemical, and biological properties of soil. The degree of nitrate inhibition was affected by soil medium composition with vermiculite and perlite, where the proportion of solid, liquid and gas space was changed (Nishiwaki et al., 1995).

It has been reported in alfalfa that the inhibition of nodulation by nitrate was reduced by medication of ethylene production inhibitor aminoethoxyvinilglycine (Ligero et al., 1991). While the exogenous ethylene inhibited nodulation on the primary and lateral roots of pea (Lee & LaRue, 1992ab). Ethylene is one of the important phytohormone regulating plant growth. Ethylene is produced through oxidative decomposition of 1-aminocyclopropane-1-carbosylic acid (ACC), and silver thiosulfate (STS) is a potent inhibitor of ethylene action in plants (Veen, 1983). Sato et al. (1999c) investigated the effect of ethylene action on soybean nodulation using ACC and STS in relation to the inhibitory mechanism of nitrate using hypernodulation mutant NOD1-3 and the parent Williams. The hypernodulation mutant of soybean NOD1-3 and its parent Williams were cultivated in culture solution with or without NO_3^-, and ACC or STS were added in the solution. The nodule dry weight was decreased by both ACC and STS treatments, however, the ratio in nodule dry weight in total plant dry weight were not significantly influenced by these treatments with or without NO_3^-. Therefore, it was concluded that the decrease in nodule dry weight by ACC or STS was caused by inferior growth. In soybean the depression of nodulation and N_2 fixation by nitrate is not mediated through ethylene action. Schmidt et al. (1999) also reported the independence of ethylene signaling on the regulation of soybean nodulation. Moreover, the nodulation of hypernodulation mutant was not specifically influenced by ACC treatments. This suggests that autoregulation of nodulation may not be involved in ethylene action or transduction pathways in soybean plants.

Recently, defective long-distance auxin transport regulation was reported in the *Medicago truncatula* super numeric nodules mutant (Van Noorden et al., 2006). However, similar trend is not observed in hypernodulation mutants of soybean. Terakado et al (2005) reported that systemic effect of brassinosteroid on nodule formation in soybean after the foliar application of brassinolide and brassinazaole, the inhibitor of brasinosteroid formation. In addition, they reported that shoot applied polyamines suppressed nodule formation in soybean (Terakado et al., 2006). Suzuki et al. reported that nodule number is controlled by the abscisic acid in *Trifolium repense* (white clover) and *Lotus japonicus* (Suzuki et al., 2004).

2. Local effect of nitrate on nodule growth and nitrogen fixation

2.1 Rapid and reversible inhibition of nodule growth and nitrogen fixation by nitrate

Short-term local effect of nitrate supply on nodule formation and nitrogen fixation was evaluated using hydroponically grown soybean plants (cultivar Williams), which were inoculated with *Bradyrhizobium japonicum*, (strain USDA110) (Fujikake et al. 2002. 2003). In the first experiment (Fujikake et al. 2002), the diameter of nodules on the upper part of nodulated soybean roots in a glass bottle was measured with a slide caliper. Nodulated soybean (cv. Williams) plants were hydroponically cultured, and various combinations of one-week culture solution with 5 mM or 0 mM nitrate were applied using 13 days old soybean seedlings during three successive weeks. The treatments were designated as 0-0-0, 5-5-5, 5-5-0, 5-0-0, 5-0-5, 0-5-5 and 0-0-5, where the three sequential numbers denote the nitrate concentration (mM) applied in the first-second-third weeks. The size of the marked individual nodules was measured periodically using a slide caliper. All the plants were harvested after measurement of the acetylene reduction activity (ARA) at the end of the treatments. In the 0-0-0 treatment, the nodules grew continuously during the treatment period. As shown in Fig. 8., individual nodule growth was immediately suppressed after 5 mM nitrate supply. However, the nodule growth rapidly recovered by changing the 5 mM nitrate solution to a 0 mM nitrate solution in the 5-0-0 and 5-5-0 treatments. In the 5-0-5 treatment, nodule growth was completely inhibited in the first and the third weeks with 5mM nitrate, but the nodule growth was enhanced in the second week with 0 mM nitrate. The nodule growth response to 5 mM nitrate was similar between small and large size nodules.

In this experiment nodule numbers are not significantly affected by nitrate treatments (Fig. 9. A), although the nodule weight was significantly affected by the period of nitrate supply (Fig. 9. B), where 5-5-5 and 0-5-5 treatments depressed nodule dry weight about 1/3 of 0-0-0 plants. After the 5-5-5, 5-0-5, 0-0-5 and 0-5-5 treatments, where the plants were cultured with 5 mM nitrate in the last third week, the acetylene reduction activity (ARA) per a plant and ARA per g nodule dry weight (DW) were significantly lower compared with the 0-0-0 treatment (Fig. 10. A,B). On the other hand, the ARA after the 5-0-0 and 5-5-0 treatments was relatively higher than that after the 0-0-0 treatment, possibly due to the higher photosynthate supply associated with the vigorous vegetative growth of the plants supplemented with nitrate nitrogen. It is concluded that both soybean nodule growth and N_2 fixation activity sensitively responded to the external nitrate level, and that these parameters were reversibly regulated by the current status of nitrate in the culture solution, possibly through sensing of the concentration of nitrate or its assimilates in roots and/or nodules.

The nitrate concentration was analyzed in each organ of soybean harvested at the end of the treatment on 34 days after planting. In the plants supplied with 5 mM nitrate during the last week in both the first and second series of treatments (5-5-5, 5-0-5, 0-5-5 and 0-0-5 treatments), the nitrate concentration was significantly high in each organ. Especially the roots and stems accumulated about 9-14 gN kg^{-1}DW and about 5-9 gN kg^{-1}DW nitrate, respectively. On the other hand, the nitrate concentration in roots (0.19 gN kg^{-1}DW), stems (0.03 gN kg^{-1}DW) and nodules (0.11 gN kg^{-1}DW) was fairly low in the 5-5-0 treatment where nitrate was not supplied during the last third week. All the accumulated nitrate during the first and second weeks was reduced and assimilated during the third week of the 0 mM nitrate treatment under the experimental conditions. The nitrate concentration in the nodules was relatively lower than that in the roots and stems, but in the 5-5-5, 0-5-5, 0-0-5 treatments, the nodules accumulated more than 1 gN kg^{-1}DW nitrate.

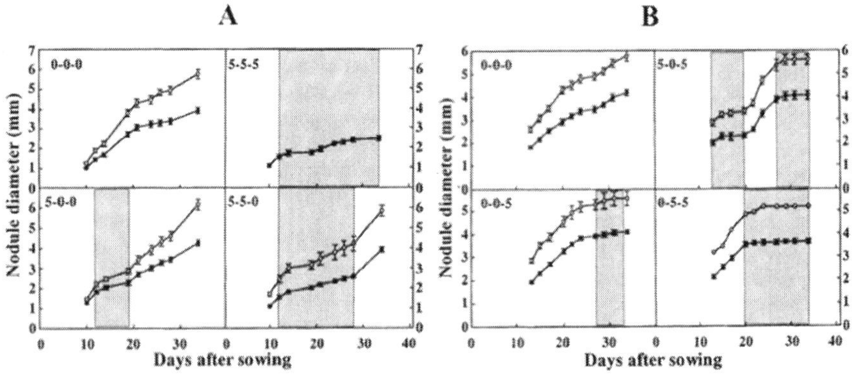

Fig. 8. Changes in nodule diameter of soybean plants with various nitrate treatments. Gray background shows the duration of 5 mM nitrate treatment, and white background shows the 0 mM nitrate. Open circle: large nodules, Closed circle: small nodules.

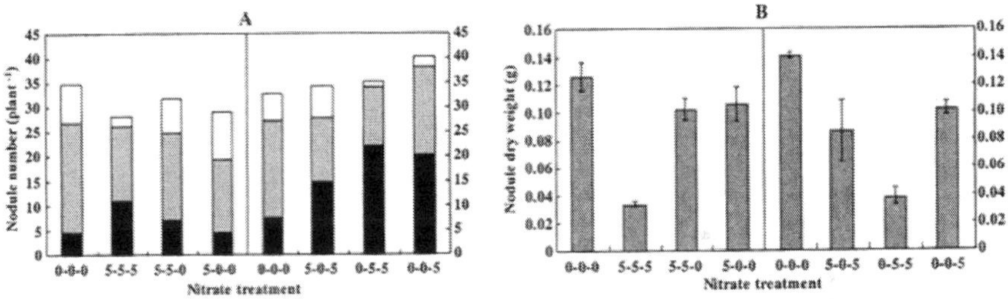

Fig. 9. Number (A) and dry weight (B) of nodules at the end on 34 days after planting. (A) nodule size were indicated by black column (3mm<), gray column (3-5 mm) or white column (<5mm).

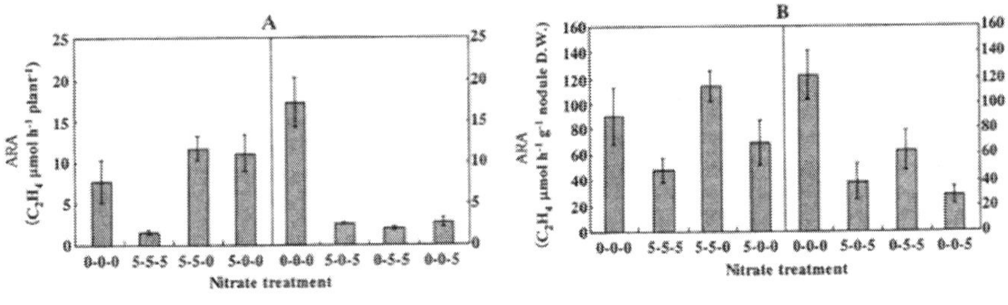

Fig. 10. Acetylene reduction activiety per a plant (A) and per nodule dry weight (B) of nitrate treatments on 34 days after planting.

In the second experiment (Fujikake et al., 2003) and the diameter of individual nodules was measured from 10-24 days after planting using a computer microscope under controlled environmental conditions (Fig. 11.). Photoes (Fig. 11A) and the diameter changes (Fig. 11B) of nodules were shown. A, nodule growth was rapid only under 0 mM nitrate conditions. The diameter of a nodule attached to the primary roots increased from 1 mm to 6 mm for 2 weeks under N free conditions (Fig. 11. Aa, Ba). The increase in nodule diameter was almost completely stopped after 1 d of supplying 5 mM NO_3^- (Fig. 11. Ab, Bb). However, nodule growth quickly returned to the normal growth rate following withdrawal of NO_3^- from the solution (Fig. 11. Ac, Bc).

The morphology of typical nodule slices of soybean observed by an optical microscope is shown in Fig. 12. (A) and the average size of infected cells, uninfected cells and inner cortex cells were measured (Fig. 12. B). It is conspicuous that the infected cells of nodules became larger under 0 mM nitrate condition (a) from 10 to 18 days after planting. On the other hand, the size of infected and uninfected cells and inner cortex cells remained small with 5 mM nitrate solution. Cell growth recovered rapidly after 2 days of 0 mM nitrate (18 days after planting) after 2 days of 5 mM nitrate (16 days after planting) (c). This result indicates that nodule growth at this stage depends on the cell expansion, rather than cell proliferation. The rapid and reversible nodule growth inhibition is caused by nodule cell growth.

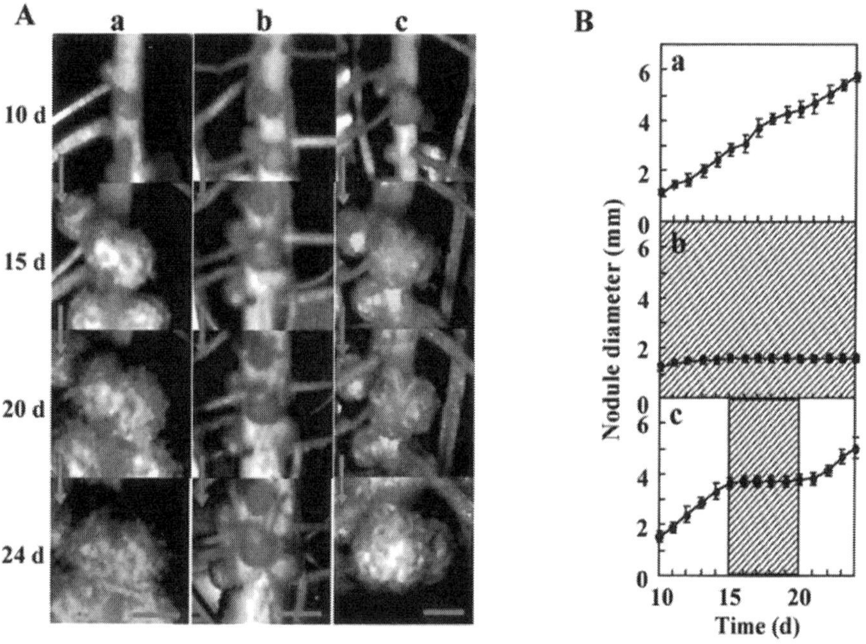

Fig. 11. (A) Growth response of soybean nodules to 0 mM nitrate (blue arrows) or 5 mM nitrate (red arrows) application in the culture solution. (B) Changes in nodule diameter with 0 mM (white background) or 5 mM (hatched background).

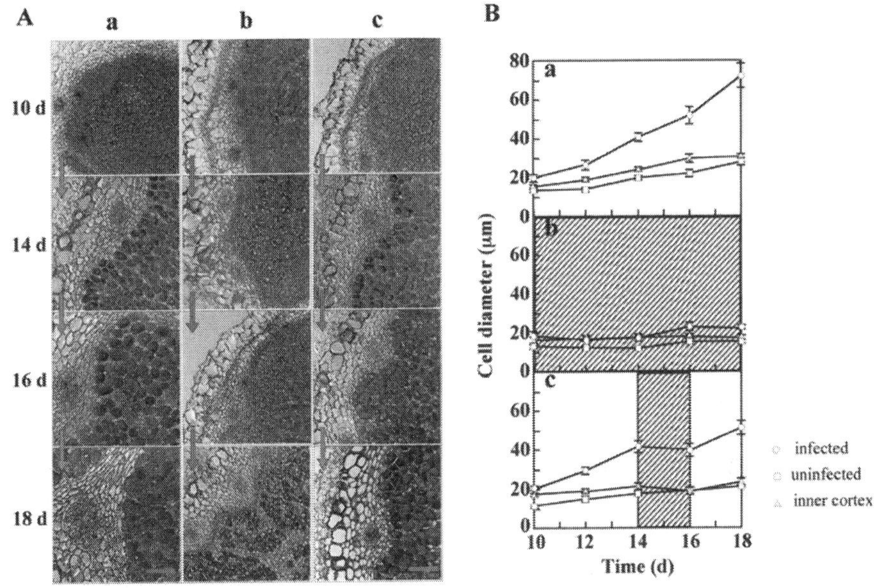

Fig. 12. Effect of 0 mM and 5 mM nitrate treatment on the size of nodule cells. (a) soybean plants were cultured in 0 mM nitrate for 10-18 days. (b) soybean plants were sultured with 5 mM nitrate for 10-18 days. (c) soybean plants were treated with 0 mM from 10-14 days and 5 mM nitrate from 14-16 days and 0 mM nitrate from 16 to 18 days after planting. The size of infected cells, uninfected cells and inner cortex cells were measured in B.

The effect of dark treatment on nodule growth was examined in combination with nitrate treatments for two days followed by normal light/dark conditions (Fig. 13.A.). Under continuous dark conditions, nodule growth maintained on the first day, but depressed on the second day. The reversible depression of nodule growth by NO_3^- was similar to the restriction of the photoassimilate supply under continuous dark conditions for 2 days. The nodule growth with 5 mM nitrate under continuous dark conditions depressed most severely among treatments. When plants were returned to the normal conditions (14 h light and 10h dark) with 0 mM nitrate, all the nodules recovered the growth rate.

The inhibitory effect of 5 mM nitrate was partially alleviated by the addition of 3% sucrose to the culture solution (Fig. 13.B.), suggesting that soybean root nodules are under carbon deficiency.

The positron emitting radioisotope $^{11}CO_2$ was supplied to the first trifoliolate leaves of 29 days after planting for 10 min, then the movement of ^{11}C was monitored by positron-emitting tracer imaging system (PETIS) (Fujikake et al., 2003). Split-root system was made by cutting the primary root of soybean seedling at 24 days after planting. Each split roots was supplied with solution containing with 0 mM or 5 mM NO_3^- for 3 days from 27-29 days after planting. Both sides of split roots were supplied with 0 mM, 0 mM (a), 5mM, 5 mM (b) or 0 mM, 5 mM (c). In the plants with 0 mM, 0 mM or 5 mM 5mM, the ^{11}C assimilated in the first trifoliate was translocated both upward to the young developing apical leaf bud and downward to the whole root system (Fig. 14. A). Very little ^{11}C was transported to the fully developed trifoliates and primary leaves. This means that fully developed leaves are not sink of photoassimilate from other leaves. Compared with the split root system supplied with 0 mM NO_3^- on one side and 5 mM NO_3^- on the other side, the transport rate of ^{11}C was faster in the split-roots supplied with 5 mM NO_3^- than those in 0 mM NO_3^- (Fig. 14. B, C). This result indicates that when NO_3^- is supplied to a part of the roots, photoassimilate flow become faster in this part.

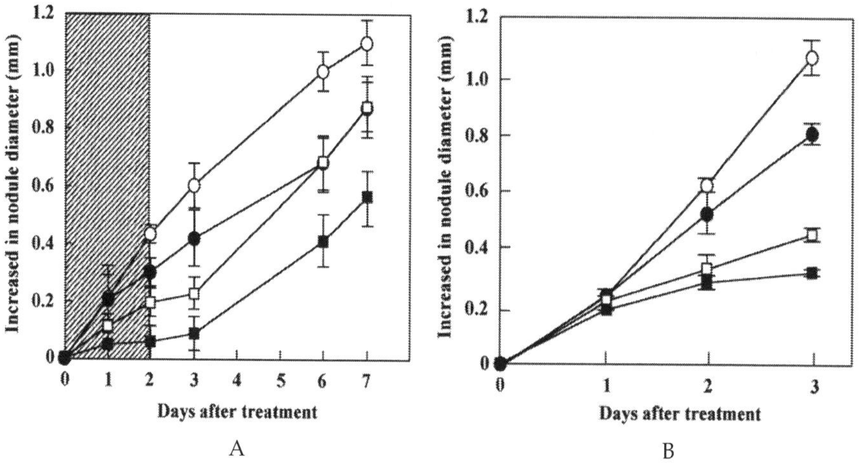

A: Open circle: 0 mM nitrate, light 14h/dark 10h. filled circle: 0 mM nitrate, continuous dark for 2 days. Open square: 5 mM nitrate, light 14h/dark 10h. filled square: 5 mM nitrate, continuous dark for 2 days. B: Effect of 3% sucrose on nodule growth (B). Open Circle: 3% sucrose + 0 mM nitrate. Filled circle: 0% sucrose + 0 mM nitrate. Open square: 3% sucrose + 5 mM nitrate. filled square: 0% sucrose + 5 mM nitrate.

Fig. 13. Nodule growth of soybean plants grown with 0 mM or 5 mM nitrate under light/dark conditions

Quantitative evaluation was conducted using ^{14}C as a tracer for the plants supplied either with 0 mM or 5 mM NO_3^- for one day before supplying $^{14}CO_2$. Whole shoot of a plant at 22 days after planting was exposed to the $^{14}CO_2$ for 120 min using circulation system. After $^{14}CO_2$ feeding, plant samples were immediately dried and the radioactivity in leaves, stems, nodules and roots were determined using Liquid Scintillation Counter. By supplying 5 mM NO_3^-, the partitioning to the underground part were almost the same in 0 and 5 mM nitrate treatments, but the ^{14}C partitioning to nodule decreased from 9.1 % to 4.3%, while that to the roots increased from 5.2 % to 9.1% (Fig. 15).

These results indicate that the decrease in photoassimilate supply to nodules may be involved in the quick and reversible nitrate inhibition of soybean nodule growth and N_2 fixation activity (Fig. 16.). The decrease in starch concentration in nodules (Vessey et al., 1988, Gordon et al., 2002) and the down-regulation of sucrose synthase transcript within 1 day of nitrate treatment (Gordon et al., 2002) may imply that NO_3^- reduces photosynthesis flow into nodules and sucrose utilization in nodules.

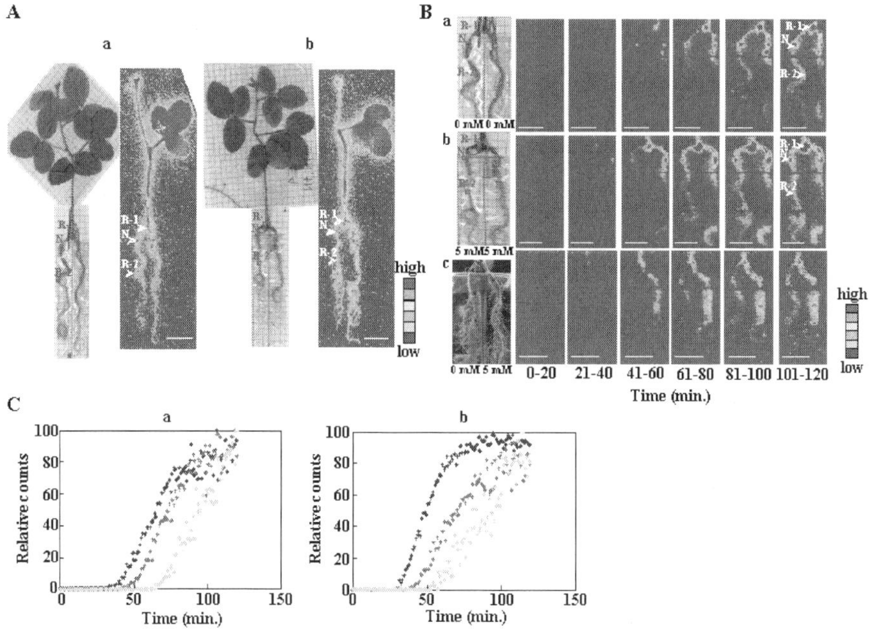

Fig. 14. ^{11}C translocation to the split root systems from first trifoliate leaves of 29 days old plants. (A) Images of the distribution of ^{11}C in soybean by Bioimaging analyzer. (B) The time course for the accumulation of radioactivity as shown by PETIS. (C) The accumulation of radioactivity for the point of R-1 (blue), N (red) and R-2 (yellow).

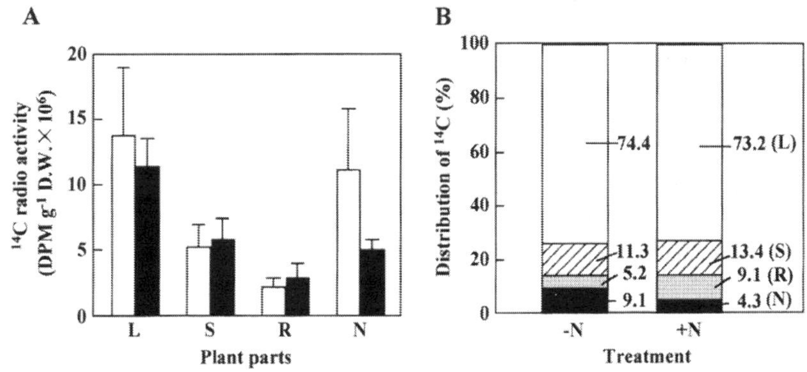

Fig. 15. Partitioning of ^{14}C labeled photoassimilate in soybean plants. (A) Radioactivity per g dry weight of each part after one day of 2 h $^{14}CO_2$ feedings to a whole shoot, with 0 mM nitrate (white column) and 5 mM nitrate (black column). (B) Distribution of ^{14}C among organs with 0 mM (-N) or 5 mM (+N) treatment for one day. (L): leaves, (S): stems, (R): roots, (N): nodules.

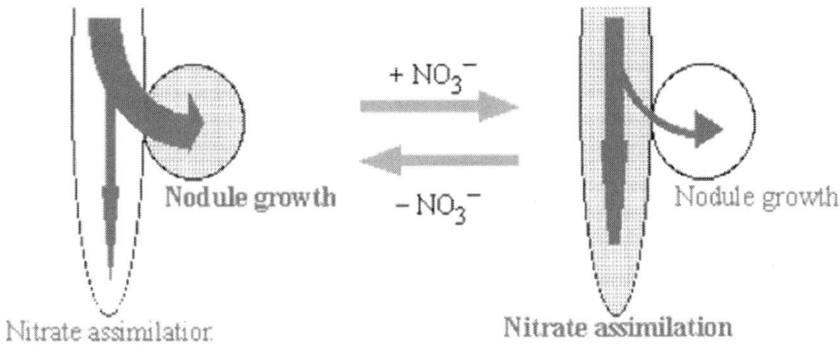

Fig. 16. A model of photoassimilate transport to nodule and roots under no nitrate or presence of nitrate.

2.2 The routes of nitrate entry into nodules

It is well known that nitrate inhibit nodule growth and nitrogen fixation activity, but the routes of nitrate entry into nodule has not fully been understood. It is postulated that there are several routes in nodules. First, NO_3^- is absorbed from the roots and transported to nodules through the xylem. However, xylem transport of NO_3^- to nodules is negligible, because nitrate accumulation is very low in separated nodules in the upper part of roots, when NO_3^- was supplied in the lower roots. In addition the role of xylem in nodule is the export of assimilated nitrogen from nodules to shoots, rather than the import of water and minerals from roots. Most of water and minerals are supplied from shoots to nodules though phloem. Second route is NO_3^- supplied via phloem. However, usually NO_3^- concentration in phloem is very low due to nitrate reduction in leaves. Third is NO_3^- is taken up from the subtending roots and transported from root cortex to the nodule cortex via symplastic pathway (Streeter 1993). Forth, NO_3^- is absorbed from nodule surface.

The nitrate transport pathway into soybean nodules were investigated using tungstate (WO_4^{2-}) and $^{15}NO_3^-$ as a tracer (Mizukoshi et al. 1995). Tungstate was used as an anion tracer as an analogue of nitrate (NO_3^-). The distribution of tungsten (W) was observed by an Electron Probe X-ray Microanalysis (EPMA). At 3 days after 1 mM tungstate treatment, a large amount of W accumulated in the root cortex, and the import of W into stele was restricted (Fig. 17A, C). It is well known that there is a barrier in endodermis between cortex and stele, where water movement is not allowed. Therefore, solutes should pass through endodermis through inside the cells via symplastic pathway. In addition, the movement of W inside the nodule was negligible (Fig. 17 C). This result suggests that the external anions cannot be readily enter into the cortex of nodules through appoplastic pathway by diffusion. In contrast, nodulated roots were treated with 1.7 mM $^{15}NO_3^-$ for one day, a relatively large amount of nitrate was accumulated in nodule cortex, although NO_3^- and ^{15}N were negligible in the symbiotic region. This result indicates that nitrate can be absorbed from the nodule surface into cytoplasm of nodule epidermis, and it is transported by symplastic pathway through plasmodesmata, and accumulated in the nodule cortex cells.

As shown in Fig. 19., the accumulation of NO_3^- in nodule cortex may be involved in the restriction of O_2 permeability into symbiotic resion, then it decreases the nitrogen fixation and nodule growth. Witty and Minchin (1990) reprted that the NO_3^- treatment decreased the air space within a oxygen barrier in inner cortex of nodules.

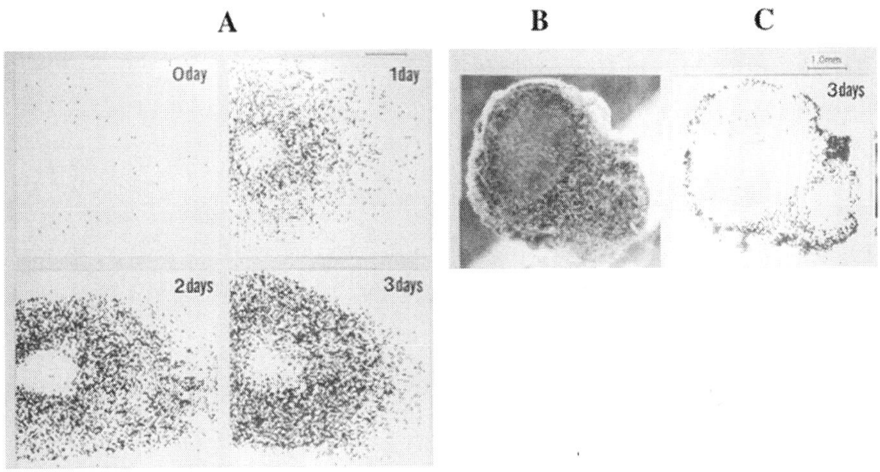

Fig. 17. Distribution of tungsten (W) in soybean root at 0, 1, 2, 3 days after 1 mM tungstate treatment (A). Distribution of tungsten (W) in soybean nodule with root at 3 days after 1 mM tungstate treatment (A). The W concentration is higher indicated by red, yellow , green, blue, and white, in this sequence.

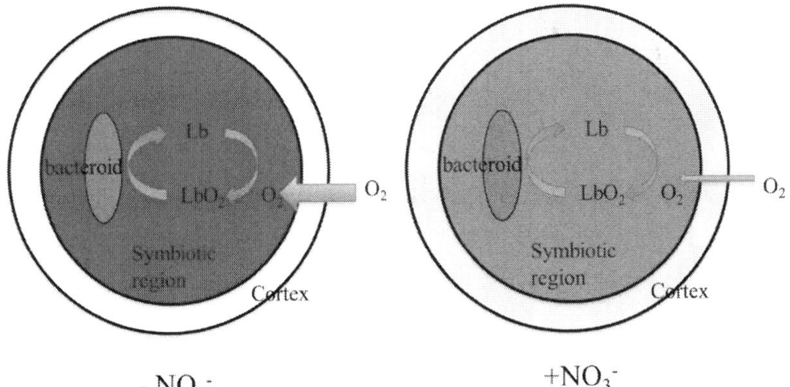

Fig. 18. Effect of nitrate accumulation in cortex of soybean nodule on oxygen permeability and nitrogen fixation activity.

Sato et al. (1999) analyzed nitrogen absorption and translocation in non-nodulated and nodulated soybean plants cultivated with 0 mM or 1 mM nitrate (Fig. 19.). The radioactivity was measured in a first trifoliate leaf after addition of $^{13}NO_3^-$ supply to the root solution. The relative radioacrivity from 0 to 40 min after $^{13}NO_3^-$ supply were similar between non-nodulated and nodulated soybean with 0 or 1 mM nitrate (Fig. 20.). This suggests that nodulation does not change the absorption and transport pattern of nitrate absorbed in the roots. However, quantitative measurement using stable isotope $^{15}NO_3^-$, total amount of ^{15}N was higher in non-nodulated soybean than nodulated soybean, especially with 1 mM $^{15}NO_3^-$, this is due to the increase in the root mass in these plants.

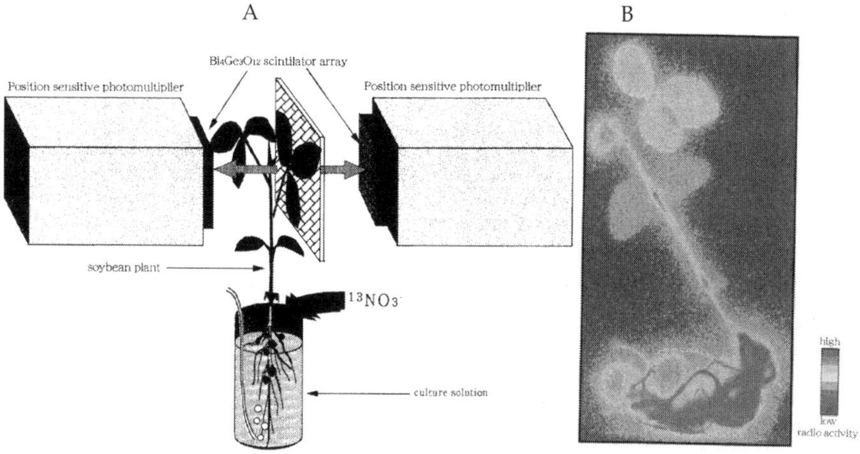

A: Positron Emitting Tracer Imaging System. B: Image of ^{13}N radioisotope in non-nodulated soybean plant (T202).

Fig. 19. Real time observation of radioactivity in the first trifoliate of soybean plant after $^{13}NO_3^-$ was supplied to the root solution.

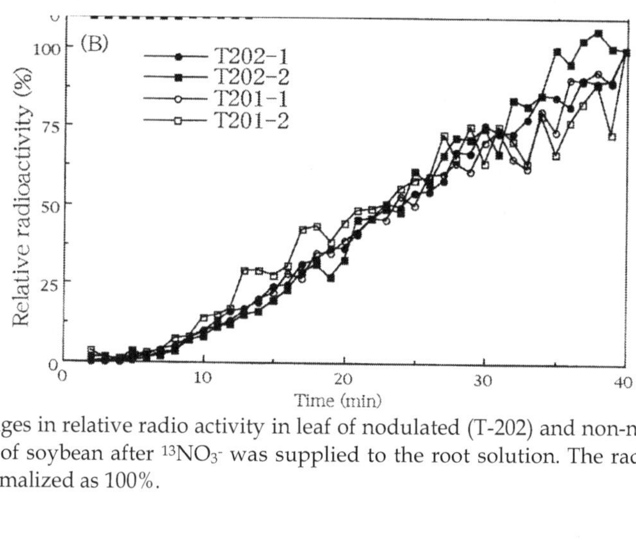

Fig. 20. Changes in relative radio activity in leaf of nodulated (T-202) and non-nodulated (T-201) isolines of soybean after $^{13}NO_3^-$ was supplied to the root solution. The radioactivity at 40 min is normalized as 100%.

3. Systemic and long-term effect of nitrate on nodule growth and nitrogen fixation

3.1 The effect of 0mM or 5 mM nitrate application in upper and lower part of roots

Local and systemic effects by nitrate on nodulation have been reported in leguminous plants. The local effect of nitrate inhibition was shown in split-root experiments where root systems had been separated into two equivalent parts. The strong and rapid nitrate inhibition of nodule growth and N_2 fixation activity is restricted in the nodules attached to the root portions that are in direct contact with nitrate; and no or milder inhibition is induced in the other part of the root system receiving no nitrate (Tanaka et al., 1985). However, some systemic inhibition of nitrate on nodulation and nitrogen fixation has also been observed with a high concentration of nitrate in clover (Silsbury et al., 1986).

We investigated the local and systemic effects of continuous supply of NO_3^- by using horizontal sprit-root system in two-layered pot system, where the lower part of roots were supplied with culture solution containing 1mM NO_3^- in the lower pot, and the upper roots were in the vermiculite medium with N-free culture solution in the upper pot (Ohyama et al., 1993). The soybean plants (cv. Williams and Norin No.2) were cultivated with 0 mM or 1 mM NO_3^- solution in the lower pot, and harvested at maturing (R7) stage (Fher et al., 1971). In this stage, there are no nodules remained in the lower part of roots. The dry weight of shoot and upper part of roots were almost the same between 0 mM and 1 mM NO_3^- supply, but nitrate treatment decreased the dry weight of nodules attached in the upper part of roots in both varieties. This result indicates that continuous long-term supply of NO_3^- may impose systemic inhibition of nodulation in soybean plants.

Systemic and local effects of long-term application of nitrate on nodule growth and N_2 fixation in soybean plants were more precisely investigated using two layered pot system. Four treatments were imposed i.e., 0/0, 0/5, 5/0 and 5/5, with the 0 mM or 5 mM NO_3^- treatment in upper pot/ lower pot, respectively. The plants were harvested at the initial flowering (R1) stage and pod setting (R4) stage, and the effect of nitrate placement on nodule number, nodule growth, and N_2 fixation activity in the upper and lower pots were elucidated (Yashima et al., 2003).

The development of the root system in the lower pots was quite different between 0 and 5 mM NO_3^- in the lower pot (Fig. 21.). The root length was longer in 0 mM treatment in lower pot (0/0, 5/0), but a bunch of short lateral roots was formed in the solution with 5 mM NO_3^- in lower pot (0/5, 5/5). In the lower pot where the nodules were in direct contact with 5 mM NO_3^-, the inhibition on the nodule number, nodule size and N_2 fixation was conspicuous. Systemic and local effect on nodule number per a plant did not occur in the upper nodules in vermiculite. On the other hand, systemic inhibition on the nodule dry weight and N_2 fixation activity in the upper pot was apparent. The 5/5 treatment depressed the nodule growth and nitrogen fixation activity in the upper nodules. Nitrate accumulation was observed only in the part of roots and nodules in direct contact with 5 mM NO_3^- either in the upper or lower pot. The concentration of total amino acids was higher in the lower roots in 0/5 treatment than those in 0/0 treatment, however, that was almost the same level in the roots and nodules of the upper pot both at R1 and R4 stage. The soluble sugar concentration in the lower roots in 0/5 treatment was lower than that in the 0/0 treatment. The similar trend was observed in the upper roots of 0/5 treatment, suggesting that the absorption of NO_3^- from the lower roots decrease sugar concentration in both lower roots in direct contact with nitrate, and the upper roots not contact with NO_3^-.

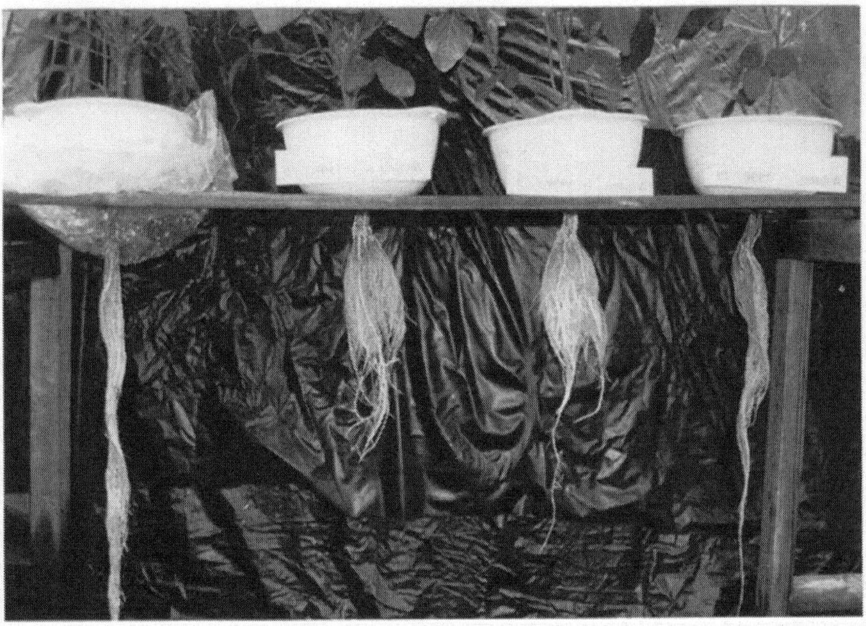

Fig. 21. Root system in the lower pot treated with 0/0 (A), 0/5 (B), 5/5 (C) and 5/0 (D) treatments (upper pot mM nitrate / lower pot mM nitrate).

3.2 Long-term effects of 0 mM, 1 mM, 5 mM nitrate application in lower part of roots

Long-term effect of NO_3^- application from the lower part of roots on the nodulation of the upper part of roots was further investigated in relation to concentration and treatment period (Yashima et al. 2005). The solution with 0 mM, 1mM or 5 mM NO_3^- was supplied from transplanting to two-layered pot system at 14 days after planting to R7 stage. Five treatments were imposed that 0-0 treatment (continuous 0 mM NO_3^-), 1-1 treatment (continuous 1 mM NO_3^-), 5-5 treament (continuous 5 mM NO_3^-), 0-5 treatment (0 mM until R3 then 5 mM NO_3^-), and 5-0 treatment (5 mM until R3 then 0 mM NO_3^-).

Total plant dry weight and seed dry weight at R7 stage was the highest in 5-5 treatment, intermediate in the 1-1, 5-0, 0-5 treatments, and lowest in the 0-0 treatment (Fig. 22.).

Fig. 23. shows the nodule number per a plant classified with nodule diameter. Nodule number in the upper pot was higher in 5-5 and 1-1 treatments than 0-0 treatment, although proportion of the small nodules under 4 mm diameter was higher in 5-5 treatment. The nitrate supply in the lower pot increased the total nodule number in the upper roots, although decreased the number of nodules in the lower roots.

The value of the nodule dry weight per a plant (Fig.24.) and N_2 fixation activity (acetylene reduction activity) per a plant (Fig. 25) and that per nodule dry weight (Fig. 26.) were lowest in the 5-5 treatment. Interestingly, the nodule dry weight in the upper roots was highest in the plants with 1-1 treatment, which exceeded the 0-0 treatment (Fig. 24.). The acetylene reduction activity per a plant of the upper nodules at R3 stage was also the highest in the 1-1 treatment (Fig. 25.). This was due to the nodule dry weight and not ARA per dry weight of nodules. These results indicated that continuous supply of low concentration of NO_3^- from the lower roots does not inhibit the nodule growth and N_2 fixation activity, but it can promote nodulation and N_2 fixation. Fig. 27. shows an example of soybean root systems cultivated with continuous supply of 1 mM nitrate at R3 stage. Nodulation was enhanced in the upper roots by supplying 1mM NO_3^- from the lower roots (Fig. 27. Ba), where the nodulation was severely depressed in the lower pot (Fig. 27. Bb).

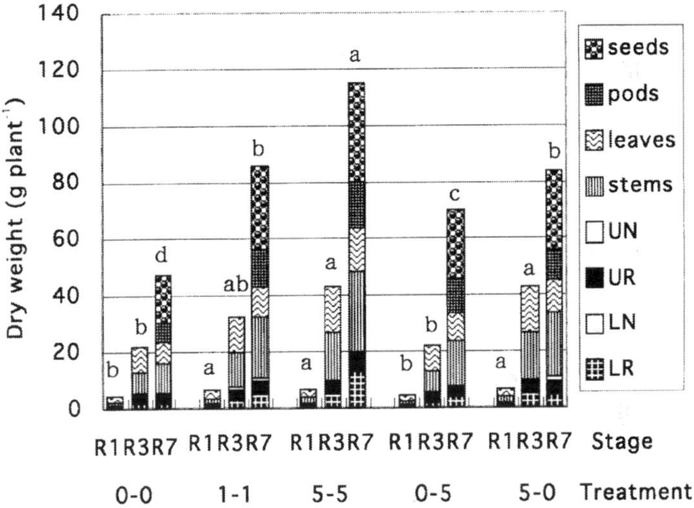

Different alphabet on the column means the statistical difference (P<0.05) between treatments.

Fig. 22. Changes in dry weight of each part of soybean plants at R1 (initial flowering stage; 40 days after planting), R3 (initial pod setting stage, 60 days after planting), R7 (maturing stage; 100 days after planting) with various nitrate treatments. 0-0, 1-1, 5-5, 0-5, 5-0 indicates mM nitrate concentration from transplanting-R3 stage, and R3-R7 stage respectively.

Withdrawal of 5 mM NO_3^- after R3 stage in 5-0 treatment markedly enhanced nodule growth and acetylene reduction activity at R7 stage when the values of both parameters decreased in the other treatments. The nitrate concentration of the nodules attached to the upper roots was very low, including continuous supply of high concentration of NO_3^- in 5-5 treatment. This result indicates that the inhibitory effect of 5 mM NO_3^-, or promotive effect of 1 mM NO_3^- was not directly controlled by nitrate itself, but was mediated through some systemic regulation.

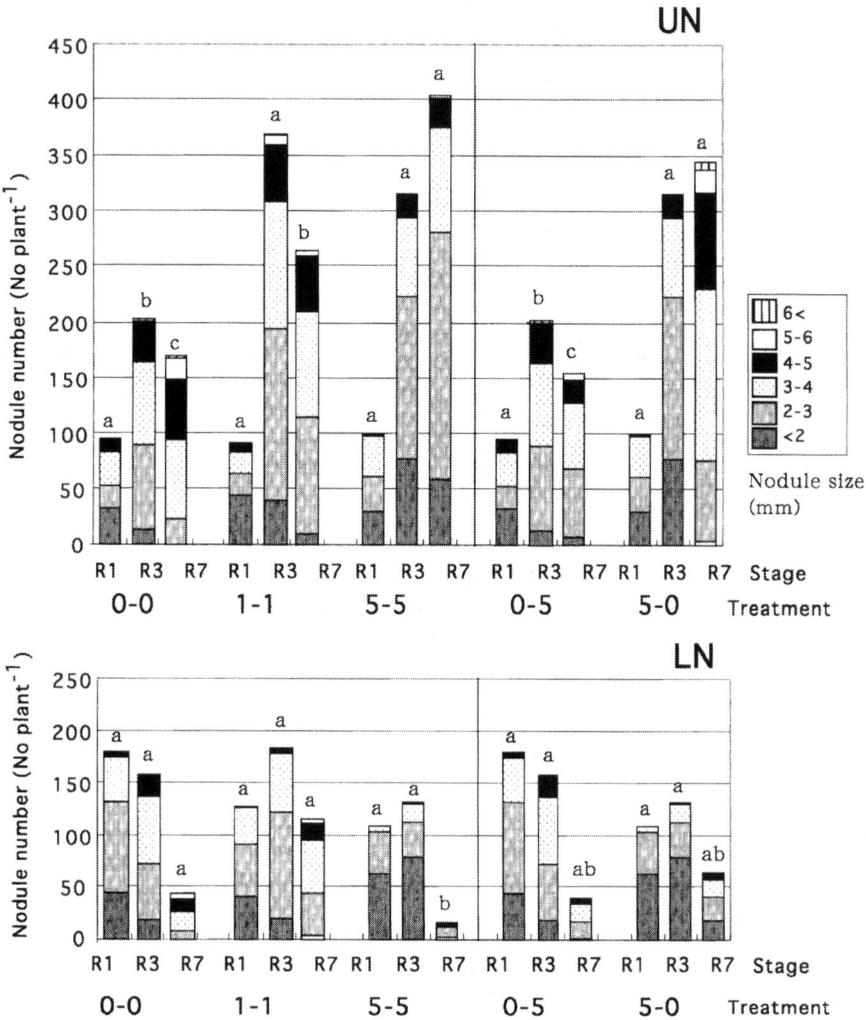

UN: upper nodules, LN: lower nodules.
Different alphabet on the column means the statistical difference ($P<0.05$) between treatments.

Fig. 23. Nodule number per a plant with various nitrate treatment in the lower pot. Size distribution is shown.

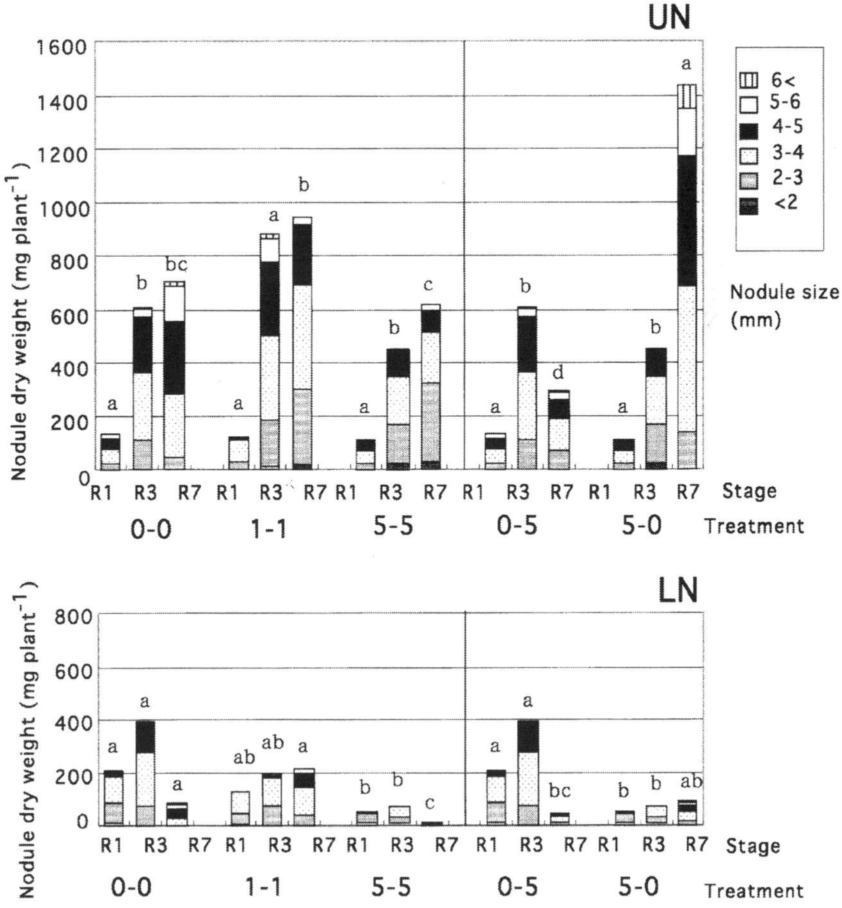

UN: upper nodules, LN: lower nodules.
Different alphabet on the column means the statistical difference ($P<0.05$) between treatments.

Fig. 24. Nodule dry weight per a plant with various treatment of nitrate in a lower pot.

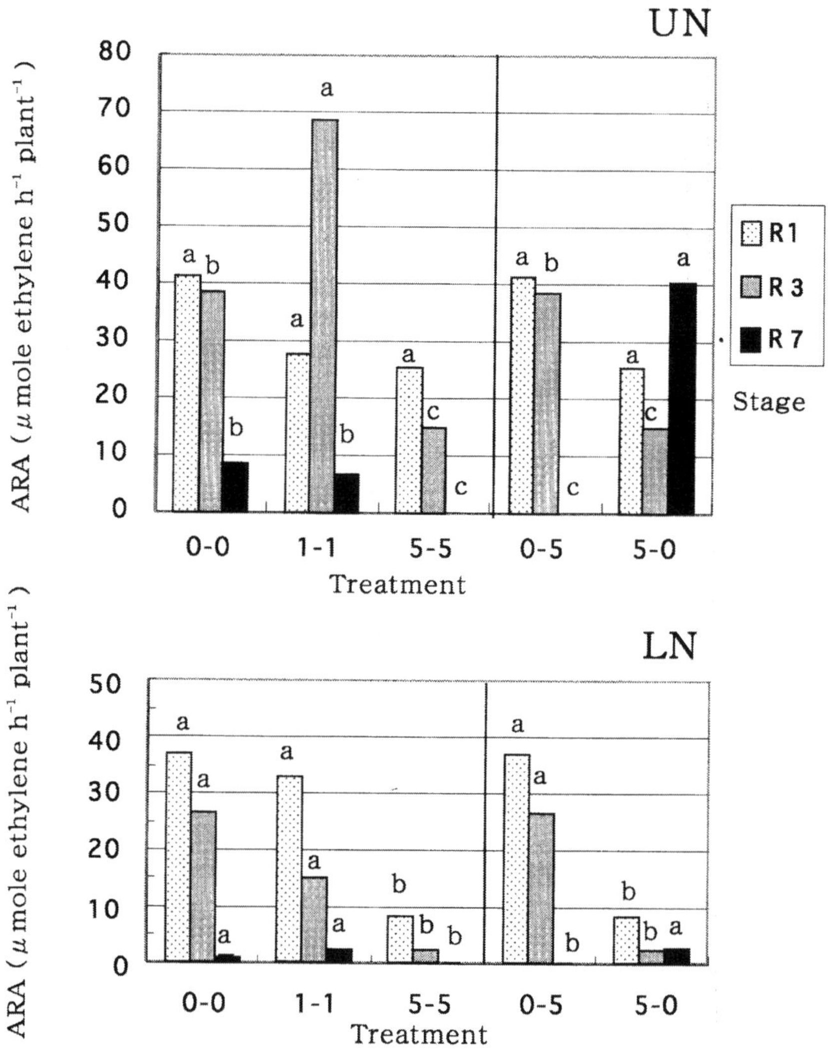

UN: upper nodules, LN: lower nodules.
Different alphabet on the column means the statistical difference ($P<0.05$) between treatments.

Fig. 25. Nitrogen fixation activity per a plant in soybean plants treated with various nitrate supply from lower pot.

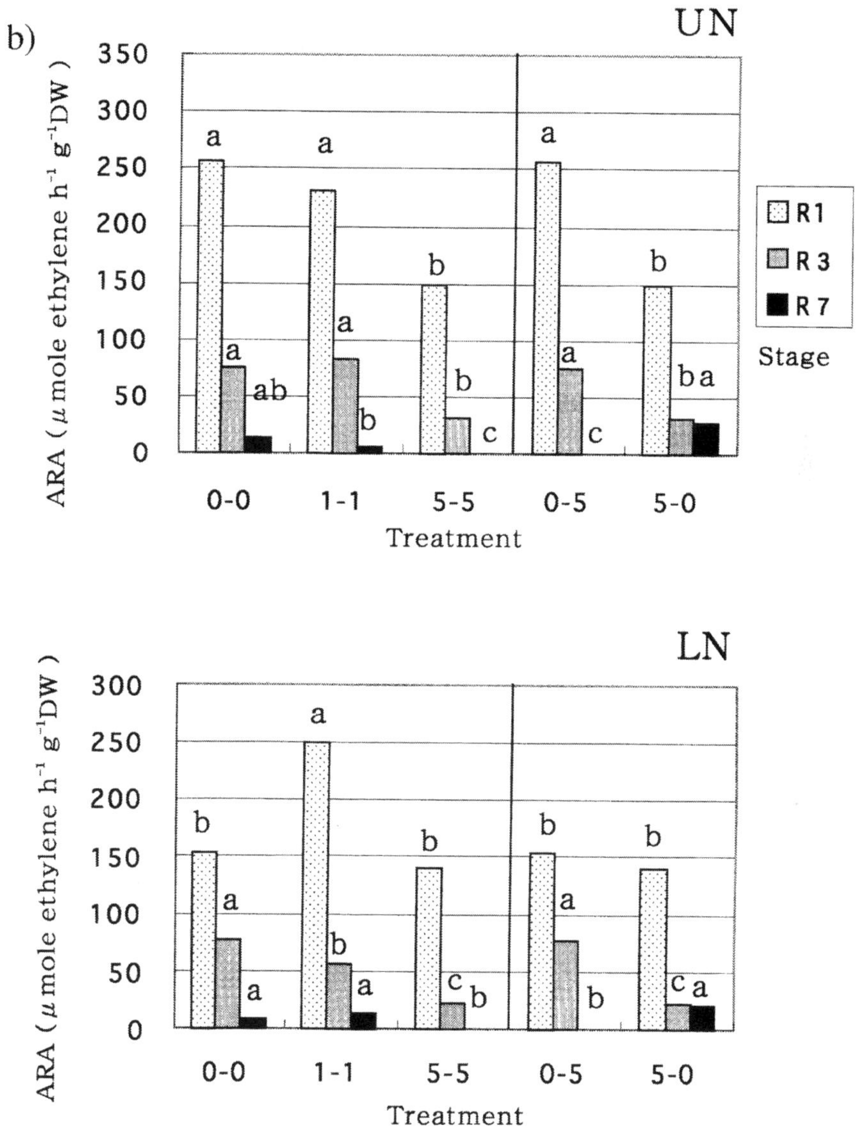

Different alphabet on the column means the statistical difference ($P<0.05$) between treatments.

Fig. 26. Nitrogen fixation activity per gram nodule dry weight in soybean plants treated with various nitrate supply from lower pot. UN: upper nodules, LN: lower nodules.

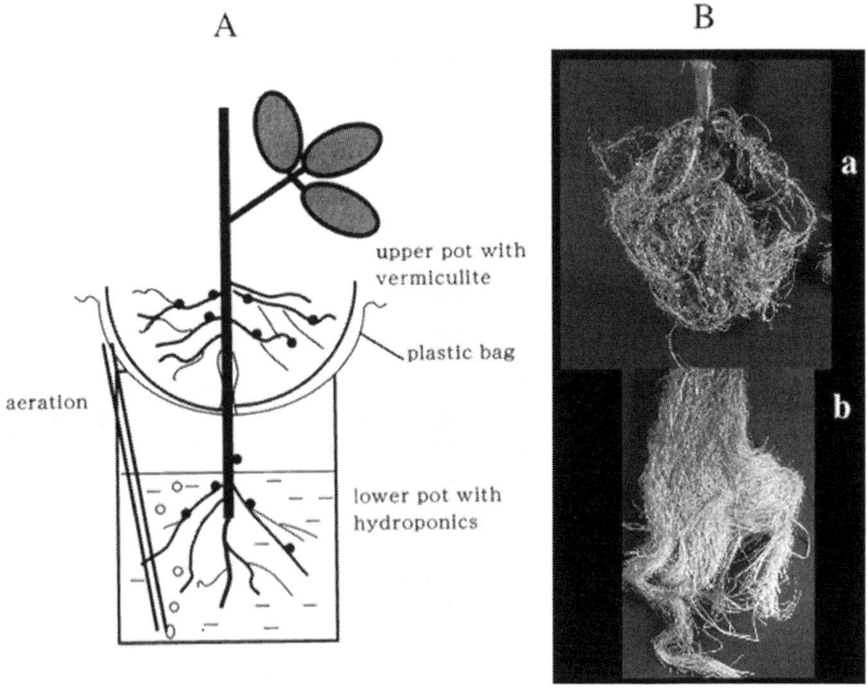

A: Apparatus for two-layered pot.
B: photograph of root system in the upper roots with 0 mM nitrate (a) and the lower roots with 1 mM nitrate(b).

Fig. 27. A vertical split root system used in long-term effect of nitrate from lower pot.

4. Conclusion

In this chapter, we focus on the effects of nitrate supply on nodule growth and nitrogen fixation in soybean plants. First, we found the rapid and reversible inhibition of 5mM nitrate on nodule growth and nitrogen fixation activity in soybean plants. When young soybean plant grown in hydroponic culture was supplied with 5 mM nitrate solution, the nodule growth was completely stopped after one day of application. The culture solution was changed back to nitrogen free solution the nodule growth and nitrogen fixation activity were recovered at one day after changing. The inhibitory effect by nitrate may be due to the changes in photoassimiate supply from nodule to the roots by the experiments exposing the shoot to positron emitting radioisotope ^{11}C or radioisotope ^{14}C labeled CO_2.

Second, the effect of long-term application of nitrate was evaluated by the vertical split root experiments using two-layered pot, which separates the upper and lower parts of the root system. Both direct and systemic inhibitions were observed for nodule dry weight and nitrogen fixation activity with long-term supply of 5 mM nitrate. Severe inhibition in the root part in direct contact with 5mM nitrate solution was observed, although milder depression was in the separate parts not in direct contact with nitrate. Nitrate accumulation in nodule was observed only in the root part in direct contact with 5 mM nitrate. When the 1mM concentration of nitrate was continuously supplied from the lower roots, the nodule growth and nitrogen fixation activity in the upper roots were promoted compared with control plants supplied with nitrogen free solution. When 5 mM nitrate supply has stopped from pod setting stage, nodule weight and nitrogen fixation activity were recovered and exceeded over control plant at maturing stage.

The routes of nitrate entry into nodules was investigated. It was suggested that most of nitrate is absorbed through the surface of the nodule, but little is transported from the lower part of roots or shoots, either via xylem or phloem. The accumulation of nitrate in the cortex of nodule may inhibit respiration by decreasing O_2 permeability or suppress the photoassimilate import to the central symbiotic region of nodules, where bacteroid reside and fix nitrogen.

From the characteristics of nitrate effects on soybean nodulation and nitrogen fixation, we have developed a new fertilizer method "deep placement of slow release nitrogen fertilizer" for promoting soybean seed yield without inhibiting nodulation and nitrogen fixation (Takahashi et al., 1991, 1992, 1999, Tewari et al., 2010).

5. References

Atkins, C.A. (1991). Ammonia assimilation and export of nitrogen from the legume nodule. In *Biology and Biochemistry of Nitrogen Fixation*, Elsevier Science Publishers, 293-319

Bacanamwo, M. & Harper, J.E. (1996). Regulation of nitrogenase activity in *Bradyrhizobium japonicum*/soybean symbiosis by plant N status as determined by shoot C:N ratio. *Physiol. Plant.*, 98, 529-538.

Bacanamwo, M. & Harper, J.E. (1997). The feedback mechanism of nitrate inhibition of nitrogenase activity in soybean may involve asparagines and/or products of its metabolism. *Physiol. Plant.* 100, 371-377.

Bloom, A.J. (2011). Energetics of Nitrogen Acquisition, *Annual Plant Reveiws*, 42, 63-81

Carroll, B.J. & Mathews, A. (1990). Nitrate inhibition of nodulation in legumes. In: Gresshoff PM, ed. *Molecular Biology of Symbiotic Nitrogen Fixation*. Boca Raton, FL: CRC press, 159-180.

Davidson, I.A. & Robson, M.J. (1986). Effect of contrasting patterns of nitrate application on the nitrate uptake, N_2-fixation, nodulation and growth of white clover. *Ann. Botany* 57, 331-338.

Fehr, W.R., Caviness, C.E., Burmood, D.T., and Pennington, J.S., 1971. Stage development description for soybean *Glycine max* (L.) Merill. *Crop Sci.*, 11, 929-931.

Fleurat-Lessard, P., Michonneau, P., Maeshima, M., Drevon, J-J. & Serraj, R. (2005). The Distribution of Aquaporin Subtypes (PIP1, PIP2 and g-TIP) is Tissue Dependent in Soybean (*Glycine max*) Root Nodules, *Annals of Botany* 96: 457–460, 2005

Fuchsman, W.H., Barton, C.R., Stein, M.M., Thomson, J.T. & Willett, R.M., (1976). Leghemoglobin: Different roles for different components?, *BBRC*, 68, 387-392

Fujikake, H., Yashima, H., Sato, T., Ohtake, N., Sueyoshi, K., and Ohyama, T., (2002) Rapid and reversible nitrate inhibition of nodule growth and N_2 fixation activity in soybean (*Glycine max* (L.) Merr.). *Soil Sci. Plant Nutr.*, 48, 211-217

Fujikake, H., Yamazaki, A., Ohtake, N., Sueyoshi, K., Matsuhashi, S., Ito, T., Mizuniwa, C., Kume, T., Hashimoto, S., Ishioka, N-S., Watanabe, S., Osa, A., Sekine, T., Uchida, H., Tsuji, A. & Ohyama, T. (2003). Quick and reversible inhibition of soybean root nodule growth by nitrate involves a decrease in sucrose supply to nodules. *Journal of Experimental Botany*, 54, 1379-1388

Fujimaki, S., Ishii, S. & Ishioka N-S. (2010). Non-invasive imaging and analyses of transport of nitrogen nutrition in intact plants using a positron-emitting tracer imaging system (PETIS). Nitrogen Assimilation in Plants, Ed. Ohyama T. & Sueyoshi K. Research Signpost, Kerala

Giannakis, C., Nicholas, D.J.D. & Wallace, W. (1988). Utilization of nitrate by bacteroids of *Bradhyrhizobium japonicum* in the soybean root nodule. *Planta*, 174, 51-58

Gibson, A.H. & Harper, J.E. (1985). Nitrate effect on nodulation of soybean by *Bradyrhizobium japonicum*. *Crop Science*, 25, 497-501.

Gordon, A.J., Skøt, L., James, C.L. & Minchin, F.R., (2002) Short-term metabolic response of soybean root nodules to nitrate. *Journal of Experimental Botany.*, 53, 423-428.

Harper, J.E. (1974). Soil and symbiotic nitrogen requirements for optimum soybean production. *Crop Sci.* 14, 255-260.

Harper, J.E. & Gibson, A.H., (1984). Differential nodulation tolerance to nitrate among legume species, *Crop Sci.* 14, 255-260

Harper, J.E. (1987). *Nitrogen metabolism. Soybeans: Improvement, production and Uses.* 2nd ed. *Agronomy Monograph* no.16. ASA-CSSA-SSSA. 497-533.

Hunt, S. & Layzell, D.B., (1993). Gas exchange of legume nodules and the regulation of nitrogenase acrivity, *Annu. Rev. Plant Physiol. and Plant Mol. Biol.*, 44, 483-511.

Ishii, S., Suzui. N., Ito, S., S-Ishioka, N., Kawachi, N., Ohtake, N., Ohyama, T. & Fujimaki, S. (2009). Real-time imaging of nitrogen fixation in an intact soybean plant with nodules using ^{13}N-labbeled nitrogen gas. *Soil Sci. Plant Nutr.* 55, 660-666

Ito, S., Ohtake, N., Sueyoshi, K. & Ohyama, T. (2006). Allocation of photosynthetic products in soybean during the early stages of nodule formation. *Soil Sci. Plant Nutr.*, 41, 438-443

Kanayama, Y. & Yamamoto, Y. (1990). Inhibition of nitrogen fixation in soybean plants supplied with nitrate I. Nitrite accumulation and formation of nitrosylleghemoglobin in nodules. *Plant Cell Physiol.* 31, 341-346

Lee, K.H. and LaRue, T.A., (1992). Ethylene as a possible mediator of light-and nitrate-indulced inhibition of nodulation of *Pisum sativum* L. cv Sparkle. *Plant Physiol.*, 100, 1334-1338

Ligero, F., Caba, J.M., Lluch, C., Oliveres, J., (1991). Nitrate inhibition of nodulation can be overcome by the ethylene inhibitor aminoethoxyvinylglycine. *Plant Physiology* 97, 1221-1225.

Mizukoshi, K., Nishiwaki, T., Ohtake, N., Minagawa, R., Ikarashi, T., & Ohyama, T. (1995). Nitrate transport pathway into soybean nodules traced by tungstate and $^{15}NO_3^-$. *Soil Sci. Plant Nutr.*, 41, 75-88

Neo, H.H. & Layzell, D.B. (1997). Phloem glutamine and the regulation of O_2 diffusion in legume nodules. *Physiologia. Plantarum* 113, 259-267.

Nishiwaki, T., Mizukoshi, K., Kobayashi, K., Ikarashi, T. & Ohyama, T. (1995). Effect of culture medium compositions on nodulation, nitrogen fixation and growth of soybean plant. *Bull. Facul. Agric. Niigata Univ.*, 47, 73-83

Ohtake, N., Nishiwaki, T., Mizukoshi, K., Minagawa, R., Takahashi, Y., Chinushi, T. & Ohyama ,T. (1995). Amino acid composition in xylem sap of soybean related to the evaluation of N_2 fixation by the relative ureide method. *Soil Sci.Plant Nutr.*, 41, 95-102.

Ohyama, T. & Kumazawa, K. (1978). Incorporation of ^{15}N into various nitrogenous compounds in intact soybean nodules after exposure to $^{15}N_2$ gas. *Soil Sci.Plant Nutr.*, 24, 525-533.

Ohyama, T. & Kumazawa, K. (1979). Assimilation and transport of nitrogenous compounds originated from $^{15}N_2$ fixation and $^{15}NO_3^-$ absorption. *Soil Sci.Plant Nutr.*, 25, 9-19

Ohyama, T. & Kumazawa, K. (1980a). Nitrogen assimilation in soybean nodules I. The role of GS/GOGAT system in the assimilation of ammonia produced by N_2 fixation. *Soil Sci.Plant Nutr.*, 26, 109-115

Ohyama, T. & Kumazawa, K. (1980b). Nitrogen assimilation in soybean nodules II. $^{15}N_2$ assimilation in bacteroid and cytosol fractions of soybean nodules. *Soil Sci.Plant Nutr.*, 26, 205-213

Ohyama, T. & Kumazawa, K. (1980c). Nitrogen assimilation in soybean nodules III. Effect of rhizaosphere pO_2 on the assimilation of $^{15}N_2$ in nodules attatched to intact plants. *Soil Sci.Plant Nutr.*, 26, 321-324

Ohyama, T. Owa N, Fujishima Y & Kumazawa, K. (1981a). Nitrogen assimilation in soybean nodules IV. Allantoin formation and transport in relation to supply with various forms of combined nitrogen. *Soil Sci.Plant Nutr.*, 27, 55-64

Ohyama, T. & Kumazawa, K. (1981b). Nitrogen assimilation in soybean nodules V. Possible pathway of allantoin synthesis in soybean nodules. *Soil Sci.Plant Nutr.*, 27, 111-114

Ohyama, T. (1983). Comparative studies on the distribution of nitrogen in soybean plants supplied with N_2 and NO_3^- at the pod filling stage. *Soil Sci.Plant Nutr.*, 29, 133-145.

Ohyama, T. (1984). Comparative studies on the distribution of nitrogen in soybean plants supplied with N_2 and NO_3^- at the pod filling stage. II. Assimilation and transport of nitrogenous constituents. *Soil Sci.Plant Nutr.*, 30, 219-229.

Ohyama, T., Mizukoshi, K., Nishiwaki, T. (1993a). Distribution of ureide originated from nodules attached to the upper roots and nitrate derived from lower roots in soybean plants cultivated by double piled pots. *Bull. Facul. Agric. Niigata Univ.* 45, 107-116.

Ohyama, T., Ohtake, N., Sueyoshi, K., Tewari, K., Takahashi, Y., Ito, S., Nishiwaki, T., Nagumo, Y., Ishii, S. & Sato, T. (2009). *Nitrogen Fixation and Metabolism in Soybean Plants*, Nova Science Publishers, Inc., New York.

Sato, T., Nishiwaki, T., Ohtake, N. & Ohyama, T. (1997). Determination of concentration of soybean nodule leghemoglobin components by capillary electrophoresis. *Jpn J. Soil Sci. Plant Nutr.*, 68, 521-526

Sato, T., Yashima, H., Ohtake, N., Sueyoshi, K., Akao, S., Harper, J.E., & Ohyama, T. (1998). Determination of leghemoglobin components and xylem sap composition by capillary electrophoresis in hypernodulation soybean mutants cultivated in the field. *Soil Sci. Plant Nutr.*, 44, 635-645

Sato, T., Yashima, H., Ohtake, N., Sueyoshi, K., Akao, S. & Ohyama, T (1999a). Possible involvement of photosynthetic supply in changes of nodule characteristics of hypernodulating soybeans. *Soil Sci. Plant Nutr.*, 45, 187-196

Sato, T., Ohtake, N., Ohyama, T., Ishioka, N.S., Watanabe. S., Osa, A., Sekine, T., Uchida, H., Tsuji, A., Matsuhashi, S., Ito, T. & Kume, T. (1999b). Analysis of nitrate absorption and transport in non-nodulated and nodulated soybean plants with $^{13}NO_3^-$ and $^{15}NO_3^-$. *Radioisotopes*, 48, 450-458.

Sato, T., Nishiwaki, T., Ohtake, N., Sueyoshi, K., & Ohyama, T. (1999c). Non-involvement of ethylene action on the nitrate inhibition of nodulation in hypernodulation soybean mutant and its parent cv. Williams. *Bull. Fac. Agric. Niigata Univ.*, 51, 121-130

Sato, T., Onoma, N., Fujikake, H., Ohtake, N., Sueyoshi, K. & Ohyama, T. (2001). Changes in four leghemoglobin components in nodules of hypernodulating soybean (*Glycine max* [L] Merr.) mutant and its parent in the early nodule developmental stage. *Plant and Soil*, 237, 129-135

Silsbury, J.H., Catchpoole, D.W.& Wallance ,W. (1986). Effects of nitrate and ammonium on nitrogenase (C_2H_2 reduction) activity of swards of subterranean clover, *Trifolium subterraneum* L. *Australian Journal of Plant Physiology* 13, 257-73.

Schimidt, J.S., Harper, J.E., Hoffman, T.K., Bent, A.F. (1999). Regulation of soybean nodulation independent of ethylene signaling. *Plant Physiology* 119, 951-959.

Schubert, K.R., (1986). Products of biological nitrogen fixation in higher plants; symbiosis, transport and metabolism. *Annu. Rev. Plant Physiol.*, 37, 539-574

Schuller, K.A., Minchin, F.R., Gresshoff, P.M. (1988). Nitrogenase activity and oxygen diffusion in nodules of soybean cv. Bragg and a supernodulating mutant: effects of nitrate. *Journal of Experimental Botany* 39, 865-877.

Serraj, R., Fleurat-Lessard, P., Jaillard, B. & Drevon, J.J. (1995). Structural changes in the inner-cortex cells of soybean root-nodules are induced by short-term exposure to high salt or oxygen concentrations. *Plant, Cell and Environment*, 18: 455–462.

Serraj, R., Frangne, N., Maeshima, M., Fleurat-Lessard, P. & Drevon, J.J. (1998). A γ-TIP cross-reacting protein is abundant in the cortex of soybean N_2-fixing nodules. *Planta* 206: 681–684

Streeter, J.G. (1988). Inhibition of legume nodule formation and N_2 fixation by nitrate. *CRC Crit. Rev. Plant Sci.*, 7, 1-23

Streeter, J.G. (1993). Translocation-A key factor limiting the efficiency of nitrogen fixation in legume nodules. *Plant Physiol.*. 87 616-623

Suzuki, A., Akune, M., Kogiso, M., Imagama, Y., Osuki, K.,Uchiumi, T., Higashi, S., Han, S., Yoshida, S., Asami, T. & Abe, M. (2004). Control of nodule number by the phytohormone abscisic acid in the roots of two leguminous species. *Plant Cell Physiol.*, 45, 914-922

Tajima, S., Nomura, M & Kouchi, H. (2004). Ureide biosynthesis in legume nodules. *Frontiers in Bioscience*, 9, 1374-1381

Takahashi, Y., Chinushi, T., Nagumo, Y., Nakano, T., and Ohyama, T., 1991. Effect of deep placement of controlled release nitrogen fertilizer (coated urea) on growth, yield and nitrogen fixation of soybean plants. *Soil Sci. Plant Nutr.*, 37, 223-231.

Takahashi, Y., Chinushi, T., Nakano, T. & Ohyama, T. (1992). Evaluation of N_2 fixation and N absorption activity by relative ureide method in field grown soybean plants with deep placement of coated urea. *Soil Sci. Plant Nutr.*, 38, 699-708.

Takahashi, Y. & Ohyama, T. (1999). Technique for deep placement of coated urea fertilizer in soybean cultivation. *JARQ*, 33, 235-242.

Tanaka, A., Fujita, K. & Terasawa, H. (1985). Growth and dinitrogen fixation of soybean root system affected by partial exposure to nitrate. *Soil Sci. Plant Nutr.* 31, 637-645.

Terakado, J., Fujihara, S., Gott S., Kutatai, R., Suzuki, Y., Yoshida, S. & Yoneyama, T., (2005). Systemic effect of a brassinosteroid on root nodule formation in soybean as revealed by the application of brassinolide and brassinozaole, *Soil Sci. Plant Nutr.*, 51, 389-395.

Terakado, J., Yoneyama, T. & Fujihara, S., (2006). Shoot-applied polyamines suppress nodule formation in soybean (*Glycine max*), *J. Plant Physiol.*, 163, 497-505.

Tewari, K., Nagumo, Y., Takahashi, Y., Sueyoshi, K., Ohtake, N. & Ohyama, T. (2010). A New Tecnology of Deep Placement of Slow Release Nitrogen Fertilizers for Promotion of Soybean Growth and Seed Yield. NOVA Science Publishers

Van Noorden, G.E., Ross, J.J., Reid, J.B., Rolfe, B.G. & Mathesius, U. (2006). Defective long-distance auxin transport regulation in the Medicago truncatula super numeric nodules mutant[1[w]]. *Plant Physiol.*, 140, 1494-1506

Vessey, J.K., Walsh, K.B. & Layzell, D.B. (1988). Can a limitation in phloem supply to nodules account for the inhibitory effect of nitrate on nitrogenase activity in soybean? *Physiol. Plant.*, 74, 137-146

Vessey, J.K. & Waterer, J. (1992). In search of the mechanism of nitrate inhibition of nitrogenase activity in legume nodues: Recent development. *Physiologia Plantarum* 84, 171-176.

Vuong, T.D., Nickell, C.D. & Harper, J.E., (1996). Genetic and allelism analyses of hypernodulation soybean mutants from two gentetic backgrounds, *Crop Sci.*, 36, 1153-1158

Witty, J.F. & Minchin, F.R. (1990) Oxygen diffusion in the legume root nodule. In *Nitrogen Fixation: Achievements and Objectives*, ed. Gresshoff, P.M., Roth, L.E., Stacy, G. and Newton, W.E., Chapman and Hal, New York, London, 285-292

Yashima, H., Fujikake, H., Sato, T., Tewari, K., Ohtake, N., Sueyoshi, K., & Ohyama, T. (2003). Systemic and local effects of long term application of nitrate on nodule growth and N_2 fixation in soybean (*Glycine max* (L.) Merr.). *Soil Sci. Plant Nutr.*, 21, 981-990

Yashima, H., Fujikake, H., Yamazaki, A., Ito, S., Sato, T., Tewari, K., Ohtake, N., Sueyoshi, K., Takahashi, Y. & Ohyama, T. (2005). Long-term effect of nitrate application from lower part of roots on nodulation and N_2 fixation in upper part of roots of soybean (*Glycine max* (L.) Merr.) in two-layered pot experiment. *Soil Sci. Plant Nutr.*, 21, 981-990

18

Causes and Consequences of the Expansion of Soybean in Argentina

S. Calvo[1], M. L. Salvador[1], S. Giancola[2], G. Iturrioz[2]
M. Covacevich[2] and , D. Iglesias[2]
[1]*National University of Córdoba*
[2]*National Institute of Agricultural Technology -INTA*
Argentina

1. Introduction

Soybean production in Argentina started in the seventies "as a productive option to provide proteins for animal feeding. This was fostered by the National Institute of Agricultural Technology (INTA) at a national level and by the Argentine Institute for the Oilseeds Development (IADO) , currently closed" (ACSOJA, 2009). In spite of this, the soybean development was very slow until the mid 90s, when two technological milestones occurred: no-till farming and the inclusion of transgenic seeds.

The development of this chain is shown in its growing competitive power. Indeed, Argentina has kept and increased its presence in the world market, being the third exporting country of soybean (13% of the total exports). It is also the first world exporting country of soybean oil and meal (55% and 50% share of the total exported). Finally, the export goods of the soybean chain represent the first exportation item of the Argentine economy (2010/2011).

This may be explained by four factors: a) the agro ecologic suitability, b) the constant improvement and technological innovation, c) the most modern and of major scope crushing facilities for soybean of the world, d) the high exportable surplus due to low domestic consumption.

In the last ten years, the development of this chain not only has favored the members of the chain -producers, manufacturers, exporters, merchants- but also the National State, which by means of the implementation of export duties[1] obtains strong revenues.

In spite of the favorable setting in which the soybean chain appears both, at a national and international level, this crop is questioned by several agents who analize how the soybean spreading is shifting the agricultural frontier. This expansion in turn is causing the substitution of agricultural and stockbreeding products, the use of not suitable lands for agriculture, the extinction of small-size producers and environmental damage among other consequences.

Therefore, the goal of this work is to describe and analize the evolution of the soybean chain in Argentina through indicators and explicative factors[2] referred to the primary, industrial

---

[1] Up to March 2010, they amount to 35% for grain and 32% for soybean oil and meal.

and exporta sectors, since the mid seventies up to the present time, with special emphasis on the last thirty years. The aim is to outline the probable future context that the direct and indirect agents connected to the soybean complex will face.

2. Primary production

At the beginning of the seventies, the soybean cultivated area reached 30,470 hectares with a total output of 26,800 tons. The low importance of this crop may be explained by the lack of attractive markets, the low international prices and the national trade policy, which strongly restricted the exportation of grain and byproducts (Civitaresi and Granato, 2003).

The expansion of the cultivated area and yields started during the seventies. The cultivated area increased 6,792 % between the 1969/1970 and the 1979/1980 seasons and 143 % between 1979/1980 and 1989/1990 whereas the production increased 12,959 % and 206 % considering the same periods. Likewise, the soybean yields increased 32.7 % between 1970/1971 (1,024 kg/hectare) and 1989/1990 (2,157 kg/ha). This outstanding growth was due, among other endogenous causes, to technological changes and its fast adoption shown in the improvement of agronomic management, agricultural mechanization and agrochemicals application (Civitaresi and Granato, 2003). In addition, the double crop wheat-soybean[3] began to develop.

Among the exogenous causes that explain the growth in this twenty years (1970/1990), highlights the increase of international prices in the mid seventies, reduction of export duties, elimination of exportation ban (1978) and modification of the policy carried out by the European Economic Community (EEC) nowadays European Union (EU). In the 70s, the EEC negotiated the entry of soybean and its byproducts with zero tariff according to a re-structural procedure of its Common Agricultural Policy in the framework of the Dillon Round of the General Agreement Tariff and Trade (GATT) (1960/1961). The strong expansion of its milk production and the subsequent need to feed its cattle urged the strong demand of vegetable proteins. The EEC tried –unsuccessfully- to cover its needs with its own cereal production by restricting the access to imported cereals (as the wheat bran from Argentina), but it could not restrict the entry of soybean, which was protected by the commercial concession in the framework of the GATT[4] (Albin and Paz, 2003).

In the last twenty years, the process of expansion of the soybean cultivated area has continued uninterrupted. The cultivated area increased 72.3 % between 1989/1990 and 1999/00 and 108.7 % between 1999/00 and the last season 2009/2010. In addition, the soybean production increased 88.2 % and 162 % for the same periods.

In Figure 1 it can be seen the expansion of the cultivated area and of the soybean production[5] as well as the leap in both indicators since the year 1996.

---

[2] It will be applied the definition by Civitaresi and Granato (2003) which classifies the factors into exogenous: that depend on macro economic policies and on international commerce, and endogenous: that derive from the strategies carried out by the protagonists of the different links of the chain.

[3] The short cycle wheat allows the sowing of wheat-soybean in the same agricultural cycle and in the same area

[4] In the eighties, it tried to "under strengthen" the soybean and its derivatives: this situation led to serious confrontations with the U.S.A.

[5] The decrease in production in 2008 and part of 2009 was due mainly to a drought and partly to the conflict between the agricultural sector and the Government. The conflict was originated by the implementation of "mobile export duties" by the Economy Minister (the duties varied according to the

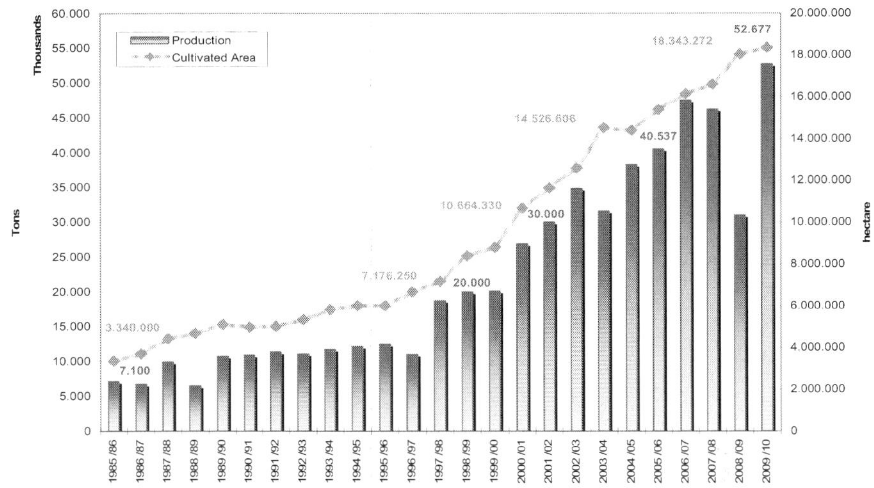

Fig. 1. Argentina soybean. Production and Cultivated Area. Years 1885/86 to 2009/2010.

Specifically in the nineties, the following endogenous factors can be featured as causes of the expansion: the genetic modification in several crops -among them the events in the soybean seed and consequently the start of the transgenic technology spreading at an international level-, and the intensification in the use of machinery and supplies, mainly the "glyphosate"[6] herbicide. In addition, a fundamental change in the plowing system took place: no-till cropping, which deserves a special paragraph.

"No-till cropping started to gain importance in the Argentine agriculture in the late eighties since in many of the most important areas of the *Pampeana* region, the cumulative effects of the soil erosion -resulting from the "agricultural cultural process" on the basis of traditional plowing practices- began to show negative operative results in the exploitation. This effect on yields, and through them, on the economic feasibility of agriculture, together with the fact that, as a result from the deregulation and opening processes of the economy, the availability of appropriate farming machinery increased, plus the reduction in direct costs (as a result of the elimination of the plowing tasks), provided an outstanding basis for the spreading of no-till cropping, which in turn had the objective of recovering, at least some of the lost productivity" (Trigo and Cap, 2006).

Fuethermore, the synergy between glyphosate tolerant soybean tolerant (RR-soybean) and the non-till system -which shortens the required time between the wheat harvest and the soybean planting- makes possible the successful use of short cycle soybean varieties as second crop and allows a double-cropping wheat-soybeans in zones where before was not possible from an agronomic point of view. According to data provided by the Argentine No Till Farmers Association (AAPRESID), at the beginning of the nineties, the area of soybean

variations in the international price of the soybean). This duties were enforced from March 2008 to July 2008 (they were derogated by a legislative decision).

[6] In the 2009/2010 season, more than 200 millon liters of glyphosate were applied in all the soybean sown area in Argentina, whereas in 1996 13 million liters were used

under no-till cropping system reached 6%, whereas in the 2001/02 season it was 74%, in 2006/07 it accounted to 85% and in 2009/10 it reached 88 %.

Analyzing the exogenous factors that fostered the development of soybean in the nineties, highlights the changes in the economic policy: tariff reduction for machinery and inputs imports, reduction and/or elimination of export duties for products of the soybean chain, investments in ports, deregulation of markets (elimination of Meat and Grain Boards) and privatizations (ports, railways) which lessen the cost of production and favor the acquisition of technology.

At an international level, relevant facts such as average international prices lower than in the previous decade , the increase of commercial barriers in the rest of the world (in spite of the final Agreement of the GATT´s Uruguay Round) have not been auspicious for the soybean chain. In this respect, the only favorable fact was the increase in the demand of grain by the Brazilian oil industry, in the frame of the MERCOSUR (Civitaresi and Granato, 2003).

Going through the last decade, the increase in the soybean cultivated area[7] is due both to the incorporation of new lands and to the substitution of other crops. The agricultural expansion towards the north of Argentina[8] is shown by a 70 % increase of the cultivated area between the last two National Agricultural Census (CNA 1988 and CNA 2002[9]), although this is a marginal area for this crop. 66 % of the lands transferred to agriculture in the north of Argentina was occupied by soybean (Giancola *et al.*, 2010).

According to Shvarzer and Tavosnanska (2007), the increase in the production for the last ten years might have several causes. One of them is the increase in international prices because of the growing world demand for soyean oil and meal. Another cause is the technological change, which involved a reduction of related costs. Likewise, between the years 2000 and 2010 it is worth mentioning: the intensification of the technological package, scale incrementes in the productive core, the suitability of the soil, the existing road infrastructure and the increase of the rainfall (Giancola *et al.*, 2010).

Complementing the previous considerations, among other fundamental causes of the soybean expansion is worth mentioning: higher financial income, lower complexity in the operation and lower risk than other crops, and farmers' knowhow, together with the use of their own seed (INTA, 2009).

As related to the topic of seeds and specially about the use of the varieties of genetically modified soybean[10] (GM), the data provided by ARGENBIO (a non governmental organization created to spread information about biotechnology) shows that in the 1996/1997 season, the area occupied with GM soybean was 6% and quickly reached 94% in 2001/2002 to cover 98% in 2004/2005.

---

[7] At present (2009/10 season) soybean occupies 18,343,272 hectares with a production of 52,677,371 tons (CIARA, 2011).

[8] The traditional region for the production of soybean is called "pampeana" and it is made up by the provinces of Córdoba, Buenos Aires, Santa Fe, Entre Ríos and La Pampa.

[9] At a primary level, 75% of the production comes from 40% of the productive units (100,000 units). The average size of the units is 170 hectares.

[10] The gene resistant to the glyphosate herbicide was initially owned by Monsanto in the United States, which granted a license to Asgrow, then this company was bought by Nidera, which introduced it in Argentina. Afterwards, when Monsanto patented the product abroad, it had already been freed by third parties to be sold in Argentina (Quaim and Traxler, 2002). This situation started a controversy between Monsanto and the Argentine government, in which the company has claimed in order to obtain royalties.

Even in Argentina, these adoption was faster than others, as for example hybrid corn or wheat with Mexican germplasma. Hybrid corn lasted 18 years to reach 70% of the acceptance that nowadays GM corn holds, and Mexican wheat reached the same percentages of adoption as soybean has nowadays (more than 90% of the market) only after 16 years (López, 2006, as cited in Trigo and Cap, 2006).
Another factor that favored the technological change is related to inputs supply for the soybean crop (Bisang, 2003) which has the following characteristics: a) technological packages: the transgenic seed -RR Soybean- works as a connector of a joint supply that includes: glyphosate, pre-emerging herbicides, insecticides and fertilizers; b) mergers and alliances with companies that offer such technological packages[11]; c) the packages of supplies offered in service centers include advice about usage techniques.
Finally, soybean share of total cultivated area and grain production of the country allows to visualize the strong soybean expansion in Argentina (Figure 2). According to production data, soybean represented 0.2% of the total cultivated area[12] (14 million hectares) in the 1969/1970 season. In the seventies, the soybean cultivated area occupied 3.1 % of the total sown, in the eighties 19%, in the nineties 33% and in the first ten years of the 21st century soybean occupied 54% (more than 18 million hectares) of the total (28 million hectares) implanted.
Regarding the importance of soybean production, and considering the total produced by the five main grains, the participation of soybean was 3.7% on total production in the seventies, 18% in the eighties, 30% in the nineties and 50% in the first ten years of the 21st century (CIARA, 2011). In the 2009/2010 season, the soybean production represented 60% of the total production whereas it was 0.12 % in the 1969/70 season.

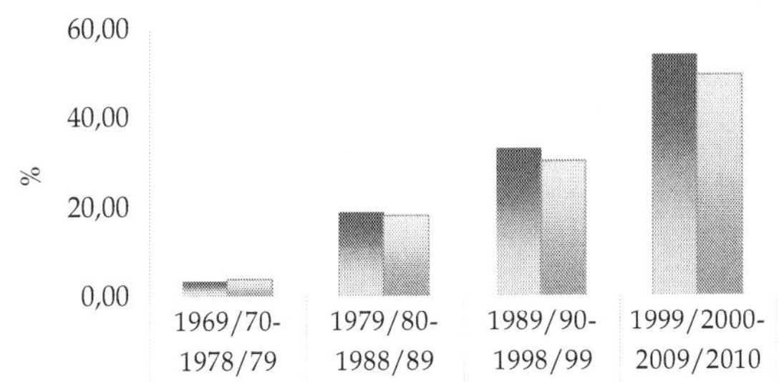

	1969/70-1978/79	1979/80-1988/89	1989/90-1998/99	1999/2000-2009/2010
Soybean /Total grain cultivated area	3,10	18,80	33,04	54,04
Soybean/Total Crop	3,68	17,84	30,00	49,34

Fig. 2. Argentina. Soybean. Production share of the total produced by the five crops and soybean cultivated area shares of total cultivated area. Periods: 1969/70 - 1978/79, 1979/80 - 1979/80 - 1988/89, 1989/90 - 1998/99 and 1999/2000 - 2009/10.

[11] Monsanto (an company originally dedicated to chemistry, which in the70s expanded to pharmaceutical chemistry) is associated with Dekalb and Cargill; the use of these seeds (specially RR Soybean from Monsanto) requires the application of glyphosate, offered by the same company with the brand "Roundup".
[12] Taking into account the five main crops: wheat, corn, sunflower, sorghum, soybean.

With respect to the geographical location, the soybean cultivated area is situated in 15 (fifteen) productive provinces. However, 85% of the cultivated area is concentrated on the traditional "pampeanas" provinces that generates 88% of the national production (Córdoba, Buenos Aires and Santa Fe provide 78% of the total production) whereas 15% of the remaining area corresponds to Northwester and Northeaster provinces. Map 1 shows the shift of the agricultural frontier[13].

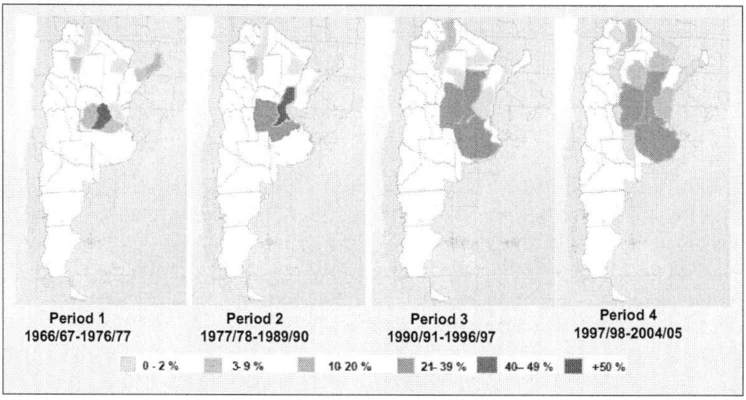

Map 1. Evolution of the space distribution of the soybean cultivated area (Brieva, 2006).

Considering the yields, Figure 3 shows that they have increased continuously, from 1.4 tons/ha in the seventies, to 2.6 tons/ha in the first decade of 2000.

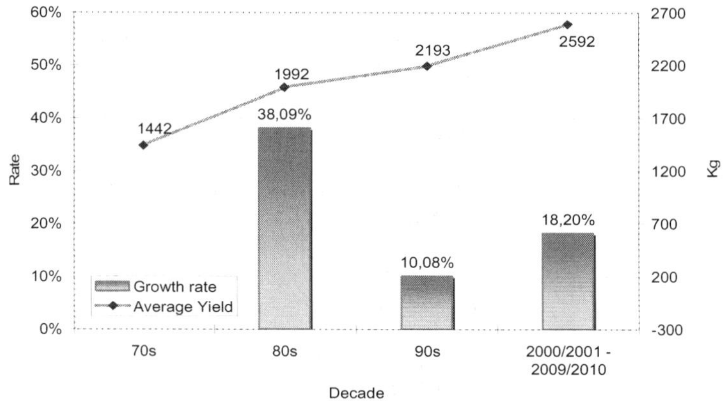

Fig. 3. Argentina Soybean. Average Yields (kg/hectare) and growth rate (%). 70, 80, 90 decades and 2000/01 - 2009/10 period.

---

[13] As mentioned, the shift of the agricultural frontier has generated the substitution of crops (sorghums, corn, cotton among others) and cattle livestock for soybean as well as the spreading of soybean over the native forest.

3. Industry and state policies

In the stage of first industrialization or milling, the productions of raw oil and meal are technologically associated, and therefore they are produced in the same industrial plants. Most part of the grain production in Argentina is oriented to milling, with a percentage that varies from 76% to 80% of the total production; approximately 18% of soybean beans is exported without processing and the rest is used for direct consumption, as seeds for sowing or for other purposes (bread, cookies, snacks, peanut butter, bird food, etc).

The industrialization process begins with the milling and extraction of oil. After going through drying - to remove humidity - and cleaning processes, the grain is broken and pressed in small sheets, which being transformed into a dough, move on to the extraction process. The remaining dough, after being dried and toasted, forms the protein meal used in the manufacture of animal feed. The gum can be used in the production of soybean lecithin or can be added to the meal in order to obtain different of protein tenors. Through a hydrogenation process, the partially refined oil can be transformed into margarine, mayonnaise and vegetable fats. Some companies integrate vertically these industrial stages.

As other grains with a low oil component, soybean oil is extracted with hexane. The solvent -a petroleum product- impregnated with oil is then separated by evaporation, and goes through a gum elimination system (degummed) to reach the stage of raw oil (Giancola *et al.*, 2010). A soybean yields 19% oil, 73% meal, 7% shell and 1% others (ashes, etc.), these values may vary depending on the drying and de-shelling degrees[14].

The manufactured products[15] are distinguished in two segments: **commodities** (for exportation and domestic market), which include raw and refined oil in bulk and meal for animals, and **speciallities** of higher added value for the final consumption of food and other uses. Some companies are present in both segments. The refined oil, apart from its direct consumption (pure or in blends) has several uses, such as margarine, mayonnaise, non edible intermediate products (candles, cosmetics, soap, paint and fine chemicals, animal food, soybean lecithin), hydrogenated vegetable fats (used in cookies, bread, ice-cream, etc), soybean derivatives for human consumption (Gutman and Lavarello, 2003). Regarding non edible uses and oriented to the chemical industry, oil and vegetable fats have similar characteristics to the petroleum ones and can be used for plastics, adhesives, solvents, lubricants, etc.

Another derived product is biodiesel. Compared to other traditional hydrocarbons, biodiesel has several advantages: it comes from a renewable product and has little environmental impact, but it also has disadvantages, being its high cost the main limitation. Biodiesel installed capacity of production is to 2.43 million tons - reaching 3 million tons by the year 2011. In 2010, 1.9 million biodiesel tons were refined and almost 1.4 tons were exported. The 500,000 remaining tons were used in the domestic market for the diesel cut.

In April 2006, Act 26093 established the rules for the regulation and promotion[16] for Production and Sustainable Use of Bio Fuels for a period of 15 years. This Law establishes that the gasoline and diesel commercialized within Argentina shall be mixed in an oil

[14] 4 tons of flour are obtained from each ton of oil.

[15] The industrial processing generates byproducts that have more restricted but dynamic markets, many of which are not developed enough in Argentina.

[16] The Law gives encouragement for investment through fiscal incentives.

distillery or refinery with 5 % - as minimum – of bio ethanol and biodiesel, respectively from 01/01/2010 on. For the year 2010, the quantity of biodiesel necessary for the 5 %[17] cut was calculated in 625,000 tons, which imply 650,000 tons of oil.

Likewise, the biodiesel industry that is developing in Argentina has a strong orientation towards the foreign market[18]. In Argentina, there are major incentives for the biodiesel exportation as a result of the way in which export duties of the soybean complex are structured. In this respect, the biodiesel exportation must pay 20 % of export duties whereas soybean oil (its main supply) and soybean pay 32 % and 35 % respectively[19]. Thus, there is a gain of competitiveness for biodiesel that comes from the efficiency in the previous links of the chain. In addition, biodiesel obtains a reimbursement of 2.5 % upon the export price, with which the competitiveness reached by the export duties structure is reinforced.

Finally, it is worth mentioning that it is currently spreading the building of small extruding facilities in farms that produce soybean. This allows to obtain oil for its use as biodiesel and protein meal for cattle feeding, adding value to thier soybean production through horizontal and vertical integration

3.1 Milling capacity

Milling is the most important destination of the primary production of soybean. While in the eigthies 71 % of the grain production was assigned to milling, in the 90 decade this percentage increased to 77 % and reached 84 % in the first 2000 decade. In the 2008 and 2009 periods, there was a 20% fall in industrialization (30 million tons average) which can be explained mainly by the decrease in the grain production due to reasons pointed earlier. In the year 2010, the milling reached 34 million tons, similar values to those of the year 2007. This volume represented 98% of the soybean supply.

In the last years, the crushing capacity grew basically in the province of Santa Fe (at an average rate of 14.4 %) while in the provinces of Buenos Aires, Córdoba and Entre Ríos it remained practically constant (CIARA, 2011). In the year 2010, soybean crushing in Argentina reached 36,824,628 tons, which implies 92% of the total milling of the country (39,898,017 tons[20]. In 1986, the soybean milling share of the total milling was 50%. (CIARA, 2011).

The increasing crushing capacity in Argentina is also reflected in the world market. In the last two decades, the world milling has grown at an annual rate of 4.5 %. In the same way, the soybean world production is destined mainly to milling (98 %) reaching 200 million tons (2009/2010)(USDA, 2010). The main countries processing soybean are also the main grain producers: U.S.A. Brazil, Argentina, China. Currently, this latter country - which specially has the goal to increase the soybean added value - contributes 21 % of the world milling. This is the reason of the strong grain importation from overseas.

---

[17] The national government established that the cut shall be 7 % since 2011.

[18] Nine plants have been authorized to export (2008), 5 of which come from investments made in the oil sector, the rest come from non agro-industrial sectors. The main refineries are: Renova, Louis Dreyfus, Unitec Bio, Patagonia Bioenergía, Ecofuel, Cargill and Aceitera General Deheza. CARBIO (Argentine Chamber of Biofuels, 2011).

[19] The export duties of grain and byproducts had been applied in previous times, as in the eighties (15 % and 41 % for grain) and nineties (3.5 % for grain).

[20] The crushing capacity in Argentina amounts to more than 57 million tons. (2010)

3.2 Oil and meal production

Soybean oil production reached 41 million tons, being the main producing countries: China, (10,3 million tons), U.S.A, Brazil and Argentina. Argentina produces 7,3 million tons (2009) and participates approximately with 18 % of the world production.

Argentina has 54 oil producing plants which are distributed in 8 provinces. Thirty-nine of these plants process soybean bean with a theoretic capacity of 160,000tons in 24 hours (CIARA, 2011). Fifty-six percent of the plants that concentrate 84 % of the processing theoretic processing capacity, operate in areas close to shipment centers in the Province of Santa Fe, since the production is oriented to exportation. The processing plants are supplied with soybean within a radius smaller than 300 km, which results in low freight cost. This closeness between primary production and transforming industry generates a major competitive advantage.

Analyzing Table 1, we can see that whereas in the 70 decade, the soybean oil production participated in 8.5% of total elaborated oil of the country, in the 2000/2004 period this participation increased to 71.88%. In 2009, soybean oil participation in Argentina reached 5,771,812 tons, thus participating 79% of the total elaborated oil (7,302,493 tons). Considering the geographical location, the province of Santa Fe elaborated more than 90% (2008/09) of soybean oil total production.

	Total Oil (T)	Soybean Oil (T)	Soybean/total (%)
Decade Averages			
1970-79	661.517	56.118	8,48
1980-89	1.735.284	571.359	32,93
1990-99	3.722.981	1.789.712	48,07
Quinquennial averages			
1985-89	2.239.749	842.542	37,62
1990-94	2.951.163	1.360.297	46,09
1995-99	4.494.799	2.219.126	49,37
2000-04	5.444.255	3.913.555	71,88

Table 1. Argentina. Total Oil and Soybean Oil Production (tons). Soybean Participation (%) in the Oil Total. Periods Selected.

The development of the Argentine oil industry was parallel to the growth of the soybean primary production. According to Obschatko (1997) and Civitaresi and Granato (2003) the main factors were: the increase[21] in the international prices of oil and meal, impelled by the increase in the world consumption and the fluent grain supply (raw material) -at an international level-, and the reduction of taxes for the exports of oil and the application of a lien differential for grain and oil -at national level. The latter measure was fundamental for the development of the industry, according to Obschatko (1997)- in order to prevent the industry from having problems for raw material supply, the National State

[21] In the nineties, international prices decreased due to the lower demand by Asian countries and higher presence of other substitute oils. All this took place in a context of strong market distortions due to protectionism mostly inin developed countries.

implemented an effective differential exchange rate for grain, oil and byproducts. Between 1976 and 1990, the differential of export duties between grain and milling products (highly praised) varied between 5.9 and 13.6 %. This measure favored the exportation of oil and meal by reducing the cost of the raw material with respect to the international market prices.

At an industrial level, the investment made in this sector as regards technology was complemented with the availability of technology from developed countries- that is the case of the Dutch enterprise (De Smet), which introduced in Argentina the technology of oil extraction by solvent. Likewise, the companies of the industrial link increased the production scale: they specialized in soybean and sunflower; they re-located in the province of Santa Fe -thus combining the closeness to the exportation ports (Rosario) and the sources of raw material supply. Finally, industry integrated forwards by building dock[22] and storage facilities. This process was intensified in the nineties. At a National State level, the tax pressure on the marketing sectors was reduced, the services related to trading activity were deregulated, public ports were privatized and new ports were built. In addition, there were investments for the sweeping and beaconing of the *Paraná* and *Río de la Plata* fluvial corridor[23] (Civitaresi and Granato, 2003).

Soybean processing is concentrated in six companies: Bunge, Cargill, Vicentín, Molinos Río de la Plata, Dreyfus, Aceitera General Deheza, which control more than 87 % of the total capacity. These companies have a modern and efficient port infrastructure of their own and an extended supply net spread throughout the entire country[24]. The oilseeds industry employ 10,000 people. In Table 2, it is identified the number of plants that each of the companies owns as well as the annual processing capacity.

Company	Number of Facilities	Total capacity		Largest capacity Facility (ton/year)
		Ton/year	%	
Bunge	5	8.220.000	17.8	2.550.000
Cargill	4	7.710.000	16.7	3.900.000
Vicentín	3	6.555.000	14.2	3.000.000
Molinos Río de la Plata	3	6.195.000	13.4	3.600.000
Dreyfus	2	6.000.000	13	3.600.000
Aceitera Gral. Deheza	4	5.700.000	12.3	2.550.000
Buyati	2	1.440.000	3.1	1.005.000
Nidera	2	1.260.000	2.7	660.000
OMHSA	2	705.000	1.5	405.000
Others	20	2.467.500	5.3	
Total	47	46.252.500	100	3.900.000

Source: Schvarzer and Tavosnasnka (2007)

Table 2. Argentina. Oil Factories by Company.

[22] In 1977, private companies were allowed to set up fluvial ports.

[23] This corridor is fundamental for the transportation of the grain production from Córdoba and Santa Fe, first and second producing provinces of soybeans.

[24] As regards the production of refined oil, 5 companies have 80 % of the production, 4 companies have 80 % of the margarine production and three companies have 80 % of the mayonnaise production.

At a world level, and according to USDA data, the production of soybean meal was approximately 177 million tons (2009/10), being the main producers: China, (25.7% of the total produced), U.S.A., Argentina (17% of the world production) and Brazil. These four countries concentrate 78% of total production (USDA, 2010).
The soybean meal share of total manufactured meal in Argentina has been higher than oil. Thus, whereas soybean meal was 20.87% of the total output in the seventies, this percentage reached 91.1% in the 2000/04 period (Table 3). In 2009, the soybean meal production was approximately 23 million tons from a total of 25,520,000 tons of manufactured meal. As in the case of oil Santa Fe province produces 93% of total soybean meal (2009).

Period	Total meal production	Soybean meal production	% soybean/ total meal
Decade average			
1970-79	1.268.956	264.807	20,87
1980-89	4.224.577	2.665.821	63,10
1990-99	10.383.969	8.124.909	78,24
Quinquennial average			
1985-89	5.674.917	3.916.431	69,01
1990-94	8.248.517	6.327.119	76,71
1995-99	12.519.422	9.922.700	79,26
2000-04	18.202.340	16.584.419	91,11

Source: CIARA (2010)

Table 3. Argentina. Total Output of meal (tons) and of soybean meal and Participation (%) in the Total Output of meal. Periods selected.

4. International trade

Regarding soybean complex, Argentina exports mainly processed products: protein meal for animal feeding and oil for human consumption. Soybean exportation began its expansion due to the demand from China, which intensified its process of importation substitution (oil).

4.1 Soybean

The world production of soybean in the 2010/2011 season reached 255.5 million tons according to USDA. The main producing countries, U.S.A. (35 %), Brazil (26 %) and Argentina (20 %), participate in 82 % of the world production. They are followed by China, with a participation of 6 % although its presence in the world market has the role of importer because of its high domestic consumption.
For a long period of time soybean main demand came from the EU. Since 1998/99, China has incremented its demand for domestic processing and since the 2004/05 harvest China turned into the main world importer of the grain (60 %) followed by the EU (14.6 %).
As related world suppliers, U.S.A, Brazil and Argentina concentrate 89 % of the total exported (98 million tons) (2009/2010). Argentina occupies the third place as soybean

exporter in the world with 14,11 % of the total[25] (see Figure 6). Taking into account the six main soybean exporting countries (U.S.A, Brazil, Argentina, Paraguay, China, Canada, Bolivia), three of them - Brazil, Argentina and Paraguay - which make up the MERCOSUR- represent half of the world exportations. This points out the importance of South America in the soybean bean production and exportation.
As seen in Figure 4, the soybean exportation has had a positive trend fostered by the increasing international prices. This indicator -grain exports-[26] shows the higher competitiveness of Argentina in the international market.
The main demand of Argentine soybean comes from China. Thus, 78 % of the exported grain had that destination. Argentina also exports to Iran, Thailand and Egypt, and in South America to Chile and Peru.

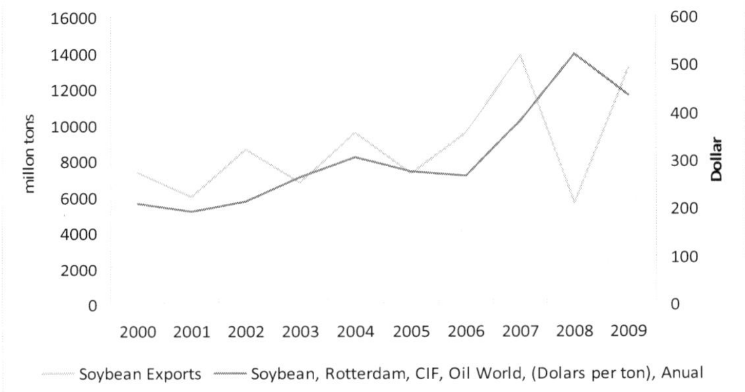

Fig. 4. Soybean. Argentine exportation (million tons) and price (U$S/ton) (CIF, Rotterdam) 2000/2009.

Argentina has also imported soybean, especially from Paraguay (98 %) (Figure 5). The need to import soybean can be explained since the processing capacity of the oil industry surpasses the local production of soybean and other oilseeds available for industrialization.
The strong increase in soybean importations since 2006 can be explained by the Decree 2147/06, which was the result of a wide negotiation between the industrial-exporting sector and the national government. This decree allows transitory importation of goods that will be subject to a process of industrial improvement. Decree 2147/06 incorporates a tax benefit upon the imported supply which allowed the exporting companies to pay export duties omly on the added value of the imported bean. Since April 2009, the Argentine Production Ministry excluded[27] the soybean from the tax benefits of temporary importations. This regulation was based on the grounds that it would promote the increase of the soybean demand in the domestic market and diminish the stock of grain stored by producers in their own farms. (Giancola *et al.*, 2010).

---

[25] U.S.A participates with 44 % and Brazil with 32 % of the total world exported of soybean.
[26] The decrease in the grain exportation is related to the drop in the production due to a drought (2008 and part of 2009).
[27] That is why no importations have been made since 2009 up to the present.

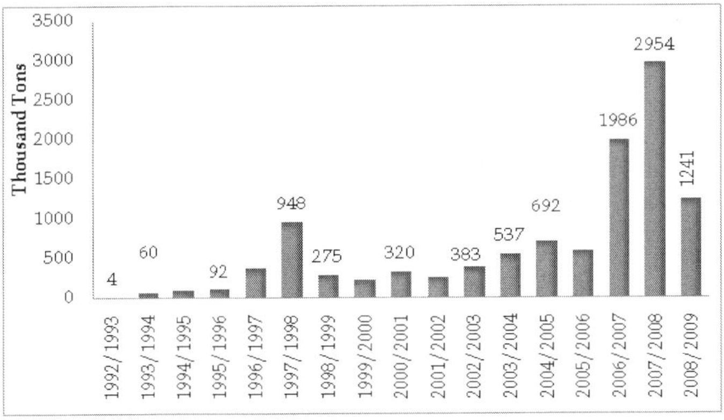

Source: USDA PSD ,2010

Fig. 5. Argentina. Importation of Soybean Bean (tons). 1992/93 - 2008/2009.

4.2 Soybean oil

The soybean oil exportation is concentrated in few countries whereas the importation is diversed. So, the first five importing countries (China, India, UE-27, Iran, Morocco) have 52% of the total imports (2010-2011).

Argentina is the main oil exporting country[28]. With 49% to 56% of the world total oil exports, followed by Brazil (16%) and the United States (13%). Considering Paraguay, the MERCOSUR is regarded as the main exporting soybean oil block. 96% of the oil exported by Argentina is raw and degummed. The refined oil incidence is very low.

The following table (4) shows Argentine apparent consumption. The data show that most of the oil production is destined to the exports.

In Figure 6 it is shown the Argentina share of total world soybean oil exported. Its main participation was in the 2005-2006 season (57%). Then, this participation decreased until it reached 49% of the 2009/2010 season. According to USDA estimates, in 2010/2011 season the Argentine exports would reach historical values (54.6%).

Year	2004	2005	2006	2007	2008	2009
Inicial stock	242.326	134.024	239.000	217.578	302.083	303.968
Production	4.569.718	5.395.724	6.161.214	6.962.675	6.024.101	5.771.812
Availability	4.812.044	5.529.718	6.400.214	7.180.253	6.326.184	6.075.780
Exports	4.588.120	4.964.180	6.086.290	6.637.770	5.125.480	4.660.400
Final stock	134.024	239.000	217.578	302.083	303.968	287.668
Apparent consumption	89.900	326.538	96.346	240.400	896.736	1.127.712

Table 4. Argentina. Apparent consumption of soybean oil.

[28] It is important to point out that since 2007, Argentine began to export biodiesel obtained from soybean.

Asian countries concentrate most of the Argentine oil demand due to the importance of their growing economy and to the fact that these countries had consumption levels per capita lower than the world average. China (45%)[29], India (14%), Bangladesh (6%), Egypt (5%), Peru (5%) are the most relevant destinations of the exported oil by Argentina (2010). Seventy eigh percent of the oil imports and 22% of the soybean imports of China have their origin in Argentina. This shows the strategic importance of soybean and its byproducts in the commercial relationship between China and Argentina.

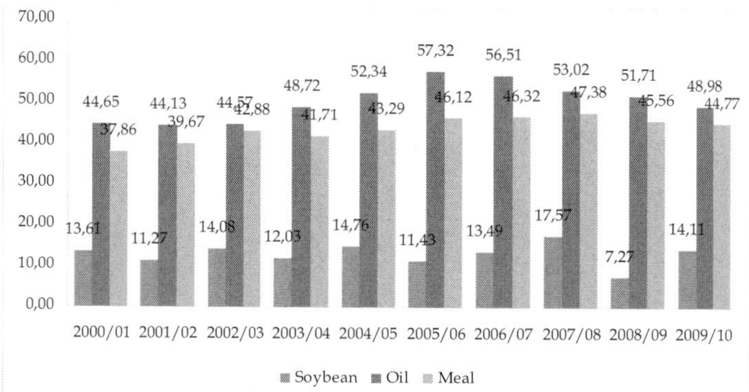

Fig. 6. Soybean. Argentina share of the world grain, oil and meal exports. (%) 2000/01-2009/10.

Finally , after a study of the soybean oil exports per company (Table 5) it comes out that the soybean exports market structure is an oligopoly[30], with foreign capital predomination (transnational companies) (Eumercopol, 2007).

Company	Soybean oil share of total oil exported (%)	Company	Soybean oil share of total oil exported (%)
Cargill	28,42	Nidera Arg.	3,12
Bunge Argentina	18,15	Aceitera C.A.	2,65
Aceitera Gral. Deheza	14,80	Oleag. Moreno	1,44
Dreyfus	10,91	Cia. Argentina de Granos	1,39
Mol. Río de la Plata	10,36		
Vicentin	3,31	A.F.A.	1,22

Source: CIARA (2011)

Table 5. Soybean Argentina. Soybean oil exporting companies. Total export share 2008.

[29] Due to the commercial restrictions imposed by China to the Argentine oil, the exports were drifted to the markets in India, Bangladesh and Iran at a lower price.

[30] The exports participation of the first eight companies (C8) is: 81% for grain, 87% for oil and 97%.for soybean meal.

4.3 Soybean meal

As in oil, Argentina is the main exporting country with 29 million tons over an exporting total of 59.4 million (Figure 7). Thus, Argentina keps -in a stable way- 44--46% (see Figure 6) of the total meal exports, followed by Brazil (23%) and the United States (14%). According to USDA for 2010/11 season, Argentina share in soybean meal export would reach 49.3%.

Argentina exports almost all its soybean meal production (between 93% and 97%), mainly to the UE-27 countries for animal feeding. The European problem with bovine spongiform encephalopathy, known as the "mad cow disease", put an end to the cattle feeding with animal proteins and intensified the demand of substitute products like the soybean meal.

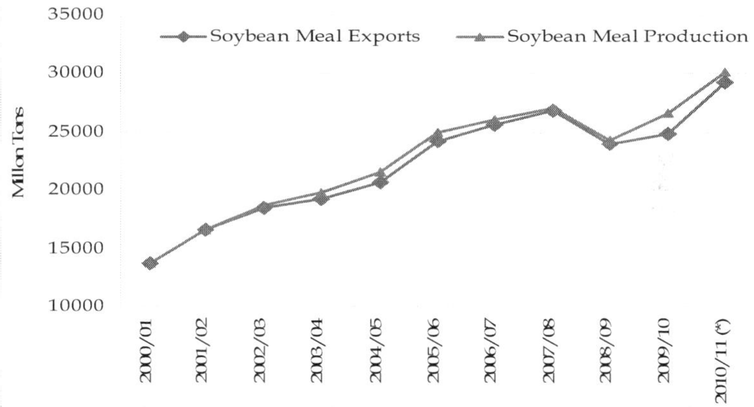

Fig. 7. Argentina. Soybean meal. Production and exports (2000/2001-2010/2011).

If the soybean meal exports by each company is examined (Table 6). The first five companies participate with 78.5% of the total national exports. It can be seen that the first four meal exporting companies are the same as the first four oil exporting companies being Cargill the first (28.4% in oil and 23% in meal, 2008).

Company	Share of total oil export (%)	Company	Share of total oil export (%)
Cargill	28,42	Vicentin	3,31
Bunge Argentina	18,15	Nidera Arg.	3,12
Aceitera Gral. Deheza	14,80	Aceitera C.A.	2,65
Dreyfus	10,91	Oleag. Moreno	1,44
Mol. Río de la Plata	10,36	Cia. Argentina de Granos	1,39

Source, CIARA (2009)

Table 6. Soybean Argentina. Exporting soybean meal companies .Participation in the total exported .2008.

5. Domestic commercialization, storage and logistics and transportation services

5.1 Comercialization. Main characteristics

Ninety five percent of the soybean harvest goes from farmers to country grain elevators (private or local cooperatives). The main final destiny is the crushing industry (oil and meal). The second destiny in importance are the exporters. The remaining activities have a small participation in the grain final destiny. Among these activities the animal feed producers and the grain purchase for own consumption (cattle producers) can be found.

The commercialization stages are divided according to the offerers, been primary operations when the seller of the goods is the producer; secondary operations when the goods are sold by any of the intermediary agents to exporters or to the industry, and finally, terciary operations which are those where exportation is involved.

In the primary stage, the "producers" are the ones who make up the supply : they can be land owners, tenants, contractors, pool producers, agricultural investments common funds, and the demand are usually country grian elevators, first grade cooperatives, livestock producers (poultry, pork, feed lot), exchange dealers[31], industries and expoterers with vertical integration strategies.

The secondary stage is focused on country grain eleators, local cooperatives, inputs providers and exchange dealers as well as exporters, industrials and livestock producers.

Brokers and institutions like the Stock Exchange, Cereal Boards and Arbitration Chambers can also take part in any of these stages. They -boards and chambers- are Arbitration Courts in case there is any kind of controversy. They set reference prices and have grain analysis laboratories. In Argentina, there are four Cereal Boards: Buenos Aires, Córdoba, Bahía Blanca and Entre Ríos.

As regards the transaction steps, the most representative ones can be divided between those made with merchandise readily available or not. The most common payment ways when the goods delivery is effectively done can be: cash on delivery, cash with insured delivery quota, cash with uncertain delivery, cash on deposit certificates or a price to be fixed. When the goods are to be delivered in the future, the payment ways can be: advance payment, exchange (with other goods), and forward payments (with a contract).

In Argentina, there are also two exchange markets where Futures and Options contracts on soybean can be made: "Rosario Futures Exchange (Rofex)" and "*Mercado Término de Buenos Aires (MATBA)*" where soybean oil contracts can be found. Although the use of these instruments is growing; it is not widely spread yet, specially among the small producers. Besides, successive economic crisis and interventions in the agricultural markets have slowed down its development. It is important to point out that this alternative is use to secure prices rather than to deliver the goods. The use of these markets is around the 28% of the total average production in the last five years (2005-2009).

The operators that take part in grain commerce have been classified according to their function into 17 categories by the National Office of Agroindustrial Commerce and Control (ONCCA, 2006). The ones involved in soybean markets are: country grain eleators (2685), grain conditioners (62), industrials (58), industrial-selector (35), laboratory (25), public scales (21), Futures Market (3), consigner without facilities (2), broker (359), importer (27), exporter (356), exchange agent (665) and delivery agent (11).

---

[31] Exchange dealers receive soybean as a form of payment for goods in general and for leasing.

Different trading channels and modalities combined with the large number of operators makes it possible to trade soybean in a very extended way. But since it's not compulsory to register the operations modality, no statistics reflect in a reliable way which the most chosen channels are. Anyway, these modalities are in general incidental and depend on the market situation.

5.2 Storage

The expansion of the storage capacity in Argentina was historically related to the volume and composition of the country grain production and also to the politicies officially developed in relation to the way of trading. In the last years, an expansion of the storage capacity is observed. This expansion had its origin, among other factors, in the sustained increase of grain production, which implied a growing need of places to keep this production. (Lopez, 2008)

	Tons	%
Grain storage companies-Cooperatives	38.204.066	53,86
Producer (fija)	15.900.000	22,42
Oil mills	7.655.511	10,79
Exports-Ports	4.759.119	6,71
Flour Mills	2.552.024	3,60
Animal Feed Proucers	785.947	1,11
Rice Mills	555.827	0,78
Conditioner	517.450	0,73
Permanent facilities	**70.929.944**	**100**
Silo bag	30.000.000	
Total	**100.929.944**	

Table 7. Argentine Storage Capacity

Producers kept the grain in the farm (when possible) or storage them in local cooperatives and country grain elevators or in the silos found in ports or factories (Table 7). National storage capacity in 2007 was nearly of 71 million tons. It must be mentioned that there's no official information in this area so the growth rate considered was similar to the one observed in former periods. (Lopez, 2008)[32]

5.2.1 The silo bag as a storage alternative

Between 1999/2000 silo bag[33] storage technique was introduced (Picture 1). It is a complementary capacity not only at a producer level but also in the case of some country grain elevators (FAO-SAGPYA, 2004). This technique consists of storaging the grains in hermetic plastic bags where the breathing process of the grain biotic components (grains, fungus, insects, etc.) consume the oxygen generating in turn carbone dioxide. This new

[32] It is important to remember information about the storage capacity in Argentina is scarce and in some cases of uncertain reliability (Lopez, 2008).

[33] The relationship installed capacity/silo bag was 2,66/1, 30% of total storage capacity. See Table 7.

atmosphere rich in carbone dioxide and poor in oxygen suppress, inactivate or reduce the reproduction capacity and/or fungus and insects development and also the proper grain activity which in turn makes it easier to store them (Casini, 2002).

Picture 1. Silo Bag

The spread of the use of this technology introduced new possibilities (it increased the retainining grain autonomy at a low cost) -although with certain limitations[34]- and also generated an additional service circuit formed by the bagging and unbagging offerers, as weel as the production of stuffing machinery (Bisang and Sztulwark, 2007). The increase observed in the use of silo bags, is an expression of a bigger storage capacity needed.
The use of this way of storing among the producers and country grain elevators allows them[35] to keep grain and make deliveries, avoiding high freight costs during the harvest time. If Soybean is stored in a dry and clean way, it can be kept between 4 and 12 months in good conditions. This allows to improve income up to 15.5 % (average, if sold in the month of January)(Ghida Daza , 2002).

5.3 Logistics and transportation services

Argentina has lower costs as regards production for most of the agricultural products where it has an important share of the global market, but it has higher costs as regards the commercialization than its most important competitors: Brazil and the United States (Dohlman *et al.*, 2001; Tavarez 2004; Nardi and Davis, 2006). The higher commercialization costs are due to, in the case of soybean, higher transport, storage and exportation tariff (Nardi and Davis, 2006). Eighty four percent of the grain production is taken to the

[34] Possibility of bag breaking and subsequent loss of stored grain quality.

[35] Other advantages would be: to activate the harvest when the production is stored in the same productive land and to obtain credits on the stored grain (Warrant)

exporting ports by truck[36], 14,5% by rail and 1,5% by barges. Grain exports are made approximately 90% by ship, 7% by truck and the rest by rail and barge.
According to data from the Stock Exchange of Rosario, the Gran Rosario area (province of Sta Fe) is the one with the highest growth as regards land cargo transportation in Argentina in the last decade with annual volumes of around 60 million tons of grain products; 8 million tons of these annual volumes come by rail[37] and 52 millons by truck. This implies that during the year an average of 5.000 trucks per day are concentrated. This number increases during the harvest months. (Giancola *et al.*, 2010).
Although there is a tariff difference between the truck and the rest of the means of transport (truck versus barge 3.25 to 1, truck versus rail[38] 2.5 to 1 (Commercial Infrastructure Area) (SAGPYA, 2007), the strong participation of the truck for domestic freight is due to its speed and flexibility to adapt to the resources and conditioning structures. The storage centers are generally located in the productive areas or within a radius of 20 km and at an average distance of 300 km to the industrial centers and port terminals[39]. The national road infrastructure has a total extension of approximately 38.000km.
Motor carrier: The trucking fleet is approximately 400.000 units. For cereal and oilseeds transport there maybe 5,000 special units, but there are around 60,000 general cargo units adaptable to this need. The average cargo capacity of a truck is 28 tons. The average age of these trucks is around 25 years, in an atomized system of car properties.
Railway transport: One of the limitations to its development as a mean of transportation is that more than 1,700 establishments which work as storage centers (83% of the total) does not have rail access (ONCCA, 2006). The present rail granted companies which are now giving cargo services are: ALL Central, ALL Mesopotamico, Ferro Expreso Pampeano, Belgrano Cargas and Nuevo Central Argentino (NCA). (Giancola *et al.*, 2010). Each company attends a different region, and all reach the Ports of Rosario. Regarding the operating capacity for the grain transport and its derived products of all the net rails, there are nowadays nearly 6.500 wagons (40 tons in each grain wagon depending on the gauge) between solid and liquid cargo and a totalnet of approximately 28,000 km (SAGPYA, 2007).
Fluvial/Maritime transport: There are 40 Argentine port terminals which have the capacity to deliver grain, oil and protein meal in bulk. Eight of these are maritime and they are located in the province of Buenos Aires. The maritime terminals deliver 19%of the cargo of this kind of products. The other 24 shipment terminals are fluvial, and from them 81% of the grain, vegetable oil and protein meal is delivered.
Most of the soybean complex products are exported from the Paraná River, in the north and south of the city of Rosario, from San Martin Port to Arroyo Seco; this section covers 70 km.

[36] Compared to Brazil and the United States, Argentina is one of the exporting countries which makes the biggest use of the truck due to the average distance between the primary productive centers and the industrial processing .
[37] With the use of the motor carrier, the rail transportation use was reduced since the former is more versatile and it does not have a fixed minimum volume or a fixed route.
[38] These differences are kept although the rail tariff has increased due to the raise in the petroleum cost since 2002.
[39] With the shift of the agricultural frontier to the North East and North West of Argentina where distances to the port of Rosario are more than 350-400 km the transport cost incidence due to the rail use would decrease making those regions more competitive .

These river port terminals deliver 76% of the cereal, oilseeds protein meal and vegetable oil exports and the rest of the river port terminals deliver 4% of global cargo.
The grain solid by products storage volume in the Paraná Hydroway ports grew from 1.5 million tons to almost 8 million. The cargo boats rhythm, in turn, grew from 23 million tons per hour to 54 million tons between 1990 and 2007. This operative capacity growth also allowed a growth of the Up-River exports from 13.5 million tons to 54 million tons as regards grains and pellets, at no extra costs of storage or delays of the cargo boats stays (SAGPYA, 2007).

6. Future context for the Argentine soybean chain

At an international level, it is considered that soybean prices will keep growing. This conclusion is drawn from the soybean world market indicators (supply, demand, stock, consumption) and ratified by the wider expansion of the world demand regarding the supply. If finantial speculation is added, the positive trend towards rising prices gets stronger. This conclusion is drawn after an analysis of the following factors (USDA, 2010):

- Grain production growth motivated specially by the larger demand of developing countries (DC) (consumption growth of food and forage for animals due to the per capita income growth and diet changes).
- Global growth more accelerated in DC (city planning processes and middle class expansion).
- Oil consumption growth in China (34% imports growth during 2011 whereas in India there will be 21% imports drop).
- Bio-fuel demand growth in the developing countries and in the DC.

It is expected that the good crop perspectives determine that the area planted remains and/ or increases, depending onthe economic perspectives of other summer crops, specially corn.
In Argentina, it is expected that in the short-medium term the discontinuous purchase of soybean oil will continue to be done by China due to the commercial barriers implemented by the government of Argentina and China, except that an agreement between both countries is reached. This has caused that the oil exports go to other countries which buy this oil at a lower price[40]. So, the Argentine oil industry -which is one of the most efficient in the world- gets lower prices and, furthermore, works under its milling capacity.
At a national level and at the primary production link, there have been and there are conflicts between the soybean protagonists (soybean producers, agrochemical companies, commercial companies; etc) and different groups that question this "soybean production system", in Argentina[41]. Among those who question this "soybean production system", it can be mention the small and medium farmers, aborigine communities and countryside workers expelled by the spreading of soybean sowing. At the same time large company

[40] The argentine soybean oil quotation suffered a discount with respect to other markets like the Brazilian one of up to U$D 75.5 per ton (average of U$D 50 per ton). These figures affect the almost 6 million soybean oil tons that Argentina would export in the current season

[41] In the mid 70s and with the use of new varieties, the sowing of two annual crops (doble -sowing) was possible. These crops are combined with the wheat production (short cycle wheat) and therefore the double crop wheat-soybean appears. This process originated a larger agricultural process (cattle is left aside) but soybean also leaves aside crops like corn and sorghum

groups[42] are consolidated and thus finance great scale farming and livestock directed by contractors who rent the fields.
Also among those that question the advance of soybean production there are some professionals and environmentalists who show the farming effects as regards the distribution of the productive and environmental resources. Some of these concerns are:

- Growing loss of food sovereignty due to the soybean growing substitution.
- The soybean monoculture generates among other consequences: productive variability fall, low genetic variability, sanitary risks (new plagues, resistence to plagues and diseases).
- Soybean takes the soil nutrients, wich demands crop rotation, and in certain soils its sowing is not advaisable. Each year, soybean takes one million tons of nytrogen and more than 227,000 tons of phosphorus. Likewise, the organic material balance is negative since the mineralized carbone cannot be compensated by the one of the soybean remains.
- The soybean monoculture produce an intense soil degradation in the long run, with a losses of between 19 and 30 soil tons per hectare depending on management , soil sloping and weather .
- The no-till cropping implies strong applications of agrochemicals. Massive use of glyphosate (160 million liters in 2004/2005) and other agrochemicals (between 20 and 25 million liters of 2-4-D (herbicide), and nearly six million liters of atrazine and 6 million liters of endosulfan; these last two, insecticides.
- The product in wich glyphosate is the active principle (Round up) has a series of coadjutant that increase its toxicity considerably, specially the polioxietil amina (surfactant), whose acute toxicity is 3 to 5 times higher than the one of the glyphosate.
- The advance of the soybean over the native forests (massive deforestation and loss of biodiversity). Between 1998 and 2002, more than 2 million hectares of native forest were deforestated[43].

Thus, there is a scenery of progressive conflict, where soybean excellent prices are signals for producers to continue with the expansion of soybean, and for other social groups (small farmers, environmentalists, professionals, etc.) to press because of the environmental and social consequences of this expansion.

7. Conclusions

Beyond the positive aspects related to the primary production (more security than other crops specially due to weather conditions, considerable finantial rentability, etc) the soybean growth has had positive effectsin Argentina, like the strong investments in new technologies (no-till cropping) biotechnology (transgenic soybean, fertilizers and herbicides), intensive use of agricultural machinery, storage (silo bags), logistics (private ports), net organitation (of services and financing), industrial machinery development, biofuel and agro food development the significant income of currency for the economy, without forgetting to mention the levying of export duties.

---

[42] "Los Grobo" Group claims that they grow 150 thousand hectares in Argentina and attemps to control 750 thousand hectares in all the Southern Cone (Paraguay, Brazil and Uruguay).
[43] Beginning of the century: 100 million hectares in forests, at the end of XX century: 33 million hectares

Without minimazing the positive effects or the conflicts brought about by the soybean expansion in Argentina, this crop has also generated a strong dependence[44] -a scarce commercial diversification –since only one product (soybean bean and its derived products) represent 25% of the total exported by Argentina (14.041 million dollars) (INDEC, 2009). This exporting dependence gets worse because the diary products, fruit, vegetable oil and meat are considered "Low Technological Level", whereas milling products are labeled as "Middle-Low Technological Level" (the technological content is based on the expenses in I+D) (CEP, 2007). At a Gross Product Level, the soybean chain (2009) represents 5.4% of the national added value (0.77% in the two-year period 73/74).

The strong soybean dependence is also seen in indicators like the soybean crop area and the soybean oil and meal production. At a primary level, soybean covers 54% of the cultivated area (more than 18 million has) over the total implanted (28 million has). For the oil and meal production numbers are more significant: the soybean oil production in Argentina participates in the 79% of all oil manufactured and in the 93% of all the meal produce.

This shows that the oil argentine industry depends basically on soybean. It must be pointed out that the oil industry milling capacity is 57 million tons; 37-38 million of which were used in the year 2010.

This idle capacity cannot be covered with the imports since the current imports regulations mekes it too costly. Thus, the industry competitiveness will be limited to the crops primary production and to the fundamental characteristics of domestic low consumption of byproducts which allow strong exportable balance. This last one, shortened with the current domestic soybean use for biodiesel production.

Finally, the National State role must be examined and also the implemented policies in relation to the agricultural production. During 2011, due to higher expenses[45], the National State do not show any sign to eliminate[46] the exporting duties[47], especially the ones which belong to the soybean complex. These obligations represent (2010) 8.7% of the total taxes. This shows the importance of the exports duties and the dependence of these assets for the expense policies. The exporting duties were 7,400 million dollars; of which 6300 million dollars were collected by the soybean complex (2011).

8. References

Ablin, E., Paz, S. (2003). El mercado mundial de la soja, la República Argentina y los organismos genéticamente modificados, In: *Las Negociaciones comerciales multilaterales de la Ronda Doha. Desafíos para la Argentina.* Centro de Economía Internacional. (7-54). Instituto del Servicio Exterior de la Nación. Ministerio de Relaciones Exteriores de Argentina. Buenos Aires. Argentina

[44] The risks of non-tariff imposition barriers for genetically modified soybean must be considered.

[45] During 2009 and 2010 the State Budget result was negative (income lower than total expenses).

[46] In the National Legislature (Commission) a project to modify the exporting duties obligations is being debated.

[47] In the year 2002 export reimbursements were implemented: the soybean oil 1.6for the raw oil in bulk; 5%for refined oil in containers of 5kg and 0,7% for the raw oil. All the above mentioned refunds were left aside (0%) since November, 2005. The raw oil but the exported volumes of refined oil represent only 3% of all the soybean oils.

ACSOJA. (2009). Importancia económica de la soja. 15/08/2010. Available from: http://www.acsoja.org.ar/ampliarII.asp?intIdSection=88&intIdContenido=217.año

Bisang, R. (2003). Apertura Económica, innovación y estructura productiva: la aplicación de biotecnología en la producción agrícola pampeana argentina. In: *Desarrollo Económico*. Vol. 43 N° 171. (413-442). Buenos Aires. ISSN 0046-001X

Bisang, R., Sztulwark, S. (2007). Tramas productivas de alta tecnología y ocupación. El caso de la soja transgénica en la Argentina. In: Estructura productiva y empleo. Un enfoque transversal. Primera edición. (181-224). ISBN 978-84-36571-68-6 (web, pdf)

Brieva, S. (2006). *Dinámica socio-técnica de la producción agrícola en países periféricos: configuración y reconfiguración tecnológica en la producción de semillas de trigo y soja en Argentina, desde 1970 a la actualidad.* Tesis doctoral. Facultad Latinoamericana de Ciencias Sociales (FLACSO). (pp.292)

Cámara Industrial de Aceites de la Republica Argentina (CIARA). (2011). *Estadísticas de soja.* 10/02/2011. Available from: http://www.ciaracec.com.ar/estadísticas

Casini, C. (2002). Guía para almacenar grano en bolsas plásticas. In: *Información técnica Proyecto eficiencia de cosecha y postcosecha de grano.* (1-4). Instituto Nacional de Tecnología Agropecuaria (INTA) Manfredi, Córdoba.

Cámara Argentina de Biocombustibles (CARBIO). (2011). *Estadísticas de biocombustibles.* 13/02/2011. Available from: www.carbio.com.ar/es/?com=bio_estadísticas

Centro de Estudios para la Producción (CEP). 2007. "*Contenido tecnológico de las exportaciones argentinas 1996-2007. Tendencias de upgrading intersectorial*". 1/08/2010. Available from:
www.cep.gov.ar/descargas_new/contenido_tecnológico_exportaciones_ 19962207.pdf

Civitaresi, H., Granato, M. (2003). El complejo oleaginoso argentino: algunos factores determinantes de su desempeño exportador. *Revista Argentina de Economía Agraria.* Vol. VI, 1, (Otoño 2003), pp. 37-47), ISSN 0327-3318

Dohlman, E., R. Schnepf., Bolling, C. (2001). Agriculture in Brazil and Argentina: Developments and Prospects for Major Field Crops. In: *Washington, DC: US Department of Agriculture, Economic Research Service. Agriculture and Trade Report. WRS-0-1-3.* 5/02/2011. Available
http://www.ers.usda.gov/publications/wrs013/wrs013g.pdf.

EUMERCOPOL (2007). EU-Mercosur. Agrifood Systems Comparison. In: Proyecto. Documento de Trabajo Work Package 4. Buenos Aires.

Facultad de Agronomía, Universidad de Buenos Aires (FAUBA). (2004). *Informe Final LART/FAUBA. Patrones espaciales y temporales de la expansión de soja en Argentina.* Universidad Nacional de Buenos Aires. Buenos Aires, Argentina.

FAO-SAGPYA. (2004). *Análisis de las cadenas de maíz y soja en Argentina, con vistas a exportación de mercaderías OVM y no OVM en el marco del art. 18.2 a) del Protocolo de Cartagena sobre seguridad de la biotecnología.* Proyecto FAO-SAGPYA. TCA/ARG 2903. Documento N° 2. Dirección Nacional de Mercados Agroalimentarios. Organización de Naciones Unidas para la Agricultura y la Alimentación. Buenos Aires, Argentina.

Fundación Producir Conservando. Capacidad de almacenamiento. (2010). Available from: http:/www.producirconservando.org.ar

Giancola, S., Salvador, L., Covacevich, M., Iturrioz, G. (2010). Análisis de la cadena de soja en la Argentina. En: *Estudios Socioeconómicos de los Sistemas Agroalimentarios y Agroindustriales Nº 3*. INTA, Buenos Aires. ISSN 1852-4605.

Ghida Daza, C. (2002). Alternativas para el almacenaje. Silo bolsa de grano seco. In: *Maíz, actualización 2002*. Información para extensionistas. INTA Marcos Juárez.

Gutman, G., Lavarello, P. (2003). La trama de oleaginosas en Argentina. In: *Estudios sobre el sector Agroalimentario. Estudio 1.EG.33.7 Componente B: Redes Agroalimentarias, Tramas B-3*. Bisang R., y G. Gutman. CEPAL-ONU, Buenos Aires.

Instituto Nacional de Estadística y Censos (INDEC). (2009). Estadísticas macroeconómicas. 20/02/2011. Available from: http://www.indec.mecon.gov.ar

Instituto Nacional de Tecnología Agropecuaria (INTA). Consejo del Centro Regional Santa Fe., (2009). Documento institucional. El avance de la soja en la Argentina y la sostenibilidad de los sistemas agrícolas. In: *INTA Centro Regional Santa Fe*. Febrero 2011. http:/www.inta.gov.ar/reconquista/crsantafe/docsoja.htm

Lopez, G., (2008). *Vamos al grano?. El Rol del Estado en el comercio granario argentino* (primera edición). Editorial Cúspide. ISBN 9789870588023. (288 pp).

Nardi, M., Davis, T., (2006). Supply chain management in Argentina and Brazil and the Effect on World Agricultural Competitiveness. *Presentation given to trading companies*. Buenos Aires, March 2006. 92 pp.

Obschatko, E., (1997). *Articulación productiva a partir de los recursos naturales. El caso del complejo oleaginoso argentino* (Primera edición). Documento de trabajo Nº 74. CEPAL, Buenos Aires. UNDP Project RLA 88/039. Buenos Aires, Argentina.

Oficina Nacional de Control Comercial Agropecuario (ONCCA). 2006. Existencia física de granos al 31 de Diciembre 2006. 6/02/2011. Available from: http://www.onnca.gov.ar

Quaim, M., Traxler, G. (2002). Roundup Ready Soybeans in Argentina: Farm Level, Environmental and Welfare Effects. *Proceedings of Conference ICABR on Agricultural Biotechnologies: New Avenues for Production, Consumption and Technology Transfer*. Ravello, Italia. Julio 2002.

SAGPyA (2007). Informe de fletes y preliminar de transporte de granos. 12/02/2011. Available from: http://www.minagri.gob.ar/publicaciones.php

Schavarzer, J., Tavosnanska, A. (2007). *El complejo sojero argentino. Evolución y perspectivas*. Centro de Estudios y Perspectivas de la Argentina (CESPA). Universidad de Buenos Aires. Facultad de Ciencias Económicas. Documento de trabajo Nº 10. Buenos Aires.

Tavarez, C. (2004). Factores críticos a Competitividade da Soja no Parana e no Mato Grosso. Companhia Nacional de Abastecimiento. Brasilia. 9 pp.

Trigo, E., Cap, E. (2006). *Diez años de cultivos genéticamente modificados en la Agricultura Argentina*. Consejo Argentino para la Información y el Desarrollo de la Biotecnología- ARGENBIO, Buenos Aires, Argentina. 53 pp

United State Department of Agriculture (USDA). (2010). USDA data base: commodities and trade. In: USDA Foreign Agricultural Service. 29/01/2011. Available from: www.fas.usda.gov/commodities.asp and /trade.asp

19

Mineral Nutrition

Vlado Kovacevic[1], Aleksandra Sudaric[2] and Manda Antunovic[1]
[1]*University J. J. Strossmayer in Osijek, Faculty of Agriculture*
[2]*Agricultural Institute Osijek*
Croatia

1. Introduction

Sixteen nutrient elements are essential for the growth and reproduction of plants. The source of carbon (C) and oxygen (O) is air, while water is source of hydrogen (H). Ninety-four percent or more of dry plant tissue is made up of C, H and O. Remaining thirteen elements, represent less than 6 percent of dry matter, are often divided in three groups (Johnson, 1987). The primary nutrients are nitrogen (N), phosphorus (P), and potassium (K). Secondary nutrients are sulfur (S), calcium (Ca) and magnesium (Mg). Micronutrients are required by the plant in very small amounts. They are iron (Fe), manganese (Mn), boron (B), zinc (Zn), copper (Cu), molybdenum (Mo) and chlorine (Cl). Nutrient removal of soybean by tone of grain and correspondingly biomass are about 100 kg N, from 23 to 27 kg P_2O_5, from 50 to 60 kg K_2O, from 13 to 15 kg CaO, from 13 to 16 kg MgO and considerable lower amounts of the other nutrients. In general, the fertilizer requirements for soybean are typically less than for other crops such as maize and wheat. Bergmann (1992) reported adequate concentrations of nutrients in dry matter of fully developed leaves at the top plant without petioles at the end of blossom as follows: 4.50-5.0 % N, 0.35-0.60 % P, 2.5-3.70 % K, 0.60-1.50 % Ca, 0.30-0.70 % Mg, 25-60 ppm Zn, 30-100 ppm Mn, 25-60 ppm B, 10-20 ppm Cu and 0.5-1.0 % Mo.

2. Nitrogen

Symbiotic nitrogen (N) fixation had important role in supplying of leguminose plants including soybean, by N. It is estimated that by this source is possible to bind from 40 to 300 kg N/ha/year (Bethlenfalvay et al., 1990). Field studies by Bezdicek et al (1978) showed that soybeans are capable of fixing over 300 kg N/ha when the soil is low in available N and effective strains of *Bradyrhizobia* are supplied in high number. Also, part of leguminose needs for N is settled by its uptake in mineral forms, mainly in NO_3^- and NH_4^+ forms. Soybean contains in mean from 1.5 to 1.6 % and from 6.5 to 7.0 % N in dry matter of aboveground part and grain, respectively (Hrustic et al., 1998). N amounts removal from soil by soybean depending on numerous external and internal factors. For forming of 1 t of grain and correspondingly vegetative mass of soybean is needed about 100 kg N.
Worldwide some 40 to 60 million metric tons (Mt) of N_2 are fixed by agriculturally important legumes annually, with another 3 to 5 million Mt fixed by legumes in natural ecosystems, providing nearly half of all the nitrogen used in agriculture (Hungria & Campo, 2004). Therefore, biological dinitrogen fixation by leguminous plants is a significant source

of available nitrogen in both natural and managed ecosystems (Galloway et al., 1995) that contributes to soil fertility and replaces the use of synthetic nitrogen fertilizer.

The host plant provides carbon substrate as a source of energy, and bacteria reduce atmospheric N_2 to NH_3 which is exported to plant tissues for eventual protein synthesis (Vincent, 1980). Nitrogen fixation occurs in different intensities in soil, during which the energy of plant assimilates is used, and because of this, bacterial activity forms unbreakable relationship with plants. The proportion of nitrogen derived from fixation varies substantially from zero to as high as 97%, and most estimates fall between 25% to 75% (Deibert et al., 1979; Keyser & Li, 1992, Russelle & Birr, 2004).

N is mobile in plants and it is quickly translocate from old to young organs. For this reason, symptoms of N deficiency (first lightgreen and later greenyellow colours of leaves) obtain on the older leaves. In the more over stages it is found falling off the flowers and pods (Vrataric & Sudaric, 2008). Excess of N had unfavorable impacts on soybean productivity, mainly due to susceptibility to diseases, low temperatures and drought. Symptoms of N oversupplies are increasing of height of plants, longer internodies and lodging incidences. Soybean is the most susceptible leguminose to nitrate oversupplies. Under these conditions inhibition of nodule forming and nitrogenase activities were found Harper & Gipson, (1984). Also, high nitrate in apoplast of soybean had effect on pH increasing, immobilization of iron and developing of iron chlorosis in soybean (Hrustic et al., 1998).

N supplies of soybean could be estimated by number and activities of bacterial nodules of genus *Rhisobium* and *Bradyrhisobium,* contents of total and mineral N in oil, nitratreductase activities etc. Inadequate N supplies are possible to correct by mineral fertilization. Activities of bacteria are reduced under good supplies of soil by N (as results either high N fertilization or favorable conditions for organic matter mineralization) and acid soil pH.

Recommendations for soybean fertilization are depended on soil test results and planned yields. Under conditions of the northern Croatia N recommended quantities are mainly in range from 60 to 90 kg N/ha mainly in spring. Using of N as urea in autumn over 100 kg/ha resulted by absence of nodule bacteria or minimizing their amounts (Vrataric & Sudaric, 2008). By testing 12 localities in fertile soils (chernozem and similar soil types) of Vojvodina (Serbia) was found that inoculation had considerable more impacts on yields of soybean compared to N fertilization and that using 90 kg N/ha was not found nodule on soybean root (Belic et al., 1987; Relic 1988 - cit Vrataric & Sudaric, 2008). Based on experiences from very fertile soils in Ohio (Johnson, 1987), soybean is not recommend for N fertilization in case of sufficient amounts of N-fixing bacteria and only in first growing of soybean on individual soil recommendation is applying 45 kg N/ha. Also, in Illinois mineral N fertilization had not effects on soybean yields even in cases of band fertilization close to soybean rows. Also, N fertilization was superfluous for maize in soybean-maize rotation (Welch et al., 1973). However, the experiences from USA are not possible to applying in less fertile soils of middle and eastern Europe.

Soil acidity is often limiting factor of the symbiotic nitrogen fixation process. Soils with low pH values lack calcium, and have surplus of toxic aluminium, so that soybean roots in acidic soils don't have mucous coating on surface which purpose is to dissolve root pectines, enables root hair curling and root hair penetration by bacteria. This is very important during the first few days after inoculation that is after sowing inoculated seed. Therefore, soils with pH value less than 5.5 (acidic soils) are not suitable for soybean growing, because they lack necessary conditions for development of useful bacteria whose growth is slowed down or completely enabled. Strains found on soybean roots in this type of soils are mostly ineffective, and when cut in half are green in colour. Situation is completely opposite in fertile neutral or mildly alkaline soils like chernozem. In these types of soil nitrogen fixing bacteria have not only good conditions for development, but also they can survive in large numbers for many years after soybean was grown. In such soils it is not necessary to perform seed bacterisation if soybean is in rotation every four years.

In case of low effects of inoculation on nodule bacteria development it is recommend top-dressing with 50 kg N/ha in form of calcium ammonium nitrate (27% N) in term close to flowering or at beginning of flowering (Vrataric & Sudaric, 2008).

Organic manures cannot alone meet the heavy demands of nutrients in intensive soybean production because of their limited availability and restricted nutrient supply. A complementary use of organic manures and mineral fertilizers may meet the goal of adequate and balanced supply of required nutrients to crops. The soybean grain yield with recommended NPK fertilization and 25 kg N/ha + 1 t neem cake/ha combinations was significantly more than the other only chemical and organic source of nutrition (Table 1).

Soybean grain quality (crude protein and oil contents) and grain yield as affected by fertilization (in. = inoculated; neem cake = n.c.; FYM = farm-yard manure 5 t/ha; NPK 20:60:40 = recommended NPK-fertilization)							
Treatments	Percent		t/ha	Treatments	Percent		t/ha
(a-e)	Protein	Oil	Yield	(f-j)	Protein	Oil	Yield
a) Control (in.)	34.42	19.03	0.69	f) b + 1 t n.c.	37.92	18.85	1.52
b) 25 kg N/ha	37.92	16.00	0.93	g) c+ 1 t n.c.	37.04	19.65	1.23
c) 50 kg N/ha	37.04	18.72	0.89	h) b + 5 t FYM	38.06	16.83	1.57
d) 5 t FYM/ha	35.73	19.23	0.80	i) c+ 5 t FYM	37.63	18.17	1.20
e) 1 t n. c/ha	35.44	17.03	1.07	j) NPK 20:60:40	38.94	18.92	1.33
LSD 5 % (a-j)	ns	1.78	0.27	LSD 5 % (a-j)	ns	1.78	0.27

Table 1. Effects of inorganic and organic sources of nutrients on grain quality and yield of soybean (Saxena et al., 2001)

3. Phosphorus

Phosphorus (P) contents in plants are in wide range, mainly from 0.1 to 0.8 % P in dry matter. Reproductive organs, especially of leguminose plants contain high levels of P about 0.6 % P. Uptake of P into plants is intensive in the early stages of growth and in period forming of generative organs (Hrustic et al., 1998). Store of P in plants, especially in grain, are mainly in form of fitine acid. P efficiency is in close connection with water and temperature regimes in soil. Under optimal soil moisture P uptake can be up to three-fold higher than in dry soil. Also, oversupplies of water, cold weather and low pH reducing P uptake in plants.

P removal by plants is mainly from 10 to 45 kg P, while by soybean is from 15 to 30 kg P/ha/year. The end of growth is the first symptom of P deficiency. Leaves are dark green and in the later stage develops chlorosis and violet color as result of increasing antociane synthesis. Necrotic spots, drying and falling of the leaves is the latest stage of P deficiency. Active nodules (dark pink center) of N-fixing bacteria are absent or few in number under conditions of P deficiencies. Also, decreasing of protein and chlorophyll synthesis was found.

Excess of P is rare. Plants reducing growth and dark frowning spots in leaves were observed. Intensity of plant development increasing and as results are the earlier flowering, grain forming and senescence. Oversupplies of P could be reason for some nutritional unbalances, for example Zn, Fe, Mn, Cu and B deficiencies.

P, mainly in combination with N and K as NPK fertilizers, can be applied broadcast and incorporated into the soil before sowing or applied as starter at sowing time. With low soil test P levels band application of fertilizer is more efficient than broadcasting. If applied as a starter, the recommend placement of the fertilizer is in band 2 inches to the side and 2 inches bellow the seed (Dahnke et al., 1992; Barbagelata et al., 2002). P materials such as triple superphosphate or from liquid or dry formulations of ammoniated phosphates are available to improve soil P status. However, organic soybean growing has restriction in P use and it is limited on rock phosphate or manures as sources of P.

Anetor and Akinrinde (2006) found that P deficiency in soil is an important growth-limiting factor in acidic alfisolof Western, Nigeria. Lime application may not be feasible for poor resourced farmers. However, the complementary benefits (liming and nutrient supply) of organic fertilizers and rock phosphates could sufficiently ameliorate acid soil conditions and greatly reduce P fertilizer cost for effective and sustainable soil fertility management.

Win et al. (2010) tested the P effects on three soybean cultivars (CKB1, SJ5 and CM60) based on the seed oil content (SOC) and the seed protein content (SPC) and to assess the physiological responses associated with changes in shoot P-utilization efficiency (SPUE). The experiment was carried out during 2008 and 2009 with a split-plot design at the Agronomy Department, Kasetsart University, Bangkok, Thailand. The main plots were for tested three P levels in a nutrient solution (0.5, 1.0 and 2.0 m*M* P), with subplots for the three soybean cultivars. The results indicated that at maturity, the P levels of 2.0 m*M* P decreased SPUE by 27% compared to that of 0.5 m*M* P (the control). SOC was not significantly affected by the P level. Relative to the control, the P nutrition levels of 1.0 and 2.0 m*M* P significantly decreased SPC by 4% and 5%, respectively. There were no significant differences in SOC between varieties. The SPC of CKB1 was 8% greater than that of SJ5 but showed no significant difference to that of CM60 (Table 2).

Zheng et al., (2010) reported effectiveness of P application in improving regional soybean yields under drought stress of the 2007 growing season in Northern China including Heilongijang, Jilin and Liaoing Provinces. Total soybean acreage of this region was around 4.5 million ha, which accounts for about 5% of the total soybean acreage in the world (FAOSTAT, 2009). Contemporary climate change is characterized by increase in frequency and intensity of drought. Total 118 soybean fields throughout Hailun County of Northern China. Regression trees analysis showed that regional soybean yield variability was mainly induced by soil available phosphorus and the amount of P applied, which explained 16.3 and 15.2% of the yield variation, respectively. The productivity of soybean over the region did not increase when P application rate reached a threshold of 55.67 kg/ha (Zheng et al., 2010).

Effects of P levels in nutrient solution and cultivars on soybean status (DM = dry matter; Sh = shoot; Prot. = protein)											
	R5 stage (Σ = Total)			**Maturity stage**				**Maturity stage**			
	Dry matter (g/plant)			**Dry weight** (g/plant)		**P-utilization*** (Q=quotient)		**Phosphorus** (mg/plant)		**Seed** %	
	Sh.	Root	Σ	Σ	Shoot	Q[a]	Eff. [b]	Sh.	Seed	Oil	Prot.
	Effects of P levels (m*M* P)in nutrient solution										
0.5	14.1c	1.35b	15.4c	25.8b	24.6b	0.213a	5.21a	118b	47.9b	16.9	40.2a
1.0	20.5a	2.36a	22.9a	36.0a	34.4a	0.139b	4.77a	258a	72.3a	18.1	38.5b
2.0	18.2b	1.96a	20.2b	33.3a	31.6a	0.117b	3.79b	292a	64.4a	17.5	38.3b
	Effects of soybean cultivars										
CKB1	22.2a	2.54a	24.7a	34.2	32.5	0.156	4.89a	236.	59.7a	16.9	40.4a
SJ5	14.2c	1.42b	15.7c	27.1	25.9	0.147	3.72b	212	52.5b	17.3	37.1b
CM60	16.4b	1.74b	18.1b	33.8	32.2	0.167	5.17a	221	72.4a	18.3	39.4a
Duncan's multiple range test (within column, means by the same letter are not significantly at 5 % level): CV %											
a	12	13	12	21	21	12	10	13	14	17	2
b	8	17	8	21	22	14	12	16	15	7	2
* [a] Shoot P-utilization quotient = plant shoot dry weight/mg P in plant shoot of P [b] Shoot P-utilization efficiency (eff.) = [(shoot DM)^2/shoot P content]											

Table 2. Effects of P on seed oil and protein contents and P use efficiency in three soybean cultivars (Win et al., 2010)

4. Potassium

Potassium (K) is essential nutrient for plant growth. K concentrations in dry matter of plants vary between 1.0 and 6.0 % and more and are generally higher than those of all other cations. The exact function of K in plant growth has not been clearly defined. By numerous investigations were found that K stimulates early growth, increases protein production, improves the efficiency of water (drought resistance), improves resistance to diseases, insects and stalk lodging (Kovacevic & Vukadinovic, 1992; Rehm & Schmitt, 1997).

Soils mainly containing enormous amounts of K, but depending on soil types, 90-98 percent of total K is unavailable. Slowly unavailable K is thought to be trapped between layers of clay minerals (Johnston, 1987; Rehm & Schmitt, 1997). K deficiency is encountered mostly on light, usually acid soils with a low cation exchange capacity or on soils with a high content of three-layered clay minerals often loess soils with illite clay (Bergmann, 1992).

Soybean requires large amounts of K and K deficiencies are easy to recognize (edge necrosis of leaves - the margins of leaflets turn light green to yellow) and correcting them is inexpensive as K is to lowest-cost major nutrient. K deficiencies as result of strong K fixation and high levels of available magnesium (Mg) were found on heavy hydromorphic soils of Sava valley area in Croatia. By ameliorative KCl fertilization yields of maize and soybean drastically increased due to improved plant nutritional status (Vukadinovic et al., 1988; Kovacevic & Vukadinovic, 1992; Kovacevic 1993; Kovacevic & Grgic, 1995¸ Kovacevic & Basic, 1997).

The K deficiency in soybeans was found on the drained gleyols which had inadequate rates of the exchangeable K and Mg (low K and high Mg status). These soil characteristics affected correspondingly K and Mg status in soybean plants (Tables 3 & 4).

Soybean (the upermost full-developed threfoliate leaf before anthesis) and soil status (means of four fields)									
The state farm (year)	Soybean (K-deficieny symptoms)					Soil status (0-30 cm of depth)			
	Leaf status (precent in dry matter)				Grain yield (kg/ha)	pH		mg/100 g (AL-method)	
	P	K	Ca	Mg		H_2O	KCl	P_2O_5	K_2O
Zupanja (1988)	0.35	0.98	1.20	0.73	1930	7.33	6.91	6.6	15.9
Vinkovci (1989)	0.57	1.05	2.22	2.14	780	7.75	6.87	28.0	12.2
Jasinje (1988)	0.38	0.87	1.30	1.11	1410	7.68	7.20	10.6	10.2
N. Gradiska (1990)	0.32	1.16	1.82	0.92	2060	7.76	6.91	7.9	16.6

Table 3. Plant and soil status (drained gleysol): symptoms of K deficiency in soybean (Kovacevic et al., 1991)

Soybean (the upermost full-developed threfoliate leaf before anthesis) nutritional status								
The state farm and date of sampling	Percent in dry matter				mg/kg (ppm) in dry matter			
	P	K	Ca	Mg	Zn	Mn	Fe	Al
	Chlorotic soybeans (K-deficiency symptoms): means of two samples/field							
Zupanja (June 19, 1987)	0.45	2.02	1.74	2.91	13.0	28.0	609	588
Vinkovci (June 13, 1986)	0.73	0.66	1.16	1.65	19.0	25.0	600	
Jasinje (June 13, 1986)	0.52	0.70	1.11	1.29	27.0	200.0	220	
Mean	0.57	1.13	1.34	1.95	19.7	84.3	476	
	Normal soybeans (oasis in the chlorotic soybeans): means of two samples/field							
Zupanja (June 19, 1987)	0.25	2.87	1.69	1.78	12.0	28.0	386	309
Vinkovci (June 13, 1986)	0.59	1.13	1.48	1.25	17.0	26.0	260	
Jasinje (June 13, 1986)	0.59	1.06	1.25	1.11	18.0	248.0	180	
Mean	0.48	1.69	1.47	1.38	15.7	100.7	275	

Table 4. Nutritional status of normal and chlorotic (K-deficiency symptoms) soybeans (Kovacevic et al., 1991)

Response of soybeans to ameliorative KCl- fertilization (the upermost full-developed threfoliate leaf before anthesis)													
Fertilization (spring 1986)			The 1986 growing season					The 1987 growing season					
			Yield	Leaf (% in dry matter)				Yield	Leaf (% in dry matter)				
N	P_2O_5	K_2O	(t/ha)	P	K	Ca	Mg	(t/ha)	P	K	Ca	Mg	
0	0	0	2.43	0.37	0.72	1.84	1.62	1.45	0.35	0.91	1.69	1.35	
120	120	180	2.40	0.36	0.87	1.87	1.49	1.48	0.32	0.99	1.37	1.35	
120	120	990	2.83	0.35	1.28	1.76	0.74	1.88	0.32	1.29	0.92	0.77	
LSD 5%			0.27	ns	0.22	ns	0.25	0.29	ns	0.16	0.30	0.31	

Table 5. Response of soybeans to potassium fertilization (Katusic et al. 1988; cit. Kovacevic & Basic, 1997)

KCl in spring 1987	Fertilization (KCl) impacts on soybean: grain yield and leaf (the upermost full-developed threfoliate leaf before anthesis) K and Mg (on dry matter basis) status– the growing seasons 1987-1989								
	1987			1988			1989		
K_2O	**Yield**	Leaf (%)		**Yield**	Leaf (%)		**Yield**	Leaf (%)	
kg/ha	kg/ha	**K**	**Mg**	kg/ha	**K**	**Mg**	kg/ha	**K**	**Mg**
150	1280	0.57	1.60	1800	0.82	1.18	780	0.60	2.16
1000	2700	1.90	0.95	2350	1.74	0.84	1470	0.75	1.79
2670	2550	2.28	0.78	2740	2.22	0.52	2530	1.17	1.41
LSD 5%	270	0.20	0.20	450	0.09	0.18	240	0.07	0.21
LSD 1%	360	0.27	0.27	600	0.13	0.24	320	0.09	0.27

Table 6. Response of soybean plants to potassium fertilization (Kovacevic & Vukadinovic 1992)

Katusic et al., (1988; cit. Kovacevic & Basic, 1997) applied increasing rates of KCl on Cerna drained gleysol. Soybean responded by yield increases for 16 % and 30 %, for the first and the second year testing, respectively. Soybean under unfertilized and usual fertilization was contained in mean 0.82 % K (acute K-deficiency with correspondingly symptoms) and 1.49 % Mg. Soybean nutritional status was considerable improved by K fertilization (mean 1.29 % K and 0.76 % Mg) - Table 5.

Kovacevic & Vukadinovic (1992) tested response of soybean and maize to increasing rates of potassium application in KCl form on silty clay gleysol developed on calcareous loess. Low levels of exchangeable K, high levels of exchangeable Ca and Mg and strong K fixation were found by the soil test (Vukadinovic et al., 1988; Kovacevic & Vukadinovic, 1992). Also, clay fraction (35.2 % of soil) composition was as follows: vermiculite/chlorite 30 %, smectite 30 %, mixed layer minerals 20 %, illite 15 % and kaolinite 5 % (Richter et al., 1990). By ameliorative K fertilization soybean yields were increased drastically (3-y means: 1286 and 2607 kg/ha, for the control and the highest rate of K) and they were in close connection with improvement of leaf K and Mg status (Table 6 and Fig. 1).

Fig. 1. Soybean status (middle of July 1989) on the control (left) and the highest rate of K (2670 kg K_2O/ha in spring 1987) application (right) - the data in Table 6 (photo V. Kovacevic)

In Ontario, Canada, studies looked at the response of soybeans to potassium fertilizer as related to K leaf tissue levels (Reid and Bohner, 2007). The data collected during that study formed the basis for updated critical and normal values for potassium in soybeans. Below a leaf K concentration of 2.0% (on dry matter basis), most of the plots showed a response to added K fertilizer. Above this level, most of the plots were unresponsive. Based on the results of these experiments and other similar studies, the critical concentration for K in soybean tissue was established at 2.0% and the maximum normal concentration from 2.5 to 3.0%. According this criterion, in our investigations under strong K-fixing conditions (Table 5) only by application of enormous K rates leaf-K concentrations were increased to normal level. However, in spite of considerable improvement of soil and plant K status, yields of high-yielding soybean cultivar were less than 3.0 t/ha (Table 5).
Long-term studies conducted on integrated nutrient management in soybean-wheat system (Singh & Swarup, 2000) revealed that continuous use of FYM along with recommended NPK for 27 crop cycle not only restricted K mining by reducing non-exchangeable K contribution to grain formation but also enchanced K uptake to the system (Table 7).

Fertilizer K added in 27 crop rotation, total K uptake, available and non-exchangeable K status in soil (maize crop was discontinued in the system since 1995)						
	Potassium (kg K /ha)				Contribution of	
Treatment	K added in 27 cycles	Available K status		Total K uptake	non-echangeable K	
		Before 1971	After 1999		kg K/ha	%
a) Control	0	370	252	3247	3129	96.4
b) 100 % N	0	370	263	4418	4311	97.6
c) 100 % NP	0	370	235	10067	9932	98.7
d) 100 % NPK	2117	370	308	11826	9647	81.6
e) d + 5 t FYM/ha	4142	370	324	14094	9906	70.3

Table 7. Removal and addition of K during 27 crops cycle of soybean-wheat-maize (fodder) cropping system (Singh and Swarup, 2000)

Morshed et al. (2009) applied six treatment of potassium (unfertilized, 50%, 70%, 100 % , 125% and 150% of recommend rate based on soil test) on equal N, P and S fertilization in Dhaka (Bangladesh) during Rabi season 2004-2005. By application of the highest K rate grain yield of soybean was increased for 83%. Slaton et al., (2009) found close connection of soybean response to K fertilization (five rates from 0 to 148 kg K/ha) and Mehlich-3-extractable soil K in eastern Arkansas. Experiments were established on silt loams at 34 site-years planted with a Maturity Group IV or V cultivar. Mehlich-3-extractable soil K ranged from 46 to 167 mg K/kg and produced relative soybean yields of 59 to 100% when no K was applied. Eleven sites had Mehlich-3-extractable K < 91 mg K/kg and all responded positively to K fertilization. Soybean grown in soil having 91 to 130 mg K/ g responded positively at nine of 15 sites. Mehlich-3 soil K explained 76 to 79% of the variability in relative yields and had critical concentrations of 108 to 114 mg K/kg, depending on the model. Based on these investigations, Mehlich-3-extractable K is an excellent predictor of soil K availability for soybean grown on silt loams in eastern Arkansas.

Gill et al. (2008) reported that imbalance and inadequate nutrient supply particularly devoid of K is main reason for low productivity and quality of soybean in India.

Yin and Vyn (2004) conducted field experiments at three locations in Ontario, Canada from 1998 through 2000 to estimate the critical leaf K concentrations for conservation-till soybean on K-stratified soils with low to very high soil-test K levels and a 5- to 7-yr history of no-till management. For maximum seed yield, the critical leaf K concentration at the initial flowering stage (R1) of development was 2.43 %. This concentration is greater than the traditional critical leaf K values for soybean that are being used in Ontario and in many U.S. Corn Belt states.

Nelson et al. (2005) compared response of soybean to foliar-applied K fertilizer and preplant application. Potassium fertilizer (K_2SO_4) was either broadcast-applied at 140, 280, and 560 kg K/ha as a preplant application or foliar-applied at 9, 18, and 36 kg K/ha at the V4, R1-R2, and R3-R4 stages of soybean development. Soybean grain yield increased 727 to 834 kg/ha when K was foliar-applied at 36 kg/ha at the V4 and R1-R2 stage of development in 2001 and 2002. Foliar-applied K at the R3-R4 stage of development increased grain yield but not as much as V4 or R1-R2 application timings. Foliar K did not substitute for preplant K in this research. However, foliar K may be a supplemental option when climatic and soil conditions reduce nutrient uptake from the soil.

Numerous studies investigated fertilization effects on soybean grain yield, but few focused on oil and protein concentrations. Haq & Mallarino (2005) determined fertilization effects on soybean grain oil and protein concentrations in 112 field trials conducted in Iowa from 1994 to 2001. Forty-two trials evaluated foliar fertilization (N-P-K mixtures with or without S, B, Fe, and Zn) at V5-V8 growth stages. Seventy trials evaluated preplant broadcast and banded P or K fertilization (35 P trials and 35 K trials). Replicated, complete block designs were used. Foliar and soil P or K fertilization increased ($P < 0.05$) yield in 20 trials. Foliar fertilization increased oil concentration in one trial and protein in one trial but decreased protein in two trials. Phosphorus fertilization increased oil concentration in two trials and protein in five trials but decreased oil in five trials and protein in two trials. Potassium fertilization increased oil in four trials and protein in two trials but decreased oil in two trials and protein in two trials. Total oil and protein production responses to fertilization tended to follow yield responses. Fertilization increased oil production in 20 trials and protein production in 13 trials. Fertilization that increases soybean yield has infrequent, inconsistent, and small effects on oil and protein concentrations but often increases total oil and protein production.

Potassium is known to play an important role in protecting the plants against drought stress. Quantity and distribution of rainfall in the major soybean regions in India is responsible for yield fluctuations about plus/minus 20% among years in comparison with national average yield of 1 t/ha. For example, K fertilization in level of 112 kg K_2O/ha resulted by soybean yield increases for 0. 2 t/ha in normal year (1980) and for 1.2 t/ha under drought stress conditions (1981). Profit from K fertilization was 44 and 259 USD/ha, for 1980 and 1981, respectively (Johnson, 1984). For this reason, K fertilization can help in curtailing the yield loss on account of drought.

There are several materials available to supply K to the soil and potassium chloride is the most economical form. However, certified organic soybean production is limited to the use of potassium sulfate or manures to supply K.

5. Secondary nutrients

Calcium, magnesium and sulfur comprise the secondary nutrient group. Documented deficiencies of these three elements are few (Council for Agricultural Science and Technology, 2009).

5.1 Calcium

Plant species differ greatly in their Ca needs. Total Ca contents in plants are mainly in range from 0.5 to 1.0% in dry matter. The Ca uptake of plants influenced by Ca status and pH value of the soil and by the concentrations of other cations, especially K and Mg. Lack of Ca in legumes prevents the development of the nodule bacteria, thus affecting N fixation. Ca containing materials are using in correction of soil pH from acid to close to neutral. Soil pH between 5.5 and 7.0 is optimal for symbiotic N fixation in soybean root nodules by *Bradyrhizobium japonicum* bacteria. Under these soil pH availability of nutrients such as N and P and microbial breakdown of crop residues are favorable. Calcium deficiency is unlikely if soil pH is maintained above 5.5 (Council for Agricultural Science and Technology, 2009).

5.2 Magnesium

The total Mg content in plants is generally between 0.1 and 0.5 % in dry matter. Mg is the central atom of chlorophyll and it is vital for photosynthesis, biological production and conversion of matter in the plant metabolism. Mg deficiency occurs on strongly leached diluvial sandy acid soils with a low cation exchange capacity. Mg deficiency can be induced not only by low Mg status but also by high concentrations of other cations, for example H^+, K^+, NH_4^+, Ca^+ and Mn_2^+(Bergmann, 1992). In Croatia were found nutritional problems of K uptake by soybean and maize induced by oversupplies of Mg and strong K-fixing (Kovacevic and Vukadinovic, 1992). Vrataric et al. (2006) reported increases of soybean yield for 5 %, contents of grain protein for 0.7% and oil for 0.7% due to foliar application of 0.5 % $MgSO_4$ (Epsom salt) solution on eutric cambisol. Importance of Mg in yield increases of field crops in Europe reviewed by Uebel (1999).

Vrataric et al. (2006) tested response of six soybean cultivars (*Kuna, Una, Nada, Ika, Lika and Tisa*) to foliar fertilization (FF) with Epsom salt ($MgSO_4.7H_2O$; 5% w/v solution in amount 400 L/ha) on Osijek eutric cambisol. The fertilization was applied on standard fertilization either once or two times (treatment designations FF 1x and FF 2x, respectively), while untreated plots were as a control (standard fertilization). The first FF was made in the soybean stage V2-V3 and the second FF ten days later before the R1 stage of soybean. The amounts of added nutrients were as follows (kg/ha): 3.2 MgO and 2.3 kg S, as well as 6.4 MgO and 4.6 kg S, for the treatment FF 1x, and FF 2x, respectively. In the growing season 1999 was by 22% higher compared to 1998. Yield of *Ika* cultivar was by 23% higher compared to *Una*. FF resulted by moderate yield increases up to 5% compared to the control. Differences of yield between FF 1x and FF 2x were non-significant. Oil contents were higher in the 1998 and 2000 (mean 21.27%) compared to 1999 and 2001 (mean 20.55%), while differences among cultivars (from 20.77% to 20.96%) were non-significant. In general, FF resulted by moderate but significant oil content increases (20.45%, 21.15% and 21.12%, for the treatment 0, FF 1x and FF 2x, respectively). Protein contents were significantly different among years from 38.53% (2000) to 39.38% (2001) and among the cultivars from 38.30%

(Lika) to 39.48% (Nada). ESFF resulted by significant increases of protein contents (38.62%, 39.11 and 39.21% for 0, FF 1x and FF 2x, respectively). Impacts of the fertilization on soybean yields were shown in the Table 8.

Cul-tivar (B)	Foliar fertilization (factor A) with Epsom salt (F1x in V2-V3 and F2x = F1x + ten days later) effects on grain properties (four year means: 1998-2001) of soybean cultivars (factor B)											
	Yield (t/ha)			X	**Oil content** (%)			X	**Protein cont.** (%)			X
	0	F1x	F2x	B	0	F1x	F2x	B	0	F1x	F2x	B
Kuna	3.70	3.98	3.99	*3.89*	20.5	21.2	21.1	*20.9*	38.0	38.7	38.9	*38.5*
Una	3.52	3.64	3.66	*3.61*	20.4	21.3	21.3	*20.9*	39.1	39.3	39.4	*39.2*
Nada	3.77	4.01	4.04	*3.94*	20.4	21.2	21.2	*21.0*	39.1	39.6	39.7	*39.5*
Ika	4.44	4.45	4.45	*4.45*	20.3	21.0	21.0	*20.8*	38.4	39.1	39.2	*38.9*
Lika	3.55	3.97	3.65	*3.63*	20.7	21.0	21.0	*20.9*	38.0	38.5	38.5	*38.3*
Tisa	3.89	4.05	4.11	*4.02*	20.4	21.2	21.2	*20.9*	39.2	39.6	39.6	*39.5*
X (A)	*3.81*	*3.97*	*3.98*		*20.5*	*21.2*	*21.1*		*38.6*	*39.1*	*39.2*	
LSD 5%	A: 0.13 B: 0.10 AB: 0.19				A: 0.36 B: ns AB: 0.46				A: 027 B: 0.31 AB: 0.46			
LSD 1%	0.35 0.13 0.31				0.84 0.66				0.50 0.48 0.66			

Table 8. Impacts of foliar fertilization with Epsom salt ($MgSO_4.7H_2O$; 5% w/v solution in amount 400 L/ha) on soybean properties - four year means (Vrataric et al., 2006)

5.3 Sulphur

Soils of humid and semi humid areas mainly contain total sulphur (S) in range from 100 to 1000 mg/kg, a range that is similar to that of total P. It is divided in inorganic and organic forms but in most soils organically bund S provides the major S reservoir. S in organic matter can be divided into two fractions, carbon bonded S and non carbon bonded S. The inorganic form of S in oil consists mainly of sulphate. In arid regions soils may accumulate high amounts of salts such as $CaSO_4$, $MgSO_4$ and $NaSO_4$. Sulphate like phosphate is adsorbed to sesquioxides and clay minerals, although the binding strength for sulphate is not a strong as that for phosphate. Under waterlogged conditions, inorganic S occurs in reduced forms such as FeS, FeS_2 and H_2S. Oxidation of S results int he formation of H_2SO_4 and is promoting factor of additional soil acidification. Sulphate acid soils are mainly extremely low pH and very rich in exchangeable Al. Soil acidification by addition of elemental S is recommend for depressing the pH of alkaline soils (Mengel & Kirkbi, 2001).

Sulphur contents in plants are mainly in range from 0.1 to 0.5 % in dry matter. S uptake by plants is in sulphate form, but plants can absorb S also in gaseous form as SO_2. Sulphate must first be reduced by the plant to sulfide before it can be incorporated mainly into S-containing amino acids methionine and cistine. S deficiencies in plants are relatively rare because of the constant inputs of sulphate with NPK fertilizers and presence of SO_2 in precipitation (acid rain). Soybeans use a considerable amount of sulfur. S deficiency is mainly occurs during cool, wet weather on highly leacheable sandy soils that are low in organic matter and in little industrialization areas. In some cases are possible damages due to S excess caused by acid rain (Bergmann, 1992).

Sarker et al., (2002) tested effects of fertilization of soybean by S and B alone or in combination up to 50 kg S/ha and u to 4.0 kg B/ha. Yield, protein and oil contents of soybean grain where significant when S and B were applied individually but their interaction were not significant. The highest biological yield and most of the yield atributes were obtained for the treatment combination of 30 kg S/ha and 1.0 kg B/ha.

6. Micronutrients (Zn, Mn, Fe, Cu, B, Cl, Mo)

6.1 Zinc

Soybean, maize and flax are the most susceptible field crops to Zn deficiency. It is often found on sandy soils low in organic matter, on high soil pH and calcaric soils, as well as on soils rich in available P. Cold and wet weather promoting Zn deficiency. N improving, while Fe and especially P, decreasing Zn uptake by plants. The first symptom of Zn deficiency in soybean is usually light green color developing between the veins on the older leaves. New young leaves will be abnormally small. Bronzing of the older leaves may occur. When the deficiency is severe, leaves may develop necrotic spots. Shortened internodes will give plants a stunted, rosetted appearance (Dahnke et al., 1992).

Zinc is essential element in metabolism of protein, carbohydrate and lipids. Zinc is compound of some enzymes (carboanhydrase, glutamat and malat-hydrogenase, alcalic phosphatase, proteinase, peptidase, etc.). Zinc has influences on auxine synthesis, intensity of respiration and uptake of Cu, Mn and especially P. Also Zn contributing to increase resistance to viruses diseases, drought and low temperature stress. Soil and leaf testing use in diagnosis of Zn status in plants. Also, important is P/Zn ratio.

Incorporation of anorganic Zn in form of $ZnSO_4.7H_2O$ (2-22 kg Zn/ha) or organic Zn in chelate form (03-6.0 kg Zn/ha), as well as foliar fertilization (0.5% solution of zinc sulfate) could be use for corrections of Zn deficiencies.

Nutritional disorders were found in soybeans grown on Osijek calcareous eutrical cambisol. Growth retardation and chlorosis were accompanied with the alkaline or a neutral soil reaction. By the foliar diagnosis zinc deficiency was found. Zinc deficiency was promoted by the excess of phosphorus or iron/aluminum in plants while K deficiency was accompanied with the excess of magnesium uptake. For example, chlorotic soybean contained in means only 16 ppm Z in dry matter (into normal soybean 27 ppm Zn). At the same time, the P:Zn ratio was 239 (the normal levels are under 180), while Fe:Zn ratio was 34 (the normal levels are under 15). The analogous values for the normal soybeans were 150 and 7, respectively (Kovacevic et al., 1991). The higher soil pH and oversupplies of plant available P are factors promoting Zn deficiency in soybean (Table 9).

Soybean: The uppermost full-developed trifoliate leaf (June 6, 1990)								Soil (0-30 cm of depth); mg/100 g = AL-method			
Percent in dry matter				mg/kg in dry matter				pH		mg/100 g	
P	K	Ca	Mg	Zn	Mn	Fe	Al	H_2O	KCl	P_2O_5	K_2O
Chlorotic and growth-retarded soybean (means of three samples)											
0.39	2.36	2.51	0.88	16.3	124	547	301	7.47	6.60	62.6	45.3
Normal soybean (oasis at the same plot: means of five samples)											
0.37	2.52	1.93	0.68	26.8	86	195	147	6.70	5.90	42.5	54.5

Table 9. Plant and soil status (eutric cambisol of Agricultural Institute Osijek): symptoms of Zn deficiency in soybean (Kovacevic et al., 1991)

Response of soybean to fertilization: pods/plant (P/P), grain/pod (G/P), 100-grain weight and grain yield															
Treatments 1-6 (kg/ha)				P/P	G/P	100gw	Yield	Treatments 7-12 (kg/ha)				P/P	G/P	100gw	Yield
N	P_2O_5	K_2O	Zn			g	t/ha	N	P_2O_5	K_2O	Zn			g	t/ha
0	0	0	0	31.6	1.96	11.5	1.46	60	80	30	0	51.4	2.03	14.1	1.89
30	0	0	0	34.7	1.98	11.6	1.64	90	40	30	0	51.9	2.12	14.4	1.90
30	40	0	0	45.1	1.99	12.6	1.74	90	60	30	0	65.9	2.14	16.6	2.23
30	40	30	0	43.7	1.99	12.6	1.79	90	80	30	0	47.2	2.07	13.5	1.77
60	40	30	0	45.1	2.01	12.7	1.76	90	80	60	0	45.0	1.84	12.3	1.71
60	60	30	0	44.5	2.05	12.9	1.82	90	80	60	25	68.9	2.14	16.1	2.48
LSD (1-12) 5 %				8.51	ns	2.29	0.34	LSD (1-12) 5 %				8.51	ns	2.29	0.34

Table 10. Effect of N, P; K and Zn application on yield attributes and grain yield of soybean (Singh et al., 2001)

Rose et al. (1981) were studied response of four soybean varieties (*Lee, Forrest, Bragg* and *Dodds*) to foliar zinc fertilization ($ZnSO_4.7H_2O$ before flowering) at three sites in central and north-west New South Wales. At Narrabri one spray of 4 kg/ha gave a yield increase of 13 %. At Trangie and Breeza, two spray each of 4 kg/ha increased yield by 57 % and 208 %, respectively. Lee was the least responsive variety at each site and Dodds and Forrest the most responsive to applied zinc. Zinc fertilizer increased plant height, leaf-Zn, oil contents (at two sites) but decreased leaf-P. Leaf-P in untreated plots was indicative of varietal sensitivity to zinc deficiency both within and between sites.

Singh et al., (2001) tested twelve nutrient combinations comprising of three levels each of nitrogen (30, 60 and 90 kg N/ha), phosphorus (40, 60 and 80 kg P_2O_5/ha) two levels of potassium (30 and 60 kg K_2O/ha) and a single level of zinc (25 kg Zn/ha) along with control. Zinc fertilization in combination with N, P and K significantly increased the growth attributes and grain yield of soybean, The highest number of pods per plant and grain yield were obtained with the joint application of N, P, K and Zn at the rates of 90, 80, 60, and 25 kg/ha, respectively (Table 10).

6.2 Iron

Leguminose plants have higher needs for Fe in comparison to cereals. Fe participating in numerous metabolic processes including protein synthesis. Under F deficiency conditions were found high levels of low-molecular N substances, especially amino acid arginine. Soybean is susceptible to Fe deficiency. Fe deficiency is a common yield limiting factor for soybean grown on high-pH, calcareous soils, as well as on some seasonally poorly drained soils. Cool and wet periods are promoting Fe deficiency. Iron may be unavailable for root absorption, not transported after absorption, or may not be utilized by the plant.

In Iowa and Minnesota, over ten million dollars in potential soybean production were lost annually due to iron chlorosis (Fleming et al., 1984). With the potential increase in alkalinity of Texas soils due to irrigation, reduced soybean production may become a problem. The problem could result from decreased yield per acre or from acreage with decreased productivity due to increased alkalinity. Iron deficiency is not easy or inexpensive to correct in the field. According to Gray et al. (1982) it would take five tons of sulfuric acid per acre to neutralize one per cent calcium carbonate in a 16.5 cm layer of soil.

Fe deficiency results in a characteristic interveinal chlorosis in new leaves and can cause substantial yield loss in soybean. In some years, developed during early growth stages and disappears as the plants mature. In more severe cases, chlorosis can persist throughout the entire season. There is wide variation in susceptibility to Fe deficiencies among soybean varieties.

Soybean in chlorotic areas had lower leaf chlorophyll concentrations, stunted growth, and poor nodule development relative to nonchlorotic plants. Also, compared to nonchlorotic areas, soil in chlorotic areas had greater soil moisture contents and concentrations of soluble salts and carbonates (Hansen et al., 2003).

Correcting Fe chlorosis often requires a combination of management practices including variety selection, application of Fe fertilizers with the seed (for example iron chelate Fe-EDDHA) or foliar treatment with 1 % solution of ferrous sulfate.

Franzen and Richardson (2000) tested soil factors affecting iron chlorosis of soybean. Total 12 sites of Red River valley of North Dakota and Minnesota were studied in the 1996-1998 period. Calcium carbonate equivalence and soluble salts were most often correlated with chlorosis symptoms.

Plant response to iron chlorosis varies between cultivars and environmental conditions (Coulombe et al., 1984; Gray et al., 1982). The reduction of iron at the root surface from Fe to Fe is an adaptive mechanism which iron efficient plants use to overcome iron deficiency. Soybean cultivars like Hawkeye have been shown to be rather effective in facilitating iron uptake by this method (Brown & Jones, 1976). Iron uptake is (1) as iron in association with chelate molecules and (2) as ionic iron after chelate splitting. Iron efficient plants have a much increased rate of iron uptake after chelate splitting during iron deficiency chlorosis (IDC)-induced stress; iron inefficient plants do not (Romheld & Marschner, 1981). Iron efficient and iron inefficient plants reportedly are distinguishable in terms of extent of iron uptake as a function of phosphorus content in the soil. Chaney & Coulombe (1982) reported that increased phosphorus inhibited the increase in iron uptake of inefficient types and slightly reduced iron uptake of efficient types.

Goos & Johnson (2003) found considerable differences of resistance of soybean varieties to iron clorosis. (Table 11) Growing of more tolerant varieties is solution for alleviation of nutritional problems induced by iron deficiency.

Soybean varieties characterizing low chlorosis score (CS <2.3)			Soybean varieties characterizing high score (CS >3.2)		
Variety	Originator	CS	Variety	Originator	CS
Trail	N.D. AES	1.7	IA 2042	Iowa AES	3.7
Danatto	N.D. AES	2.0	IA 2041	Iowa AES	3.7
MN 0201	Minn. AES	2.0	MN 1103SP	Minn. AES	3.7
92 M10	Pioneer	2.0	MN 101SP	Minn. AES	3.7
IA 1005	Iowa AES	2.0	Minnatto	Minn. AES	3.5
Jim	N.D. AES	2.2	IA 2050	Iowa AES	3.3
MN 0203 SP	Minn. AES	2.2	IA 2033	Iowa AES	3.3
Mn 0302	Minn. AES	2.2	MN 2101SP	Minn. AES	3.3
Nornatto	N.D. AES	2.2	IA 2050	Iowa AES	3.3
MK 0649	Richland Organics	2.2	Parker	Minn. AES	3.3
CV = 31.7; LSD 5% = 1.0					

Table 11. Chlorosis scores of soybean varieties in Minnesota 2003 (Goos & Johnson, 2003): score 1.0 = no chlorosis, 5 = most severe chlorosis (choice 20 extremely of 104 tested genotypes)

Silman and Motto (1990) tested under greenhouse conditions in nutrient solutions influences zinc on the growth and composition of an Fe-efficient (*Hawkeye*) and Fe-inefficient (*PI-54619-5-1*) soybean genotypes in various levels of Fe. In general, increased Zn levels resulted in growth reduction in both genotypes with the Fe-inefficeint plants being more sensitive to Zn level. The Fe-efficient genotype had a higher Fe content than the Fe-inefficient at corresponding treatment levels.

6.3 Manganese

Plants vary considerably in their Mn requirements and levels only 20 ppm often being sufficient for normal plant growth. Levels of Mn in plants vary between species and soil properties more than those of other nutrients. Plants generally absorb Mn from the soil as Mn_2^+. Mn is important in plant metabolism because of its redox properties and thus its ability to control oxidation, reduction and carboxilation reactions in the carbohydrate and protein metabolism.

Mn deficiency causes soybean plants to be stunted. The leaves are yellow to whitish but with green veins. Mn deficiency is most pronounced in cool weather on alkalic and slow alkalic soils rich in organic matter. Soil pH is the most important factor affecting Mn availability because it is extremely soluble at low pH and insoluble at high pH levels.

Foliar fertilization of young and moderately young crops with 8-15 kg $MnSO_4$/ha as 1-2 % solution (2-3 applications) is recommendation for prevention of Mn deficiency on soils with a high pH value.

Manganese deficiencies in oats and soybeans were reported by Willis (1928), primarily in spots in the coastal plain of North Carolina. This problem was associated with very high soil pH, and thus this observation was likely the first evidence of "overliming." The soils in the coastal plain are inherently low in manganese, especially the more poorly drained ones, since manganese can be reduced and leached in the soil-forming process (Cox, 1965). Interveinal chlorosis is a clear symptom of manganese deficiency. Cox (1968) used both extractable manganese by the Mehlich-1 extract and soil pH and developed a yield response prediction and manganese soil test interpretation for soybeans. Extractable concentrations at the critical level, which varied from 3 to 9 with Mehlich-1 depending on pH, were sufficient to be measured readily. Critical levels of manganese in soybean leaves at various growth stages and effective rates of fertilization for correcting manganese deficiency in soybeans reported by Mascagni & Cox (1985a, 1985b).

6.4 Copper

Copper (Cu) is seldom deficient in soil. Only on soils high in organic mater and under conditions pH above 6.0 would Cu likely be deficient. The color of legume and forage plants deficient in Cu tends to be grayish-green, blue-green or olive green. The internodes become shortened to produce a bushy type of plants (Sauchelli, 1969). Soybean has low requirements for Cu.

Williams (1930) noted that crops grown on muck soils in North Carolina often responded to copper application. This observation was researched in detail by Willis (1937), Willis & Piland (1936) and these researches centered on the aspect that copper may be a catalyst in oxidation-reduction processes in soils.

6.5 Boron

Total boron (B) contents in soils are into range of 20 to 200 mg/kg dry weight, most of which is unavailable for plants. The available, hot water soluble fraction in soils adequately supplied with B ranges from 0.5 to 2.0 mgB/L. Soluble B consists mainly of boric acid which under most soil pH conditions (ph 4-8) is undissociated. In soils of arid and semi arid regions, B may accumulate to toxic concentrations in the upper soil layer because lack of drainage and the reclamation of such soils requires about three times as much water as that of saline soils. Soil organic matter is closely associated with the accumulation and availability of B in soils (Mengel & Kirkby, 2001).

B deficiency leads to disturbance of growth and development of plants. B is known to influence carbohydrate metabolism, sugar transport, the nucleic acid and protein household, N metabolism, flower formation and pollen germination, water household, energetic processes of phosphorylation and dephosphorylation, etc. Conditions that favour B shortage are high pH (7.0-8.0), soils low in organic matter, drought, high concentrations of iron and aluminium hydroxide. Difference between adequate and toxic concentrations of B is very small. Soybean is very susceptible to B toxicity. Alfalfa and sugar beet have high requirements in B. Broadcasting and incorporation of 0.5 to 1.0 kg B/ha (for example, the most commonly used borax) is satisfied for needs of crops in rotation for a few years (Berrgmann, 1992; Mengel & Kirkby, 2001).

6.6 Chlorine

Chlorine (Cl) is in group of elements which can have a beneficial effects on plant growth. Plant tissues usually contain substantial amount of Cl often in range of 2 to 20 mg/kg dry weight. Soils considered low in Cl are below 2 mg water soluble Cl/kg soil which is rare. The effects of excess Cl in plants are more serious problem. Crops growing on salt affected soils often show symptoms of Cl toxicity. These include burning of leaf tips too margins, bronzing, premature yellowing and abscission of leaves. Plant species differ in their sensitivity to Cl. Some leguminous species are very prone to Cl toxicity and using of sulfate instead of chlorine fertilizers is recommend (Mengel and Kirkby, 2001).

6.7 Molybdenum

Molybdenum (Mo) is an essential plant nutrient. The concentrations of Mo may vary from less than 0.1 to more than 300 ppm. Roots contain a greater proportion of Mg than aboveground part or seed. Molybdenum is needed by the soybean and other leguminose plant itself and also by the nitrogen-fixing *Rhizobia* bacteria in the soil. In contrast to other micronutrients, Mo availability increases with soil pH. Seldom is there Mo deficiencies with soil pH above 6.0. Since the element is critical for nitrogen fixation, the pale green or yellow plants are identical to a nitrogen deficiency. In this case, leaves generally begin to yellow first on the lower leaves. Needless to say, symptoms usually do not occur on soils high enough in nitrogen to make up for lack of nodule fixation (Holshouser, 1997). Efficiency of symbiotic N_2 fixation can be limited by micronutrient deficiencies, especially of molybdenum. Soybean generally responds positively to fertilization with Mo in soils of low fertility and in fertile soils depleted of Mo due to long-term cropping.

Sodium or ammonium molybdate are mainly used for correction of Mo deficiency either as a solid to the soil or by spraying on the foliage or by treating the seed. The first step, however, is always to establish the proper soil pH. The micronutrient can be supplied by seed treatment, however toxicity of Mo sources to *Bradyrhizobium* strains applied to seed as inoculant has been observed, resulting in bacterial death and reductions in nodulation, N2 fixation and grain yield. Therefore, use of seeds enriched in Mo could be a viable alternative to exterior seed treatment. Campo et al., (2009) demonstrated the feasibility of producing

Mo-rich seeds of several soybean cultivars, by means of two foliar sprays of 400 g Mo/ha each, between the R3 and R5 stages, with a minimum interval of 10 days between sprays (Table 12). In most cases, Mo-rich soybean seeds did not require any further application of Mo-fertilizer (Campo et al., 2009).

Soybean variety	Foliar fertilization by Mo (R3 = beginning of pod formation; R5 = beginning of pod filling; 2x = two spraying at R3+R5 with ½ dose)										
	0	400 g Mo/ha			800 g Mo/ha			1600 g Mo/ha			CV
		R3	R5	2x	R3	R5	2x	R3	R5	2x	%
	Mean contents of molybdenum in seed of soybean (ug/g)										
Embrapa 48	3g	23e	17f	31d	36c	25e	43b	39c	39c	61a	8.8
BRS 133	4f	22e	18e	35c	33d	26d	50b	50b	39c	82a	18.7
BRS 156	4f	19e	20e	35d	33d	35d	56b	52b	43c	81a	13.2

Table 12. Impacts of foliar fertilization on Mo contents in soybean grain (Campo et al., 2009)

7. Harmful elements (Cd, Cr, Hg and Pb) and heavy metal toxicities

Heavy metals are the intrinsic component of the environment. It is usually accumulated due to unplanned municipal waste disposal, mining and use of extensive pesticides. Other agro-chemicals uses as chemical fertilizer are the significant cause of elevation in environment.
Shute et al. (2006) reported results of greenhouse study regarding Cd and Zn accumulation in soybean. The highest dose of Cd (100 mg/kg) reduced plant height and dry weight (down to 40 % and 34 % of control, respectively), while the analogical data for the highest dose of Zn (2000 mg/kg) were 55 % and 70 %, respectively. With both metals present, the plants were approximately the same size as those treated with cadmium only. When both metals were added to the soil, 80-100 % of the cadmium and 46-60 % of the zinc were bioavailable. Concentrations of both metals were highest in root tissue (10-fold higher for Cd and up to 2-fold higher for Zn). Although relatively little Cd was translocated to pods and seeds, the seeds of all plants (including those from control and zinc-treated plants) had concentrations of cadmium 3-4 times above the limit of 0.2 mg/kg set by the Codex Alimentary Commission. This was surprising given that Cd in the soil was only 1 mg /kg well below the maximum allowable amount for agricultural soil.
The heavy metal content of municipal and industrial sewage sludge and swine manure lagoon sludge are quite high in Cu and Zn and cause a buildup of the elements in the soil and for this reason have potential toxicity to the environment. (King, 1986; King & Hajjar 1990). Physiological effects of zinc toxicity in soybean elaborated Fontes (1992) and Fontes & Cox, 1995, 1998). Borkert & Cox (1999) evaluated the effects of high concentrations of both Zn and Cu on soybean status. Miner (1997) looked at soil factors affecting plant concentrations of these elements in sludge-amended soils. When concentrations of heavy metals are high, knowledge of their solubility becomes important.
As soybeans are one of the principle sources of dietary intake in the Japanese population, the Codex Committee on Food Additives and Contaminants has proposed an upper limit of 0.2 mg/kg for cadmium concentration in soybean grain with aim of protection dietary uptake of harmful quantities of Cd (Arao et al., 2003).
Arao et al. (2003) tested Cd uptake and distribution of Cd in 17 soybean varieties grown in pots (three soils: Mid-Cd Soil, High-Cd Soil, Low-Cd Soil) and under field conditions in un-

polluted soil (low-Cd field). The sources of cadmium pollution were thought to be mine waste in the case of the Mid-Cd Soil, and refining plant waste in the High-Cd Soil. The seed cadmium concentration was lowest for the *En-b0-1-2* soybean variety, and highest for *Harosoy*. The seed cadmium levels of *Tohoku 128*, a cross between *Enrei* and *Suzuyutaka*, were intermediate between those of the parents (Table 13). For four soil types, containing from 0.2 to 6.5 mg kg^{-1} extractable cadmium, the ranking of soybean genotypes based on seed cadmium level was similar, indicating that there is a genetic factor involved in the varietal differences in cadmium concentration. The lower levels of cadmium found in the seeds of certain varieties of soybean could be result from the combination of lower initial uptake and retention of higher levels of cadmium in the roots, thus limiting its translocation to the shoot.

Different actions can be undertaken in order to reduce the absorption of Cd by plants. The addition of amendments such as calcium carbonate, zeolite, and manganese oxide can reduce Cd uptake in plants. With that regard, zeolite was more effective in suppressing Cd uptake by plants than calcium carbonate or manganese oxide (Chen et al., 2000; Putwattanaa et al., 2010). Also, organic amendment such as farmyard manure and compost which contains a high proportion of humified organic matter can decrease the bioavailability of Cd and other heavy metals in soil (Li et al. 2006, Pichtel & Bradway, 2008; Tordoff et al., 2000). Shamsi et al. (2010) tested effects of potassium supplementation on alleviation of Cd toxicity in hydroponics experiment. K supplementation at a rate of 380 mg/l in combination either with Cd addition (1 ug Cd) or without Cd. K supplementation alleviated the reduction of growth, photosynthesis and nutrient uptake in Cd-treated soybean plants. It was concluded that Cd toxicity could be alleviated through enhanced K nutrition in soybean.

Soybean cultivars show significant differences in seed cadmium concentrations, primarily because of genetic rather than environmental factors. One-six of the total soybean produced in Japan exceeded 0.2 mg Cd/kg, the international standard proposed by the Codex Alimentarius Commission. Further, the soybean crops had considerably higher Cd contents than other field crops MAFFJ (2002). Sugiyama & Noriharu (2009) investigated the seed Cd concentrations in four soybean cultivars (*Suzuyutaka, Hatayutaka, Enrei* and *Kantou 100*) in pot experiment on Cd-polluted soil. In *Suzuyutaka,* which had high Cd concentrations in the seeds, the concentrations of Cd distributed from the shoots to the leaves was 67% and that distributed from the shoots to the seeds was 13%. In *Kantou 100* which had low Cd concentrations into seeds, 57% Cd was distributed from the shoots to the leaves and 21% from the shoots to the seeds. These results suggest that cultivars that have a low capacity for Cd accumulation in the roots have a mechanism that prevents Cd accumulation into seeds by promoting its accumulation in the leaves (Sugiyama & Noriharu, 2009).

Chromium (Cr) is a nonessential and toxic element to plants. Chromium interferes with several metabolic processes, causing toxicity to plants as exhibited by reduced seed germination or early seedling development (Sharma et al., 1995), root growth and biomass, chlorosis, photosynthetic impairing and finally, plant death (Scoccianti et al., 2006). Normal range of Cr is from 10 to 50 mg/kg depending on the parental material (Pandey & Pandey, 2008). Researchers have demonstrated experiments with plants associated with high levels of Cr. Thus, 1-5 ppm Cr present in the available form in the soil solution, either as Cr (III) of Cr (VI), is the critical level for a number of plant species. Increased Cr (VI) concentration of 10-800 mg/l in culture medium led to the detection of inhibited growth parameters. There was a reduction in growth, dry weight and vigour index in four soybean genotypes of soybean at 5 -200 mgl-1 concentrations of chromium, according to control application (Ganesh et al., 2009).

Testing of 17 soybean cultivars				High-Cd soil (66 days after sowing)		
Soybean Cultivar (1-17)	Pot experiment		Field	Soybean cultivar	Cadmium	
	High-Cd soil	Mid-Cd soil	Low -Cd		ppm	ug/plant
	Seed Cd (ppm in dry matter)			Leaves		
En-b0-1-2	1.43 a	0.46a	0.08 a	En-b0-1-2	5.5a	67.6a
Tamahomare	2.52 abc	0.70abc	0.10 ab	Tohoku 128	12.2b	152.8b
En-b0-01	1.96 abc	0.82bcd	0.10 ab	Suzuyutaka	12.9b	86.9a
Goyoukuromame	1.99 abc	1.16ef	0.10 ab	LSD 5%	1.6	27.1
Hayagin	2.22 abc	0.91cde	0.11 ab	Stem		
Enrei	2.09 abc	0.89cde	0.11 ab	En-b0-1-2	4.3a	48.0a
En-b2-110	2.06 abc	0.91cde	0.11 ab	Tohoku 128	10.3b	130.1b
Dewamusume	5.24 d	1.05def	0.12 b	Suzuyutaka	20.3c	120.1b
Tachiyutaka	3.29 c	1.47g	0.12 b	LSD 5%	5.1	51.2
En-N0-2	4.94 d	1.91h	0.13 b	Pod		
Tachinagaha	2.88 bc	1.17f	0.13 bc	En-b0-1-2	2.1a	8.4a
Nattousyouryuu	2.90 bc	0.59ab	0.13 bc	Tohoku 128	5.5ab	8.3a
Getenshirazu 1	1.72 ab	0.78bcd	0.13 bc	Suzuyutaka	13.7b	8.6a
EN 1282	5.33 d	2.21i	0.16 c	LSD 5%	9.4	5.9
Tohoku 128	2.83 bc	0.97cde	0.22 d	Total		
Suzuyutaka	7.46 e	1.50g	0.31 e	En-b0-1-2		124.0a
Harosoy	12.68 f	2.68j	0.40 f	Tohoku 128		291.3b
Average 1 - 17	3.61	1.14	0.15	Suzuyutaka		215.6a
LSD 5 %	1.35	0.28	0.03	LSD 5%		83.8

Table 13. Seed Cd concentrations of soybean varieties grown in pots (choice of High-Cd soil and Mid-Cd soil) and under field conditions (Arao et al., 2003)

Mercury (Hg) and his compounds are among the strongest phytotoxic substances and are also extremely dangerous to human and animals. It is a constituent of many crop protection agents. Non-contaminated soils contain only 0.003 to 0.03 mg Hg/kg. Hg levels of about 0.04 mg/kg in dry matter can be considered normal in plants. Maximum tolerance limit of 0.05 mg/kg in fresh matter is proposed for foodstuffs. Mercury uptake in plants is very slight because it is strongly sorbet in the soil, mainly by complexation with organic matter. Apart from growth inhibition, the symptoms of Hg toxicity include chlorosis, necrotic lesions and death. These are mainly results of severe root damage and the consequent inhibition of nutrient and water uptake. Since little Hg is translocated out of the root, there is a little danger of its entering to the food chain through the soil. The mobility of Hg and its uptake by plants can be greatly reduced by liming (Bergmann, 1992).

Lead (Pb) is major chemical pollutant of the environment, and is highly toxic for man. The major source of Pb pollution arises from petrol combustion. This source accounts for about 80% of the total Pb in the atmosphere. Pb is toxic because it mimics many aspects of the metabolic behavior of Ca and inhibits many enzyme systems. There is evidence that Pb pollution can induce brain damage in man and aggressive behavior in animals. Pb toxicity interferes with Fe metabolism and the formation of haem. The total Pb concentrations of agricultural soils lie between 2 to 200 mg/kg soil. Pb contamination very clearly follows the motorway areas. Vegetation at the side of the road may have levels of 50 mg/kg dry matter but in distance of only 150 m away from the motorway the level is normally about 2 to 3 mg/kg. Contamination occurs only on the outer part of plant seed or leaves and stem, and high proportion can be removed by washing (Mengel and Kirkby, 2001).

8. Genetic aspects of mineral nutrition of soybean

Plant varieties of the same species differ in absorption and utilization of nutrients from the environment. Varietal differences in the uptake of individual nutrients can be used as basis for the testing of both commercial varieties and selection materials under unfavorable soil conditions. An adequate distribution of soybean varieties based on their tolerance or susceptibility to less favorable conditions could contribute to better utilization of their yield potential (Saric, 1981; Saric & Loughman, 1983). According to Epstein (1976) agricultural intervention in the process of nature has two corresponding strategies: selection and genetic manipulation of the organism and modification of the environment. Many crops in Brazil have their yield improved thanks to the selection and breeding, especially in the large savanna (cerrado) region of Central Brazil. Soybean cultivation in the low-latitude acidic soils of Brazilian Savanach has become a reality since 1970's. Great contribution for this success has been achievements in soil science and plant breeding. There are however, constraints for sustainable production like high-Al and low-Ca in the deep layers of the soil. Measures can be taken to reduce the negative effects of acidity on plant growth are liming and selection of more tolerant genotypes.

Kastori (1978) found that Ca uptake was higher in *Corsoy* than in *Stella* and *Wilkin*. Later research showed that K uptake was the highest at the variety *Corsoy* (Kastori et al., 1979). Keoght et al., (1977) found lower N uptake in the varieties *Hill, Lee* and *Bragg* than in *Hood*. Queiroz et al. (1980) tested residual effects of P fertilizers on the yield of three soybean varieties over four years. The variety *Bossier* increased grain yield by 320 kg/ha, *Parana* by 640 kg/ha, whereas variety *Vicoja* did not show any response to P fertilization. Saric and Krstic (1982) tested ten soybean varieties 30 days in N-deficient nutrient solution. The variety *Yoslie Kataya 2* showed the lowest and the variety *Traverse* the highest N contents.

Kovacevic and Krizmanic (1987) tested 12 soybean genotypes of maturity group I (*Corsoy* and *Hodgson* as standard varieties and remaining ten are experimental lines from the F8 generation) under calcareous soil conditions. Grain yield of soybean genotypes ranged from 2.1 4 to 3.11 t/ha. The highest yield of *Vuka* and the lowest yield of *Os 155/82* on this soil may be due to the lowest Ca status by the former and the highest Ca status by the latter genotype (Table 14).

Spehar (1995a, b; 1999) studied genetic differences in the accumulation of nutrients in leaves and seeds of tropical soybean cultivars from diallel crosses with the cultivars *IAC-9, IAC-2, UFV-1, IAC-5, IAC-8, Vx5-281, IAC-7, Biloxi* and *Cristalina* under high and low Al-stress. The diallel analysis indicated that an additive-dominance model could explain the genetic differences among those genotypes for nutrient accumulation in leaves and seeds. The diallel analysis, although not conclusive, indicated that the mechanisms of mineral element accumulation in the leaves are not fully associated to those of accumulation in the seeds of soybeans. The expression of these characters is, however, dependent on mineral plant-stress.

Yield (t/ha) and leaf composition of 12 soybean genotypes (the uppermost full-developed trifoliate leaf at beginning of flowering) on calcareous soil (pH in KCl 7.35; $CaCO_3$ 7.93 %)											
Soybean genotype	t/ha	Leaf (% in dry matter)				Soybean genotype	t/ha	Leaf (% in dry matter)			
		N	K	Ca	Mg			N	K	Ca	Mg
Corsoy	3.10	5.20	2.05	0.91	0.55	***Os 8***	2.30	5.20	1.85	0.76	0.41
Hodgson	2.89	5.78	1.98	0.90	0.44	***Os 9***	3.00	4.82	2.18	0.85	0.36
Vuka	3.11	5.58	2.01	0.76	0.35	***Os 45***	2.41	5.07	1.99	0.82	0.35
Podunavka	2.74	5.73	2.06	0.98	0.51	***Os 89***	2.51	5.60	2.15	0.88	0.41
Sava	2.71	5.59	2.12	0.80	0.40	***Os155/82***	2.14	4.91	1.99	1.12	0.61
Os 5	2.82	5.48	2.15	0.95	0.39	***Os442/83***	2.45	5.46	2.24	0.86	0.50
LSD 5%	0.26	0.12	0.08	0.16	0.09	LSD 5%	0.26	0.12	0.08	0.16	0.09
LSD 1%	0.35	0.17	0.11	0.22	0.12	LSD 1%	0.35	0.17	0.11	0.22	0.12

Table 14. Yield and nutritional status of 12 soybean genotypes on Osijek calcareous soil (Kovacevic & Krizmanic, 1987)

Sudaric et al. (2008) reported about the effectiveness of biological nitrogen fixation in soybean linked to genotype for four growing season in eastern Croatia. Fields study involved eight cultivars in two treatments (control - without ionoculation and inoculation by *Bradyrhizobium japonicum*). The obtain results suggested on significant positive effect of rhizobial inoculation on both nitrogen fixation indicators and grain yield at all tested soybean cultivars (Table 15). Significant differences among tested cultivars in each measured trait indicate genetic diversity of tested material in both potential of biological nitrogen fixation and compatibility cultivar by *B. japonicum* strain, as well. Tested cultivars with the best potential for nitrogen fixation (OS-1-00, OS-3-0, OS-3-I) had the highest grain yield increasing (14.4%, 14.3% and 14.0%, respectively). These results indicate that the cultivars with the favorable performances of biological nitrogen fixation could be used as the parents for development new cultivars that are able to accomplish high grain yield with lower nitrogen level in soil.

Phosphorus is a major limiting factor for crop production of many tropical and subtropical soils. In Brazilian soils, high productivities of soybean are achieved by soil amendment techniques, using lime and fertilizers, supplying the nutrients required for best crop performance. The yield potential is an intrinsic factor and depends on plant germplasm characters that can be modified by selection and breeding (Furlani et al., 2002). Differences in grain yield among soybean cultivars under field conditions for P-, K- and N-efficiencies , were also reported by Raper and Barber (1970), De Mooy et al. (1973), Sabbe & Delong (1998), Sarawgi & Tripathi (1998), Hanumanthappa et al. (1998; 1999) and Ogburia et al. (1999).

Plant efficiency for phosphorus uptake and utilization may contribute to improve crop yield potential in situations of low P availability. Furlani et al., (2002) evaluated and classified twenty nine soybean cultivars in relation to the response to phosphorus (P) levels in nutrient solution. P uptake and use efficiency were estimated by the variables: shoot and root dry matter (DM) yield, P-concentrations and contents in plant parts and P-efficiency index (EI). The experiment was conducted in a greenhouse, during 1999, at Campinas, State of São Paulo, Brazil. The experimental design consisted of randomized complete blocks, arranged in split-plots, with three replications. The main plots were the P levels in the nutrient solution (64.5; 129; 258 and 516 mmol L^{-1}), and the subplots were the twenty-nine soybean cultivars, grouped according days to maturity. Multivariate analysis showed high correlation among the variables shoot-DM, total-DM and shoot P-concentration and P-efficiency index (EI). Cultivars were classified in efficient-responsive (ER)¾ *IAC-1, IAC-2, IAC-4, IAC-5, IAC-6, IAC-9, Sta. Rosa* and *UFV-1*; efficient-non-responsive (ENR) ¾ *IAC-7, IAC-11, IAC-15, S. Carlos* and *Cristalina*; inefficient-responsive (IR) ¾ *IAC-8, IAC-10, IAC-14, Bossier* and *Foscarin*; and inefficient-non-responsive (INR) ¾ *IAC-12, IAC-13, IAC-16, IAC-17, IAC-18, IAC-19, IAC-20, IAC-22, Paraná, IAS-5* and *BR-4*. The efficient-responsive soybean cultivars showed the highest values for shoot and total DM and EI, and the lowest shoot P-concentrations.

Soybean cultivar	Properties of uninoculated (- = control) and inoculated (+) soybeans							
	Nodule number/plant		Above-ground part of plant				Grain yield t/ha	
			Dry matter weight (g)		Nitrogen (% N)			
	-	+	-	+	-	+	-	+
OS-1-00	0	44.6	4.29	6.19	1.84	2.65	3.54	4.05
OS-2-00	0	37.8	3.41	4.91	1.69	2.44	3.39	3.85
OS-1-0	0	38.2	3.34	4.80	1.71	2.46	3.53	4.02
OS-2-0	0	35.8	3.21	4.62	1.67	2.40	3.36	3.81
OS-3-0	0	49.2	4.24	6.11	1.84	2.65	3.42	4.00
OS-1-I	0	31.9	2.64	3.80	1.46	2.11	3.57	4.03
OS-2-I	0	36.3	3.24	4.67	1.65	2.38	3.83	4.33
OS-3-I	0	40.2	4.17	6.00	1.82	2.62	3.71	4.23
LSD 5%	4.0		1.13		0.22		0.31	
LSD 1%	5.3		2.02		0.41		0.36	

Table 15. Mean values of nitrogen fixation indicators and grain yield of 8 soybean cultivars (2004-2007; Osijek, Croatia) (Sudaric et al., 2008)

Ojo et al. (2010) tested 55 soybean genotypes under acid soil conditions in area of Umudike, Nigeria for two growing seasons. Highly significant differences in genotypic effects were observed for all the traits (days to 50% flowering, plant height at maturity, number of pods/plant, 100-seed weight and grain yield). Eight acid tolerant varieties were found (*Conqvista, TGX 1896-3F, TGX 1897-17F, TGX 1866-7F, TGX 1805-31F, Milena, Doko* and *TGX 1844-18E*) with a higher grain yield of >1.80tons/ha compared to <1.45tons/ha in the previously recommended varieties (*TGX 1485- 1D* and *TGX 1440-1E*). The result also showed the potential of the EMBRAPA genotypes in upgrading the TGX varieties for higher productivity. The eight identified acid tolerant varieties could therefore be explored in the development of improved high yielding soybean genotypes for production on acid soils of Nigeria.

9. Agronomic management practice and nutritional status of soybean

9.1 Fertilization

In general, mainly fertilization of soybean with nitrogen, phosphorus and potassium are common agronomic practice. Additional using the other nutrients are more exception than rule. In some cases application of the higher P and K rates are needed for achieving of satisfied yields of soybean.

Phosphorus and potassium are limiting factor of field crops yield on some hydromorphic soils in Croatia (Kovacevic, 1993; Kovacevic et al., 2007; 2011; Rastija et al., 2006). By application of the ameliorative rates of NPK fertilizer up to 3748 kg/ha level soybean yields were increased up to 32 %. Protein contents in soybean grain were independent on the fertilization, while oil contents were increased up to 0.66% compared to the control (Rastija, et al., 2006). In the second experiment, P and K applied separately up to 1500 kg/ha either P_2O_5 or K_2O and in their combination (1000 + 1000 kg/ha). Yields of soybean were increased up to 21% (influences of P), 17% (influences of K) and 30% (PK influences). However, protein and oil contents in grain were independent on fertilization (Kovacevic et al., 2007).

Soybean is generally responsive to fertilization with inadequate nutrient supplies. For example, grain yields of soybeans were increased by 40% and 34% as affected by the K and P fertilization, respectively (Table 16). According to status of the uppermost full-developed trifoliate leaf (Jones, 1967, cit. Bergmann and Neubert, 1976; Bergman, 1992) the adequate P, and high Ca and Mg status as well as low K status was found in the soybean leaves when ordinary fertilization was applied (Table 16). However, nutritional status of soybean was considerably improved when affected by the ameliorative fertilization. Calcium uptake by soybean leaves was high and it was practically independent on the fertilization. Also, the K fertilization influenced the Mg status in soybean leaves: it was decreased in relative amount by about 30 % compared to ordinary fertilization. More favorable relationship between K and Mg was associated with K fertilization: 1.13 and 3.20 for ordinary fertilization and the highest rate of added K, respectively (Table 16).

Rastija et al. (2006) applied four rate of ameliorative PK-fertilization on acid soil. As affected by the fertilization grain yields of soybean were increased up to 32%. However, yield differences among three ameliorative treatments were non-significant. Protein contents in soybean grain were independent on the fertilization, while oil contents were increased up to 0.66% compared to the control (Table 17).

Fertilization (March 22, 1990) by P and K rates on equal (kg/ha: 90 N + 137 P_2O_5+ 132 K_2O) NPK fertilization and soybean propreties (the growing season 1990: the uppermost full-developed trifoliate leaf before anthesis)											
K_2O kg/ha	Yield t/ha	Leaf concentrations (% in dry matter)				P_2O_5 kg/ha	Yield t/ha	Leaf concentrations (% in dry matter)			
		P	K	Ca	Mg			P	K	Ca	Mg
132	2.13	0.32	1.17	1.80	1.04	195	2.86	0.33	1.24	1.86	1.05
433	2.69	0.32	1.49	1.82	0.92	325	2.71	0.35	1.39	1.89	1.04
735	2.98	0.32	1.65	1.77	0.83	585	2.57	0.38	1.52	1.94	0.98
1337	2.81	0.33	2.01	1.66	0.79	1105	2.52	0.49	1.80	1.92	0.91
2532	2.82	0.33	2.37	1.85	0.74	585	2.60	0.41	2.01	1.80	0.81
LSD 5%	0.49	0.01	0.13	0.19	0.08	LSD 5%	0.49	0.01	0.13	0.19	0.08
LSD 1%	0.66	0.02	0.17	0.26	0.11	LSD 1%	0.66	0.02	0.17	0.26	0.11

Table 16. Response of soybean to ameliorative P and K fertilization (Kovacevic, 1993)

Kovacevic et al. (2007) tested response of soybean to ameliorative P and K fertilization alone or in their combination. As affected by applied fertilization soybean yields were increased up to 21% (influences of P), 17% (influences of K) and 30% (PK influences). However, protein and oil contents in grain were indenpendent on fertilization (Table 18).

Fertilization (April 23, 2004)			Soybean properties (the 2005 growing season)				
	kg/ha		(t/ha)	Percent in dry matter			
Treatment	P_2O_5	K_2O	Grain	Grain		Leaves*	
			yield	Protein	Oil	P	K
Control	125	82	3.88	41.92	20.33	0.530	2.67
PK-1	375	248	4.87	40.89	20.80	0.537	2.71
PK-2	625	414	4.73	41.42	20.62	0.571	2.85
PK-3	875	582	4.98	40.64	20.99	0.487	2.73
PK-4	1125	746	5.14	41.94	20.73	0.501	2.69
LSD 5%			0.72	n.s.	n.s.	n.s.	n.s.
			* the uppermost full-developed threefoliate leaf before anthesis				

Table 17. Residual impact of PK-fertilization on soybean properties (Rastija et al., 2006)

Kovacevic et al. (2011) reported residual impacts of increasing rates of PK-fertilization up to 1000 kg P_2O_5/ha and 672 K_2O/ha in spring 2004 and liming by granulated fertdolomite (24.0 % CaO + 16.0 % MgO + 3.0 % N + 2.5 % P_2O_5 + 3.0 % K_2O) in autumn 2007 on soybean status in the growing season 2010. As affected by liming yields of soybean were increased for 18 % (means 3279 and 3854 kg/ha, for unlimed and limed plots, respectively). Also, grain quality parameters were improved by liming (thousand grain weight were 151.8 and 168.3 g; protein contents were 35.24 and 39.06 %, respectively), while oil contents were decreased (23.84 and 22.62 %, respectively). However, impact of P and K fertilization was considerably lower in comparison with liming (Table 19).

Fertilization (April 23, 2004)*				Soybean properties (the 2005 growing season)				
		kg/ha		kg/ha	Percent in dry matter			
Treatment		P_2O_5	K_2O	Grain	Grain		Leaves*	
				yield	Protein	Oil	P	K
a	Control	125	82	3600	41.27	20.74	0.537	2.81
b	P-1	625	82	3580	40.76	20.83	0.513	2.86
c	P-2	1125	82	3460	41.57	20.64	0.593	3.15
d	P-3	1625	82	4360	41.74	20.52	0.603	2.88
e	K-1	125	582	4010	41.13	21.20	0.520	2.87
f	K-2	125	1082	4200	40.21	21.10	0.547	2.95
g	K-3	125	1582	4080	40.59	21.10	0.573	3.29
h	P2K2	1125	1082	4670	40.28	21.14	0.593	3.05
			LSD 5%	370	n.s.	n.s.	0.055	0.30
			LSD 1%	510			n.s.	n.s.
* for next year: 80 N + 125 P_2O_5 + 82 K_2O				* the uppermost full-developed trifoliate leaf before anthesis				

Table 18. Residual influences of NPK-fertilization on soybean properties (Kovacevic et al., 2007)

Residual effects of fertilization and liming on soybean (cultivar *Lucija*) grain yield and grain quality in the 2010 growing season											
Factor B*: Fertilization (April 2004)			Factor A**: Lime(t/ha) (Oct. 2007)		Mean B	Factor A**: Lime t/ha) (Oct. 2007)		Mean B	Factor A**: Lime (t/ha) (Oct. 2007)		Mean B
	kg/ha		0	10		0	10		0	10	
	P_2O_5	K_2O	Grain yield (kg/ha)			Protein contents (%)			Oil contents (%)		
a	0	0	3141	3730	3422	36.12	37.68	37.68	23.70	22.49	23.10
b	250	168	3231	3837	3534	34.92	36.84	36.84	23.58	22.61	23.09
c	500	336	3285	4047	3666	35.73	37.59	37.59	23.58	22.29	22.94
d	750	504	3352	3826	3589	34.59	36.67	36.76	24.32	22.90	23.61
e	1000	672	3387	3860	3624	34.83	36.88	36.88	24.02	22.79	23.41
	Mean A		3279	3854		35.24	37.15		23.84	22.62	
	LSD 5% LSD 1%		A: 209 482	B: 156 ns	AB: ns	A: 0.63 1.46	B: ns	AB: ns	A: 0.53 1.23	B: ns	AB: ns
* Ameliorative fertilization by NPK 10:30:20 (a-e) on ordinary fertilization; N added by NPK-fertilizer were equalized with CAN (calcium ammonium nitrate: 27% N); ** ferdolomite (24.0 % CaO + 16.0 % MgO + 3.0 % N + 2.5 % P_2O_5 + 3.0 % K_2O) 10 t/ha											

Table 19. Residual effects of PK-fertilization (April 2004) and liming (Oct. 2007) on soybean (Kovacevic et al., 2011)

Response of soybean to phosphorus and potassium fertilization on yield and at the uppermost trifoliate leaves status									
Impacts of P (triple superphosphate) fertilization					Impacts of K (KCl form) fertilization				
P rate	Soybean yield (kg/ha)		Leaf P (%) at R2 stage		K rate	Soybean yield (kg/ha)		Leaf K (%) at R2 stage	
kg/ha	1997	1998	1997	1998	kg/ha	1997	1998	1997	1998
0	615	700	0.23	0.28	0	914	1180	1.51	2.27
5	814	890	0.26	0.35	9	973	1280	1.58	2.87
10	826	960	0.44	0.35	18	1092	1299	2.34	2.64
20	925	1338	0.44	0.57	36	1188	1320	2.36	2.54
30	1188	1667	0.46	0.51	54	1559	2236	2.66	2.64
40	1585	2433	0.46	0.40	72	2294	2725	2.62	2.62
50	2443	2731	0.44	0.35	90	2246	2773	2.71	2.70
60	2598	2814	0.50	0.35	108	2544	3164	2.25	2.65
70	2713	2938	0.50	0.38	135	2520	2815	2.21	2.61
P treatments received a blanket application of 108 kg K/ha; K treatments received a blanket application of 60 kg P/ha; Sufficiency ranges at the uppermost threefoliate leaves R2 stage: 0.26-0.50 % P and 1.71-2.50 % K									

Table 20. Response of soybean to P and K fertilization (Casanova, 2000)

Casanova (2000) reported that the primary nutritional limitation for successful soybean production under savanna soils in Venezuela (Guarico state) are soil acidity and deficiencies of P, K, N and Ca. Other nutrients as Mg, S and Zn become limiting when the soil is cultivated for several years. Application of triple superphosphate up to 70 kg P/ha and potassium chloride up to 135 kg K/ha on fixed rate of 108 kg K/ha for P plots and 60 kg P/ha for K plots resulted by considerable yield increases. The treatment combination of 60 kg P/ha and 108 kg K/ha produced the best grain yield of 3.16 t/ha (Table 20).
Mottaghian et al. (2008) applied in a silty loam soil in Mazandaran province, Iran, for soybean eight fertilization treatments as follows: 20 and 40 t/ha of organic fertilizers (municipal solid waste compost, vermicompost and sewage sludge) enriched with 50% of anorganic fertilizers need by the soil), only inorganic fertilizers (potassium sulphate and triplephosphate 75 kg/ha) and control (unfertilized). Mixture of 40 t/ha sewage sludge and inorganic fertilizers produced the highest yield and micronutrient (Mn, Cu, Zn and Fe) grain concentrations.

9.2 Liming

Acid soils occupy 3.95 billion ha (about 30 % of the world's ice-free land area (von Uexkull and Mutert, 1995). The poor production of crops grown in acid soils is due to combinations of toxicity (Al, Mn, Fe, H) and deficiencies (N, P, Ca, Mg, K, Fe, Zn). Soil acidity is certainly one of the most damaging soil conditions affecting the growth of most crops. Many factors are involved, but Al toxicity is of outermost significance because of damaging root growth and therefore reduces water and nutrient uptake.
Poor growth of soybean in acid soils as been attributed to a number of factors that include: low pH, high level of Al, Mn, and H, low levels of Ca, Mg, P, K, micronutrients like B, Zn etc. (Fageria, 1994), low population of beneficial micro-organisms like rhizobia, vesicular arbuscular (VAM) fungi and inhibition of root growth (Maddox and Soileux, 1991).
Management practices, such as acidificatying effects of acid-forming N fertilizers, removal of cations by harvested crops, increased leaching and leguminous crops (N_2-fixation), have resulted in the lowering of natural soil pH (Baligar and Fageria, 1997).
Plant growth in mineral acid soils can be restricted by complex of factors. Malnutrition of plants on acid soils is mostly the results of limited soil nutrient availability, often strengthened by impaired uptake capability of the root. Based for over 5000 observations for soybeans worldwide, the optimum pH value for soybeans indicated by this approach lies between ph 5.7 and 6.0 (Sumner, 1997).
An optimum liming regime should achieve the reduction plant available Al and Mn concentrations to levels which allow optimal production of a particular crops, supplying adequate levels of plant available Ca and Mg for optimum root growth and crop performance, creating conditions for optimal performance of beneficial soil fauna and flora particularly int he rhizosfere, and in case of legumes, creating environment which promotes infection and nodulation of root with effective N-fixing rhizobia. (Keltjens, 1997).
For successful soybean production, large quantities of lime and phosphorus fertilizers may be required (Fageria *et al.*, 1995). Liming improves microbiological activities of acid soils, which in turn increases N fixation by legumes, and also promotes mineralization of organic materials. However, over liming may reduce crop yield by inducing P and micronutrient deficiencies (Fageria, 1984).
Unfortunately, over 50% of the world's potential arable land surface is composed of acid soils mostly distributed in developing countries (von Uexküll and Mutert, 1995; Kochian et al., 2005). This restricts the production of soybeans and other legumes due to their

sensitivity to acid soil infertility. The growth of leguminous crops and development of symbiosis on acid soils are generally affected by deficiencies of Ca, K, P, Mg, S, Zn and Mo and/or toxicities of Al, Mn and Fe (Foy,1984; Clark et al., 1988).

Liming has been used to ameliorate the problem of aluminium toxicity and low pH in soils. Liming the top soil, however, remains a temporary solution due to subsoil acidity. Restriction in root growth due to subsoil acidity reduces plant nutrient acquisition and access to subsoil water which culminates in the reduction of crop yield (Ferrufino et al., 2000). Moreover, the cost of liming particularly in developing countries is prohibitive and does not justify such huge investment given the return on investment from grain yield of soybeans. The input cost of the recommended quantity of 0.5 to 1.00 tons/ha of liming material (Yusuf and Idowu, 2001), is about the expected total revenue from the current average yield of 0.7 tons/ha in the South-East and South-South regions of Nigeria. The identification of acid stress tolerant cultivars of soybeans, therefore, remains a viable alternative.

Opkara et al., (2007) conducted field experiments in Southeastern Nigeria, in the 2003 and 2004 growing seasons to assess the effect of liming on the performance of four high yielding soybean varieties (early maturing TGX 1485-1D, TGX 1799-8F, TGX 1805-8F and medium maturing TGX 1440-1E). Five lime rates of 0, 0.5, 1.0, 1.5 and 2.0 t/ha were applied to the main plots while the four soybean varieties were planted in the sub-plots. Liming significantly increased soil pH, number of nodules and number of pods per plant and grain yield, especially in 2004 but did not significantly influence plant height, shoot dry matter, days to 50% flowering and 100-seed weight. The 1.0 t/ha lime rate proved to be optimum and is thus recommended for high grain yield in soybean. Mean grain yield at 1.0 t/ha lime rates was higher than the yield in the control (no lime) by 66%. The medium maturing TGX 1440-1E gave, on the average, significantly higher number of leaves and number of pods per plant and grain yield than other varieties.

Kovacevic et al., (1987) tested response of maize, soybean and wheat on liming by hydrated lime to level 20 t/ha. The field experiment was conducted in triplicate for maize- soybean-wheat rotation. Depending on the year grain yield of soybean ranged between 2.59 and 4.03 t/ha. Liming with 10 t of lime increased soybean yield by 17 %. Increased lime rates did not affect grain yield. Low grain yield were obtained under dry and warm weather conditions in 1983 and cold and wet weather conditions in 1984. Liming with 20 tons of lime per hectare increased soil pH from 4.0 to 6.4 at end of the first year of testing (Table 21).

	Hydrated lime (t/ha: autumn 1980)							LSD	
Year	0	1	5	10	15	20	Mean	5%	1%
	Grain yield of soybean in maize-soybean-wheat rotation (t/ha)								
1981	2.67	2.72	2.80	2.81	2.95	2.85	2.80	0.11	0.15
1982	3.48	3.75	4.06	4.28	4.29	4.32	4.03	0.11	0.15
1983	2.24	2.39	2.60	2.72	2.77	2.83	2.59	0.08	0.11
1984	2.27	2.32	2.49	3.08	3.03	3.02	2.70	0.15	0.20
1985	3.37	3.43	3.58	3.61	3.59	3.59	3.53	0.09	0.12
Mean	2.81	2.92	3.11	3.30	3.33	3.32			
	Soil pH (1n KCl) status at end of the growing season								
1981	4.03	4.21	5.03	5.44	5.94	6.42			
1984	4.40	4.36	4.93	5.46	6.02	6.53			

Table 21. Response of soybean to liming on Fericanci acid soil (Kovacevic et al., 1987)

Loncaric et al. (2007) applied liming with carbocalk up to 20 t/ha (spring 2003) and three degrees of fertilization (every year for 4-year period) on Donji Miholjac dystric luvisol. Soybean was grown on the experimental field in the fourth growing season (2006). Depending on the treatment, soybean yields were in range from 2.7 t/ha (unlimed and unfertilized plots) to 4.4 t/ha (treatment kg/ha: 140 N + 300 P_2O_5 + 300 K_2O) and phosphorus removals (P_2O_5/ha) by soybean were 56 and 78, respectively.

9.3 Soil tillage

Different tillage techniques affect the root absorption of nutrients. Lavado et al. (2001) tested effects of conventional and zero tillage (CT and ZT) on nutritional status of soybean, wheat and maize with emphasis on heavy metals. The field experiments were conducted in area far from contaminated sources in Buenos Aires Province, Argentina. The effects of tillage were limited for nutrient concentrations, but significant for heavy metals. Soybean appeared to be more sensitive than cereals to the apparent effect of soil tillage. Grain composition of soybean was independent on soil tillage. Under CT conditions leaves and stem N as well as root Cu were significantly higher, while root -Zn, -Pb and -Ni were significantly lower in comparison with ZT (Table 22).

Soil tillage treatments (ZT = zero tillage; CT = conventional tillage) and soybean nutritional status (G = grain; L+S = leaves and stems; R = root) of soybean under field conditions (mg/kg)*												
	ZT	CT	ZT	CT	ZT	CT	ZT	CT	ZT	CT	ZT	CT
	Nitrogen		*Phosphorus*		*Potassium*		*Sulfur*		*Copper*		*Zinc*	
G	33800a	46900a	4500a	1900a	19800a	18000a	2200a	2600a	17.10a	20.83b	44.85a	43.50a
L+S	9500a	14400b	2100a	1500a	15000a	15000a	900a	800a	10.93a	13.45a	21.48a	18.70a
R	7000a	8600a	2900a	900b	1300a	900a	800a	800a	18.70a	27.98b	64.73a	41.78a
	Boron		*Molybdenum*		*Lead*		*Nickel*		*Cadmium*		*Chromium*	
G	5.77a	6.15a	2.95a	1.71a	0.85a	0.80a	4.30a	4.26a	<0.05a	<0.05a	0.93a	1.20a
L+S	4.03a	4.60a	1.49a	1.70a	0.69a	0.63a	1.55a	2.08a	<0.05a	<0.05a	1.74a	2.36a
R	5.10a	6.00a	1.15a	1.36a	3.51a	2.41b	9.46a	6.77b	<0.05a	<0.05a	10.80a	11.93a
	* Means with different letter in each row are significantly different between treatments at LSD 5 %											

Table 22. Impacts of soil tillage on nutrient and heavy metal status of soybean (Lavado et al., 2001)

Jug et al (2006) reported about soil tillage impacts on nutritional status of soybean on chernozem soil for four growing seasons (stationary field experiment from 2002 to 2005). Three treatment of soil tillage were applied as follows: a) conventional tillage, b) reduced tillage (DH = diskharrowing instead of ploughing) and c) no-till (NT). In general, the characteristics of growing season (the factor „year") were more influencing factor of soybean nutritional status (aerial part in stage of full-developed pods) in comparison with the soil tillage. In this study, low influences of applied soil tillage treatments on nutritional status of soybean were found because significant differences on soybean composition were found only for four (Cu, Cr, Sr and Ba) from total 20 analysed elements. For example, conventional tillage resulted by the higher plant Cu (by 15% and 18% in comparison with DH and NT, respectively), and the lower plant Sr (by 12% and 16%, respectively) and Ba (by 26% and 23%, respectively), while under DH conditions by 22% lower plant Cr was found. Main nutrient status were independent on soil tillage (Table 23). For this reason, usual fertilization practice is recommended for possible application of soil tillage reduction under conditions of calcareous chernozem.

Stipesevic et al. (2009) reported response of winter wheat and soybean to different soil tillage systems on chernosem soil for four years. Three applied soil tillage treatments were applied as follows: a) CT - conventional soil tillage, based on mouldboard ploughing, b) DH - soil tillage based on diskharrowing instead of ploughing; and c) NT - no-tillage. Both crops showed decreasing concentration of Zn within the plant tissue as a result of the soil tillage reduction in the order CT>DH>NT, presumably due to the limited roots growth in lesser disturbed soil at DH and NT treatments. Winter wheat recorded generally lower than optimal Zn concentrations and higher P:Zn ratios at reduced soil tillage treatments, as a result of lower Zn uptake. The recommendation for the winter wheat production by reduced soil tillage is additional Zn fertilization, whose exact amounts and way of application shall follow further research.

The year (the factor A) and soil tillage (ST = the factor B: conventional = CT; diskharrowing = DH; no-till = NT) and composition of soybean (cultivar Tisa)**											
Year (A)	ST (B)	The aerial part of soybean at full-developed pods stage*									
		Percent on dry matter basis					mg/kg on dry matter basis				
		P	**K**	**Ca**	**Mg**	**S**	**Zn**	**Mn**	**Fe**	**Cu**	**B**
2002		0.316	1.85	1.67	0.636	0.171	21.9	38.6	158	9.4	37.6
2003		0.297	1.56	1.91	0.639	0.212	22.4	111.1	385	8.7	46.7
2004		0.407	2.49	1.75	0.470	0.265	34.3	69.2	229	10.3	59.2
2005		0.397	2.66	1.74	0.515	0.223	24.0	84.3	282	9.5	44.2
LSD A 5%		0.047	0.47	n.s.	0.101	0.034	3.7	26.5	127	n.s.	6.0
	CT	0.370	2.18	1.68	0.538	0.226	25.3	71.7	272	10.5	46.8
	DH	0.366	2.31	1.75	0.565	0.215	26.7	75.0	222	9.1	47.3
	NT	0.327	1.93	1.87	0.592	0.212	25.0	80.7	296	8.9	46.7
LSD B 5%		n.s.	n.s.	n.s.	n.s.	n.s.	n.s.	n.s.	n.s.	1.3	n.s.
Mean		0.354	2.14	1.77	0.565	0.218	25.7	75.8	264	9.5	46.9
		mg/kg on dry matter basis					mg/kg on dry matter basis				
		Mo	**Co**	**Ni**	**Cr**	**Sr**	**Ba**	**Al**	**Pb**	**Cd**	**Na**
2002		0.148	0.055	2.81	0.851	25.0	6.59	112	0.376	0.030	44.6
2003		0.198	0.160	1.83	0.609	21.1	6.64	312	0.336	0.042	29.3
2004		0.126	0.163	1.84	0.406	16.4	5.30	171	0.307	0.028	34.8
2005		0.313	0.174	2.44	0.450	21.3	9.48	215	0.503	0.086	59.2
LSD A 5%		0.090	0.040	n.s.	0.193	4.1	2.52	123	0.097	0.034	9.6
	CT	0.267	0.149	2.56	0.626	19.1	5.75	214	0.374	0.051	39.5
	DH	0.175	0.120	2.14	0.485	21.7	7.80	163	0.394	0.043	41.6
	NT	0.146	0.145	1.99	0.626	22.1	7.46	230	0.374	0.045	44.9
LSD B 5%		n.s	n.s.	n.s.	0.126	2.5	1.70	n.s.	n.s.	n.s.	n.s.
Mean		0.196	0.138	2.23	0.579	21.3	9.48	215	0.381	0.047	42.0
		* under detectable levels (mg/kg): **Se** (<0.60), **Hg** (<0.12), **As** (<0.40)									

Table 23. Influences of the growing season and soil tillage on nutritional status of soybean (Jug et al., 2006)

10. Mineral nutrition in function of diseases and pest control

To control diseases and pest the farmers have several options as follows: genetics (cultivation of less susceptible or even resistant to diseases and pest), biological control (utilization of predators), chemical control (using correspondingly pesticides), plant and soil management practices (creating optimal growth conditions of the cultivated crops and /or to eradicate those conditions, which are favorable for multiplication of diseases and pest) and plant nutrition.

Nutrition of plant has a substantial impact on the predisposition of plants to be attacked or affected by diseases and pests. The ratio between nitrogen and potassium plays obviously a particular role in the host/pathogen relationship. However, unbalanced fertilization is wead spread Developing countries apply nitrogenous and potassic fertilizers at a ratio of 1: 0.2, the situation in developed countries is slightly better with a NK ratio of 1:0.4 (Krauss, 2001).

Generally, potassium tends to improve plant health (Perrenoud, 1990).

Useof potassium decreased the incidence of fungal diseases in 70% of the cases. Simultaneously, K increased yield of plants infested with fungal diseases by 42% (Perrenoud, 1990). Mondal et al. (2001) found a negative correlation between K contents in soybean with incidence and positive correlation with their respective yield.

Insects activately select plants best suited as a food source by, among other factors, appearance, stage of development and composition oft he plant. A precondition for successful infestation is the coincidence of certain developmental stages of both host and pathogen. The use of fertilizers can affect this coincidence by either accelerating or slowing down the development of the host plant relative to that of pathogen. A good example is the control of stem cancer (*Diaporthe phaseolorum*) in soybean by potash use, because the fungus can attack soybean only at a particular phenological stage. Earliness due to balanced fertilization provides the possibility to escape (Ito et al., 2001).

Rodrigues et al., (2009) found that spraying of soybean by potassium silicate (Psi) solutinon reduced the intensity of soybean rust. Soybean rust severity at the highest applied KSi rate in level 60 g/L (pH 5.5) was 70% less than the control (plant spraying with water). This finding may be valuable in areas where soybean is grown as a monoculture, and where high yielding but susceptible cultivars cannot be grown because of occurence of frequent severe epidemics. However, Duarte (2009) reported that there was no effects of KSi on rust control in susceptible soybean cultivar *Monarca*.

11. References

Anetor, M. O. & Akinrinde, E. A. (2006). Response of Soybean [*Glycine max* (L.) Merrill] to Lime and Phosphorus Fertilizer Treatments on an Acidic Alfisol of Nigeria. *Pakistan Journal of Nutrition*, Vol. 5, No. 3, pp. 286-293, ISSN 1680-5194

Arao, T., Ae, N., Sugiyama, M. & Takahashi, M. (2003). Genotypic differences in cadmium uptake and distribution in soybeans. *Plant and Soil,* Vol. 251, pp. 247–253.

Baligar,V. C. & Fageria, N. K. (1997). Nutrient use efficiency in acid soil, nutrient management and plant use efficiency. *Plant-Soil Interactions at Low pH,* A.C. Moniz et al. (Ed.), pp. 75-95, Brazilian Soil Science Society, Campinas/Vicosa.

Barbagelata, P., Melchiori, R. & Paparotti, O. (2002). Phosphorus feertilization of soybeans in clay soils of Entre Rios province. *Better Crops International,* Vol. 16, No 1, pp. 3-5.

Bergman, W. (1992). *Nutritional disorders of plants - development, visual and analytical diagnosis,* Gustav Fischer Vela Jena, Stuttgart, New York.

Bergman, W. & Neubert, P. (1976). *Pflanzendiagnose und pflanzenanalyse,* WEB Gustav Fisher, Jena.

Bethlenfalvay, J. K., Franson, L. R. & Brown, S. M. (1990). Nutrition of mychorrhizal soybean evaluated by the diagnosis and recommendation integrated system (DRYS). *Agronomy Journal,* Vol. 82, pp. 302-403.

Bezdicek, D.F., Evans, D.W., Adebe, B. & Witters, R.E. (1978). Evaluation of peat and granular inoculum for soybean yield and N fixation under irrigation. *Agronomy Journal*, Vol. 70, pp. 865-868.

Borkert, C.M., & Cox, F.R. (1999). Effects of acidity at high soil zinc, copper, and manganese on peanut, rice, and soybean. *Commun Soil Sci Plant Anal*, Vol. 30, pp. 1371-84.

Brown, J.C. & Jones, W.E. (1976). A technique to determine iron efficiency in plants. *Soil Sci. Soc. Am. J.*, Vol. 40, pp.398-405.

Casanova, E. F. (2000). Phosphorus and potassium fertilization and mineral nutrition of soybean in Guarico state. *Better Crops International*, Vol. 14, No 2, pp. 6-9.

Clark, R.B., Flores, C.I. & Gourley, L.M. (1988). Mineral element concentrations in acid soil tolerant and susceptible Sorghum genotypes. *Communications in Soil Science and Plant Analysis*, Vol. 19, pp. 1003 - 1017.

Campo, R. J., Araujo, R. S. & Hungria, M. (2009). Molybdenum-enriched soybean seeds enhance N accumulation, seed yield, and seed protein content in Brazil. *Field Crops Research*, Vol. 110, pp. 219-224.

Chaney, R.L. & Coulombe, B.A. (1982). Effect of phosphate on regulation of Fe-stress response in soybean and peanut. *Journal of Plant Nutrition*, Vol. 5, pp. 469-487.

Chen, H.M., Zheng, C.R., Tu, C. & She, Z.G. (2000). Chemical methods and phytoremediation of soil contaminated with heavy metals. *Chemosphere*, Vol. 41, pp. 229–34.

Coulombe, B.A., Chaney, R.L. & Wiebold, W.J. (1984). Bicarbonate directly induces iron chlorosis in susceptible soybean cultivars. *Soil Sci. Soc. Am. J.*, Vol. 48, pp. 1297-1301.

Council for Agricultural Science and Technology (2009). *Sustainable of U.S. Soybean Production*. Conventional, Transgenic and Organic Production Systems. Special publication No 30, Ames, Iowa, USA, pp. 24-30.

Cox, F.R. (1965). Factors affecting the Mn content and yield of soybeans. *NC Soil Sci Soc Proc*, Vol. 8, pp. 57-67.

Cox, F.R. (1968). Development of a yield response prediction and manganese soil test interpretation for soybeans. *Agronomy Journal*, Vol. 60, pp. 521-524.

Dahnke, W. C., Fanning, C. & Cattanach, A. (1992). Fertilizing soybean, Available from: www.ag.ndsu.edu/pubs/plantsci/soilfert/sf719w.htm (10. 03. 2011).

Deibert, E.J., Bijeriego, M. & Olson, R.A. (1979). Utilization of ^{15}N fertilizer by nodulating and non-nodulating soybean isolines. *Agronomy Journal*, Vol. 71, pp. 717-723.

De Mooy, W. C., Pesek, J. & Spaldon, E. (1973). Mineral Nutrition. In: *Soybeans. improvement, production, and uses* , Caldwell, B.E. (Ed.), p.267-352, Madison.

Duarte, H. S. S. (2009). Potassium silicate, acilbenzolar-S-methyl and fungicides on the control of soybean rust. *Cienc. Rural* (online), Vol. 39, No 8, pp. 2271-2277. ISSN 0103-8478. Available from http://www.scielo.br (8. 03. 2011.)

Epstein, E. (1976). Genetic potentials for solving problems of soil mineral stress, adaptation of crops to salinity. In: *Plant Adaptatation to Mineral Stress in problem Soils.* M. J. Wright, pp.73-82, Cornell Univ. Agric. Exp. Station, Ithaca NY.

FAOSTAT (2009). Statistics database. Available from htpp.//faostat.fao.org/default.aspx (verified 23 June 2009).

Fageria, N.K., Zimmermann, F.J.P. & Baligar, V.C. (1995). Lime and P interaction on growth and nutrient uptake by upland rice, wheat, common bean and corn in an oxisol. *Journal of Plant Nutrition*, Vol. 18, pp. 2519-2532.

Fageria, N.K. (1984). Response of rice cultivars to liming in certado soil. *Pesq. Agro spec., Bras. Brasilia*, Vol. 19, pp. 883-889.

Fageria, N.K. (1994). Soil acidity affects availability of Nitrogen, Phosphorus and Potassium. *Better crops International,* Vol. 10, pp. 8-9.

Ferrufino, A., Smyth, T.J., Israel, D.W. & Carter Jr. T.E. (2000). Root elongation of soybean genotypes in response to acidity constraints in a subsurface solution compartment. *Crop Science*, Vol. 40, pp. 413 - 421.

Fleming, A.L., Chaney, R. L. & Coulombe, B. A.(1984). Bicarbonate inhibits Fe-stress response and Fe uptake -translocation of chlorosis-susceptible soybean cultivars. *Journal of Plant Nutr*ition, Vol. 7, pp. 699-714.

Fontes, R.L.F. (1992). *Zinc toxicity in soybean as affected by plant iron and sulfur* [dissertation]. Raleigh (NC). North Carolina State University.

Foy, C. D. (1984). Physiological effects of hydrogen, aluminum and manganese toxicities in acid soils. In. *Soil Acidity and Liming* (Adams F. ed.) pp. 57-97. Am. Soc. Agron. Madison, Wisconsin.

Fontes, R.L.F. & Cox, F.R. (1995). Effects of sulfur supply on soybean plants exposed to zinc toxicity. *Journal of Plant Nutrition,* Vol. 18, pp. 1893-1906.

Fontes, R.L.F. & Cox, F.R. (1998). Zinc toxicity in soybean grown at high iron concentration in nutrient solution. *Journal of Plant Nutrition,* Vol. 21, pp. 1723-30.

Franzen, D. W. & Richardson, J. L. (2000). Soil factors affecting iron chlorosis int he Red River Valley of North Dakota and Minnesota. *Journal of Plant Nutrition,* Vol. 23, No. 1, pp. 67-78.

Furlani, A. M. C., Furlani, P. R., Tanak, R. T., Mascarenhas, H. A. A. & Delgado, M. D. P. (2002). Variability of soybean germplasm in relation to phosphorus uptake and use efficiency for dry matter yield. In: *Plant nutrition – food security and sustainability of agro-ecosystems,* Horst W. J. et al., Developments in Plant and Soil Sciences, Vol. 92, pp. 74-75. Kluwer Academic Publishers Dordrecht/Boston/London.

Galloway, J.N., Schlesinger, W.H., Levy, H., Micaels, A. & Schnoor, J.L. (1995). Nitrogen fixation: Atmospheric enhancement-environmental response. *Global Biogeochem Cycles*, Vol. 9, pp. 235-252.

Ganesh, K.S., Baskaran, L., Chidambaram, A.A. & Sundaramoorthy, P. (2009). Influence of chromium stress on proline accumulation in soybean (*Glycine max* L. Merr.) genotypes. *Global J. Environ. Res.*, Vol. 3, No 2, pp.106-108.

Gill, M. S., Singh, V. K. & Shukla, A. A. (2008). Potassium Management for Enhancing Productivity in Soybean-based Cropping System. *Indian J. of Fertilizers,* Vol. 4, pp. 13-22.

Gray, C., Pennington, H. D. & Matocha, J. (1982). Identifying and correcting iron deficiency in field crops. *Texas Agric. Ext. Ser.* L-723.

Goos, R. J. & Johnson, B. (2003). Screening soybean varieties for resistance to iron chlorosis. Available from: http://ww.docstock.com/docs/32446064/107. (10. 03. 2011).

Hansen, N. C., Schmitt, M. A., Anderson, J. E. & Strock, J. S. (2003). Iron deficiency of soybean int he Upper Meadwest and associated soil properties. *Agronomy Journal,* Vol. 95, pp. 1595-1601.

Hanumanthappa, M., Hosmani, S. A. & Sreeramulu, K. J. R. (1998). Effect of different levels of phosphorus on yield and yield components of soybean varieties. *Agricultural Science Digest,* Vol. 18, pp. 140-142.

Hanumanthappa, M. , Sreeramulu, K. J. R. & Naik, R. G.(1999). Influence of phosphorus levels on dry matter production and yield in soybean varieties. *Journal of Maharashtra Agricultural Universities*, Vol. 23, pp.195-196.

Haq, M. U. & Mallarino, A. P. (2005). Response of Soybean Grain Oil and Protein Concentrations to Foliar and Soil Fertilization. *Agronomy Journal*, Vol. 97, pp. 910-918.

Harper, J. E. & Gipson, A. H. (1984). Differential nodulation tolerance to nitrate among legume species. *Crop Science*, Vol. 24, pp. 797-801.

Holshouser, D. (1997). Micronutrient deficiencies. Available from http://webipm.ento.vt.edu/ipm-www/nipmn/VA-IPM/updates/micronu.html. (10. 03. 2011)

Hrustic, M., Vidic, M. & Jockovic, Dj. (1998). *Soja / Soybean.* Institut za ratarstvo i povrtarstvo Novi Sad, Soja protein d.d. Becej, Novi Sad - Becej, Serbia.

Hungria, M. & Campo, R.J. (2004). Economical and environmental benefits of inoculation and biological nitrogen fixation with soybean:situation in South America. *Proceedings of World Soybean Research Conference VII,* pp.488-498, ISBN 85-7033-004-9, Brasil, March 2004, Foz do Iguassu.

Ito, M. F., Mascayrenhas, H. A. A., Tanaka, R. T., Martins, A. L. M., Otsuk, I. P., Carmello, Q. A. C. & Muraoka T. (2001). Control of stem cancer in soybeans by liming and potassium fertilizer. *Rev. de Agricultura, Piracicaba,* Vol. 76, fasc. 2, pp. 307-316.

Johnson, J. (1984). Fertilizer reduces yield loss from weather stress. *Potash Review,* Subject 6, Suit 49, No 6, pp.1-3.

Johnson, J. (1987). Soil fertility and crop nutrition. In: *The soybean in Ohio.* Ohio Cooperative Extension Service, The Ohio State University, pp. 34-41.

Jug, I., Jug, D., Kovacevic, V., Stipesevic, B. & Zugec, I. (2006). Soil tillage impacts on nutritional status of soybean. *Cereal Research Communications,* Vol. 34, No 1, pp. 537-540.

Kastori, R. (1978). Change of different forms of Ca in soybean varieties during growth season. *Zbornik za prirodne nauke,* Matica srpska Novi Sad, 54, 85-93 (in Serbian).

Kastori, R., Belic, B., Molnar, I., Petrovic, N. & Dzilitov, S. (1979). Dynamics of content, accumulation and distribution of N, P, K, Ca and Mg during growth of some soybean varieties. *Savremena poljoprivreda,* Novi Sad, Vol. 27, pp. 433-446. (in Serbian)

Keltjens, W. G. (1997). Plant adaptation and tolerance to acid soil; it is possible Al avoidance - a review. In: *Plant-Soil Interactions at Low pH,* A.C. Moniz et al. (Ed.), p. 109-11. Brazilian Soil Science Society Campinas/Vicosa.

Keogh, J., Sable, W. E. & Caviness, C. E. (1977). Leaf nutrient concentration in selected soybean cultivars as affected by fertilization, stage of growth and year. *Report Series, Agr. Exp. Station, University of Arkansas,* N0 234.

Keyser, H.H. & Li, F. (1992). Potential for increasing biological nitrogen fixation in soybean. *Plant and Soil,* Vol. 141, pp. 119-135.

King, L.D. (1986). Agricultural use of municipal and industrial sludges in the southern United States. Raleigh (NC). *North Car olina State University. Southern cooperative series; Bulletin 314.*

King, L.D. & Hajjar, L.M. (1990). The residual effect of sewage sludge on heavy metal content of tobacco and peanut. *Journal Environ Quality,* Vol. 19, pp. 738-48.

Kovacevic, V. (1993). Phosphorus and potassium deficiency in soybeans on gleysol of Croatia. *Rostlinna Vyroba,* Vol. 39, No 4, pp. 345-352.

Kovacevic, V. & Basic, F. (1997). The soil potassium resources and the efficiency of potassium fertilizers in Croatia (Country Report 10). *International Potash Institute, Coordinator Central/Eastern Europe,* CH-4001 Basel/Switzerland.

Kovacevic, V. & Grgic, D. (1995). Response of soybeans to potassium fertilization on a high potassium fixing soil. *Rostlinna Vyroba,* Vol. 41, No 4, pp. 24-248.

Kovacevic, V., Juric, I., Zugec, I. (1987). Soybean response to lime and phosphorus under the growing conditions of eastern Croatia. *Eurosoya* Vol. 6. pp. 58-61.

Kovacevic, V. & Krizmanic, M. (1987). Nutrient uptake by soybean genotypes on calcareous soil in eastern Croatia. *Eurosoya,* Vol. 6, pp. 71-73.

Kovacevic, V., Seput, M., Andric, L. & Sostaric, J. (2007). Response of maize and soybeans to fertilization with phosphorus and potassium on acid soil. *Cereal Research Communications,* Vol. 35, No. 2, pp. 645-648

Kovacevic, V., Sudaric, A., Sudar, R., Rastija M. & Iljkic D. (2011). Residual impacts of liming and fertilization on soybean yield and quality. *Növénytermelés,* Vol. 60, No.2 (suppl.), pp.259-262.

Kovacevic, V. & Vukadinovic, V. (1992). The potassium requirements of maize and soybean on a high K-fixing soil. *South African Journal of Plant and Soil,* Vol. 9, pp.10-13.(reprint. *Potash Review,* Subject 6, Cereals, 8th suite, No 1, 1992, International Potash Institute Basel, Schwitzerland).

Kovacevic, V., Vukadinovic, V. & Komljenovic, I. (1991). Tipovi kloroze soje uslijed debalansa ishrane na tlima istočne Hrvatske (The types of soybean chlorosis induced by nutritional debalances on soils of eastern Croatia). *Znan. Prak. Poljopr. Tehnol,* Vol. 21 (Suppl.), pp. 15-23.

Krauss, A. (2001). Potassium and biotioc stress. *Proceedings of 1st FAUBA-FERTILIZAR-IPI Workshop, Potassium in Argenina`s agricultural systems,* November 20-21 Buenos Aires, Argentina.

Lavado, R. S., Porcelli, C. A. & Alvarez, R. (2001). Nutrient and heavy metal concentration and distribution in corn, soybean and wheat as affected by different tillage systems int he Argentine Pampas. *Soil & Tillage Research,* Vol. 62, pp. 55-60.

Li, S., Liu, R.,.Wang, M.,.Wang, X., Shen, H. & Wang, H. (2006). Phytoavailability of cadmium to cherry-red radish in soils applied composed chicken or pig manure. *Geoderma,* Vol. 136, pp. 260–271.

Loncaric Z., Popovic B., Karalic K., Rastija D., Engler M. (2007). Phosphorus fertilization and liming impact on soil properties. *Cereal Research Communications* Vol. 35, No. 2, pp. 733-746.

Maddox, J.S. & Soileux J. M. (1991). Effects ofphosphate fertilizer, lime amendment and innoculation with VA-mychorrhizal fungi on soybean in an acid soil. In: *Plant and Soil interraction at low pH* . Wright *et al.* (Ed.) pp: 215-225. Kluwer Academic Publishers.

Mascagni, H.J. Jr & Cox, F.R. (1985). Critical levels of manganese in soybean leaves at various growth stages. *Agronomy Journal,* Vol. 77, pp. 373-5.

Mascagni, H.J. Jr & Cox, F.R. (1985). Effective rates of fertilization for correcting manganese deficiency in soybeans. *Agronomy Journal,* Vol. 77, pp. 363-366.

Mengel, K. & Kirkby, E. A. (2001). *Principles of Plant Nutrition.* Kluwer Academic Publishers , Dordrecht/Boston/London. ISBN 0-7923-7150-X.

Miner, G.S., Gutierrez, R. & King, L.D. (1997). Soil factors affecting plant concentrations of cadmium, copper, and zinc on sludge-amended soils. *Journal of Environment Quality,* Vol. 26, pp. 989-994.

Mondal, S. S., Pramanik, C. K. & Das, J. (2001). Effects of nitrogen and potassium on oil yield, nutrient uptake and soil fertility in soybean - sesame intercropping system. *Indian J. Agric. SCi.,* Vol. 71, pp. 44-46.

Morshed, R. M., Rahman, M. M., Rahman, M. A. & Hamidullah, A. T. M. (2009). Effect of Potassium on Growth and Yield of Soybean. *Bangladesh J. Agric. and Environ.,* Vol. 5, No 2 , pp. 35-42.

Mottaghian, A., Pirdashti, H., Bahmanyar, M. A. & Abbasian, A. (2008). Leaf and seed micronutrient accumulation in soybean cultivars in response to integrated organic and chemical fertilizers application. *Pakistan Journal of Biological Sciences,* Vol. 11, No 9, pp. 1227-1233.

Nelson, K. A., Motavally, P.P. & Nathan, M. (2005). Response of No-Till Soybean [*Glycine max* (L.) Merr.] to Timing of Preplant and Foliar Potassium Applications in a Claypan Soil. *Agronomy Journal,* Vol. 97, pp. 832-838.

Ogburia, M. N., Atabaeva, H. N. & Hassanshin, R. U. (1999). Evaluation of varietal response of soybean (*Glycine max* L. Merril) to nitrogen (N) fertilization in Tashkent, Central Asia. *Acta Agronomica Hungarica,* Vol. 47, pp. 329-333.

Ojo, G. O. S., Bello, L. L. & Adeyemo, M. O. (2010). Genotypic variation for acid stress tolerance in soybean in the humid rain forest acid soil of south Eastern Nigeria. *Journal of Applied Biosciences,* Vol. 36, pp. 2360- 2366. ISSN 1997–5902. Available from www.biosciences.elewa.org. (10. 03. 2011).

Opkara, D. A., Muoneke, C. O. & Ihediwa, N. (2007). Influence of liming on the performance of high-yielding soybean varieties in southeastern Nigeria. *Journal of Agriculture, Food, Environment and Extension,* Vol. 6, No 2, ISSN 1119-7455 URL: http://www.agrosciencejournal.com/. (11. 03. 2011).

Pandey, S.K. & Pandey, S.K. (2008). Germination and Seedling growth of Field Pea *Pisum sativum* Malviya Matar-15 (HUDP-15) and Pusa Prabhat (DDR-23) under varying level of Copper and Chromium. *J. Am. Sci.,* Vol. 4, No. 3, pp. 28-40.

Perrenoud, S. (1990). Potassium and plant health. *IPI Research Topics No. 3, 2nd rev. Edition,* Basel, Switzerland.

Pichtel, J. & Bradway, D. (2008). Conventional crops and organic amendments for Pb, Cd and Zn treatment at a severely contaminated site. *Bioresource Tech,* Vol. 99, pp. 1242–1251.

Putwattanaa, N., Kruatrachueb, M., Pokethitiyooka, P. & Chaiyaratc, R. (2010). Immobilization of cadmium in soil by cow manure and silicate fertilizer, and reduced accumulation of cadmium in sweet basil (Ocimum basilicum). *ScienceAsia,* Vol. 36, pp. 349-354. doi: 10.2306/scienceasia1513-1874.2010.36.349

Queiroz, E. F., Garcia, A. A., Sfredo, G. J. & Gordeiro, D. S. (1980). Resposta de cultivares de soya a niveis de abudacao fosfatada. *Resultados de pesquisa de soya 1979/80,* Londrina, Brazil, pp. 203-205.

Raper, J. R. & Barber, S. A. (1970). Rooting System of Soybeans. II. Physiological effectiveness as nutrient absorption surfaces. *Agronomy Journal,* Vol. 62, pp. 585-588.

Rastija, M., Kovacevic, V., Vrataric, M., Sudaric, A. & Krizmanic, M. (2006). Response of maize and soybeans to ameliorative fertilization in Bjelovar-Bilogora County. *Cereal Research Communications,* Vol. 34 (part 2), pp. 641-644.

Rehm, G. & Schmitt, M. (1997). Potassium for crop production. (www.extension.umn.edu/distribution/cropsystem/DC6794.html). (08. 03. 2011).

Reid, K. & Bohner, H. (2007). Interpretation of Plant Analysis for Soybeans. Available from: http:// www.omafra.gov.on.ca/english/crops/facts/soybean_analysis.htm.

Richter, D., Kovačević, V. & Flossmann, R. (1990). Ergebnisse und Untersuchung von K-fixierenden Boden Jugoslawiens. *Richtig dungen-mehr ernten,* KALI-BERGBAU Handelsgesellsch. mbH Berlin, Vol. 14, No 2, pp. 1-6.

Rodriguez, F. A., Duarte, H. S. S., Domiciano, G. P., Souza, C. A., Korndorfer, G. H. & Zambolim, G. (2009). Foliar application of potassium silicate reduces the intensity of soybean rust. *Australasian Plant Pathology,* Vol. 38, No 4, pp. 366-372.

Romheld, V. & Marschner, H. (1981). Effect of Fe stress on utilization of Fe chelates by efficient and inefficient plant species. *Journal of Plant Nutrition,* Vol. 3, pp. 551-560.

Rose, I. A., Felton, W. L. & Banks, L. W. (1981). Response of four soybean varieties to foliar zinc fertilizer. *Australian Journal of Experimental Agriculture and Animal Husbandry,* Vol. 21 (109), pp. 236-24.

Russelle, M.P. & Birr, A.S. (2004). Biological nitrogen fixation: Large Scale Assessment of Symbiotic Dinitrogen fixation by Crops: Soybean and Alfalfa in the Mississippi River Basin. *Agronomy Journal,* Vol. 96, pp. 1754-1760.

Sabbe, W. E. & Delong, R. E. (1998). Influence of phosphorus plus potash fertilizer and irrigation on grain yields of soybean cultivars. *Arkansas Experiment Station Research Series* 459.

Sarawgi, S. K. & Tripathi, R. S. (1998). Yield and economics f soybean *(Glycine max* L. Merril) varieties in relation to phosphorus level. *Journal of Oilseeds Research,* Vol. 15, pp. 363-365.

Saric, M. R. (1981). Genetic spewcifity in relation to plant mineral nutrition. *Journal of Plant Nutrition,* Vol. 3, No 5, pp. 743-766.

Saric M. & Krstic B. (1982).Genetisch bedingte Unterschiede im Stickstoffgehalt verschiedener Sojasorten. Archiv fuer Acker- und Pflanzenbau und Bodenkunde, Vol. 26, No 12, pp. 755-761.

Saric, M. & Loughman, B. C. (eds.) (1983). *Genetic aspects of plant mineral nutrition.* Martinus Nijhoff / Dr. W. Junk Publishers, The Hague, The Netherlands.

Sarker, S. K., Chowdhury, M. A. H. & Zakir, H. M. (2002). Sulphur and boron fertilization on yield quality and nutrient uptake by Bangladesh Soybea-4. *Journal of Biological Sciences,* Vol. 2, pp. 729-733.

Sauchelli, V. (1969). *Trace elements in agriculture.* Nostrand Reinhold Company, New York.

Saxena, S. C., Manral, H. S. & Chandel, A. S. (2001). Effects of inorganic and organic sources of nutrients on soybean. *Indian Journal of Agronomy,* Vol. 46, No 1, pp. 135-140.

Scoccianti, V., Crinelli, R., Tirillini, B., Mancinelli, V. & Speranza, A. (2006). Uptake and toxicity of Cr(III) in celery seedlings. *Chemosphere,* Vol. 64, pp. 695-1703.

Sharma, D.C., Chatterjee, C. & Sharma, C.P. (1995). Chromium accumulation by barley seedlings (*Hordeum vulgare* L.). *Journal of Experimental Boany,* Vol. 25, pp. 241-251.

Shute, T. & Macfie, S. M. (2006). Cadmium and zinc accumulation in soybean: A treat to food safety? *Sceince of the Total Environment* , Vol. 371, pp. 63-73. Availabvle from: www.sciencedirect.com. (08. 03. 2011).

Shamsi, H. I., Jiang, L., Wei, K., Jilani, G., Hua, S. & Zhang, G. P. (2010). Alleviation of cadmium toxicity in soybean by potassium supplementation. *Journal of Plant Nutrition,* Vol. 33, pp. 1926-1938.

Silman, Z.T. & Motto, H. L. (1990). Differential response of two soybean genotypes to zinc induced iron deficiency chlorosis. *Journal King Saud. Univ.*, Vol. 2, Agric. Scie (1), pp. 71-80.

Singh, G., Singh, H., Kolar, J. S., Singh, G. & Singh, H. (2001). Response of soybean (*Glycine max* (L) Merrill.) to nitrogen, phosphorus, potassium and zinc fertilization. *Journal of Research Punjab Agricultural University*, Vol. 38, No 1-2, pp. 14-16.

Singh, G. B. & Swariup, A. (2000). Lessons from long-term fertility experiemnts. *Fertilizer News*, Vol. 45, No 2, pp. 13-24.

Slaton, N. A., Golden, B. R., DeLong, R. E. & Mozaffari, M. (2009). Correlation and Calibration of Soil Potassium Availability with Soybean Yield and Trifoliolate Potassium. *Soil Sci. Soc. Am. Journal*, Vol. 74, No. 5, pp. 1642-1651.

Spehar, C.R. (1995a). Genetic differences in the accumulation of mineral elements in seeds of tropical soybeans (*Glycine max* (L.) Merril). *Pesquisa Agropecuária Brasileira*, Vol. 30, pp. 89-94.

Spehar, C.R. (1995b). Diallel analysis for mineral element absorption in tropical adapted soybeans (*Glycine max* L. Merril). *Theoretical and Applied Genetics*, Vol. 90, pp. 707-713.

Spehar, C.R. (1999). Diallel analysis for grain yield and mineral absorption rate of soybeans grown in acid Brazilian Savannah soil. *Pesquisa Agropecuária Brasileira*, Vol 34, pp.1003-1009.

Stipesevic, B., Jug, D., Jug, I., Tolimir, M., & Cvijovic, M. (2008). Winter wheat and soybean zinc uptake in different soil tillage systems. *Cereal Research Communications*, Vol. 37, No. 2, pp. 305-310.

Sugiyama, M. & Noriharu, A. E. (2009). Differences among soybean cultivars with regard to the cadmium-accumulation patterns in various organs. *Proceedings of the XVI International Plant Nutrition Colloquium, UC Davis, California*. Available from http://escholarship.org/uc/item/71n1j212. (08. 03. 2011.)

Sudaric, A., Vratarić, M., Duvnjak, T. Majic, I. & Volenik, M. (2008). The effectiveness of biological nitrogen fixation in soybean linked to genotype and environment. *Cereal Research Communications*, Vol. 36, pp. 67-70.

Sumner, M. E. (1997). Procedures used for diagnosis and correction of soil acidity, a critical review. *Plant-Soil Interactions at Low pH*, A.C. Moniz et al. (Ed.), pp. 195-204. Brazilian Soil Science Society Campinas/Vicosa.

Tordoff, G.M., Baker, A.J.M. & Willis A.J. (2000). Current approaches to the revegetation and reclamation of metalliferous mine wastes. *Chemosphere*, Vol. 41, pp. 219–28

Uebel E. (1999). Znacenje magnezija u povecanju prinosa i kvalitete usjeva - rezultati poljskih pokusa u nekim europskim zemljama / Yield and quality increases by magnesium fertilization - results of field experiments in some European countries. Poljoprivreda Vol. 5 (99), No. 1, pp. 47-53.

Vincent, J.M. (1980). Factors controllling the legume- Rhizobium symbiosis. *Nitrogen fixation*, Vol. 2, Eds. W.E. Newton & W.H. Orme-Johnson, pp. 103-129. University Park Press, Baltimore, MD.

Von Uexkull & Mutert, E. (1995). Global extent, development and economic impact of acid soils. *Plant-Soil Interactions at Low pH: Principles and Management.* (Date R. A., et al. Ed.) pp. 5-19. Kluwer Academic Publisher, Dordrecht, The Netherlands.

Vrataric, M., Sudaric, A. (2008). *Soja / Soybean*. Sveuciliste J.J. Strossmayera u Osijeku i Poljoprivredni institut Osijek, Hrvatska.

Vrataric, M., Sudaric, A., Kovacevic, V., Duvnjak, T., Krizmanic, M. & Mijic, A. (2006). Response of soybean to foliar fertilization with magnesium sulfate (Epsom salt). *Cereal Research Communications,* Vol. 34, No 1, pp. 709-712.

Vukadinović, V., Bertić, B. & Kovačević, V. (1988). Kalium- und Phosphorverfuegbarkeit auf den Boeden in Gebiet von Posavina. *Tagungsbericht - Akademie der Landwirtschaftswissenschaften der DDR,* No. 267, pp.73-80.

Welch, L. F., Boone, L. V., Chambliss, C. G., Christiansen, A. T., Mulvaney, DF. L., Oldham, M. G. & Pendleton, J. W. (1973). Soybean yields with direct and residual nitrogen fertilization. *Agronomy Journal,* Vol. 65, pp. 547-550.

Williams, C.B. (1930). Factors influencing the productivity of muck soils. Raleigh (NC) *North Carolina Agricultural Experiment Station. Report 53.* pp 46-47.

Willis, L.G. (1928). Response of oats and soybeans to manganese on some coastal plain soils. *North Carolina Agric. Exp. Station Bulletin 257.*

Willis, L.G. (1937). Evidences of the significance of oxidation-reduction equilibrium in soil fertility problems. *Soil Science Society of America Proceeding,* Vol. 1, pp. 291-297.

Willis, L.G & Piland, J.R. (1936). The function of copper in soils and its relation to the availability of iron and manganese. *Journal of Agricultural Research,* Vol. 52, pp. 467-76.

Win, M., Nakasathien, S. & Sarobol, E. (2010). Effects of phosphorus on seed oil and protein contents and phosphorus use efficiency in some soybean varieties. *Kasetsart Journal (Nat. Sci.),* Vol. 44 , pp. 1-9.

Yin, X. Y. & Vyn, T. J. (2004). Critical Leaf Potassium Concentrations for Yield and Seed Quality of Conservation-Till Soybean. *Soil Science Society of America Journal,* Vol. 68, pp. 1626-1634.

Yusuf, I.A. & Idowu, A. A. (2001). Evaluation of four soybean varieties for performance under different lime regimes on the acid soil of Uyo, Nigeria. *Tropical Oilseeds Journal,* Vol. 6, pp. 65 - 70.

Zheng, H., Chen, L., Han, X., Ma, Y. & Zhao, X. (2010). Effectivenss of phosphorus application in improving regional soybean yields under drought stress. A multivariate regression three analysis. *African Journal of Agricultural Research,* Vol. 5 (23), pp. 3251-3258.

20

Up-Regulation of Heme Oxygenase by Nitric Oxide and Effect of Carbon Monoxide on Soybean Plants Subjected to Salinity

Guillermo Noriega, Carla Zilli, Diego Santa Cruz, Ethel Caggiano, Manuel López Lecube, María Tomaro and Karina Balestrasse
University of Buenos Aires, Consejo Nacional de Investigaciones Científicas y Técnicas
Argentina

1. Introduction

Reactive oxygen species (ROS) are generated in small amounts in the normal metabolism of the cells and in increased amounts under many conditions of altered cell physiology; they are responsible for many kinds of cell injuries (Sies 1993) and have been shown to induce a significant reprogramming of gene expression (Colburn 1992).

Salt stress is one of the most important abiotic stresses that adversely affects soybean growth and causes significant crop loss worldwide. Salinity has always been considered a serious constraint on agricultural productivity (Hay & Porter 2006) and affects plant's physiology. Salt stress is a complex phenomenon that involves morphological and developmental changes. Two major components have been identified in this insult, osmotic stress and ion toxicity (Darwish et al. 2009). Higher plants have multiple protective mechanisms against salt stress including ion homeostasis, osmolyte biosynthesis, ROS scavenging, water transport, and transducers of long-distance response coordination. It is generally accepted that many stresses, including salinity, induce an overproduction of ROS, such as H_2O_2, O_2 •-, and HO·, and these species are thought to be responsible for the oxidative damage associated with plant stress (Zilli et al. 2009). To counteract the toxicity of ROS, defense systems that scavenge cellular ROS have been developed in plants to cope with oxidative stress via the non-enzymatic and enzymatic systems (Demiral & Turkan 2005; Mandhania et al. 2006)

Nitric oxide (NO) acts as a signaling molecule and mediates multiple physiological processes in plants (Leitner et al. 2009). In addition, it has been implicated in responses to biotic and abiotic stresses, such as disease resistance, salinity , drought, heat stress, among others (Beligni & Lamattina 1999; Romero-Puerta et al. 2004; Corpas et al. 2009). There are several sources of NO in plants, but mainly it can be enzymatically produced by nitrate reductase and nitric oxide synthase-like enzymes (Wilson et al. 2008 and Corpas et al. 2009). NO is a reactive nitrogen species and, depending on its concentration, it produces either protective or toxic effects. A low dose of NO modulates superoxide anion formation and inhibits lipid peroxidation, resulting in an antioxidant function during stress (Boveris et al. 2000 and Santa Cruz et al. 2010). Moreover, microarray studies have shown that NO induces a large number of genes at transcriptional level, among them those of antioxidant enzymes (Parani et al. 2004). It has also been reported that Nitric oxide gives rise to signaling pathways mediating responses of specific genes to ultraviolet-B (UV-B) radiation, such as chalcone synthase and phenylalanine ammonia lyase (Mackerness et al. 2001). However, information about the role that NO plays in regulation of antioxidant enzymes to counteract salt-induced oxidative stress is rather limited.

Nitric oxide is believed to act as a signal molecule mediating responses to both biotic and abiotic stresses in plants (reviewed in Xuan et al. 2010 and Nürnberger & Scheel 2001) and its presence has been shown to induce seed germination (Liu et al. 2010), to affect growth and development of plant tissue (Beligni & Lamatina 2001, to increase iron homeostasis (Martin et al. 2009), to regulate plant maturation and senescence (Yaacov et al. 1998 and Jasid et al. 2009) to mediate abscisic acid-induced stomatal closing (Garcia-Mata & Lamattina, 2007). Recently, a few studies suggested that NO can play a role in protecting plants from oxidative stresses (Shantel et al. 2008) and NO-donor treatment protected plants from damage by increasing the activity of antioxidative enzymes.

Heme oxygenase catalyzes the oxidative degradation of heme and has well-known antioxidant properties in mammals by mean of its products biliverdin IXα and carbon monoxide (CO) (Kikuchi et al. 2005). One of the three known mammalian isoforms, heme oxygenase-1 (HO-1), is induced in animal tissues by many factors including its own substrate heme, heavy metals, UV-A radiation among others (Tomaro & Batlle 2002). While earlier studies pointed to plant HO as a source of phytochrome chromophore (Terry et al. 2002), more recent works showed that HO synthesis increases in soybean plants subjected to oxidative stress conferring resistance to a subsequent insult (Noriega et al. 2004; Balestrasse et al. 2005). Moreover, we have recently demonstrated that ROS are involved in HO-1 up-regulation in soybean leaves subjected to UV-B radiation (Yannarelli et al. 2006 and Santa Cruz et al. 2010). We hypothesized that NO may also participate in this process, as it regulates the oxidative status and mediates other UV-B responses.

The aim of the present study was to investigate whether NO or CO could protect soybean against salt-induced oxidative stress through the modulation of HO activity. Soybean plants were subjected to salt stress after pre-treatments with different concentrations of sodium nitroprussiate (SNP), a well-characterized NO-donor or CO. Overall, our results indicate that in soybean plants NO is involved in the signaling pathway leading to HO-1 up-regulation under salinity, and that a balance between NO and ROS is important to trigger the antioxidant response against oxidative stress. On the other hand pretreatment with CO did not provoke any change.

2. Materials and methods

2.1 Plant material and treatments

Surface sterilized soybean seeds (*Glycine max.* L.) (A6445RG) were germinated for 10 days in plastic pots containing vermiculite in controlled environmental chambers, with a photoperiod of 16 h, photon flux density of 175 μmol m^{-2} s^{-1}, and a day/night regime of 25/20°C. Afterwards, they were pretreated hydroponically with different sodium nitroprusiate concentrations (250-750 μM) for 72 h and then with NaCl (200 mM) for 48 h.

Carbon monoxide was generated from H_2SO_4 and formic acid (HCOOH). Stock solution was prepared by bubbling CO in a Hoagland solution for 40 min and was immediately diluted (50%) to perform analysis.

Plants were then harvested. When the effect of Zn-protoporphyrin IX (ZnPPIX) was investigated, roots were pretreated with 22 μM ZnPPIX during 4 h before addition of NaCl.

Controls were incubated in buffer. For fresh weight determination, plants were filtered, washed three times with distilled water, kept on filter paper for a few minutes to remove of excess liquid and weighed. Three different experiments were performed, with three replicated measurements for each parameter assayed

2.2 Thiobarbituric acid reactive substances (TBARS) determination

Lipid peroxidation was measured as the amount of TBARS determined by the thiobarbituric acid (TBA) reaction as described by Heath and Packer (1968). Fresh control and treated roots (0.3 g) were homogenized in 3ml of 20% (w/v) trichloroacetic acid (TCA). The homogenate was centrifuged at 3,500 x g for 20 min. To 1 ml of the aliquot of the supernatant, 1ml of 20% TCA containing 0.5% (w/v) TBA and 100 ml 4% butylated hydroxytoluene (BHT) in ethanol were added. The mixture was heated at 95ºC for 30min and then quickly cooled on ice. The contents were centrifuged at 10,000 x g for 15 min and the absorbance was measured at 532 nm. Value for non-specific absorption at 600 nm was substracted. The concentration of TBARS was calculated using an extinction coefficient of 155 $mM^{-1}cm^{-1}$

2.3 Heme oxygenase preparation and assay

Roots (0.3 g) were homogenized in a Potter-Elvehejm homogenizer using 4 vol. of ice-cold 0.25M sucrose solution containing 1mM phenylmethyl sulfonyl fluoride, 0.2mM EDTA and 50mM potassium phosphate buffer (pH 7.4). Homogenates were centrifuged at 20,000 x g for 20min and supernatant fractions were used for activity determination. Heme oxygenase activity was determined as previously described with minor modifications (Muramoto et al. 2002). The standard incubation mixture in a final volume of 500ml contained 10mmol potassium phosphate buffer (pH 7.4), 60 nmol NADPH, 250ml HO (0.5mg protein), and 200 nmol hemin. Incubations were carried out at 378ºC during 60min. Activity was determined by measuring biliverdin formation, which was calculated using the absorbance change at 650 nm employing an 1 value of 6.25$mM^{-1}cm^{-1}$ (vis_{max} 650 nm)

2.4 Glutathione determination

Non-protein thiols were extracted by homogenizing 0.3 g of roots in 3.0 ml of 0.1 N HCl (pH 2.0), and 1 g PVP. After centrifugation at 10,000 x g for 30 min at 4°C, the supernatants were used for analysis. Total glutathione (GSH plus GSSG) was determined in the homogenates spectrophotometrically at 412 nm, after precipitation with 0.1 N HCl, using yeast-glutathione reductase, 5,5' dithio-bis-(2-nitrobenzoic acid) (DTNB) and NADPH. GSSG was determined by the same method in the presence of 2-vinylpyridine and GSH content was calculated from the difference between total glutathione and GSSG (Anderson, 1985).

2.5 Classical antioxidant enzymes

Extracts for determination of catalase (CAT), ascorbate peroxidase (APX) and glutathione reductase (GR) activities were prepared from 0.3 g of roots homogenized under ice-cold conditions in 3 ml of extraction buffer, containing 50 mM phosphate buffer (pH 7.4), 1 mM EDTA, 1 g PVP, and 0.5% (v/v) Triton X-100 at 4 °C. The homogenates were centrifuged at 10,000 × g for 20 min and the supernatant fraction was used for the assays.

CAT activity was determined in the homogenates by measuring the decrease in absorption at 240 nm in a reaction medium containing 50 mM potassium phosphate buffer (pH 7.2) and 2 mM H_2O_2. The pseudo-first order reaction constant ($k' = k[CAT]$) of the decrease in H_2O_2

absorption was determined and the catalase content in pmol mg−1 protein was calculated using $k = 4.7 \times 10^7 M^{-1}s^{-1}$.
APX activity was measured immediately in fresh extracts and was assayed as described by Nakano and Asada (1981), using a reaction mixture (1 ml) containing 50 mM K-phosphate buffer (pH 7.0), 0.1 mM H_2O_2, 0.5 mM Na-Ascorbate and 0.1 mM EDTA. The hydrogen peroxide-dependent oxidation of Ascorbate was followed by a decrease in the absorbance at 290 nm (ε: 2.8 $mM^{-1} cm^{-1}$). One unit of APX forms 1 μmol of ascorbate oxidized per minute under the assay conditions.
GR activity was measured by following the decrease in absorbance at 340 nm due to NADPH oxidation. The reaction mixture contained tissue extract, 1 mM EDTA,0.5 mM GSSG, 0.15 mM NADPH and 50 mM Tris–HCl buffer (pH 7.5) and 3 mM $MgCl_2$ (Schaedle and Bassham 1977).

2.6 Histochemical analysis

In order to analyze H_2O_2 generation roots were excised and immersed in a 1% solution of 3,3'-Diaminobenzidine (DAB) in Tris-HCl buffer (pH 6.5), vacuum-infiltrated for 5 min and then incubated at room temperature for 16 h in the absence of light. Roots were illuminated until appearance of brown colors characteristic of the reaction of DAB with H_2O_2.
In the same way to show $O_2^{\cdot-}$ production roots were excised and immersed in a 0.1% solution of NBT in K-phosphate buffer (pH 6.4), containing 10 mM Na-azide, and were vacuum-infiltrated for 5 min and illuminated until appearance of dark spots, characteristic of blue formazan precipitate.

2.7 Isolation of RNA and RT-PCR analysis

Total RNA was extracted from soybean roots by using the Trizol reagent (Gibco BRL). Four micrograms of total RNA were treated with RNase-free DNase I (Promega, CA, USA) and then 1.0 μg was reversed transcribed into cDNA using random hexamers and M-MLV Superscript II RT (Invitrogen, CA, USA). PCR reactions were carried out using *Glycine max* HO-1 and 18S specific primers, as previously described (Yannarelli and others, 2006). The PCR profile was set at 94°C for 1 min and then 29 cycles at 94°C for 0.5 min, 54°C for 1 min, and 72°C for 1 min, with a final extension at 72°C for 7 min. Each primer set was amplified using an optimized number of PCR cycles to ensure the linearity requirement for semi-quantitative RT-PCR analysis. The amplified transcripts were visualized on 1.5% agarose gels with the use of ethidium bromide. Gels were then scanned (Fotodyne Incorporated, WI, USA) and analyzed using Gel-Pro Analyzer 3.1 software (Media Cybernetics, MD, USA).

2.8 Protein determination

Protein concentration was evaluated by the method of Bradford (1976), using bovine serum albumin as a standard.

2.9 Statistics

Values in the text, figures and tables indicate mean values ± SEM. Differences among treatments were analyzed by one-way ANOVA, taking $p<0,05$ as significant according to Tukey's multiple range test.

3. Results

3.1 Growth parameter

Experiments were carried out in the presence of different SNP concentrations ranging from 200 to 750 μM. Root length was measured as a parameter to asses the optimal condition. Figure 1 shows that 250 μM SNP brought about a 45% increase in root length, whereas a diminution was observed under the other concentrations. Depending on its dose, NO can promote or inhibit root growth. According to these result, 250 μM SNP was chosen as the concentration to be used in pretreatment.

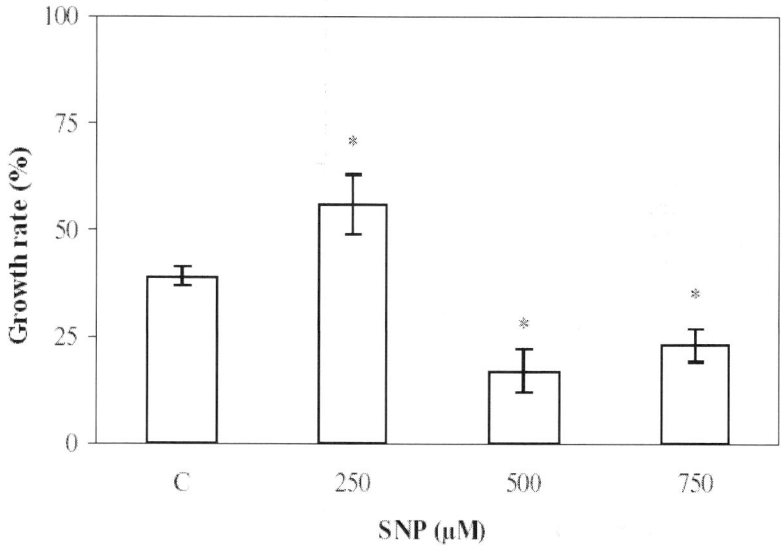

Fig. 1. Effect of different SNP concentrations on root growth. * Significant difference ($p < 0.05$) according to Tukey's test.

3.2 Lipid peroxidation

Increment in TBARS is a good reflection of oxidative damage to membrane lipids and other vital molecules such as proteins, DNA and RNA. Figure 2 shows that TBARS levels increased 75% respect to controls under salt treatment which is in agreement with results of other studies (Deng et al. 2010).

To complete this analysis, the effect of SNP pre-treatment was evaluated. Figure 2 indicates that in this case, membrane damage was more moderated, as indicated by a 14% augmentation respect to controls. Treatment with SNP alone did not show any difference respect to controls.

3.3 Glutathione content

GSH is a leading substrate for enzymatic antioxidant functions and it is also a known radical scavenger. Previous reports from our laboratory demonstrated that oxidative stress induces the formation of oxidant species and therefore affects GSH content in soybean plants

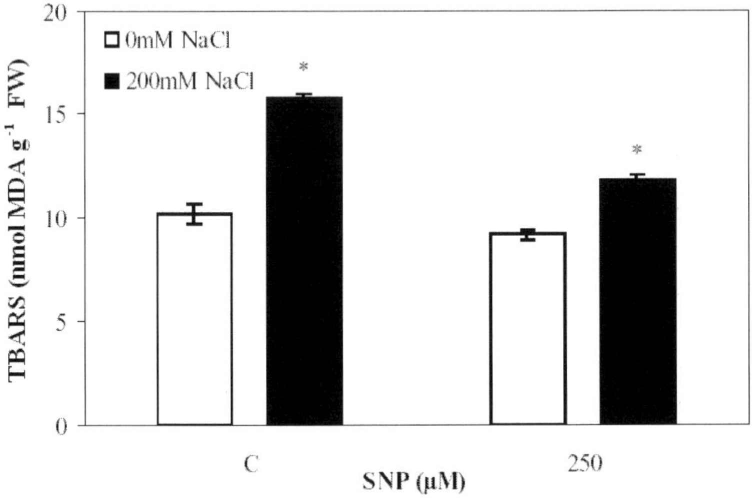

Fig. 2. Effect of salinity on TBARS formation and SNP regulation. * Significant difference ($p<0.05$) according to Tukey test.

(Balestrasse et al. 2001 and Noriega et al. 2004). Surprisingly, data in Figure 3 show that GSH concentration in soybean roots treated with NaCl was enhanced 3.5-fold respect to controls. Pre-treatment with SNP brought about a 4-fold augmentation respect to controls. Moreover, SNP alone provoked a 2-fold increase respect to controls.

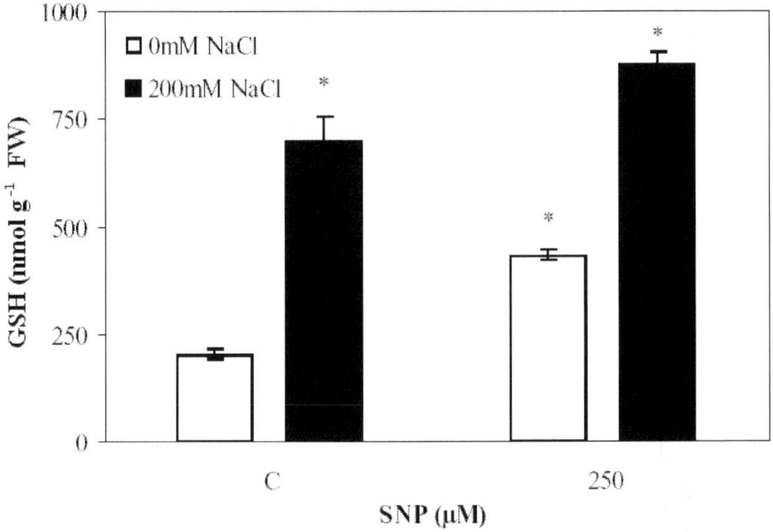

Fig. 3. Effect of salinity on GSH levels and SNP regulation. * Significant difference ($p<0.05$) according to Tukey test.

3.4 H_2O_2 and $O_2^{\cdot-}$ localization *in situ*

Accumulation of H_2O_2 and $O_2^{\cdot-}$ were also evaluated *in situ* by histochemical methods as shown in Figure 4a NaCl produced 32% H_2O_2 spots area versus total root area, while pretreatment with 250 μM SNP prevented this effect and spot area was similar to controls (Figure 4a). Data in Figure 4b showed that roots treated with NaCl produced 41% $O_2^{\cdot-}$ spots area versus total root area. Pretreatments with 250 μM SNP completely prevented the $O_2^{\cdot-}$ production induced by NaCl.

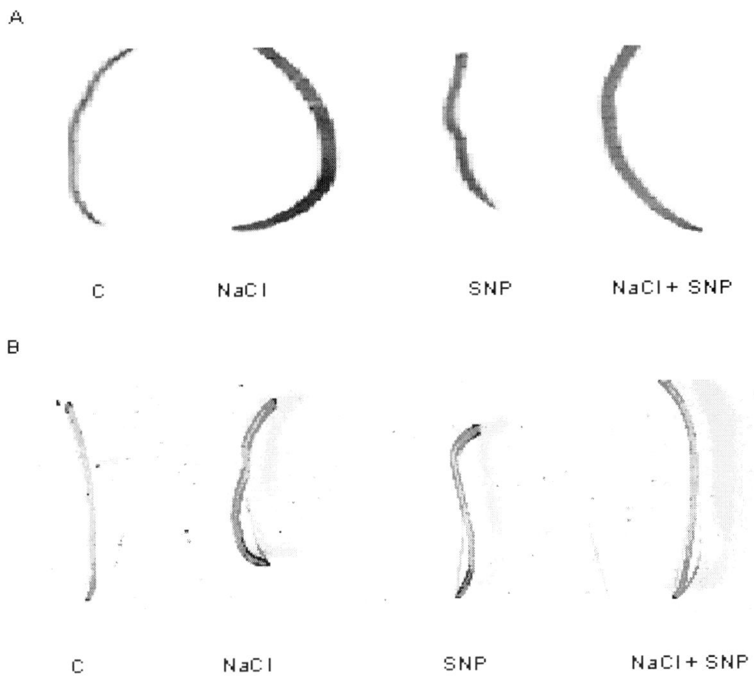

Fig. 4. Histochemical detection of H_2O_2 (A) and $O_2^{\cdot-}$ (B) in soybean roots. Experiments were performed as described in Materials and Methods. Pictures are representative of three different experiments with three replicated measurements for each treatment.

3.5 Effect of NO on antioxidant enzyme activities

We also investigated whether NO can modulate the activities of classical antioxidant enzymes such as CAT and APX. These are the main H_2O_2-scavenging enzymes that control ROS-mediated responses under biotic and abiotic stresses (Mittler 2002). CAT and APX activities were significantly affected by NaCl (Table 1). They were increased by 47% and 33% in NaCl-treated plants compared to controls, respectively. Moreover, CAT activity significantly augmented up to 24% with respect to controls of SNP-treated plants, whereas APX only showed a mild increase (19%). Heme oxygenase behavior was similar to that found for CAT (Table 1).

Treatment	HO-1 (U/mg protein)	CAT (pmol/mg protein)	APX (U/mg protein)
Control	0.065 ± 0.001[a]	120 ± 12[a]	0.0040 ± 0.0010[a]
NaCl	0.073 ± 0.001[b]	176 ± 9[b]	0.0053± 0.0012[a]
SNP	0.079 ± 0.002[c]	149± 2[c]	0.0047 ± 0.0010[a]
SNP+ NaCl	0.083 ± 0.004[c]	138 ± 14[c]	0.0050 ±0.0010[a]

Table 1. Antioxidant enzyme activities in soybean roots subjected to 200mM NaCl and 250 μM SNP pretreatment. Enzymatic activities were assayed as described in Materials and Methods. Different letters within columns indicate significant differences ($P < 0.05$) according to Tukey's multiple range test.

3.6 Heme oxygenase-1 activity and gene expression

Previous findings from our group demonstrated the protective role that HO-1 plays against oxidative stress in soybean plants (Noriega et al. 2004 and Balestrasse et al. 2005). Figure 5 indicates that salt stress caused HO-1 mRNA induction (13%, respect to controls). This enhancement is positively correlated with enzyme activity (Table 1). Pretreatment with 250 μM SNP brought about an augmentation of gene expression in control plants (21%), as well as salt treated plants (27%) (Figure 5). Once again, this behavior was also found when enzyme activity was determinated (Table 1) .These results indicate on one hand, that NO

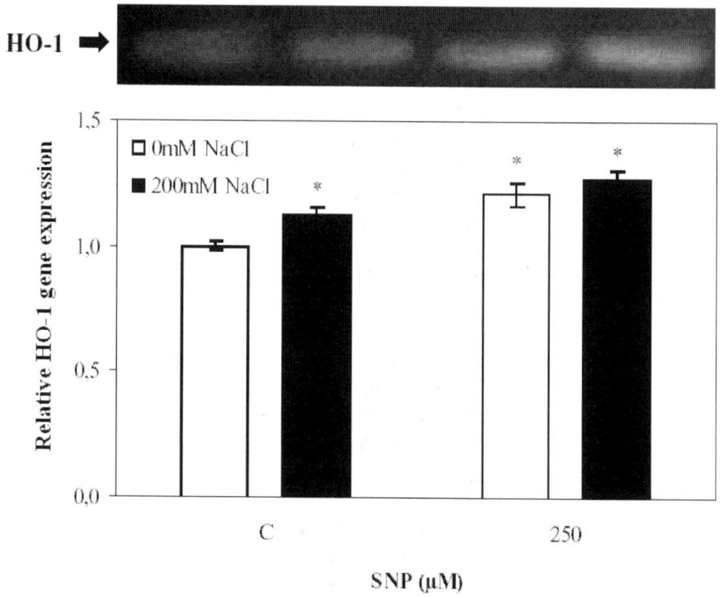

Fig. 5. HO-1 mRNA expression was analyzed by semi-quantitative RT-PCR as described in Materials and Methods. The 18S amplification band is shown to confirm equal loading of RNA and RT efficiency. Relative HO-1 transcript expression taking control as 1 U. Data are means of three independent experiments and bars indicate SE. *Significant differences (P < 0.05 according to Tukey test).

induces HO-1 more efficiently than NaCl, and on the other hand, both compounds have a synergic effect on this induction. To asses whether HO-1 is involved in the protection against NaCl exerted by NO, experiments were carried out in plants treated with ZnPPIX, a well known irreversible HO-1 inhibitor. Plants with inhibited HO-1 activity can not cope with NaCl insult (data not shown). We can assume that protection exerted by SNP may be due to the augmentation of the activity of this antioxidant enzyme.

3.7 Effect of NO and CO

3.7.1 Glutathione content

As already stated, there is a positive relationship between NO content and GSH levels (Figure 3). This result prompted us to investigate whether HO is involved in the regulation of this tripeptide. To fulfill this purpose, experiments were carried out in plants treated with ZnPPIX and then subjected to NO (HO inductor) or CO (HO reaction product) for 48 h before salt stress. Afterwards, GSH content (Figure 6) as well as HO-1 gene expression (Figure 8) was determinated.

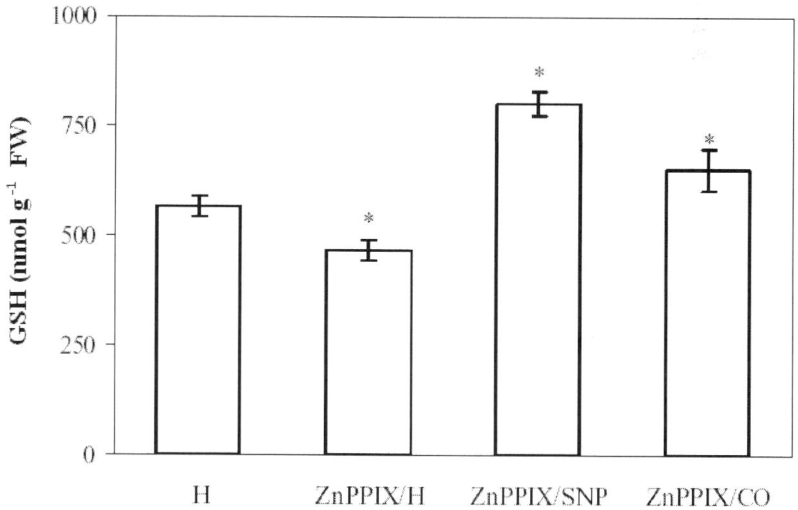

Fig. 6. Effect of NO and CO on GSH content. (H) Control plants, (ZnPPIX/H) plants pretreated with ZnPPIX and then with H; (ZnPPIX/SNP) plants pretreated with ZnPPIX and then with SNP; (ZnPPIX/CO) plants pretreated with ZnPPIX and then with CO as described in Materias and Methods. * Significant difference ($p<0.05$) according to Tukey's test.

In plants pretreated with ZnPPIX for 72 h before Hoagland (H) treatment (ZnPPIX/H), GSH level diminished 20% respect to controls (H). Figure 6 shows that NO (ZnPPIX/SNP) as well as CO (ZnPPIX/H) enhanced GSH levels (40% and 15%, respectively).

3.7.2 Glutathione reductase activity

Taking into account the fact that GSH synthesis is affected by HO-1 inhibition and NO pretreatment GR activity was determinated under the same conditions. Figure 7 indicates a positive relationship between GSH levels and GR activity. Enzyme activity (GR) diminished 22% respect to controls when HO was inhibited, but an increase was detected in plants treated with NO and CO (33% and 26%, respectively).

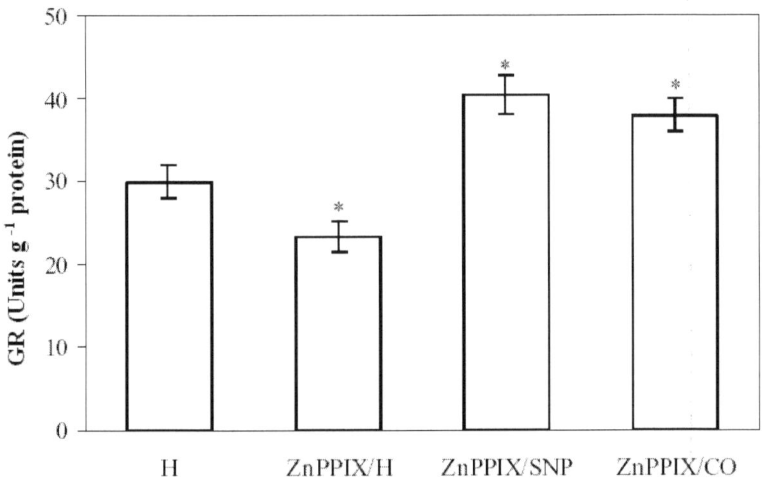

Fig. 7. Effect of NO or CO on GR activity. (H) Control plants, (ZnPPIX/H) plants pretreated with ZnPPIX and then with H; (ZnPPIX/SNP) plants pretreated with ZnPPIX and then with SNP; (ZnPPIX/CO) plants pretreated with ZnPPIX and then with CO as described in Materias and Methods. * Significant difference ($p<0.05$) according to Tukey's test.

3.7.3 HO-1 gene expression

Figure 8 shows HO-1 gene expression under different conditions. ZnPPIX/SNP treatment brought about a 20% augmentation respect to controls. This increase is positively correlated with GSH content and GR activity. On the other hand, CO did not show any effect. It is interesting to note that the enhancement of GSH content is not related to oxidative stress, since TBARS levels in roots of SNP and CO treated plants do not differ from controls. In contrast, HO inhibition brought about an enhancement (28%) in TBARS levels.

4. Discussion

In a previous work, we found that SNP pre-treatment ameliorates Cd-induced oxidative stress and modulates HO-1 gene expression in soybean plants (Noriega et al. 2007). Taking into account the fact that NO is involved in various signaling pathways, in the present study we evaluated whether this molecule could enhance HO activity conferring a major protection against salt stress.

Our data demonstrated that, depending on its concentration, NO can improve the plant antioxidant response against salinity. This model was appropriate to determine the beneficial effect of exogenously added NO. While the lower dose of SNP did not reduce the oxidative damage (data not shown), the application of 500 or 750μM SNP showed a deleterious effect suggesting a pro-oxidant behavior of NO at these concentrations (Figure 1).

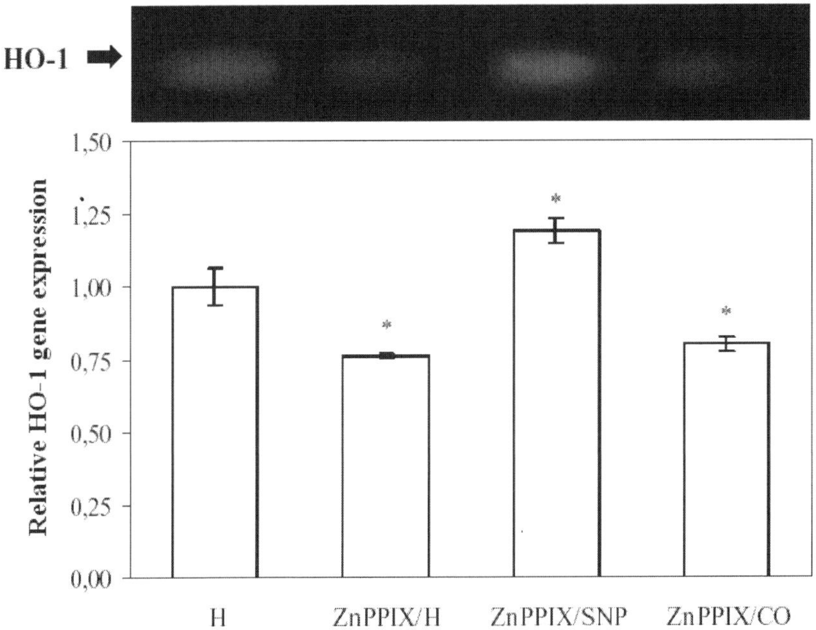

Fig. 8. Effect of NO or CO on HO-1 gene expression. (H) Control plants, (ZnPPIX/H) plants pretreated with ZnPPIX and then with H; (ZnPPIX/SNP) plants pretreated with ZnPPIX and then with SNP; (ZnPPIX/CO) plants pretreated with ZnPPIX and then with CO as described in Materias and Methods. * Significant difference ($p<0.05$) according to Tukey's test.

The pre-treatments with 250 μM SNP effectively ameliorated NaCl-induced oxidative stress, as indicated by the decrease in H_2O_2 and $O_2^{\cdot -}$ formation (Figure 4), preventing TBARS formation (Figure 2) and enhancing GSH content (Figure 3). The activities of classical antioxidant enzymes, such as CAT and APX were also augmented by SNP treatment, instead of the drastically diminution observed with salinity alone (Table 1). These data are in agreement with reports showing a protective effect of NO in plants subjected to other stresses (Zhao et al. 2004; Shi et al. 2005 and Noriega et al. 2007). Nevertheless, the molecular mechanism that mediates NO enhancement of antioxidant enzyme activities is not completely understood. Interestingly, we found that HO and CAT activities had a similar behaviour with respect to SNP pre-treatment under salinity (Table 1). A recent study showed that the time-course of induction of those enzymes in soybean nodules subjected to Cd stress is related (Balestrasse et al. 2008). These results suggest a close relationship between the signal transduction pathways involved in the response of HO and CAT after oxidative stress generation and support the antioxidant role of HO. In addition, there was a positive correlation between HO-1 transcript levels and enzyme activity (Figure 5 and Table 1). Previous reports have also demonstrated that the enhancement of HO activity is associated with an increase in HO-1 transcript levels and protein content (Yannarelli et al. 2006 and Balestrasse et al. 2008). Although this mechanism can account for the changes observed in HO activity, the incidence of post-translational modifications or different HO isoforms under stress conditions needs to be addressed. Experiments carried out in plants treated with SNP in the absence of NaCl showed that NO itself can up-regulate HO-1 mRNA expression, but to a lesser extent (Figure 5). This observation indicates that a certain balance between NO and ROS is required to trigger the full response. Interestingly, a recent report found that the ROS-NO ratio is important to elicit ROS-activated stress responses and cell death regulation in plant leaves during ozone exposure (Ahlfors et al. 2009). Moreover, new evidence suggests that plastids and peroxisomes are important regulators of NO levels in plants (Corpas et al. 2009 and Gas et al. 2009).

Biliverdin, one of the products of the HO, is an efficient scavenger of ROS and it can account for the antioxidant properties of this enzyme both in animals and plants (Otterbein et al. 2003 and Noriega et al. 2004). More recently, it has been shown that CO released by HO is an important signal molecule for the tolerance mechanisms against cadmium and salt stress (Han et al. 2008). It would be interesting to determine whether CO could also play a role in the defense against salinity in soybean plants.

Pretreatment with ZnPPIX decreased HO-1 expression (Figure 8) and increased parameters of oxidative stress. When the inhibitor was added before NO or CO treatment, HO-1 expression as well as GSH content (Figure 6) and GR activity were increased (Figure 7). These results let us suppose that a close relationship between HO-1 induction and GSH content could exist. Taking together, these data provide evidence of one of the possible roles that NO, as well as CO could play against oxidative insult.

5. Conclusion

The present study together with previous results (Balestrasse et al. 2008 and Zilli et al. 2008) support the protective role of HO in soybean plants against salinity. Data here reported let us understand the mechanisms involved in HO response in NaCl-treated soybean plants. This model proposes that NO is implicated in the HO signaling pathway and, together with ROS, modulates the activity of this enzyme under salinity. In plants treated with ZnPPIX, CO did not induce HO-1, but an augmentation of GSH levels as well as GR activity was observed. On the other hand, NO not only caused a more important enhancement in GSH content and GR activity, but also brought about the induction of HO-1. Moreover, NO can enhance the antioxidant system allowing an improved plant defense to the subsequent oxidative insult. Interestingly, while NO may directly potentiate NaCl-induced HO-1 transcription, pre-treatment with SNP followed by salinity stress may protect and enhance by inducing free radical scavenging enzymes and GSH. An appropriate balance of ROS-NO is necessary to trigger the full HO response. In contrast to other stress conditions, induction of HO-1 occurs together with an enhancement of GSH levels and GR activity. In conclusion, the present study provides new insights into the molecular response of soybean plants to salinity and also evidences that HO plays an important role during stress conditions.

6. Acknowledgments

We thank Dr T. Kohchi for kindly providing the Arabidopsis HO-1 antibodies. This work was supported by grants from the Universidad de Buenos Aires (Argentina) and from Consejo Nacional de Investigaciones Científicas y Técnicas (CONICET) (Argentina). M.L.T and K.B.B. are career investigators from CONICET.

7. References

Ahlfors, R.; Broschél, M.; Kollist, H.; Kangasjärvil, J. (2009). Nitric oxide modulates ozone-induced cell death, hormone biosynthesis and gene expression in Arabidopsis thaliana. *Plant Journal*, Vol.58, No.1, pp.1-12, ISSN 0960-7412.

Anderson, E. (1985). Determination of glutathione and glutathione disulfide in biological samples, *Methods in Enzymology*. Vol.113, pp. 548-554, ISSN 0076-6879

Balestrasse, K.; Gardey, L.; Gallego, S. & Tomaro, M. (2001). Response of antioxidative defense system in soybean nodules and roots subjected to cadmium stress. *Australian Journal of Plant Physiology*, Vol.28, pp.497–504. ISSN 0310-784.

Balestrasse, K.; Noriega, G.; Batlle; A. & Tomaro, M. (2005). Involvement of Heme Oxygenase as Antioxidant Defense in Soybean Nodules, *Free Radical Research*, Vol.39, No.2, pp.145-51. ISSN 1071-5762.

Balestrasse, K.; Zilli, C. & Tomaro, M. (2008). Signal transduction pathways and haem oxygenase induction in soybean leaves subjected to salt stress, *Redox Report*,Vol.13; No. 6, (December 2008), pp. 255-262, ISSN 1351-0002.

Balestrasse, K.;Gardey, L.; Gallego, S. & Tomaro, M. (2001). Response of antioxidant defence system in soybean nodules and roots subjected to cadmium stress, *Australian Journal of Plant Physiology*. Vol.28, pp.497–504, ISSN 0310-7841.

Beligni, M. & Lamattina, L. (1999). Is nitric oxide toxic or protective?. *Trends in Plant Science*, Vol.4, No.8, (August 1999), pp. 299-300, ISSN.

Beligni, M. & Lamattina,L. (2001). Nitric oxide: a non-traditional regulator of plant growth, *Trends in Plant Science*, Vol6, No. 11, (November 2001), pp. 508-509, ISSN 1360-1385.

Boveris, A. & Poderoso, J. (2000). Regulation of Oxygen Metabolism by Nitric Oxide, *Nitric Oxide*, Vol.51, pp. 355-368, ISSN 1089-8306.

Bradford, M. (1976). A rapid and sensitive method for the quantitation of microgram quantities of protein utilizing the principle of protein-dye binding, *Analytical Biochemistry*, Vol.72, pp.248-254, ISSN 0003-2697.

Colburn, NH.(1992). Gene regulation by active oxygen and other stress inducers, Spatz L, Bloom AD (eds) *Biological consequences of oxidative stress*, pp 121–137, ISBN 10: 0195072960, Oxford University Press, New York.

Corpas, F.; Palma, J.; Del Río, L. & Barroso, J. (2009). Evidence supporting the existence of l-arginine-dependent nitric oxide synthase activity in plants, *New Phytologist*, Vol.184, pp.1–3, ISSN *0028-646X*.

Darwish, E.; Testerink, C.; Khalil, M.; El-Shihy, O. & Munnik, T. (2009). Phospholipid Signaling responses in salt-Stressed Rice Leaves. *Plant and Cell Physiology*, Vol.50, No.5, pp.986-99, ISSN 0032-0781.

Demiral, T. & Turkan, I. (2005). Comparative lipid peroxidation, antioxidant defense systems and proline content in roots of two rice cultivars differing in salt tolerance, *Environmental and Experimental Botany* , Vol.53, pp. 247-257, ISSN 0098-8472.

Garcia-Mata, C. & Lamattina, L. (2007). Abscisic acid (ABA) inhibits light-induced stomatal opening through calcium- and nitric oxide-mediated signaling pathways *Nitric Oxide*, Vol.17, No.3-4, (November-December 2007), pp.143-151, ISSN 1089-8306.

Harminder, P.; Shalinder Kaur, S.; Batish, D.; Sharma, V.; Sharma, N. & Kohli, R. (2009). Nitric oxide alleviates arsenic toxicity by reducing oxidative damage in the roots of *Oryza sativa* (rice), *Nitric Oxide*, Vol.20, No.4, (June 2009), pp. 289-297, ISSN 1089-8306.

Hay, R. & Porter, J. (2006). *The physiology of crop yield*, Wiley-Blackwell Publishing, ISBN 1405108592, Singapore.

Heath, R. & Packer, L. (1968). Photoperoxidation in isolated chloroplasts. I Kineties and stoichiometry of fatty acid peroxidation, *Archives of Biochemistry and Biophysics*, Vol.125, pp.189-198, ISSN 0003-9861.

Jasid, S.; Galatro, A.; Villordo, J.; Puntarulo, S. & Simontacchi, M. (2009).Role of nitric oxide in soybean cotyledon senescence, *Plant Science*, Vol. 176, No. 5, (May 2009), pp.662-668, ISSN 0168-9452.

Kikuchi, G.;Yoshida, T. & Noguchi, M. (2005). Heme oxygenase and heme degradation, *Biochemistry and Biophysical Research Communication*, Vol. 338, pp.558–567, ISSN 0006-291X.

Leitner , M.; Vandelle, E.; Gaupels, F.;Bellin,D.& Delledonne, F. (2009). NO signals in the haze Nitric oxide signalling in plant defence, *Current Opinion in Plant Biology*, Vol.12, pp.451–458. ISSN 1369-5266.

Liu, Y.; Xu, S.; Ling, T.; Xu, L. & Shen, W. (2010). Heme oxygenase/carbon monoxide system participates in regulating wheat seed germination under osmotic stress involving the nitric oxide pathway, *Journal of Plant Physiology* , Vol.167, pp. 1371–1379, ISSN 0176-1617.

Mackernessa, S.; Johna, F.; Jordan, B. & Thomasa, B. (2001). Early signaling components in ultraviolet-B responses: distinct roles for different reactive oxygen species and nitric oxide, *FEBS Letters*, Vol.489, pp.237-242, ISSN 0014-5793.

Mandhania, S.; Madan, S. & Sawhney, V. (2006). Antioxidant defense mechanism under salt stress in wheat seedlings, *Biologia Plantarum*, Vol.227, pp. 227- 231, ISSN 0006-3134.

Martin, M.; Rodríguez Colman,M.; Gómez-Casati, D.; Lamattina, L. & Julián, E. Zabaleta. (2009). Nitric oxide accumulation is required to protect against iron-mediated oxidative stress in frataxin-deficient Arabidopsis plants, *FEBS Letters*, Vol.583, No.3, (February 2009), pp. 542-548, ISSN 0014-5793.

Mittler, R. (2002) Oxidative stress, antioxidants and stress tolerance, *TRENDS in Plant Science*, Vol.7, No.9, (September 2002),pp.521-546, ISSN 1360-1385.

Muramoto, T.; Tsurui, N.; Terry, M.; Yokota, A. & Kohchi, T. (2002). Expression and biochemical properties of a ferredoxin-dependent heme oxygenase required for phytochrome chromophore synthesis. *Plant Physiology*, Vol.130, pp.1958-1966, ISSN 0032-0889.

Nakano, Y. & Asada, K. (1981). Hydrogen peroxide is scavenged by ascorbate-specific peroxidase in spinach chloroplast. *Plant and Cell Physiology*, Vol.22, pp.867-880, ISSN 0032-0781.

Noriega, G.; Balestrasse, K.; Batlle, A & Tomaro, M. (2004). Heme oxygenase exerts a protective role against oxidative stress in soybean leaves. *Biochemical and Biophysical Research Communications*. Vol.323, pp. 1003-1008, ISSN 0006-291X.

Noriega, G.; Yannarelli, G.; Balestrasse, K.; Batlle, A. & Tomaro, M. (2007). The effect of nitric oxide on heme oxygenase gene expression in soybean leaves, *Planta*, Vol. 226, pp.1155–1163, ISSN 0032-0935.

Nürnberger, T. & Scheel, D. (2001) Signal transmission in the plant immune response, *TRENDS in Plant Science*, Vol.6, No.8 (August 2001), pp. 551-579, ISSN 1360-1385.

Otterbein, L.; Soares, M.; Yamashita, K. & Bach, F. (2003). Hemeoxygenase-1:unleashing the protective properties of heme. *Trends Immunology,* Vol.24, pp.449-455, ISSN 1471-4906.

Parani, M. ; Rudrabhatla. S.; Myers, R.; Weirich, H.; Smith, B.; Leaman, B. & Goldman, S. (2004). Microarray analysis of nitric oxide responsive transcripts in *Arabidopsis, Plant Biotechnology Journal,* Vol.2, pp.359–366, ISSN 14677644.

Romero-Puertas, M.; Rodríguez-Serrano, M.; Corpas, F.; Gómez, M.; Del Río, L. & Sandalio, L. (2004). Cadmium-induced subcellular accumulation of $O_2^{\cdot-}$ and H_2O_2 in pea leaves, *Plant and Cell Environment,* Vol.27, pp.1122–1134, ISSN 0140-7791.

Santa-Cruz, D.;Pacienza, N.; Polizio A., Balestrasse K.;Tomaro M. & Yannarelli, G. (2010). Nitric oxide synthase-like dependent NO production enhances heme oxygenase up-regulation in ultraviolet-B-irradiated soybean plants. *Phytochemistry,* Vol.71, pp.1700–1707, ISSN.

Shaedle, M. & Bassham, J. (1977). Chloroplast glutathione reductase, *Plant Physiology,* Vol. 59, pp.1011–1012, ISSN 0031-9422.

Shantel, A.; Fowler, R.; Virgen, A.; Gossett, D.; Banks, S. & Rodriguez, S. (2008). Opposing roles for superoxide and nitric oxide in the NaCl stress-induced upregulation of antioxidant enzyme activity in cotton callus tissue, *Environmental and Experimental Botany,* Vol.62, No.1, (January 2008), pp. 60-68, ISSN 0098-8472.

Shi, S.; Wang, G.; Wang, Y.; Zhang, L. & Zhang, L. (2005). Protective effect of nitric oxide against oxidative stress under ultraviolet-B radiation, *Nitric Oxide,* Vol.13, pp.1–9, ISSN 1089-8603.

Sies, H. (1993). Damage to plasmid DNA by singlet oxygen and its protection. *Mutation Research/Genetic Toxicology,* Vol. 299, No.3-4, May 1993, pp 183-191, ISSN 1383-5718.

Terry, M.; Linley, P. & Kochi, T. (2002). Making light of it: the role of plants heme oxygenases in phytochrome chromophore synthesis, *Biochemical Society Transactions,* Vol. 30, pp. 604–609, ISSN 0300-5127.

Terry, M.; Linley, P. & Kochi, T. (2002). Making light of it: the role of plants heme oxygenases in phytochrome chromophore synthesis, *Biochemical Society Transactions,* Vol. 30, pp.604–609, ISSN 0300-5127.

Tomaro, M., Batlle, A. (2002). Bilirubin: its role in cytoprotection against oxidative stress. *International Journal of Biochemistry and Cell Biology,* Vol.34, pp.216–220. ISSN 1357-2725.

Wilson, I.; Neill, S. & Hancock, J. (2008).Nitric oxide synthesis and signalling in plants, *Plant Cell and Environment,* Vol.31, pp.622–631, ISSN 0140-7791.

Xuan, Y.; Zhou, S.; Wang, L.; Cheng, Y. & Zhao L. (2010). Nitric Oxide Functions as a Signal and Acts Upstream of AtCaM3 in Thermotolerance in Arabidopsis Seedling, *Plant Physiology.* Vol.153, pp.1443-1444, ISSN 0032-0889.

Yaacov,Y.; Leshem, R.;Wills, B. & Veng-Va Ku, W. (1998). Evidence for the function of the free radical gas – nitric oxide (NO•) – as an endogenous maturation and senescence regulating factor in higher plants, *Plant Physiology and Biochemistry,* Vol. 36, No. 11, (November 1998), pp. 825-833, ISSN 0981-9428.

Yannarelli, G.; Noriega, G.; Batlle, A. & Tomaro, M. (2006). Heme oxygenase up-regulation in ultraviolet-B irradiated soybean plants involves reactive oxygen species, *Planta,* Vol.224, pp.1164–1172, ISSN 0032-0935.

Zhao, L.; Zhang, F.; Guo, J.; Yang, Y.; Li, B. & Zhang, L. (2004). Nitric oxide functions as a signal in salt resistance in the calluses from two ecotypes of reed (*Phragmites communis Trin.*). *Plant Physiology*, Vol.134, pp.849–857. ISSN 0032-0889.

Zilli, C.; Santa-Cruz, D.; Yannarelli, G.; Noriega, G.; Tomaro, M. & Balestrasse, K. (2009). Heme Oxygenase Contributes to Alleviate Salinity Damage in *Glycine max* L. Leaves, *International Journal of Cell Biology*. Published online (September 2009), PMCID: PMC2809017.

21

Regulation of Leaf Photosynthesis Through Photosynthetic Source-Sink Balance in Soybean Plants

Minobu Kasai
Department of Biology, Faculty of Agriculture and Life Science, Hirosaki University
Japan

1. Introduction

Plant photosynthesis is the basis for matter production needed for all living organisms. In the future, plant photosynthesis would be more important, since environmental problems such as climatic warming due to increasing environmental CO_2 concentration and problems of food and energy shortages due to increasing populations may be severer (von Caemmerer & Evans, 2010; Raines, 2011). Increasing plant leaf photosynthesis and thereby increasing plant matter production would be expected as a realistic way to resolve the problems. There is, however, a well-known hypothesis that in plants leaf photosynthesis can be down regulated through accumulated photosynthetic carbohydrates in leaf under excessive photosynthetic source capacity, which also means sink limitation, although the detailed mechanism is not clear (see Kasai, 2008). Actually, for example, there is evidence for the excessive photosynthetic source capacity causing down regulation of photosynthesis in crop plants under field conditions (Okita et al., 2001; Smidansky et al., 2002, 2007). Therefore, for the better improvement of leaf photosynthesis in plants, it is important to elucidate the regulatory mechanism for leaf photosynthesis under excessive photosynthetic source capacity and thereby clarify way of the improvement of leaf photosynthesis.
To elucidate the regulatory mechanism for leaf photosynthesis under excessive photosynthetic source capacity, experimental construction of the excessive photosynthetic source capacity is important. Excising sink organs such as pods, fruits or flowers from plant materials is a way to construct excessive photosynthetic source capacity, and it has often been conducted to study the regulatory mechanism of photosynthetic source-sink balance in plants (see Kasai, 2008). However, the way excising sink organs results not directly but indirectly in excessive photosynthetic source capacity by diminishing sink capacity, and can give some damages to plant materials. Recent studies using transgenic plants have shown that overexpression of Calvin cycle enzymes (sedoheptulose-1,7-bisphosphatase and fructose-1,6-bisphosphatase) or leaf plasma membrane CO_2 transport protein increases the leaf photosynthetic rate significantly (Raines, 2003, 2006). Therefore, the use of the transgenic plants with improved higher leaf photosynthetic rate may be useful to study the regulatory mechanism for leaf photosynthesis under excessive photosynthetic source capacity, since the higher photosynthetic rate is likely to result in excessive photosynthetic source capacity. However, it seems difficult to analyze the down regulation of

photosynthesis that may hide in the improved photosynthetic rate. Actually, down regulation of photosynthesis that is associated with excessive photosynthetic source capacity has not been analyzed in the transgenic plants with improved higher photosynthetic rate. Exposure to high CO_2 or continuous exposure to light of plant materials is thought as the other way to construct excessive photosynthetic source capacity. It is well known that leaf photosynthetic rate, especially, in C_3 plants does not reach the saturation at the present atmospheric CO_2 concentration and thus the rate increases initially under high CO_2 conditions (Ward et al., 1999). Therefore, in C_3 plants, exposure to high CO_2 is expected to result in excessive photosynthetic source capacity. However, the way of exposure to high CO_2 may be not suitable to analyze the regulatory mechanism for leaf photosynthesis under excessive photosynthetic source capacity, because of the same reason described for the transgenic plants with improved photosynthetic rate and well-known action of high CO_2 to decrease stomatal aperture (Bredmose & Nielsen, 2009). In contrast, continuous exposure to light of plant materials, which prolongs photosynthetic period, can result in excessive photosynthetic source capacity without affecting directly the sink organs, leaf photosynthetic rate and stomatal aperture and giving direct damage to the plant materials.

Soybean plants, although it is single-rooted soybean leaves, have largely contributed to study the regulatory mechanism for leaf photosynthesis under excessive photosynthetic source capacity through the experimental system using continuous exposure to light. Single-rooted soybean leaves are source-sink model plants with a simple organization of a leaf, a short petiole and roots developed from the petiole in individuals and were developed by Sawada et al. (1986) using the primary leaves of intact soybean plants (*Glycine max* L. Merr. cv. Tsurunoko). Studies using single-rooted soybean leaves have shown that treating the plants with continuous light results in accumulation of photosynthetic carbohydrates (sucrose and starch) in the leaf and decrease in the leaf photosynthetic rate, which correlates with the increase in leaf carbohydrate (sucrose or starch) content (Sawada et al., 1986, 1989, 1990, 1992). Also, it has been shown in the single-rooted soybean leaves that deactivation of Rubisco, a CO_2-fixing enzyme is caused by the treatment of continuous exposure to light (Sawada et al., 1990, 1992). As continuous exposure to light of single-rooted soybean leaves also increased the leaf phosphorylated intermediates' contents (Sawada et al., 1989), and there have been findings that in vitro, inorganic phosphate promotes activation of Rubisco by enhancing the affinity of uncarbamylated inactive Rubisco to CO_2 (Bhagwat, 1981; McCurry et al., 1981; Anwaruzzaman et al., 1995), the studies using single-rooted soybean leaves have suggested that there is a regulatory mechanism of leaf photosynthetic rate through deactivation of Rubisco, which is associated with accumulation of photosynthetic carbohydrates in leaf under excessive photosynthetic source capacity, and that the deactivation of Rubisco may be caused by limitation of inorganic phosphate (Sawada et al., 1990, 1992). Data from a study using single-rooted soybean leaves demonstrate that the plants do not change the leaf area and leaf dry weight other than the weights of major photosynthetic carbohydrates (sucrose and starch) and grow only the roots during experimental period, irrespective of whether light conditions are normal (daily light/dark periods of 10/14 h) or continuous without darkness (Sawada et al., 1986). Although the source-sink model plants with simple source-sink organization have been developed from various plant species, only the single-rooted soybean leaves have been demonstrated to show almost no growth in the source organ (Sawada et al., 2003). No growth of the source organ and the simple organization of source and sink in the single-rooted soybean leaves are attractive characteristics to analyze comprehensively the regulatory mechanism of photosynthetic source-sink balance in plants, including the regulatory mechanism for leaf photosynthesis under excessive photosynthetic source capacity. Actually, as mentioned above, various analyses have been conducted in the single-rooted soybean leaves, especially in studies for elucidating the regulatory mechanism for leaf photosynthesis under excessive photosynthetic source capacity. Therefore, the single-rooted soybean leaves are important plant materials to elucidate further the regulatory mechanism for leaf photosynthesis under excessive photosynthetic source capacity. However, the plants are made artificially, and do not exist in nature, and in addition, as already mentioned, the plant leaf originates from only the primary leaf in intact soybean plants (Sawada et al., 1986). Therefore, there is the possibility that properties of single-rooted soybean leaves may not reflect those of the original, intact soybean plants or the other intact plants. Thus, it is important to examine the regulatory mechanism for leaf photosynthesis under excessive photosynthetic source capacity using the original, intact soybean plants.

The present study used the original intact soybean plants, and it was analyzed how continuous exposure to light affects the leaf photosynthetic rate and related characteristics, such as leaf stomatal conductance and intercellular CO_2 concentration, contents of water, chlorophyll, major photosynthetic carbohydrates (sucrose and starch), total protein and Rubisco protein in leaf, and activity and activation ratio (ratio of initial to total activity) of Rubisco and amount of protein-bound ribulose-1,5-bisphosphate (RuBP) in leaf extract, which were analyzed to evaluate the amount of uncarbamylated inactive Rubisco (Brooks & Portis, 1988). The same series of analyses have not been conducted together in studies that have performed the experiment of continuous exposure to light using plants.

2. Materials and methods

2.1 Plant materials

Soybean (*Glycine max* L. Merr. cv. Tsurunoko) seeds were sown in plastic pots (13.5 cm in height, 12.5 cm in diameter) containing almost equal volumes of vermiculite and sand that had been mixed, and were grown in growth chambers (Koitotoron, HNL type; Koito Industries Ltd., Tokyo, Japan) under daily light/dark periods of 10/14 h, day/night temperatures of 24/17°C and relative humidity of 60 %. After 8 weeks, plants were divided into two groups, and one group was grown for 3 days with continuous light, and another group was grown for 3 days under daily light/dark periods of 10/14 h as controls. Nutrients were supplied once a week with a 1000-fold diluted solution of Hyponex [6-10-5 type (N:P:K = 6:10:5); Hyponex Co., Osaka, Japan], and tap water was supplied in sufficient amounts. Light was supplied with incandescent lamps at an intensity of 480 μmol photons m^{-2} s^{-1} (400-700 nm) at the middle height of plants grown for 8 weeks.

2.2 Leaf photosynthetic rate, stomatal conductance and intercellular CO_2 concentration

Leaf photosynthetic rate, stomatal conductance and intercellular CO_2 concentration were determined in fully expanded fourth trifoliate leaves at a light intensity of 1000 μmol photons m^{-2} s^{-1}, air flow rate of 200 ml min^{-1}, air temperature of 25 °C, relative humidity of 60 % and CO_2 concentration of 350 ppm on day 3 after treating plants with continuous light using a portable photosynthetic analyzer (Cylus-1; Koito Industries Ltd.). After measurements, leaf disks (1.79 cm^2) were cut off from fourth trifoliate leaves, immediately frozen in liquid nitrogen and stored at -80 °C until used for the other analyses described below.

2.3 Other analyses

The activity of Rubisco in leaf extract was determined at 25 °C as described previously (Kasai, 2008). For the initial activity, 20 µl of a leaf extract obtained by homogenizing a leaf disk with ice-cold buffer (100 mM HEPES-KOH, pH 7.8, 2 ml) was added to a cuvette containing 1.98 ml of assay medium [100 mM Bicine-KOH (pH 8.2), 20 mM $MgCl_2$, 20 mM $NaHCO_3$, 5 mM creatine phosphate, 1 mM ATP, 0.2 mM NADH, 20 units creatine kinase, 20 units 3-phosphoglycerate kinase and 20 units glyceraldehyde-3-phosphate dehydrogenase], immediately followed by the addition of RuBP (final concentration 0.6 mM) and mixed well. For total activity, RuBP was added 5 min later after 20 µl of the leaf disk extract was immediately combined with the assay medium. The change in absorbance at 340 nm was monitored using a spectrophotometer (Model U-2000; Hitachi Co., Tokyo, Japan).
The amount of protein-bound RuBP in leaf extract was determined as described previously (Kasai, 2008). A leaf extract (800 µl) obtained by homogenizing a leaf disk with an ice-cold buffer (100 mM HEPES-KOH, pH 7.8, 1 ml) was centrifuged (100 g, 1 min) after loading onto a column containing Sephadex G-50 (bed volume before centrifugation, 4 ml) that had been equilibrated with the same buffer. The eluent (500 µl) from the column lacking free RuBP was centrifuged (10,000 g, 10 min) after mixing with an acidic solution (5.5 M $HClO_4$, 50 µl) to precipitate protein in the eluent. The resulting supernatant was centrifuged (10,000 g, 10 min) after neutralizing to pH 5.6 with K_2CO_3, and RuBP in the supernatant was determined in the assay medium for determining Rubisco activity using purified spinach Rubisco (0.5 units).
Leaf Rubisco content was determined as described by Makino et al. (1986). Leaf total protein was extracted as described by Makino et al. (1986) and quantified by the method of Bradford (1976). Leaf chlorophyll content was determined according to the method of Mackinney (1941). Leaf sucrose and starch contents were determined as described by Sawada et al. (1999). Leaf water content was analyzed by measuring fresh weight and dry weight of leaf disks. Leaf disks were dried for 2 days at 75 °C.

3. Results

Analyzed leaf photosynthetic rate was significantly lower in intact soybean plants grown for 3 days with continuous light than in control plants grown under daily light/dark periods of 10/14h (Fig. 1).
Leaf stomatal conductance was also significantly lower in continuous light-treated plants than in control plants (Fig. 2). Leaf intercellular CO_2 concentration did not differ significantly between control and continuous light-treated plants (Fig. 2).
When activation ratio (percentage of initial activity to total activity) of Rubisco in leaf extract was calculated from analyzed initial and total activities of Rubisco in leaf extract, the ratio was significantly lower in continuous light-treated plants than in control plants (Fig. 3). The ratios in control and continuous light-treated plants were 74.2% and 56.6%, respectively.

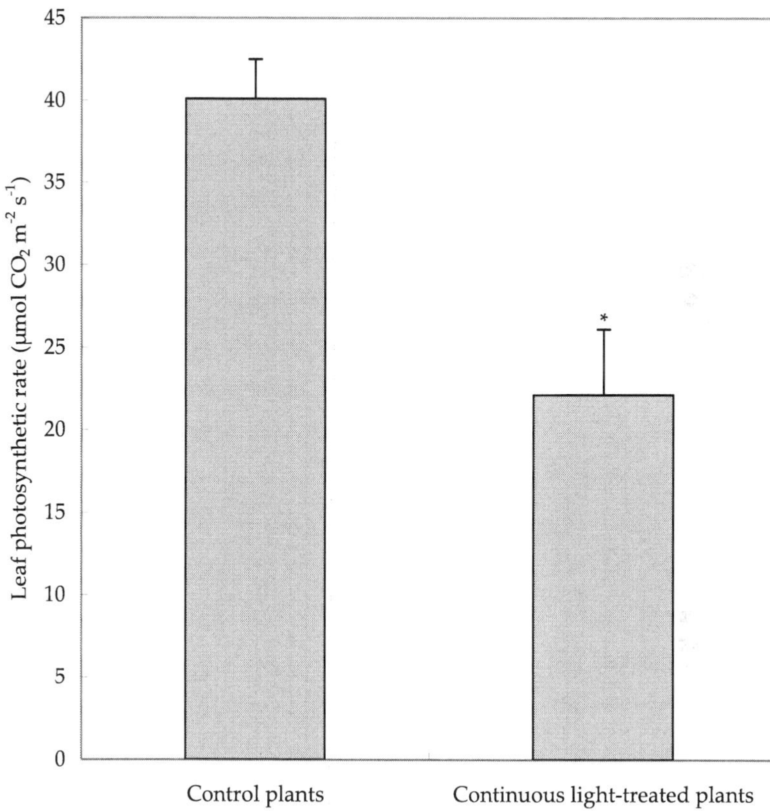

Fig. 1. Leaf photosynthetic rate in soybean plants on day 3 after continuous exposure to light. Control plants were grown under daily light/dark periods of 10/14h for 3 days. Vertical bars indicate S.D. (n=4). *P<0.01 (t-test) when compared with control plants.

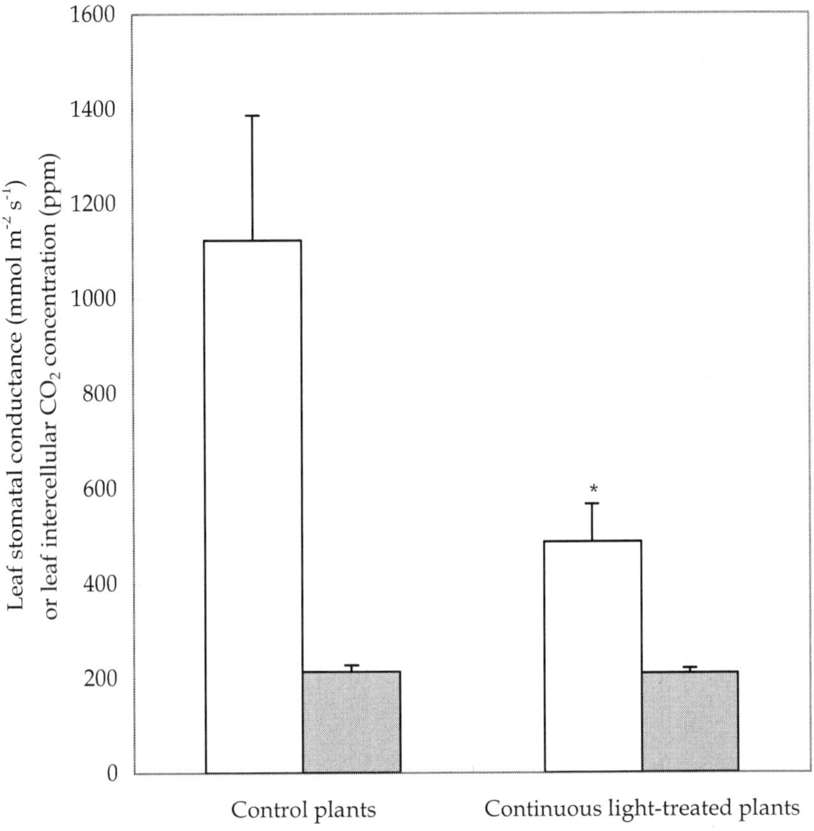

Fig. 2. Leaf stomatal conductance and leaf intercellular CO_2 concentration in soybean plants on day 3 after continuous exposure to light. Control plants were grown as described in Fig. 1. Open bar, leaf stomatal conductance; closed bar, leaf intercellular CO_2 concentration. Vertical bars indicate S.D. (n=4). $^{*}P<0.01$ when compared with control plants. The intercellular CO_2 concentration did not differ significantly ($P>0.05$) between control and continuous light-treated plants.

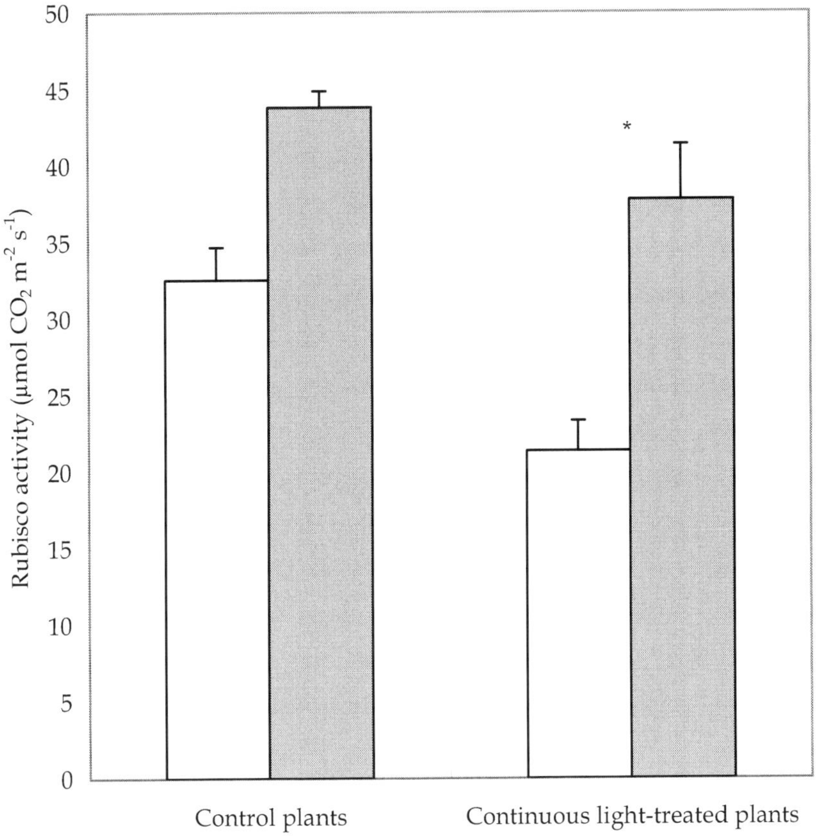

Fig. 3. Initial and total activities of Rubisco in leaf extract from soybean plants on day 3 after continuous exposure to light. Control plants were grown as described in Fig. 1. Open bar, initial activity; closed bar, total activity. Vertical bars indicate S.D. (n=4). In comparison with control plants of the activation ratio of Rubisco calculated as a percentage of the initial activity to total activity, $^*P<0.01$.

In a study investigating the light activation of Rubisco using *Arabidopsis thalian*, it was demonstrated that the amount of protein-bound RuBP in leaf extract reflects the amount of uncarbamylated inactive Rubisco (Brooks & Portis, 1988). When the amount of protein-bound RuBP was analyzed, the amount was significantly more in continuous light-treated plants than in control plants (Fig. 4).

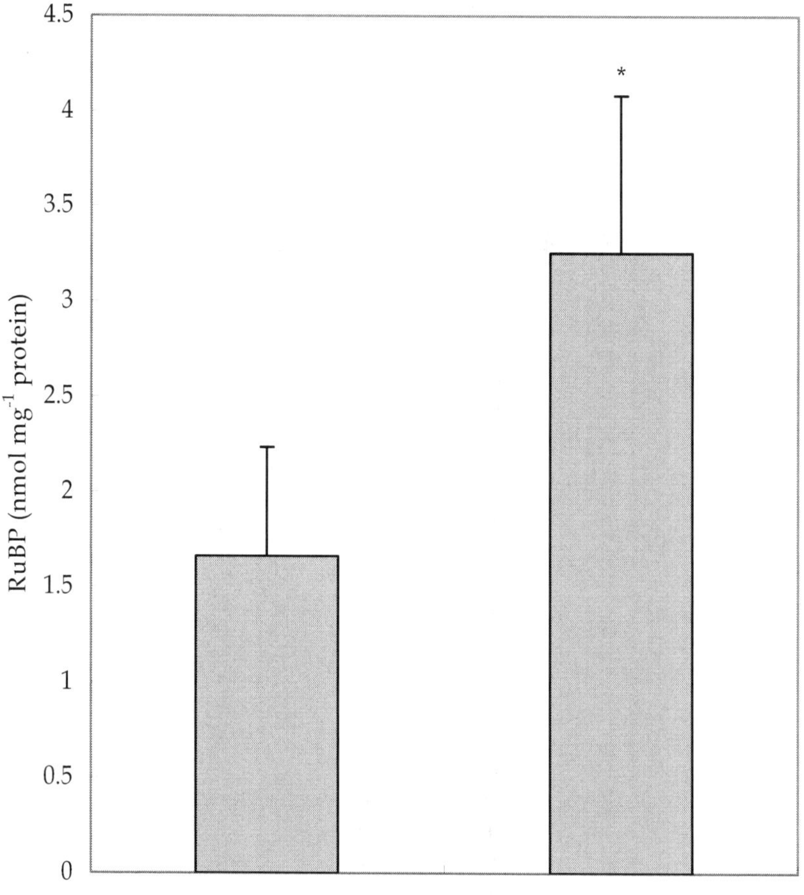

Fig. 4. Amount of protein-bound RuBP in leaf extract from soybean plants on day 3 after continuous exposure to light. Control plants were grown as described in Fig. 1. Vertical bars indicate S.D. (n=4). *$P<0.05$ when compared with control plants.

Contents of sucrose and starch, which are the major photosynthetic carbohydrates, in leaf were both significantly higher in continuous light-treated plants than in control plants (Fig. 5).

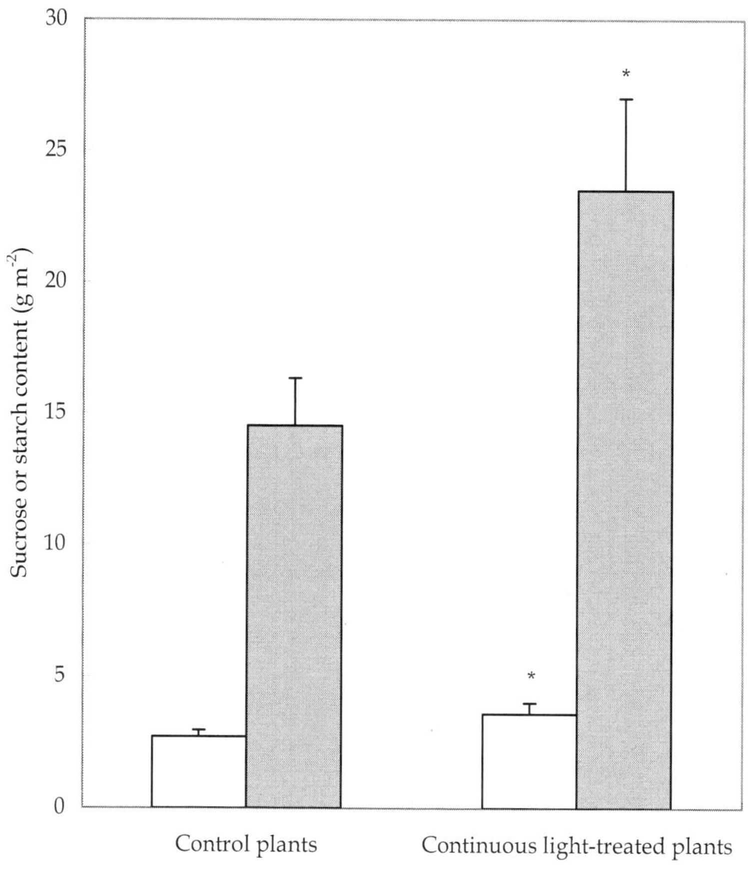

Fig. 5. Leaf sucrose or starch content in soybean plants on day 3after continuous exposure to light. Control plants were grown as described in Fig. 1. Open bar, sucrose content; closed bar, starch content. Vertical bars indicate S.D. (n=4). $^{*}P<0.05$ when compared with control plants.

Analyzed contents of chlorophyll, water, total protein and Rubisco protein in leaf did not differ significantly between control and continuous light-treated plants (Fig. 6 and 7).

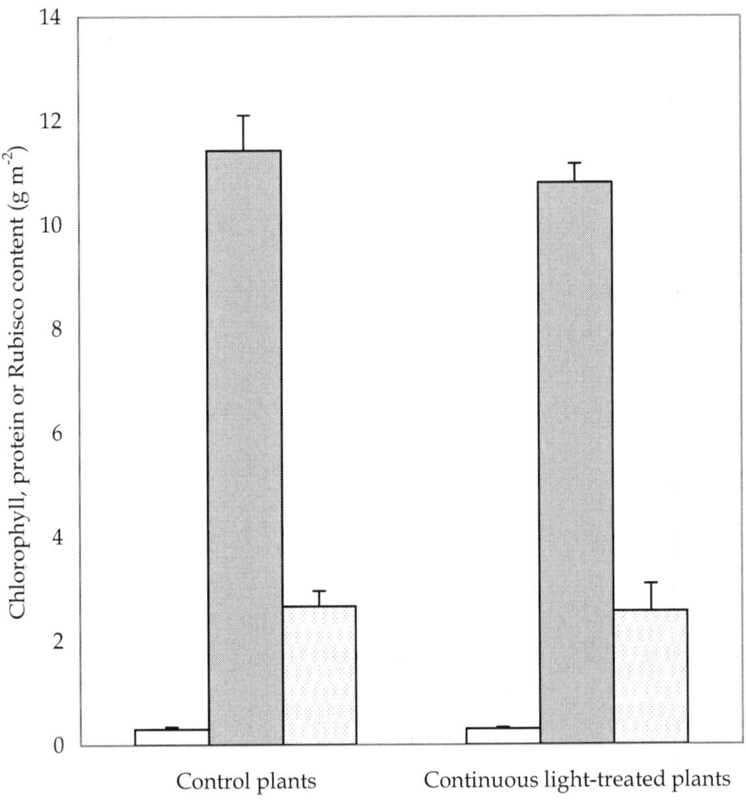

Fig. 6. Leaf chlorophyll, total protein or Rubisco content in soybean plants on day 3 after continuous exposure to light. Control plants were grown as described in Fig. 1. Open bar, chlorophyll content; closed bar, total protein content; dotted bar, Rubisco content. Vertical bars indicate S.D. (n=4). The chlorophyll, total protein and Rubisco contents did not differ significantly ($P>0.05$) between control and continuous light-treated plants.

Analyzed leaf dry weight other than the weights of sucrose and starch was heavier a little in continuous light-treated plants than in control plants (Fig. 5 and 7). The mean dry weights after subtracting the weights of sucrose and starch in control and continuous light-treated plants were 49.0 g m^{-2} and 57.5 g m^{-2}, respectively.

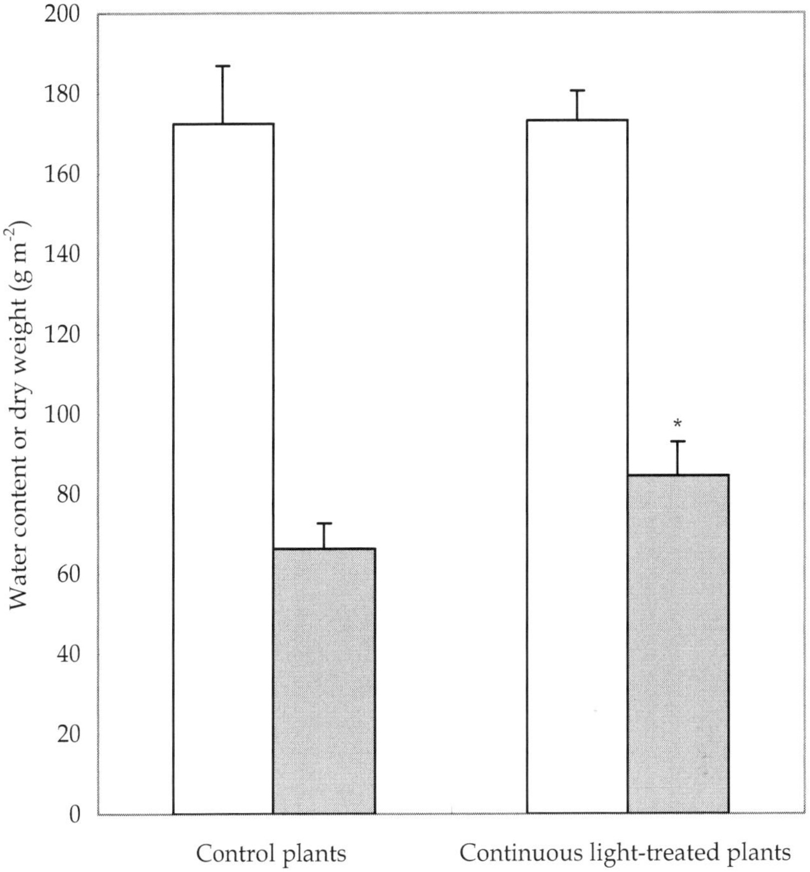

Fig. 7. Leaf water content and leaf dry weight in soybean plants on day 3 after continuous exposure to light. Control plants were grown as described in Fig. 1. Open bar, leaf water content; closed bar, leaf dry weight. Vertical bars indicate S.D. (n=4). $^{*}P<0.05$ when compared with control plants. The leaf water content did not differ significantly ($P>0.05$) between control and continuous light-treated plants.

4. Discussion

The present study was conducted to examine the regulatory mechanism for leaf photosynthesis under excessive photosynthetic source capacity in intact soybean plants. The experimental construction of excessive photosynthetic source capacity was conducted by treating the plants with continuous light for 3 days. The data show that the treatment of continuous exposure to light for intact soybean plants decreased significantly the leaf photosynthetic rate (Fig. 1). Since the light treatment also decreased the leaf stomatal conductance in soybean plants (see Fig. 2), it is thought that the decrease in leaf photosynthetic rate caused by treatment of continuous exposure to light might have resulted from stomatal limitation of CO_2 diffusion. However, the treatment of continuous exposure to light did not affect significantly leaf intercellular CO_2 concentration (see Fig. 2), implicating that the light treatment decreased CO_2 incorporation by leaf photosynthetic cells, as it affected leaf stomatal conductance. In addition, the light treatment decreased activation ratio of Rubisco in leaf extract and did not affect significantly leaf Rubisco content (see Fig. 3 and 6). Furthermore, the light treatment increased the amount of protein-bound RuBP in leaf extract (see Fig. 4). The decrease in activation ratio of Rubisco and increase in the amount of protein-bound RuBP in leaf extract (Brooks & Portis, 1988) strongly suggest an increase in the amount of uncarbamylated inactive Rubisco in leaf. Therefore, it is suggested that the decrease in leaf photosynthetic rate caused by treatment of continuous exposure to light is likely to be due to deactivation of Rubisco in leaf. Treatment of continuous exposure to light for intact soybean plants also increased significantly both the contents of sucrose and starch, which are the major photosynthetic carbohydrates, in leaf (see Fig. 5), indicating that the light treatment could result in an excessive photosynthetic source capacity in the plants. The present study also shows that analyzed leaf chlorophyll, total protein and water contents were not affected significantly by the treatment of continuous exposure to light (see Fig. 6 and 7). Therefore, results obtained in the present study strongly suggest that the decrease in leaf photosynthetic rate in intact soybean plants caused by treatment of continuous exposure to light is unlikely to be due to simple damages such as the breakdown of cellular compartments, but is likely to be due to deactivation of Rubisco, which is associated with accumulation of photosynthetic carbohydrates (sucrose and starch) in leaf under excessive photosynthetic source capacity.

As described in the Introduction, single-rooted soybean leaves have quite been helpful to study the regulatory mechanism for leaf photosynthesis under excessive photosynthetic source capacity, since the plants have simple source-sink organization and have excellent characteristics [growing only the sink organs (roots) without growing source organ (leaf)], which have not been found in other plants (Sawada et al., 1986, 2003). However, as already mentioned, as the plant leaf is constituted from only the primary leaf in intact soybean plants, there is the possibility that properties of single-rooted soybean leaves may not reflect those of the original intact soybean plants or the other intact plants. However, results obtained in the present study of the changes in leaf photosynthetic rate, initial activity and activation ratio of Rubisco in leaf extract, and contents of major photosynthetic carbohydrates (sucrose and starch) and chlorophyll in leaf caused by treatment of continuous exposure to light corresponded with results from studies that have performed similar experiments of continuous exposure to light using single-rooted soybean leaves (Sawada et al., 1986, 1990, 1992). Leaf intercellular CO_2 concentration, amount of protein-bound RuBP in leaf extract and leaf Rubisco content have not been analyzed in the single-rooted soybean leaves. As already mentioned, the present study used the original intact soybean plants from which single-rooted soybean leaves can be made. Therefore, the correspondence of data from original intact soybean plants and those from single-rooted soybean leaves highlights that properties of single-rooted soybean leaves and those of original intact soybean plants are very similar, thus suggesting that properties of single-rooted soybean leaves and those of original intact soybean plants can reflect each other. As described in the Introduction, studies using single-rooted soybean leaves have implicated that there is a regulatory mechanism of leaf photosynthetic rate through deactivation of Rubisco, which is associated with accumulation of photosynthetic carbohydrates in leaf under excessive photosynthetic source capacity (Sawada et al., 1986, 1989, 1990, 1992, 1999, 2003). Data from the present study using the original intact soybean plants have also suggested the same regulatory mechanism of leaf photosynthetic rate. Therefore, the suggested regulatory mechanism of leaf photosynthetic rate may be a common mechanism in plants. With respect to the excellent characteristic of single-rooted soybean leaves that do not change the leaf dry weight other than the weights of major photosynthetic carbohydrates (sucrose and starch) (Sawada et al., 1986), a little change (increase) of leaf (fourth trifoliate leaves) dry weight other than the weights of major photosynthetic carbohydrates (sucrose and starch) was observed by treatment of continuous exposure to light in the original intact soybean plants (see Fig. 5 and 7). Although the present study conducted various analyses to examine the regulatory mechanism for leaf photosynthesis under excessive photosynthetic source capacity, the same series of analyses have not been conducted together in other studies that have performed the treatment of continuous exposure to light using plants.

Treatment of continuous exposure to light for plants results, in most cases, in accumulation of photosynthetic carbohydrate(s) in leaf and decrease in leaf photosynthetic rate. However, in addition to these effects of the light treatment, there are other effects of the light treatment that are different from those indicated by the present study. In tomato, egg plant, peanut and potato, treatment of continuous exposure to light has been shown to result in leaf decolorization (Bradley & Janes, 1985; Globig et al., 1997; Murage et al., 1996, 1997; Rowell et al., 1999; Wheeler & Tibbitts, 1986; Tibbitts et al., 1990). In young leaves of potato and *Arabidopsis*, the continuous light treatment has been shown to accelerate expressions of photosynthetic genes, pigments and proteins, and subsequent declines of the expressions (Cushman et al., 1995; Stessman et al., 2002). In a study using young apple, a decrease in leaf photosynthetic rate caused by treatment of continuous exposure to light was suggested to be due to stomatal limitation of CO_2 diffusion rather than a reduction of Rubisco activity, although, in the study, leaf water content, which is likely to affect stomatal aperture (Brodribb & McAdam, 2011), was not analyzed (Cheng et al., 2004). Therefore, leaf photosynthetic rate may also be regulated through changes in expressions of photosynthetic genes, pigments and proteins and through a regulation of stomata under excessive photosynthetic source capacity in plants.

Other ways, which indirectly construct excessive photosynthetic source capacity as described in the Introduction, have also been shown to result in accumulation of photosynthetic carbohydrate(s) in leaf and decrease in leaf photosynthetic rate. With respect to the cause(s) of why leaf photosynthetic rate declines under the excessive photosynthetic source capacity, for example, data from photosynthetic carbohydrate-feeding or high CO_2 treatment experiments suggest that decreased expressions of photosynthetic genes, including genes for chlororphyll-related protein and Rubisco protein can be causes (Paul & Foyer, 2001; Martin et al., 2002; Paul & Pellny, 2003). However, there is also evidence from high CO_2 treatment experiments using various C_3 plants that decreased Rubisco activity in leaf rather than changes in leaf Rubisco content is likely to be a main cause (Sage et al., 1989). Data from experiments conducting excisions of sink organs (pods or flower buds and flowers) or petiole girdling suggest that a decrease of stomatal conductance or Rubisco activity or Rubisco content in leaf, or both decreases of Rubisco activity and Rubisco content in leaf can be responsible for the decrease in leaf photosynthetic rate under excessive photosynthetic source capacity (Mondal et al., 1978; Setter & Brun, 1980; Setter et al., 1980; Wittenbach, 1982, 1983; Xu et al., 1994; Crafts-Brandner & Egli, 1987; Cheng et al., 2008). As described in the Introduction, excising sink organs or high CO_2 treatment can have side effect(s) other than inducing excessive photosynthetic source capacity. In the present study using intact soybean plants in which excessive photosynthetic source capacity was constructed by treatment of continuous exposure to light, visible damages such as leaf decolorization and wilt were not observed. Treatment of continuous exposure to light did not affect significantly leaf chlorophyll, total protein and water contents analyzed. However, as mentioned above, totally, the effects of indirectly constructed excessive photosynthetic source capacity on leaf carbohydrate status, photosynthetic rate, stomatal conductance, Rubisco activity and photosynthetic gene expressions including Rubisco gene expression are similar to those of excessive photosynthetic source capacity that is constructed by treatment of continuous exposure to light.

Regarding the detailed mechanism(s) of why leaf photosynthetic rate declines under excessive photosynthetic source capacity, recent studies using transgenic plants show that hexokinase could be involved in carbohydrate-mediated repression of photosynthetic gene expression (Jang et al., 1997; Dai et al., 1999; Moore et al., 2003). Other recent study shows that protein kinases (KIN10 and KIN11) may be involved in governing the entirety of carbohydrate metabolism, growth and development in response to carbohydrates in plants (Baena-Gonzalez et al., 2007). Data from a study investigating the effect of chilling stress on leaf photosynthetic rate suggest that H_2O_2, a reactive oxygen species can induce deactivation of Rubisco (Zhou et al., 2006). As described in the Introduction, inorganic phosphate has been found to promote activation of Rubisco by enhancing the affinity of uncarbamylated inactive Rubisco to CO_2 (Bhagwat, 1981; McCurry et al., 1981; Anwaruzzaman et al., 1995). Data from a more recent study suggest that pH within the chloroplasts can be an important factor affecting leaf photosynthetic rate, since the study has demonstrated that pH can affect distribution of Rubisco activase within the chloroplasts by affecting binding of the enzyme to the thylakoid membranes (Chen et al., 2010). Distribution of Rubisco activase within the chloroplasts can affect activation state of Rubisco, since Rubisco activase plays a role in promoting the activation of Rubisco by dissociating RuBP from uncarbamylated inactive Rubisco (Crafts-Brandner & Salvucci, 2000), which tightly binds RuBP (Jordan & Chollet, 1983). Since ATP is needed for the catalytic action of Rubisco activase (Crafts-Brandner & Salvucci, 2000) and it is well known that ATP is needed for regeneration of RuBP, a substrate for Rubisco in Calvin cycle (see Kasai, 2008), it is evident that ATP is also an important factor affecting leaf photosynthetic rate. However, the precise mechanism of how hexokinase and protein kinases exercise regulation of photosynthetic carbohydrate metabolism including the carbohydrate-mediated repression of photosynthetic gene expression is not yet clear. In addition, effects of excessive photosynthetic source capacity on the levels of H_2O_2, inorganic phosphate, pH and ATP within the chloroplasts in which central photosynthesis is performed have not been analyzed in intact plants at real times under light. A main reason seems to be the lack of appropriate methods. Therefore, further researches including those following the development of new methods are important to elucidate further the regulatory mechanism for leaf photosynthesis under excessive photosynthetic source capacity.

Recent studies using transgenic plants have shown that overexpression of Calvin cycle enzymes (sedoheptulose-1,7-bisphosphatase and fructose-1,6-bisphosphatase) or leaf plasma membrane CO_2 transport protein increases the leaf photosynthetic rate and the biomass production (Raines, 2003, 2006). Increasing plant leaf photosynthesis and thereby increasing plant matter (biomass) production seems to be an effective way to resolve the serious problems such as climatic warming and food and energy shortages. However, data obtained in the present study and those from other studies strongly suggest that excessive photosynthetic source capacity decreases the efficiency of leaf photosynthetic matter production. This means that under excessive photosynthetic source capacity, efficiency of plant matter (biomass) production decreases. There is also evidence for the excessive photosynthetic source capacity causing down regulation of photosynthesis in plants under field conditions (Okita et al., 2001; Smidansky et al., 2002, 2007). Therefore, it is strongly suggested that for the efficient improvement of plant matter (biomass) production, well-balanced improvement of source and sink would be essential. Further studies are desired for deeper and more comprehensive understanding of the regulatory mechanism of photosynthetic source-sink balance including the regulatory mechanism for leaf photosynthesis under excessive photosynthetic source capacity. Soybean plants (*Glycine max* L. Merr. cv. Tsurunoko) used in the present study from which single-rooted soybean leaves can be made are one of the important experimental materials.

5. Conclusion

Studies using single-rooted soybean leaves, each of which is constituted from a primary leaf, a short petiole and roots developed from the petiole, have implicated that there is a regulation of leaf photosynthesis through deactivation of Rubisco, which is associated with accumulation of photosynthetic carbohydrates in leaf under excessive photosynthetic source capacity. The present study using intact soybean plants from which single-rooted soybean leaves can be made has also suggested the same regulatory mechanism for leaf photosynthesis under excessive photosynthetic source capacity. It is therefore concluded that for efficient improvement of plant matter (biomass) production, well-balanced improvement of source and sink would be essential. Further studies are desired for more complete understanding of the regulatory mechanism for leaf photosynthesis under excessive photosynthetic source capacity and its application.

6. References

Anwaruzzaman, Sawada, S., Usuda, H., Yokota, A. (1995). Regulation of ribulose-1,5-bisphosphate carboxylase/oxygenase activation by inorganic phosphate through stimulating the binding of the activator CO_2 to the activation sites. *Plant & Cell Physiology*, 36: 425-433.

Baena-Gonzalez, E., Rolland, F., Thevelein, J.M., Sheen, J. (2007). A central integrator of transcription networks in plant stress and energy signaling. *Nature*, 448: 938-942.

Bhagwat, A.S. (1981). Activation of spinach ribulose-1,5-bisphosphate carboxylase by inorganic phosphate. *Plant Science Letters*, 23: 197-206.

Bradford, M.M. (1976). A rapid and sensitive method for the quantitation of microgram quantities of protein utilizing the principle of protein-dye binding. *Analytical Biochemistry*, 72: 248-254.

Bradley, F.M., Janes, H.W. (1985). Carbon partitioning in tomato leaves exposed to continuous light. *Acta Horticulturae*, 174: 293-302.

Bredmose, N.B., Nielsen, K.L. (2009). Controlled atmosphere storage at high CO_2 and low O_2 levels affects stomatal conductance and influences root formation in kalanchoe cuttings. *Scientia Horticulturae*, 122: 91-95.

Brodribb, T.J., McAdam, S.A.M. (2011). Passive origins of stomatal control in vascular plants. *Science*, 331: 582-585.

Brooks, A., Portis, A.R. (1988). Protein-bound ribulose bisphosphate correlates with deactivation of ribulose bisphosphate carboxylase in leaves. *Plant Physiology*, 87: 244-249.

Chen, J., Wang, P., Mi, H.L., Chen, G.Y., Xu, D.Q. (2010). Reversible association of ribulose-1,5-bisphosphate carboxylase/oxygenase activase with the thylakoid membrane depends upon the ATP level and pH in rice without heat stress. *The Journal of Experimental Botany*, 61: 2939-2950.

Cheng, Y., Arakawa, O., Kasai, M., Sawada, S. (2008). Analysis of reduced photosynthesis in the apple leaf under sink-limited conditions due to girdling. *Journal of the Japanese Society for Horticultural Science*, 77: 115-121.

Cheng, Y., Arakawa, O., Sawada, S. (2004). A mechanism of photosynthetic inhibition in sink-limited young apple trees by continuous light. *Horticultural Research*, 3: 393-398.

Crafts-Brandner, S.J., Egli, D.B. (1987). Sink removal and leaf senescence in soybean. *Plant Physiology*, 85: 662-666.

Crafts-Brandner, S.J., Salvucci, M.E. (2000). Rubisco activase constrains the photosynthetic potential of leaves at high temperature and CO_2. *Proceedings of the National Academy of Sciences U.S.A.*, 97: 13430-13435.

Cushman, K.E., Tibbitts, T.W., Sharkey, T.D. (1995). Constant-light injury of potato: temporal and spatial patterns of carbon dioxide assimilation, starch content, chloroplast integrity, and necrotic lesions. *Journal of the American Society for Horticultural Science*, 120: 1032-1040.

Dai, N., Schaffer, A., Petreikov, M., Shahak, Y., Giller, Y., Ratner, K., Levine, A., Granot, D. (1999). Overexpression of *Arabidopsis* hexokinase in tomato plants inhibits growth, reduces photosynthesis, and induces rapid senescence. *The Plant Cell*, 11: 1253-1266.

Globig, S., Rosen, I., Janes, H.W. (1997). Continuous light effects on photosynthesis and carbon metabolism in tomato. *Acta Horticulturae*, 418: 141-151.

Jang, J.C., Leon, P., Zhou, L., Sheen, J. (1997). Hexokinase as a sugar sensor in higher plants. *The Plant Cell*, 9: 5-19.

Jordan, D.B., Chollet, R. (1983). Inhibition of ribulose bisphosphate carboxylase by substrate ribulose-1,5-bisphosphate. *The Journal of Biological Chemistry*, 258: 13752-13758.

Kasai, M. (2008). Regulation of leaf photosynthetic rate correlating with leaf carbohydrate status and activation state of Rubisco under a variety of photosynthetic source/sink balances. *Physiologia Plantarum*, 134: 216-226.

Mackinney, G. (1941). Absorption of light by chlorophyll solutions. *The Journal of Biological Chemistry*, 140: 315-322.

Makino, A., Mae, T., Ohira, K. (1986). Colorimetric measurement of protein stained with Coomassie Brilliant Blue R on sodium dodecyl sulfate-polyacrylamide gel electrophoresis by eluting with formamide. *Agricultural and Biological Chemistry*, 50: 1911-1912.

Martin, T., Oswald, O., Graham, I.A. (2002). *Arabidopsis* seedlings growth, storage mobilization, and photosynthetic gene expression are regulated by carbon: nitrogen availability. *Plant Physiology*, 128: 472-481.

McCurry, S.D., Pierce, J., Tolbert, N.E., Orme-Johnson, W.H. (1981). On the mechanism of effector-mediated activation of ribulose bisphosphate carboxylase/oxygenase. *The Journal of Biological Chemistry*, 256: 6623-6628.

Mondal, M.H., Brun, W.A., Brenner, M.L. (1978). Effects of sink removal on photosynthesis and senescence in leaves of soybean (*Glycine max* L.) plants. *Plant Physiology*, 61: 394-397.

Moore, B., Zhou, L., Rolland, F., Hall, Q., Cheng, W.H., Lui, Y.X., Hwang, J., Jones, T., Sheen, J. (2003). Role of the *Arabidopsis* glucose sensor HXK1 in nutrient, light, and hormonal signaling. *Science*, 300: 332-336.

Murage, E.N., Watashiro, N., Masuda, M. (1996). Leaf chlorosis and carbon metabolism of eggplant in response to continuous light and carbon dioxide. *Scientia Horticulturae*, 67: 27-37.

Murage, E.N., Watashiro, N., Masuda, M. (1997). Influence of light quality, PPFD and temperature on leaf chlorosis of eggplants grown under continuous illumination. *Scientia Horticulturae*, 68: 73-82.

Okita, T.W., Sun, J., Sakulringharoj, C., Choi, S.B., Edwards, G.E., Kato, C., Ito, H., Matsui, H. (2001). Increasing rice productivity and yield by manipulation of starch synthesis. *Novartis Found Symp*, 236: 135-146 (discussion 147-152).

Paul, M.J., Foyer, C.H. (2001). Sink regulation of photosynthesis. *The Journal of Experimental Botany*, 52: 1383-1400.

Paul, M.J., Pellny, T.K. (2003). Carbon metabolite feedback regulation of leaf photosynthesis and development. *The Journal of Experimental Botany*, 54: 539-547.

Raines, C.A. (2003). Minireview The calvin cycle revisited. *Photosynthesis Research*, 75: 1-10.

Raines, C.A. (2006). Transgenic approaches to manipulate the environmental responses of the C_3 carbon fixation cycle. *Plant Cell & Environment*, 29: 331-339.

Raines, C.A. (2011). Increasing photosynthetic carbon assimilation in C_3 plants to improve crop yield: Current and future strategies. *Plant Physiology*, 155: 36-42.

Rowell, T., Mortley, D.G., Loretan, P.A., Bonsi, C.K., Hill, W.A. (1999). Continuous daily light period and temperature influence peanut yield in nutrient film technique. *Crop Science*, 39: 1111-1114.

Sage, R.F., Sharkey, T.D., Seemann, J.R. (1989). Acclimation of photosynthesis to elevated CO_2 in five C_3 species. *Plant Physiology*, 89: 590-596.

Sawada, S., Arakawa, O., Muraki, I., Echigo, H., Miyashita, M., Iwafune, M., Kasai, M. (1999). Photosynthesis with single-rooted *Amaranthus* leaves. I. Changes in the activities of ribulose-1,5-bisphosphate carboxylase and phosphoenolpyruvate carboxylase and the amounts of intermediates in photosynthetic metabolism in response to changes in the source-sink balance. *Plant & Cell Physiology*, 40: 1143-1161.

Sawada, S., Hasegawa, Y., Kasai, M., Sasaki, M. (1989). Photosynthetic electron transport and carbon metabolism during altered source/sink balance in single-rooted soybean leaves. *Plant & Cell Physiology*, 30: 691-698.

Sawada, S., Hayakawa, T., Fukushi, K., Kasai, M. (1986). Influence of carbohydrates on photosynthesis in single, rooted soybean leaves used as a source-sink model. *Plant & Cell Physiology*, 27: 591-600.

Sawada, S., Sato, M., Kasai, A., Yaochi, D., Kameya, Y., Matsumoto, I., Kasai, M. (2003). Analysis of the feed-forward effects of sink activity on the photosynthetic source-sink balance in single-rooted sweet potato leaves. I. Activation of RuBPcase through the development of sinks. *Plant & Cell Physiology*, 44: 190-197.

Sawada, S., Usuda, H., Hasegawa, Y., Tsukui, T. (1990). Regulation of ribulose-1,5-bisphosphate carboxylase activity in response to changes in the source/sink balance in single-rooted soybean leaves: the role of inorganic orthophosphate in activation of the enzyme. *Plant & Cell Physiology*, 31: 697-704.

Sawada, S., Usuda, H., Tsukui, T. (1992). Participation of inorganic orthophosphate in regulation of the ribulose-1,5-bisphosphate carboxylase activity in response to changes in the photosynthetic source-sink balance. *Plant & Cell Physiology*, 33: 943-949.

Setter, T.L., Brun, W.A. (1980). Stomatal closure and photosynthetic inhibition in soybean leaves induced by petiole girdling and pod removal. *Plant Physiology*, 65: 884-887.

Setter, T.L., Brun, W.A., Brenner, M.L. (1980). Effect of obstructed translocation on leaf abscisic acid, and associated stomatal closure and photosynthesis decline. *Plant Physiology*, 65: 1111-1115.

Smidansky, E.D., Clancy, M., Meyer, F.D., Lanning, S.P., Blake, N.K., Talbert, L.E., Giroux, M.J. (2002). Giroux, Enhanced ADP-glucose pyrophosphorylase activity in wheat endosperm increases seed yield. *Proceedings of the National Academy of Sciences U.S.A.*, 99: 1724-1729.

Smidansky, E.D., Meyer, F.D., Blakeslee, B., Weglarz, T.E., Greene, T.W., Giroux, M.J. (2007). Expression of a modified ADP-glucose pyrophosphorylase large subunit in wheat seeds stimulates photosynthesis and carbon metabolism. *Planta*, 225: 965-976.

Stessman, D., Miller, A., Spalding, M., Rodermel, S. (2002). Regulation of photosynthesis during *Arabidopsis* leaf development in continuous light. *Photosynthesis Research*, 72: 27-37.

Tibbitts, T.W., Bennett, S.M., Cao, W. (1990). Control of continuous irradiation injury on potatoes with daily temperature cycling. *Plant Physiology*, 93: 409-411.

von Caemmerer, S., Evans, J.R. (2010). Enhancing C_3 photosynthesis. *Plant Physiology*, 154: 589-592.

Ward, J.K., Tissue, D.T., Thomas, R.B., Strain, B.R. (1999). Comparative responses of model C_3 and C_4 plants to drought in low and elevated CO_2. *Global Change Biology*, 5: 857-867.

Wheeler, G.M., Tibbitts, T.W. (1986). Growth and tuberization of potato (*Solanum tuberosum* L.) under continuous light. *Plant Physiology*, 80: 801-804.

Wittenbach, V.A. (1982). Effect of pod removal on leaf senescence in soybeans. *Plant Physiology*, 70: 1544-1548.

Wittenbach, V.A. (1983). Effect of pod removal on leaf photosynthesis and soluble protein composition of field-grown soybeans. *Plant Physiology*, 73: 121-124.

Xu, D.Q., Gifford, R.M., Chow, W.S. (1994). Photosynthetic acclimation in pea and soybean to high atmospheric CO_2 partial pressure. *Plant Physiology*, 106: 661-671.

Zhou, Y.H., Yu, J.Q., Mao, W.H., Huang, L.F., Song, X.S., Nogues, S. (2006). Genotypic variation of Rubisco expression, photosynthetic electron flow and antioxidant metabolism in the chloroplast of chilled-exposed cucumber plants. *Plant & Cell Physiology*, 47: 192-199.

22

Soybean Cyst Nematode (*Heterodera glycines* Ichinohe)

Qing Yu
Eastern Cereal and Oilseed Research Centre
Agriculture and Agric-Food Canada, Ottawa, ON
Canada

1. Introduction

Soybean cyst nematode (SCN) (*Heterodera glycines* Ichinohe) is the most serious nematode pest on soybean in the world, infests most of the soybean producing countries of the world with the exception of west European countries and Oceania countries and causes up to 1.5 billions US$ economical loss according to some estimates (Wrather et al., 2001). The cyst nematode is also an international quarantined pest. Although it was first discovered and described in Japan in 1952, it is now widely believed originated in China as its host soybean was. In America, since the first report in North Carolina, USA in 1955, it has spread to 26 out of 28 soybean producing states, to the province of Ontario, Canada, and to the soybean producing countries of South America. Race of the nematode was recognized in 1954, and a total of 14 races were reported and widely distributed, especially in the USA, which has created a series of problems for developing resistant cultivars. As the climate change intensifies, it is likely that this nematode pest is going to spread to new soybean producing areas. Many resistant cultivars have been developed, especially in USA where resistant cultivars were developed using resistant parents selected from resistant plant introductions (PI) of the exotic accessions in the USDA soybean germplasm collection. These resistant PIs were collected from the oriental countries. Since the 80s, international seed companies like Pioneer, and Monsanto have been the driving force for the resistant cultivar development and marketing. SCN resistant cultivars alone used to be the solution for the control of the nematode pest for soybean production in USA. Because of the new emerging races and the shifting between existing races, resistant cultivars in many cases lose their usefulness dramatically. With the new realization that the agriculture biodiversity plays an essential role for pest management, new control methods have been developed and tested such as rotating with nonhost crops, planting multiline cultivars mixtures, using biological control agents, and applying green manure. There are a lot of literatures related with SCN, one book by Schmitt et al, 2004, and a review by Noel 1993 nevertheless are the excellent sources of information. A few non scientific aspects are also used for the synthesis of this paper. China, where the nematode is believed originated, becomes economically integrated into the world system at a pace never seen before. Soybeans there as a crop are shifting from vegetable to pulse, and to oil seeds. Soybean seeds are increasingly produced, controlled and marketed by a few international companies, as the result, fewer cultivars (more monocultures) are planted at a given year compared with the time when farmers used to get their seeds from all sorts of channels, and productions are in large scales. The climate change, especially the global warming caused by human activities is inevitably impacting the soybean production, and the soybean cyst nematode.

2. Taxonomic position

When the SCN was discovered, it was believed as a race of *Heterodera schachtii,* the sugar beet cyst nematode, since these 2 species are closely related biologically. Around that time, cyst nematodes were generally considered races of *H. schachtii.* In 1940 Franklin's comparative morphological studies led to many "races" being elevated to species. It was morphologically consistent that the morphological distinctions of the SCN led Ichinohe to elevate the race to a new species in 1952.

Phylum Nematoda,
- Class Secernentea,
  - Order Tylenchida,
    - Suborder Hoplolaimina,
      - Superfamily Hoplolaimoidea,
        - Family Heteroderidae,
          - Genus *Heterodera,*
            - Species *Heterodera glycines* (Ichinohe 1952)

3. Morphology and identification

H. glycines is a typical cyst forming nematode within the family of Heteroderidae: characterized by sexual dimorphism: male is vermiform, while female lemon shaped (Figure 1). The brown cyst is the dead female with viable eggs inside. The second stage juvenile (J2) is vermiform, much smaller than the male. The length of the stylet, and the hyaline tail terminus of J2, and the characters of the vulva cone of the cyst are the most important characters for the identification (Table 1). More than one reference descriptions are usually required for comparison because there are variance among isolates from different crops and locations (Mulvey & Golden 1983, Wouts 1985, Golden 1986, Burrows & Stone 1985, Tylor 1975, Hesling 1978, Graney & Miller 1986).

Fig. 1. Morphology of *Heteridera glycines*: A: male; B: J2; C: head of J2; D: cyst; E: hyaline tail terminus

Character		Measurment (μ)	
Cyst		Average	Range
Fenestra	length	55	30-70
	width	42	25-60
Vulval slit	length	53	43-60
J2			
Body	length	440	375-540
Stylet	length	23	22-24
Tail	length	50	40-61
	hyaline length	27	20-30

Table 1. Measurement of *Heterodera glycines* (after Tylor, 1975, Graney & Miller, 1982)

4. Biology and life cycle

After the death of the female, the eggs are retained inside the hardened body (cyst), until suitable conditions arrive. The cysts can remain viable for several years in the soil. The eggs hatch to juveniles at stage 2, stimulated by exudates from the roots (Masamune et al., 1982). The 2nd stage juveniles (J2) of *H. glycines* are the only stage that the nematode can penetrate the root near the tip. The J2 once inside the root, become sedentary and establish a syncytic feeding site (Moore, 1984). J2 swells, and moults to J3, J4, and become adults (Wyss and Zunke, 1992). The life cycle is usually from 21 to 24 days (Fig. 2). Time required for the nematode to complete its life cycle is usually from 20-25 days at 20-24 °C, the lower the temperature is, the longer the time it takes to finish its life cycle (Melton et al., 1986).

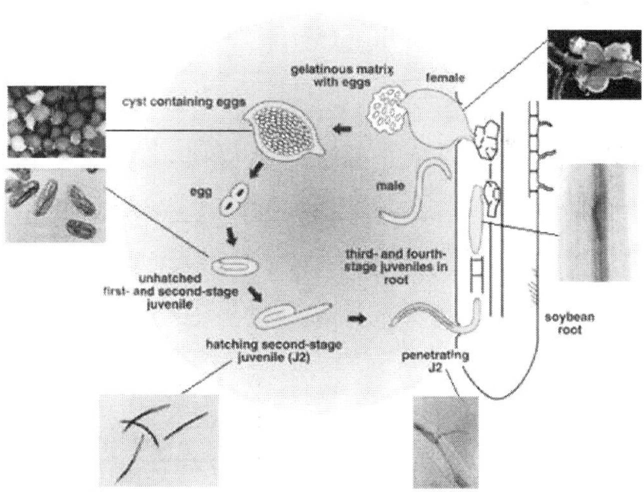

Fig. 2. Life cycle of *Heterodera glycines* (courtesy of Dirk Charlson, Iowa State University)

5. Distribution

It is impossible to know exactly the source(s) of infestation of the nematode species. More and more people believe that *H. glycines* is very likely a native of China, for two compelling evidences: 1. One of the most important hosts for the nematode the soybean originated (domesticated 5,000 years ago, the early stage of the Chinese civilization) (Qiu et al., 2011, Liu et al., 1997) in China, and 2. Most of the resistant cultivars used today have their roots in cultivars from China (Bernard et al., 1988). Probably the spreading and the pathway of the nematode followed footsteps of its host soybean (Fig. 3). The soybean was introduced to Japan, Korea around 300 AD, the nematode was discovered in 1915 (Hori), and later was described by Ichinohe in 1952 with the type locality in Hokkaido. It was first found in the United States in 1954 (Winstead et al., 1955) and spread with the expansion of soybean growing areas such as in Canada (Anderson et al., 1988). The nematode was also found in Colombia in the 1980s, and more recently in the major soybean producing areas in Argentina and Brazil (Mendes & Dickson 1992). SCN has also been reported from Iran and Italy (Fig. 3).

Africa: Egypt (unconfirmed)

Asia: China (Anhui, Hebei, Hubei, Heilongjiang, Henan, Inner Mongolia, Jiangsu, Jilin, Liaoning, Shanxi, Shandong), Indonesia (Java), Korean peninsula, Japan, Taiwan (unconfirmed), Russia (Amur District in the Far East).

North America: Canada (Ontario), USA (Alabama, Arkansas, Delaware, Florida, Georgia, Illinois, Indiana, Iowa, Kansas, Kentucky, Louisiana, Maryland, Minnesota, Michigan, Mississippi, Missouri, Nebraska, New Jersey, North Carolina, North Dakota, Ohio, Oklahoma, Pennsylvania, South Carolina, South Dakota, Tennessee, Texas, Virginia and Wisconsin).

South America: Argentina, Brazil, Chile, Columbia, Ecuador.

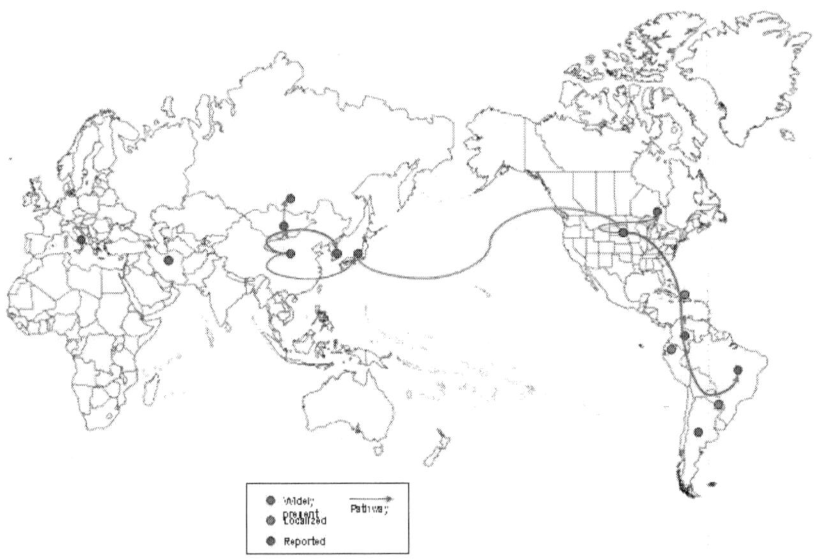

Fig. 3. Distrubition map of *Heterodera glycines* in the World.

6. Disease symptoms

"Yellow dwarf" is a good description of the above ground symptoms when soybeans are infested with the soybean cyst nematode. When soybeans are heavily infested, the plants usually become stunt (Fig. 4). Low level infestation usually does not produce obvious symptoms above ground. Belowground symptoms include poorly developed and darkened roots, reduced nodule formation.

Fig. 4. Above ground symptoms of soybeans infested by soybean cyst nematode

7. Spread

7.1 Canada

Since it was first identified in 1987, SCN has been identified in 12 counties in Ontario, Canada. Infected counties include Essex, Kent, Lambton, Elgin, Perth, Haldimand-Norfolk, Middlesex, Glengarry, Prescott, Stormont, Huron and Oxford. It is obvious that *H. glycines* has been spreading north and northeast wards (Fig. 5). It would be difficult to exclude the possibility that climate change is the cause, or at least one of the causes. This finding proves what Boland et al. 2004 has predicated. It is likely to spread along the St. Lawrence seaway towards the Maritime provinces as new cultivars suitable for the cold climate being developed and planted in the region.

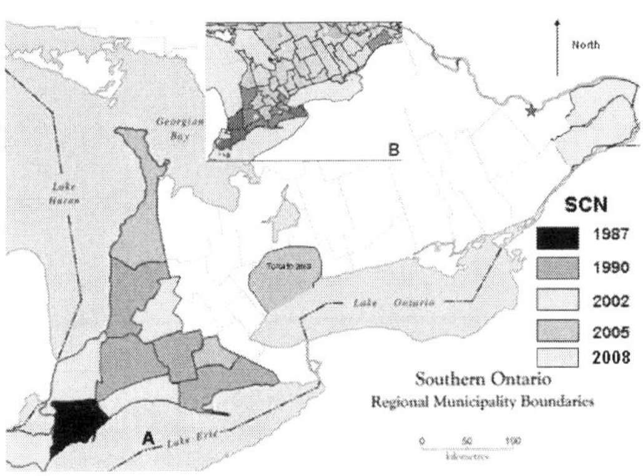

Fig. 5. Distribution of *Heterodera glycines* in Ontario, Canada from 1987 to 2007 (insert B: soybean growing area in Ontario).

7.2 USA

Since *H. glycines* was first discovered in North Caroline in 1955. Within the next 6 years, it had been reported in 7 states along the Mississippi river (Riggs, 2004). Today, it has spread to 29 soybean producing states in USA, as far northeast as the state of New Jersey (Riggs, 2004) (Fig. 6).

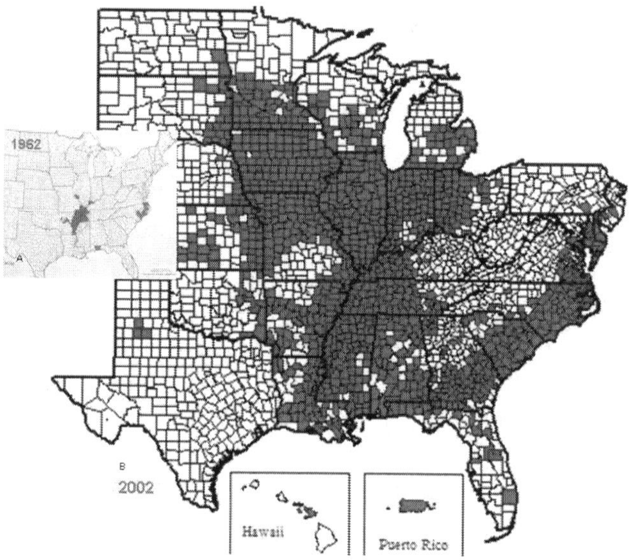

Fig. 6. Distribution map of *Heterodera glycines* in USA: A: distribution in 1962; B: distribution in 2002.

8. Host

The soybeans are the most economic important host for *H. glycines*. It has a broad host range, up to 100 plant species world wide, some selected common hosts are listed in Table 2 (Baldwin & Mundo-Ocampo, 1991), especially some legumes including beans, vetch, clover, and pea. It also attacks many species of weeds (Riggs & Hamblen, 1966). These weeds in the fields must be taken into consideration for any management measures. Some nonhosts are also listed that may be used in rotation.

9. Race

Variability in virulence among populations of soybean cyst nematode was recognized by researchers soon after the nematode was discovered in USA. The term race has been adopted by an *ad hoc* committee of the Society of Nematologists to designate the SCN populations differ in their ability to develop on a set of differential soybean cultivars (Table 3). This scheme was developed in 1970 (Golden et al., 1970) and later was refined by (Riggs & Schmitt 1988). The scheme defines 16 possible races. In 2002, HG type was introduced to more adequately define the diversity in virulence phenotypes (Niblack et al., 2002).

Host species	Non-host
Soybeans	Alfalfa
Green beans	Barley
Snap beans	Canola
Dry beans	Red clover
Red beans	White clover
Lima beans	Ladino clover
Mung beans	Oats
Bush beans	Rye
Adzuki beans	Sorghum
Graden peas	Wheat
Cowpeas	
Corn	
Vetch (common, hairy)	
Lespedeza (common, Korean, round bush, sericea)	
Birdsfoot-trefoil	
Sweetclover	
White lupines	

Table 2. A list of hosts and nonhosts for *Heterodera glycines*.

RACE	Pickett	Peking	PI 88788	PI 90763
1	-	-	+	-
2	+	+	+	-
3	-	-	-	-
4	+	+	+	+
5	+	-	+	-
6	+	-	-	-
7	-	-	+	+
8	-	-	-	+
9	+	+	-	-
10	+	-	-	+
11	-	+	+	-
12	-	+	-	+
13	-	+	-	-
14	+	+	-	+
15	+	-	+	+
16	-	+	+	+

A "+" rating is given if the number of females produced by a soybean cyst nematode population on the soybean cultivar is equal to or greater than 10% of the number produced on the susceptible cultivar Lee. If it is less than 10%, a "-" rating is given.

Table 3. Race classification scheme for *Heterodera glycines*.

Races 1, 2, 3, 4 were described in 1970 (Golden et al., 1970), race 5 in 1979 from Japan (Inagaki, 1979), and race 7 from China (Chen et al., 1988). Today a total 14 races have been reported (Table 4). Only races 12 and 16 have not been reported. USA has the most races. Race 3 is the most common race in the world.

With relative short history of growing soybean (100-200 years), the fact that USA has the highest number of races for the nematode indicates that the races were probably the results of its widespread planting of the resistant cultivars. Argentina (Doucet et al., 2008) and Brazil (Dias et al., 1998) both have a very short history of growing soybean (less than 50 years) have 6, and 9 races respectively. Very likely these races have been imported through seed stocks and on used machinery from several sources in USA. In comparison, the oriental countries, Korea (Kim et al., 1998), Japan (Ichinohe, 1988) and China (Liu et al., 1998) where soybean has been growing for several thousands years, have relative fewer races, because the modern practice of resistant cultivar selection and development is relatively new. Races 8, 11, 13, and 15 have only been found in USA. Lack of race 4 in both Japan and Korean was a surprise.

Country	Races	Total Races	Dominate Race
Argentina	1,3,5,6,9,14	6	3
Brazil	1,2,3,4,5,6,9,10,14	9	3
Canada	1,2,3,5,6	5	3
China	1,2,3,4,5,6,7,14	8	3 Northern, 4 Southern
Japan	1,3,5	3	3
Korea	1,3,5,6	4	3
USA	1,2,3,4,5,6,7,8,9,10,11,13,14,15	14	3

Table 4. Distribution of races of *Heterodera glycines* in the world

In USA, the most prevalent race for the northern states is the race 3, while the race 6 is the most common race. That difference of variability was noted along the 35 ^{0} N latitude line. Niblack and Riggs (2004) postulated that is likely caused by the history of using different resistant cultivars.

In central China lies between the Yellow and Yangtze rivers which include the province of Shandong, Anhui, Jiangsu, and Henan, and Shanxi, where the soybean may have originated, where the soybeans are planted in the summer and harvested in the fall (short growing season), the most common race is race 4 (Lu et al., 2006), race 3 has not been reported in that region. In northern China includes the province of Heilongjiang, Jilin, Liaoning, and Inner Mongolia where the soybeans are sown in spring and harvested in fall (long growing season) the prevalent race is race 3 and race 4 has only been reported in one case (Dong et al. 2008) (Fig. 7). China is only country that has these 2 races geologically separated. In the

Fig. 7. The race 3 prevalent region and the race 4 prevalent region of *Heterodera glycines* in China

race classification scheme, the race 3 and the race 4 are opposite to each other, all the 4 cultivars are resistant to the race 3 but susceptible to the race 4. The cultivars have their resistant genes rooted in the cultivars collected from the northern region of China where the race 3 is the prevalent. It is likely that these 2 races are the 2 original races and are native to these 2 regions respectively in China, with race 4 is the ancestral to the race 3, since the cradle of the ancient Chinese civilization happened to be in the race 4 region, and ancient Chinese civilization was an agricultural civilization. The north region (race 3 region) was not an ancient agricultural area, rather a nomadic region.

10. Management

10.1 Resistant cultivar

A search for sources of SCN resistance led to the evaluation of large number of the plant introduction (PI) from among exotic accession in the USDA soybean germplasm collection. Five accessions were selected as parents. The first SCN resistant cultivar "Picket" with Peking as the source for resistance, was breed and released in 1966 in USA (Brim & Ross, 1966), hundreds have been developed for all the Maturity Groups. In 1970, field populations were found readily reproduce on Picket, and Peking. That finding led to the search for another source of resistance. Chosen from USDA soybean germplasm collection, PI 88788 was used for the development of "Bedford" (Hartwig & Epps, 1978). With the widespread deployment of the SCN resistant cultivars, new races emerged. This has been the story for SCN resistant cultivar development in USA. Since the 80s, the soybean seed breeding has been transferred from public institutes to private company. At much later stage, some accessions of the Chinese soybean germplasm collection were identified have resistance. At present, Roundup ready cultivars developed by Monsanto are the most widely used cultivars in USA, and elsewhere.
Relying on few resistant cultivars alone for SCN control had been proven misguided, as the high number and shifting of the races in USA indicated (Young, 1992).
In North America, the basic management tactics of planting resistant cultivars at different fashions, and rotating with non-hosts will continue to be the main methods to manage the SCN problem, even though the tactics face great challenges of the shitting of the nematode races, and of the uneconomical of the non-hosts (Niblack & Chen, 2004).

10.2 Using mutiline cultivars for SCN management

Probably, it is hard to argue against the fact that monoculture farming has been one of causes for disease and pest epidemics, the best example is without doubt the Irish Potato Famine caused by the potato late blight disease (*Phytophthora infestans*). Recent cases of other invasive alien species such as the Dutch Elm disease also remind us that biodiversity is very important in fighting pests and diseases. The crop biodiversity used to be a norm practice before the modern agriculture (few cultivars, and a few pesticides), each farmer had to grow different kinds of crops for all household needs (grains, vegetable, and others). The usefulness of mixture of multiline cultivars and cultivars mixtures for disease control has been well documented (Mundt, 2002, Wolfe, 1985). The recent successful cases of using multiline cultivars or cultivar mixtures for controlling diseases, such as potato late blight on potato (Garrett & Mundt, 1999), on barley (Wolfe et al. 1981), on rice in China (Zhu et al., 2000) demonstrated that the practical difficulties associated with the mixtures have been overestimated. This concept has not been carefully tested for SCN control. In the few tested cases, the mixtures were not superior to the resistant cultivars in terms of their yield increasing

(Young & Hartwig, 1988). More studies are recommended. As Mundt (2002) demonstrated that for biodiversity to be functional, there must be an appropriate match between the resistant genes in a mixture and the virulence genes present in the target pathogens or parasites.

10.3 Cover crop

Cover crops are commonly used to prevent soil erosion. These crops are usually planted in rotation with primary crops. When the cover crops are incorporated into the soil at the certain stage of the growing season, this practice is being referred as green manure. A major benefit obtained from green manures is the addition of organic matter to the soil, which increases the food supply for macro, and micro organisms in the soil resulting increased biodiversity in soil. There is a lot of information on the benefit effects of soil biodiversity on disease control (Brussaard et al., 2007).

This agriculture practice with certain crops which contain nematicidal compounds is especially interesting. Marigold, especially French marigold (*Tagetes patula*) has been shown reduced the populations in soil of several root-knot nematodes, and root lesion nematodes (Motsinger et al., Ploeg, 2000, Pudasaini, 2007). Castor beans, sesame, Sudan grass, sorghum, and Crucifers have all shown are toxic against plant parasitic nematodes. Among them, plants from Brassica have received considerable attention for their possibility in controlling plant parasitic nematodes by incorporating them into soil (Mojtahedi et al., 1993, Potter et al., 1998). The principle reason is that glucosinolates which exist in these plants convert upon decomposition to isothiocyanates, a group of chemicals proven to have a wide spectrum of biological activities, including nematicidal activity (Brown & Morra 1997), a few these chemicals are volatile, the practice has been referred "Biofumigation". Among these converted isothiocynates, allyl isothiocyanate (AITC) has been proven as being the most toxic against *H. glycines* (Lazzeri et al. 1993). AITC is the decomposition product of allyl glucosinolate (generally called sinigrin), which exists in plants of *Armoracia lapathifolia, Brassica carinata, B. juncea, B. napus, B. oleracea,* and *Peltaria alliacea* (Brown & Morra, 1997). Among them, mustards have been cited most promising, especially the oriental mustard (Brassica juncea) which contains highest concentration of Ally isothiocynate (AITC) in plant (Tsao et al., 2000). AITC toxicity was found highly selective, was highly toxic against J2 of *H. glycines*, but less toxic on

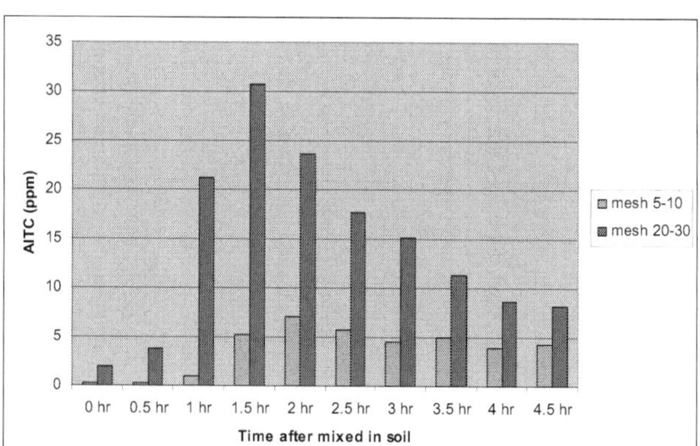

Fig. 8. Effect of particle size of mustard materials on AITC releasing in soil

free-living nematodes, AITC also inhibited the egg hatching of the nematode (Yu et al., 2005). Some materials from this oriental cultivar have been demonstrated effective in reducing population of *Pratylecnhus penetrans* in soil (Yu et al., 2007 a, 2007b).

Recently using mustard such as oil radish, or other mustard related crops as a cover crop for controlling *H. glycines* have tested, but the results have not been conclusive. The potential factors that caused the inconsistency includes: 1) targeted nematode species; 2) mustard varieties; and 3) environmental factors. In another study we found that the particle size had dramatic effect on releasing the AITC in to the soil (Fig. 8). It is likely that with a mustard variety of high AITC concentration, and plant tissue macerated to very fine particles, mustard crops as cover crops for the SCN control can be an effective method.

11. Concluding remarks

The soybean cyst nematode is more likely going to spread to new soybean growing areas around the world as the climate change intensifies, and as the world becomes more integrated. The soybean seeds are more than ever developed and marketed by a few international companies, the soybean farming practices in the world will become more and more uniform, less diverse unfortunatly. New races could emerge. This creates greater challenges for managing the SCN.

The whole genome sequencing project of *H. glycines* has been completed by Monsanto Company and Divergence. The sequencing information although has been submitted to Genbank, it remains inaccessible to the public. DOE Join Genomic Institute led by Kris Lambert and Matthew E. Hudson (Univ. of Illinois at Urbana-Champaign) is in the process of sequencing the pest as well, in a hope that it will lead us to learn more about the races, and to find new ways for the controlling of the pest.

There are a few soybean germplasm collections in the world with the USDA collection being the largest, the holding information is accessible to the public. The Chinese soybean germplasm collection holds 6644 accessions of *Glycine soja*, a potential rich pool of source of resistance. Collaborations between the collections such as sharing information and germ lines are essential. The management of SCN must not rely on a few resistant cultivars. An integrated approach involving several cultivars, rotating with nonhosts, and cultivation practices that encourage biodiversity in the soil must be the future.

12. Acknowledgment

Appreciation goes to Dr. W. Ye of Nematode Assay Section, Agronomic Division, North Carolina Department of Agriculture & Consumer Services, NC, USA for providing some of the pictures.

13. References

Anderson, T. R., Welacky, T.W., Olechowski, T. H., Ablett, G., & Ebsary, B.A. (1988). First report of *Heterodera glycines* in Ontario, Canada. *Plant Disease* 72: 453.

Baldwin, J. G. & Mundo-Ocampo, M. (1991). Heteroderinae, cyst-and non-cyst- forming nematodes, In: *Manual of Agricultural Nematology*, WR Nickle, (Ed.), pp. 275-362, New York: Marcel Dekker Inc.

Boland, G.J., Melzer, M.S., Hopkin, A., Higgins, V., & Nassuth A. (2004). Climate change and plant disease in Ontario. *Canadian Journal of Plant Pathology* 26: 335-350

Brim, C.A., & Ross, J.P. (1966). Registration of Pickett soybeans. *Crop Science* 6: 305.

Brown P. D. & Morra, M.J. (1997). Control of soil-borne plant pests using glucosinolate-containing plants. *Advances in agronomy* 61:167-231.

Brussaard, L., de Ruiter, P., & Brown, G.G. (2007). Soil biodiversity for agricultural sustainability. *Agriculture, Ecosystems and Environment* 121:233-244.

Burrows, P.R. & Stone, A.R. (1985). *Heterodera glycines. CIH Descriptions of Plant- Parasitic Nematodes* No. 118. CAB International, Wallingford, UK.

CABI, http://www.cabi.org/isc/?compid=5&dsid=27027&loadmodule=datasheet&page=481&site=144.

Chen, P.H., Zhang, D.S. & Chen, S.Y. (1988). First report on a new physiological race (race 7) of soybean cyst nematode (*Heterodera glycines*). *Journal of the Chinese Academy of Agricultural Sciences* 20: 94.

Dias, W.P., Silva, J.F.V, & Pereira, J.E. (1998). Survey of *Heterodera glycines* races in Brazil, harvest 1997/98. *Nematologia Brasileira* 22: 33.

Dong, L.M., Xu, Y.L., Li, C.J., Pan, F. J., Xie, Y.J., Han, Y.B., Teng, W.L., & Li, W.B. (2008). Cyst density and subspecies identification of soybean cyst nematode in Heilongjiang province. *Chinese Journal of Oil Crop Sciences* 30:108-111.

Doucet, M. E., Paola L. & Coronel, N. (2008). The soybean cyst nematode *Heterodera glycines* Ichinohe in Argentina, In: *Integrated Management and Biocontrol of Vegetable and Grain Crops Nematodes*, A. Ciancio and K. G. Mukerji (ed.),. Springer the Netherland.

Garrett, K.A., & Mundt, C. C. (1999). Epidemiology in mixed host populations. *Phytopathology* 89: 984-990.

Golden, A. M. (1986). Morphology and identification of cyst nematodes, In: *Cyst Nematodes* Lamberti F, Taylor CE, (Ed.), pp. 23–45, Plenum press, New York, USA.

Golden A.M., Epps J.M., Riggs R.D., Duclos L.A., Fox J.A., & Bernard R.L. (1970). Terminology and identity of infraspecific forms of the soybean cyst nematode (*Heterodera glycines*). *Plant Disease Reporter* 54: 544-546.

Graney, L.S.O. & Miller, L.I. (1982) Comparative morphological studies of *Heterodera schachtii* and *H. glycines*. In: *Nematology in the Southern region of the United States* Riggs, R.D. (Ed.) pp 96-107, Southern Cooperative Series Bulletin No. 276

Inagaki, H. (1979). Race statues of five Japanese populations of *Heterodera glycines*. *Japanese Journal of Nematology* 9: 1-4.

Hartwig, E.E. & Epps, J.M. (1978). Registration of Bedford soybeans. *Crop Science* 18:915.

Hesling, J.J. (1978). Cyst nematodes: morphology and identification of *Heterodera, Globodera* and. *Punctodera*. In: *Plant nematology*, Southey, J.F. (Ed.), pp 125-155, HMSO, London, UK.

Hori, S. (1915). Phytopathological notes: sick soil of soybean caused by a nematode (in Japaneses). *Byotyugai-Zasshi* 2: 927-930.

Ichinohe, M. (1952). On the soy bean nematode, *Heterodera glycines* n. sp., from Japan. *Oyo-Dobutsugaku-Zasshi* 17: 4.

Ichinohe M. (1988). Current research on the major nematode problems in Japan. *Journal of Nematology* 20: 184-190.

Kim, D.G., Lee, J.K., & Lee, Y.K. 1999. Distribution of races of soybean cyst nematode in Korea. *Korea Journal of Applied Entomology* 38: 249-253.

Liu, X.Z., Li, J.Q., & Zhang, D.S. (1997). History and statues of soybean cyst nematode in China. *International Journal of Nematology* 7: 18-25.

Lu, W.G., Gai, J.Y., & Li, W.D. (2006). Sampling survey and identification of races of soybean cyst nematode (*Heterodera glycines*) in Huang-Huai valley. *Scientia Agricultura Sinica* 39: 306-312.

Masamune, T., Anetai, M., Takasugi, M. & Katsui, N. (1982). Isolation of a natural hatching stimulus, glycinoeclepin A, for the soybean cyst nematode. *Nature* 297:495-496.

Melton, T. A. , Jacobsen, B. J. , & Noel, G. R. (1986). Effects of Temperature on Development of *Heterodera glycines* on *Glycine max* and *Phaseolus vulgaris*. *Journal of Nematology* 18: 468-474.

Mendes, M.L., & Dickson, D.W. (1992). *Heterodera glycines* found on soybean in Brazil. *Journal of Nematology* 24: 606.

Mojtahedi, H., Santo, G. S., Wilson, J. H. & Hang, A. N. (1993). Managing *Meloidogyne chitwoodi* on potato with rapeseed as green manure. *Plant Disease* 77:42-46.

Morre, W. F. (1984). *Soybean cyst nematode*. Mississippi Cooperative Extension Service.

Motsinger, R. E., Moody, E. H. & Gay, C. M. (1977). Reaction of certain French marigold (*Tagetes patula*) cultivars to three *Meloidogyne* spp. *Journal of Nematology* 9: 278.

Mulvey, R.H. & Golden, A.M. (1983). An illustrated key to the cyst-forming genera and species of Heteroderidae in the Western Hemisphere with species morphometrics and distribution. *Journal of Nematology* 15: 1-59.

Mundt, C.C. (2002). Use of multiline cultivars and cultivar mixtures for disease management. *Annual Review of Phytopathology* 40: 381-410.

Niblack, T.L., Arelli, P. R., Noel, G.R., Opperman, C.H., Orf, J., Schmitt, D.P., Shannon, J.G., & Tylka, G.L. (2002). A revised classification scheme for genetically diverse populations of *Heterodera glycines*. *Journal of Nematology* 34:279-288.

Niblack, T. L., & Chen S. Y. (2004). Cropping systems and crop management practices, In: *Biology and Management of the Soybean Cyst Nematode* pp. 181–206, D. P. Schmitt, J. A. Wrather, & R. D. Riggs, (Eds.), Schmitt & Associates of Marceline. Marceline, Missouri, USA.

Noel, G. (1993). *Heterodera glycines* in Soybean. *Nematologia Brasileira* 17: 103-121.

Ploeg, A. T. (2000). Effects of amending soil with *Tagetes patula* cv. Single Gold on *Meloidogyne incognita* infestation on tomato. *Nematology* 2: 489-493.

Potter, M. Davies, J., K. & Rathjen, A. J. (1998). Suppressive impact of glucosinolates in Brassica vegetable tissues on root lesion nematode *Pratylenchus neglectus*. *Journal of Chemical Ecology* 24:67-80.

Pudasaini, M.P., Viaene, N., & Moens, M. 2006. Effect of marigold (Tagetes patula) on population dynamics of *Pratylenchus penetrans* in a field, Nematology, Vol. 8 (4): 477-484.

Qiu, L.J., Chen, P.Y., Liu, Z. X., Li, Y. H., Guan, R. X., Wang, L. H., & Chang, R. Z. (2011). The worldwide utilization of the Chinese soybean germplasm collection. *Plant Genetic Resources: Characterization and Utilization* 9: 109-122.

Riggs, R.D. (2004). History and distribution. In: *Biology and Management of the Soybean Cyst Nematode* pp. 9-40, D. P. Schmitt, J. A. Wrather, & R. D. Riggs, (Eds.), Schmitt & Associates of Marceline. Marceline, Missouri, USA.

Riggs, R. D. , & Hamblen, M. L. (1966). Further studies on the host range of the soybean cyst nematode. *Arkansas Agriculture Experiment Station Bulletin* 718:1-19.

Riggs, R.D. & Schmitt, D.P. (1988). Complete characterization of the race scheme for *Heterodera glycines. Journal of Nematology* 20: 392-395.

Schmitt, D. P., Wrather, J. A. & Riggs, R.D. (2004). Biology and Management of the soybean Cyst Nematode (Second Edition). Schmitt & Associates of Marceline, Marceline, Missouri, USA

Shannon, J.G., Arelli, P.R., & Young, L.D. (2004). Breeding resistance and tolerance. In: *Biology and Management of the Soybean Cyst Nematode* pp. 155-180, D. P. Schmitt, J. A. Wrather, & R. D. Riggs, (Eds.), Schmitt & Associates of Marceline. Marceline, Missouri, USA.

Taylor, A.L. (1975). Identification of soybean cyst nematodes for regulatory purposes. Proceedings. *Soil and Crop Science Society of Florida* 34: 200-206

Tsao, R., Yu, Q., Friesen, I., Potter, J., & Chiba, M. (2000). Factors affecting the dissolution and degradation of oriental mustard-derived sinigrin and allyl isothiocyanate in aqueous media. *Journal of Agriculture and Food Chemistry* 48: 1998-1902.

Young, L.D. (1992). Croping sequencing effects on soybean and *Heterodera glycines. Plant Disease* 76:78-81.

Young, L.D., & Hartwig, E.E. (1988). Selection pressure on soybean cyst nematode from soybean cropping sequences. *Crop Science* 28:845-847.

Winstead, N.N., Skotland, C.B. & Sasser, J.N. (1955). Soybean cyst nematode in North Carolina. *Plant Disease Reporter* 39: 9-11.

Wolfe, M.S. (1985). The current status and prospects of multiline cultivars and variety mixtures for disease resistance. *Annual Review of Phytopathology* 23: 251-271.

Wolfe, M.S., Barrett, J.A., & Jenkin, J.E.E. (1981). The use of cultivar mixtures for disease control. In: *Strategies for the control of cereal diseases,* J.F., Jenkyn, R.T., Plumb (Eds.), pp. 73-80 Blackwell Scientific Publications, Oxford, UK.

Wouts, W. M. (1985). Phylogenetic classification of the family Heteroderidae (Nematoda: Tylenchida). *Systematic parasitology* 7:295-328.

Wrather, J. A., Anderson, T. R., Arsyad, D. M., Tan, Y., Ploper, L. D., Porta-Puglia, A., Ram, H. H., & Yorinori, J. T. (2001). Soybean disease loss estimates for the top ten soybean-producing countries in 1998. *Canadian Journal of Plant Pathology* 23:115-121.

Wyss, U., & U. Zunke. (1992). Observations on the feeding behaviour of *Heterodera schachtii* throughout development, including events during moulting. *Fundamental and Applied Nematology* 15:75-89.

Yu, Q., Tsao, R., Chiba, M. & Potter, J. (2005). Selective nematicidal activity of allyl isothiocyanate. *Journal of Food, Agriculture and Environment* 3: 218-221.

Yu, Q., Tsao, R., Chiba, M., & Potter, J. (2007a). Elucidation of the nematicidal activity of bran and seed meal of Oriental mustard (*Brassica juncea* L.) under controlled conditions. *Journal of Food, Agriculture and Environment* 5 (3&4): 374-379.

Yu, Q., Tsao, R., Chiba, M., & Potter, J.W. (2007b). Oriental mustard bran reduces Pratylenchus penetrans on sweet corn. *Canadian Journal of Plant Pathology* 29:421-426.

Zhu, Y. Y., Chen, H. R., Fan, J. H., Wang, Y. Y., Li, Y., Chen, J. B., Fan, J. X., Yang, S. S., Hu, L. P., Leung, H., Mew, .T W., Teng, P. S., Wang, Z. H., & Mundt C C. (2000). Genetic diversity and disease control in rice. Nature 406: 718-722.

23

The Asian Soybean Rust in South America

Gustavo B. Fanaro and Anna Lucia C. H. Villavicencio
Instituto de Pesquisas Energéticas e Nucleares (IPEN)
Brazil

1. Introduction

Soybean is infected by two species of fungi that cause the rust: the *Phakopsora meibione* (Arth.) Arth. (American soybean rust), which is native from American continent, existing from Puerto Rico to southern Brazil and not cause concerns for farmers and the *Phakopsora pachyrhizi* Sydow & Sydow (Asian soybean rust), a serious disease which causes a high yield losses. The differentiation of these two species is only possible through DNA testing (Yorinori & Lazzarotto, 2004).

The *P. meibione* is the less aggressive soybean rust species and was reported in the western hemisphere, South and Central America and Caribbean. It was reported in Puerto Rico in 1913, Mexico in 1917 and Cuba in 1926 on hyacinth bean and some other leguminous species, but only in Puerto Rico in 1976 was related on soybean (Pivonia & Yang, 2004). This species occurs under mild temperatures (average below 25 °C) and high relative humidity (Yorinori & Lazzarotto, 2004).

The *P. pachyrhizi* was described as a pathogen on the legume *Pachyrhizus erosus* (L.) Urb. (well-know as jacatupé in Brazil; jícama or pois patate in France; jícama, yam and mexican turnip in English language and jícama, pipilanga, yacón or nabo mexicano in Spanish language) in Taiwan, published by Sydow & Sydow in 1914 and can infect many leguminous species in numerous orders of the family Leguminosae (Deverall et al., 1977).

The *P. pachyrhizi* was first identified in Japan in 1902, and then was detected in India (1906), Australia (1934), China (1940), in Southeast Asia (1950s) and Russia (1957). For many years it remained confined to Asia and Australia, until to be found in Hawaii in 1994 and in Africa continent (from Uganda to South Africa) in 1997 (Begenisic et al., 2004).

P. pachyrhizi was first identified in the America continent in March 2001 in Paraguay, which caused yield reduction of 1,100 kg/hectare. In May, it was also found in Paraná (Brazil). In 2001/02 harvest, the disease recurred throughout Paraguay and was also found in Argentina, Bolivia and in several states of Brazil. In the worst hit places, the reductions in grain yield were estimated between 10% and 80% (Yorinori, 2002).

The fungal inoculum, for the initial outbreak in South America, is thought to originated from southern Africa where soybean rust has been observed since the late 1990s (Scherm et al., 2009). Since 1994, the disease has been identified by several countries, damaging up to 40% of crops in Thailand, 90% in India, 50% in the south of China and 40% in Japan (Hartman et al., 1991; Mendes et al., 2009). In the United States, this disease was first reported at the Louisiana State University AgCenter Research Farm in 2004, but yield loss was not as high as those reported from other countries (Cui et al., 2010).

Soybean plants are susceptible to the fungus at all growth stages. As a general rule, the earlier a crop is attacked, the higher will be the loss (Mendes et al., 2009), however, if the attack occurs at flowering and pod filling stage, which is commonly observed in soybean fields, the yield reducing can be higher than in others stages (Kawuki et al., 2004).

2. Contamination

The Asian soybean rust is one of the most destructive diseases of soybean because it produces a high amount of airborne spores that can infect large areas of soybeans and cause significant yield loss. The fungal spores (uredospores) are deposited on leaves in the lower region of the canopy through the rain or wind transport from nearby plants during the growing season (Huber & Gillespie, 1992). The figure 1 shows a soybean leave contaminated with Asian soybean rust.

The *P. pachyrhizi* is highly moisture dependent, requiring at least 6 hours of free moisture on the lower trifoliolates to starts the contamination. Warm temperature is ideal, but not limiting, since the disease can be established between 15 °C and 30 °C (Embrapa, 2005). These moist conditions can be achieved through any form of wetness as drizzle, mist, fog or dew and the minimum duration of wetness is dependent on the ambient temperature after spore deposition (Schmitz & Grant, 2009).

Fig. 1. (Godoy et al. 2009). Soybean leaf contaminated with Asian soybean rust.

Early rust symptoms are characterized by small dots of 1 to 2 mm of diameter, darker than the healthy leaf tissue with a greenish to greenish gray coloration. On the local corresponding at the dark spot, there is initially a tiny lump, like a bubble formed by burning, showing the early formation of the fruiting structure of fungi. As soon as the death of infected tissues, the blemishes increase in size and acquire a reddish brown color. The uredospores, initially has a crystalline color, become beige and accumulate around the pores or are carried by the wind and the number of uredias per point can vary from one to six. The uredias that no longer sporulete shows the pustules with open pores, which allows distinguish them from bacterial pustule, which often causes confusion comparison (Bromfield, 1980; Embrapa, 2004).

The infection causes rapid browning and premature leaf fall, preventing the full grain formation. The earlier the defoliation, smaller is the grain size and lower is the yield and quality. In severe cases, when the disease reaches the stage of the soybean pod formation, it can cause abortion and drop of the pods, resulting in a total loss of income (Constamilan, 2002; Godoy & Canteri, 2004; Soares et al., 2004).

The life cycle is typical of the majority of other rust fungi (Fig. 2) and their uredospores are easily transported by air currents and disseminated hundreds of kilometers in few days (Tremblay et al., 2010).

Once germination occurs, the uredospore produces a single germ tube (GT) that grows across the leaf surface until it reaches an appropriate surface where an appressorium (AP) forms. This penetration occurs between 7-12h after the spore lands on the leaf adaxial surface. Appressoria form over anticlinal walls or over the center of epidermal cells, but rarely over stomata, in contrast to the habit of many other rusts. Thus, penetration is direct rather than through natural openings or through wounds in the leaf tissue. Approximately twenty hours after the spore landing, the *penetration hyphae* (PH), stemming from the appressorium cone, pass through the cuticle to emerge in the intercellular space where a septum is formed to produce the *primary infection hypha* (IH). This IH grows between palisade cells to reach the spongy mesophyll cells where it forms the haustorium (H) (Tremblay et al., 2010).

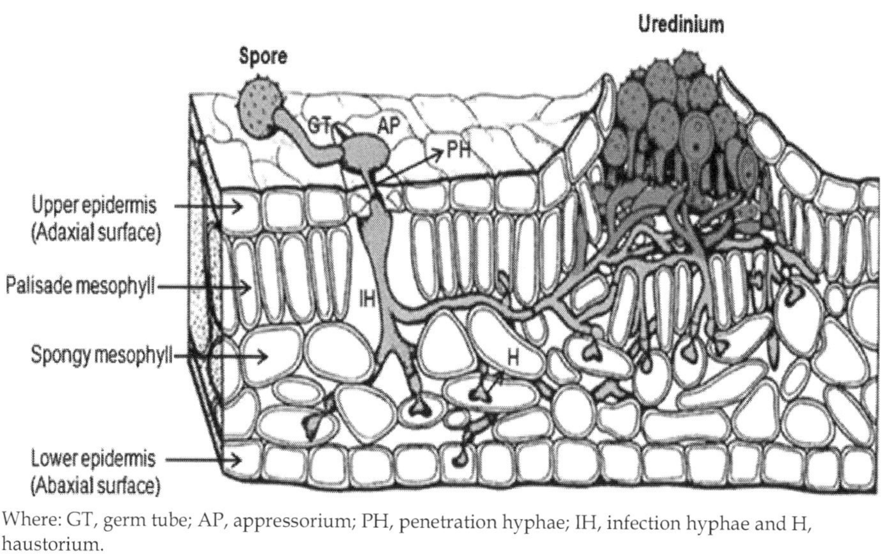

Where: GT, germ tube; AP, appressorium; PH, penetration hyphae; IH, infection hyphae and H, haustorium.

Fig. 2. (Hahn, 2000 apud Tremblay et al., 2010). Internal structure of a typical dicotyledon leaf showing the different cell layers and infection by a rust fungus.

Once this first stage has been reached, about 4 days after spore landing, additional hyphae emerge and spread through the entire spongy mesophyll layer of cells where many other haustoria are formed. At approximately 6 days after infection, some necrosis of epidermal cells occurs which is visible at the adaxial surface of the leaves (Fig. 3a). Hyphae aggregate and a uredinium arise in the spongy mesophyll cell layer. Uredinia can develop 6-8 days after spore landing and development might extend up to 4 weeks (Tremblay et al., 2010).

The first uredospores produced by the uredinium emerge at the abaxial leaf surface in 9-10 days after spore landing and spore production can be observed for up to 3 weeks. High rate of sporulation is typical of a susceptible reaction where lesions on the upper surface of the leaf are tan (Fig. 3b). Plants classified as resistant develop a dark, reddish-brown lesion with few or no spores (Fig. 3c) (Tremblay et al., 2010).

Soybean rust diagnosis is usually performed by experienced plant pathologists or plant disease diagnosticians, but nowadays, several technologies are being performed as crop health sensor, either optical or electronic or bio-electronic based to improve the perform crop disease diagnosis (Cui et al., 2010).

A useful tool that can be used as many by experienced professionals as amateurs is the diagrammatic scale to assess the severity of rust. It is very important once provides data on the severity of contamination in the plantation and the result should be informed when the competent organs are contacted, as well as help to define the goals for fungicides treatment. There are several types of scales such as developed by Godoy et al. (2006) (Fig. 4a) and Martins et al. (2004) (Fig. 4b) witch is highly recommended to be used together.

(3a)

(3b)

(3c)

Fig. 3. (Tremblay et al., 2010). Symptoms observed on soybean leaves. (a) Yellow mosaic discoloration (b) Tan lesions and (c) reddish-brown lesions.

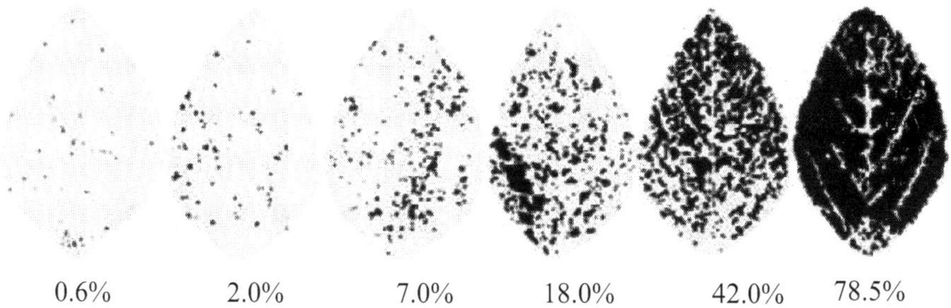

(4a)

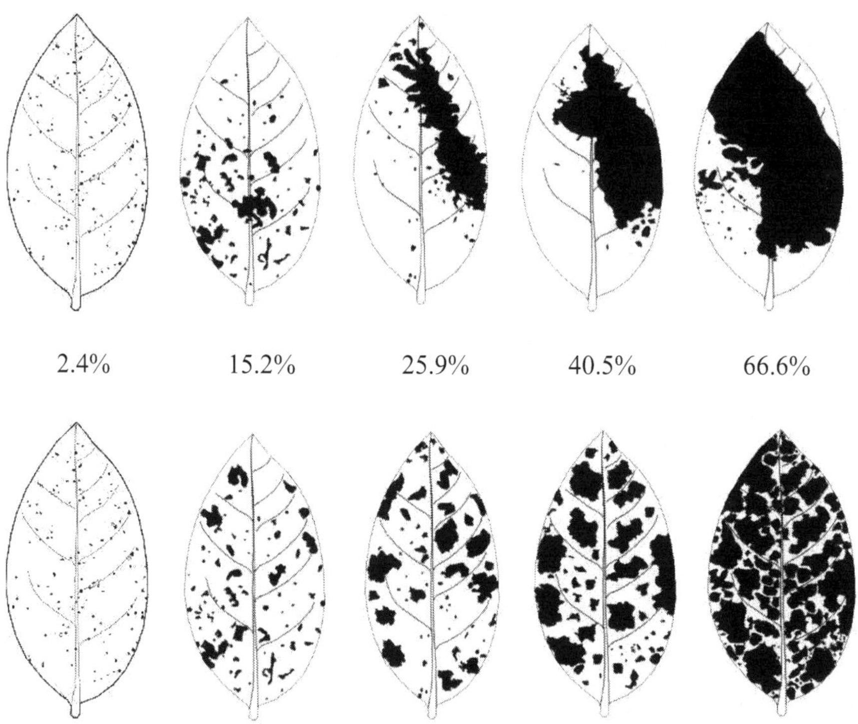

(4b)

Fig. 4. Two diagrammatic scales of Asian soybean rust severity with percentage that represents the area of disease contamination.

3. The Asian soybean rust in South America

3.1 Argentina

The soybean crop has been converted from the middle of the nineties in the main seasonal crop of Argentina, both in area planted and in its total production (CAS, 2008). During 2004/05 season, the area devoted to soybeans was 14.4 million hectares (51% of the total planted with cereals and oilseeds), producing 38 million tones. In the season 2005/06, the soybean planting grew up to 15 and 15.3 million hectares (PNR, 2011a). In the 2006/07 season, reached a volume of 47.5 million tonnes, representing 50% of total the country grain production estimated at 94.4 million tones (CAS, 2008).

The Asian soybean rust first appeared in Argentina at the end of the 2001/02 season, in a test group in the town of Alem, province of Misiones. The infected plants samples were sent to the United States for identification by molecular analysis techniques and confirmed that the pathogen *P. pachyrhizi* was present. This finding coincided with the species identified in Brazil and Paraguay (Begenisic et al., 2004).

In the following season (2002/03), the soybean rust was detected at the end of the cycle, by a group of technicians from "Instituto Nacional de Tecnología Agropecuaria - INTA" (National Institute of Agricultural Technology) in test samples from the town of Cerro Azul, province of Misiones and two commercial lots located in the town of Gobernador Virasoro, in the province of Corrientes. Although this pathogen has penetrated the national territory, it was far from the main producing areas of the country. Because of the history of the disease, there were high producers and technicians concern about the losses that could result in the coming years (Begenisic et al., 2004).

As a result, at the beginning of the 2003/04 season, bearing in mind that rust had caused heavy losses in neighboring countries and displaying that it could become a serious concerns for Argentina, despite until that moment the rust had a little history in the country, the Ministry of Agriculture launched the "Programa Nacional de Roya de la Soja - PNRS" (National Program for Soybean Rust), coordinating activities with various agencies and public and private institutions in order to minimize the possible impact of the disease in the country (PNRS, 2011a).

The implementation of the PNRS was the first opportunity in which all public institutions have joined forces, in a cooperative manner, carrying out activities in a coordinated way and contributing in those components of the program according to their specific duties incumbent on each institution (PNRS, 2011a).

During crop season of 2004/05, the disease was detected in 13 provinces, including provinces witch the disease was detected for the first time, representing the advance of the disease significantly when compared with the last season. In most of the contaminated provinces, the rust did not cause economic losses due to its late appearance in the crop, with the exception of the province of Entre Rios where the disease was much more severe, reporting significant yield losses up to 30% (PNRS, 2011b).

These low levels of contamination can be explained for (PNRS, 2011b):

- Drought conditions and high temperatures occurring in some regions of Brazil, Paraguay and Bolivia, neighborhood to Argentina, that were not conducive to serious infections and consequently to the high seed production regional;
- The drought in Argentina during the first months of 2005 that prevented the infection conditions for generating a second "weather barrier";
- The limited survival of the fungus because the Argentina has not double cropping of soybeans in the year as in Brazil or Bolivia;

- The high level of adoption of fungicide use in Brazil, Paraguay and Bolivia.

The fungicides recommended by Argentina are the strobilurin, triazoles and their mixtures. The decision to apply is at the first signs and/or when was possible to anticipated the diagnosis in the field or when they are found in areas close to their lots and recorded favorable environmental conditions to ensure at least 7-10 hours of wet leaf and average temperatures of 22°C.

The Argentina has an official monitoring system that allow to analyze a large number of samples for the detection and disease monitoring through the website "www.sinavimo.gov.ar" (in Espanish).

3.2 Bolivia

Bolivia is the eighth country in soybeans production and is the fourth in the South America, after Brazil, Argentina and Paraguay and is one of the most important and is the successful of the national economy, due to growth in primary production, processing and export during the last fifteen years (CAS, 2008).

In Bolivia, the Asian soybean rust was first detected in the winter crop season of 2003 at Ichilo, City of Yapacani and is currently distributed throughout in all soybean crops of Santa Cruz and affecting the crop of Tarija (Yacuiba). Before the advent of Asian rust, the number of applications of fungicides for control of diseases ranging from 0 to 1, however, today the value has increased from 3 to 5 applications, increasing the costs of fungicides from 10 to 70 US$ respectively. This disease, year after year is responsible for at least 30 to 50% of loss in total production area, which in economic terms is between US$100 and US$150 million witch concerns to the use of agrochemicals and total yield loss per year (Condori, 2009).

In the 2007 winter cropping season the problems to control the rust emerged from a series of technical and climatic factors as (Condori, 2009):

- Soybean planted between harvests (April-May) generated the Asian rust inoculum that infected soybean fields planted in June until early winter season;
- The prolonged period of drought and the continuous moisture in the months of August and September stressed cultivation, focusing directly into the beginning of flowering that occurred at 65 to 70 days;
- The fungicides applied were exposed to critical climatic conditions as high temperatures (30-35°C) and low relative humidity (50-40%) which affected the residual effect and effectiveness of their control.

Those factors explain that this season (2007/08) was one of the most catastrophic, mainly in the north and east of the Santa Cruz de La Sierra due to continuous rains that prevented raising the winter planting crop, to perform the planting on summer and, the most important fact, the delay of fungicide applications, generating a "explosion" of the rust, forced farmers to make up 7 applications of fungicides per hectare. The economic losses quantified by the "Asociación de Productores de Oleaginosas y Trigo - ANAPO" (Association of Producers of Oilseeds and Wheat), exceed US$ 150 million, for the past two seasons (summer 2007/08 and winter 2008) (Condori, 2009).

The Bolivia are implementing the sanitary break in cities located in the Integrated Zone (Andrés Ibáñez Province, Warnes, Ichilo, Sara, Bishop Santiestevan and Guarayos), Expansion Area (Andres Ibanez, Chiquitos, Ñuflo Chavez, Guarayos) of the Santa Cruz state. This project will benefit more than 14,000 small, medium and large producers, of various nationalities and a planting area between the 700,000 to 1,000,000 hectares. Training, dissemination and sharing of technical and legal measures, through different media

available as workshops, seminars, television and radio messages is also referred in this project (Condori, 2009).
Fungicides recommended for control of Asian rust are bencimidazoles, triazols, triazol + triazol, triazol + benzimidazol and triazol + estrobilurina products. Those products were chose based on research results conducted by different agricultural companies and research institutions (Condori, 2009).

3.3 Brazil

Brazil is the second largest producer of soybeans. In the 2006/07 season, the culture occupied an area of 20.69 million hectares, which totaled a production of 58.4 million tons. The United States, the worldwide producer, accounted for the production of 86.77 million tons of soybean. The yield of soybeans in Brazil is 2,823kg per hectares, reaching about 3,000 kg/ha in Mato Grosso, the largest state producer (EMBRAPA, 2011).
The soybean is the crop which has the higher development in Brazil in the last three decades and accounts for 49% of grains area planted in the country. The grain is an essential component in the manufacture of animal feeds and the growing use food is increasing (MAPA, 2011). Data from the Ministry of Development show that soy has a major share of Brazilian exports. In 2006, were US$ 9.3 billion, representing 6.77% of total exported (EMBRAPA, 2011).
The Asian soybean rust was identified in Brazil in May of 2001 and spread quickly to the main producing regions, becoming a major problem for the national soybean producers. To propose solutions was created in September of 2004, the "Consórsio Antiferrugem - CAF" (Antirust Consortium). The consortium constituents are representative institutions of various soybean segments as foundations, universities, research institutes, representatives of entities of inputs manufacturers and farmer cooperatives. One of the aims of the Consortium is to bring the farmer all available information about the disease and enable him to handle it (Farias, 2009).
The CAF main information and communication vehicle is the consortium website: "www.consorcioantiferrugem.net" (in Portuguese) where the laboratories accredited update information about the disease outbreaks in all producing regions of Brazil during a season. In the system are recorded and presented a map of Brazil, the city of occurrence, date of detection, the fenological phases of culture and type of area (warning unit, commercial field, irrigated area etc). Thus, epidemics of soybean rust have been monitored and the spread of the disease are presented in real time at the consortium website, describing it as the main source of data for the record of events and the spread of the disease in Brazil (Spolti et al., 2009).
When the disease arrives, both farmers and technicians were not prepared to identify soybean rust. Factors such as dry climate, the symptoms likely with other diseases of end of cycle and because it was a new disease in the Americas, their identification was difficult and there was no species resistant to fungus attack (Constamilan, 2002 (2005)). It is estimated that over 60% of soya production in Brazil has been contaminated in the season of 2001/02, causing grains losses estimated at 569.2 thousand of tons or the equivalent of US$ 125.5 million (US$ 220.50/t) (Yorinori, 2004).
In the season 2002/03, the occurrence was different from the last season. In localities where the disease was severe in 2001/02, the high temperatures prevented, despite the high amount of rain, the development of the disease, except in Rio Grande do Sul and Santa Catarina, where the late cultivars were affected. But where the rust had not been reported earlier, favorable

climatic conditions and a new strain of *P. pachyrhizi* caused major losses. The states of Bahia, Goiás, Minas Gerais and Mato Grosso were severely affected (Yorinori, 2004).
However, despite the intensive campaigns to alert and guidance on methods for identification and control, held in 2002 and in January and February of 2003, through lectures, publications and other means of dissemination, the technical assistance and most producers were not prepared to control the rust. In many crops, the fungicide application was delayed due to lack of product and/or excessive rain which precluded the spraying (Yorinori, 2004). In this season the damage caused by the rust (amounting the grain losses, control expenses and revenues falling) were approximately US$ 1.29 billion (Soares et al. 2004).
The beginning of 2003/04 season was characterized by irregular rainfall and high temperatures, which probably not favored the outbreak of rust as expect. Moreover, the experience of loss in the previous crop left farmers in the areas previously affected readiness and "armed" for the chemical control. However in the southern region, the beginning of the harvest was characterized by mild temperatures and frequent rainfall, which favored the early emergence of *P. pachyrhizi*. The total damage caused by rust, in this year, adding the grain losses, control spending and falling revenue was approximately US$ 2.28 billion (Yorinori, 2004).
Among the crops of 2005/06 to 2008/09 were recorded, respectively, 1,369; 2,766; 2,107 and 2,880 reports of the occurrence of soybean rust. While there is an increase in the number of reports over the years, it is not possible to assert that the severity of epidemics is related to the number of outbreaks, since it is observed only presence of disease in crops, once in 2006/07, when they were registered the greatest losses in productivity caused by rust due to the higher disease severity, the number of reported outbreaks was lower than in 2008/09 when, according to regional information, the attack of the disease was not as severe as that year (Spolti et al., 2009).
Since the disease monitoring, the Asian rust was not observed before the month of October, whether in the commercial field or in units of alert. The progress of the number of reports of disease presents a sigmoid pattern with a logarithmic phase and a stationary phase when approaches the end of growing season. The maximum rate of increase, indifferent to the season, was observed between January and March (90 to 150 days after October 1st), this period can be defined as critical in the epidemics development, being responsible for the differentiation of the final number of focus reported in the cycle (Spolti et al., 2009).
At present, around 70 fungicides are registered in the Brazilian Ministry of Agriculture for managing soybean rust and many of these have been evaluated annually since 2003/2004 in a nationwide network of standardized coordinated by Embrapa Soja, a research unit of the Brazilian Agricultural Research Corporation (Godoy et al., 2010; Scherm et al., 2009). The fungicides registered for control of Asian soybean rust belong into two main groups: Triazoles and strobilirins (Godoy & Flausino, 2008).

3.4 Paraguay

The soybean in Paraguay is the main agricultural export item, with a market of 70% of national output in the form of grain. This is due to high charges imposed by the European Union, the main buyer, for other soy subproducts such as soybean oil. Today, Paraguay is the sixth soybean production in the world (preceded by USA, Brazil, Argentina, China and India) and has a weighted average of 2,600kg/ha, performance similar to Argentina and Brazil (CAS, 2008).

Since the appearance of Asian soybean rust in 2001 in Paraguay, there have been major changes in the soybean production system and it also contributed to better care for the crop, getting even better yields by protecting against various diseases of economic importance appellant in soybean. The productivity losses were very important in years when climatic conditions were favorable for the disease, especially during the breeding season and when constant rainfall recorded during the months of January and February, considered the most critical for the development of an epidemic Paraguay (Morel & Bogado, 2009).

In the first year where the disease was recorded yield losses were estimated, in the cultivars most affected, at more than 60%. On the next seasons, 2001/02 and 2002/03, the severity of the disease was not very important because of the drought but the late sown soybean crop showed severe losses of more than 50% of performance. This epidemic is especially observed when the rains season from the month of March to May (Morel & Bogado, 2009).

In all the years that the rust has been detected early, it was observed in plants of 30-35 days from sentinel plots in the region of Pirapó, considered an endemic area once is possible to detect the strong presence of the volunteer plant, being a fungus host, named Kudzu (*Pueraria lobata*), but the severity level ever has thrived in the vegetative phase. This demonstrates the importance of the survival of the disease during the winter, which has made a strong campaign of awareness among farmers aiming the elimination of inoculum source in areas with no winter crop, in order to avoid the primary infection in an early period of soybean cultivation (Morel & Bogado, 2009).

The crop of 2005/06, was the largest epidemic in the normal planting season, resulting in a loss of more than US$ 400 million. This strong impact due to multiple factors, neglect of producers, the time control and problems in application of the technology. The number of fungicide applications was a maximum of 5 and a minimum of 2. In the season of 2008/09, the rust incidence was reported again in a very early (second fortnight of October) in crops planted in September, but the severe drought that affected the whole area of soybean production allowed the progress of the disease (Morel & Bogado, 2009).

In rare cases and in regions where rains started to become evident, controls measures have been made, but around the country more than 1 application of fungicides should be done for each producer. This drought was so important that did not allow the progress of the disease throughout the production area, except in some regions of late-sown soybean (Morel & Bogado, 2009).

4. Disease control

Once commercial soybean cultivars used in the major soybean producing countries are susceptible to soybean rust, management of the disease is done using fungicides, although some cultural and crop management practices also may decrease disease risk at field and regional scales. Early research in Asia indicated that mancozeb and, to a more limited extent, the benzimidazole fungicides suppressed soybean rust but required three to five applications to be effective. The disease control was significantly improved after the introduction of the triazole fungicides (Scherm et al., 2009).

Scherm et al. (2009) studying the efficacy of several fungicides on a soybean crop in Brazil showed that triazole fungicides had significantly efficiency than strobilurins classes. The combination of triazoles with strobilurins improved disease control and yield gain compared with triazoles or strobilurins alone. However the combination of triazoles with a benzimidazole fungicide did not improve the desease control when compared with triazoles alone. They either conclude that the two fungicides with the best disease control efficacy were combinations of two active ingredients as flusilazole + carbendazim and azoxystrobin + cyproconazole.

The triazoles group acts to inhibit the ergosterol biosynthesis and have the primary site of action the C-14 demethylation and the strobilurins interfere with mitochondrial respiration by blocking electron transfer by the cytochrome bc1 complex, formulated alone or in mixtures (Godoy & Flausino, 2008).

Besides the fungicides application, other measures could be taken as the use of earliest varieties, seed at the beginning of recommended time for each region, avoid prolonging the period of sowing, inspect crops and verify if there are temperature and high humidity favorable to the pathogen (Reunião de Pesquisa da Soja da Região Sul, 2002).

One way to anticipate the presence of this fungus before it reaches the crop would be the establishment of sentinel plots in one or more locations, depending on the area of the property. These traps, seeded with 15 to 20 days in advance of the first crops are intensively monitored to identify the first symptoms. Once detected the presence of disease, the traps must be destroyed or heavily treated with an effective fungicide. From this initial detection, the commercial areas should be treated or monitored more carefully for making treatment decisions to be made (Yorinori & Lazzarotto, 2004).

Another method for disease control is the adoption of absence of living plants in the field of this culture denominated sanitary break. This technician aiming to reduce the amount of uredospores in the environment on off-season and then, inhibit the early attack to soybean plants, trough the smaller inoculum presence (Seixas & Godoi, 2007) and is adopted in many countries. The general rule is that all regions are forbidden to cultivate soybean in the period established and the remaining plants from the last crop should be eradicated with chemicals or other means. The producer who does not obey the sanitary break will be required to pay large fines. Another caution that the producer should be is to remove the soy plants that may grow due to grain that fell in the soil and germinate during the harvest.

Also, the kudzu (*Pueraria lobata*), a leguminous plant witch is highly susceptible to the Asian soybean rust, founded in Paraguay and Brazil, shown to be an efficient source of inoculum, presenting the first symptoms and fungal growth before the first crops of soybean (Yorinori & Lazzarotto, 2004). In those countries where kudzu is found, control policies of the rust also include the control of this plant.

5. References

Begenisic, F.; Ploper, L.D.; Ivancovich, A. (2004). *Roya de la soja: Características de la enfermedad*. Documento de Trabajo N° 1, Programa Nacional de la Roya de la Soja

Bromfield, K.R.; Melching, J.S.; Kingsolver, C. H. (1980). Virulence and aggressiveness of *Phakopsora pachyrhizi* isolates causing soybean rust. *Phytopathology*, Vol. 70, pp. 17-21

CAS - Consejo Agropecuario del Sur. (2008). *El Mercado de la soja en los países del Consejo Agropecuario del Sur*. Grupo II. Sistema de Información de Mercados y Pronósticos de Cosecha

Condori, M. (2009). Roya asiatica de la soya en Bolivia (Santa Cruz de La Sierra), In: *Reunião do Consórcio Antiferrugem Safra 2008-09*, Godoy, C.V.; Seixas, C.D.; Soares, R.M., pp. 47-54, Embrapa Soja, Londrina, Paraná, Brazil, Documentos 315

Constamilan, L.M.; Bertagnolli, P.F.; Yorinori, J.T. (2002). *Comunicado Técnico 96*, 10 fev. 2005, Available from: <www.embrapa.com.br>

Cui, D.; Zhang, Q.; Li, M.; Hartman, G. L.; Zhao, Y. (2010). Image processing methods for quantitatively detecting soybean rust from multispectral images. *Biosystems Engineering*, Vol. 107, pp. 186- 193

Deverall, B. J.; Keogh, R. C.; McLeod, E. (1977). Responses of soybean to infection by, and to germination fluids from, urediniospores of *Phakopsora pachyrhizi*. *Transactions of the British Mycological Society*, Vol. 69, No. 3, pp. 411-415, 1977.

EMBRAPA - Empresa Brasileira de Pesquisa Agropecuária. *A soja*, 14 mar. 2011, Available from: <www.cnpso.embrapa.br/index.php?op_page=22&cod_pai=16>

Embrapa. 12 fev. 2005, Available from: <www.embrapa.com.br>

Embrapa. *Sistemas de produção*, 17 mai. 2004, Available from: <http://sistemasdeproducao.cnptia.embrapa.br/FontesHTML/Soja/SojaCentralBrasil2003/doenca.htm>.

Farias, J.R.B. (2009). Apresentação. In: *Reunião do Consórcio Antiferrugem Safra 2008-09*, Godoy, C.V.; Seixas, C.D.; Soares, R.M., Embrapa Soja, Londrina, Documentos 315

Godoy, C.V.; Canteri, M.G. (2004). Efeitos protetor, curativo e erradicante de fungicidas no controle da ferrugem da soja causada por *Phakopsora pachyrhizi*, em casa de vegetação. *Fitopatologia brasileira*, Vol. 29, pp. 97-101

Godoy, C.V.; Flausino, A.M. (2007). *Eficiência de fungicidas para o controle da ferrugem asiática da soja em Londrina e Tamarana, PR, na safra 2007/08*, Circular Técnica 57, Embrapa Soja, Londrina, Paraná, Brazil

Godoy, C.V.; Utiamada, C.M.; Silva, L.H.C.P. (2010). *Eficiência de fungicidas para o controle da ferrugem asiática da soja, Phakopsora pachyrhizi, na safra 2009/10*: resultados sumarizados dos ensaios cooperativos, Circular Técnica 80, Embrapa Soja, Londrina, Paraná, Brazil

Hartman, G.; Wang, T.; & Tschanz, A. (1991). Soybean rust development and the quantitative relationship between rust severity and soybean yield. *Plant Disease*, Vol. 75, No. 6, pp. 596-600

Huber, L.; & Gillespie, T.J. (1992). Modeling leaf wetness in relation to plant disease epidemiology. *Annual Review of Phytopathology*, Vol. 30, pp. 553-577

Kawuki, R.S.; Tukamuhabwa, P.; Adipala, E. (2004). Soybean rust severity, rate of rust development, and tolerance as influenced by maturity period and season. *Crop Protection*, Vol. 23, pp. 447-455

MAPA - Ministério da Agricultura, Pecuária e Abastecimento. *Soja*, 14 jan. 2011, Available from: <www.agricultura.gov.br >

Mendesa, R.K.; Carvalhal, R.F.; Stach-Machado, D.R.; Kubota, L.T. (2009). Surface plasmon resonance immunosensor for early diagnosis of Asian rust on soybean leaves. *Biosensors and Bioelectronics*, Vol. 24, pp. 2483-2487

Morel, W.; Bogado, N. (2009). Relato de la situación de la roya de soja en Paraguay. In: *Reunião do Consórcio Antiferrugem Safra 2008-09*, Godoy, C.V.; Seixas, C.D.; Soares, R.M., pp. 45-46, Embrapa Soja, Londrina, Paraná, Brazil, Documentos 315

Pivonia, S.; Yang, X. B. (2004). Assessment of the potential year-round establishment of soybean rust throughout the world. *Plant Disease*, Vol. 88, No. 5, pp. 523-525

Programa Nacional de la Roya de la Soja. *Acciones previstas para la campaña 2005/06*, 13 jan. 2011a, Available at: <www.minagri.gob.ar/SAGPyA/agricultura/roya_soja>

Programa Nacional de la Roya de la Soja. *Roya asiática de la soja: Escenario pasado, actual y perspectivas futuras en Argentina*, 13 jan. 2011b, Available from: <www.minagri.gob.ar/SAGPyA/agricultura/roya_soja>

Reunião de pesquisa de soja da região sul. (2002). *Indicações técnicas para a cultura da soja no Rio Grande do Sul e em Santa Catarina 2002/2003*, Fundacep/Fecotrigo, Cruz Alta, Rio Grande do Sul, Brazil

Scherm, H.; Christiano, R.S.C.; Esker, P.D.; Del Ponte, E.M.; Godoy, C.V. (2009). Quantitative review of fungicide efficacy trials for managing soybean rust in Brazil. *Crop Protection*, Vol. 28, pp. 774-782

Schmitz, H.F.; Grant, R. H. (2009). Precipitation and dew in a soybean canopy: Spatial variations in leaf wetness and implications for *Phakopsora pachyrhizi* infection. *Agricultural and Forest Meteorology*, Vol. 149, pp. 1621-1627

Seixas, C.D.S.; Godoy, C.V. (2007). Vazio sanitário: panorama nacional e medidas de monitoramento, In: Simpósio brasileiro de ferrugem asiática da soja. *Anais do simpósio brasileiro de ferrugem asiática da soja*, Embrapa Soja, Londrina, Paraná, Brazil

Soares, R.M.; Rubin, S.A.L.; Wielewicki, A.P.; Ozelame, J.G. (2004). Fungicidas no controle da ferrugem asiática (*Phakopsora pachyrhizi*) e produtividade da soja. *Ciência rural*, Vol. 34, No. 4, pp. 1245-1247

Spolti, P.; Godoy, C.V.; Del Ponte, E.M. (2009). Sumário da dispersão em larga escala das epidemias de ferrugem asiática da soja no Brasil em quatro safras (2005/06 a 2008/09). In: *Reunião do Consórcio Antiferrugem Safra 2008-09*, Godoy, C.V.; Seixas, C.D.; Soares, R.M., pp. 11-19, Embrapa Soja, Londrina, Paraná, Brazil, Documentos 315

Tremblay, A.; Hosseini, P.; Alkharouf, N.W.; Li, S.; Matthews, B.F. (2010). Transcriptome analysis of a compatible response by *Glycine max* to *Phakopsora pachyrhizi* infection. *Plant Science*, Vol. 179, pp. 183-193

Yorinori, J.T. (2002). Ferrugem asiática da soja (*Phakopsora pachyrhizi*): ocorrência no Brasil e estratégias de manejo. In: *II Encontro Brasileiro sobre Doenças da Cultura da Soja*, Resumos de palestras, pp. 47-54, Passo Fundo, Rio Grande do Sul, Brazil

Yorinori, J.T.; Lazzarotto, J.J. (2004). *Situação da ferrugem asiática da soja no Brasil e na América do Sul*, Documentos 236, Embrapa, Londrina, Paraná, Brazil